TECHNICAL DRAWING AND DESIGN

TECHNICAL DRAWING AND DESIGN

David L. Goetsch
John A. Nelson

Delmar Publishers Inc.

DEDICATION

From David L. Goetsch . . . .

. . . . To Savannah Day, Toby, Dustin, and Clifford Jay

From John A. Nelson . . . .

. . . . To my wife, Joyce

Cover ELLE STUDIO

Delmar staff
  Administrative editor: Mark W. Huth
  Production editor: Frances Larson

For information, address Delmar Publishers, Inc.
2 Computer Drive West, Box 15-015
Albany, New York 12212-9985

Printed in the United States of America
Published simultaneously in Canada
by Nelson Canada,
a division of International Thomson Limited

10 9 8 7 6 5 4 3 2 1

Library of Congress Cataloging in Publication Data

Goetsch, David L.
  Technical drawing and design.

  Includes index.
  1. Mechanical drawing.   I. Nelson, John A.,
1935-   II. Title.
T353.G63 1986   604.2'4   85-15859
ISBN 0-8273-2223-2

Brief Contents

CONTENTS vi

PREFACE ix

SECTION ONE • **BASICS** 1

Introduction 2   **1** Drafting Instruments and Their Use 23   **2** Lettering, Sketching, and Line Techniques 66   **3** Geometric Construction 96

SECTION TWO • **TECHNICAL DRAWING FUNDAMENTALS** 147

4 Multiview Drawings 148   **5** Sectional Views 189   **6** Auxiliary Views 219   **7** Descriptive Geometry 235   **8** Patterns and Developments 266   **9** Dimensioning and Notation 299

SECTION THREE • **COMPUTER-AIDED DRAFTING** 339

10 Computer-Aided Drafting Technology 340   **11** Computer-Aided Drafting Operations 357

SECTION FOUR •**DESIGN DRAFTING APPLICATIONS** 377

12 Geometric Dimensioning and Tolerancing 378   **13** Fasteners 403 **14** Springs 452   **15** Cams 468   **16** Gears 483   **17** Assembly and Detail Drawings 501   **18** Pictorial Drawings 529

SECTION FIVE • **RELATED TECHNOLOGY** 569

19 Welding 570   **20** Shop Processes 594   **21** Shortcuts 628   **22** Media and Reproduction 642   **23** Mechanical Drafting Mathematics 649

GLOSSARY 678

APPENDIX CONTENTS 684

INDEX 739

Contents

SECTION ONE • **BASICS**   1

Introduction   2

drawings described • types of drawings • types of technical drawings • purpose of technical drawings • applications of technical drawings • regulation of technical drawings • what students of technical drawing and drafting should learn

Chapter 1 • **Drafting instruments and their use**   23

conventional and CAD/CAM drafting equipment • conventional drafting requisites • drawing sets • scales • measuring • ink tools • technical pens • mechanical lettering/sets • butterfly-type scriber • airbrush • paper sizes • whiteprinters • files • open-end typewriter • care of drafting equipment

Chapter 2 • **Lettering, sketching, and line techniques**   66

freehand lettering • freehand lettering techniques • line work • types of sketches • sketching materials • sketching techniques

Chapter 3 • **Geometric construction** 96

geometric nomenclature • geometric forms • elemental construction principles • polygon construction • circular construction • supplementary construction

SECTION TWO •   **TECHNICAL DRAWING FUNDAMENTALS**   147

Chapter 4 • **Multiview drawings**   148

orthographic projection • planning the drawing • sketching procedure • centering the drawing • rounds and fillets • runouts • treatment of intersecting surfaces • curve plotting • cylindrical intersections • incomplete views • aligned features • how to represent holes • conventional breaks • visualization • first-angle projection • industry print

Chapter 5 • **Sectional views**   189

sectional views • cutting-plane line • direction of sight • section lining • multisection views • kinds of sections • sections through ribs or webs • holes, ribs and webs, spokes and keyways • aligned sections • fasteners and shafts in section • intersections in section • industry print

Chapter 6 • **Auxiliary views**   219

auxiliary views defined • how to draw an auxiliary view • how to plot an irregular curved surface • secondary auxiliary views • partial views • auxiliary section • half auxiliary views • industry print

Chapter 7 • **Descriptive geometry** 235

descriptive geometry projection • steps used • notations • fold lines • projecting a line into other views • projecting points into other views • determining the true length of a line • determining the point view of a line • finding the true distance between a line and a point in space • determining the true distance between two lines in space • projecting a plane surface in space • developing an edge view of a plane surface in space • determining the true distance between a plane surface and a point in space • determining the true angle between plane surfaces in space

Chapter 8 • **Patterns and developments** 266

developments • parallel line development • radial line development • triangulation development • true-length diagram • notches • bends • industry print

Chapter 9 • **Dimensioning and notation** 299

specifying the scale • dimensioning systems • dimension components • general rules of dimensioning • specific dimensioning techniques • locational dimensioning systems • notation • rules for applying notes on drawings • industry print

SECTION THREE • **COMPUTER-AIDED DRAFTING** 339

Chapter 10 • **Computer-aided drafting technology** 340

overview of CAD • computer-aided drafting systems • CAD hardware • CAD software • CAD users • modern CAD system configurations • advantages of CAD

Chapter 11 • **Computer-aided drafting operations** 357

general system operation • input commands • manipulation commands • output commands

SECTION FOUR • **DESIGN DRAFTING APPLICATIONS** 377

Chapter 12 • **Geometric dimensioning and tolerancing** 378

general tolerancing • geometric dimensioning and positional tolerancing defined • modifiers • datums • feature control symbol • true position • flatness • straightness • circularity (roundness) • cylindricity • angularity • parallelism • perpendicularity • profile • runout • industry print

Chapter 13 • **Fasteners** 403

classifications of fasteners • threads • screw thread forms • tap and die • threads per inch (TPI) • pitch • single and multiple threads • right-hand and left-hand threads • thread representation • thread relief (undercut) • screw, bolt and stud • rivets • keys and keyseats • grooved fasteners • spring pins • fastening systems • retaining rings • industry print

Chapter 14 • **Springs** 452

spring classification • terminology of springs • required spring data • how to draw compression springs • how to draw an extension spring • other spring design layout, including torsion springs • standard drafting practices • section view of a spring • isometric views

Chapter 15 • **Cams** 468

cam principle • basic types of followers • cam mechanism • cam terms • cam motion • laying out the cam from the displacement diagram • how to draw a cam with an offset follower • how to draw a cam with a flat-faced follower • timing diagram • dimensioning a cam • industry print

Chapter 16 • **Gears**   483

kinds of gears • gear ratio • pitch diameter • gear blank • backlash • basic terminology • diametral pitch • pressure angle • center-to-center distances • measurements required to use a gear tooth caliper • required tooth-cutting data • rack • bevel gear • worm and worm gear • gear train • materials • design and layout of gears • industry print

Chapter 17 • **Assembly and detail drawings**   501

the engineering department • drawing revisions • invention agreement • title block • checking procedure • numbering system • parts list • personal technical file • the design procedure • working drawings • pattern drawings • computer drawings (see Chapters 10 and 11) • industry print

Chapter 18 • **Pictorial drawings**   529

types of pictorial drawings • axonometric drawings • oblique drawings • perspective drawing • isometric principles • nonisometric lines • hidden lines • offset measurements • center lines • box construction • irregularly shaped objects • isometric curves • isometric circles or arcs • isometric arcs • isometric knurls • isometric screw threads • isometric spheres • isometric intersections • isometric rounds and fillets • dimensioning • isometric templates • perspective drawing procedures

SECTION FIVE • **RELATED TECHNOLOGY**   569

Chapter 19 • **Welding**   570

welding processes • basic welding symbol • size of weld • length of weld • placement of weld • intermittent welds • process reference • contour symbol • field welds • welding joints • types of welds • multiple reference line • spot weld • projection welds • seam welds • welding template

Chapter 20 • **Shop processes**   594

shop processes • casting • forging • extruding • stamping • machining • special workholding devices • heat treatment of steels

Chapter 21 • **Shortcuts**   628

inking techniques • use of appliques • use of burnishing plates • typewritten text • overlay drafting • scissors drafting techniques

Chapter 22 • **Media and reproduction**   642

drafting vellum • polyester drafting film • sepia diazo intermediate paper • diazo print paper • sensitized polyester intermediate film • graphite pencil lead • plastic lead • erasers • reproduction of drawings • diazo printing • high-speed printing

Chapter 23 • **Mechanical drafting mathematics**   649

mathematics for drafters • rounding decimal fractions • expressing common fractions as decimal fractions • millimetre-inch equivalents (conversion factors) • evaluating formulas • ratio and proportion • arithmetic operations on angles expressed in degrees, minutes, and seconds • degrees, minutes, seconds - decimal degree conversion • types of angles • angles formed by a transversal • types of triangles • common polygons • definitions of properties of circles • geometric principles of circle circumference, central angles, arcs, and tangents • geometric principles of angles formed inside a circle • geometric principles of angles formed on a circle and angles formed outside a circle • internally and externally tangent circles • trigonometry: trigonometric functions • trigonometry: basic calculations of angles • trigonometry: basic calculations of sides • trigonometry: common drafting applications • trigonometry: oblique triangles - law of sines and law of cosines

GLOSSARY   678

APPENDIX CONTENTS   684

INDEX   739

Preface

Purposes *Technical Drawing and Design* is intended for use in such courses as basic and advanced technical drawing, basic and advanced drafting, engineering graphics, descriptive geometry, mechanical drafting, machine drafting, tool and die design and drafting, and manufacturing drafting. It is appropriate for those courses offered in comprehensive high schools, area vocational schools, technical schools, community colleges, trade and technical schools, and at the freshman and sophomore levels in universities.

Prerequisites No prerequisites are necessary. The text begins at the most basic level and moves step-by-step to the advanced levels. It is as well suited for students who have had no previous experience with technical drawing as it is for students with a great deal of prior experience.

Innovations An advantage of the text is that it evolved during a time when the world of technical drawing and design is undergoing a period of major transition from manual to automated techniques. Computer-aided drafting (CAD) is slowly but steadily gaining a foothold. This transitional period will last at least until the turn of the century, with CAD gaining greater acceptance every year.

This transition has created a need for a major text that deals with both traditional knowledge and skills and CAD-related knowledge and skills. *Technical Drawing and Design* fills this need. Even when the world of technical drawing and design has become fully automated, drafters and designers will still need to know the traditional basics and technical drawing fundamentals. These basic factors will not change. Therefore, the traditional fundamentals are treated in depth in this text.

What is changing, and will continue to change, is the way that drafters and designers prepare technical drawings. For this reason, CAD is also treated in depth, and many of the drawings and illustrations were prepared on various CAD systems. Along with this treatment, *Technical Drawing and Design* offers students and teachers a special blend of the manual and automated knowledge and techniques that are needed now through the turn of the century, and even beyond.

Another advantage of the text is that it was written after the latest update of the most frequently used drafting standard — ANSI Y14.5. This standard was updated with major revisions in 1982, and is now ANSI Y14.5M - 1982. Consequently, all dimensioning and tolerancing material in *Technical Drawing and Design* is based on this most recent edition of the standard.

Grouping of Chapters The authors combine more than 32 years of classroom experience and 18 years of industrial experience. When using various textbooks, they were never satisfied with the grouping or sequencing of chapters. To solve this problem and, in turn, to make *Technical Drawing and Design* a more effective teaching tool, the authors came up with five major sections for the book: Basics, Technical Drawing Fundamentals, Computer-aided Drafting, Design Drafting Applications, and Related Technology.

All chapters are sequenced under one of the major section headings in such a way as to correspond with normal teaching and learning patterns. The basics are treated first. Technical drawing fundamentals that should be learned using traditional manual tools and techniques are discussed next. The third section deals with CAD. This section allows teachers the option of progressing into advanced chapters using the manual approach, the CAD approach, or a combination of both. Section Four contains the advanced technical drawing and design chapters for those programs that go beyond the fundamentals. All of the non-drawing or reference chapters are grouped together in Section Five, thus allowing teachers and students to turn to any one comprehensive reference section as needed.

Comprehensive Listing of Contents The detailed Contents listing is designed to give students and teachers an all-inclusive but easy to follow "roadmap" for locating material in the text. In addition to the chapter titles, the Contents contains a complete breakdown of the major topical headings for each chapter so that students and teachers can turn readily to information that is needed without having to guess as to its location in the text. A brief Contents is also provided for quick reference to subject matter.

Simply Written Students often find the language of technical drawing books verbose, abstract, and complicated. Many of the principles covered in such books ARE complicated, but this does not mean that their explanations must be. This text is written in such a way as to communicate plainly and simply the most complex principles and concepts so that students can concentrate on what they are learning, rather than trying to unscramble overly complicated explanations.

Well Illustrated The text is copiously illustrated for all of the concepts and principles presented. A second color has been used freely to better illustrate important points. The color not only enhances the book's appearance, making it more readable, but it also simplifies some of the more complicated concepts presented.

Review Sections Each chapter contains a Review designed to test the students' knowledge of all theories, principles, and concepts presented. In order to answer the reviews, students must comprehend the material that has been presented. Teachers may use the reviews as homework or student self-tests, or for classroom discussions.

Application Problems Each drawing chapter contains numerous drawing problems that range from the simple to the intermediate to the complex in terms of the material covered. Chapter 17 contains major drawing projects that can be used to challenge and stretch the stamina of the students' attention span.

CAD Drawing and Industry Prints This text contains many drawings prepared on various CAD systems. These drawings allow students to become accustomed to the differences between manually prepared and computer-aided drawings. In addition, 10 chapters contain an industry print, each of which was prepared by the design and drafting departments of specific industrial firms. These prints are accompanied by questions for students to answer. This activity gives students experience in dealing with "real world" drawings while still in the classroom, and is designed to allow students to serve an in-school internship.

Instructor's Guide An Instructor's Guide is provided to make the teacher's job easier. The guide contains learning objectives for each chapter, answers to the reviews, answers to the questions accompanying the "real world" drawings, and four test drawings with solutions for each chapter that involves drawing. Instructors may use end of chapter drawings for learning activities, skills development, and practice drills, while reserving the test drawings in the Instructor's Guide for conducting evaluations.

Acknowledgments The authors acknowledge the efforts of several people without whose assistance this project would not have been completed. Special acknowledgment is made to Edward G. Hoffman, author of Chapter 20, "Shop Processes"; and Robert D. Smith, author of Chapter 23, "Mechanical Drafting Mathematics." We thank Ray Adams, Dana Welch, Susan Wilkinson, and Ron Ryals for their assistance with illustrations. We thank Deborah M. Goetsch for her assistance with photography and typing. We thank Joyce Nelson for her assistance with typing. We also thank Mark Huth, Administrative Editor for Delmar Publishers, and our editor on several other projects. Mark gave rise to the idea that became this book. He did what a successful editor must do — he saw a need and filled it.

The following individuals reviewed the manuscript and made valuable suggestions to the authors. We and the publisher greatly appreciate their contributions to this textbook.

John G. Boyden, Jr.
Canton Agricultural and
　Technical College
Canton, NY

J. Douglas Frampton
The Ohio State University
Columbus, OH

Robert F. Franciose
Chairman Y14
American National
　Standards Institute

Steven F. Horton
San Jacinto College
Pasadena, TX

Mark Knott
Texas State Technical
　Institute
Harlingen, TX

Richard Latimer
The University of
　Alabama
University, Alabama

Dave Price
Orange Coast College
Costa Mesa, CA

George Pruitt
Ridgecrest, CA

Paul Salvucci
Boston Gear
Quincy, MA

Edwin B. Thomas
Grambling State
　University
Grambling, LA

James R. Vandervest
Gulf Coast Community
　College
Panama City, FL

David L. Goetsch
John A. Nelson

ABOUT THE AUTHORS David L. Goetsch is Director of Occupational Education, Head of Drafting and Design, and Professor of Computer-aided Design and Drafting of Okaloosa-Walton Junior College in Niceville, Florida. His drafting and design program has won national acclaim for its pioneering efforts in the

John A. Nelson has a strong background in industry and the classroom. Before entering education full time, Nelson spent 11 years in drafting and design in the private sector, beginning as a detailer and work-

John A. Nelson

David L. Goetsch

area of computer-aided drafting (CAD). In 1984, his school was selected as one of only ten schools in the country to earn the distinguished Secretary's Award for an Outstanding Vocational Program. Goetsch is a widely acclaimed teacher, author, and lecturer on the subject of drafting and design. He won Outstanding Teacher of the Year honors in 1976, 1981, 1982, 1983, and 1984. He entered education full time after a successful career in design and drafting in the private sector where he spent more than eight years as a Senior Drafter and Designer for a subsidiary of Westinghouse Corporation. This is his 12th book.

ing up to designer. He currently continues to freelance in this field. He has more than 20 years of teaching experience in the classroom at both the high school and college levels, and was New Hampshire's Vocational Teacher of the Year in 1982. He holds the associate in arts, bachelor of science, and master of education degrees. This is his 16th book.

United States Department of Education
=== *Secretary's Award* ===

The Secretary of Education
recognizes Okaloosa-Walton Junior College
for an outstanding Vocational Education Program
in 1984, Drafting and Design Technology

T. H. Bell
United States Secretary of Education

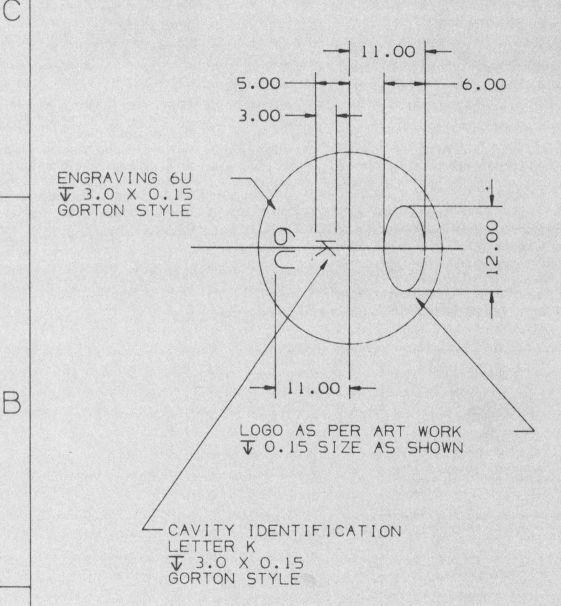

ENGRAVING 6U
∇ 3.0 X 0.15
GORTON STYLE

11.00

5.00   6.00

3.00

12.00

11.00

LOGO AS PER ART WORK
∇ 0.15 SIZE AS SHOWN

CAVITY IDENTIFICATION
LETTER K
∇ 3.0 X 0.15
GORTON STYLE

C

B

A

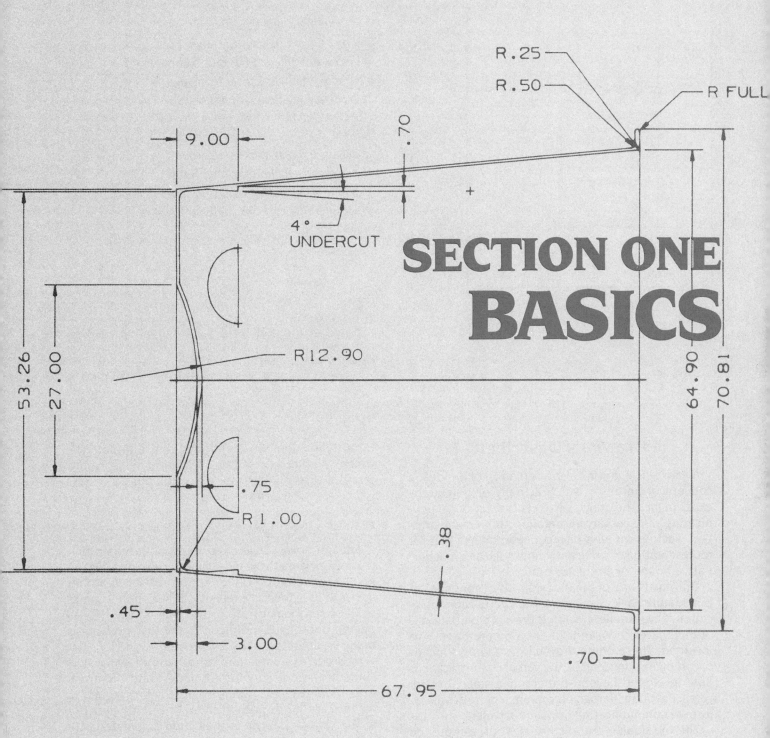

R.25

R.50

R FULL

9.00

.70

4°
UNDERCUT

+

SECTION ONE
BASICS

R12.90

53.26

27.00

64.90

70.81

.75

R1.00

.38

.45

3.00

.70

67.95

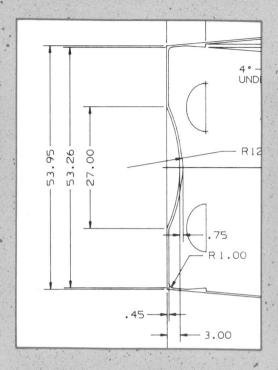

This introduction to Technical Drawing and Design presents the concept of technical drawing and traces its evolution from primitive manual techniques to modern computer-aided drafting (CAD) techniques. Major topics covered include: drawings described; types of drawings; types of technical drawings, their purpose, applications, and regulation; and a checklist of what students of technical drawing and drafting should learn.

Introduction

Drawings Described

A *drawing* is a graphic representation of an idea, a concept or an entity which actually or potentially exists in life. The drawing itself is 1) a way of communicating all necessary information about an abstraction, such as an idea or a concept; or 2) a graphic representation of some real entity, such as a machine part, a house, or a tool, for example.

Drawing is one of the oldest forms of communication, dating back even farther than verbal communication. Cave dwellers painted drawings on the walls of their caves thousands of years before paper was invented. These crude drawings served as a means of communicating long before verbal communications had developed beyond the grunting stage. In later years, Egyptian hieroglyphics were a more advanced form of communicating through drawings.

The old adage "one picture is worth a thousand words" is still the basis of the need for technical drawings.

Types of Drawings

There are two basic types of drawings: artistic and technical. Some experts believe there are actually three types: the two mentioned and another type which combines these two. The third type is usually referred to as an illustration or rendering.

Artistic Drawings

Artistic drawings range in scope from the most simple line drawings to the most famous paintings. Regardless of their complexity or status, artistic drawings are used to express the feelings, beliefs, philosophies or abstract ideas of the artist. This is why the lay person often finds it difficult to understand what is being communicated by a work of art.

In order to understand an artistic drawing, it is sometimes necessary to first understand the artist. Artists often take a subtle or abstract approach in communicating through their drawings. This gives rise to the various interpretations often associated with artistic drawings.

Technical Drawings

The technical drawing, on the other hand, is not subtle or abstract. It does not require an understanding of its creator; only an understanding of technical

Figure I-1 Technical
drawing (mechanical)

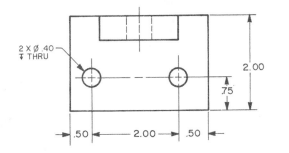

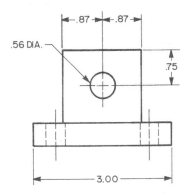

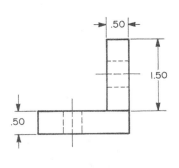

MATERIAL: STAINLESS STEEL

MANUFACTURERS NOTE:

FINISH ALL OVER

drawings. A *technical drawing* is a means of clearly and concisely communicating all of the information necessary to transform an idea or a concept into reality. Therefore, a technical drawing often contains more than just a graphic representation of its subject. It also contains dimensions, notes, and specifications.

The mark of a good technical drawing is that it contains all of the information needed by individuals for converting the idea or concept into reality. The con-

version process may involve manufacturing, assembly, construction, or fabrication. Regardless of the process involved, a good technical drawing allows the conversion process to proceed without having to ask designers or drafters for additional information or clarification.

Figures I-1 and I-2 contain samples of technical mechanical drawings which are used as guides by the people involved in various phases of manufactur-

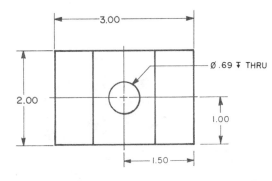

MATERIAL: STAINLESS STEEL

MANUFACTURERS   NOTE:

FINISH ALL OVER

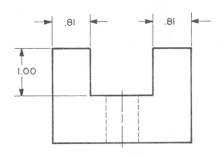

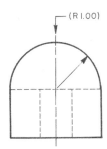

Figure I-2 Technical
drawing (mechanical)

Figure I-3 Rendering

Figure I-4 Rendering

Figure I-5 Mechanical illustration (*Courtesy Ken Elliott*)

ing the represented parts. Notice that the drawings contain a graphic representation of the part, dimensions, material specifications, and notes.

Illustrations or Renderings

Illustrations or renderings are sometimes referred to as a third type of drawing because they are not completely technical, neither are they completely artistic; they combine elements of both, as shown in Figures I-3, I-4, I-5, and I-6. They are technical in that they are drawn with mechanical instruments or on a computer-aided drafting system, and they contain some degree of technical information. However, they are also artistic in that they attempt to convey a mood, an attitude, a status or other abstract, nontechnical feelings.

Types of Technical Drawings

Technical drawings are based on the fundamental principles of projection. A *projection* is a drawing or representation of an entity on an imaginary plane or planes. This projection plane serves the same purpose in technical drawing as is served by the movie screen in a theater.

As can be seen in Figure I-7, a projection involves four components: 1) the actual object that the drawing or projection represents, 2) the eye of the viewer looking at the object, 3) the imaginary projection plane (the viewer's drawing paper or the graphics display in a computer-aided drafting system), and 4) imaginary lines of sight called *projectors*.

Two broad types of projection, both with several subclassifications, are parallel projection and perspective (converging) projection.

Parallel Projection

Parallel projection is subdivided into the following three categories: orthographic, oblique, and axonometric projections.

Orthographic projections are drawn as multiview drawings which show flat representations of principal views

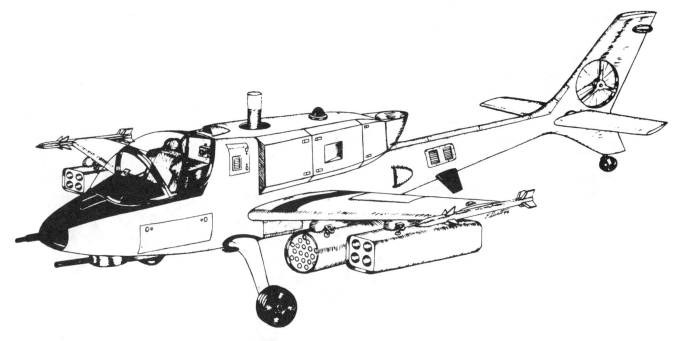

Figure I-6 Mechanical illustration (*Courtesy Ken Elliott*)

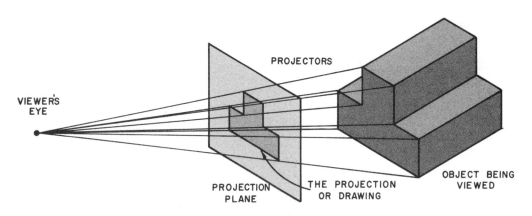

Figure I-7 The projection plane

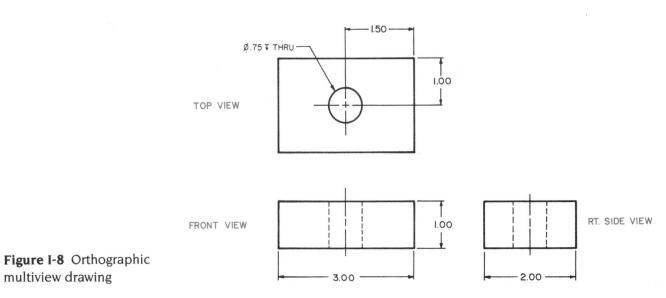

Figure I-8 Orthographic
multiview drawing

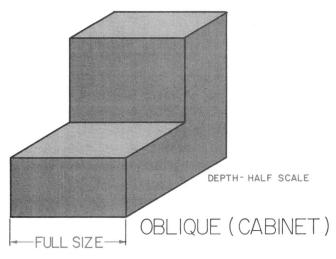

DEPTH- HALF SCALE

OBLIQUE (CABINET)

├── FULL SIZE ──┤

Figure I-9 Oblique projection (cabinet)

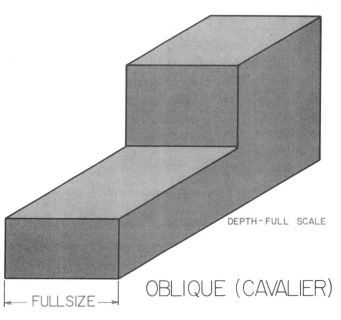

DEPTH-FULL SCALE

OBLIQUE (CAVALIER)

├── FULLSIZE ──┤

Figure I-10 Oblique projection (cavalier)

of the subject, Figure I-8. *Oblique projections* actually show the depth of the subject, and are of two varieties: *cabinet* (half scale) or *cavalier* (full scale) projections, Figures I-9 and I-10. *Axonometric projections* are three-dimensional drawings, and are of three different varieties: isometric, dimetric, and trimetric, Figures I-11, I-12, and I-13.

perspective projections: one-point, two-point, and three-point projections, Figures I-14, I-15, and I-16.

Perspective Projection

Perspective projections are drawings which attempt to replicate what the human eye actually sees when it views an object. That is why the projectors in a perspective drawing converge. There are three types of

Purpose of Technical Drawings

To appreciate the need for technical drawings, one must understand the design process. The design process is an orderly, systematic procedure used in accomplishing a needed design.

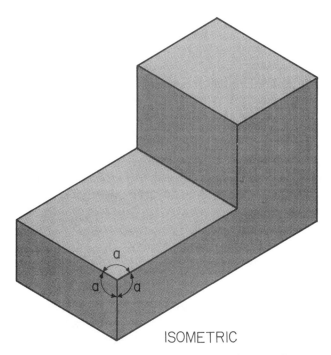

ISOMETRIC

Figure I-11 Axonometric projection (isometric)

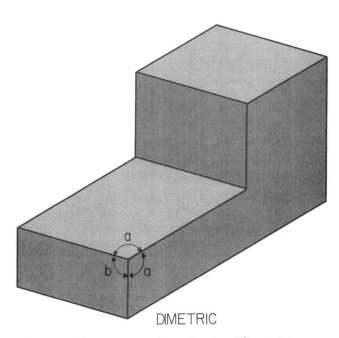

DIMETRIC

Figure I-12 Axonometric projection (dimetric)

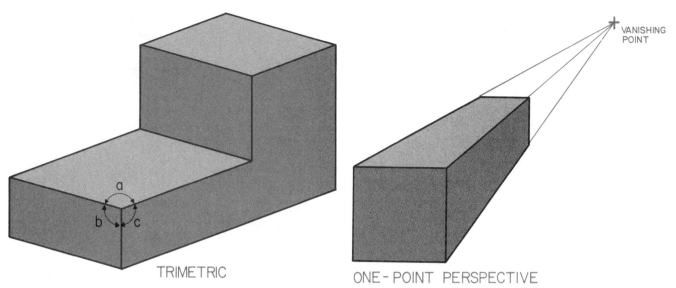

Figure I-13 Axonometric projection (trimetric)

TRIMETRIC

ONE-POINT PERSPECTIVE

Figure I-14 One-point perspective projection

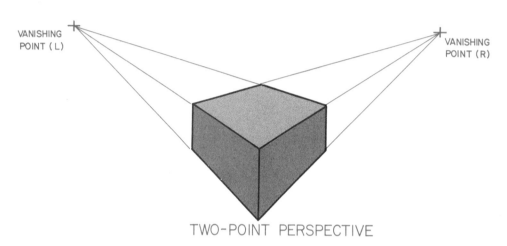

VANISHING POINT (L)

VANISHING POINT (R)

TWO-POINT PERSPECTIVE

Figure I-15 Two-point perspective projection

Any product that is to be manufactured, fabricated, assembled, constructed, built or subjected to any other type of conversion process must first be designed. For example, a house must be designed before it can be built. An automobile must be designed before it can be manufactured. A printed circuit board must be designed before it can be fabricated.

The Design Process

The *design process* is an organized, step-by-step procedure in which mathematical and scientific principles, coupled with experience, are brought to bear in order to solve a problem or meet a need. The design process has five steps. Traditionally, these steps have been 1) identification of the problem or a need, 2) development of initial ideas for solving the problem, 3) selection of a proposed solution, 4) devel-

opment and testing of models or prototypes, and 5) developing working drawings, Figure I-17.

The age of computers has altered the design process slightly for those companies which have converted to computer-aided design and drafting. For these companies, the expensive, time-consuming fourth step in the design process — the making and testing of actual models or prototypes — has been substantially altered, Figure I-18. This fourth step has been replaced with three-dimensional computer models which can be quickly and easily produced on a CAD system using the data base built-up during the first three phases of the design process, Figure I-19.

Whether in the traditional design process or the more modern computer version, in either case, working drawings are an integral part of the design process from start to finish.

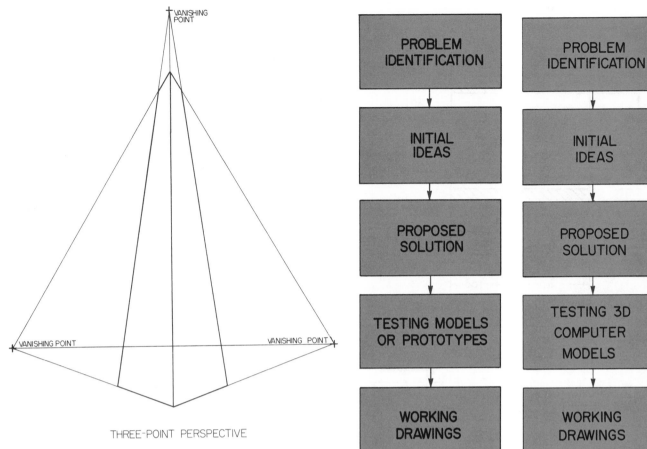

Figure I-16 Three-point perspective projection

Figure I-17 The design process (manual)

Figure I-18 The design process (CAD)

The purpose of technical drawings is to document the design process. Creating technical drawings to support the design process is called *drafting*. People who do drafting are known as *drafters* or *drafting technicians*. The words "draftsman" or "draughtsman" are no longer used.

In the first step of the design process, technical drawings are used to help clarify the problem or the need. The drawings may be old ones on file or they may be new ones created for the purpose of clarification. In the second step, technical drawings — often in the form of sketches or preliminary drawings — are used to document the various ideas and concepts formed. In the third step, technical drawings — again, usually preliminary drawings — are used to communicate the purposed solution.

If the traditional fourth step in the design process is being used, preliminary drawings and sketches from the first three steps will be used as guides in constructing models or prototypes for testing. If the more modern fourth step is being used, the data base built-up during documentation of the first three steps can be used in developing three-dimensional computer models. In both cases, the final step is the development of complete working drawings for guiding

individuals involved in the conversion process. Figure I-20 is a working drawing documenting the design of a simple mechanical part. The drawing was produced manually. Figure I-21 is a similar drawing produced on a CAD system.

Figure I-19 Three-dimensional computer model
(Courtesy Terak Corporation)

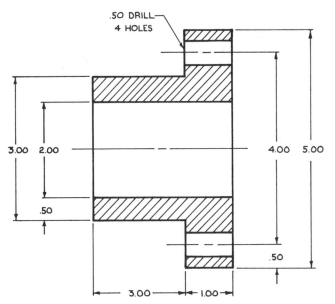

Figure I-20 Simple mechanical drawing (manual)

Applications of Technical Drawings

Technical drawings are used in many different applications. They are needed in any setting which involves design, and in any subsequent form of conversion process. The most common applications of technical drawings can be found in the fields of manufacturing, engineering, architecture and construction, and all of their various related fields.

Architects use technical drawings to document their designs of residential, commercial, and industrial buildings, Figures I-22 and I-23. Structural, electrical, and mechanical [heating, ventilating, air conditioning (HVAC) and plumbing] engineers who work with architects also use technical drawings to document those aspects of the design for which they are responsible, Figures I-24, I-25, and I-26.

Surveyors and civil engineers use technical drawings to document such work as the layout of a new

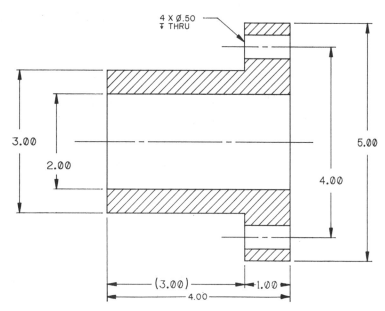

Figure I-21 Simple mechanical drawing (CAD)

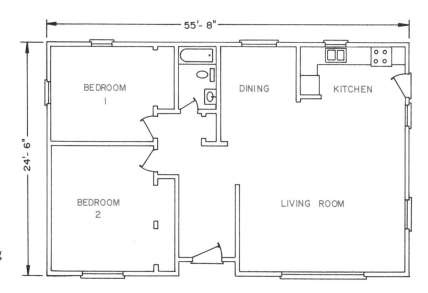

Figure I-22 Technical drawing (architectural)

Figure I-23 Technical drawing (architectural)

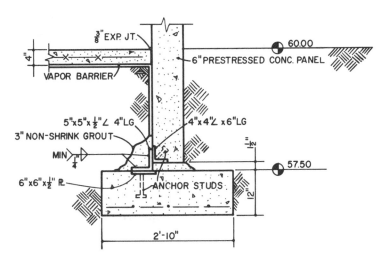

TYPICAL 6" WALL BASE CONN FTG DETAIL

Figure I-24 Technical drawing (structural)

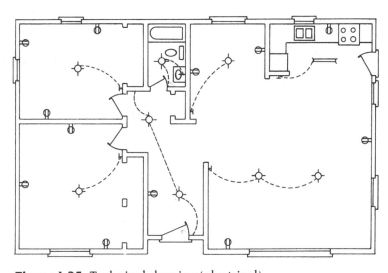

Figure I-25 Technical drawing (electrical)

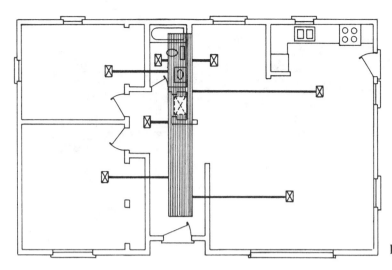

Figure I-26 Technical drawing (HVAC)

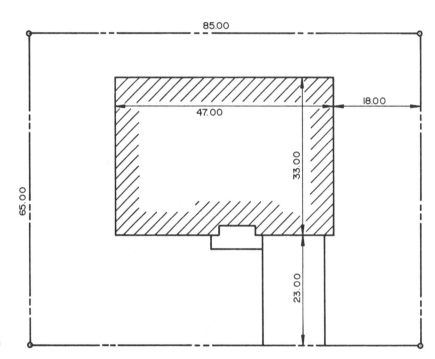

Figure I-27 Technical drawing (civil)

subdivision, or the marking-off of the boundaries for a piece of property, Figure I-27. Contractors and construction personnel use technical drawings as their blueprints in converting architectural and engineering designs into reality, Figures I-28 and I-29.

Technical drawings are equally important to engineers, designers, and various other individuals working in the manufacturing industry. Manufacturing engineers use technical drawings to document their designs. Technical drawings guide the collective efforts of individuals who are concerned with the same common goal, Figures I-30 and I-31, pages 15 and 16.

Regulation of Technical Drawings

Technical drawing practices must be regulated because of the diversity of their applications. Just as the English language must have certain standard rules of grammar, the graphic language must have certain rules of practice. This is the only way to ensure that all people attempting to communicate using the graphic language are speaking the same language.

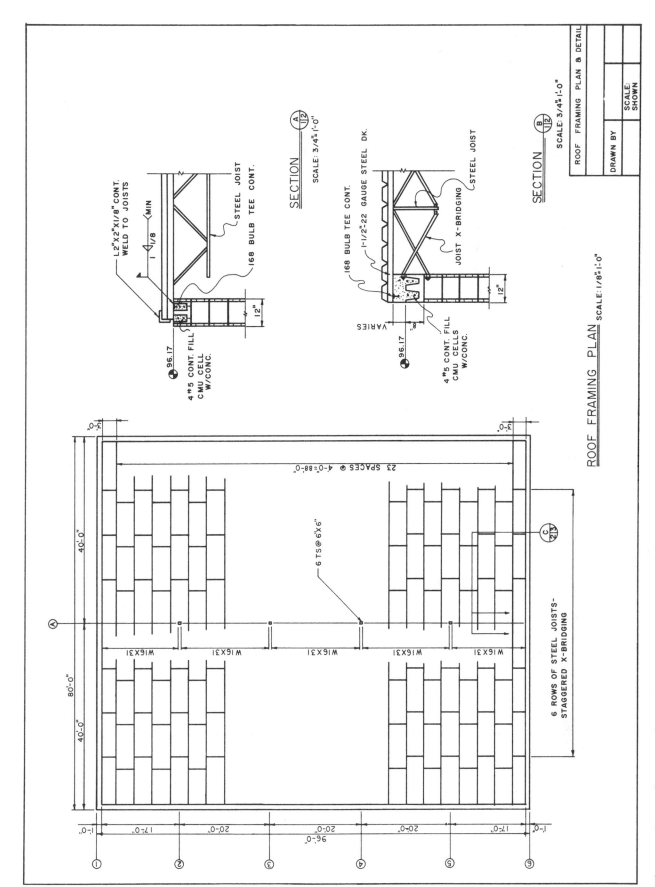

Figure I-28 Architectural/Engineering drawing

SECTION A
SCALE: 3/4"=1'-0"

L2"X2"X1/8" CONT.
WELD TO JOISTS

1 1/8 MIN

STEEL JOIST

168 BULB TEE CONT.

96.17

4 #5 CONT. FILL
CMU CELL
W/CONC.

12"

SECTION B
SCALE: 3/4"=1'-0"

168 BULB TEE CONT.

1-1/2"22 GAUGE STEEL DK.

STEEL JOIST

JOIST X-BRIDGING

96.17

VARIES

8"

4 #5 CONT. FILL
CMU CELLS
W/CONC.

12"

ROOF FRAMING PLAN SCALE:1/8"=1'-0"

ROOF FRAMING PLAN & DETAIL

SCALE: 3/4"=1'-0"

DRAWN BY

SCALE:
SHOWN

23 SPACES @ 4'-0"=88'-0"

40'-0"

80'-0"

40'-0"

3'-0"

3'-0"

6 TS@6'x6'

W16X31

W16X31

W16X31

W16X31

W16X31

A

C
2/3

6 ROWS OF STEEL JOISTS-
STAGGERED X-BRIDGING

1'-0" 17'-0" 20'-0" 20'-0" 20'-0" 17'-0" 1'-0"
96'-0"

1 2 3 4 5 6

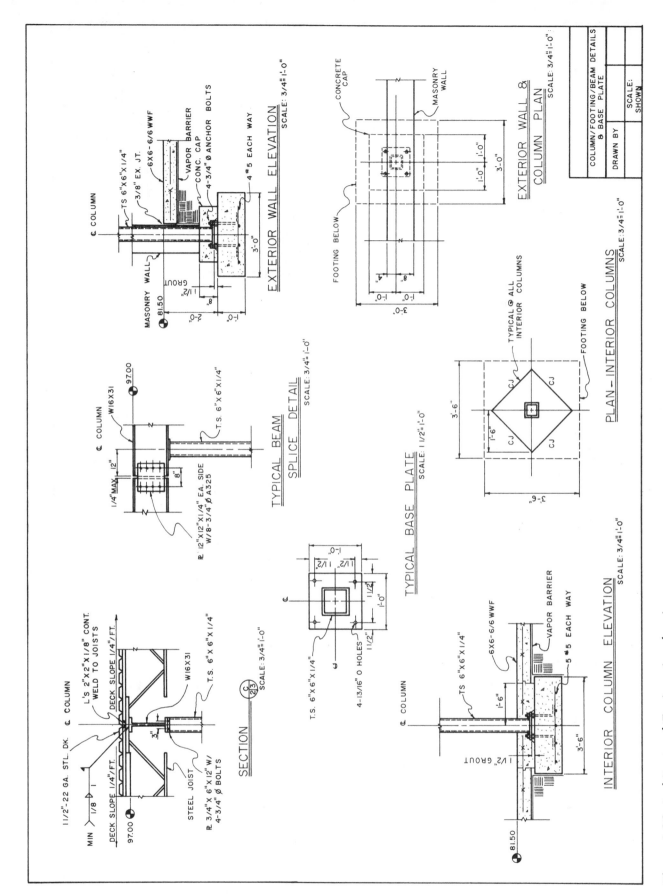

Figure I-29 Architectural/Engineering drawing

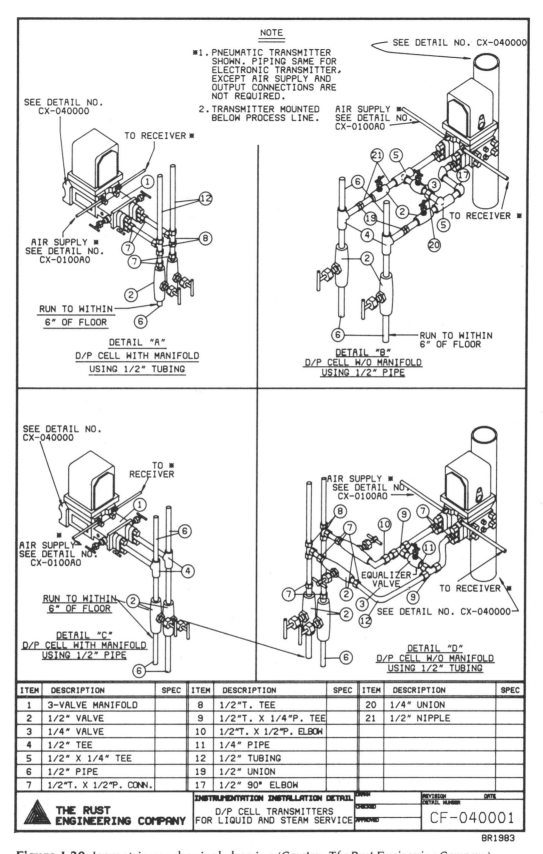

Figure I-30 Isometric mechanical drawing (*Courtesy The Rust Engineering Company*)

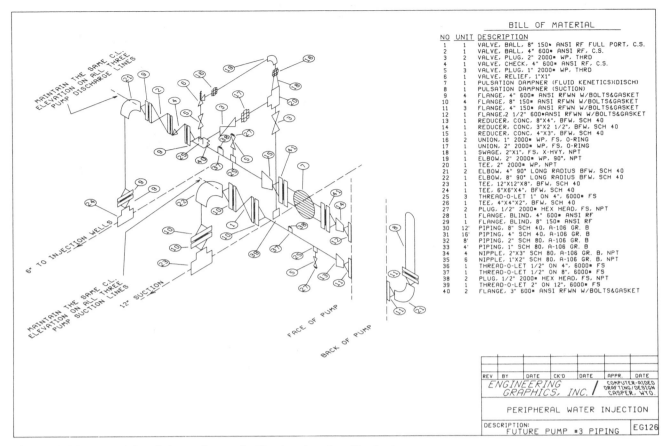

Figure I-31 Isometric piping schematic (*Engineering Graphics, Inc.*)

Standards of Practice

A number of different agencies have developed standards of practice for technical drawing. The most widely used standards of practice for technical drawing and drafting are those of the U.S. Department of Defense (DOD), the U.S. Military (MIL), and the American National Standards Institute (ANSI).

The American National Standards Institute does not limit its activities to the standardization of technical drawing and drafting practices. In fact, this is just one of the many fields for which ANSI maintains a continuously updated set of standards.

Standards of interest to drafters, designers, checkers, engineers, and architects are contained in the "Y" series of ANSI standards. Figure I-32 contains a list of ANSI standards frequently used in technical drawing and drafting specifications.

What Students of Technical Drawing and Drafting Should Learn

Many people in the world of work use technical drawings in various forms. Engineers, designers, checkers, drafters, and a long list of related occupations use technical drawings as an integral part of their jobs. Some of these people must be able to actually make drawings; others are only required to be able to read and interpret drawings; some must be able to do both.

What students of technical drawing and drafting should learn depends on how they will use technical drawings in their jobs. Will they make them? Will they read and interpret them? This textbook is written for

SIZE AND FORMAT_____Y14.1

LINE CONVENTIONS AND LETTERING__Y14.2

PROJECTIONS_____Y14.3

PICTORIAL DRAWING_____Y14.4

DIMENSIONING AND TOLERANCING_____Y14.5M

SCREW THREADS_____Y14.6

GEARS, SPLINES AND SERRATIONS____Y14.7

GEAR DRAWING STANDARDS_____Y14.7.1

MECHANICAL ASSEMBLIES_____Y14.14

Figure I-32 Sample list of drafting standards

LEARNING CHECKLIST
FOR
STUDENTS OF TECHNICAL DRAWING

FUNDAMENTAL KNOWLEDGE AND SKILLS	RELATED KNOWLEDGE	ADVANCED KNOWLEDGE AND SKILLS
DRAFTING EQUIPMENT	MATH	DEVELOPMENT
FUNDAMENTAL DRAFTING TECHNIQUES	WELDING	GEOMETRIC DIMENSIONING AND TOLERANCING
SKETCHING	SHOP PROCESSES	THREADS AND FASTENERS
GEOMETRIC CONSTRUCTION	MEDIA AND REPRODUCTION	SPRINGS
MULTIVIEW DRAWING		CAMS
SECTION VIEWS		GEARS
GRAPHICAL DESCRIPTIVE GEOMETRY		MACHINE DESIGN DRAWING
AUXILIARY VIEWS		PICTORIAL DRAFTING
GENERAL DIMENSIONING		DRAFTING SHORT-CUTS
NOTATION		CAD TECHNOLOGY
		CAD OPERATION

Figure I-33 Checklist for students of technical drawing

Figure I-34 Modern CAD technology (*Courtesy Auto-Trol Technology Corporation*)

students in the fields of engineering, design, drafting, and architecture, among others, who must be able to make, read, and interpret technical drawings. These students should develop a wide range of knowledge and skills, Figure I-33.

The learning required of technical drawing students can be divided into three categories: fundamental knowledge and skills, related knowledge, and advanced knowledge and skills.

In the "fundamentals" category, students of technical drawing and drafting should develop knowledge and skills in the areas of drafting equipment; such fundamental drafting techniques as line work, lettering, scale use, and sketching; geometric construction; multiview drawing; sectional views; descriptive geometry; auxiliary views; general dimensioning; and notation.

In the "related knowledge" category, students of technical drawing and drafting should develop a broad knowledge base in the areas of related math, welding, shop processes, and media and reproduction.

In the "advanced" category, students of technical drawing and drafting should develop knowledge and skills in the areas of development, geometric dimensioning and tolerancing, threads and fasteners, springs, cams, gears, machine design drawing, pictorial drafting, drafting shortcuts, and CAD/CAM technology and operations. The latter area represents a significant change in techniques used to create, maintain, update, and store technical drawings, Figures I-34 and I-35.

Figures I-36 through I-40 contain examples of several different kinds of technical drawings taken from the "real world" of drafting.

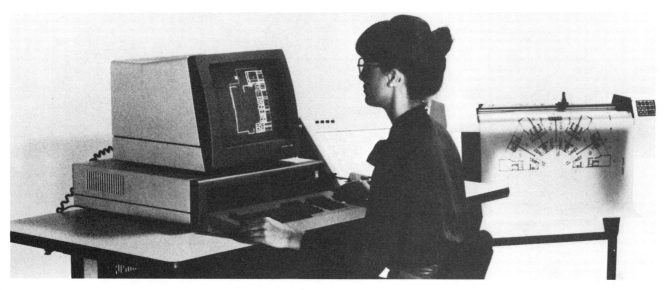

Figure I-35 Modern CAD system (*Courtesy Bausch & Lomb, Inc.*)

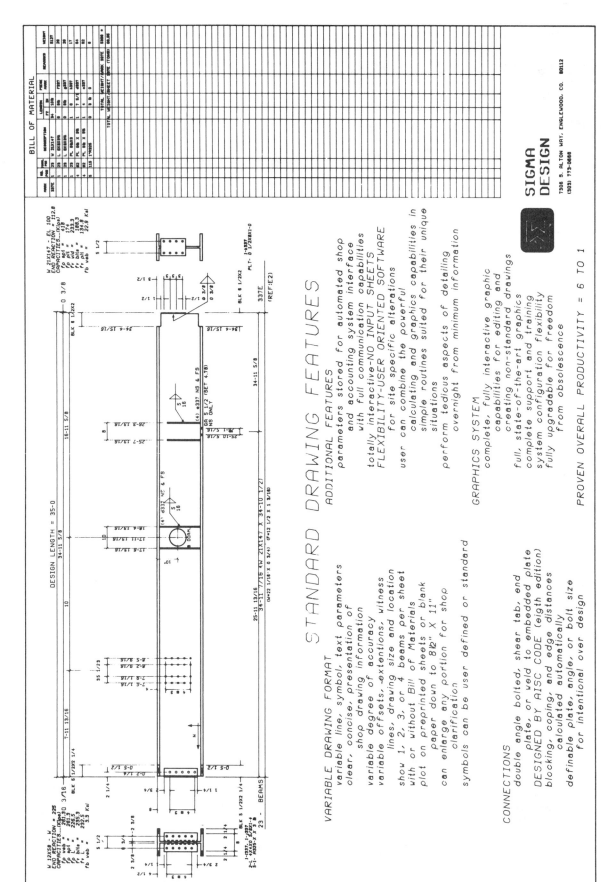

Figure I-36 Structural steel drawing *(Courtesy Sigma Design)*

STANDARD DRAWING FEATURES

VARIABLE DRAWING FORMAT
variable line, symbol, text parameters
clear, concise, presentation of
shop drawing information
variable degree of accuracy
variable offsets,-extentions, witness
lines, drawing size and location
show 1, 2, 3, or 4 beams per sheet
with or without Bill of Materials
plot on preprinted sheets or blank
paper down to 8½" X 11"
can enlarge any portion for shop
clarification
symbols can be user defined or standard

CONNECTIONS
double angle bolted, shear tab, end
plate, or weld to embedded plate
DESIGNED BY AISC CODE (eigth edition)
blocking, coping, and edge distances
calculated automatically
definable plate, angle, or bolt size
for intentional over design

ADDITIONAL FEATURES
parameters stored for automated shop
and accounting system interface
with full communication capabilities
totally interactive-NO INPUT SHEETS
FLEXIBILITY-USER ORIENTED SOFTWARE
user can combine the powerful
calculating and graphics capabilities in
simple routines suited for their unique
situations
perform tedious aspects of detailing
overnight from minimum information

GRAPHICS SYSTEM
complete, fully interactive graphic
capabilities for editing and
creating non-standard drawings
full, state-of-the-art graphics
complete support and training
system configuration flexibility
fully upgradable for freedom
from obsolescence

PROVEN OVERALL PRODUCTIVITY = 6 TO 1

SIGMA DESIGN
7306 S. ALTON WAY, ENGLEWOOD, CO. 80112
(303) 773-0666

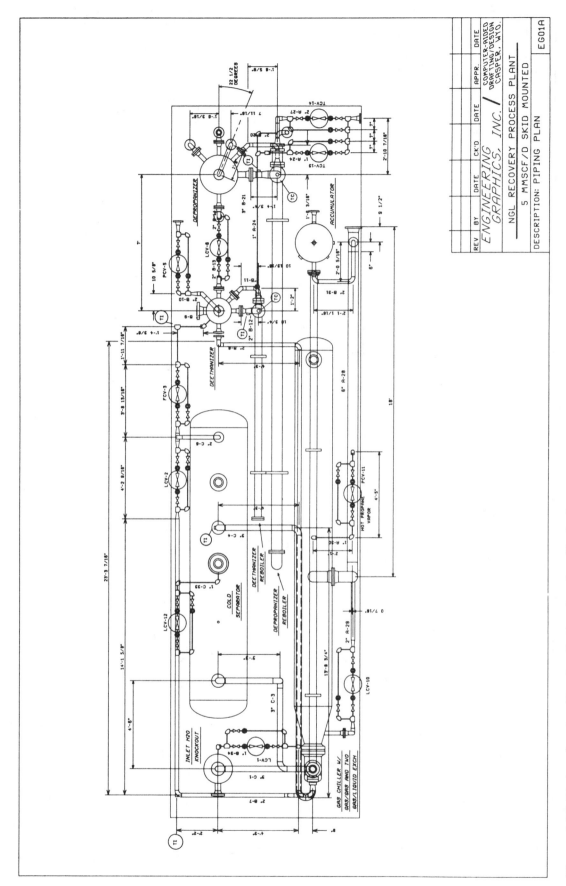

Figure I-37 Piping drawing (*Courtesy Engineering Graphics, Inc.*)

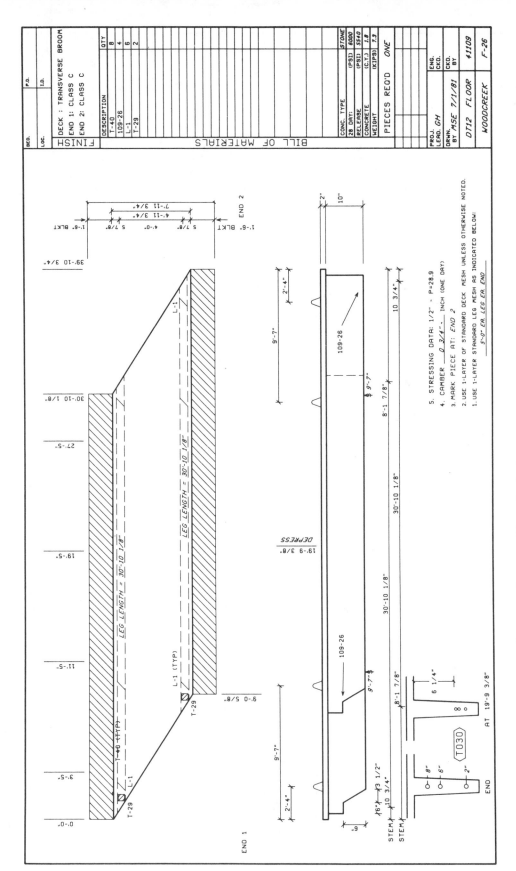

Figure I-38 Prestressed concrete drawing (*Courtesy Sigma Design*)

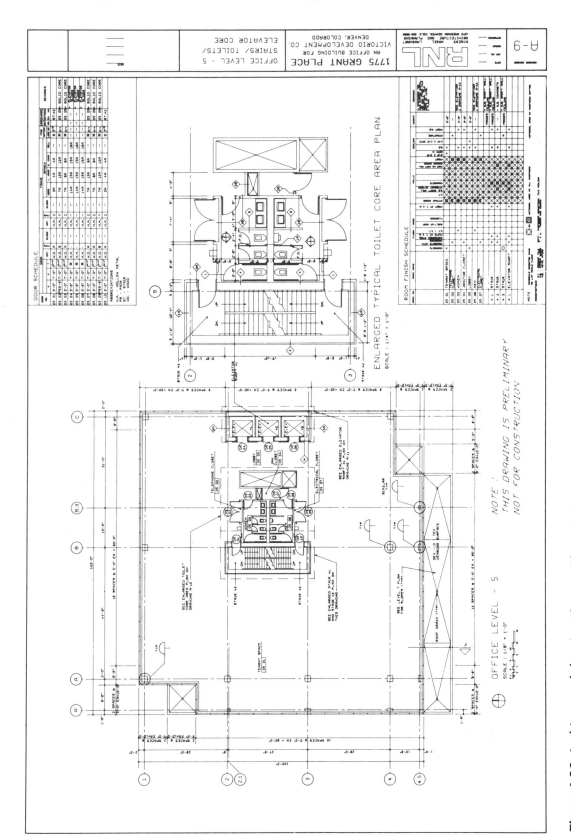

Figure I-39 Architectural drawing (*Courtesy Sigma Design*)

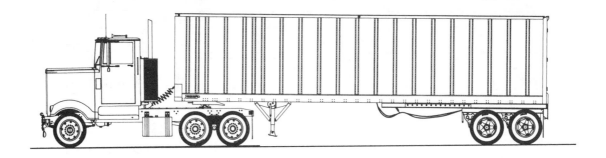

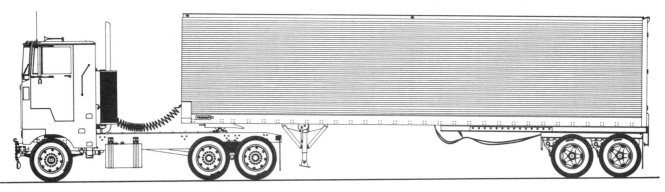

Figure I-40 Mechanical pictorial drawing (*Courtesy Fruehauf Corporation*)

Review

1. What is a drawing?

2. What old adage explains the basis of the need for technical drawings?

3. What are the two basic types of drawings?

4. Explain the major difference between the two basic types of drawings.

5. What are the four components of a projection?

6. Name the three subdivisions of parallel projection.

7. Name the three types of axonometric projection.

8. Name the three types of perspective projection.

9. What are the five steps in the design process?

10. Name four fields in which technical drawings are used extensively.

11. Explain how technical drawings differ from artistic drawings.

12. Name three organizations that regulate technical drawing practices.

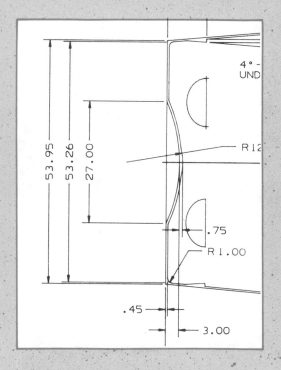

This chapter discusses in detail the use of conventional drafting instruments and equipment, including their proper care; touches on CAD/CAM equipment, which is described more fully in Section 3; line-making methods; drafting media; copying equipment; measuring devices; scales; and many other basic drafting requirements and techniques.

CHAPTER ONE
Drafting Instruments and Their Use

Conventional and CAD/CAM Drafting Equipment

In drafting, no lines are made freehand. Each and every line is drawn using some kind of a drafting tool. It is up to the drafter to own a complete set of standard drafting tools in order to be fully functional.

When purchasing conventional drafting equipment, care must be taken to obtain quality equipment from a reliable dealer. It is advisable to consult with an experienced drafter, a drafting instructor, or a reputable dealer. The following is a list of the minimum required drafting equipment. Special templates and special equipment must be added to this list, depending upon the field of drafting and, in some cases, the actual product manufactured by the company. Each of the following pieces of equipment is illustrated in this chapter.

- Drawing board — 24" x 36" (60 cm x 90 cm) minimum size
- T-square (parallel straightedge or drafting machine) to suit board
- 45° triangle — 8" (20 cm) size
- 30°-60° triangle — 8" (20 cm) size
- Triangular scale (depending upon the field of drafting)
- Center wheel bow compass — 6" (15 cm) with extension bar)
- Drop bow compass (recommended)
- Irregular curves (two or three different configurations)
- Dividers — 6" (15 cm)
- Drafting brush
- Mechanical drafting pencils with lead
- Protractor or adjustable triangle
- Erasing shield
- Eraser
- Lead pointer with steel cutting wheel (pencil)
- Lead sandpaper or flat file (for compass lead)
- Circle template
- Ellipse template
- Drafting tape or drafting dots
- Calculator
- Dry cleaning pad (optional)

The following is a list of the required inking supplies.

- Technical pen #2 1/2 (for lettering scriber)
- Technical pen #1 (for lettering scriber)
- Technical pen #0 (for lettering scriber)

Figure 1-1 A computer-aided design system (*Courtesy Bausch & Lomb, Inc.*)

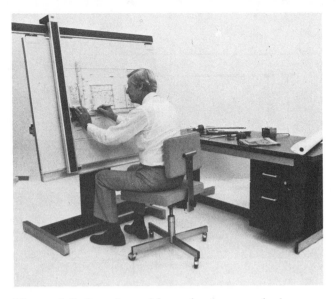

Figure 1-2 Drawing table and reference desk (*Courtesy Keuffel & Esser Co.*)

- India ink — black
- Pen cleaning solvent
- Lettering scriber
- Lettering template #100
- Lettering template #175
- Lettering template #290
- Compass adaptor for pen

Note that some specialized or expensive equipment is often furnished by the company.

CAD/CAM equipment is another drafting tool and is covered more fully in Chapters 10 and 11 in Section 3 of this text. For a very basic description, refer to Figure 1-1. The photo shows the following equipment, from left to right:

Printer — The printer is an output device used for producing printouts of alphanumeric data.

Workstation — (Text display, trackball, auxiliary keyboard, push-button menu board).

  a. *The text display*
    An output device for displaying user prompts which tell the user what to do and how to do it.

  b. *The trackball*
    A baseball-shaped input device used for creating horizontal, vertical, and diagonal lines, and for positioning the cursor.

  c. *Auxiliary keyboard*
    An input device for entering commands, text, annotation, and dimensions.

  d. *Push-button menu board*
    An input device for entering commands from a menu, and calling up stored data.

Graphics display — An ouput device for displaying drawings and other graphic data as they are being worked on.

Figure 1-3 Small drawing table (*Courtesy Stacor Corp.*)

Digitizing table with puck — An input device used for entering graphic data into the system. Graphic data is converted to digital data (digitized) and entered into the system as X-Y coordinates.

Conventional Drafting Requisites

Drawing Table

Drawing tables are available in a variety of styles. Most are adjustable up and down, and can tilt to almost any angle from vertical 90° to horizontal, Figures 1-2 and 1-3.

Drawing Surface

The drawing surface, whether it is a drawing table-top or drawing board, must be flat, smooth and large enough to accommodate the drawing and some drafting equipment. If a T-square is used on a drawing board, at least one edge of the board must be absolutely true. Most quality drafting boards have a metal edge to ensure against warpage and against which to hold the T-square securely.

Standard drafting boards range in size from small, 12" x 17" (30 cm x 43 cm), to large, 31" x 42" (78 cm x 105 cm). Standard drafting tabletops range in size from 31" x 42" (78 cm x 105 cm) to 37½" x 60" (94 cm x 150 cm).

Lighting It is important that the drawing surface be fully lighted without any shadows, Figure 1-4.

Top Cover The drafting board should have a top cover which protects the board surface and provides a perfect drawing surface A good top cover actually seals over holes made by compass points, and it is easily cleaned, Figure 1-5.

Efficiency

To be fully efficient at drafting, all equipment must be clean, correctly adjusted and/or sharpened, and stored in a convenient location, ready for use at all times. It is good drafting practice to store each piece of equipment in a specific location and return it to its location after use. An organizer, such as the one shown in Figure 1-6, aids in keeping all equipment in its place.

Left-handed Drafters

Most drafting equipment is designed for the right-handed drafter, although left-handed types of drafting machines can be purchased. The T-square is simply placed on the right side of the drafting board by left-handed drafters, but everything else is right handed. The left-handed drafter has to adapt. The lettering scriber is especially difficult to manipulate.

T-Square

While T-squares are not used today in industry, they do provide a parallel straightedge for the beginning drafter. The T-square is used to draw horizontal lines. Draw lines only against the upper edge of the blade. Make sure the head is held securely against the left

Figure 1-4 Fluorescent lamp for drawing surface
(*Courtesy Waldmann Lighting Co.*)

Figure 1-5 Drafting board top cover (*Courtesy Modern School Supplies, Inc.*)

Figure 1-6 Equipment organizer (*Courtesy Stacor Corp.*)

Figure 1-7 T-square in use (*Courtesy Teledyne* Post)

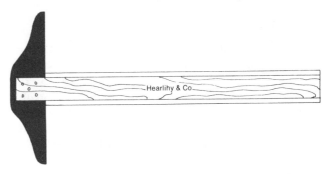

Figure 1-8 Parts of the T-square (*Courtesy Hearlihy and Co.*)

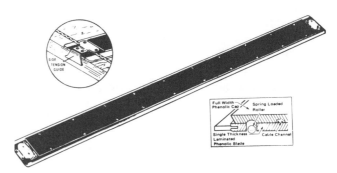

Figure 1-9 Parallel straightedge (*Courtesy Modern School Supplies, Inc.*)

edge of the drawing board to guarantee parallel lines, Figure 1-7.

The T-square is composed of two parts: the head and the blade, Figure 1-8. The two parts are fastened together at an exact right angle. The blade must be straight and free of any nicks or imperfections. A transparent acrylic edge is recommended since this allows the drafter to see the drawing underneath the edge. T-squares can be purchased with adjustable heads for drawing specific angles.

Parallel Straightedge

The parallel straightedge is always parallel, regardless of where it is placed upon the drawing surface. Parallel control is accomplished by a system of cords and pulleys, Figure 1-9. The parallel straightedge replaced the T-square in industry, and is still used somewhat today. Most straightedges come with a transparent acrylic edge, and some have rollers for a smooth gliding action. Some have a locking brake that permits the straightedge to be locked in any position.

Drafting Machines

A drafting machine is a device that attaches to the drafting table and replaces both the T-square and the parallel straightedge. There are two basic kinds of drafting machines. One is the arm type, Figure 1-10, and the other, the newer, is the track type, Figure 1-11. On both types, a round head holds two straightedges at right angles to each other. The head can be rotated to set the straightedges at any angle. Most machines are available with interchangeable straightedges marked with different scales along their edges.

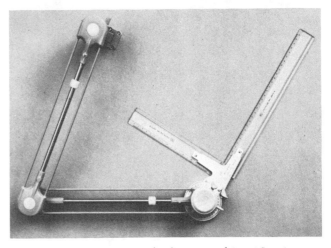

Figure 1-10 Arm-type drafting machine (*Courtesy Teledyne Post*)

Figure 1-11 Track-type drafting machine (*Courtesy Vemco, Inc.*)

Most drafting machines have a protractor and a vernier which permit readings to 5 minutes of an arc, Figure 1-12. Notice that zero on the protractor is in line with the zero on the vernier. The venier is graduated in 5-minute increments from zero to 60 minutes. To read the vernier, first read the protractor, Figure 1-13. In this example, the zero on the vernier points between 18° and 19°. On the vernier, notice that the only line that lines up with a line on the protractor is the 45; thus, this is read as 18°-45'. Some drafting machine heads simplify this process by adding a digital readout, see Figure 1-14.

Drafting machine straightedges come in sizes of 9" (23 cm), 12" (30 cm) and 18" (45 cm), graduated or ungraduated, in both transparent plastic and aluminum scale, Figure 1-15.

A drafting machine, although a precision instrument, should be checked for accuracy at least once a week. The instructions for checking and adjusting a drafting machine are included with the manufacturer's information.

Drafting machines replace straightedges, scales, triangles, and protractors. They increase accuracy and greatly reduce drafting time. A drafting machine is one of the few tools that can be purchased either as a right-handed or a left-handed instrument.

Drawing Sets

Typical drawing sets include compasses, dividers, and a ruling pen, Figure 1-16. Many sets include a variety of tools not normally used by the drafter. It is recommended that only those tools actually needed be purchased.

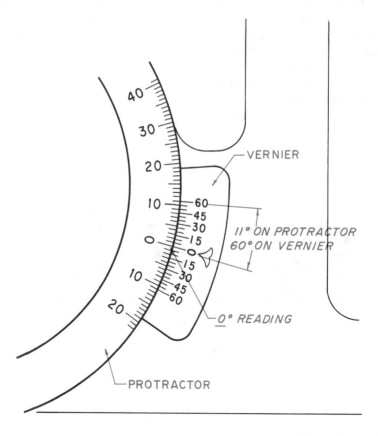

Figure 1-12 Protractor with vernier

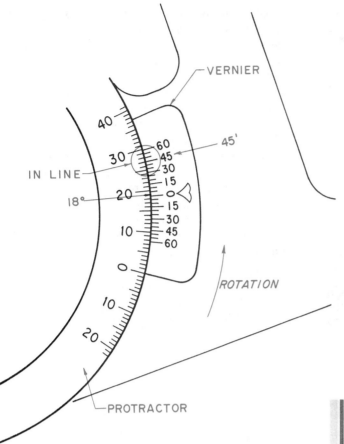

- VERNIER
- IN LINE
- 18°
- 45'
- ROTATION
- PROTRACTOR

Figure 1-13 Reading the vernier

Compasses

There are two main types of compasses: the friction-joint type, and the spring-bow type. The friction-joint type is still widely used for lightly laying out pencil drawings which will be inked. The disadvantage of this type of compass is that the setting may slip when strong pressure is applied to the lead.

The spring-bow type of compass, Figure 1-17, is best for pencil drawings and tracings as it retains its

Figure 1-14 Drafting machine head with digital readout (*Courtesy Consul & Mutoh, Ltd.*)

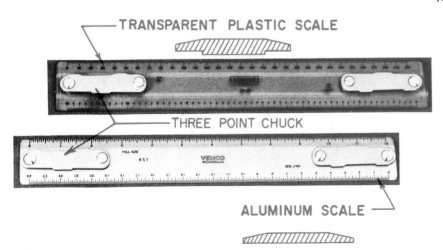

TRANSPARENT PLASTIC SCALE

THREE POINT CHUCK

ALUMINUM SCALE

Figure 1-15 Transparent and metal drafting machine straightedges with scale (*Courtesy Vemco, Inc.*)

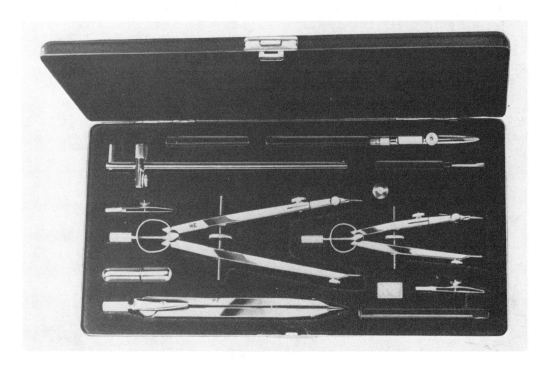

Figure 1-16 Complete drafting set (*Courtesy Keuffel & Esser Co.*)

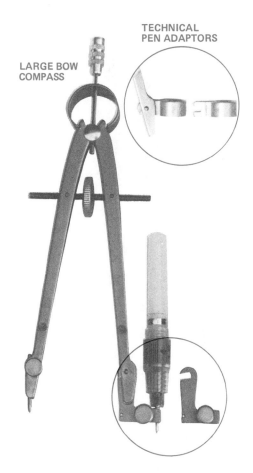

LARGE BOW
COMPASS

TECHNICAL
PEN ADAPTORS

Figure 1-17 Spring bow compass with pen adaptor (*Courtesy Vemco, Inc.*)

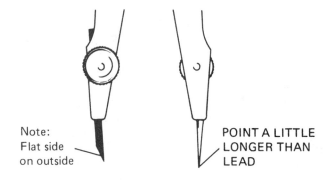

Note:
Flat side
on outside

POINT A LITTLE
LONGER THAN
LEAD

Figure 1-18 Bow compass lead and metal points

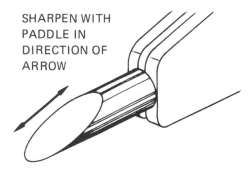

SHARPEN WITH
PADDLE IN
DIRECTION OF
ARROW

Figure 1-19 Bow compass lead point shape (*Courtesy Drafting for Trades and Industry, Basic Skills, Nelson, Delmar Publishers Inc.*)

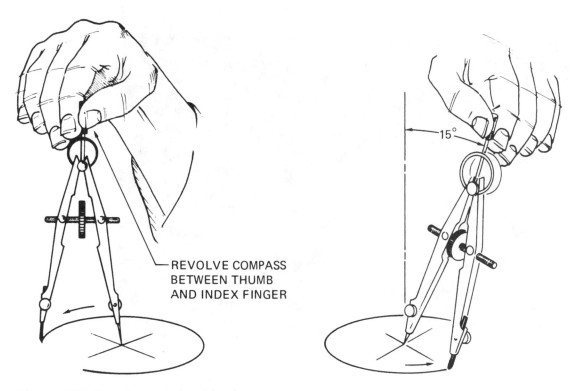

Figure 1-20 Drawing a circle with a bow compass

setting even when strong pressure is applied to obtain dark lines. The spring, located at the top of the compass, holds the legs securely against the adjusting screw. The adjusting screw is used to make fine adjustments.

Compass leads should extend approximately 3/8" (0.9 cm). The metal point of the compass extends slightly more than the lead to compensate for the distance the point penetrates the paper, Figure 1-18. The lead is sharpened with a sandpaper paddle to produce clean, sharp lines. The flat side of the lead faces *outward* in order to produce circles of very small diameter, Figure 1-19. Sharpen with paddle in direction of arrow, as shown, in order to keep the lead sharp longer.

To draw a circle with the bow compass, the compass is revolved between the thumb and the index finger, Figure 1-20. Pressure is applied downward on the metal point to prevent the compass from jumping out of the center hole.

Drop Bow Compass The drop bow compass, Figure 1-21, is used for circles of .03" (0.08 cm) to .50" (1.3 cm) diameter. The compass is adjusted to the required radius. The center point is located on the circle or arc swing point and held in place with the index finger. Rotate the knurled head of the compass between the thumb and second finger.

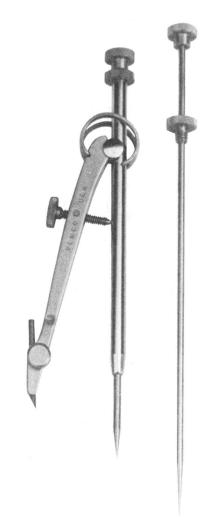

Figure 1-21 Drop bow compass (*Courtesy Vemco, Inc.*)

Figure 1-22 Bow compass with special lead clutch (*Courtesy B. Carter Lykins*)

Figure 1-23 Inking bow compass (*Courtesy Koh-I-Noor Rapidograph*)

Figure 1-24 Inking drop bow compass (*Courtesy Koh-I-Noor Rapidograph*)

Bow Compass with Lead Clutch In order to eliminate the process of sharpening compass leads, some drafters use a compass with a lead clutch of 0.5-mm lead, Figure 1-22. This compass saves time, and ends messy lead sharpening. Special compasses are designed only for inking, Figures 1-23 and 1-24. An adaptor to attach to a standard compass to draw ink lines is illustrated in Figure 1-25.

Beam Compass A beam compass, Figure 1-26, is used to draw large circles or arcs. Fine line adjustments can be obtained and locked in place. Beam compasses come in sizes from 13" (33 cm) bars and upwards.

Adjustable Curve An adjustable curve, Figure 1-27, has a locking knob, and is used to draw any radius from 6.75" to 200" (17 cm to 500 cm). This tool takes over where the ordinary compass leaves off, and eliminates the beam compass.

Divider

A divider is similar to a compass except that it has a metal point on each leg. It is used to lay off distances and to transfer measurements, Figure 1-28.

Proportional Dividers Proportional dividers are used to enlarge or reduce an object in scale. This tool has a sliding, adjustable pivot which varies the proportions of the tips of each leg, Figure 1-29.

Triangle

Two standard triangles are used by drafters. One is a 30-60-degree triangle, usually written as 30°-60° triangle. The other is a 45-degree triangle, written as 45° triangle. The 45° triangle consists of two 45-degree angles, and one 90-degree angle, Figure 1-30A. The 30°-60° triangle contains a 30-degree angle, a 60-degree angle, and a 90-degree angle, Figure 1-30B.

Figure 1-25 Standard compass ink
adaptor (*Courtesy
Koh-I-Noor Rapidograph*)

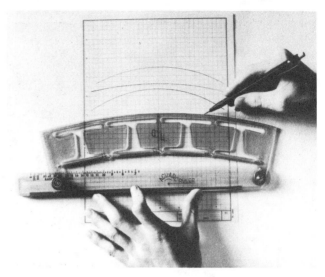

Figure 1-27 Adjustable curve (*Courtesy Hoyle Products,
Inc.*)

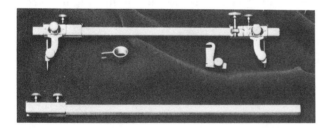

Figure 1-26 Beam compass (*Courtesy Vemco, Inc.*)

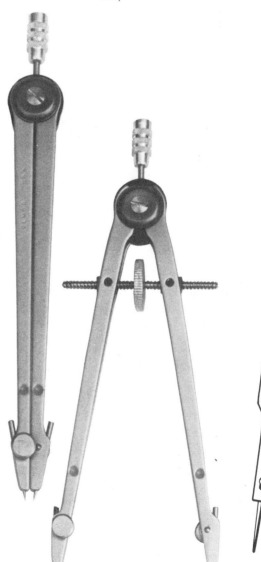

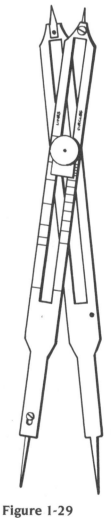

Figure 1-28 Divider
(*Courtesy Vemco, Inc.*)

Figure 1-29
Proportional dividers
(*Courtesy Modern
School Supplies, Inc.*)

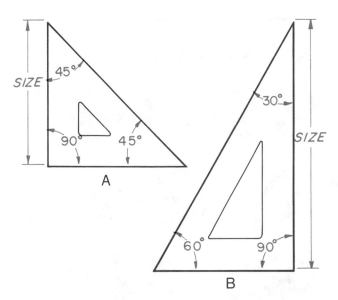

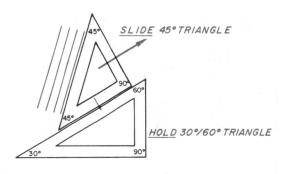

Figure 1-31 Drawing parallel angular lines

Figure 1-30 (A) 45° triangle, and (B) 30°-60° triangle

Triangles are made of plastic and come in a variety of sizes other than those mentioned. When laying out lines, triangles are placed firmly against the upper edge of the straightedge. Pencils are placed against the left edge of the triangle and lines drawn upwards, away from the straightedge. Parallel angular lines are made by moving the triangle to the right after each new line has been drawn, Figure 1-31.

Adjustable Triangle An adjustable triangle may take the place of both the 30°-60° and 45° triangles, Figure 1-32. It is recommended, however, that this tool be used only for drawing angles that cannot be made with the two standard triangles. The adjustable triangle is set by eye and thus is not as accurate as the solid triangle.

Template

A template is a thin, flat piece of plastic containing various cutout shapes, Figures 1-33, 1-34, and 1-35. It is designed to speed the work of the drafter and to make the finished drawing more accurate. Templates are available for drawing circles, ellipses, plumbing fixtures, bolts and nuts, screw threads, electronic symbols, springs, gears, and structural metals, to name just a few uses.

Templates come in many sizes to fit the scale being used on the drawing. A template should be used wherever possible to increase accuracy and speed. It is preferable to purchase templates that are stamped and not molded, as molded templates become brittle in time and break.

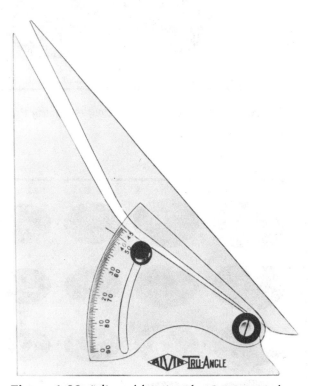

Figure 1-32 Adjustable triangle (*Courtesy Modern School Supplies, Inc.*)

French Curves

French curves are thin, plastic tools that come in an assortment of curved surfaces, Figure 1-36. They are used to produce curved lines that cannot be made with a compass. Such lines are referred to as *irregular curves*. Most good French curves are actually segments of such geometric curves as ellipses, parabolas, hyperbolas, and the like.

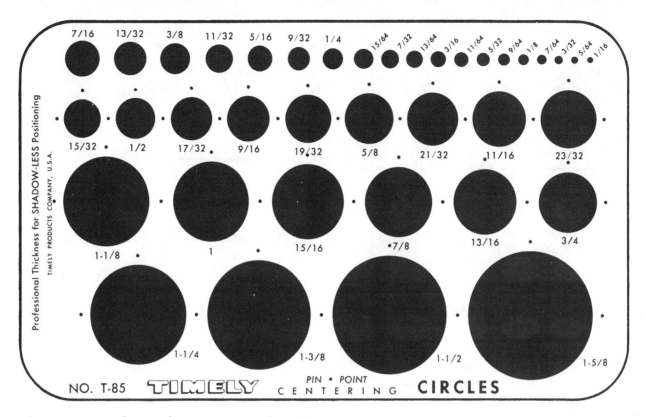

Figure 1-33 Circle template (*Courtesy Timely Products Co.*)

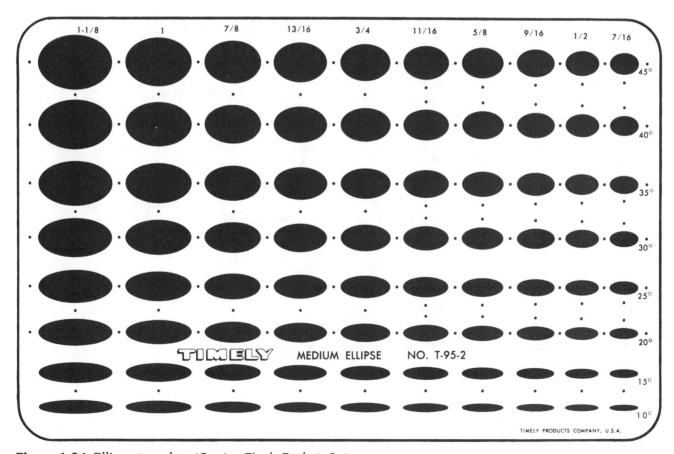

Figure 1-34 Ellipse template (*Courtesy Timely Products Co.*)

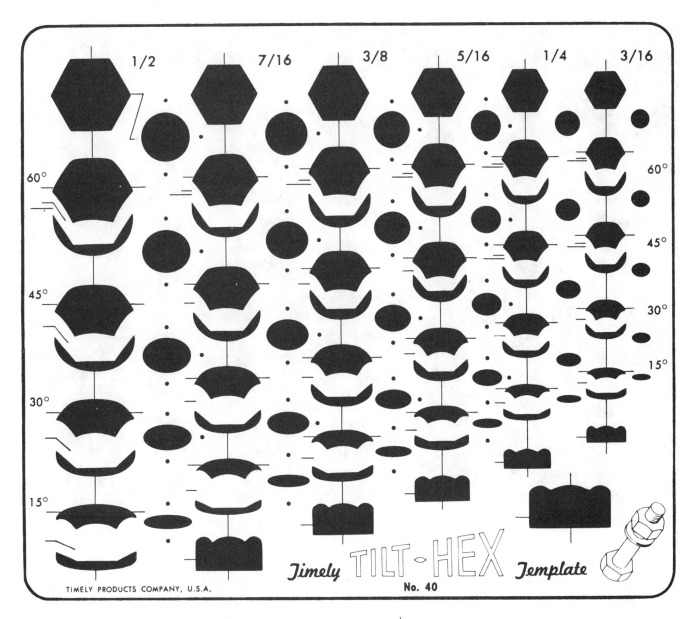

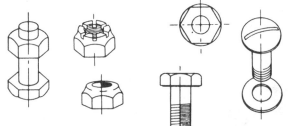

Figure 1-35 Bolt and nut template (*Courtesy Timely Products Co.*)

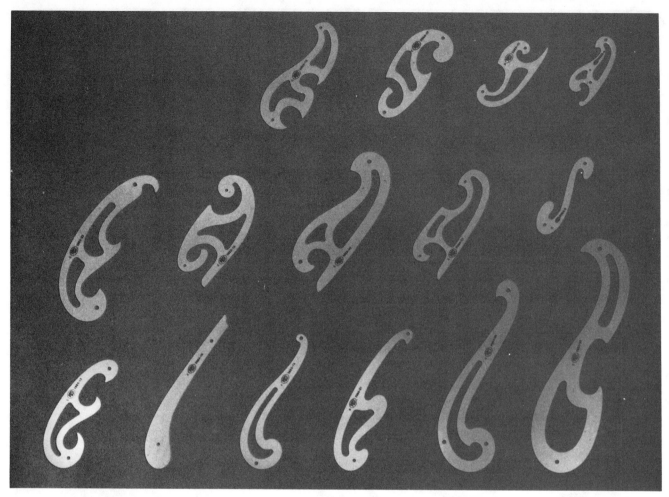

Figure 1-36 Assortment of French curves (*Courtesy Modern School Supplies, Inc.*)

Using a French Curve To use a French curve, the irregular curve must be defined by a series of dots. *Lightly* connect straight lines to get a general idea of where the curved line is going. If the line makes an abrupt turn, a line *lightly* sketched in place of the straight

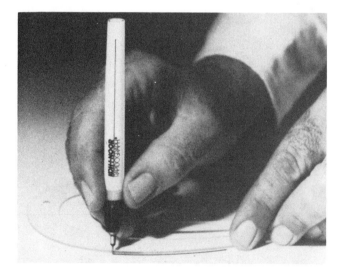

Figure 1-37 Drawing an irregular curve (*Courtesy Koh-I-Noor Rapidograph*)

lines may be more useful. Starting from one side or the other, line up the French curve along as many points as possible and draw a dark line connecting these points, Figure 1-37. Readjust and align the French curve along all additional points, and continue drawing the curved line. Proceed in this manner until the line is completed.

Adjustable Curve Adjustable curves form smooth curves. Figure 1-38 shows a flexible steel measuring tape that measures the perimeter of the curve to be drawn. The curve is held by friction between many layers of interlocking channels.

Protractor

A protractor is used to measure and lay out angles, Figure 1-39. It can be used in place of a drafting machine or an adjustable triangle.

To use the protractor, place the center point (located at the lower edge of the protractor) on the corner point of the angle. Align the base of the protractor along one side of the angle. The degrees are read along the semicircular edge.

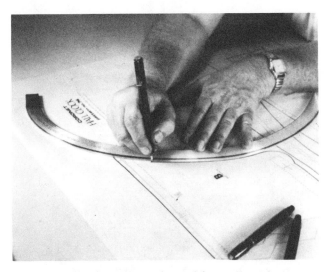

Figure 1-38 Using an adjustable curve (*Courtesy Hoyle Products, Inc.*)

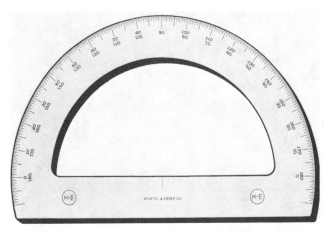

Figure 1-39 Protractor (*Courtesy Keuffel & Esser Co.*)

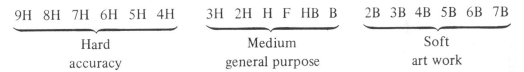

9H 8H 7H 6H 5H 4H	3H 2H H F HB B	2B 3B 4B 5B 6B 7B
Hard	Medium	Soft
accuracy	general purpose	art work

Figure 1-40 Grades of lead

Pencils and Leads

Lead for a mechanical drafting pencil comes in 18 degrees of hardness, ranging from 9H, which is very hard, to 7B, which is very soft, Figure 1-40. For drafting purposes, the scale of hardness is as follows: 4H lead is recommended for layout work, and 2H lead is recommended for all other lines. Experiment with various leads to determine which ones give the best line thickness. This varies depending upon the pressure applied to the point while drawing lines. Figure 1-41 shows leads for a mechanical drafting pencil.

Regular pencils are sharpened with a pencil sharpener. It is important that enough wood is removed to ensure that the lead, not the wood, of the pencil comes in contact with the straightedge or triangle edges.

Figure 1-41 Drafting lead (*Courtesy Staedtler Mars*)

Figure 1-42 Lead holder (*Courtesy Teledyne Post*)

Lead Holders and Leads

Lead holders hold sticks of lead, Figure 1-42. The leads designed for lead holders come in the same range of hardness as those for regular mechanical pencils, and are used for the same purposes. The main advantage is that they are more convenient to use. Leads are usually sharpened in a lead pointer. Electric lead pointers are fully automatic. A slight downward pressure of the lead starts the motor

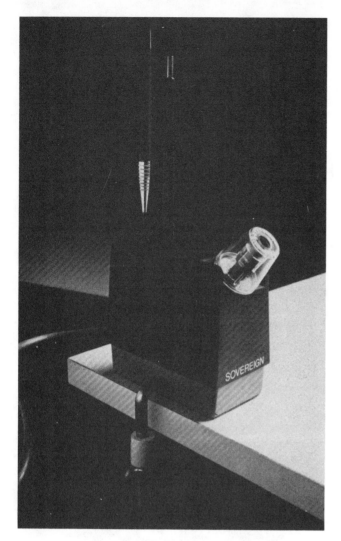

Figure 1-43 Lead pointer (*Courtesy Rotex Co.*)

Figure 1-44 Sandpaper paddle (*Courtesy Keuffel & Esser Co.*)

Figure 1-45 Block-type eraser (*Courtesy Staedtler Mars*)

Figure 1-46 Pencil-type eraser with clutch (*Courtesy Staedtler Mars*)

action, Figure 1-43. This machine produces a perfectly tapered point, and eliminates all loose clinging graphite.

Sandpaper Paddle A sandpaper paddle consists of several layers of sandpaper attached to a small wooden holder, Figure 1-44. The sandpaper is used to sharpen compass leads only. Do not sharpen leads over a drawing as the graphite will smear the drawing surface.

Erasers

Various kinds of erasers are available to a drafter. One of the most commonly used is a soft, white block-type eraser, Figure 1-45. Figure 1-46 shows a pencil-type eraser with an adjustable clutch. By developing good drawing habits, erasing can be kept to a minimum.

Electric Eraser An electric eraser speeds up corrections. Some models take a 7" (17.5 cm) long eraser strip.

The model illustrated in Figure 1-47 has a slip clutch to hold the eraser strip in place.

A cordless erasing machine can be used with or without the standard electric cord, and uses rechargeable NiCad batteries, Figure 1-48. As the eraser is placed in the stand, the batteries are recharged.

Electric erasers do save time, but care must be taken not to burn through the drawing paper. This can be avoided by using an erasing shield and placing a thick sheet of paper beneath the drawing to cushion it.

Erasing Shield An erasing shield restricts the erasing area so that correctly drawn lines will not be disturbed during the erasing procedure. It is made from a thin, flat piece of metal with variously sized cutouts, Figure 1-49. The shield is used by placing it over the line to be erased and erasing through the cutout.

Drafting Brush

The drafting brush is used to remove loose graphite and eraser crumbs from the drawing surface, Figure 1-50. Do not brush off a drawing surface by hand as this tends to smudge the drawing. Drafting brushes

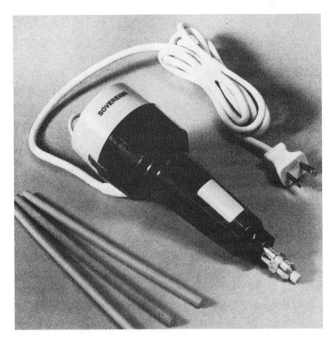

Figure 1-47 Electric eraser with slip clutch (*Courtesy Rotex Co.*)

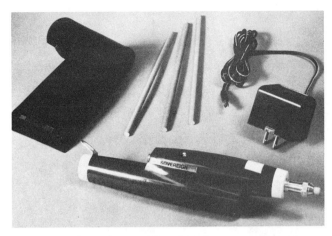

Figure 1-48 Rechargeable electric eraser (*Courtesy Rotex Co.*)

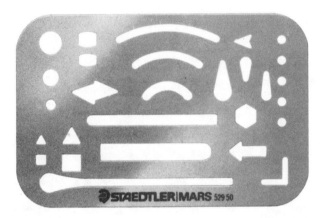

Figure 1-49 Erasing shield (*Courtesy Staedtler Mars*)

come in various sizes from 10½" to 14" (26 cm to 35 cm). The bristles can be either horsehair or nylon, and they can be cleaned with warm, soapy water.

Dry Cleaning Powder

Cleaning powder is used to help keep drawings clean, to avoid smearing, and to speed up the drafting process. Cleaning powder comes in a can or as pads, Figure 1-51, and is sprinkled over the drawing before starting. It is imperative that all cleaning powder is removed before placing the original drawing into a whiteprinter as the powder tends to stick to the roller. If good drafting habits are followed, the dry cleaning powder is not necessary.

Scales

Various kinds of scales are used by drafters, Figure 1-52. A number of different scales are included on each instrument. Scales save the drafter the work of computing new measurements every time a drawing is made larger or smaller than the original.

Scales come open divided and full divided. A *full-divided scale* is one in which the units of measurement are subdivided throughout the length of the scale. An *open-divided scale* has its first unit of measurement subdivided, but the remaining units are open or free from subdivision.

Figure 1-50 Drafting brush (*Courtesy Hearlihy & Co.*)

Mechanical Engineer's Scale

Mechanical engineer's scales are divided into inches and parts to the inch. To lay out a full-size measurement, use the scale marked 16. This scale has each inch divided into 16 equal parts or divisions of 1/16 inch. It is used by placing the 0 on the

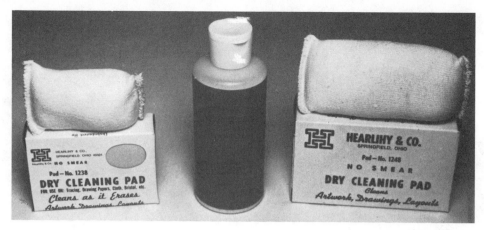

Figure 1-51 Dry cleaning powder (*Courtesy Hearlihy & Co.*)

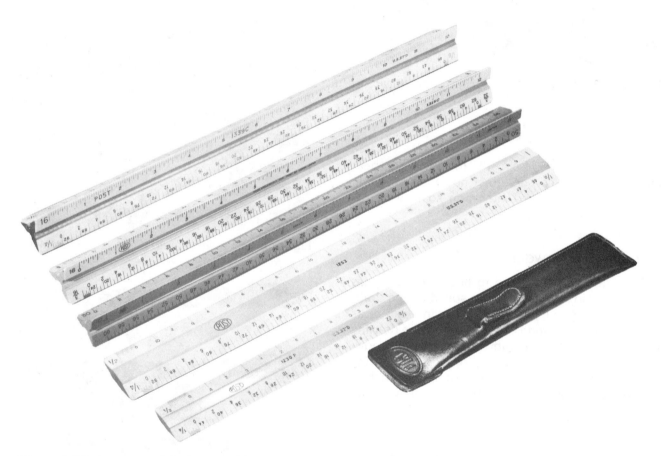

Figure 1-52 A variety of drafting scales (*Courtesy Teledyne Post*)

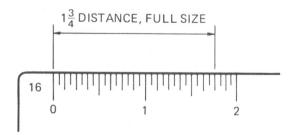

Figure 1-53 Regular full-size scale

point where measurement begins, and stepping off the desired length, Figure 1-53.

To reduce a drawing 50 percent, use the scale marked 1/2. The large 0 at the end of the first subdivided measurement lines up with the other unit measurements that are part of the same scale. The large numbers crossed out in Figure 1-54 go with the 1/4 scale starting at the other end. These numbers are ignored while using the 1/2 scale. To lay out 1

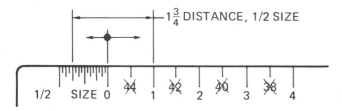

Figure 1-54 Half-size scale

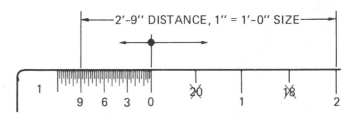

Figure 1-55 Quarter-size scale

3/4 inches at the 1/2 scale, read full inches to the right of 0 and fractions to the left of 0.

The 1/4 scale is used in the same manner as the 1/2 scale. However, measurements of full inches are made to the left of 0 and fractions to the right, because the 1/4 scale is located at the opposite end of the 1/2 scale, Figure 1-55.

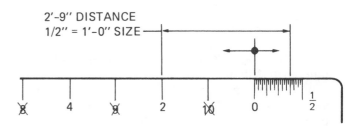

Figure 1-56 Architectural scale (full size)

Architect's Scale

The architect's scale is used primarily for drawing large buildings and structures. The full-size scale is used frequently for drawing smaller objects. Because of this, the architect's scale is generally used for all types of measurements. It is designed to measure in feet, inches, and fractions of an inch. Measure full feet to the right of 0; inches and fractions of an inch to the left of 0. The numbers crossed out in Figure 1-56 correspond to the 1/2 scale. They can be used, however, as 6 inches as each falls halfway between full-foot divisions. Measurements from 0 are made in the opposite direction of the full scale, because the 1/2 scale is located at the opposite end of the scale, Figure 1-57.

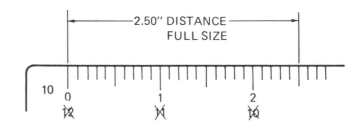

Figure 1-57 Architectural scale (half size)

Civil Engineer's Scale

A civil engineer's scale is also called a *decimal-inch scale*. The number 10, located in the corner of the scale in Figure 1-58, indicates that each graduation is equal to 1/10 of an inch or .1". Measurements are read directly from the scale. The number 20, located in the corner of the scale shown in Figure 1-59, indicates that it is 1/2 scale. To read 1/4 scale, the number 40 scale (not shown) would be used.

Using the same scale for civil drafting, one inch equals two hundred feet, Figure 1-60, and one inch equals one hundred feet, Figure 1-61.

A metric scale is used if the millimetre is the unit of linear measurement. It is read the same as the decimal-inch scale except that it is in millimetres, Figure 1-62.

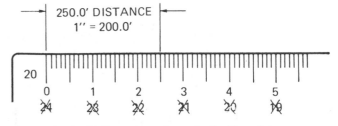

Figure 1-58 Civil engineer's scale

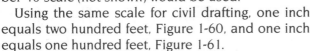

Figure 1-59 Civil engineer's scale (half scale)

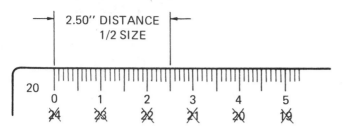

Figure 1-60 Mechanical scale (half size)

Figure 1-61 Civil scale

Figure 1-62 Metric scale

Pocket Steel Ruler

The drafter should make use of a pocket steel ruler. The pocket steel ruler is the easiest of all measuring tools to use. The inch scale, Figure 1-63, is six inches long, and is graduated in 10ths and 100ths of an inch on one side and 32nds and 64ths on the other side.

The metric scale is 150 millimetres long (approximately six inches) and is graduated in millimetres and half millimetres on one side, Figure 1-64. Sometimes metric pocket steel rulers are graduated in 64ths of an inch on the other side.

Measuring

The metric system uses the metre (m) as its basic dimension. A metre is 3.281 feet long or about 3 3/8 inches longer than a yardstick. Its multiples, or parts, are expressed by adding prefixes. These prefixes represent equal steps of 1000 parts. The prefix for a thou-

sand (1000) is *kilo*; the prefix for a thousandth (1/1000) is *milli*. One thousand metres (1000 m), therefore, equals one kilometre (1.0 km). One thousandth of a metre (1/1000 m) equals one millimetre (1.0 mm). Comparing metric to English then:

One millimetre (1.0 mm) = 0.001 metre (0.01 m) = .03937 inch

One thousand millimetres (1000 mm) = 1.0 metre (1.0 m) = 3.281 feet

One thousand metres (1000 m) = 1.0 kilometre (1.0 km) = 3281.0 feet

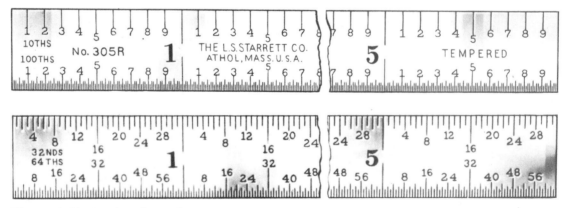

Figure 1-63 Steel scale (inch) (*Courtesy* L. S. Starrett Co.)

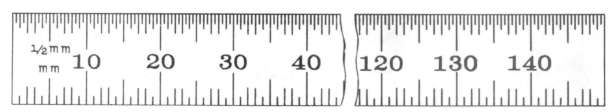

Figure 1-64 Steel scale (metric) (*Courtesy* L. S. Starrett Co.)

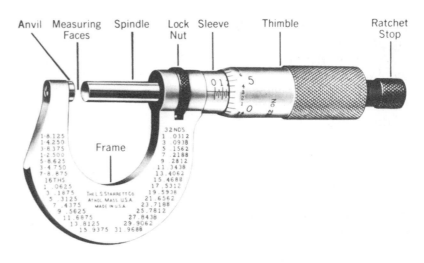

Figure 1-65 Micrometer
(*Courtesy* L. S. *Starrett Co.*)

Anvil Measuring Spindle Lock Sleeve Thimble Ratchet
Faces Nut Stop

Frame

How To Read a Micrometer Graduated in Thousandths of an Inch (.001″)

A micrometer consists of a highly accurate ground screw or spindle which is rotated in a fixed nut, thus opening or closing the distance between two measuring faces on the ends of the anvil and spindle, Figure 1-65. A piece of work is measured by placing it between the anvil and spindle faces, and rotating the spindle by means of the thimble until the anvil and spindle both contact the work. The desired work dimension is then found from the micrometer reading indicated by the graduations on the sleeve and thimble, as described in the following paragraphs.

Since the pitch of the screw thread on the spindle is 1/40″ or 40 threads per inch in micrometers that are graduated to measure in inches, one complete revolution of the thimble advances the spindle face toward or away from the anvil face precisely 1/40 or .025 inch.

The reading line on the sleeve is divided into 40 equal parts by vertical lines that correspond to the number of threads on the spindle. Therefore, each vertical line designates 1/40 or .025 inch and every fourth line, which is longer than the others, designates hundreds of thousandths. For example: the line marked "1" represents .100″, the line marked "2" represents .200″, and the line marked "3" represents .300″, and so forth.

The beveled edge of the thimble is divided into 25 equal parts, with each line representing .001″ and every line numbered consecutively. Rotating the thimble from one of these lines to the next moves the spindle longitudinally 1/25 of .025″ or .001 inch; rotating two divisions represents .002″, and so forth.

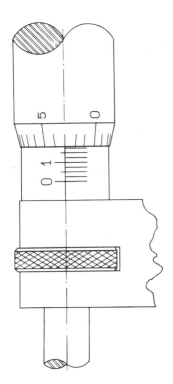

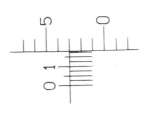

THIMBLE

SLEEVE

READING .178″

Figure 1-66 Reading a micrometer
(*Courtesy* L. S. *Starrett Co.*)

Twenty-five divisions indicate a complete revolution: .025 or 1/40 of an inch.

To read the micrometer in thousandths, multiply the number of vertical divisions visible on the sleeve by .025", and to this add the number of thousandths indicated by the line on the thimble which coincides with the reading line on the sleeve.

Example (See Figure 1-66):

The 1" line on the sleeve is visible, representing .100"

Three additional lines are visible, each representing .025". Thus, 3 x .025" = .075"

The third line on the thimble coincides with the reading line on the sleeve, each line representing .001". Thus, 3 x .001" = .003"

The micrometer reading is .100" + .075" + .003" = .178"

An easy way to remember how to read a micrometer is to think of the various units as if you were making change from a ten dollar bill. Count the figures on the sleeve as dollars, the vertical lines on the sleeve as quarters, and the divisions on the thimble as cents. Add up your change and put a decimal point instead of a dollar sign in front of the figures.

Micrometers come in both English and metric graduations. They are manufactured with an English size range of 1 inch through 60 inches, and a metric size range of 25 millimetres to 1500 millimetres. The micrometer is a very sensitive device and must be treated with extreme care.

Vernier Caliper

Vernier calipers have the capability of measuring both the outside and the inside measurements of an object, Figures 1-67 and 1-68. Use the bottom scale when measuring an outside size. Use the top scale when measuring an inside size.

Microfinish Comparator

The microfinish comparator is a handy tool for the drafter to approximate surface irregularities. Various kinds of microfinish comparators are available. Figure 1-69 illustrates a comparator for cast surfaces.

Ellipses Instrument

Two unique instruments are used to draw large ellipses. An ellipsograph is shown in Figure 1-70A. The OvalCompass is shown in Figure 1-70B. With these tools, the height and width of the ellipse are measured, locked-in, and quickly drawn. A template is used to draw small ellipses.

Ink Tools

Some fields of drafting, such as civil (map) drafting, require that all drawings be done in ink. Some companies ink their drawings so that they can be reduced and filed on film. All artwork that is to be reproduced

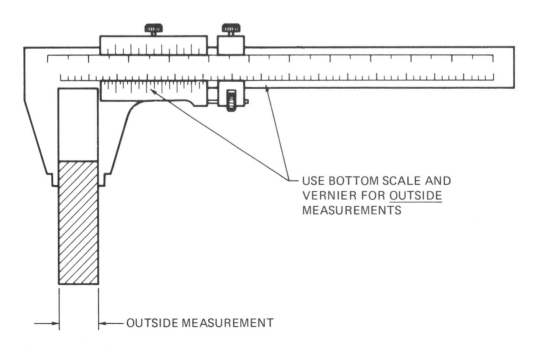

USE BOTTOM SCALE AND VERNIER FOR <u>OUTSIDE</u> MEASUREMENTS

OUTSIDE MEASUREMENT

Figure 1-67 Caliper—outside measurement

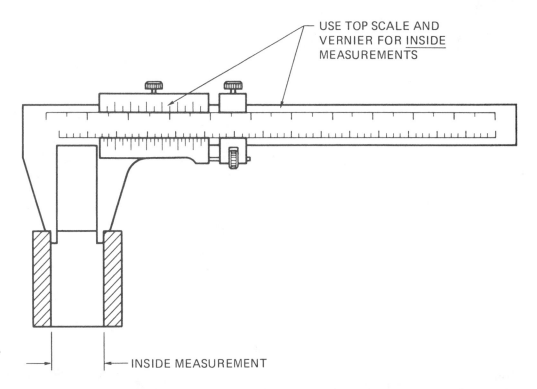

Figure 1-68 Caliper—
inside measurement

INSIDE MEASUREMENT

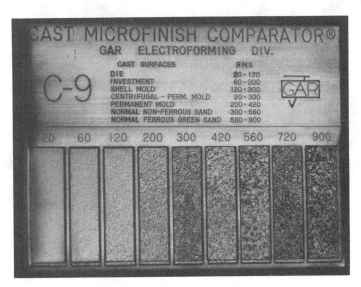

Figure 1-69 Microfinish comparator (*Courtesy* GAR
Electroforming Div., Mite Corp.)

by camera, such as in the field of technical illustration, is done in ink. Ink drawing is no more difficult than pencil drawing, Figure 1-71.

Technical Pens

The key to successful inking is a good technical pen, Figure 1-72. Technical pens are produced in two styles. Notice the ends of the two pens in the figure; one has a tapered end, the other a straight end. The tapered pen is used primarily for artwork; the straight end is used for drafting and mechanical lettering. Pens are available in various sizes and styles of pen-holder sets, Figures 1-73 and 1-74.

Technical pen points are manufactured of stainless steel, tungsten or jewels. The stainless steel point is chromium plated for use on tracing paper or vellum. Tungsten points are long wearing for use on abrasive, coated plotting film or triacetate. Jewel points are used on a plotter that has a controlled pen force.

Figure 1-70A Ellipsograph
(*Courtesy Omicron Co.*)

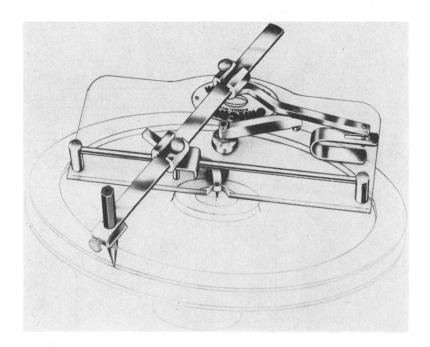

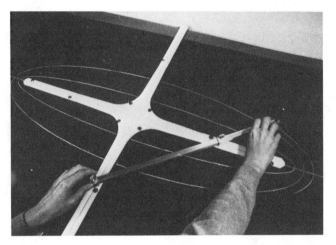

Figure 1-70 B Oval compass (*Courtesy Oval Compass*)

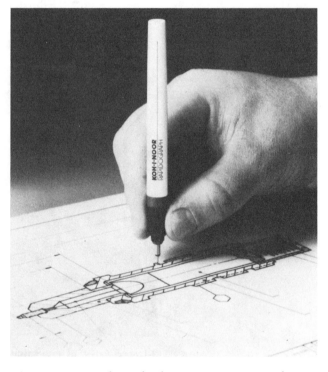

Figure 1-71 Technical inking pen (*Courtesy Koh-I-Noor Rapidograph*)

Figure 1-72 Drafting and art technical pens
(*Courtesy Koh-I-Noor Rapidograph*)

Figure 1-73 Revolving pen holder (*Courtesy Koh-I-Noor Rapidograph*)

Figure 1-74 Flat pack pen holder (*Courtesy Koh-I-Noor Rapidograph*)

.13/5x0 .18/4x0 .25/3x0 .30/00 .35/0 .45/1 .50/2 .70/2½ .80/3 1.0/3½ 1.2/4 1.4/5 2.0/6

Figure 1-75 Pen sizes (*Courtesy Staedtler Mars*)

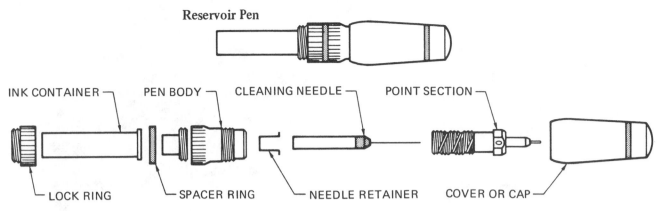

Reservoir Pen

INK CONTAINER ── PEN BODY ── CLEANING NEEDLE ── POINT SECTION

LOCK RING ── SPACER RING ── NEEDLE RETAINER ── COVER OR CAP

Figure 1-76 Internal parts of a technical inking pen

Pen points are available in thirteen standard sizes of varying widths, Figure 1-75. For general drafting inking, numbers .45/1 and .70/2½ are recommended.

Cleaning Technical Pens

Pens should be cleaned when they get sluggish or before storing them for long periods of time. The parts of most technical inking pens are similar to those shown in Figure 1-76. When not in use, technical pens should be kept in a storage clamp or else capped to prevent ink from drying in the point. If a pen does get clogged, remove the point and hold it under warm tap water. This normally softens the ink. If the ink has dried, use an ultrasonic cleaner or a mild solvent. If the pens will not be used for a week or more, all ink should be removed and the pens stored empty and clean. Care must be taken when removing and replacing the cleaning needle. An ultrasonic cleaner is used quickly and efficiently to clean technical pens, Figure 1-77.

When pens are to be cleaned by hand, use the following recommended steps:

Cleaning. Pens can be ruined by improper cleaning. Study Steps 1 through 5 and follow them closely when cleaning pens. (Refer again to Figure 1-76.)

Step 1. Remove the cap and the ink container.

Step 2. Soak the body of the pen in hot water. The ink container should also be soaked if ink has dried in it.

Figure 1-77 Technical pen ultrasonic cleaner (*Courtesy Keuffel & Esser Co.*)

Figure 1-78 Drawing ink

Step 3. After soaking, remove the pen body from the water. Hold the knurled part of the body with the top downward. Unscrew and remove the point section. Remove the end of the cleaning needle weight. Do not bend the cleaning needle or it will break.

Step 4. Immerse all body parts in a good pen-cleaning fluid or hot water mixed half with ammonia.

Step 5. Dry and clean.

Filling. To fill the pen, follow Steps 1 through 5:

Step 1. Unscrew and remove the knurled lock ring.

Step 2. Remove the ink container. Leave the spacer ring in place.

Step 3. Fill the ink container with lettering ink. Do not fill it more than ¾ inch from top.

Step 4. Hold the filled container upright and insert the pen body into the container.

Step 5. Replace the knurled lock ring.

Ink

A high-quality, fast-drying ink must be used in technical pens for the best results. The ink must be black and erasable, and it must not crack, chip or peel, Figure 1-78. Keep inks out of extremely warm or cold temperatures. The bottles or jars should be kept airtight, and the excess ink should be cleaned from the neck of the container to keep it from drying in the cap. Inks in large containers should be transferred to smaller bottles or directly into pens, *away from working areas to avoid the possibility of spillage.*

Mechanical Lettering Sets

Lettering sets come in a variety of sizes and templates, Figure 1-79. All sets contain a scriber, and various pen sizes and templates.

Scriber Templates

Scriber templates consist of laminated strips with engraved grooves which are used to form letters. A

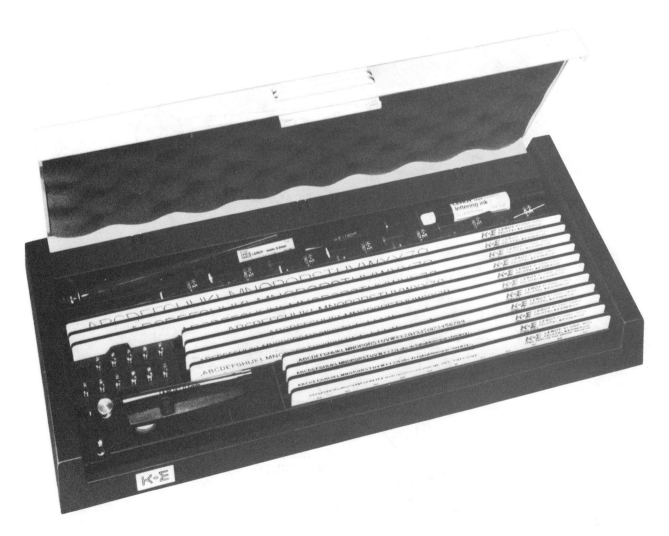

Figure 1-79 Lettering set (*Courtesy Keuffel & Esser Co.*)

Figure 1-80 Forming letters with a scriber (*Courtesy Koh-I-Noor Rapidograph*)

tracer pin moving in the grooves guides the scriber pen (or pencil) in forming the letters, Figure 1-80.

Guides for different sizes and kinds of letters are available for any of the lettering devices. Different point sizes are made for special pens so that fine lines can be used for small letters and wide lines for large letters. Scribers may be adjusted to form vertical or slanted letters of several sizes from a single guide by simply unlocking the screw underneath the scriber and extending the arms, Figure 1-81.

One of the principal advantages of lettering guides is that they maintain uniform lettering. This is especially useful where many drafters are involved. Another important use is for the lettering of titles, and note headings and numbers on drawings and reports.

Letters used to identify templates are:

U = Uppercase
L = Lowercase
N = Numbers

Thus, a template identified as 8-ULN means it is 8/16 inch high (1/2"), and has uppercase letters and lowercase letters, and numbers.

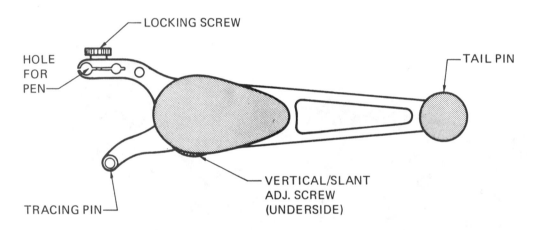

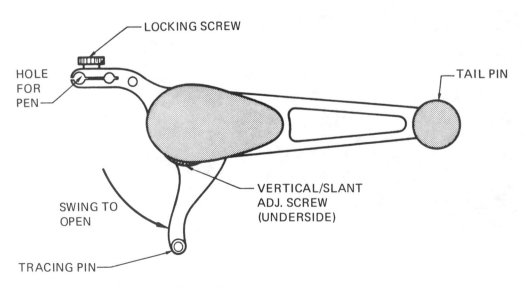

Figure 1-81 Scribers are adjustable

Tracing Pin Better, more expensive scribers use a double tracing pin, Figure 1-82. The blunt end is used for single-stroke lettering templates or very large templates that have wide grooves. The sharp end is used for very small lettering templates, double-stroke letters or script-type lettering using a fine groove. Most tracing pins have a sharp point, but some do not. Always screw the cap back on the unused end after turning the tracing pin to the desired tip. Be careful with the points as they will break if dropped and can cause a painful injury if mishandled.

Standard Template

Learning to form mechanical letters requires a great deal of practice. Figure 1-83 shows a template having three sets of uppercase and lowercase letters. Practice forming each size letter and number until they can be made rapidly and neatly. Use a very light, delicate touch so as not to damage the template, scriber or pen.

Size of Letters The size or height of the lettering on a template is called out by the number used to identify each set. Sizes are in thousandths of an inch. A #100 is .100 inch high, or slightly less than an eighth of an inch; a #240 is .240 inch high, or slightly less than a quarter of an inch.

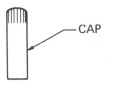
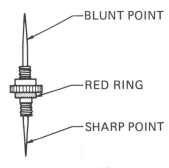

Figure 1-82 Double tracing pen

Another system to determine template size uses simple numbers. These numbers are placed above the number 16 to indicate the fraction height of the letter. For instance, the number 3 placed above the number 16 would read as 3/16 inch in height.

Pens

There are two types of pens: the regular pen and the reservoir pen, Figure 1-84. The regular pen must

Figure 1-83 Lettering template

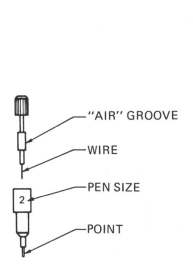

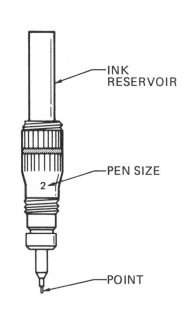

Figure 1-84 Ink pens used for lettering

be cleaned after each use. The reservoir pen should be cleaned when it gets sluggish or before being stored for long periods of time. This procedure is the same as it is for the cleaning of technical pens as described previously.

Butterfly-Type Scriber

Basic Parts

The butterfly-type scriber shown in Figure 1-85 is a delicate, precision tool that does its job without requiring any adjustments, repairs or maintenance. The clear plastic base of the scriber bears the setting chart used in adjusting the pen arm for enlargements, reductions, verticals, and slants to be produced by tracing the engraved letters of a letter guide template.

The pen arm of the scriber holds the pen accessories for the various jobs to be performed. The pen and the arm have a thumb-tightening screw device for securing the pen being used, and an adjustable pressure post screw with locking nut for controlling the amount of pressure at which the pen is set. The pressure post rides on the surface of the work when in use, and is used only in conjunction with the swivel knife. The bull's-eye setting marker at the opposite end of the pen arm offers a concise, accurate means of setting the scriber for the various percentages and angles desired.

The tracing pin is the hardened tool steel point used in tracing the template letter. The tail pin serves as the pivot point for the triangular action of the scriber. This pin travels in the center groove of the template.

Operation of the Butterfly Scriber

The butterfly scriber, a precision lettering tool, is the key to producing clean, sharp, controlled lettering. The setting chart, using the bull's-eye at the end of the pen arm for a marker, begins at the outer edge with a starting line marked "vertical." In this position, the scriber produces a vertical letter of normal size with the template being used. To enlarge this letter, set the bull's-eye at a position above the 100 percent intersection. At 120 percent, the scriber produces a letter 20 percent greater in height than it does at 100 percent. A reduction can be produced by setting the bull's-eye at a position below the 100 percent intersection. Variations in height range from 100 up to 140 percent and down to 60 percent. The extreme settings produce condensed letters, and the intermediate settings produce headings, subheadings or large or small letters.

Slants in all sizes are easily produced by setting the bull's-eye on a line other than the vertical line. Normal slants or italics are produced in all height adjustments by setting the bull's-eye on either the 15-degree or 22 1/2-degree line, and at the desired percent of height of the letter on the template.

Variations may be produced in slants ranging from 0 degree to 50 degrees forward. Tracing the engraved template letter requires a very light and delicate touch. This results in more accurately traced letters and less wear on the equipment. Each lettering application requires its own specific pen, and will place at the fingertips of the drafter the very best in standard typeface and hand-lettered alphabets for fast, easy rendering.

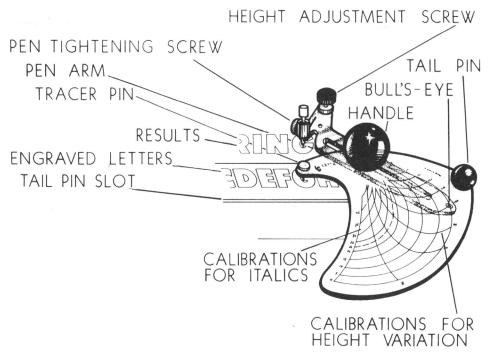

Figure 1-85 Butterfly-type scriber (*Courtesy Letterguide Inc.*)

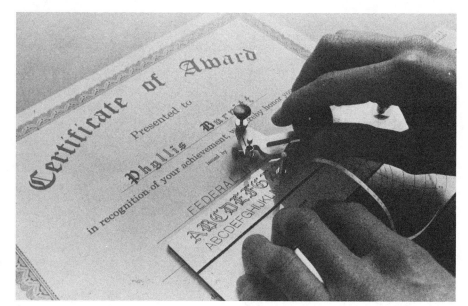

Figure 1-86 Adjustable scriber creates special effects (*Courtesy Letterguide Inc.*)

Figure 1-87 Sample lettering styles (*Courtesy Letterguide Inc.*)

Figure 1-88 Additional special effects (*Courtesy Letterguide Inc.*)

Special Effects By using one's imagination many special effects can be achieved, Figures 1-86, 1-87 and 1-88.

Airbrush

Airbrush guns are used for such purposes as production designing, pictorial rendering, portrait figure rendering, architectural rendering, and technical illustration. There are two kinds of airbrushes: the single-action type and the double-action type. In the single-action airbrush, the trigger controls the flow of air only. The fluid control is adjusted in front by the nozzle. In the double-action airbrush, the trigger controls both the flow of air and the amount of fluid to be sprayed, Figure 1-89.

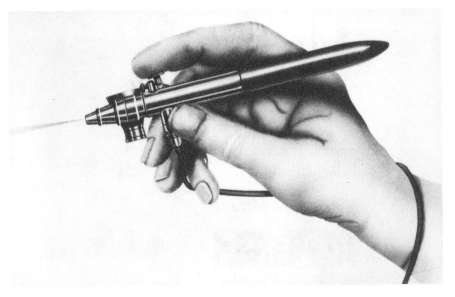

Figure 1-89 Airbrush (*Courtesy Badger Airbrush Co.*)

INCHES			MILLIMETRES	
SIZE	DIMENSIONS		SIZE	DIMENSIONS
A	8 1/2 x 11	9 x 12	A-4	210 x 297
B	11 x 17	12 x 18	A-3	297 x 420
C	17 x 22	18 x 24	A-2	420 x 594
D	22 x 34	24 x 36	A-1	594 x 841
E	34 x 44	36 x 48	A-0	841 x 1189

Figure 1-90 Paper sizes

Paper Sizes

Two basic standard paper sizes are 8 1/2 x 11 inches and 9 x 12 inches. The basic standard metric size, A-4, is 210 x 297 millimetres. See Figure 1-90. Examples of paper folded to A-size are shown in Figure 1-91.

Borders

The location of the borders varies with each size sheet of paper, Figure 1-91A. This chart indicates the various standard borders used today. A standard horizontal border is shown in Figure 1-91B. A standard vertical border is shown in Figure 1-91C.

Zoning

Zoning is used to pinpoint a particular detail on a drawing. The exact rectangular zone is located by the use of numbers running horizontally and letters running vertically in the margins. By extending these imaginary lines, the exact rectangular zone, Zone 7-C, is located as shown in Figure 1-91A. See the corresponding symbol below the chart. The number at the left (1) indicates the page number, the number at the top right (7) indicates the corresponding number on the horizontal margin. The letter at the lower right (C) indicates the corresponding letter in the vertical margin.

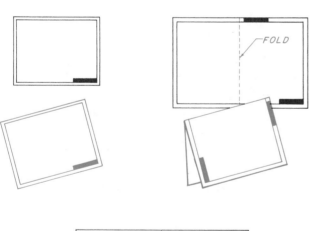

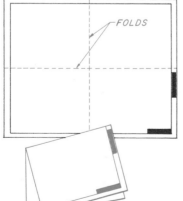

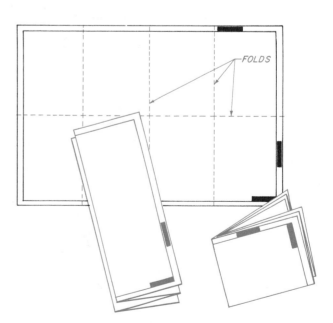

Figure 1-91 Paper folded to A-size

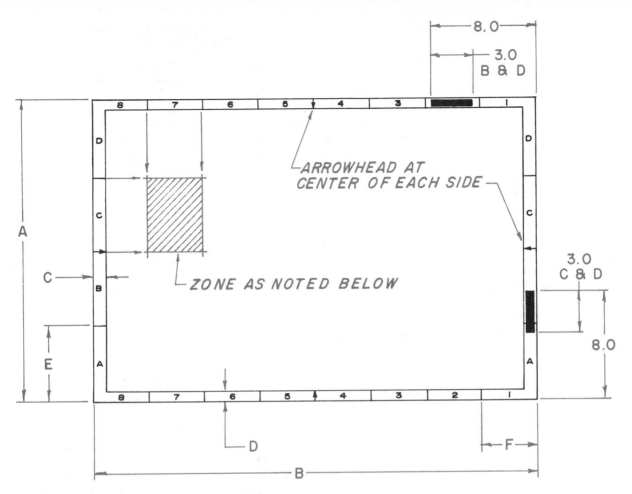

STANDARD BORDER SIZES

DRAWING SIZE		A	B	C	D	E	F
A	HORIZONTAL	8.5	11.0	.25	.38	2 AT 4.25	2 AT 5.50
A	VERTICAL	11.0	8.5	.38	.25	2 AT 5.50	2 AT 4.25
B		11.0	17.0	.62	.38	4 AT 2.75	4 AT 4.25
C		17.0	22.0	.50	.75	4 AT 4.25	4 AT 5.50
D		22.0	34.0	1.00	.50	4 AT 5.50	8 AT 4.25

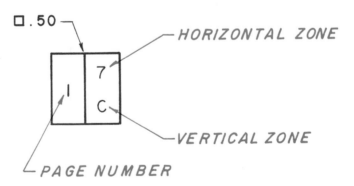

ZONE IDENTIFICATION
SEE ZONE ABOVE

Figure 1-91A Standard border sizes

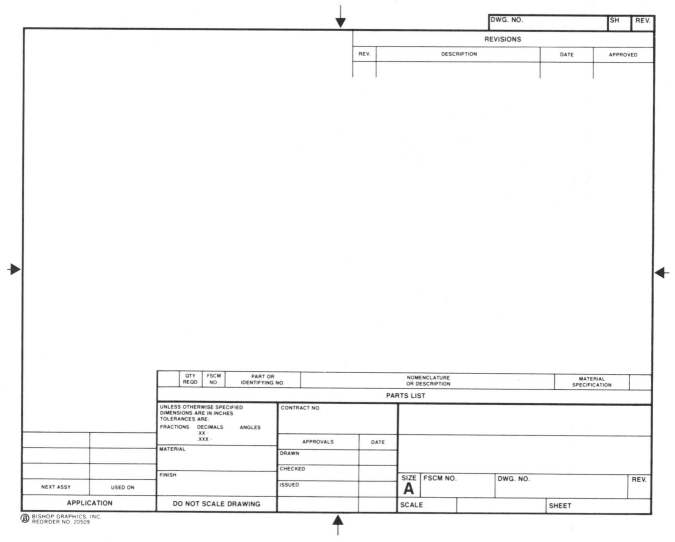

Figure 1-91B Standard horizontal border (*Courtesy Bishop Graphics Co.*)

Whiteprinter

Many types of whitepapers are available for use in drafting rooms. A *whiteprinter*, Figure 1-92, reproduces a drawing through a chemical process. Most of these machines work on the same basic principle, Figure 1-93. A bright light passes through the translucent original drawing and onto a coated whiteprint paper. The light breaks down the coating on the whiteprint paper, but wherever lines have been drawn on the original drawing, no light strikes the coated sheet.

Then the whiteprint paper is passed through ammonia vapor for developing. This chemical developing causes the unexposed areas — those that were shaded by lines on the original — to turn blue or black.

Most whiteprinters have controls to regulate the speed and flow of the developing chemical. Each type of machine requires different settings and has different controls. Before operating any whiteprinter, read all of the manufacturer's instructions.

Today, with the advent of new technology, copies are made on an outprint printer, Figure 1-94.

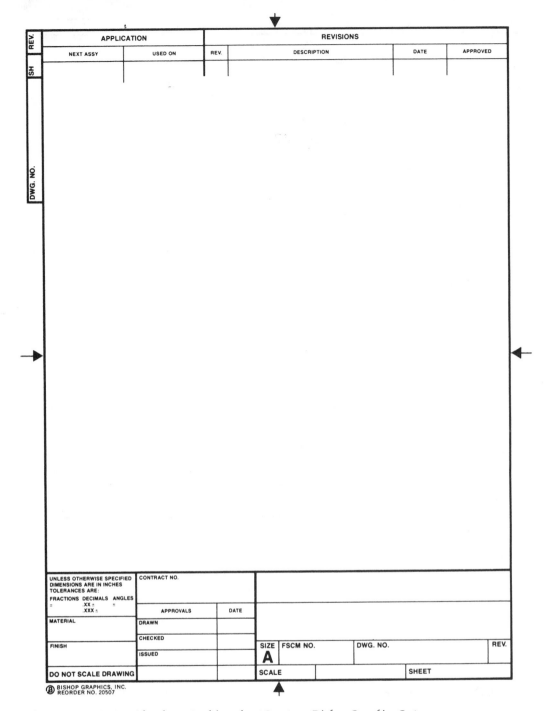

Figure 1-91C Standard vertical border (*Courtesy Bishop Graphics Co.*)

Figure 1-92 Whiteprinter (*Courtesy Blu-Ray Inc.*)

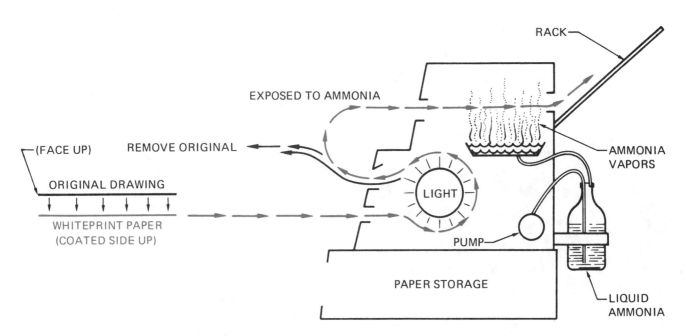

RACK

EXPOSED TO AMMONIA

AMMONIA
VAPORS

(FACE UP)

REMOVE ORIGINAL

ORIGINAL DRAWING

LIGHT

WHITEPRINT PAPER
(COATED SIDE UP)

PUMP

PAPER STORAGE

LIQUID
AMMONIA

Figure 1-93 Whiteprinter process

Figure 1-94 Copier (*Courtesy J. S. Staedtler Inc.*)

Files

A finished drawing represents a great deal of valuable drafting time and is, therefore, a costly investment. Drawings must be stored flat in a clean storage area, Figure 1-95. Vertical drawing storage is provided by hangers, Figure 1-96. Most engineering firms keep their files in fireproof and theftproof vaults.

Open-End Typewriter

A word processor-equipped, open-end typewriter is used to speed up the lettering process on a large drawing, Figure 1-97.

Care of Drafting Equipment

Drafting tools are precision instruments, and the proper care will ensure that they last a lifetime.

Plastic Tools

Plastic drafting tools, such as T-squares, parallel straightedges, templates and triangles, should be wiped immediately after use with a damp cloth to

Figure 1-95 Drawing file system (*Courtesy Safco Products Inc.*)

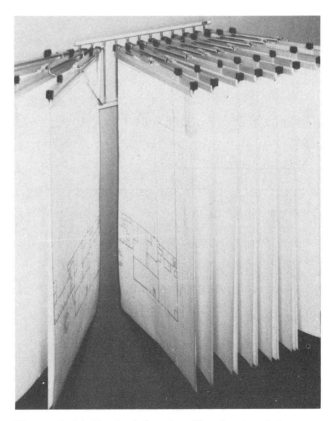

Figure 1-96 Vertical drawing file (*Courtesy Safco Products Inc.*)

Figure 1-97 Open-end typewriter
(*Courtesy Diagram Corp.*)

remove ink or graphite that may stain the tools or be carried to the next drawing. Once a plastic instrument is stained, a mild soap or ammonia solution will dissolve many water- and oil-based inks. Be careful *not* to use a solvent such as paint thinner, lacquer thinner or alcohol.

Plastic drafting tools should be kept out of direct sunlight and away from warm surfaces to prevent them from becoming brittle, cracked, and warped. They should be stored in a flat position with cloth or paper between them to reduce scratching the surface.

A great number of plastics are used in drafting instruments. Most are made from either styrene or acrylic plastic. Styrene is a more flexible and softer plastic than acrylic. Although acrylic instruments are harder, they are more prone to chipping. Because both types of plastic are relatively soft, plastic drafting instruments should never be used for a cutting edge.

Compasses

Almost all compasses are made of brass that is chrome- or nickel-plated. To clean these instruments, use a mild solution of soap and water to remove residue and dirt.

Compasses should not normally need oiling, unless they are kept in a damp area which could cause rust. If a compass is oiled unnecessarily, there is a risk of soiling the next drawing on which it is used.

Tables and Chairs

Wooden drafting furniture is cared for in the same manner as any other wooden furniture. It may be polished or waxed with ordinary products. Do not polish the insides of drawers or cabinets. These areas may retain the wax, which can then be transferred to drawings.

Steel furniture can be cleaned with soap and water, and then waxed.

The gears and joints on adjustable drafting tables are lubricated at the factory, and generally do not require further oiling. Additional oiling increases the risk of getting oil or grease on a drawing.

The tops of most drafting tables are coated with a vinyl film such as melamine or a phenol-laminate material. A glass cleaner or mild ammonia solution is used to clean these surfaces.

Review

1. Among the pencil lead grades from hard to soft, which two are recommended for drafting?

2. Describe the whiteprinter process used in today's drafting rooms.

3. List three kinds of drafting triangles.

4. Explain how to read a micrometer.

5. Why are the drawing surface and top cover so important?

6. List the various kinds of drafting tools used to draw horizontal lines, and explain which is the best and why.

7. List the standard paper sizes used in industry today.

8. What is the difference between an open-divided scale and a full-divided scale?

9. Which kind of compass is recommended and why?

10. For what is dry cleaning powder used, and why must it be removed from the drawing?

11. Why is a dusting brush used in drafting?

12. Explain how to read a vernier protractor. Why is it used?

13. Figures 1-98 through 1-111 contain graphic illustrations of various scales with distances indicated. Examine each scale, and record the distance indicated.

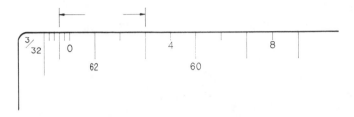

Figure 1-98 Reading the 3/32" scale

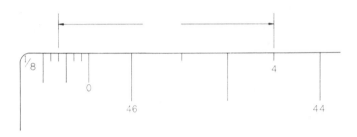

Figure 1-99 Reading the 1/8" scale

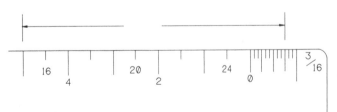

Figure 1-100 Reading the 3/16" scale

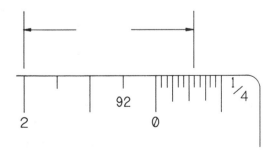

Figure 1-101 Reading the 1/4" scale

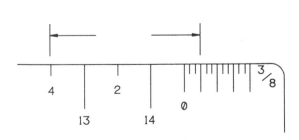

Figure 1-102 Reading the 3/8" scale

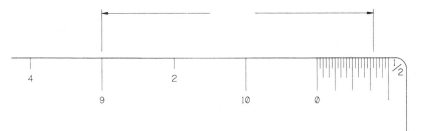

Figure 1-103 Reading the 1/2" scale

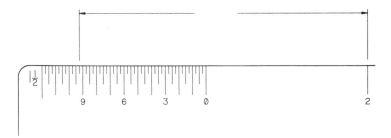

Figure 1-104 Reading the 3/4" scale

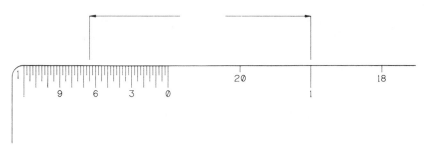

Figure 1-105 Reading the 1" scale

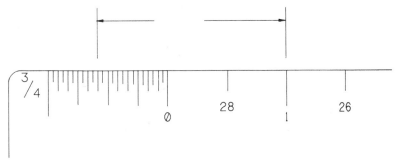

Figure 1-106 Reading the 1 1/2" scale

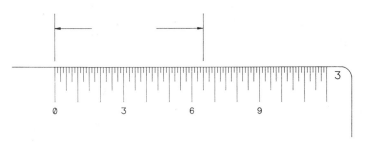

Figure 1-107 Reading the 3" scale

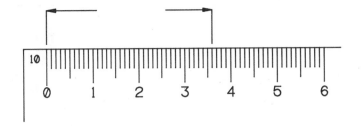

Figure 1-108 Reading the full scale on the engineer's scale

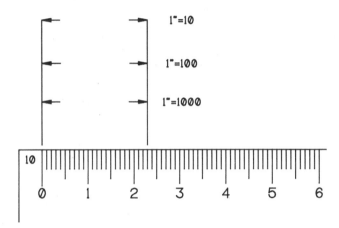

Figure 1-109 Reading the 10 scale

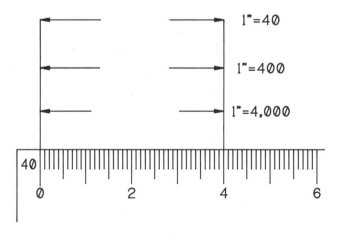

Figure 1-110 Reading the 40 scale

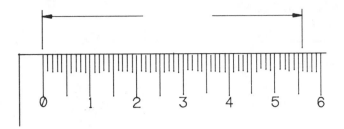

Figure 1-111 Reading the metric scale

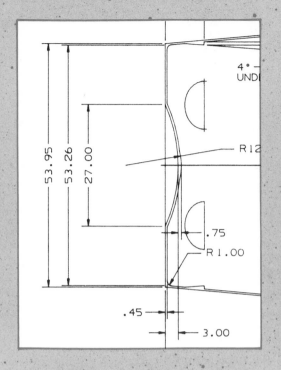

This chapter covers three of the basics that are needed in preparing all types of technical drawings. All three concepts represent manual drafting techniques. Computer-aided drafting (CAD) techniques are covered in later chapters. The major topics covered in this chapter are freehand lettering, freehand lettering techniques, line work, and sketching.

CHAPTER TWO

Lettering, Sketching, and Line Techniques

Freehand Lettering

Text is an important part of a technical drawing. Not all information required on technical drawings can be communicated graphically; the most obvious data being dimensions. *Text* on technical drawings consists of dimensions, notes, legends, and other data that are best conveyed using alphanumeric characters, Figure 2-1.

Several different ways are used to create text on technical drawings. The traditional method is by freehand lettering. Other methods include such mechanical lettering techniques as scriber templates, typewritten notation, and typed lettering generated by computer-aided drafting systems. This chapter focuses on freehand lettering. Other methods are described elsewhere in this text.

Lettering Styles

There are numerous different lettering styles or fonts, Figure 2-2. The standard style for freehand lettering on technical drawings, as established in American National Standards document Y14.2-1973, is single-stroke Gothic lettering. Vertical, single-stroke Gothic letters are the most universally used of the various styles available to drafters, Figure 2-3.

Some modifications of the standard Gothic configuration of letters are often made, without actually changing from the Gothic style of lettering. One way is through the use of uppercase and lowercase letters, Figure 2-4, but this is seldom acceptable on technical drawings. Another method is to condense or extend the letters, Figure 2-5.

The most common way of modifying Gothic letters is by inclining them slightly to the right, Figure 2-6. Inclined lettering is easier to make as it lends itself to a natural direction of wrist action. The correct angle of the inclined elements is a two-unit incline to the right for each five units of letter height. Errors are not as detectable with inclined letters as they are with vertical elements. As the inclined elements are longer, they are easier to read. However, inclined lettering is not universally accepted, and caution must be exercised to not conflict with customary drafting styles. A backhanded or left-leaning inclination is never an acceptable modification.

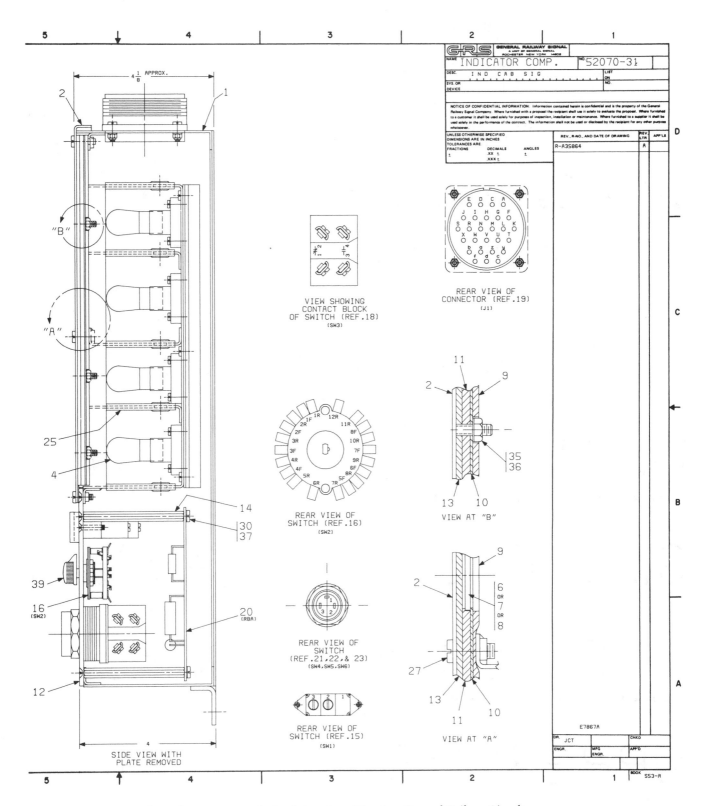

Figure 2-1 Examples of text on a technical drawing *(Courtesy General Railway Signal)*

THIS IS BLOCK FONT

THIS IS FAST FONT

THIS IS FUTURA FONT

THIS IS LEROY FONT

THIS IS OLD ENGLISH FONT

THIS IS RIVERA FONT

THIS IS TIMES FONT

THIS IS HELVET FONT

BAUSCH & LOMB ▼ FONTS

Figure 2-2 Sample lettering fonts (*Courtesy Bausch & Lomb, Inc.*)

SINGLE-STROKE GOTHIC

LETTERING SAMPLE

Figure 2-3 Single-stroke Gothic lettering

UPPERCASE GOTHIC

lowercase Gothic

Figure 2-4 Uppercase and lowercase Gothic lettering

EXTENDED VARIATION

CONDENSED VARIATION

Figure 2-5 Extended and condensed variations of Gothic lettering

SAMPLE OF INCLINED LETTERING

Figure 2-6 Inclined Gothic lettering

SLOPPY LETTERING IS DIFFICULT TO READ

Figure 2-7 Sloppy lettering is difficult to read

Characteristics of Good Lettering

Good freehand lettering, regardless of whether it is uppercase or lowercase, condensed or extended or vertical or inclined, must have certain characteristics. These requisites include neatness, uniformity, stability, proper spacing, and speed.

Neat lettering is important so that the information being conveyed can be easily read. Few things detract from the appearance and quality of a technical drawing more than sloppy lettering, Figure 2-7.

For uniformity, all letters should be the same in height, proportion, and inclination. A necessary tactic for maintaining uniformity is the use of guidelines, Figure 2-8. The customary heights of characters in technical drawing are 1/8" (6 mm) for regular text, and 3/16" (9 mm) for headings and titles.

The proper stability or balance of letters is an important characteristic in freehand lettering. Each letter should appear balanced and firmly positioned to the human eye. Top-heavy letters are not balanced because they appear about to topple over, Figure 2-9.

The proper spacing of letters and words is important, and it takes a lot of practice to accomplish. A good rule of thumb to follow in terms of spacing is to use close spacing within words, and far spacing between words, Figure 2-10. The proper positions of letters relative to one another in words is accomplished by spacing the letters in the word equally in the area, not by trying to equalize the spacing between letters. This becomes automatic if the drafter concentrates on the word being lettered, not on each letter. Another rule of thumb for spacing is to allow the width of one round letter, such as O, C, Q or G, between words. Figure 2-11 illustrates how this type of spacing can make the lettering much easier to read.

In the modern drafting room, because time is money, speed in freehand lettering is critical. Typically, freehand lettering is one of the slowest, most time-consuming tasks drafters must perform. It takes many hours of practice to develop freehand lettering that is neat, uniform, balanced, properly spaced,

UNIFORMITY ——GUIDELINES

Figure 2-8 Guidelines improve uniformity

BEFR382   TOP HEAVY

BEFR382   CORRECT FORM

Figure 2-9 Top-heavy letters are not balanced

Figure 2-10
The proper
spacing of
letters and
words

SPACING WITHIN WORDS
SHOULD BE CLOSE.

SPACING BETWEEN WORDS
SHOULD BE FAR.

THIS IS A PROPERLY SPACED SENTENCE.

THISISANIMPROPERLYSPACEDSENTENCE.

Figure 2-11
Spacing between
words is important

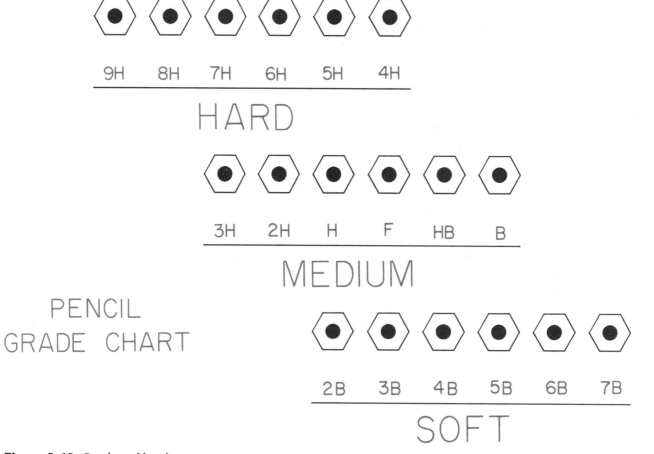

HARD

MEDIUM

PENCIL
GRADE CHART

SOFT

Figure 2-12 Grades of lead

TASK	LEAD
CONSTRUCTION LINES	3H, 2H
GUIDE LINES	3H, 2H
LETTERING	H, F, HB
DIMENSION LINES	2H, H
LEADERLINES	2H, H
HIDDEN LINES	2H, H
CROSSHATCHING LINES	2H, H
CENTERLINES	2H, H
PHANTOM LINES	2H, H
STITCH LINES	2H, H
LONG BREAK LINES	2H, H
VISIBLE LINES	H, F, HB
CUTTING PLANE LINES	H, F, HB
EXTENSION LINES	2H, H
FREEHAND BREAK LINES	H, F, HB

Figure 2-13 Lead-lines chart

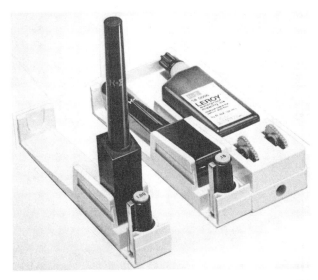

Figure 2-14 Modern technical pens (*Courtesy Keuffel & Esser Co.*)

and fast. Some drafters never reach this goal. Those who do, reach it through constant practice, coupled with continual efforts to improve.

Freehand Lettering Techniques

Freehand lettering techniques are learned by knowing what grades of lead to use, how to make the basic lettering strokes, and how to use guidelines; and by constantly practicing and trying to improve.

Figure 2-12 catagorizes the various grades of lead available to drafters. Figure 2-13 matches the lead grades with typical line styles. The actual choice of lead for a given situation in technical drawing is up to the user. The personal preference of one technician will differ from that of another. The specifications set forth in Figure 2-13 should be viewed as guidelines, rather than hard-and-fast rules.

Lettering in ink has been greatly simplified in recent years. Old-fashioned tools, such as adjustable nib ruling pens and speedball pens, have been replaced by the less cumbersome, easier-to-use technical pen, Figure 2-14. When lettering in ink, drafters still use light guidelines made with pencil lead. The actual lettering is done with the desired pen point size. Commonly used pen points for lettering in ink are sizes 0, 1, and 2, which are standard American sizes. In metrics, these point sizes represent line widths of 0.35 mm, 0.50 mm, and 0.60 mm.

All letters and numbers are created using six basic strokes, Figure 2-15. The first stroke is a single stroke made downward and to the right at approximately 45 degrees. The second stroke is made downward and to the left at approximately 45 degrees. The third stroke is vertical, and is made from top to bottom. Stroke number four is horizontal, and is made from left to right. The fifth stroke is a half-circular stroke to the left, made from top to bottom. The sixth stroke is a half-circular stroke to the right, made from top to bottom. All alphanumeric characters can be created using combinations of these six strokes. Figures 2-16 and 2-17 give examples of how these strokes are used for making selected characters.

Lettering Guidelines

Guidelines are a critical part of freehand lettering. Uniformity, neatness, and stability cannot be achieved without using guidelines. Guidelines can be made using a small plastic device called a lettering guide, Figure 2-18. The lettering guide shown is a device with a vertical side on the left, and a 68-degree inclined side on the right, the 2-to-5 incline needed for slanted lettering. In the center is a movable circular component containing three rows of variously spaced holes. The holes are used for the spacing of guidelines. A row of numbers on the bottom of the circular component allows drafters to set the guide for various heights of letters. The numbers — 2, 3, 4, 5, 6, 7, 8, and 10 — indicate 32nds of an inch. A setting of 4 means 4/32nds of an inch, or guidelines that will yield letters 1/8th of an inch in height.

A complete set of guidelines consists of three lines: a top line, a middle line, and a bottom line, Figure

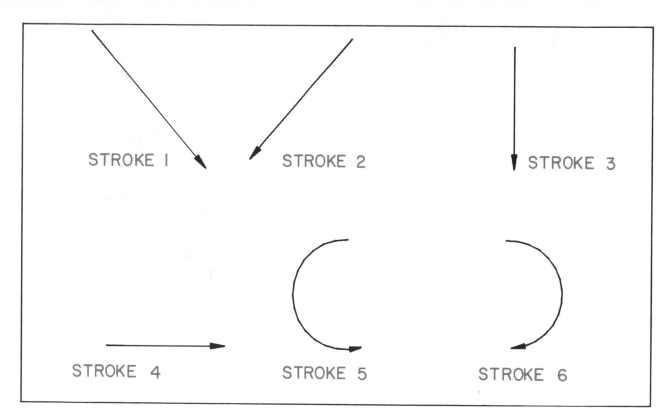

Figure 2-15 Six basic strokes are used for lettering

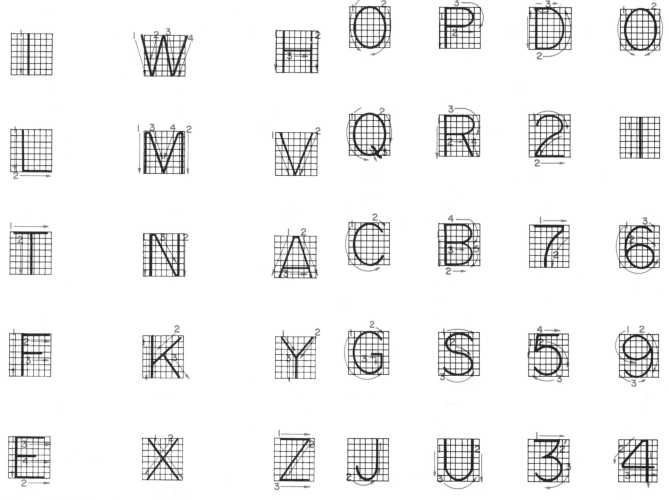

Figure 2-16 Forming straight letters

Figure 2-17 Forming curved letters

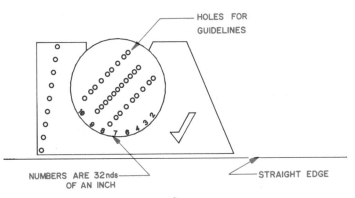

Figure 2-18 Lettering guide

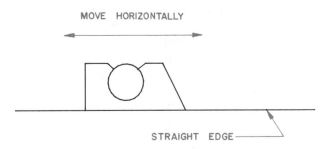

Figure 2-19 A complete set of guidelines

2-19. The middle guideline helps drafters space middle elements of such letters as A, B, E, F, G, H, P, R, S, and Y.

The center row of holes on the lettering guide, being equally spaced, is the one used most frequently for freehand Gothic lettering. The two outer rows of holes are used when lowercase letters are desired. The outer row of holes on the left produces guidelines for lowercase letters, the bottom portions of which are 3/5ths of the letter's total height. The outer row of holes on the right produces guidelines for lowercase letters, the bottom portions of which are 2/3rds of the total height of the letters.

Horizontal guidelines are made by placing the pencil point in the appropriate holes and sliding the guide across the top of the parallel bar, Figure 2-20. Guidelines for vertical lettering are produced by using the guide together with a triangle or vertical straightedge, Figure 2-21.

Guidelines for letters, numbers, and fractions are created in the same way. However, when lettering fractions, drafters should be careful to leave a visible distance between the numerator, denominator, and the fraction crossbar, which should be parallel, Figure 2-22.

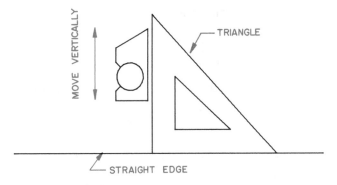

Figure 2-20 Making horizontal guidelines

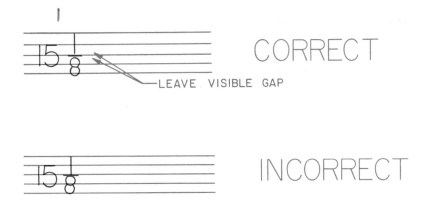

Figure 2-21 Making vertical guidelines

Figure 2-22 Correct method for lettering fractions

Line Work

Line work is the generic term given to all of the various techniques used in creating the graphic data on technical drawings. Mechanical line work is made using either mechanical pencils or technical pens. Such devices as parallel bars, drafting machines, triangles, scales, and numerous other tools are used to guide the line-making. Since inking is dealt with in the first chapter, this chapter focuses on pencil line work.

Characteristics of Lines

Twelve basic types of lines are used in manual drafting. Each has its own individual characteristics. The *visible* line is thick and dark. The *hidden* line is a series of short dashes separated by even shorter breaks. The hidden line is thinner than the visible line. *Dimension* lines and *extension* lines are solid, thin lines of approximately the same width as hidden lines. *Dimension* lines should be broken for dimensions, and have arrowheads for terminations.

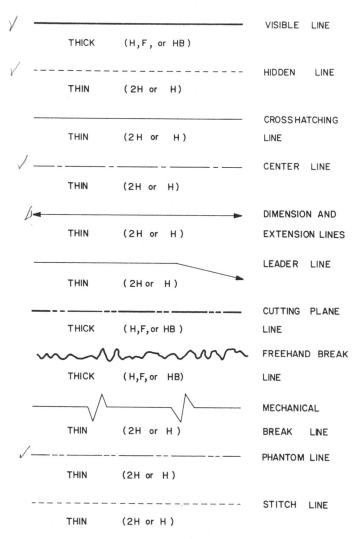

Figure 2-23 Line types used on technical drawings

The *center* line is broken with one short dash in its center. It is the same width as the hidden, dimension, and extension lines. The *phantom* line is just like the center line except that it has two dashes. The dashes are repeated approximately every two inches. The *cutting plane* line is thick like the visible line and consists of a series of long, equally spaced dashes. All lines used on technical drawings should closely match those in Figure 2-23.

Horizontal and Vertical Lines

Horizontal lines are formed by pressing the straight-edge (T-square, parallel bar, drafting machine scale, and so forth) against the worksheet with one hand and moving the pencil with the other. Uniformity of line widths and weights can be achieved by holding the pencil at approximately 60 degrees from the drawing surface, maintaining an even pressure downward, and slowly revolving the pencil axially as it moves across the drawing surface, Figure 2-24. This keeps the lead tip symmetrical.

Vertical lines are created according to the same principles, except that the drafter's hand moves upward rather than from left to right. The angle of inclination, the amount of pressure, and the rotating motion are the same as they are for horizontal lines.

Angular Lines

Many modern devices are available to assist drafters in making angular lines. These include protractors, adjustable triangles, and adjustable arms on drafting machines. However, most angular lines can be created simply by using the standard 30°-60° and 45° triangles alone, and in various combinations, Figures 2-25, 2-26, 2-27, and 2-28. These standard tools create angles of 15°, 30°, 45°, 60°, and 75°.

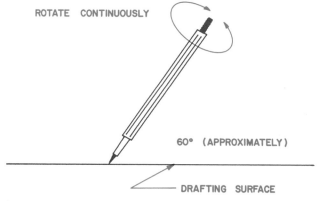

Figure 2-24 Maintaining uniformity of lines

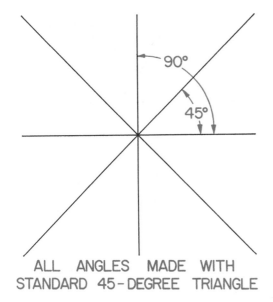

ALL ANGLES MADE WITH
STANDARD 45-DEGREE TRIANGLE

Figure 2-25 Making angular lines with the 45° triangle

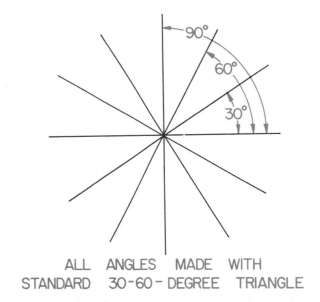

ALL ANGLES MADE WITH
STANDARD 30-60-DEGREE TRIANGLE

Figure 2-26 Making angular lines with the 30°-60° triangle

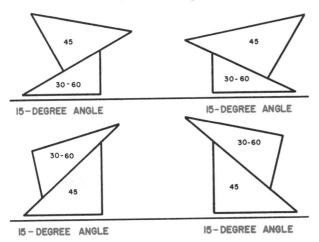

Figure 2-27 Making 15° angles

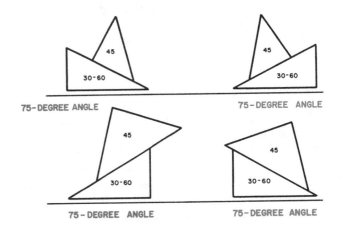

Figure 2-28 Making 75° angles

Parallel Lines

Parallel lines can be created in a number of different ways. Vertical (and horizontal) parallel lines are made by simply moving the straightedge the required distance and making each successive line, Figure 2-29.

Parallel lines at angles can be created by using the 30°-60° and 45° triangles in combination much the same as they are used for making angular lines. When using triangles to create angular lines, the first line is created at the desired angle. Aligning one edge of a

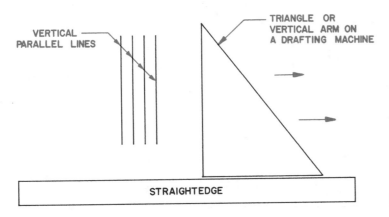

Figure 2-29 Making vertical parallel lines

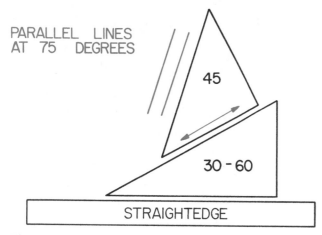

Figure 2-30 Creating successive parallel lines

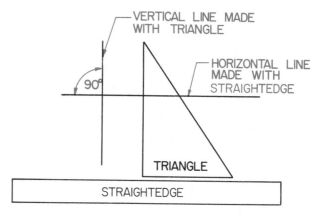

Figure 2-31 Making perpendicular lines

triangle to the line, register any side of the second triangle against one of the nonaligned edges of the first triangle. Holding the second triangle to prevent it from moving, and sliding the first triangle along the engaged edge of the second triangle, will reposition the originally aligned edge to any desired parallel position. Successive parallel lines are created in the same way, Figure 2-30.

Perpendicular Lines

Drawing perpendicular lines can be accomplished in a manner similar to drawing parallel lines. Horizontal and vertical perpendicular lines can be created using a straightedge and a triangle, Figure 2-31.

Creating a line perpendicular to a nonhorizontal or nonvertical line is accomplished by using triangles in conjunction with a straightedge, Figure 2-32.

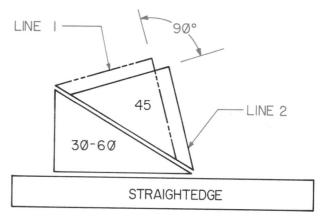

Figure 2-32 Creating a line perpendicular to a nonvertical or nonhorizontal line

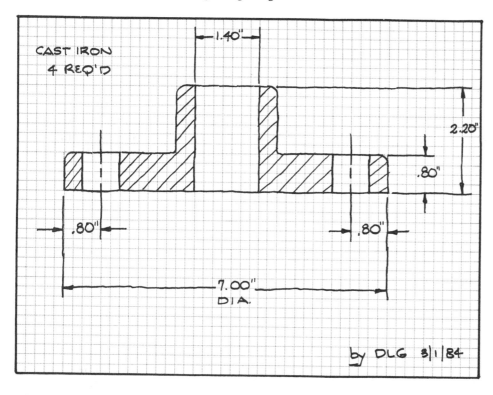

Figure 2-33
Typical design sketch

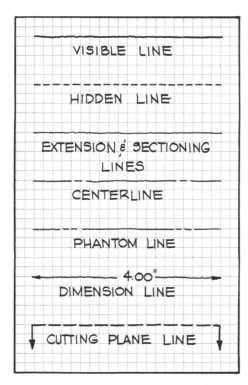

Figure 2-34 Lines used in sketching

Line 1 in this figure is drawn first. Then the 45° triangle is slid along the 30°-60° triangle and the perpendicular line is created with the opposite side of the 45° triangle.

Sketching

Even in the world of high technology and computers, sketching is still one of the most important skills for drafters and designers. Sketching is one of the first steps in communicating ideas for a design, and it is used in every step thereafter. It is common prac-

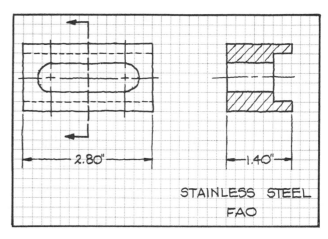

Figure 2-35 Sample design sketch

tice for designers to prepare sketches that are turned over to drafters for conversion to finished working drawings. Figure 2-33 is an example of a typical design sketch.

Sketching Lines

The lines used in creating sketches closely correspond to those used in creating technical drawings except, of course, that they are not as sharp and crisp. Figure 2-34 illustrates the various types of lines used in making sketches.

The basic line types are: visible line, hidden line, center line, dimension line, sectioning line, extension line, and cutting plane line. These lines represent the various lines available for creating sketches. The character of each line, as illustrated in Figure 2-35, should be closely adhered to when making sketches.

Types of Sketches

The types of sketches correspond to the types of technical drawings. There are four types of sketching: orthographic, axonometric, oblique, and perspective, Figures 2-36, 2-37, 2-38 and 2-39.

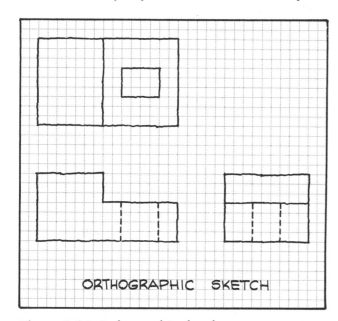

Figure 2-36 Orthographic sketch

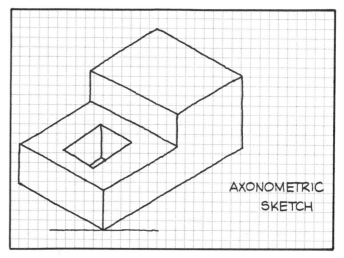

Figure 2-37 Axonometric sketch

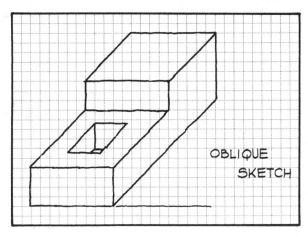

Figure 2-38 Oblique sketch

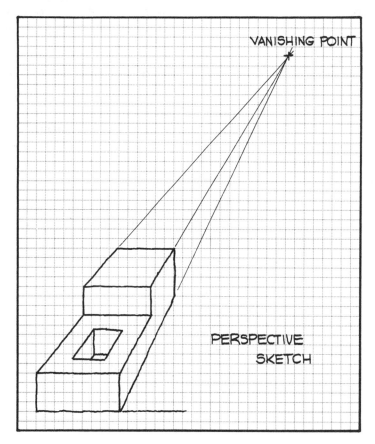

Figure 2-39 Perspective sketch

Orthographic sketching relates to flat, graphic facsimiles of a subject showing no depth. Six principal views of a subject may be incorporated in an orthographic sketch: top, front, bottom, back, right side, and left side, Figure 2-40. The views selected for use in a sketch depend on the nature of the subject and the judgment of the sketcher.

Axonometric sketching may be one of three types: isometric, dimetric, or trimetric, Figure 2-41. The type most frequently used is *isometric*, in which length and

width lines recede at 30° to the horizontal and height lines are vertical, Figure 2-42. In sketching, the use of these terms is academic as they relate to proportional scales and angle positions of length, width, and depth, which are only estimated in sketching.

Oblique sketching involves a combination of a flat, orthographic front surface with depth lines receding at a selected angle (usually 45°), Figure 2-43.

Perspective sketching involves creating a graphic facsimile of the subject. Consequently, depth lines must

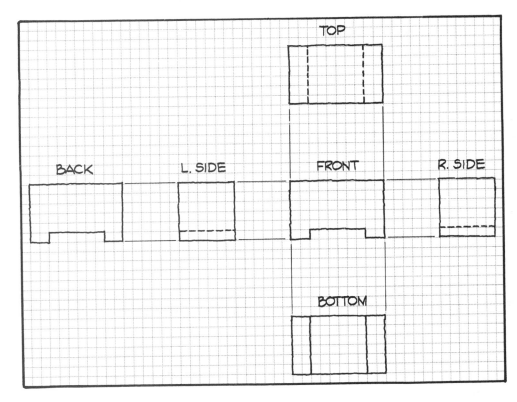

Figure 2-40
Six principal views in orthographic sketching

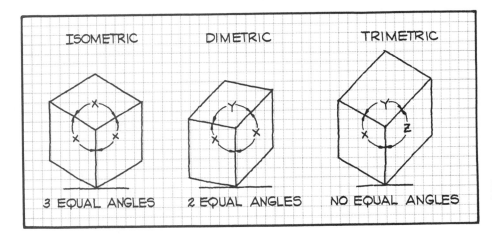

Figure 2-41
Three types of axonometric sketches

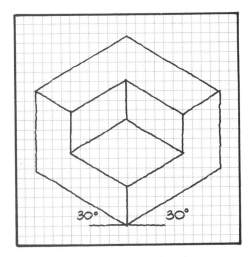

Figure 2-42 Isometric sketch

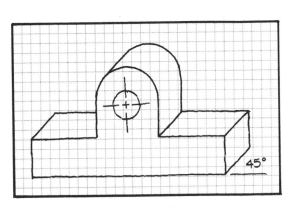

Figure 2-43 Oblique sketch

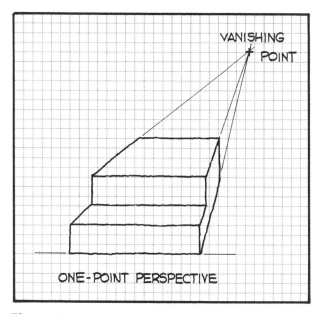

Figure 2-44 One-point perspective sketch

recede to a hypothetical vanishing point (or points), Figures 2-44 and 2-45. In fact, all pictorial sketches naturally tend to assume characteristics of perspective sketches as a result of how the eye views the apparent relative proportions of objects. This is not necessarily undesirable.

Sketching Materials

An advantage of sketching is that it requires very few material aids. Whereas drafters must have a complete collection of tools, equipment, and materials in order to do working drawings, sketching requires nothing more than a pencil and a piece of paper. It is not uncommon for a sketch to be drawn on a paper napkin during a hurried luncheon meeting.

Sketching done in an office environment requires three basic materials: pencil, media (paper or graph paper), and an eraser. Graph paper simplifies the sketching process considerably, especially for students just learning, and should be used freely.

Sketching Techniques

Sketches, as with drawings, consist of straight and curved lines. With practice, drafters can become skilled in creating neat, sharp, clear examples — straight or curved — of all the various lines types introduced previously. When sketching, the following general rules apply.

1. Hold the pencil firmly, but not so tightly as to create tension or hand fatigue.

2. Grip the pencil approximately one inch to one and one-half inches up from the point.

3. Maintain a comfortable angle between the pencil and the sketching strokes.

4. Draw horizontal lines from left to right using short, slightly overlapping strokes.

5. Draw vertical lines from top to bottom using short, slightly overlapping strokes.

6. Draw curved lines using short slightly overlapping strokes.

In addition to these general rules, some specific techniques are used in making the various line types for sketching.

Sketching Straight Lines and Curves

Making straight lines on graph paper is a simple process of guiding the pencil using the existing lines.

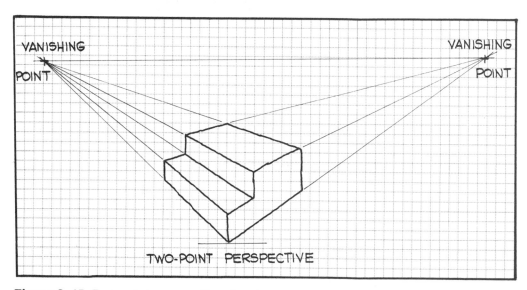

Figure 2-45 Two-point perspective sketch

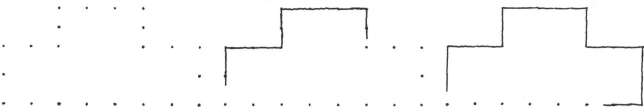

Figure 2-46 Dots used as guides in sketching

Figure 2-47 Using dots in sketching lines

Figure 2-48 Continuing the sketch by connecting the dots

If graph paper is not available, pencil dots can be positioned to plot the path of the line, Figure 2-46. In this figure, the sketcher enters a series of pencil dots on the paper which provide a basic outline as to the shape of the object. Then, using a series of short, slightly overlapping strokes, the pencil dots are connected, Figures 2-47, 2-48, and 2-49. This technique is also used for curved lines, Figure 2-50.

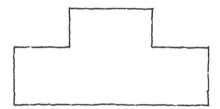

Figure 2-49 Completing the sketch made by using dots as guides

Sketching Circles

Figure 2-51 illustrates a series of six steps that can be used for sketching a circle. Vertical and horizontal center lines are sketched, which positions the center of the circle (Step 1) and the radial distances of the desired circle size are marked on each of these lines, equidistant from the center (Step 2). A square is drawn symmetrically around the center, with the sides located at the radial line marks (Step 3). On the diagonals of the square (Step 4), the radial distances are again marked off from the center (Step 5). This provides four positions for the circumference to pass through at the sides of the square, and four more positions on the diagonals of the square. The right half of the circle is sketched in from top to bottom using short, slightly overlapping strokes, and then the left half is sketched in the same manner (Step 6).

Sketching Ellipses

A similar technique is used for sketching ellipses, except that the square becomes a rectangle, Figure 2-52. Ellipses are oriented on the object being sketched as shown in the diagram in Figure 2-53.

Proportion in Sketching

Sketches are not done to scale, but it is important that they are made proportionately accurate. All of the various components of a sketch should be kept in proportion to those of the actual object. This technique takes a great deal of practice to master.

Some methods for achieving proportion recommend using a pencil or a strip of paper as a simulated scale. These techniques are not only unrealistic in terms of the real world, they defeat the very purpose of sketching. A skilled sketcher must learn to maintain proportion without the use of tools and aids. The best device for accomplishing proportion in sketching is the human eye. With practice, the drafter can become proficient in maintaining proportion without the use of extraneous, time-consuming devices. The following general rules relating to proportion will also help.

Step 1. In sketching, use graph paper whenever possible.

Step 2. Examine the object to be sketched and mentally break it into its component parts.

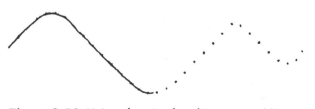

Figure 2-50 Using dots in sketching curved lines

Figure 2-51 Sketching a circle

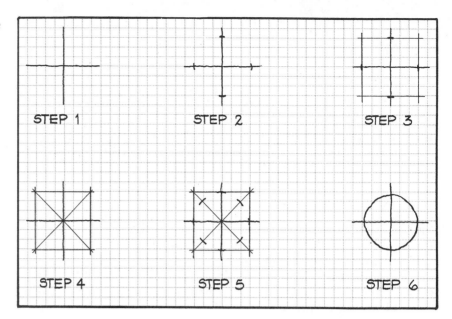

Figure 2-52
Sketching an ellipse

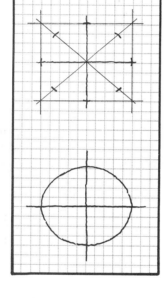

Figure 2-53 Orienting an ellipse on an object

Step 3. Beginning with the largest components (length and width), estimate the proportion, such as the length is 4/3 times the width or 5/2 times the width, and so on.

Step 4. Lay out the largest component according to the proportions decided upon in Step 3. Use

construction line squares and rectangles to block in irregularly shaped components, Figure 2-54.

Repeat Steps 3 and 4 until the entire object is finished.

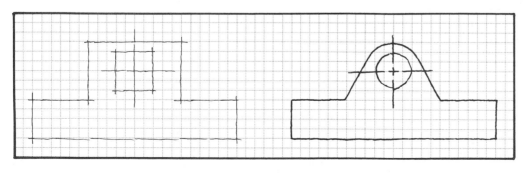

Figure 2-54 Blocking in components

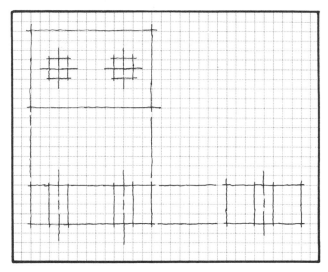

Figure 2-55 Blocking in an orthographic sketch

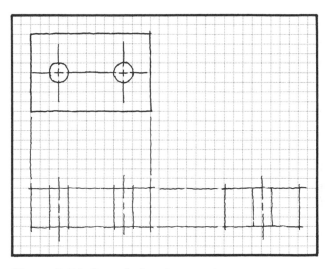

Figure 2-56 Completing the top view

Orthographic Sketching

Orthographic sketching may involve sketching any combination of the six principal views of the subject. The top, front, and right-side views are normally selected for representing an object in an orthographic sketch. However, these views are not always appropriate. The sketcher must learn to choose the most appropriate views. These are the views that show the most detail and the fewest hidden lines. A good rule of thumb to use in selecting views is to select the views which would give you all of the information you would need if you had to make the object yourself.

Once the views have been selected, the orthographic sketch may be laid out using the techniques set forth earlier in this chapter. To ensure that the sketched views align, the entire sketch should be blocked in before adding details, Figure 2-55. Once

the layout is blocked in, the details can be added one view at a time, Figures 2-56, 2-57, and 2-58.

Axonometric Sketching

As was mentioned earlier, there are three types of axonometric projection: isometric, dimetric, and trimetric. Isometric projection is used in sketching. Dimetric and trimetric projection have little application in sketching, due to the difficulty in proportioning scale values of length, width, and height. Isometric views have the same scaling value in all three directions, eliminating the need to vary proportions among the three directions. In an isometric sketch, height lines are vertical, and width and length lines recede at approximately 30° and 150° (180°−30°) from the horizontal.

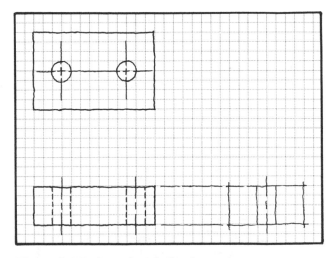

Figure 2-57 Completing the front view

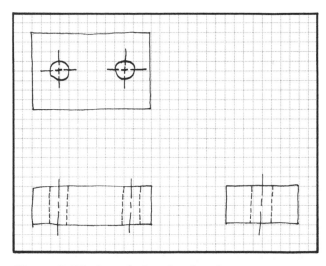

Figure 2-58 Completing the sketch

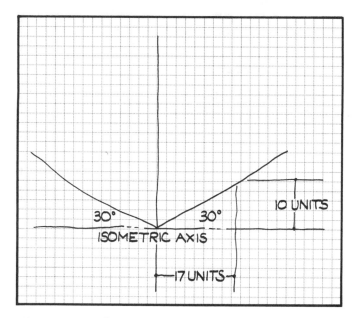

Figure 2-59 The isometric axis in sketching

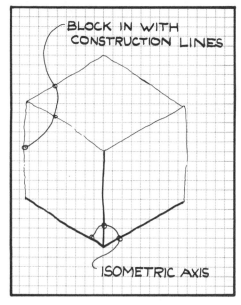

Figure 2-60 Blocking in the object

The first step in creating an isometric sketch is to lay out the isometric axis, Figure 2-59. All normal lines will be parallel to one of the axis lines. The next step is to block in the object using construction lines, Figure 2-60. Five steps in the development of an isometric sketch are shown in Figures 2-61, 2-62, 2-63, 2-64 and 2-65.

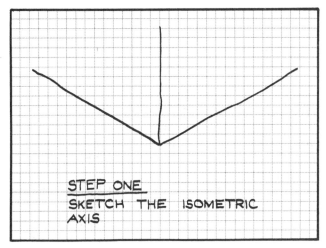

Figure 2-61 Step one in making an isometric sketch

Oblique Sketching

Oblique sketching involves laying out the front view of an object, and showing the depth lines receding at an angle (usually 45°) from the horizontal. Oblique sketching is particularly useful for dealing with an object having round components. Oblique sketching allows round components to be drawn round, rather than elliptical.

Using the blocking in method, the flat front surface of the object is laid out. The depth is then blocked in using parallel lines, and the sketch is completed by outlining the exposed profile of the rear surface. Figures 2-66, 2-67, and 2-68 illustrate three steps in creating an oblique sketch.

Perspective Sketching

Perspective sketching closely approximates how the human eye actually sees an object. Two common types of perspective sketches are one-point and two-point perspectives.

A one-point perspective sketch is similar to an oblique sketch, except that depth lines recede to a

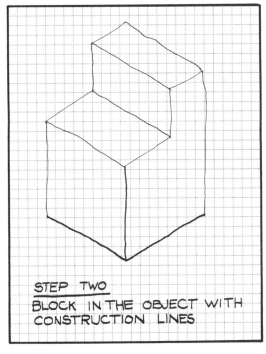

Figure 2-62 Step two in making an isometric sketch

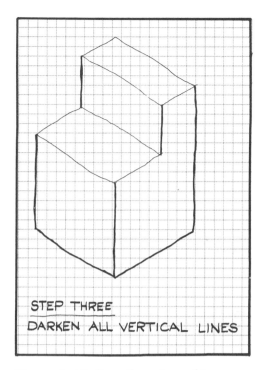

STEP THREE
DARKEN ALL VERTICAL LINES

Figure 2-63 Step three in making an isometric sketch

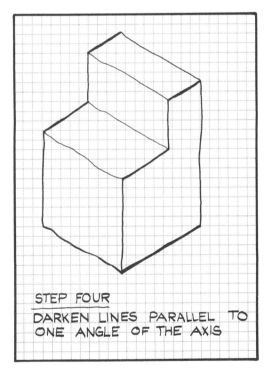

STEP FOUR
DARKEN LINES PARALLEL TO ONE ANGLE OF THE AXIS

Figure 2-64 Step four in making an isometric sketch

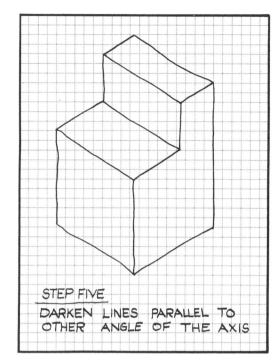

STEP FIVE
DARKEN LINES PARALLEL TO OTHER ANGLE OF THE AXIS

Figure 2-65 Step five in making an isometric sketch

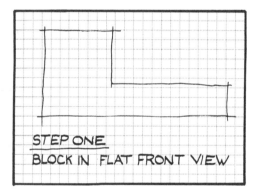

STEP ONE
BLOCK IN FLAT FRONT VIEW

Figure 2-66 Step one in making an oblique sketch

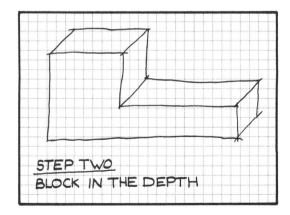

STEP TWO
BLOCK IN THE DEPTH

Figure 2-67 Step two in making an oblique sketch

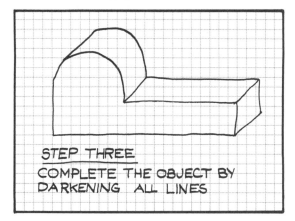

Figure 2-68 Step three in making an oblique sketch

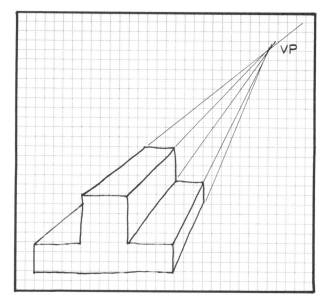

Figure 2-69 One-point perspective sketch

vanishing point instead of receding parallel to one another, Figure 2-69. In constructing a one-point perspective, the following procedures apply.

Step 1. Lay out the flat front surface of the object using the blocking in method, Figure 2-70.

Step 2. Select and mark a single vanishing point. Project all points on the front surface back to the vanishing point, Figure 2-71.

Step 3. Estimate the depth of the object and mark it off on all line projectors, Figure 2-72.

Step 4. Complete the sketch by outlining the exposed profile of the rear surface, Figure 2-73.

A two-point perspective resembles an isometric sketch, except that width and depth lines recede to the left and right vanishing points rather than receding in parallel, Figure 2-74.

In constructing a two-point perspective, the following procedures apply.

Step 1. Lay out the two-point perspective frame which consists of the vertical height line, the vanishing point left, the vanishing point right, and the receding lines (all estimated locations), Figure 2-75. The horizon line when positioned below the view provides a view of the bottom of the object, and when positioned above shows the top. The vanishing points must be on the horizon.

Step 2. Block in the object, estimating the length and width for proportion, Figure 2-76.

Step 3. Lay out the details, lightly giving special attention to proportion, Figure 2-77.

Step 4. Complete the two-point perspective sketch, Figure 2-78.

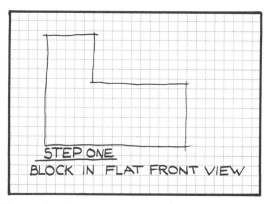

Figure 2-70 Step one in making a one-point perspective sketch

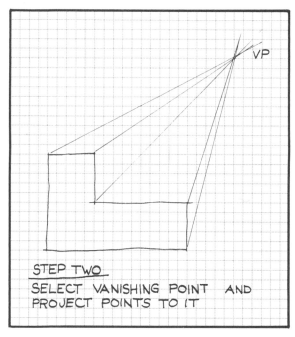

Figure 2-71 Step two in making a one-point perspective sketch

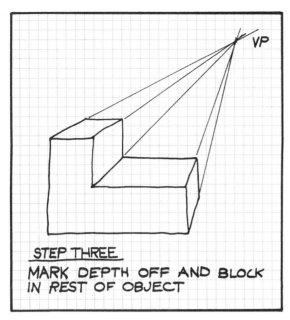

STEP THREE
MARK DEPTH OFF AND BLOCK
IN REST OF OBJECT

Figure 2-72 Step three in making a one-point perspective sketch

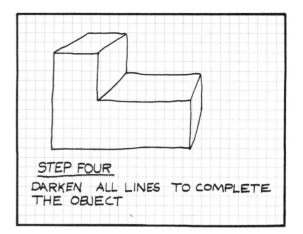

STEP FOUR
DARKEN ALL LINES TO COMPLETE
THE OBJECT

Figure 2-73 Step four in making a one-point perspective sketch

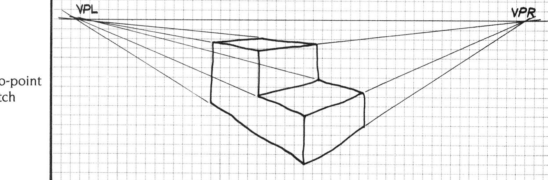

Figure 2-74 Two-point perspective sketch

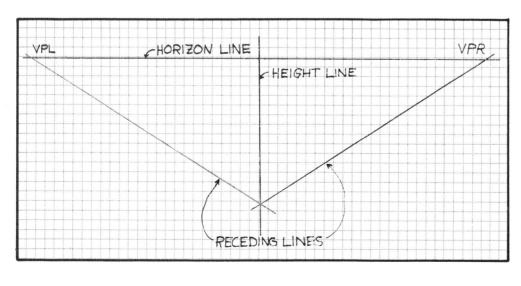

Figure 2-75 Laying out the two-point perspective frame

Figure 2-76
Blocking in the object

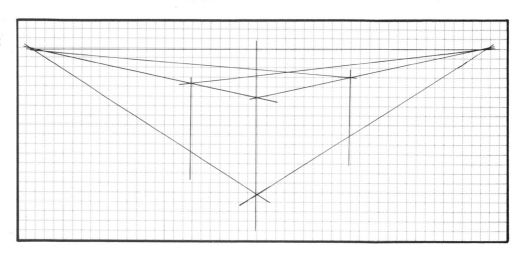

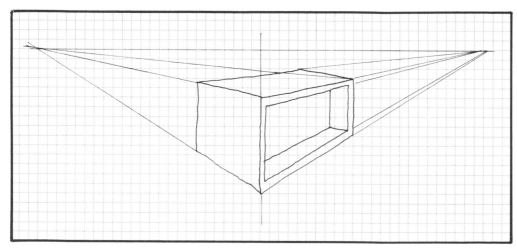

Figure 2-77
Laying out the details

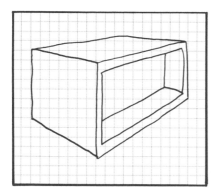

Figure 2-78 Completing the
two-point perspective sketch

Review

1. What is the standard style of freehand lettering on technical drawings?

2. What is the slope of slanted lettering?

3. What are the five characteristics of good
   freehand lettering?

4. How many strokes are required to make a
   letter B?

5. How high should fractions be?

6. How much space should be left between
   words?

7. What are two grades of lead commonly used
   for freehand lettering?

8. What are guidelines?

9. Why are guidelines important?

10. A setting of 8 on a lettering guide will
    produce letters of what height?

11. Define the term *line work*.

12. Name ten basic lines types used on technical
    drawings.

13. What are two advantages of sketching over
    mechanical drawing?

14. What are the six principal views of ortho-
    graphic projection?

15. Which axonometric projection is preferred
    for sketching?

16. What kind of projection exhibits a circle as an
    elliptic shape?

Chapter Two Problems

The following problems are intended to give the beginning drafter practice in using the various lettering, line work, and sketching techniques that are fundamental to drafting. Each problem should be studied carefully, while following the individual instructions.

Problem 2-1

Lay out an A-size sheet for lettering practice, according to the figure shown. Use 1/2" borders. Use 5/16" guidelines, using a lettering guide if available, and letter 3 rows of characters in the manner shown, under each respective row. Leave 1/8" space between each row.

Problem 2-2

Repeat Problem 2-1, but use 1/4" guidelines.

Problem 2-3

Repeat Problem 2-1, but use 3/16" guidelines.

Problem 2-4

Repeat Problem 2-1, but use 1/8" guidelines.

A B C D E F G H I

J K L M N O P Q R S T

U V W X Y Z 0 1 2 3 4 5 6 7 8 9

Problems 2-1 through 2-4

Lay out an A-size sheet for lettering practice, according to the figure shown. Use 1/2" borders. Use 1/4" guidelines, using a lettering guide if available, and letter 4 rows of text in the manner shown, under each respective row. Leave a 1/8" space between each row.

Repeat Problem 2-5, but use 3/16" guidelines.

Repeat Problem 2-5, but use 1/8" guidelines.

GOOD LETTERING TAKES PRACTICE

IMPROVEMENT REQUIRES EFFORT

PATIENCE AND HARD WORK

Problems 2-5 through 2-7

Lay out an A-size sheet for lettering practice according to the figure shown. Use 1/2" borders. Use 1/4" guidelines, using a lettering guide if available. Letter one line of the first row of text. Copy two rows of the second line of text, four rows of the third line, and five rows of the last line.

Using the estimation marks provided for proportion only, sketch the top, front, and right-side views of the objects specified by Problems 2-9 through 2-13.

SINGLE-STROKE GOTHIC IS THE STANDARD

LETTERING STYLE FOR TECHNICAL DRAWING

1'-0"   1.267'   300.06'   27'-6"

3/16"   1'-7 3/4"   62'-7 1/4"   106'-2 5/8"

Problem 2-8

Problem 2-9

Sketch the figure shown.

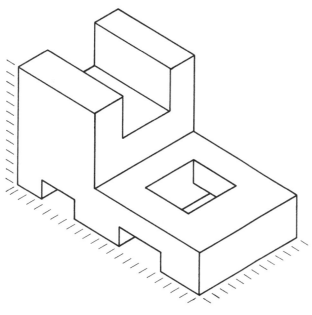

Problem 2-9

Problem 2-11

Sketch the figure shown.

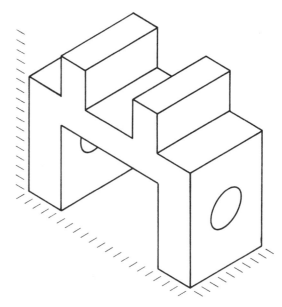

Problem 2-11

Problem 2-10

Sketch the figure shown.

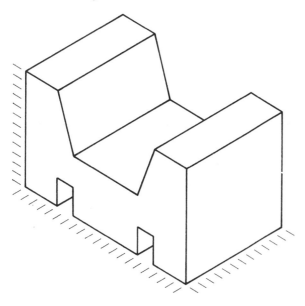

Problem 2-10

Problem 2-12

Sketch the figure shown.

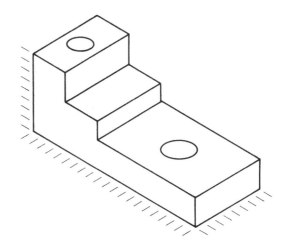

Problem 2-12

Problem 2-13

Sketch the figure shown.

Using the estimating scale provided for proportion only, make isometric sketches of the objects specified by Problems 2-14 through 2-17.

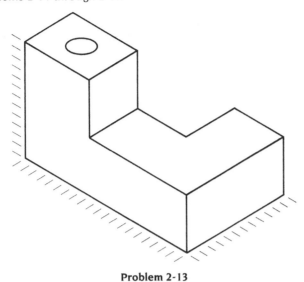

Problem 2-13

Problem 2-14

Sketch the figure shown.

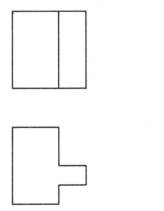

Problem 2-14

Problem 2-15

Sketch the figure shown.

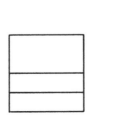

Problem 2-15

Problem 2-16

Sketch the figure shown.

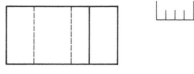

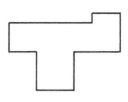

Problem 2-16

Problem 2-17

Sketch the figure shown.

Using the estimating scale provided for proportion only, make oblique sketches of the objects specified by Problems 2-18 through 2-21.

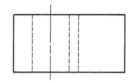

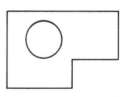

Problem 2-17

Problem 2-18

Sketch the figure shown.

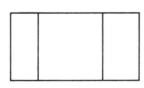

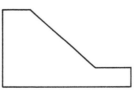

Problem 2-18

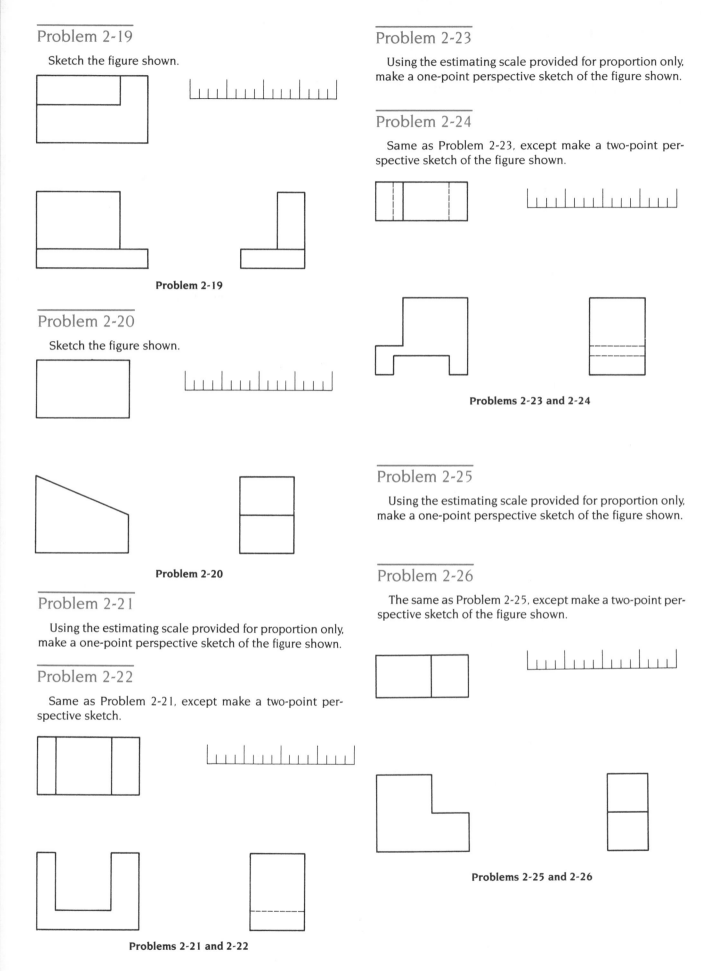

Problem 2-19

Sketch the figure shown.

Problem 2-19

Problem 2-20

Sketch the figure shown.

Problem 2-20

Problem 2-21

Using the estimating scale provided for proportion only, make a one-point perspective sketch of the figure shown.

Problem 2-22

Same as Problem 2-21, except make a two-point perspective sketch.

Problems 2-21 and 2-22

Problem 2-23

Using the estimating scale provided for proportion only, make a one-point perspective sketch of the figure shown.

Problem 2-24

Same as Problem 2-23, except make a two-point perspective sketch of the figure shown.

Problems 2-23 and 2-24

Problem 2-25

Using the estimating scale provided for proportion only, make a one-point perspective sketch of the figure shown.

Problem 2-26

The same as Problem 2-25, except make a two-point perspective sketch of the figure shown.

Problems 2-25 and 2-26

Problems 2-27 through 2-31

Practice your line work technique by duplicating the objects specified by Problems 2-14 through 2-18 on A-size sheets. Try to match all line characteristics as exactly as possible. Use the scale provided for determining line lengths.

Problems 2-32 through 2-36

Practice your line work technique by duplicating the objects specified by Problems 2-19 through 2-25 on A-size sheets. Try to match all line types as exactly as possible. Use the scale provided for determining line lengths.

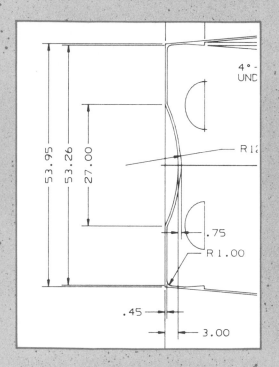

The basic techniques associated with geometric construction must be thoroughly mastered. The various procedures discussed in this chapter will be used in solving all drawing problems throughout this book, as well as later on the job. By using these geometric construction techniques, drawings will be of professional quality and accomplished in the least amount of time. It is important for the beginning drafter to thoroughly know and understand these techniques and, more importantly, to know when and where to apply them.

CHAPTER THREE

Geometric Construction

To be truly proficient in the layout of both simple and complex drawings, the drafter must know and fully understand the many geometric construction methods used. These methods are illustrated in this chapter, and are basically simple principles of pure geometry. These simple principles are used to actually develop a drawing with complete accuracy, and in the fastest time possible, without wasted motion or any guesswork. Applying these geometric construction principles give drawings a finished, professional appearance.

In laying out the various geometric constructions, it is important to use a very sharp, 4-H lead and to be extremely accurate at all times. Always draw light construction lines that can hardly be seen when held at arm's length. These light construction lines should not be erased as this takes up valuable drawing time and they can also be reused to check layout work if necessary.

Geometric Nomenclature

Points in Space

A *point* is an exact location in space or on a drawing surface, Figure 3-1. A point is actually represented

on the drawing by a crisscross at its exact location. The exact point in (drawing) space is where the two lines of the crisscross intersect. When a point is located on an existing line, a light, short dashed line or crossbar is placed on the line at the location of the exact point. Never represent a point on a drawing by a dot, except for sketching locations. This is not accurate enough, and is considered to be poor drafting practice.

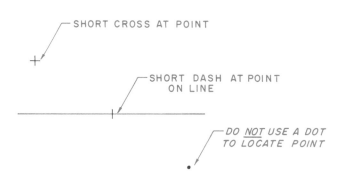

Figure 3-1 Points in space or on a surface

Figure 3-2 A straight line is the shortest distance between two points

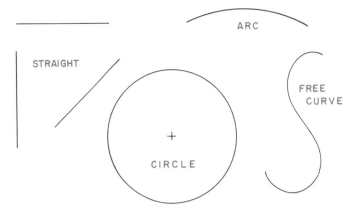

Figure 3-3 Kinds of lines

Line

A straight line is the shortest distance between two points, Figure 3-2. It can be drawn in any direction. A *line* can be straight, an arc, a circle or a free curve, as illustrated in Figure 3-3. If a line is indefinite, and the ends are not fixed in length, the actual length is a matter of convenience. If the end points of a line are important, they must be marked by means of small, mechanically drawn crossbars, as described by a point in space.

Straight lines and curved lines are considered to be parallel if the shortest distance between them remains constant. The symbol used for parallel lines is //. Lines that are tangent and at 90° are considered perpendicular. The symbol for perpendicular lines is ⊥ (singular), Figure 3-4, and ⊥'s (plural). The symbol for an angle is ∠ (singular) and ∠'s (plural). To draw an angle, use the drafting machine, a triangle, or a protractor. For extra accuracy, use the vernier on the drafting machine or a vernier protractor.

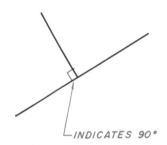

Figure 3-4 Perpendicular lines

Angle

An *angle* is formed by the intersection of two lines. There are three major kinds of angles: right angles, acute angles, and obtuse angles, Figure 3-5. The *right angle* is an angle of 90°. An *acute angle* is an angle at less than 90°, and an obtuse angle is an angle at more than 90°. Note that a straight line is 180°. Figure 3-5 also illustrates the complementary and supplementary angles of a given angle.

There are 360 degrees (360°) in a full circle. Each degree is divided into 60 minutes (60'). Each minute is divided into 60 seconds (60"). Example: 48°28'38" is read as 48 degrees, 28 minutes, and 38 seconds.

To convert minutes and seconds to decimal degrees, divide minutes by 60 and seconds by 3600.

Example:

$$21°18'27" = 21° + (18/60)° + (27/3600)° = 21.3075°$$

To convert decimal degrees to degrees, minutes, and seconds, multiply the degree decimal by 60 to obtain minutes, and the minute decimal by 60 to obtain seconds.

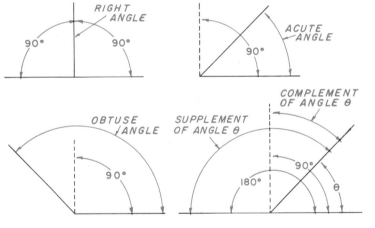

Figure 3-5 Kinds of angles

Example:

$$77.365° = 77° + (.365 \times 60)' = 77°21.9'$$
$$77°21.9' = 77°21' + (.9 \times 60)" = 77°21'54"$$

A vernier may be used to measure and read off minutes and seconds of a degree. The vernier scale is discussed in Chapter 1.

Triangle

A *triangle* is a closed plane figure with three straight sides and three interior angles. The sum of the three

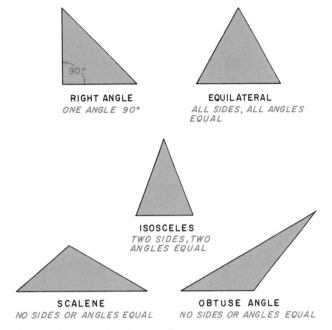

RIGHT ANGLE
ONE ANGLE 90°

EQUILATERAL
ALL SIDES, ALL ANGLES EQUAL

ISOSCELES
TWO SIDES, TWO ANGLES EQUAL

SCALENE
NO SIDES OR ANGLES EQUAL

OBTUSE ANGLE
NO SIDES OR ANGLES EQUAL

Figure 3-6 Kinds of triangles

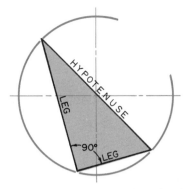

Figure 3-7 Any triangle inscribed within a semicircle is a right triangle

internal angles is always exactly 180°, half of the 360° in a full circle. Figure 3-6 shows the various kinds of triangles: a right angle, an equilateral triangle, an isosceles triangle, a scalene triangle, and an obtuse angle triangle.

A *right triangle* is a triangle having a right angle or an angle of 90°. The two sides forming the right angle are called legs, and the third side (the longest) is the hypotenuse. Any triangle inscribed in a semicircle is a right triangle, Figure 3-7.

An *equilateral triangle*, as its name implies, is a triangle with all sides of equal length. All of its interior angles are also equal. An *isosceles triangle* has two sides of equal length and two equal interior angles. A *scalene* *triangle* has no equal sides or angles. An *obtuse angle triangle* is a triangle having an obtuse angle greater than 90°, with no equal sides.

Polygon

A *polygon* is a closed plane figure with three or more straight sides, Figure 3-8. More specifically, shown in the figure are regular polygons, meaning that all sides are equal in each of these examples. The most important of these polygons as they relate to drafting are probably the triangle with three sides, the square with four sides, the hexagon with six sides, and the octagon with eight sides.

Quadrilateral

A *quadrilateral* is a plane figure bounded by four straight sides. When opposite sides are parallel, the quadrilateral is also considered to be a *parallelogram*, Figure 3-9.

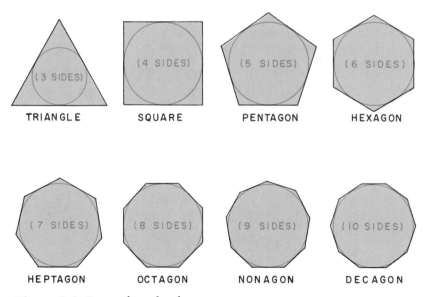

TRIANGLE (3 SIDES)   SQUARE (4 SIDES)   PENTAGON (5 SIDES)   HEXAGON (6 SIDES)

HEPTAGON (7 SIDES)   OCTAGON (8 SIDES)   NONAGON (9 SIDES)   DECAGON (10 SIDES)

Figure 3-8 Examples of polygons

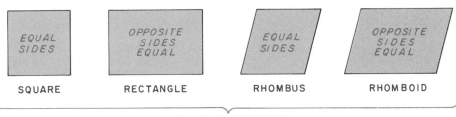

SQUARE   RECTANGLE   RHOMBUS   RHOMBOID

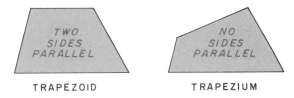

PARALLELOGRAMS

TRAPEZOID   TRAPEZIUM

Figure 3-9 Quadrilaterals

Circle

A *circle* is a closed curve with all points on the circle at the same distance from the center point. The major components of a circle are the diameter, the radius, and the circumference, Figure 3-10.

The *diameter* of a circle is the straight distance from one outside curved surface through the center point to the opposite outside curved surface. The diameter of a circle is twice the size of the radius.

The *radius* of a circle is the distance from the center point to the outside curved surface. The radius is half the diameter, and is used to set the compass when drawing a diameter.

The *circumference* of a circle is the distance around the outer surface of the circle. To calculate the circumference of a circle, multiply the value of π (use the approximation 3.1416) by the diameter. A chart similar to the one found in the appendix of this text may be used.

Other important parts of a circle are the central angle, the sector, the quadrant, the chord, and the segment, Figure 3-11.

A *central angle* is an angle formed by two radial lines from the center of the circle.

A *sector* is the area of a circle lying between two radial lines and the circumference.

A *quadrant* is a sector with a central angle of 90°, and usually with one of the radial lines oriented horizontally.

A *chord* is any straight line whose opposite ends terminate on the circumference of the circle. (A diameter is a chord passing through the center of the circle.)

A *segment* is the smaller portion of a circle separated by a chord.

Concentric circles are two or more circles with a common center point, Figure 3-12.

Eccentric circles are two or more circles without a common center point, Figure 3-12.

A *semicircle* is half of a circle.

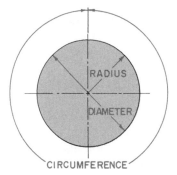

NOTE, THE COMPASS IS SET TO THE *RADIUS*

Figure 3-10 The major components of a circle

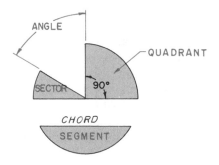

Figure 3-11 Other parts of a circle

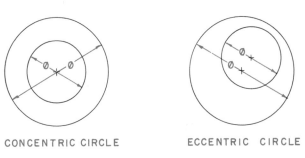

CONCENTRIC CIRCLE   ECCENTRIC CIRCLE

Figure 3-12 Concentric and eccentric circles

Figure 3-13 Polyhedrons (solids)

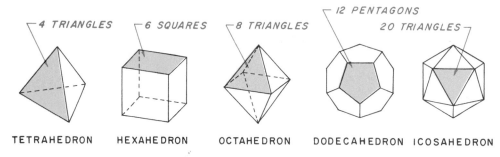

4 TRIANGLES   6 SQUARES   8 TRIANGLES   12 PENTAGONS   20 TRIANGLES

TETRAHEDRON   HEXAHEDRON   OCTAHEDRON   DODECAHEDRON   ICOSAHEDRON

REGULAR SOLIDS

Polyhedron

A *polyhedron* is a solid object bounded by plane surfaces. Each surface is called a face. If the faces are equal, regular polygons, the solid figure is a regular polyhedron, Figure 3-13.

Prism

A *prism* is a solid having ends that are parallel matched polygons, and sides that are parallelograms, Figure 3-14. This definition also applies to round or circular objects, such as a cylinder. When the polygon on one end of a prism is not parallel to the other end, it is said to be *truncated*. The *altitude* of a prism is the perpendicular distance between its end polygons (or bases).

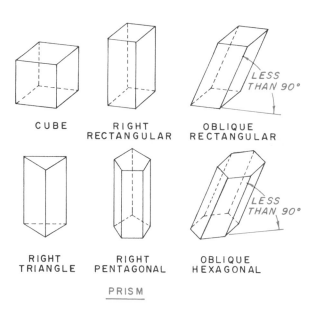

CUBE   RIGHT RECTANGULAR   OBLIQUE RECTANGULAR

LESS THAN 90°

RIGHT TRIANGLE   RIGHT PENTAGONAL   OBLIQUE HEXAGONAL

LESS THAN 90°

PRISM

Figure 3-14 Prism

Pyramid and Cone

A *pyramid* is a polyhedron having a polygon as its base. Three or more triangles form its lateral sides which meet at a common vertex, Figure 3-15. A *cone* is a pyramid with a central axis, and an infinite number of sides which form a continuous curved lateral surface. When the vertex of a pyramid or cone has been removed by a plane that intersects all the lateral sides (which forms a new polygon), the pyramid or cone is said to be *truncated*, Figure 3-16.

Sphere

A *sphere* is a closed surface, every point on which is equal from a common point or center, Figure 3-17. If a sphere is cut into two equal parts, the parts are called *hemispheres*. *Poles* are two reference measuring positions on the surface of the sphere on opposite sides of its center.

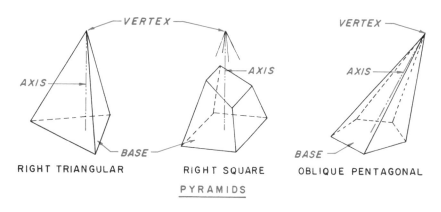

VERTEX

AXIS   AXIS

BASE

RIGHT TRIANGULAR   RIGHT SQUARE

PYRAMIDS

VERTEX

AXIS

BASE

OBLIQUE PENTAGONAL

Figure 3-15 Pyramid

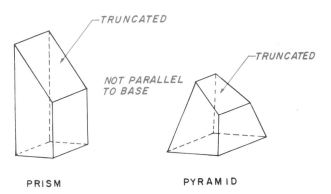

PRISM　　　　　　PYRAMID

Figure 3-16 Truncated pyramid

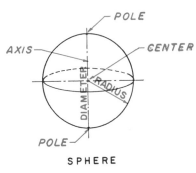

SPHERE

Figure 3-17 Sphere

Elemental Construction Principles

The remaining portion of this chapter is devoted to illustrating step-by-step the many geometric construction principles used by the drafter to develop various geometric forms. It is important that each step be fully understood and followed. As the beginning drafter uses these geometric construction principles, various shortcuts will become evident, thus reducing the drawing time and increasing accuracy even more. At the end of the chapter, each of these techniques is incorporated or used in some way to complete the various given problems.

How To Bisect a Line

To *bisect* a line means to divide it in half or to find its center point. In the given process, a line will also be constructed at the exact center point at exactly 90°.

Given: Line A-B, Figure 3-18A.

Step 1. Set the compass approximately two-thirds of the length of line A-B and swing an arc from point A, Figure 3-18B.

Step 2. Using the exact same compass setting, swing an arc from point B, Figure 3-18C.

Step 3. At the two intersections of these arcs, locate points X and Y, Figure 3-18D.

Step 4. Draw a straight line connecting point X with point Y. Where this line intersects line A-B is the bisect of line A-B, Figure 3-18E. Line X-Y is also perpendicular to line A-B at the exact center point.

GIVEN:

A ├─────────────┤ B

Figure 3-18A How to bisect a line

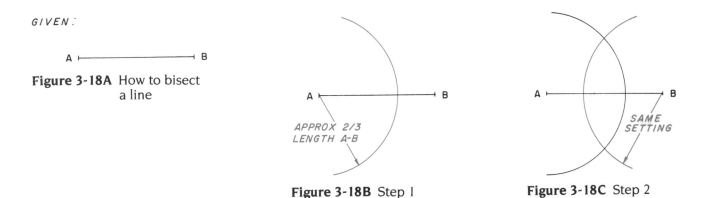

Figure 3-18B Step 1

Figure 3-18C Step 2

Figure 3-18D Step 3

Figure 3-18E Step 4

Figure 3-19A How to bisect an angle

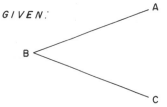
GIVEN:

Figure 3-19B Step 1
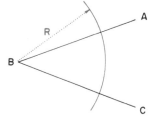

Figure 3-19C Step 2

Figure 3-19D Step 3

How To Bisect an Angle

To *bisect* an angle means to divide it in half or to cut it into two equal angles.

Given: Angle ABC, Figure 3-19A.

Step 1. Set the compass at any convenient radius and swing an arc from point B, Figure 3-19B.

Step 2. Locate points X and Y on the legs of the angle, and swing two arcs of the same identical length from points X and Y, respectively, Figure 3-19C.

Step 3. Where these arcs intersect, locate point Z. Draw a straight line from B to Z. This line will bisect angle ABC and establish two equal angles: ABZ and ZBC, Figure 3-19D.

How To Draw an Arc or Circle (Radius) through Three Given Points

Given: Three points in space at random: A, B, and C, Figure 3-20A.

Step 1. With straight lines, lightly connect points A to B, and B to C, Figure 3-20B.

Step 2. Using the method outlined for bisecting a line, bisect lines A-B and B-C, Figure 3-20C.

Step 3. Locate point X where the two extended bisectors meet. Point X is the exact center of the arc or circle, Figure 3-20D.

Step 4. Place the point of the compass on point X and adjust the lead to any of the points A, B, or C (they are the same distance), and swing the circle. If all work is done correctly, the arc or circle should pass through each point, as shown in Figure 3-20E.

GIVEN:

Figure 3-20A How to draw an arc or circle (radius) through three given points

Figure 3-20B Step 1

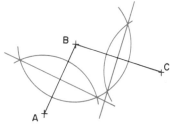

Figure 3-20C Step 2

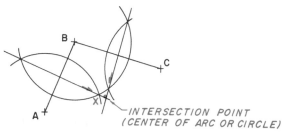

INTERSECTION POINT
(CENTER OF ARC OR CIRCLE)

Figure 3-20D Step 3

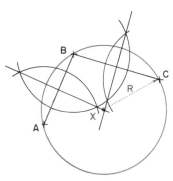

Figure 3-20E Step 4

GIVEN:

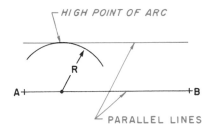

Figure 3-21B Step 1

Figure 3-21A How to draw a line parallel
to a straight line at a
given distance

Figure 3-21C Step 2

How To Draw a Line Parallel to a Straight Line at a Given Distance

Given: Line A-B, and a required distance to the parallel line, Figure 3-21A.

Step 1. Set the compass at the required distance to the parallel line. Place the point of the compass at any location on the given line, and swing a light arc whose radius is the required distance, Figure 3-21B.

Step 2. Adjust the straightedge of either a drafting machine or an adjustable triangle so that it lines up with line A-B, slide the straightedge up or down to the extreme high point of the arc, then draw the parallel line, Figure 3-21C.

Note: The distance between parallel lines is measured on any line that is perpendicular to both.

How To Draw a Line Parallel to a Curved Line at a Given Distance

Given: Curved line A-B, and a required distance to the parallel line, Figure 3-22A.

Step 1. Set the compass at the required distance to the parallel line. Starting from either end of the curved line, place the point of the compass on the given line, and swing a series of light arcs along the given line, Figure 3-22B.

Step 2. Using an irregular curve, draw a line along the extreme high points of the arcs, Figure 3-22C.

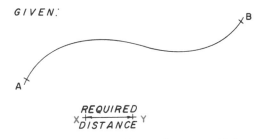

Figure 3-22A How to draw a line parallel to a curved line at a given distance

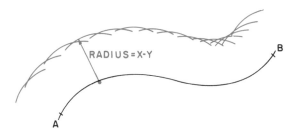

Figure 3-22B Step 1

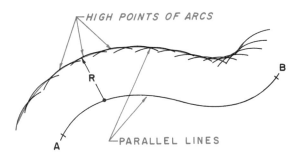

Figure 3-22C Step 2

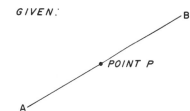

Figure 3-23A
How to draw a
perpendicular to a line at a point (method 1)

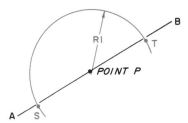

Figure 3-23B
Step 1

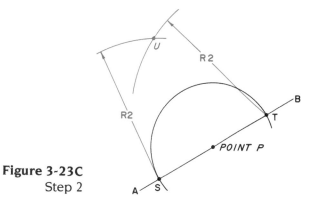

Figure 3-23C
Step 2

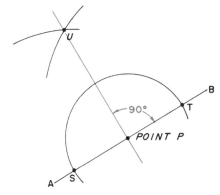

Figure 3-23D
Step 3

How To Draw a Perpendicular to a Line at a Point (Method 1)

Given: Line A-B with point P on the same line, Figure 3-23A.

Step 1. Using P as the center, make two arcs of equal radius or one continuous arc (R1) to intercept line A-B on either side of point P, at points S and T, Figure 3-23B.

Step 2. Swing larger but equal arcs (R2) from each of points S and T to cross each other at point

U, Figure 3-23C.

Step 3. A line from U to P is perpendicular to line A-B at point P, Figure 3-23D.

How To Draw a Perpendicular to a Line at a Point (Method 2)

Given: Line A-B with point P on the line, Figure 3-24A.

Step 1. Swing an arc of any convenient radius whose center O is at any convenient location

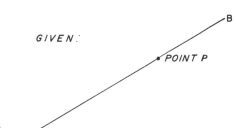

Figure 3-24A
How to draw a
perpendicular to a line at a point (method 2)

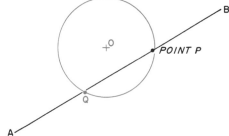

Figure 3-24B
Step 1

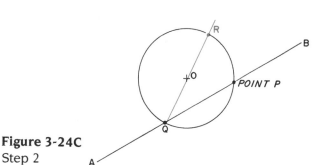

Figure 3-24C
Step 2

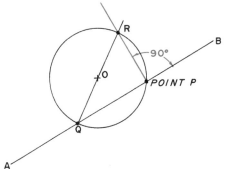

Figure 3-24D
Step 3

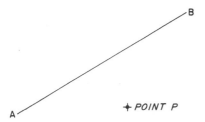

Figure 3-25A How to draw a perpendicular to a line from a point not on the line

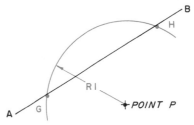

Figure 3-25B Step 1

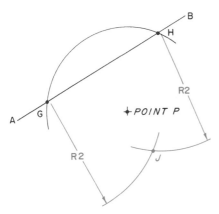

Figure 3-25C Step 2

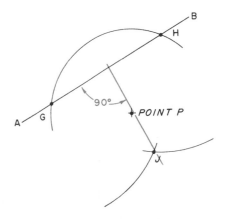

Figure 3-25D Step 3

NOT on line A-B, but positioned to make the arc cross line A-B at points P and Q, Figure 3-24B.

Step 2. A line from point Q through center O intercepts the opposite side of the arc at point R, Figure 3-24C.

Step 3. Line R-P is perpendicular to line A-B. (A right triangle has been inscribed in a semicircle.) See Figure 3-24D.

How To Draw a Perpendicular to a Line from a Point Not on the Line

Given: Line A-B and point P, Figure 3-25A.

Step 1. Using P as a center, swing an arc (R1) to intercept line A-B at points G and H, Figure 3-25B.

Step 2. Swing larger, but equal length arcs (R2) from each of the points G and H to intercept each other at point J, Figure 3-25C.

Step 3. Line P-J is perpendicular to line A-B, Figure 3-25D.

How To Divide a Line into Equal Parts

Given: Line A-B, Figure 3-26A.

Step 1. Draw a line 90° from either end of the given line. A 90° line from point B is illustrated in Figure 3-26B, but either end or either direction will work as well.

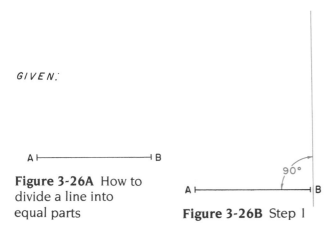

Figure 3-26A How to divide a line into equal parts

Figure 3-26B Step 1

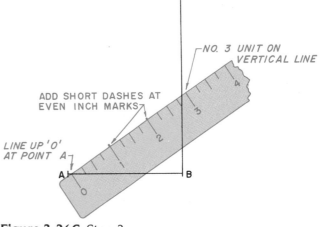

Figure 3-26C Step 2

Figure 3-26D Step 3

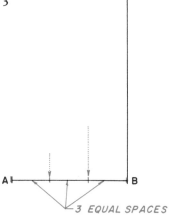

3 EQUAL SPACES

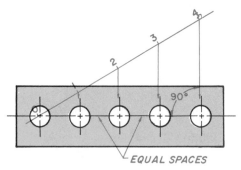

EQUAL SPACES

Figure 3-26E An example of equal spacing within a given length

Step 2. Place a scale with its zero at point A of the given line. If three equal parts are required, pivot the scale until the three-inch measurement (or any length representing three equal units of measure: 30, 60 and 90 mm, for example) is on the perpendicular line drawn in Step 1. Place a short dash at these points. The example in Figure 3-26C shows these dashes at the 1-inch and 2-inch marks.

Step 3. Project lines downward from these points and add short hash marks where these projected lines cross the original given line A-B. This divides the line A-B into three exact equal parts. Check all work, comparing your final step with Figure 3-26D. An example of equal spacing, where equally spaced holes are required within a given length, is illustrated in Figure 3-26E.

How To Divide a Line into Proportional Parts

Given: Line A-B, Figure 3-27A. Problem: Locate point X at 2/3 of the distance from point A to point B.

Step 1. Draw a line 90° from either end of the line. A 90° line from point B is illustrated in Figure 3-27B.

Step 2. Place a scale with its zero on point A of the given line. Because a 2/3 proportion is required, pivot the scale until any multiple of three units of measure intersects the perpendicular line drawn in Step 1. In this example, 6 is used, representing three 2-unit increments. The 2/3 position of this length is two 2-unit increments, or the 4 position, where a hash mark is made, as shown in Figure 3-27C. Projecting this point downward to line A-B, it becomes point X, which is 2/3 the overall distance from point A.

GIVEN :

A ├─────────────┤ B

Figure 3-27A How to divide a line into proportional parts

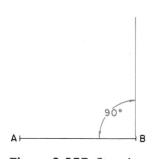

Figure 3-27B Step 1

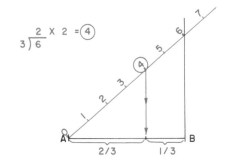

$$3\overline{)6} \quad \frac{2}{3} \times 2 = \textcircled{4}$$

2/3   1/3

Figure 3-27C Step 2

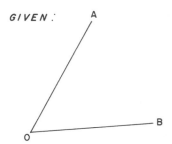

GIVEN :

Figure 3-28B Given: Point 0′

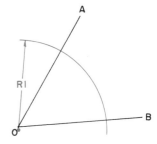

Figure 3-28C Step 1

Figure 3-28A How to transfer an angle

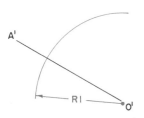

Figure 3-28D Transferred angle

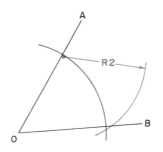

Figure 3-28E Step 2

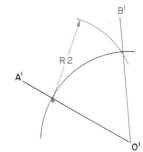

Figure 3-28F Step 3

How To Transfer an Angle

Given: An angle formed by two straight lines, 0A-0B, Figure 3-28A, and one location of where the transferred angle begins (point 0′), Figure 3-28B.

Step 1. Refer to Figure 3-28C. Draw an arc through both legs of a given angle (R1) and then duplicate this radius at the transferred angle location, Figure 3-28D.

Step 2. Transfer the chord length between the two angle legs at the intersection of the arc (R2) to the arc at the transfer angle location.

Step 3. A line from the arc center to the intersection of arc and chord length forms the second line, forming an angle equal to the original, Figures 3-28E and 3-28F.

How To Transfer an Odd Shape

Given: Triangle ABC, Figure 3-29A.

Step 1. Letter or number the various corners and point locations of the odd shape in counterclockwise order around its perimeter. In this example, place the compass point at point A of the original shape and extend the lead to point B. Refer back to Figure 3-29A. Swing a light arc at the new desired location, Figure 3-29B. Letter the center point as A′ and add letter B′ at any convenient location on the arc. It is a good habit to lightly letter each point as you proceed.

Step 2. Place the compass point at letter B of the original shape, Figure 3-29C, and extend the compass lead to letter C of the original shape.

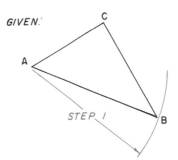

GIVEN:

STEP 1

Figure 3-29A How to transfer an odd shape

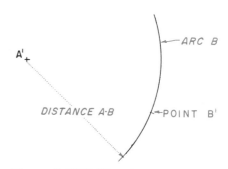

Figure 3-29B Step 1

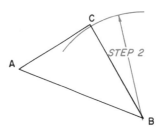

STEP 2

Figure 3-29C Step 2

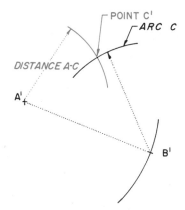

Figure 3-29D Step 3

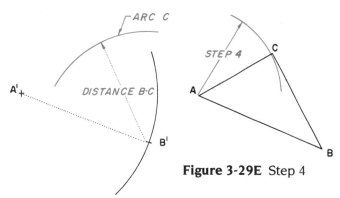

Figure 3-29E Step 4

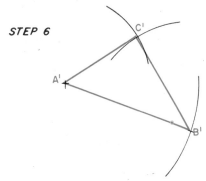

Figure 3-29F Step 5

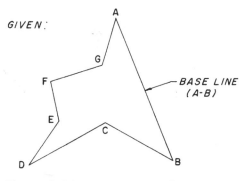

Figure 3-29G Step 6

Step 3. Transfer this distance, B-C, to the layout, Figure 3-29D.

Steps 4 and 5. Going back to the original object, place the compass point at letter A, Figure 3-29E, and extend the compass lead to letter C. Transfer the distance A-C as illustrated in Figure 3-29F. Locate and letter each point.

Step 6. Connect points A', B', and C' with light, straight lines. This completes the transfer of the object, Figure 3-29G. Recheck all work and, if correct, darken lines to the correct line weight.

How To Transfer Complex Shapes

A complex shape can be transferred in exactly the same way by reducing the shape into simple triangles and transferring each triangle using the foregoing method.

Given: An odd shape, A, B, C, D, E, F, G, Figure 3-30A. Letter or number the various corners and point locations of the odd shape in clockwise order around the perimeter, Figure 3-30A. Use the longest line or any convenient line as a starting point. Line A-B is chosen here as the example.

Step 1. Lightly divide the shape into triangle divisions, using the baseline if possible. Transfer each triangle in the manner described in Figures 3-29A through 3-29G. Suggested triangles to be used in example Figure 3-30A are ABC, ABD, ABE, ABF, and ABG, Figure 3-30B.

Step 2. This completes the transfer. Check all work and, if correct, darken in lines to correct line thickness. See Figure 3-30C.

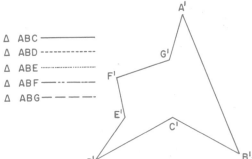

Figure 3-30A How to transfer complex shapes

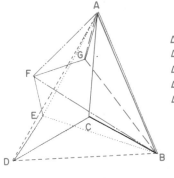

Figure 3-30B Step 1

Figure 3-30C Step 2

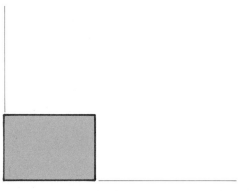

Figure 3-31A How to proportionately enlarge or reduce a shape

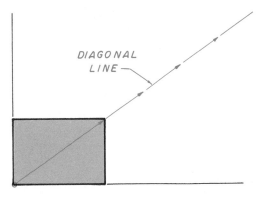

Figure 3-31B Step 1

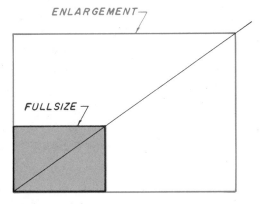

Figure 3-31C Step 2

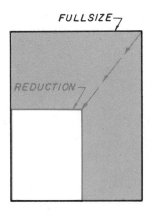

Figure 3-31D Step 1

How To Proportionately Enlarge or Reduce a Shape

Given: A rectangle, Figure 3-31A. *Problem*: To enlarge or reduce its size proportionately.

Step 1. Draw a line from corner to corner diagonally, and extend it as shown in Figure 3-31B.

Step 2. The rectangle is enlarged to any size proportionately if the vertical and horizontal sides are located from the extended diagonal line, Figure 3-31C.

Step 1. The rectangle is reduced proportionately if the vertical and horizontal lines are located on the unextended diagonal line, Figure 3-31D.

Polygon Construction

How To Draw a Triangle with Known Lengths of Sides

Given: Lengths 1, 2, and 3, Figure 3-32A.

Step 1. Draw the longest length line, in this example length 3, with endpoints A and B, Figure 3-32B. Swing an arc (R1) from point A whose radius is either length 1 or length 2; in this example, length 1.

Step 2. Using the radius length *not* used in Step 1, swing an arc (R2) from point B to intercept the arc swung from point A at point C, Figure 3-32C.

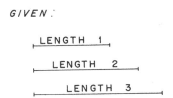

Figure 3-32A How to draw a triangle with known lengths of sides

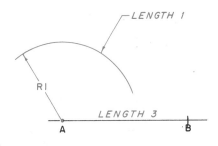

Figure 3-32B Step 1

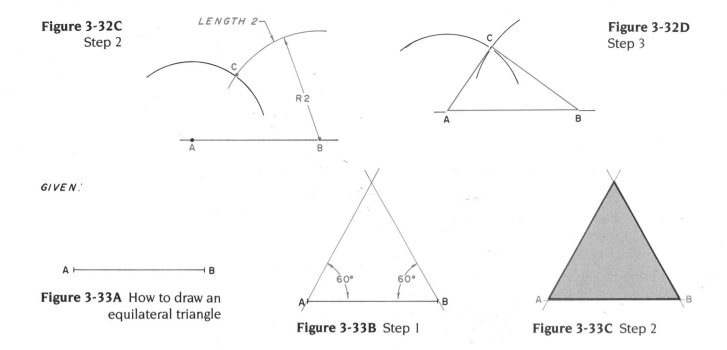

Figure 3-32C
Step 2

LENGTH 2

C

R2

A   B

Figure 3-32D
Step 3

C

A   B

GIVEN:

A |————————————| B

Figure 3-33A How to draw an
equilateral triangle

60°   60°

A |————————————| B

Figure 3-33B Step I

A |————————————| B

Figure 3-33C Step 2

Step 3. Connect A to C and B to C to complete the triangle, Figure 3-32D.

How To Draw an Equilateral Triangle

Given: A baseline and the given length of each side, Figure 3-33A.

Step 1. Either adjust the drafting machine angle to 60° or use a 30°-60° triangle and project from points A and B at 60° using light lines, Figure 3-33B. Allow light construction lines to cross as shown; do not erase them.

Step 2. Check to see that there are three equal sides and, if so, darken in the actual triangle using correct line thickness, Figure 3-33C. Care should be exercised in constructing the sharp corners. Again, do *not* erase the light construction lines.

How To Draw a Square

Given: The location of the center and the required distance across the "flats" of a square, Figure 3-34A.

Step 1. Lightly draw a circle with a diameter equal to the distance across the "flats" of the square. Set the compass at half the required diameter, Figure 3-34B.

Step 2. Using a triangle, lightly complete the square by constructing tangent lines to the circle. Allow the light construction lines to project from the square, as shown, without erasing them, Figure 3-34C.

Step 3. Check to see that there are four equal sides and, if so, darken in the actual square using the correct line thickness, Figure 3-34D. Care should be exercised in constructing the sharp corners. Again, do *not* erase the light construction lines.

GIVEN:

LOCATION OF CENTER

DISTANCE
ACROSS "FLATS"

Figure 3-34A How to draw a square

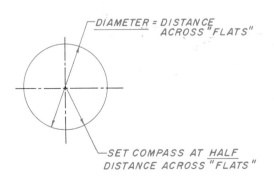

DIAMETER = DISTANCE
ACROSS "FLATS"

SET COMPASS AT HALF
DISTANCE ACROSS "FLATS"

Figure 3-34B Step I

Figure 3-34C Step 2

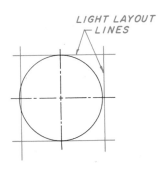

LIGHT LAYOUT
LINES

SHARP
CORNERS

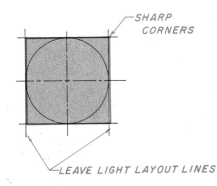

Figure 3-34D
Step 3

LEAVE LIGHT LAYOUT LINES

How To Draw a Pentagon (5 Sides)

Given: The location of the pentagon center, and the diameter that will circumscribe the pentagon, Figure 3-35A.

Step 1. Locate point A at the top-center of the circle and, using a drafting machine, position an angle of 72° (360/5) from the horizontal (or 18° from the vertical) through point A to locate point B where the angle crosses the circumference of the circle, Figure 3-35B.

Step 2. Draw a horizontal line from point B to locate point C on the circumference of the circle on the opposite side, Figure 3-35C.

Step 3. Set the compass at the distance from point B to point C, and swing this distance from the points as illustrated in Figure 3-35D to locate points X and Y.

Step 4. Lightly connect the points. Check to see that there are five equal sides and, if correct, darken in the actual pentagon taking care to construct five sharp corners, Figure 3-35E.

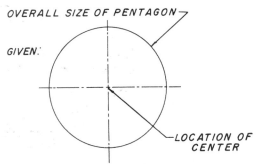

OVERALL SIZE OF PENTAGON

GIVEN.

LOCATION OF
CENTER

Figure 3-35A How to draw a pentagon

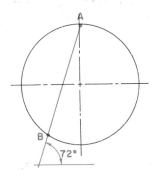

Figure 3-35B Step 1

72°

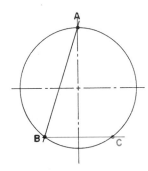

Figure 3-35C Step 2

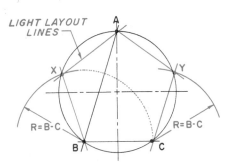

LIGHT LAYOUT
LINES

R=B-C   R=B-C

Figure 3-35D
Step 3

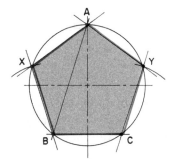

Figure 3-35E
Step 4

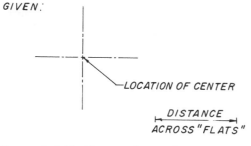

Figure 3-36A How to draw a hexagon

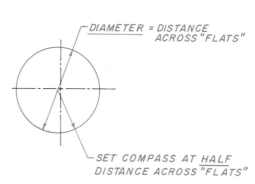

Figure 3-36B Step I

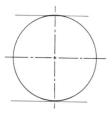

Figure 3-36C Step 2

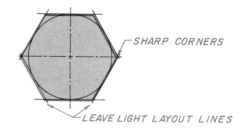

Figure 3-36D Step 3

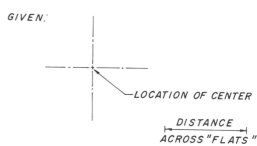

Figure 3-36E Step 4

How To Draw a Hexagon (6 Sides)

Given: The location of the required center and the required distance across the "flats" of a hexagon, Figure 3-36A.

Step 1. Lightly draw a circle with a diameter equal to the distance across the "flats" of the hexagon. Set the compass at half the required diameter, Figure 3-36B.

Step 2. Draw two horizontal lines tangent to the curve, or two vertical lines if the hexagon is to be oriented at 90° to the illustrated position, Figure 3-36C.

Step 3. Using a 30°-60° triangle, lightly complete the hexagon by constructing tangent lines to the circle, Figure 3-36D. Allow the light construction lines to extend as shown; do not erase them.

Step 4. Check to see that there are six equal sides and, if so, darken in the actual hexagon using correct line thickness and taking care to construct six sharp corners, Figure 3-36E. Again, do *not* erase the light construction lines.

How To Draw an Octagon (8 Sides)

Given: The location of the required center and the required distance across the "flats" of an octagon, Figure 3-37A.

Step 1. Lightly draw a circle with a diameter equal to the distance across the "flats" of the octagon. Set the compass at half the required diameter, Figure 3-37B.

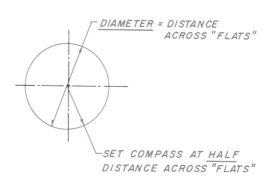

Figure 3-37A How to draw an octagon

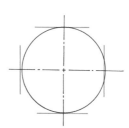

Figure 3-37B Step I

Figure 3-37C Step 2

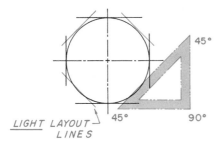

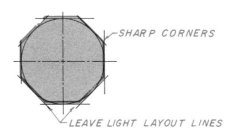

Figure 3-37D Step 3

Figure 3-37E Step 4

Step 2. Lightly draw two horizontal lines and two vertical lines tangent to the circle, as illustrated in Figure 3-37C.

Step 3. Using a 45° triangle, lightly complete the octagon by constructing tangent lines to the circle, Figure 3-37D. Allow the light lines to extend.

Step 4. Check that there are eight equal sides and, if so, darken in the actual octagon using correct line thickness and taking care to construct eight sharp corners, Figure 3-37E. Again, do not erase the light construction lines.

Circular Construction

How To Locate the Center of a Given Circle

Given: A circle without a center point, Figure 3-38A.

Step 1. Using the drafting machine or T-square, draw a horizontal line across the circle approxi-

mately halfway between the *estimated* center of the given circle and the uppermost point on the circumference. Label the end points of the chord thus formed as A and B, Figure 3-38B.

Step 2. Draw perpendicular lines (90°) downward from points A and B. Locate points C and D where these two lines pass through the circle, Figure 3-38C.

Step 3. Carefully draw a straight line from point A to point D and from point C to point B. Where these lines cross is the exact center of the given circle. Place a compass point on the center point; adjust the lead to the edge of the circle and swing an arc to check that the center is accurate, Figure 3-38D.

Alternatively, the intersection of perpendicular bisectors of any two nonparallel chords serve to locate the center. This is a modification of the previous construction for passing a circle through any three nonaligned points.

GIVEN:

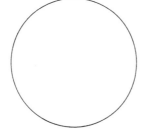

Figure 3-38A How to locate the center of a given circle

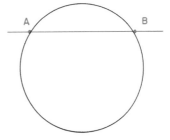

Figure 3-38B Step 1

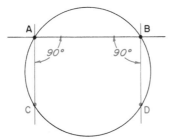

Figure 3-38C Step 2

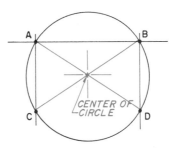

Figure 3-38D Step 3

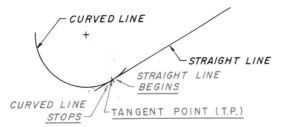

Figure 3-39 Tangent points

Tangent Points

A *tangent point* is the exact location or point where one line stops and another line begins. Tangent also means to "touch." As an example, a tangent point is the exact point where a curved line stops and a straight line begins, Figure 3-39.

How To Locate Tangent Points

The tangent point is a point 90° from the straight line to the swing point of the curved line. Place a light hash mark at each tangent point. It is a good habit to always find *all* tangent points on the object in all views before darkening in the drawing.

The tangent points for a right angle bend are illustrated in Figure 3-40A. The tangent points for an acute angle bend are illustrated in Figure 3-40B. The tangent points for an obtuse angle bend are illustrated in Figure 3-40C.

In each preceding example, a light line is constructed 90° from the straight line to the exact swing point of the radius. Find all the tangent points before darkening in the final work. Always darken in all compass work first, followed by the straight lines.

The tangent point between two arcs or circles is the exact point where one arc or circle ends and the next arc or circle begins. The tangent point could also be where one arc or circle touches another arc or circle, Figure 3-40D.

The tangent point is found by drawing a straight line from the swing point of the first arc or circle to the swing point of the second arc or circle, Figure 3-40E. Place a short light dash at the exact tangent point.

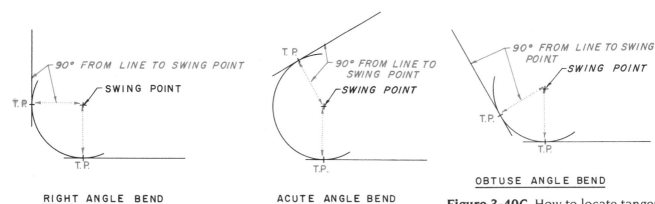

RIGHT ANGLE BEND

Figure 3-40A How to locate tangent points on a right angle

ACUTE ANGLE BEND

Figure 3-40B How to locate tangent points on an acute angle

OBTUSE ANGLE BEND

Figure 3-40C How to locate tangent points on an obtuse angle

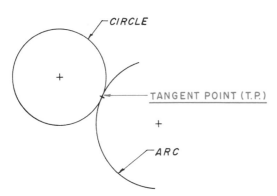

Figure 3-40D How to locate tangent points between arcs or circles

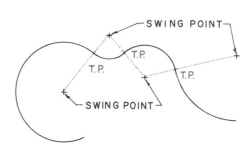

Figure 3-40E Tangent points between arcs or circles

GIVEN.

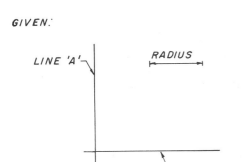

Figure 3-41A How to construct an arc tangent to a right angle

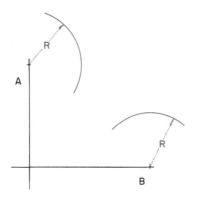

Figure 3-41B Step 1

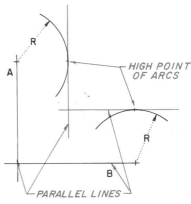

Figure 3-41C Step 2

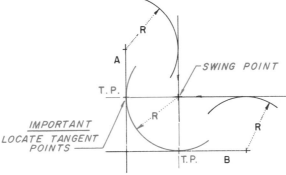

Figure 3-41D Step 3

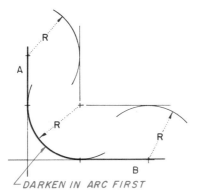

Figure 3-41E Step 4

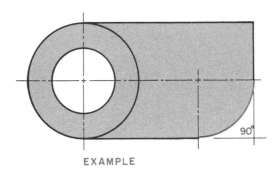

EXAMPLE

Figure 3-41F Example of an arc tangent to two lines at a right angle

How To Construct an Arc Tangent to a Right Angle (90°)

Given: A right angle (90°), lines A and B and a required radius, Figure 3-41A.

Step 1. Set the compass at the required radius and, out of the way, swing a radius from line A and one from line B, as illustrated in Figure 3-41B.

Step 2. From the extreme high points of each radius, construct a light line parallel to line A and another line parallel to line B, Figure 3-41C.

Step 3. Where these lines intersect is the exact location of the required swing point. Set the compass point on the swing point and lightly construct the required radius, Figure 3-41D. Allow the radius swing to extend past the required area, as shown in the figure. It is important to locate all tangent points (T.P.) before darkening in.

Step 4. Check all work, and darken in the radius using the correct line thickness. Darken in connecting straight lines as required. Always construct compass work first, followed by straight lines. Leave all light construction lines. See Figure 3-41E.

An example of an arc tangent to two lines at a right angle (90°) is illustrated in Figure 3-41F.

How To Construct an Arc Tangent to an Acute Angle (Less Than 90°)

Given: An acute angle, lines A and B, and a required radius, Figure 3-42A. Follow the exact same procedure as outlined in the preceding example.

Step 1. Set the compass at the required radius and, out of the way, swing a radius from line A and one from line B, as illustrated in Figure 3-42B.

Step 2. From the extreme high points of each radius, construct a light line parallel to line A and another line parallel to line B, Figure 3-42C.

Step 3. Where these lines intersect is the exact location of the required swing point. Set the compass point on the swing point and lightly construct the required radius, Figure 3-42D. Allow the radius swing to extend past the required area, as shown in Figure 3-42D.

Step 4. Check all work, and darken in the radius using the correct line thickness. Darken in connecting straight lines as required. Always construct compass work first, followed by straight lines. Leave all light construction lines. See Figure 3-42E.

An example of an arc tangent to two lines at an acute angle (22°) is illustrated in Figure 3-42F.

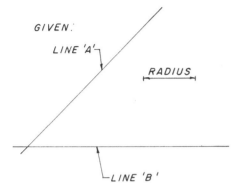

Figure 3-42A How to construct an arc tangent to an acute angle

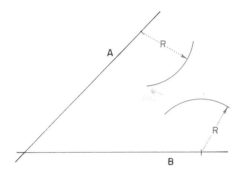

Figure 3-42B Step 1

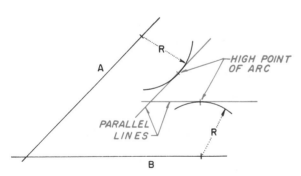

Figure 3-42C Step 2

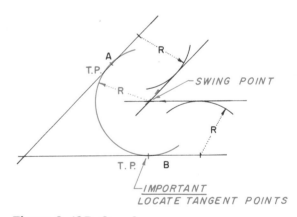

Figure 3-42D Step 3

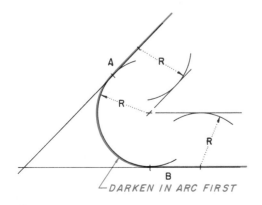

Figure 3-42E Step 4

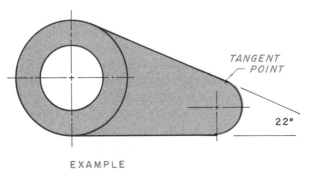

Figure 3-42F Example of an arc tangent to two lines at an acute angle

How To Construct an Arc Tangent to an Obtuse Angle (More Than 90°)

Given: An obtuse angle between lines A and B, and a required radius, Figure 3-43A. Follow the exact same procedure as outlined in the two preceding examples.

Step 1. Set the compass at the required radius and, out of the way, swing a radius from line A and one from line B, as illustrated in Figure 3-43B.

Step 2. From the extreme high points of each radius, construct a light line parallel to line A and one parallel to line B, Figure 3-43C.

Step 3. Where these lines intersect is the exact location of the required swing point. Set the compass point on the swing point and lightly construct the required radius. Allow the radius swing to extend past the required area, as shown in Figure 3-43D.

Step 4. Check all work, darken in the radius using the correct line thickness. Darken in connecting straight lines as required. Always construct compass work first, followed by straight lines. Leave all light construction lines. See Figure 3-43E.

An example of an arc tangent to two lines at an obtuse angle (120°) is illustrated in Figure 3-43F.

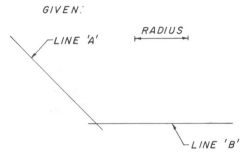

Figure 3-43A How to construct an arc tangent to an obtuse angle

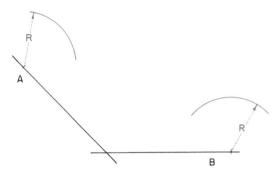

Figure 3-43B Step 1

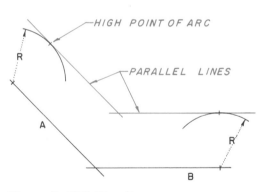

Figure 3-43C Step 2

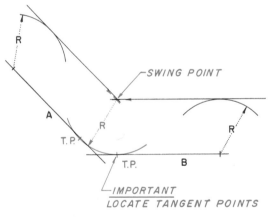

Figure 3-43D Step 3

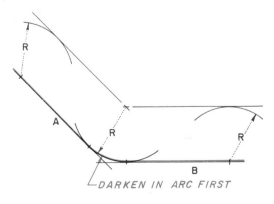

Figure 3-43E Step 4

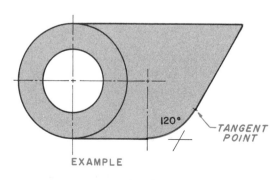

Figure 3-43F Example of an arc tangent to two lines at an obtuse angle

How To Construct an Arc Tangent to a Straight Line and a Curve

Given: Straight line A, an arc B with a center point, and a required radius, Figure 3-44A.

Step 1. Set the compass at the required radius and, out of the way, swing a radius from the given arc B and one from the given straight line A, Figure 3-44B.

Step 2. From the extreme high points of each radius, construct a light straight line parallel to line A, and construct a radius outside the given arc B equal to the required radius, as illustrated in Figure 3-44C.

Step 3. Where these lines intersect is the exact location of the required swing point. Set the compass point on the swing point and lightly construct the required radius, Figure 3-44D. Allow the radius swing to extend past the required area as shown. Locate all tangent points (T.P.) before darkening in.

Step 4. Check all work, darken in the radius using the correct line thickness. Darken in the arcs first and the straight line last, Figure 3-44E.

An example of an arc tangent to a straight line and a curve is illustrated in Figure 3-44F.

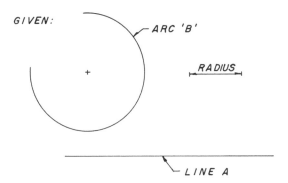

Figure 3-44A How to construct an arc tangent to a straight line and a curve

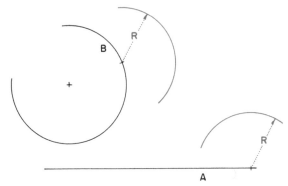

Figure 3-44B Step 1

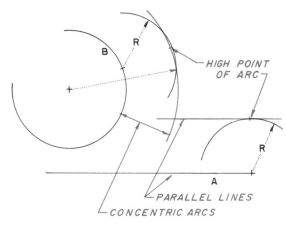

Figure 3-44C Step 2

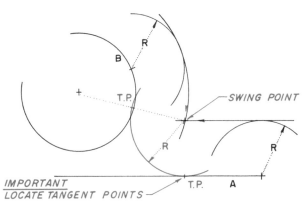

Figure 3-44D Step 3

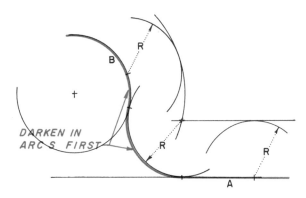

Figure 3-44E Step 4

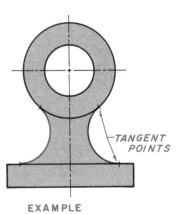

Figure 3-44F Example of an arc tangent to a straight line and a curve

How To Construct an Arc Tangent to Two Radii or Diameters

Given: Diameter A and arc B with center points located, and the required radius, Figure 3-45A.

Step 1. Set the compass at the required radius and, out of the way, swing a radius of the required length from a point on the circumference of given diameter A. Out of the way, swing a required radius from a point on the circumference of given arc B, Figure 3-45B.

Step 2. From the extreme high points of each radius, construct a light radius outside of the given radii A and B, as illustrated in Figure 3-45C.

Step 3. Where these arcs intersect is the exact location of the required swing point. Set the compass point on the swing point and lightly construct the required radius, Figure 3-45D. Allow the radius swing to extend past the required area as shown. Before darkening in, it is important to locate all tangent points (T.P.), Figure 3-45D.

Step 4. Check all work, darken in the radii using the correct line thickness. Darken in the arcs or radii in consecutive order from left to right or from right to left, thus constructing a smooth connecting line having no apparent change in direction, Figure 3-45E.

An example of an arc tangent to two radii is illustrated in Figure 3-45F.

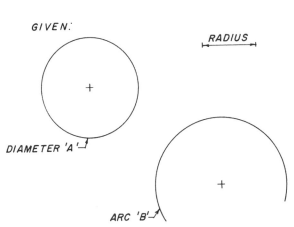

Figure 3-45A How to construct an arc tangent to two radii or diameters

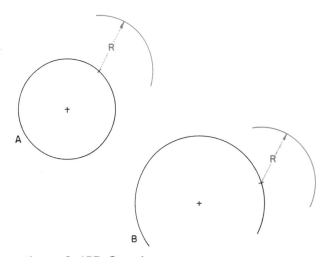

Figure 3-45B Step 1

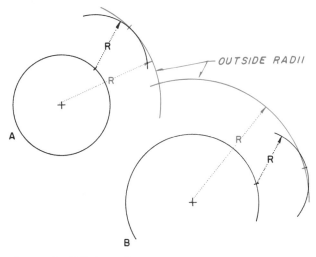

Figure 3-45C Step 2

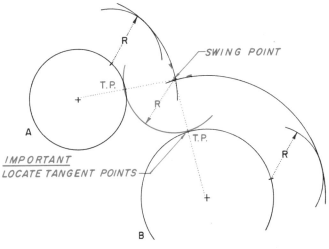

Figure 3-45D Step 3

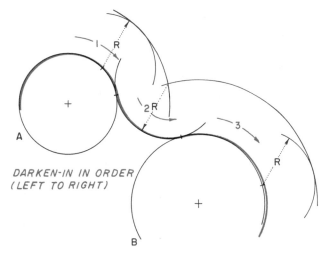

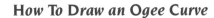
DARKEN-IN IN ORDER
(LEFT TO RIGHT)

Figure 3-45E Step 4

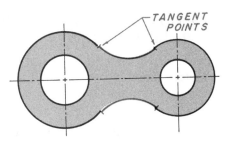
EXAMPLE

Figure 3-45F Example of an arc
tangent to two
radii

GIVEN:

Figure 3-46A How to draw an ogee curve

How To Draw an Ogee Curve

An *ogee curve* is used to join two parallel lines. It forms a gentle curve that reverses itself in a neat symmetrical geometric form.

Given: Parallel lines A-B and C-D, Figure 3-46A.

Step 1. Draw a straight line connecting the space between the parallel lines. In this example, from point B to point C, Figure 3-46B.

Step 2. Make a perpendicular bisector to line B-C to establish point X, Figure 3-46C.

Step 3. Make perpendicular bisectors to the lines B-X and X-C, Figure 3-46D.

Step 4. Draw a perpendicular from line A-B at point B to intersect the perpendicular bisector of B-X, which locates the first required swing center. Draw a perpendicular from line C-D at point C to intersect the perpendicular bisector of C-X which locates the second required swing center, Figure 3-46E.

Step 5. Place the compass point on the first swing point and adjust the compass lead to point B, and swing an arc from B to X. Place the compass point on the second swing point and swing an arc from X to C, Figure 3-46F. This completes the ogee curve.

Note: point X is the tangent point between arcs. Check and, if correct, darken in all work.

An example of an ogee curve is illustrated in Figure 3-46G.

Figure 3-46B Step 1

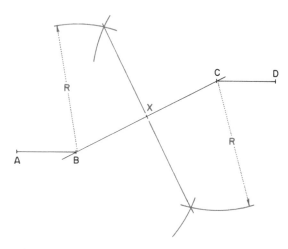

Figure 3-46C Step 2

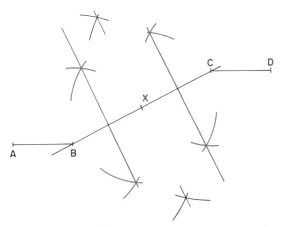

Figure 3-46D Step 3

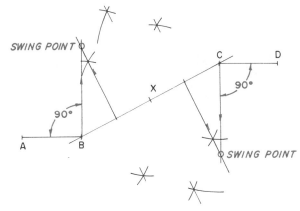

Figure 3-46E Step 4

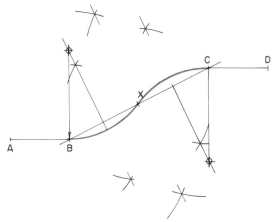

Figure 3-46F Step 5

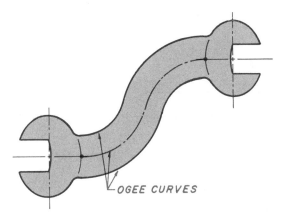

Figure 3-46G Step 6

Conic Sections

A *conic section* is a section cut by a plane passing through a cone. These sections are bounded by various kinds of shapes. Depending upon where the section is cut, the various shapes can be a triangle, a circle, an ellipse, a parabola or a hyperbola, Figures 3-47A through 3-47E.

Triangle, Figure 3-47A. This shape results when a plane passes through the apex of the cone.

Circle, Figure 3-47B. This shape results when a plane passes through the cone, parallel with the base and perpendicular to the axis.

Ellipse, Figure 3-47C. This shape results when a plane passes through the cone inclined to the axis.

Parabola, Figure 3-47D. This shape results when a plane passes through the cone parallel to one element.

Hyperbola, Figure 3-47E. This shape results when a plane passes through the cone parallel with the axis of the cone.

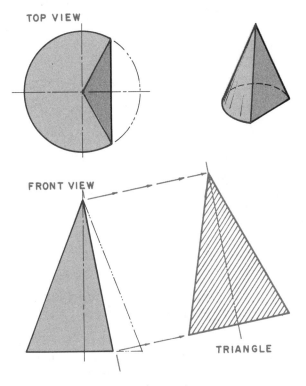

Figure 3-47A Triangle section

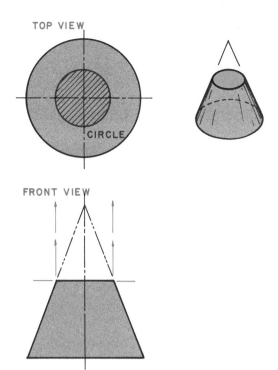

Figure 3-47B Circle section

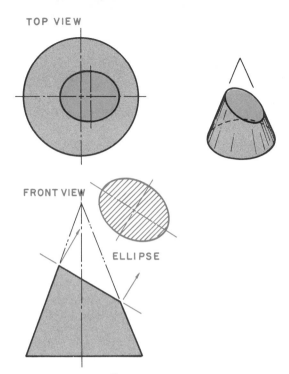

Figure 3-47C Ellipse section

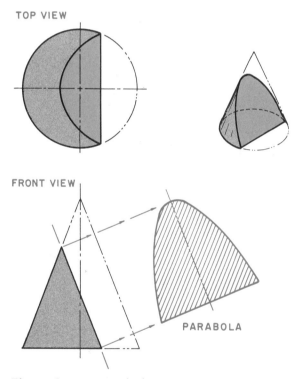

Figure 3-47D Parabola section

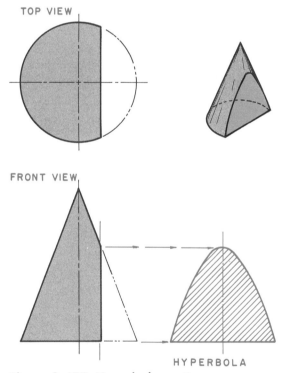

Figure 3-47E Hyperbola section

How To Draw an Ellipse, Concentric Circle Method

Given: The location of the center point, the major diameter, and the minor diameter, Figure 3-48A.

Step 1. Lightly draw one circle equal to the major diameter and another circle equal to the minor diameter, Figure 3-48B.

Step 2. Divide both circles into 12 equal divisions by passing lines through the center at every 30 degrees. Number points 1 through 5 in clockwise consecutive order on the major diameter,

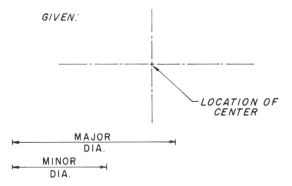

Figure 3-48A How to draw an ellipse (concentric method)

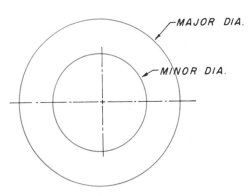

Figure 3-48B Step 1

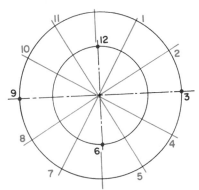

Figure 3-48C Step 2

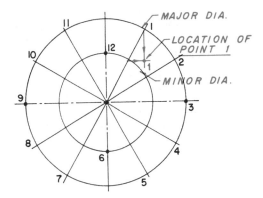

Figure 3-48D Step 3

positioning the 3 location on the right horizontal axis. Number points 7 through 11 in clockwise consecutive order, positioning the 9 location on the left horizontal axis. Point 6 is on the lower vertical axis at the minor diameter, and point 12 is on the upper vertical axis at the minor diameter, Figure 3-48C. If the ellipse were to be positioned at 90° from this example, the point locations would be rotated accordingly.

Step 3. Project down from point 1 on the major diameter, and to the right from point 1 on the minor diameter. Where these lines intersect is the exact location where point 1 will be on the ellipse, Figure 3-48D.

Step 4. Project down from point 2 on the major diameter, and to the right from point 2 on the minor diameter. Where these lines intersect is the exact location where point 2 will be on the ellipse, Figure 3-48E.

Step 5. This completes the first quadrant of the ellipse. Continue around the circle to locate points 4, 5, 6, 7, 8, 9, 10, and 11 in the same manner except in reverse, Figure 3-48F.

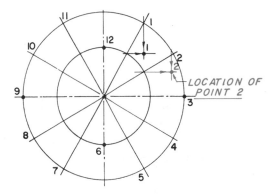

Figure 3-48E Step 4

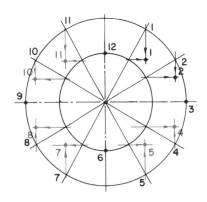

Figure 3-48F Step 5

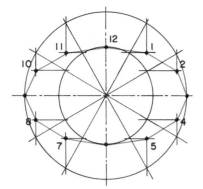

Figure 3-48G Step 6

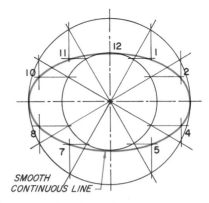

Figure 3-48H Step 7

Step 6. Lightly draw straight lines connecting points 1 through 12 in order to get a general idea of the ellipse outline, Figure 3-48G.

Step 7. Darken in the ellipse using an irregular curve. Carefully connect all the points with a smooth continuous line of the correct thickness, Figure 3-48H. It is sometimes helpful to divide into 10° or 15° spaces the two ends where the ellipse curves the fastest in order to have more points around the extreme ends. In this example, this is done between points 2-4 and 8-10.

Examples of ellipses are illustrated in Figure 3-48I, showing how a rotated circle would look at 0°, 15°, 45°, 60°, and 90°.

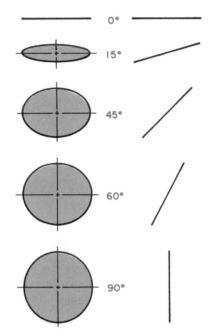

Figure 3-48I Example of ellipses at various degrees

How To Draw a Trammel Ellipse

Given: A major or minor axis of a required ellipse, and any located point through which the ellipse must pass, Figure 3-49A. If a major axis is given, mark the end points A and B. If the minor axis is given, as is shown in the figure, mark the end points C and D. Mark as P the known location of where the ellipse must pass.

Step 1. Draw a perpendicular bisector of the given axis to locate the unknown axis position, crossing the known axis at a point that will become 0, Figure 3-49B.

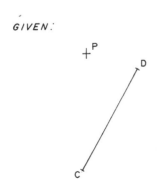

Figure 3-49A Trammel ellipse

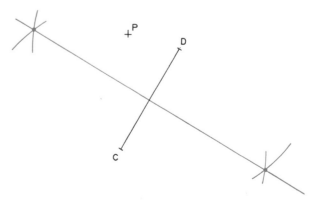

Figure 3-49B Step 1

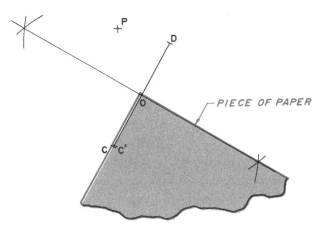

Figure 3-49C Step 2

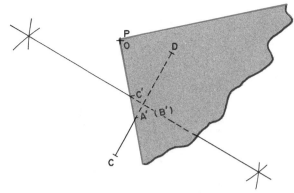

Figure 3-49D Step 3

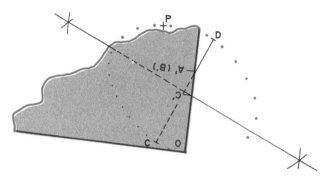

Figure 3-49E Step 4

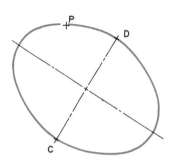

Figure 3-49F Step 5

Step 2. Mark an end corner of a separate strip of paper as point 0, and mark along the edge of the strip one-half of the distance A-B or C-D, whichever is known. Mark this point on the strip as A' or C' to correspond to the known axis, Figure 3-49C. Position the strip so that point 0 lies on the known point through which the ellipse must pass. Position the other strip mark on the axis that is not represented by the half-length on the strip, at the position (of two possible positions) that is nearest to the ellipse center.

Step 3. With the strip positioned as specified in Step 2, mark the strip at the location of where it crosses the other axis. The length of end point 0 to this new position represents one-half of the unknown axis length, and should correspondingly be labeled as A' or C', Figure 3-49D.

Step 4. Using the two marked strip locations, reposition the strip so that the mark representing one-half of the major axis always lies on the minor axis, and the mark representing one-half of the minor axis always lies on the major axis. Repeat as needed to establish enough points for constructing the ellipse, Figure 3-49E.

Step 5. Using a French curve, pass a smooth line through each of these established points, Figure 3-49F.

How To Draw a Foci Ellipse

Given: Major and minor axes of a required ellipse. The end points of the major axis are labeled A and B, the end points of the minor axis are labeled C and D, and the ellipse center is labeled 0, Figure 3-50A.

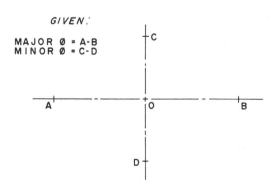

Figure 3-50A How to draw a foci ellipse

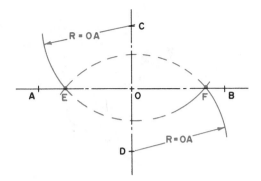

Figure 3-50B Step 1

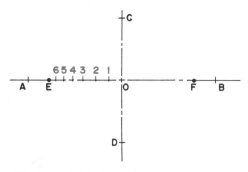

Figure 3-50C Step 2

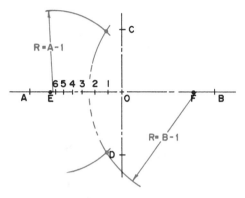

Figure 3-50D Step 3

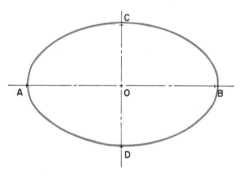

Figure 3-50E Step 4

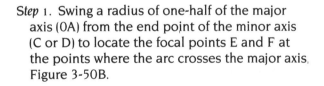

Figure 3-50F Step 5

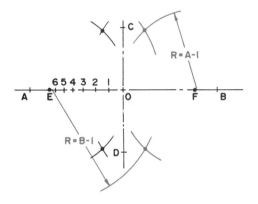

Figure 3-50G Step 6

Step 1. Swing a radius of one-half of the major axis (0A) from the end point of the minor axis (C or D) to locate the focal points E and F at the points where the arc crosses the major axis, Figure 3-50B.

Step 2. Select a random array of points on the major axis between one focal point and the ellipse center (space those at the ends more closely). Number these in consecutive order, Figure 3-50C.

Step 3. Swing an arc whose radius is the distance from point B to number 1. Swing this dis-

tance from focal point F. Intersect this arc with an arc whose radius is the distance from point A to the same numbered point, number 1 in this example, but whose center is at focal point E, Figure 3-50D.

Step 4. Repeat Step 3, but reverse the focal point positions of the arc centers, Figure 3-50E.

Step 5. Repeat Steps 3 and 4 for each point selected in Step 2, Figure 3-50F.

Step 6. Connect the points into a smooth curve, using a French curve, Figure 3-50G.

126   Section 1

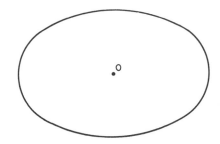

Figure 3-51A How to locate the major and minor axes of a given ellipse

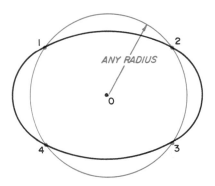

Figure 3-51B Step 1

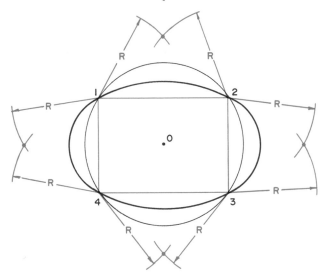

Figure 3-51C Step 2

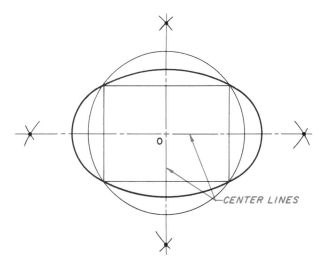

Figure 3-51D Step 3

How To Locate the Major and Minor Axes of a Given Ellipse with a Located Center

Given: An ellipse, with center 0 located, Figure 3-51A.

Step 1. From the ellipse center 0, draw a circle at a radius that allows intersecting the ellipse at any four locations. Label these points in successive order 1, 2, 3, and 4, moving clockwise around the ellipse center, Figure 3-51B.

Step 2. Draw a rectangle by connecting each of the labeled points in successive order. As shown in Figure 3-51C, the major and minor axes are parallel to these sides, found by drawing perpendicular bisectors of two consecutive sides.

Step 3. Draw the major and minor axes parallel to sides 1-2 and 2-3 through point 0, Figure 3-51D.

How To Draw a Tangent to an Ellipse at a Point on the Ellipse

Given: An ellipse, and a point P on its perimeter, Figure 3-52A. If not provided, locate the major

GIVEN:

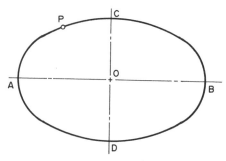

Figure 3-52A How to draw a tangent to an ellipse at a point on the ellipse

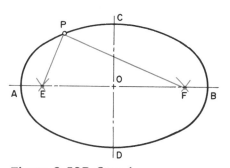

Figure 3-52B Step 1

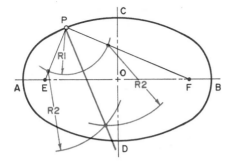

Figure 3-52C Step 2

axis A-B and minor axis C-D, and the focal points E and F.

Step 1. Draw E-P and F-P, Figure 3-52B.

Step 2. Bisect the angle EPF, Figure 3-52C.

Step 3. Draw a perpendicular to the angle bisector of angle EPF at point P. This is the required tangent, Figure 3-52D.

How To Draw a Tangent to an Ellipse from a Distant Point

Given: An ellipse and a distant point P, Figure 3-53A.

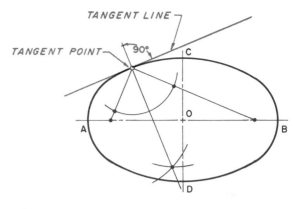

Figure 3-52D Step 3

Step 1. If not provided, locate the major axis A-B and focal points E and F, Figure 3-53B.

Step 2. Swing an arc R1 from point P to pass through either focal point. F is selected for this example, Figure 3-53C.

Step 3. Swing an arc whose radius is equal to the major axis, from the other focal point (E in this example). Intersect the arc from Step 2 at points R and S, Figure 3-53D.

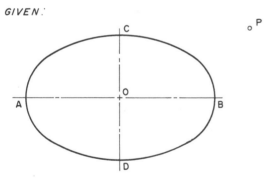

Figure 3-53A How to draw a tangent to an ellipse from a distant point

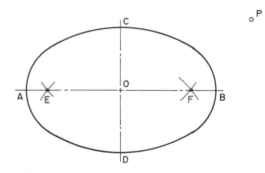

Figure 3-53B Step 1

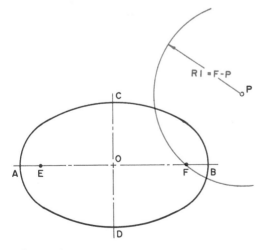

Figure 3-53C Step 2

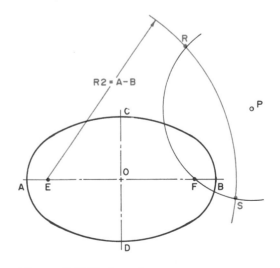

Figure 3-53D Step 3

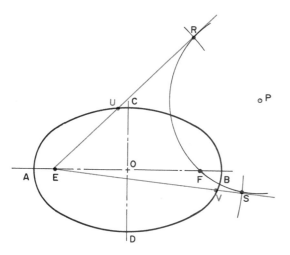

Figure 3-53E Step 4

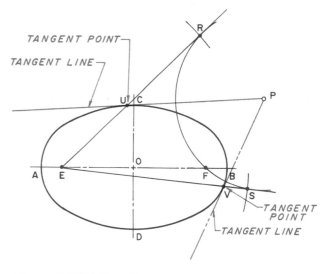

Figure 3-53F Step 5

Step 4. Extend the lines from E to each of points R and S, to cross the ellipse at U and V, Figure 3-53E.

Step 5. The two possible points of tangency of the two possible lines from point P are U and V. Select and draw either, or both, Figure 3-53F.

How To Draw a Parabola (Method 1)

Given: The front view and plan view of cone ABC with cutting plane X-Y, Figure 3-54A.

Step 1. Locate points Y' and Y'' in the plan view. In the front view, place the point of the compass on point Y and position the lead on point X. Swing point X to the extended baseline and down into the plan view, as illustrated in Figure 3-54B.

Step 2. Locate a point along line X-Y (in this example, point D) and, using point Y as a swing point, swing point D to the extended baseline and down into the plan view. In the front view, reset the compass to a distance equal to the horizontal distance from the cone axis to the outer edge (identified as length 1), illustrated

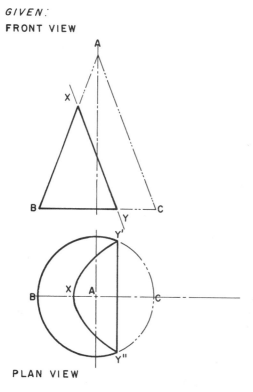

Figure 3-54A How to draw a parabola (method 1)

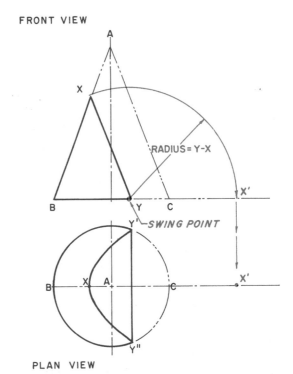

Figure 3-54B Step 1

FRONT VIEW

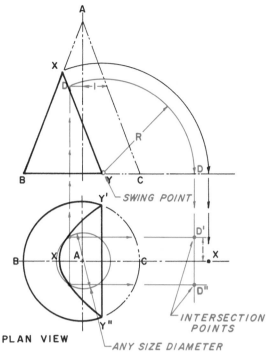

PLAN VIEW

Figure 3-54C Step 2

FRONT VIEW

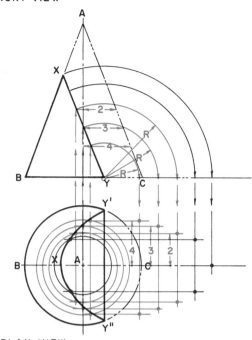

PLAN VIEW

Figure 3-54D Step 3

FRONT VIEW

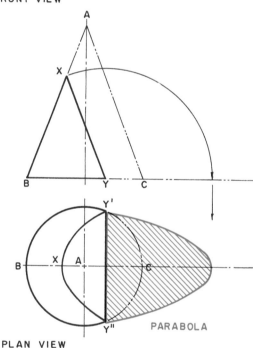

PLAN VIEW

Figure 3-54E Step 4

in Figure 3-54C. Transfer this distance to the plan view. Project point D down into the plan view and, where point D intersects the arc, locate points D′ and D″. Project these points to the right, intersecting with point D from above at the actual intersection points D′ and D″.

Step 3. Choose other various points along line X-Y in the front view, and project them over and down as described in Step 2, Figure 3-54D.

Step 4. Using an irregular curve, connect all points. This completes the parabola, Figure 3-54E.

How To Draw a Parabola (Method 2)

Given: Required rise A-B, and required span A-C. *Problem*: Construct a parabola within the required rise and span, Figure 3-55A.

Step 1. Divide half the span distance A-0 into any number of equal parts. (In this example, 4 equal parts are used.) See Figure 3-55B. Divide half the rise A-B into equal parts amounting to the square of the equal parts in Step 1 (in this example, $4^2 = 16$ equal parts).

Step 2. From line A-0, each point on the parabola is offset by a number of units equal to the square of the numbers of units from point 0, see Figure 3-55C. For example, point 1 projects 1 unit; point 2 projects 4 units; point 3 projects 9 units, and so forth.

Step 3. Using an irregular curve, connect the points to form a smooth parabolic curve, Figure 3-55D.

How To Find the Focus Point of a Parabola

Given: A parabolic curve with points A, 0, and B, Figure 3-56A. *Problem*: Find the focus point of the parabola A0B.

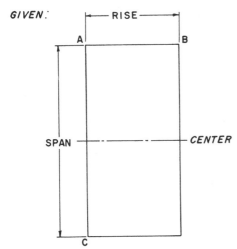

Figure 3-55A How to draw a parabola (method 2)

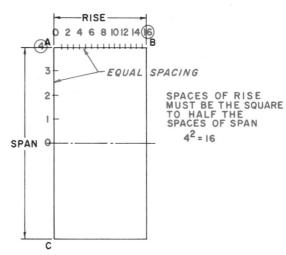

SPACES OF RISE
MUST BE THE SQUARE
TO HALF THE
SPACES OF SPAN

$4^2 = 16$

Figure 3-55B Step 1

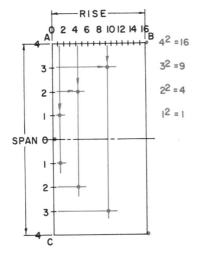

$4^2 = 16$

$3^2 = 9$

$2^2 = 4$

$1^2 = 1$

Figure 3-55C Step 2

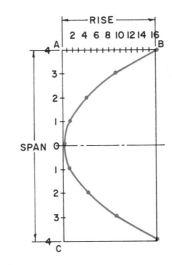

Figure 3-55D Step 3

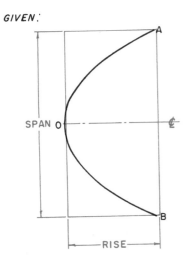

Figure 3-56A How to find a focus point of a parabola

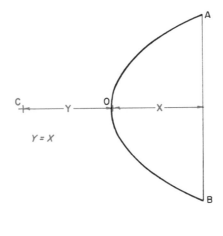

$Y = X$

Figure 3-56B Step 1

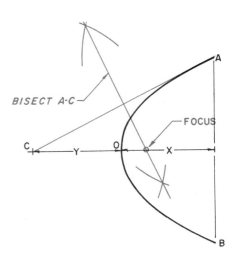

BISECT A-C

FOCUS

Figure 3-56C Step 2

Step 1. Draw a line from point A to point B. Continue the center line to a point equal to length X, distance Y, to find point C, Figure 3-56B.

Step 2. Draw a line from point C to point A and bisect it. The intersection of the bisect line and the axis is the focus of the parabola, Figure 3-56C.

How To Draw a Hyperbola (Method 1)

Given: The front view and plan view of cone ABC with cutting plane X-Y, Figure 3-57A.

Step 1. Locate points Y' and Y'' in the plan view. In the front view, place the point of the compass on point Y and set the lead to point X. Swing point X to the extended baseline and down to the plan view, as illustrated in Figure 3-57B.

Step 2. Locate a point along line X-Y (in this example, point D). Using Y as a swing point, swing point D to the extended baseline and down into the plan view. In the front view, reset the compass to a distance equal to the horizontal distance from the cone axis to the outer edge (length 1), as illustrated in Figure 3-57C. Transfer the arc to the plan view; where the arc intersects with the cutting plane line X-Y, project to the right, and where these points inter-

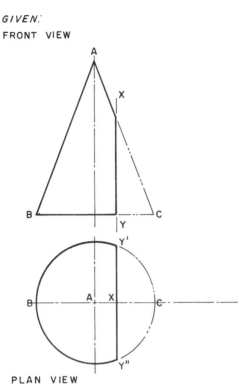

Figure 3-57A How to draw a hyperbola (method 1)

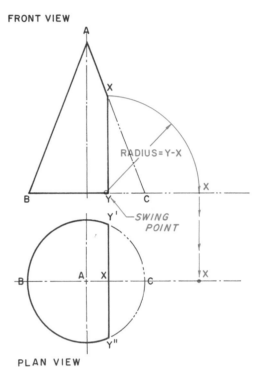

Figure 3-57B Step 1

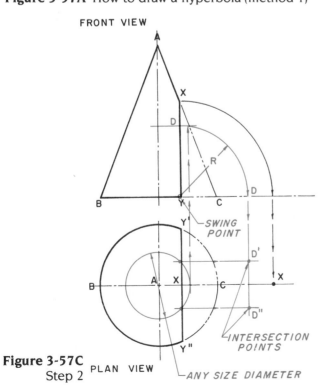

Figure 3-57C Step 2

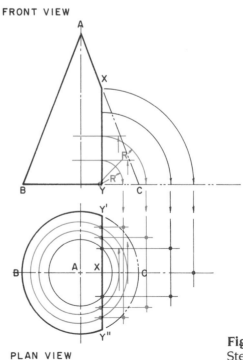

Figure 3-57D Step 3

132   Section 1

FRONT VIEW

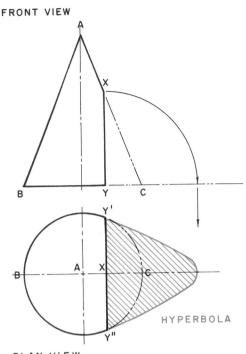

PLAN VIEW

Figure 3-57E Step 4

sect with point D from above is the actual location of points D' and D''.

Step 3. Choose other points along line X-Y in the front view and project them over and down as described in Step 2, Figure 3-57D.

GIVEN:

SQUARE EQUAL TO THE TRANSVERSE AXIS WITH POINTS A-B AS SHOWN

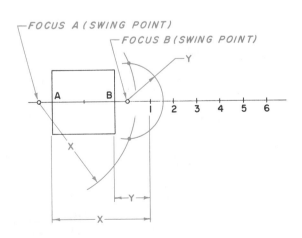

□ EQUAL TO TRANSVERSE AXIS

Figure 3-58A How to draw a hyperbola (method 2)

Figure 3-58C Step 2

Step 4. Using an irregular curve, connect all points. This completes the parabola, Figure 3-57E.

How To Draw a Hyperbola (Method 2)

Given: Coordinates, a square whose sides equal the transverse axis, and points A and B, Figure 3-58A.

Step 1. With the center at the intersection of the diagonal lines from opposite corners of the square, place the compass lead at the corner of the given square and swing arcs to intersect the horizontal line through the center. This locates focus A and focus B. Progressing outward along the horizontal line, mark off equal spaces of arbitrary length from the focus points, Figure 3-58B.

Step 2. Set the compass at dimension X (A-1) and swing an arc from focus A. Set the compass at dimension Y (B-1) and swing an arc from focus B. See Figure 3-58C.

Step 3. Follow the same procedure for as many points as are required (in this example, 6 points). Set the compass at distance A-2 and swing from focus A. Set the compass at distance B-2 and swing from focus B, and so on, Figure 3-58D.

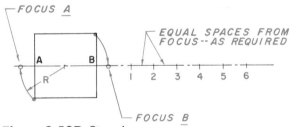

Figure 3-58B Step 1

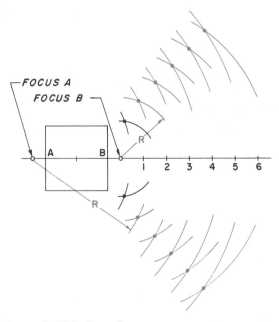

Figure 3-58D Step 3

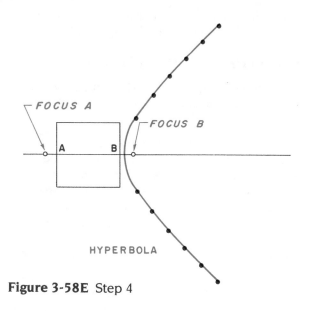

Figure 3-58E Step 4

Step 4. Using an irregular curve, carefully complete the hyperbola curve, Figure 3-58E.

How To Join Two Points by a Parabolic Curve

Given: Points X and Y with 0 an assumed point of tangency, Figures 3-59A, 3-59B, and 3-59C.

Step 1. Divide line X-0 into an equal number of parts; divide 0-Y into the exact same number of equal parts. See Figures 3-59D, 3-59E, and 3-59F.

Step 2. Connect corresponding points, Figures 3-59G, 3-59H, and 3-59I. Using an irregular curve, draw the parabolic curve as a smooth flowing curve.

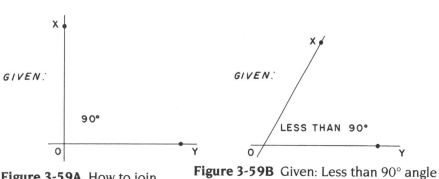

Figure 3-59A How to join two points by a parabolic curve (given: 90° angle)

Figure 3-59B Given: Less than 90° angle

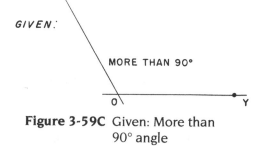

Figure 3-59C Given: More than 90° angle

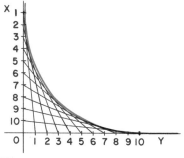

Figure 3-59D Step 1

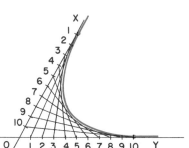

Figure 3-59E Step 1

Figure 3-59F Step 1

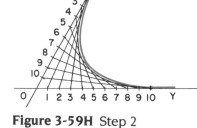

Figure 3-59G Step 2

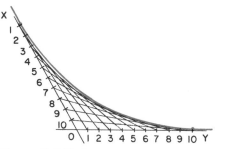

Figure 3-59H Step 2

Figure 3-59I Step 2

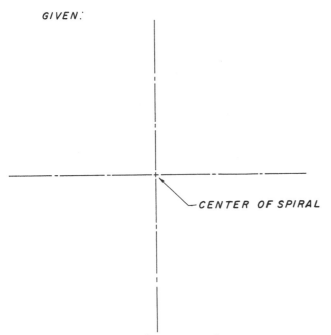

Figure 3-60A How to draw a spiral

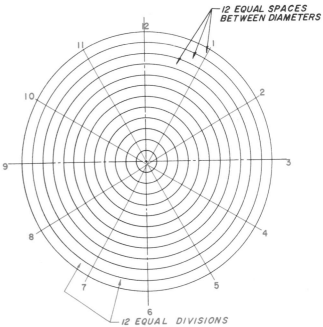

Figure 3-60B Step 1 (30° APART)

Supplementary Construction

How To Draw a Spiral

Given: Crossed axes, Figure 3-60A.

Step 1. Divide the circle into equal angles (in this example, 30°). Set the compass at a required radius and draw various diameters to suit, evenly spaced, as illustrated in Figure 3-60B.

Step 2. Starting any place along the angles, step over one angle and up one diameter until the required spiral is completed, Figure 3-60C.

How To Draw a Helix

A *helix*, a form of spiral, is used in screw threads, worm gears, and spiral stairways, to mention but a few uses. A helix is generated by moving a point around and along the surface of a cylinder with uniform angular velocity about its axis. A *cylindrical helix* is simply known as a helix, and the distance measured parallel to the axis traversed by a point in one revolution is called the *lead*.

Given: The top and front views of a cylinder with a required lead, Figure 3-61A.

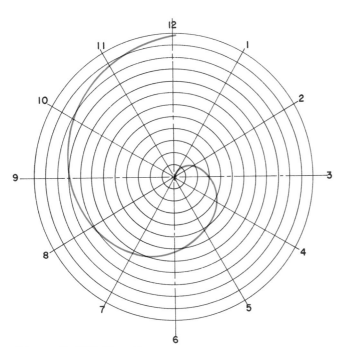

Figure 3-60C Step 2

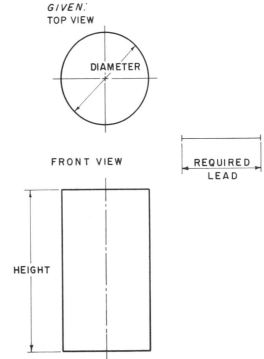

Figure 3-61A How to draw a helix

Step 1. Divide the top view into an equal number of spaces (12 in this example). Draw a line equal to one revolution and/or circumference, project the lead distance from the end, and label all points per illustration 3-61B.

Step 2. Divide the circumference into the same amount of equal spaces, and project each from the inclined line to the corresponding point projected from the front view, Figure 3-61C.

Step 3. Connect all points of the helix, and draw as hidden lines the lines that would disappear from view on an actual helix form, Figure 3-61D.

Notice that a right-hand helix is illustrated. To construct a left-hand helix, simply project all the points in the opposite direction, Figure 3-61E.

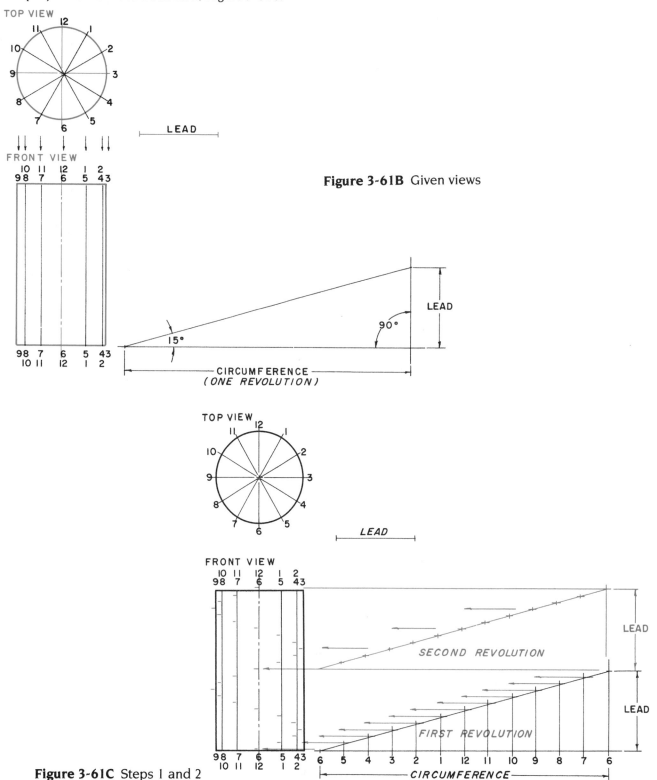

Figure 3-61B Given views

Figure 3-61C Steps 1 and 2

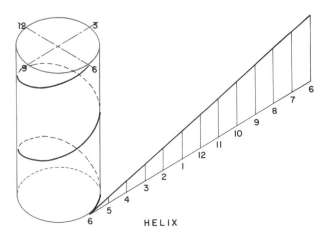

Figure 3-61D Step 3

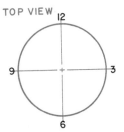

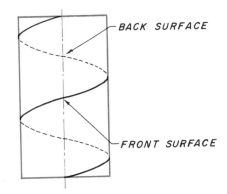

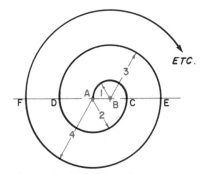

Figure 3-61E Step 4

How To Draw an Involute of a Line

The curved path of a point on a string as it unwinds from a line, a triangle, a square or a circle is an *involute*. The involute is used to construct involute gears. The involute forms the face and part of the flank of the teeth of the gear.

Given: Line A-B, Figure 3-62A.

Step 1. Extend line A-B, as illustrated in Figure 3-62B. Set the compass at the length A-B, and, using point B as the swing point, swing semicircle A-C. With the compass set at length A-C, and, using point A as the swing point, swing semicircle A-D. Continue in this manner, alternating between points A and B until the required involute is completed, as shown in the figure.

Figure 3-62A
How to draw an involute of a line

Figure 3-62B Step 1

How To Draw an Involute of a Triangle

Given: Triangle ABC, Figure 3-63A.

Step 1. Extend straight lines from the triangle, as illustrated in Figure 3-63B. Set the compass at length A-C, and, using point A as the swing point, swing semicircle A-C. With the compass set at length B-D, and, using point B as the swing point, swing semicircle B-E. Continue similarly around the triangle until the required involute is completed, as shown in the figure.

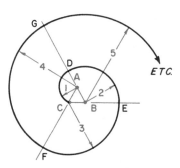

Figure 3-63A
How to draw an involute of a triangle

Figure 3-63B Step 1

How To Draw an Involute of a Square

Given: Square ABCD, Figure 3-64A.

Step 1. Extend straight lines from the square, as illustrated in Figure 3-64B. Set the compass at length A-B, and using point A as the swing point, swing semicircle A-E. With the compass set at length B-E, and, using point B as the swing point, swing semicircle B-F. Continue in the same manner around the square until the required involute is completed, as shown in the figure.

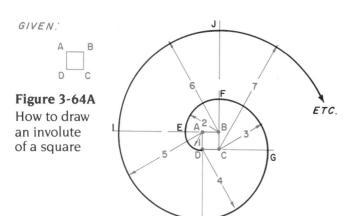

Figure 3-64A
How to draw
an involute
of a square

Figure 3-64B Step I

Figure 3-65A
How to draw
an involute
of a circle

Figure 3-65B Step I

How To Draw an Involute of a Circle

Given: Circle A, Figure 3-65A.

Step I. Divide the circle into a number of equal parts and number each point clockwise around the circle (in this example, 12 equal parts). Project a line perpendicular to the radius at each point in a clockwise direction, Figure 3-65B.

Step 2. Set the compass at length 1-2, and, using point 2 as the swing point, swing semicircle 1-2. With the compass set at length 1-2 plus 2-3, and, using point 3 as the swing point, swing semicircle 1-3. Continue in the same manner around the circle until the required involute is completed, as shown in Figure 3-65C.

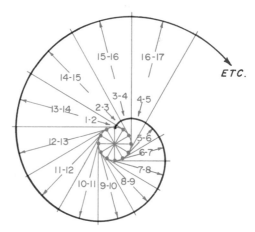

Figure 3-65C Step 2

How To Draw a Cycloidal Curve

Given: A required span of a cycloid.

Step I. As the span represents the rolling distance of one revolution of a diameter, divide the span length by pi to find the required diameter. Divide the span into an equal number of divisions. Twelve is a convenient number, Figure 3-66A.

Step 2. Draw the rolling diameter tangent to the given span at point 0/12. Divide this diameter into the same number of equal spaces as done in the Step I span division. Consecutively number the radial end points of these divisions to make them meet their corresponding number of the span divisions as the diameter rolls along the span, Figure 3-66B.

Step 3. At each span division draw the rolling diameter tangent to the span in the rolled position, with the division numbers of span and diameter matching at their contact point.

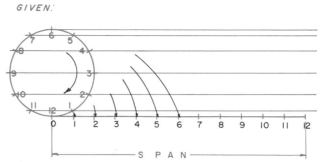

Figure 3-66A How to draw a cycloidal curve
(Step I)

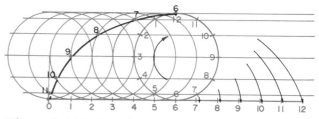

Figure 3-66B Step 2

Locate the point 0/12 position on the diameter at each of these locations, marked with a small cross, Figure 3-66C.

Step 4. Connect all the point 1 diameter division positions with a smooth curve to complete the cycloidal curve, Figure 3-66D.

Note: If the given span is a concave arc, a *hypocycloid* will be drawn, Figure 3-67. If the span is convex, an *epicycloid* will be drawn, Figure 3-68.

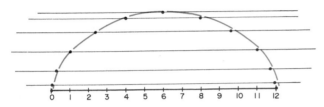

Figure 3-66C Step 3

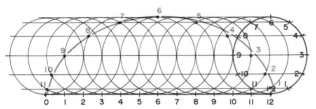

Figure 3-66D Step 4

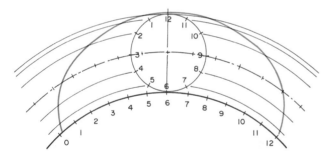

Figure 3-67 A hypocycloid

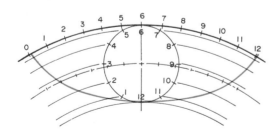

Figure 3-68 An epicycloid

Review

1. Explain the following terms: right angle, acute angle, obtuse angle

2. List six polygons.

3. Why are geometric construction procedures so important to the drafter?

4. What is meant by the term *tangent*, and why is it important to find all tangent points before darkening in a drawing?

5. Explain the following parts of a circle: central angle, sector, quadrant, segment

6. Explain the difference between a concentric circle and an eccentric circle.

7. How is an actual point in space represented on a drawing?

8. Completely describe an angle.

9. List five kinds of triangles.

10. What is meant by the circumference of a circle, and how is it calculated?

11. What is the sum of the three internal angles of a triangle?

12. Define a *line*.

Chapter Three Problems

The following problems are intended to give the beginning drafter practice in using the many geometric construction techniques used to develop drawings. Accuracy, line work, neatness, speed, and centering are stressed. It is recommended that dimensions not be added to problems at this time. The beginning drafter should practice using drafting instruments and correct line thicknesses, and should concentrate on developing good drafting habits.

The steps to follow in laying out all drawings throughout this book are:

Step 1. All geometric construction work should be made very accurately using a sharp 4-H lead.

Step 2. All geometric construction work must be laid out very lightly.

Step 3. Do not erase construction lines. If constructed lightly, they will not be seen on the whiteprint copy.

Step 4. Make a rough sketch of each problem before beginning to calculate the overall shape.

Step 5. Lightly draw each problem completely, first.

Step 6. Locate *all* tangent points as you proceed. Make short light dashes at each tangent point.

Step 7. Try to center the problem in the work area.

Step 8. Check each dimension for accuracy.

Step 9. Darken in the drawing using the correct line thickness and the following steps:

- Locate and draw all center lines with a thin black line.
- Darken in all diameters using a compass.
- Darken in all radii using either a compass or a circle template.
- Darken in all horizontal lines, either from right to left or from left to right.
- Recheck all dimensions.
- Check all lines for correct thickness.
  Fill out the title block using light guidelines and neat lettering.

Problem 3-1

Divide the work area into four equal spaces. In the upper left-hand space, draw an inclined line 3.25 long. Bisect this line. Show all construction lines lightly.

In the upper right-hand space, draw an acute angle with intersecting lines, each approximately 2.75 long. Bisect this angle. Show all light construction lines.

In the lower left-hand space, draw an inclined line 2.88 long. Divide this line into 5 spaces. Show all light construction lines.

In the lower right-hand space, draw an inclined line 3.125 long. Divide this line into three proportional parts 2, 3, and 4. Show all light construction lines.

Problem 3-2

Divide the work area into four equal spaces. In the center of the upper left-hand space, draw an equilateral triangle having sides of 2.0. Bisect the interior angles. Show all light construction lines.

In the center of the upper right-hand space, draw a square having sides of 2.0. Show all light construction lines.

In the center of the lower left-hand space, draw a hexagon having the distance of 2.0 across the flats. Show all light construction lines.

In the center of the lower right-hand space, draw an octagon having the distance of 2.0 across the flats. Show all light construction lines.

Problem 3-3

Divide the work area into four equal spaces. In the upper left-hand space, draw two lines intersecting at 60° and draw an arc with a .88 radius tangent to the two lines. Show all light construction lines.

In the upper right-hand space, draw two lines intersecting at 120° and draw an arc with a .625 radius tangent to the two lines. Show all light construction lines.

In the lower left-hand space, draw two lines intersecting at 90° and draw an arc with 1.25 radius tangent to the two lines. Show all light construction lines.

In the lower right-hand space, draw an ellipse with a major diameter of 3.00 and a minor diameter of 1.75.

Problem 3-4

Divide the work area into four equal spaces. In the upper left-hand space, center a line 10 mm long and label it line A-B. Construct a straight line involute with five arcs.

In the upper right-hand space, center a triangle with sides equal to .25 and label the triangle ABC. Construct a triangular involute with five arcs.

In the lower left-hand space, center a square with sides equal to 10 mm and label the square ABCD. Construct a square involute with five arcs.

In the lower right-hand space, center a circle with a diameter of 1.00. Construct a circle involute with as many arcs as space will allow.

Problem 3-5

Divide the work area into four equal spaces. In the center of the upper left-hand space draw a spiral with 30° spaces, at 4 mm increments, for one 360° rotation.

In the center of the upper right-hand space, draw two lines at right angles intersecting a point 0. One line is to be 3.5 long, and the other 2.5 long. Draw a parabolic curve between these two lines.

In the center of the lower left-hand space, draw a rectangle 2.0 x 4.50 in size. Label the 2.0 side as the rise, and the 4.50 as the span. Construct a parabola (Method 2), and locate the focus point.

In the center of the lower right-hand space, draw a line 80 mm long, and label its end points as A and B. Locate a point on line A-B at 5/8 of the line's length from point A. Label the point X.

Problems 3-6 through 3-19

Using the art for problems 3-6 through 3-19, center each object within the work area using correct line thickness. Do not erase all light construction lines.

Locate and draw a light short dash at all tangent points. Do not dimension objects.

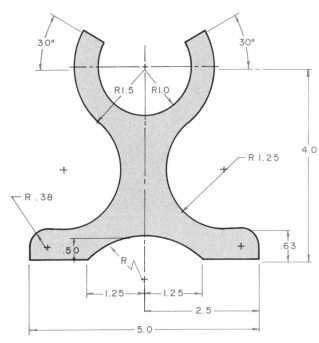

Problem 3-7

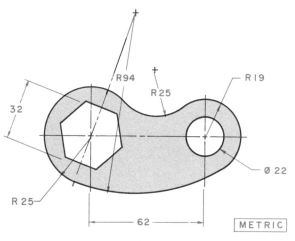

Problem 3-8

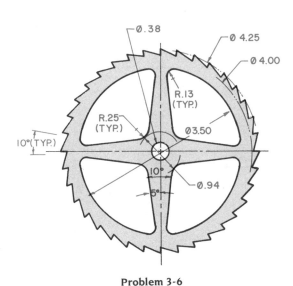

Problem 3-6

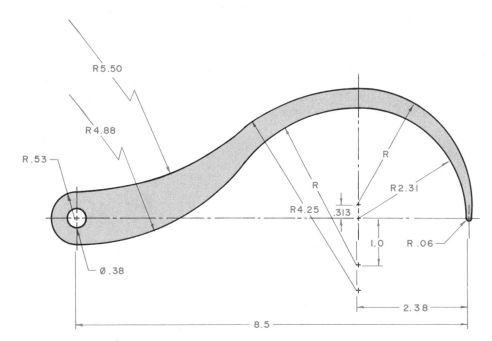

Problem 3-9

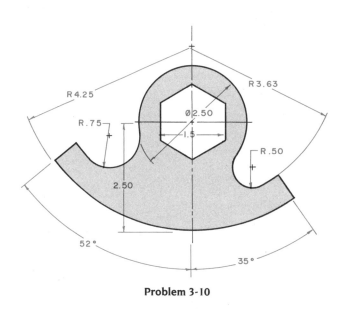

Problem 3-10

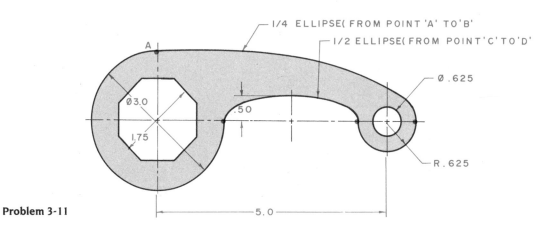

1/4 ELLIPSE(FROM POINT 'A' TO 'B'

1/2 ELLIPSE(FROM POINT 'C' TO 'D'

Ø.625

R.625

Ø3.0

1.75

.50

5.0

Problem 3-11

142   Section 1

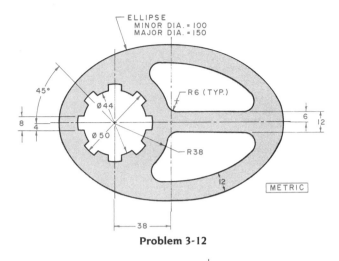

ELLIPSE
MINOR DIA. = 100
MAJOR DIA. = 150

45°

Ø44

Ø50

R6 (TYP.)

R38

8
4

6
12

12

38

METRIC

Problem 3-12

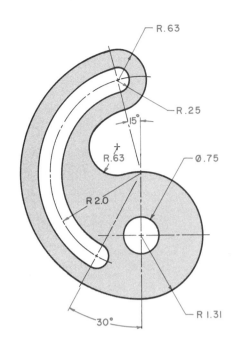

R.63

R.25

15°

R.63

R 2.0

Ø.75

30°

R 1.31

Problem 3-15

45°

30°

R 3.0

R3.12

R.25
(TYP.)

Ø1.25
(TYP.)

2X Ø.56

.63

R1.25
R1.63

HEX .68 ACROSS FLATS

Problem 3-13

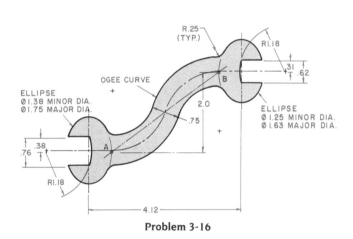

R.25
(TYP.)

R1.18

OGEE CURVE

.31
.62

ELLIPSE
Ø1.38 MINOR DIA.
Ø1.75 MAJOR DIA.

B

2.0

.75

ELLIPSE
Ø 1.25 MINOR DIA.
Ø 1.63 MAJOR DIA.

.38
.76

A

R1.18

4.12

Problem 3-16

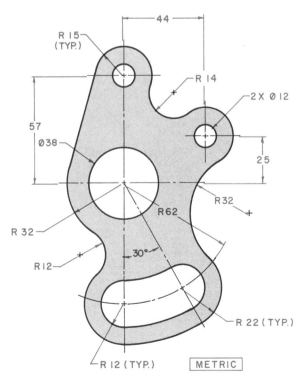

44

R 15
(TYP.)

R 14

2X Ø 12

57

Ø38

25

R 32

R32

R62

30°

R 12

R 22 (TYP.)

R 12 (TYP.)

METRIC

Problem 3-14

R 15 (TYP.)

82

R56

HEX. 14
(ACROSS FLATS)

68

B

OGEE CURVE
FROM POINT 'A'TO 'B'

A

5X Ø12, EVENLY SPACED

METRIC

Problem 3-17

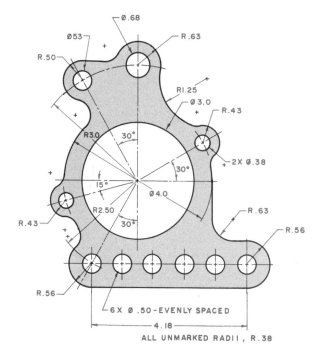

Problem 3-18

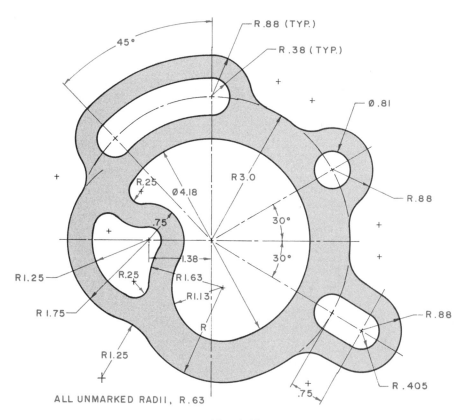

Problem 3-19

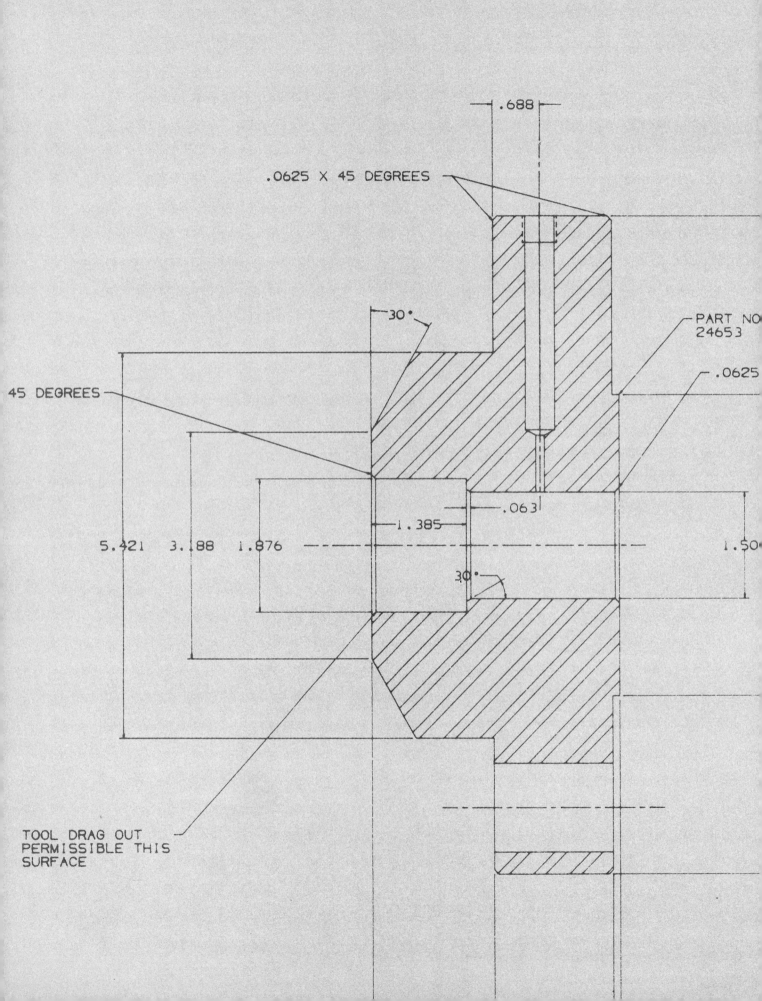

.688

.0625 X 45 DEGREES

30°

PART NO
24653

.0625

45 DEGREES

.063

1.385

5.421   3.188   1.876

1.500

30°

TOOL DRAG OUT
PERMISSIBLE THIS
SURFACE

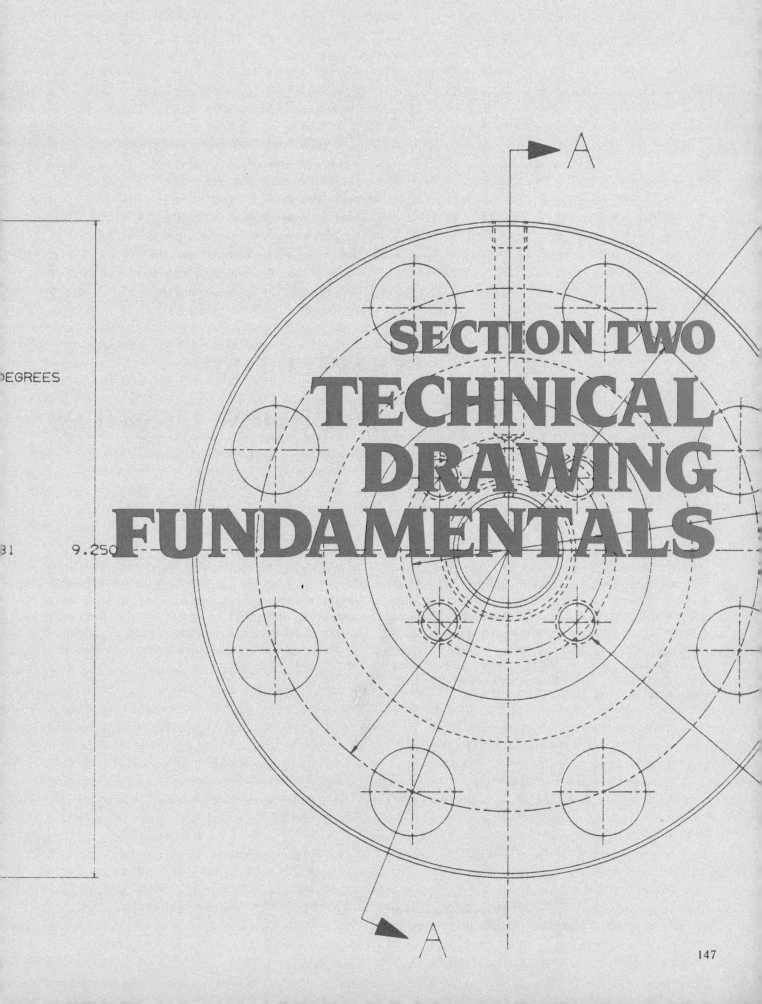

SECTION TWO
TECHNICAL DRAWING FUNDAMENTALS

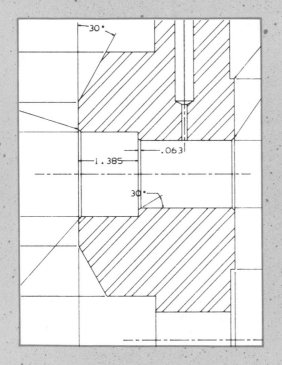

Most drawings produced and used in industry are multiview drawings of one, two, three or more views. The concepts of multiview projection, that is, taking a three-dimensional object and drawing it on a two-dimensional sheet of paper, must be fully understood and mastered by the student. All drafting practices are fully illustrated in this chapter, and must be learned by the beginning drafter.

CHAPTER FOUR

Multiview Drawings

Providing accurate shape descriptions of objects by drawing methods requires that three-dimensional object information be presented in a flat, two-dimensional drawing space. This is usually done by drawing images of the object from multiple directions. Commonly shared methods and interpretations are essential for all who make or use such drawings. Technical drawings seldom use more than lines to outline an object's features. Visual qualities, such as color and texture, are more accurately specified by written requirements.

Viewing an object by eye creates a depth distortion. This phenomenon is recreated in a field of drawing called *perspective projection*, Figure 4-1. This distortion occurs because the visual rays used to create the image in the viewer's eye are not parallel. This distortion, while useful in providing the illusion of distance, results in a loss of image accuracy and fails to provide the information needed for most detailed technical communication.

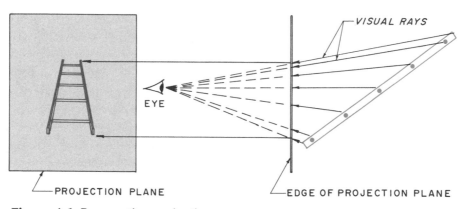

Figure 4-1 Perspective projection

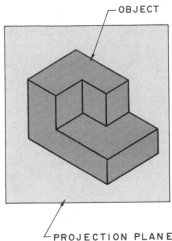

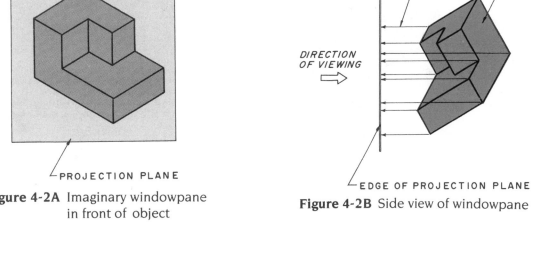

Figure 4-2A Imaginary windowpane in front of object

Figure 4-2B Side view of windowpane

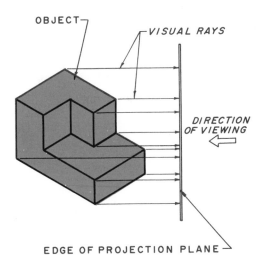

Figure 4-2C Reversed projection

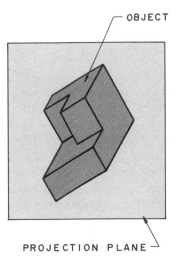

Figure 4-2D Reversed projection

Orthographic Projection

Orthographic projection is the most accurate method of shape description wherein an undistorted image of the object appears in a flat, transparent, but imaginary projection plane. A *projection plane* may be thought of as an imaginary pane of window glass, with an object underneath or behind it, Figure 4-2A. Viewed from the side, the pane of glass appears as a line, Figure 4-2B. Assume that light beams emitting from the surfaces of the object are projected to the projection plane, and that each light beam from each exposed surface is directed toward the viewing plane, Figure 4-2C. The image shown in Figure 4-2A traces the path of each light beam as it intersects the viewing plane. This image is called a *projection*, and illustrates the reflected features of the exposed surfaces of the object. Note, however, that Figure 4-2B is an image

created in the same manner from Figure 4-2A. This reversal in shown in Figures 4-2C and 4-2D. Figure 4-2C shows the object's image resulting from the surface projectors striking the viewing plane, as illustrated in Figure 4-2D.

Normal Surfaces

The surface images of the figures just described do not represent the size and shape of their corresponding real surfaces. Rectangular surfaces appear as parallelograms. The lengths of edges are shorter in the image than their actual lengths (a phenomenon known as *foreshortening*). However, if the viewing plane were positioned parallel to some of the object's surfaces, as shown in Figure 4-3A, then the image appearing in that viewing plane would provide the actual size and shape of those surfaces, as shown in

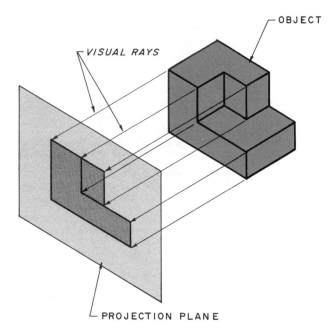

Figure 4-3A Viewing plane parallel to surfaces

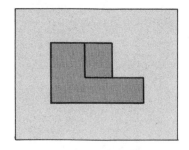

Figure 4-3B Viewing plane provides actual size and shape of surfaces

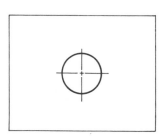

ONE–VIEW DRAWING

Figure 4-4 Simple one-view drawing

Figure 4-3B. A surface that is parallel to a viewing plane is said to be *normal* to the viewing plane.

A simple object, such as the round cannonball in Figure 4-4, needs only one view to describe its true size and shape. An object such as the flat gasket in Figure 4-5 needs only one view and a callout stating its required thickness. The callout can be noted on the drawing, as illustrated, or listed in the title block under "material." Figure 4-6A shows an object with a viewing plane placed in a normal position relative to some surfaces of the object, resulting in the orthographic views shown in Figure 4-6B. The sizes and shapes of the object's normal surfaces are shown in perfect outline in the orthographic view, but the viewer still cannot determine other features of the object. For example, Figures 4-6C, 4-6D, and 4-6E are different objects that provide exactly the same projected view. Furthermore, not all surfaces of the projected views of Figures 4-6C and 4-6E are shown in their real size and shape, as they are not normal to the viewing plane. There is no way to fully distinguish these without additional information.

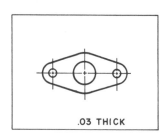
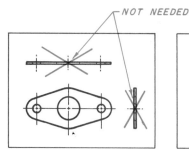

ONE–VIEW DRAWING

Figure 4-5 Simple one-view drawing

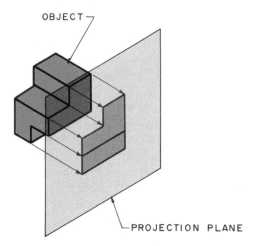

Figure 4-6A Viewing plane in a normal position

Two Orthographic Views

In order to present the images of each viewing plane in a flat drawing area, one of the viewing planes must be repositioned to lie in the same plane of the drawing as the other. The procedure used to do this is to create a fold line where the imaginary perpendicular orthographic viewing planes meet each other along the straight edge of intersection, as shown in

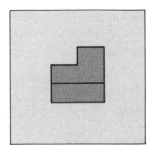

Figure 4-6B Orthographic view of the object

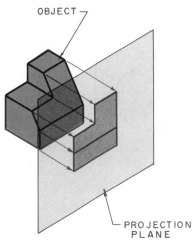

Figure 4-6C Different object with the same orthographic view

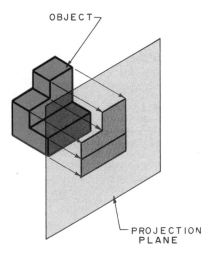

Figure 4-6D Another object with the same orthographic view

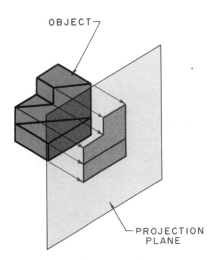

Figure 4-6E Another object with the same orthographic view

Figure 4-7A. The figure shows the object in Figure 4-6A with a second viewing plane positioned. The fold line acts as a hinge line around which the intersecting viewing planes are swung into the same plane, Figure 4-7B. The resulting orthographic views are shown in Figure 4-7C, each of which contains the image of a 90° rotated view or projection of the other. Note that the distance from the viewing plane to the object's surfaces in each image can be determined by locating from the fold line the position of the same surface in the adjoining image. In this procedure, the features and surfaces are always aligned to each other and thus can be located. The same two orthographic views are shown in Figures 4-7D, 4-7E, and 4-7F in haphazard relationship to each other, putting them each in error and making it impossible to determine surface positions or the real shape of the object.

Labeling Two Views

Figure 4-8A shows the positioning of two adjoining orthographic viewing planes to describe the object shown in Figure 4-6C. The orthographic view that presents the most characteristic shape of the object is

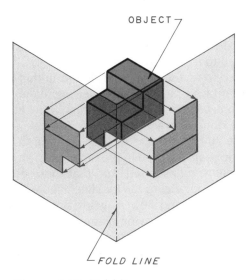

Figure 4-7A Fold line

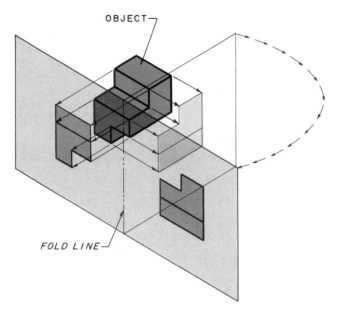

Figure 4-7B Fold line acts as a hinge

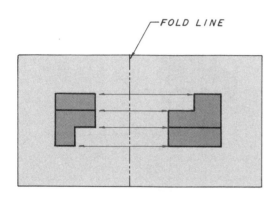

Figure 4-7C Projecting features from one view to the next view

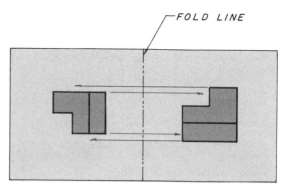

Figure 4-7D Incorrect positioning of views

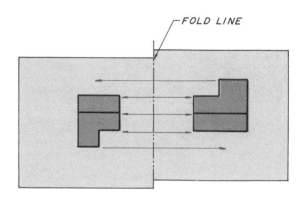

Figure 4-7E Incorrect positioning of views

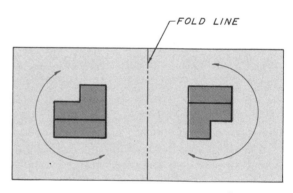

Figure 4-7F Incorrect positioning of views

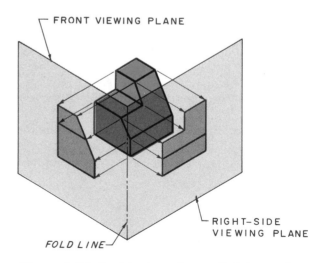

Figure 4-8A Positioning of two adjoining orthographic viewing planes

usually selected and identified as the front view. If this is not feasible, the normal surface containing the most visible details is usually the next best choice. Once the front view is selected, the orthographic view that is projected to the right of the front view is identified as the right-side view. The term *front view* is often referred to as the *front elevation* and the *right-side view* is referred to as the *right profile*. To aid in this example, the normal surfaces to each viewing plane in Figure 4-8B are flattened out by shading, but this should not be done in practice.

Note that the edge view of each of the shaded surfaces appears as a line in the other orthographic view from where it is shaded, and that those edge views are always parallel to the fold line. Figure 4-9A shows viewing planes and view images to accurately describe Figure 4-6D. The viewing planes were earlier identified as imaginary; therefore, the fold line is imaginary also. The finished orthographic drawing should not include the borders of a viewing plane, which are usually a matter of choice. This omission is shown

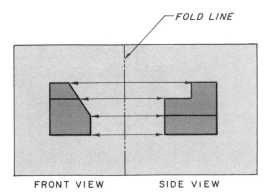

Figure 4-8B Two adjoining orthographic viewing planes flattened out

FRONT VIEW SIDE VIEW

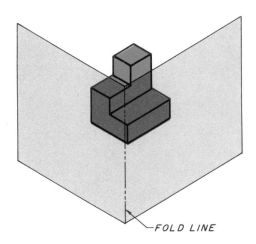

Figure 4-9A Viewing planes and image

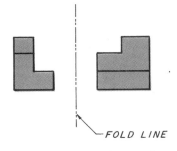

Figure 4-9B Orthographic drawing omitting viewing planes

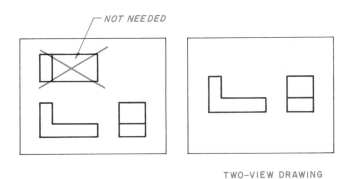

TWO-VIEW DRAWING

Figure 4-10 Only two views are required

in Figure 4-9B. Often, there is a need to draw the fold line for view constructions, as is discussed later, but it is preferable to remove it from the final drawing. The distance of the image from the fold line identifies the object's location from the viewing plane and is selected by the drafter. This decision is often based on the amount of space available for the drawings, dimensions, and notes.

Many objects can be drawn with only two views. A third view would only duplicate the same information. Figure 4-10 shows an example of a drawing requiring only two views. Do not use more views than needed to draw an object. Too many views will only complicate the drawing, and waste a great deal of drawing time.

Multiple Orthographic Views

Figure 4-11A shows two orthographic views that do not completely describe the object. Multiple interpretations of the object are possible, causing the viewer to be misled by inadequate information. Correctly interpreted, the object appears as shown in Figure 4-11B. Incorrect interpretations are shown in Figures 4-11C, 4-11D, and 4-11E.

A third viewing plane is commonly used to ensure adequate definition of the represented object. The third viewing plane is usually above the object, and perpendicular to the other two. It is called the *top view* or is sometimes identified as the *plan view* in construction-related drawings. This addition to the views is shown in Figure 4-12. The selection of views may be based upon an attempt to present the object's shape using the most efficient use of drawing area,

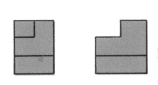

Figure 4-11A Orthographic views of an object

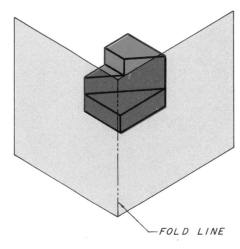

Figure 4-11B Correct interpretation

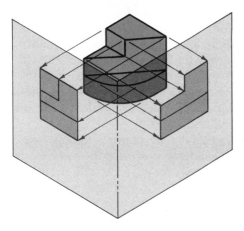

Figure 4-11C Incorrect interpretation

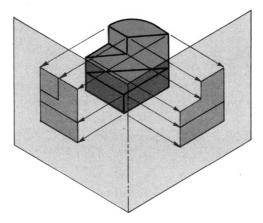

Figure 4-11D Incorrect interpretation

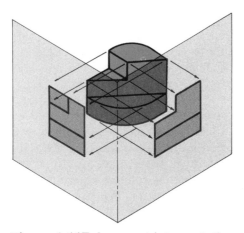

Figure 4-11E Incorrect interpretation

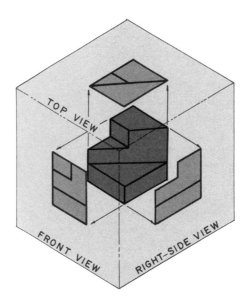

Figure 4-12
Third viewing plane added

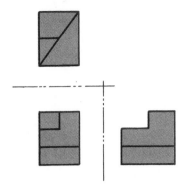

Figure 4-13A Orthographic view
of an object

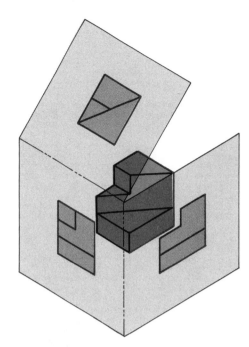

Figure 4-13B Three principal planes

Figure 4-13A, or to maintain the orientation of the object that is most easily understood by the reader, as illustrated in Figure 4-13B. Recall that the front view should be an attempt to show the most characteristic shape of an object. It can be disconcerting to read a view of a building that is drawn sideways.

The three viewing planes shown in Figure 4-12 are called *principal viewing planes*, as they are all perpendicular to one another, beginning with the orientation of the front view. There are three other principal planes. When combined with the front, top, and right side, they form the sides of a transparent box that completely surrounds the object. Figure 4-14A illustrates such a box, and Figure 4-14B shows some of the many possible fold line selections that would result in the multiple orthographic views of the object shown in Figure 4-14C.

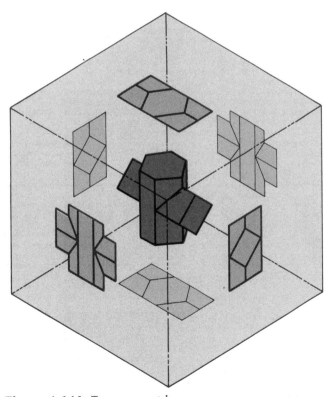

Figure 4-14A Transparent box

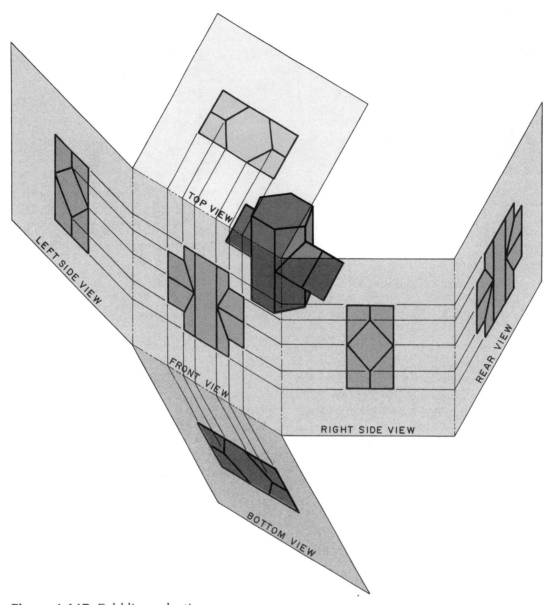

Figure 4-14B Fold line selections

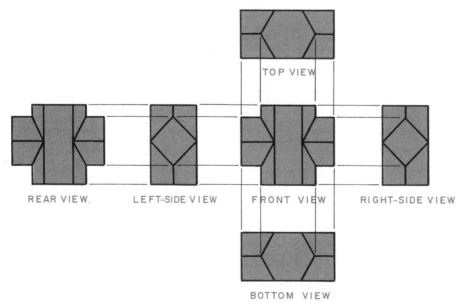

Figure 4-14C Resulting multiple views

156   Section 2

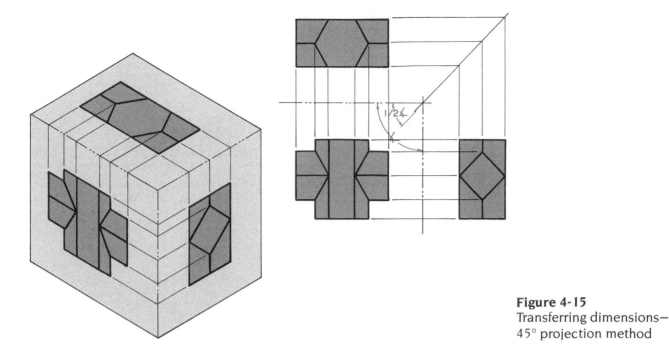

Figure 4-15
Transferring dimensions—
45° projection method

Object Description Requirements

All six views of Figure 4-14C are not necessary; three will provide sufficient information. Either the front view or the rear view may be selected as the front view, as each provides the same information. The common arrangement of top, front, and right-side view then follows. In Figure 4-15, the top view aligns directly above the front view and the right-side view aligns directly to the right of the front view. Both the top view and right-side view images are behind the front view projection plane by the same distance.

Dimension Transfer Methods

Figure 4-15 shows a method of transferring dimensions between the top and right-side views. This method employs a miter line at an angle that is at half of the angle between the intersection of the view fold lines in the drawing space.

In Figure 4-16, using a compass, the distance from the front view fold line to the image in the top view is rotated to the right-view fold line to the image in the right-side view.

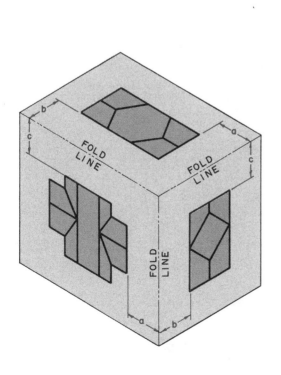

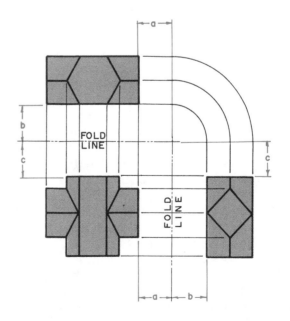

Figure 4-16 Transferring dimensions—
compass arc method

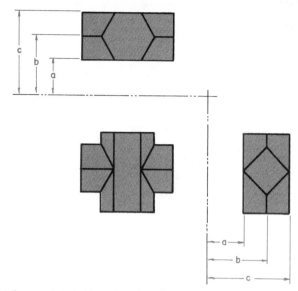

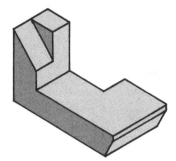

Figure 4-18 Isometric view
of an object

Figure 4-17 Transferring dimensions—transfer
method

Figure 4-17 shows the transfer method between top and front views. This method is potentially the most accurate. If the front and top views are drawn first, a fold line is arbitrarily positioned between them, but recall that a fold line must remain perpendicular to the projectors, or projection lines between views. A second fold line is drawn perpendicular to projectors or projection lines to the right of the front view *at any distance from the front view.* The distances from the fold line to the top view feature, dimensions a through c, are simply transferred directly to corresponding dimensions a through c, using a scale or dividers. If the front and right-side view are drawn first, the reverse order of fold line selection and distance transfers would occur.

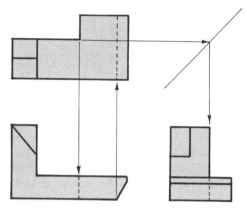

Figure 4-19 Orthographic views of
the same object

Hidden Lines

Figure 4-18 is an isometric view of an object. The orthographic top, front, and right-side views of this same object are illustrated in Figure 4-19. H*idden lines* are used to represent feature outlines whose visual rays must pass through some obstruction before reaching the viewing plane. Hidden lines mark the real but invisible features in each of the viewing planes where they are used.

A hidden line is often covered by a visible (solid) object line in the same view which hides the hidden line. This concept is shown in Figures 4-20A and 4-20B. Surface B aligns with corner A, as shown in Figure 4-20B. This causes part of the hidden line representing surface B in the right-side view of Figure 4-20B to "hide" behind the solid line representing edge B. Considering that each intersection of flat surfaces of an object in orthographic projection is represented by either a solid line or a hidden line, then almost every solid line in Figure 4-20B is covering

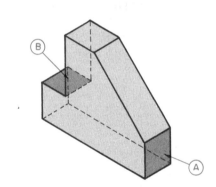

Figure 4-20A Isometric view
of an object

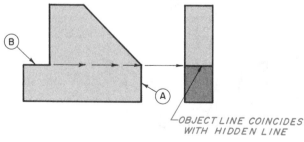

OBJECT LINE COINCIDES
WITH HIDDEN LINE

OBJECT LINE TAKES
PRECEDENCE OVER HIDDEN
LINE

Figure 4-20B Object line takes precedence over
hidden line

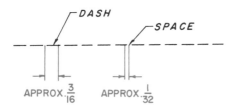

Figure 4-21 Average hidden line construction

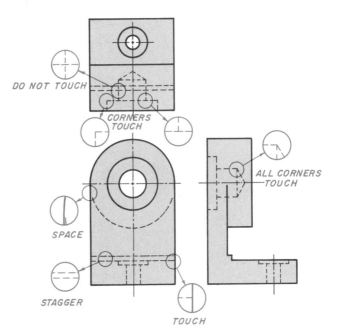

Figure 4-22 Hidden line drafting practices

one or more hidden lines. The ability to mentally perceive visible and hidden lines from orthographic views can provide the reader with an understanding of the object's actual configuration. This process is called *visualization.*

The lengths of dashes and dash spacing in hidden line construction are allowed to vary according to the size of the features being drawn. The dashes vary from 1/8 inch to 1/4 inch long (3.2 mm to 6.4 mm) and the spacing is correspondingly varied from 1/32 inch to 1/16 inch (0.8 mm to 1.6 mm). Little variation in size should be made from the average hidden line construction shown in Figure 4-21 for most drawing. The technical drawing trainee should concentrate on maintaining some consistency of dash sizes and spacing by estimation. Measuring dash lengths for preciseness is time consuming and does little to improve the learner's recognition time. However, there is some value for beginning drafters to measure their hidden lines at first to gain an awareness of the desired size and spacing. Recall that the line width of a hidden line should be of a medium width, perceptibly narrower than the width of the solid object line. Figure 4-22 illustrates the various drafting practices used in drawing hidden lines.

Center Lines

Center lines are used primarily as origin locations of circular, cylindrical or spherical features, but they can also be used to specify locations of other principal symmetries as well. As with hidden lines, the relative size of long and short dashes can be varied slightly with the size of the image being drawn. Crossed perpendicular center lines indicate a two-coordinate location of where an axis exists, and a single center line indicates the path of an axis. With the exception of noncritical surface radius contours of 1/8 inch or less, the axis of all circular features should be identified in all views.

Surface Categories

Recall that *normal surfaces* are parallel to one of the principal projection planes of the imaginary transparent box surrounding the object. In Figure 4-23, these surfaces are identified as the top, front, and

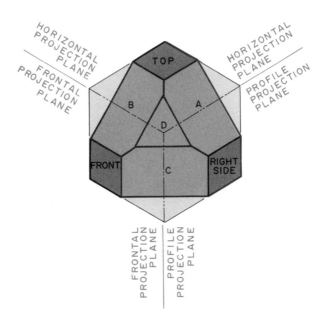

Figure 4-23 Surface categories

right side. The viewing planes containing these images are identified as the horizontal, frontal, and profile projection planes, respectively. A normal surface's size and shape appears in one of the principal views, Figure 4-24, but the surface appears as an edge view or line in the other principal views.

Surfaces A, B, and C are *inclined surfaces;* they are not parallel to any of the principal viewing planes, but each inclined surface is perpendicular to one of them. As with the normal surfaces, a surface that is perpendicular to a principal viewing plane will appear as an edge or line in that plane's image of the object.

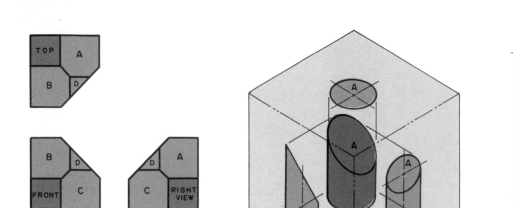

Figure 4-24 Orthographic views

Figure 4-25 Object inside transparent box

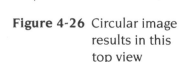

Figure 4-26 Circular image results in this top view

Surface A, for example, is perpendicular to the frontal viewing plane and accordingly appears as a line. It does appear in distorted and smaller size than its real size and shape in both the top and right-side views because of the foreshortening phenomenon. The maximum amount of foreshortening occurs when a surface is perpendicular to the viewing plane, which causes that surface to become so small as to condense it into a single straight line. Surfaces B and C are configured similarly to surface A, with B perpendicular to the right side (or *profile*) principal projection plane, and C perpendicular to the top (or *horizontal*) principal projection plane. Surface D is not perpendicular to any of the principal projection planes and is, therefore, categorized as an *oblique* viewing plane. It appears in a partially foreshortened condition in all of the principal projection planes.

The transparent box of principal projection planes that surrounds cylindrical surfaces uses the cylindrical axis as the line of orientation, Figure 4-25. If the cylindrical axis is made perpendicular to one of the principal viewing planes, then the cylindrical surface is foreshortened into a single-line circular arc. Figure 4-25 shows a cylinder whose axis is perpendicular to the top horizontal principal projection plane, and the circular image that results in the top view is shown in Figure 4-26. The lateral surface of the cylinders cannot be classified as normal, inclined or oblique, as the surface has a quality of continuous change in orientation. Surface A in Figure 4-26 is inclined to the horizontal viewing plane, creating elliptic contour with the lateral side. If surface A is 45° from the horizontal viewing plane, a circular image of the surface will appear in the right-side view.

Planning the Drawing

Before beginning a three-view drawing of any object, make a sketch of the object. Figure 4-27B shows preliminary sketches of the object in Figure 4-27A. The

sketches were used to aid in selecting the best front view and its best position. Sketches reduce the possibility of errors in the finished drawing, and are helpful in selecting the required views and their positions.

Follow these basic steps before beginning a three-view drawing:

Step 1. Visualize the object. Be sure you have a good mental picture of exactly what the object actually is.

Step 2. Decide which view to use as the front view by sketching it in various positions. See Figure 4-27B. Keep in mind:

- the front view is the most important view.
- the front view should show the most basic shape in profile.
- the front view should be drawn so that it appears in a stable position. To accomplish this, always place the heavy part at the bottom surface of the view.
- the front view should be placed in such a position that the other views have as few hidden edges as possible. This may take a little practice, but is very important.
- the front view should show the most detail.

Step 3. Decide how many views are needed to completely illustrate the object without question.

Step 4. Decide in which position to place the front view. Figures 4-28, 4-29, and 4-30 illustrate *poor* positioning. Study each figure carefully to understand why each is poor practice. Figure 4-31 illustrates the *best* position for this object. Do not measure distances or use a straightedge when making a sketch. A sketch should be nothing more than its name implies.

Step 5. Make sure the views are neat and centered within the work area.

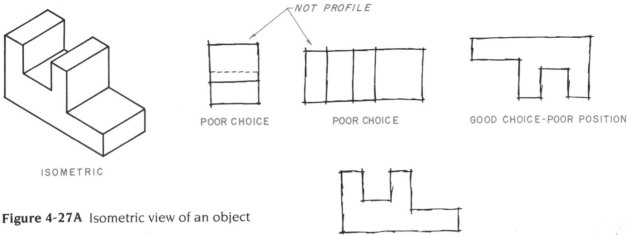

Figure 4-27A Isometric view of an object

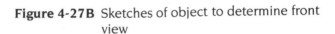

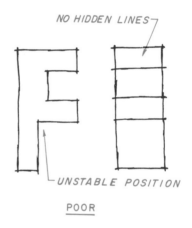

Figure 4-27B Sketches of object to determine front view

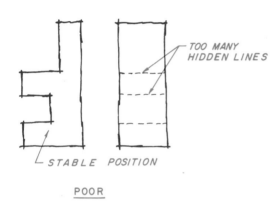

Figure 4-28 Poor front view—too many hidden lines

Figure 4-30 Poor front view—unstable

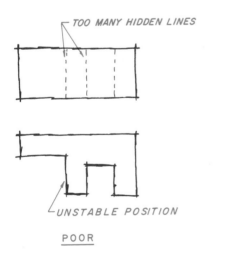

Figure 4-29 Poor front view—unstable, too many hidden lines

Positioning the Views within the Work Area

The example shown on the left of the Figure 4-32 is poorly centered. It has wasted space on both sides of the top view and front view, and both views are too close to the top and bottom of the paper. Objects should never be drawn within ½ inch (12 mm) of the border lines. The example shown on the right of Figure 4-32 is better than the first example, because it appears balanced and well centered. A rule to remember is to try to keep the white space evenly distributed around the view, if possible.

Sketching Procedure

Sketching should be done freehand, quickly, and only to an approximate scale. Do not take the time to make fancy sketches, and do not use any straightedges or compasses. Select the front view, and, using

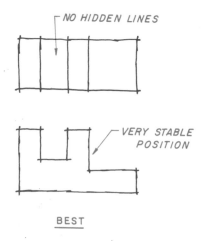

Figure 4-31 Best position- front view

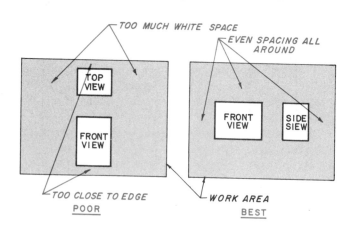

Figure 4-32 Positioning the views within the work area

the criteria previously outlined for the most important view, sketch it in position. Project upwardly to make the top view, and horizontally to make the right-side view. Lightly draw the basic shape of each view first, and then add the details to each view.

Centering the Drawing

The drawing must be neatly centered within the work area of the paper or within the border, if one is provided. A full one-inch (25 mm) space should be placed between all views drawn, regardless of which scale is used. This space may be adjusted with increased experience, and as the demands of dimensioning are introduced. Figures 4-33 and 4-34 show

the procedures used to center a drawing within a specified work area. Given is an isometric view of the object with all dimensions added. In this example, the total *horizontal* distance of the views (front view and side view) is determined by adding 4.0, the width of the front view, plus the 1.0 space between views, plus the 2.0 depth of the side view, for a total of 7.0 inches. See Figure 4-33. To center these two views horizontally, subtract 7.0 from 11.0, the width of the example work area. The answer 4.0 represents available extra space. This answer, or 4.0, is divided in two in order to have equal spacing on either side; refer to dimension D.

To center the drawing vertically, the same basic procedure is followed. The total *vertical* distance of the

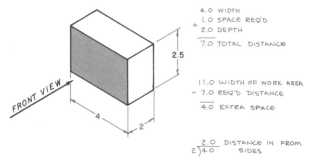

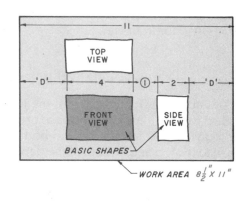

Figure 4-33 Centering views horizontally

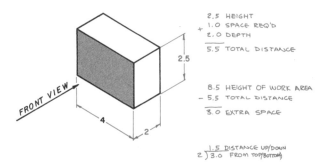

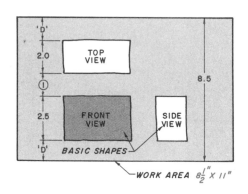

Figure 4-34 Centering views vertically

THINK OF EACH LINE AS,

A ———————————————— B

TWO POINTS IN SPACE

A •   B •

Figure 4-35 A line is made up of two points in space

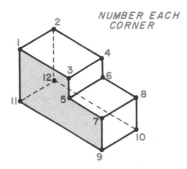

NUMBER EACH CORNER

Figure 4-36 Given: an isometric view of an object

two views is determined by adding the 2.5 height of the front view, plus the 1.0 space between views, plus the 2.0 depth of the side view, for a total of 5.5. See Figure 4-34. To center these two views vertically, subtract 5.5 from 8.5, the width of the example work area. This answer, 3.0, represents the available extra space. This space is divided by two to distribute it equally. One half of 3 or 1.5 inches is placed at the bottom of the work area, and the other half or 1.5 inches is placed above the top view. This centers the required view vertically. The same process is followed each time a drawing is to be centered, regardless of the drawing size or available work area.

Numbering Drawings

Many times, a multiview drawing is so complicated that the drafter is not quite sure of how some part of a view will look. Numbering points of various features of the more difficult drawings makes them easier to visualize, and helps to ensure that the final drawing is correct. Think of a drawing as a group of points in space joined together by lines. For instance, think of each line as two points in space, Figure 4-35. If the ends of a line can be found, all that has to be done to complete a line is to connect the ends. Once the ends have been found and numbered in one view, the same can be done in other views by simple projection. Figure 4-36 is an isometric view of a simple object. Each end of each line has been assigned a number.

Use the following steps to add numbers to a difficult drawing:

Step 1. Lightly draw the basic shape of the required views. Draw the light projection line also.

Step 2. Assign a number to each corner of the view where the profile is found. In this example, the front view has been used as a starting point. See Figure 4-37 and refer to the isometric view in Figure 4-36.

Step 3. Project these points up to the top view. Number the points in the top view, as shown in Figure 4-38.

Step 4. Project these points across from the front view to the right-side view, Figure 4-39. Project

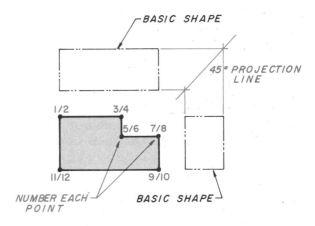

BASIC SHAPE

45° PROJECTION LINE

NUMBER EACH POINT

BASIC SHAPE

Figure 4-37 Number one view

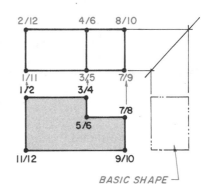

BASIC SHAPE

Figure 4-38 Project numbers into the top view

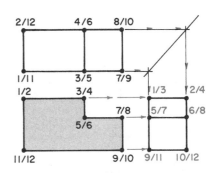

Figure 4-39 Project numbers into the right-side view

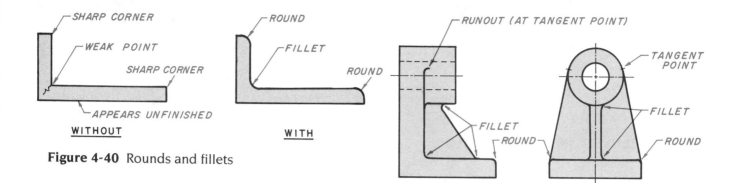

Figure 4-40 Rounds and fillets

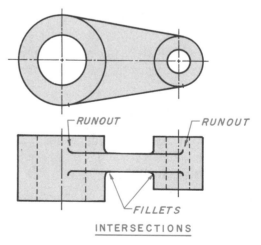

Figure 4-41 Runouts

the same points from the top view, over to the 45° line. Where the lines meet the 45° line, project down to the right-side view. The points where these lines cross are the exact locations of each point. Number all points, as shown in Figure 4-39. Other methods may be employed. Review Figures 4-15, 4-16, and 4-17.

Step 5. Connect points together, for example, point 1 to point 2, point 3 to point 4, and so on until the figure is completed in each view. Refer to the isometric view if necessary, Figure 4-36.

Note: If a point can be found in any two views, regardless of which two views, it can be projected into the third view.

Rounds and Fillets

It is very difficult to manufacture objects with absolutely square corners, especially with processes that require the object's material to flow into position. *Rounds* replace sharp external edges on cast objects, and are rounded outside corners.

Fillets are the opposite of rounds; they are inside rounded corners. Cast objects tend to crack due to the strain placed on the metal during the cooling process. Fillets distribute the strain and prevent

cracking. Rounds and fillets also enhance the appearance of an object. Examples of rounds and fillets are shown in Figure 4-40.

Fillets and rounds

- remove sharp corners from the object.
- add strength to the object.
- enhance the appearance of the object, giving it a "finished" look.

Runouts

Runouts are curved surfaces formed where a flat and curved surface meet, Figure 4-41. To find the exact intersection where a runout will occur

- locate the tangent points of the curved surface, Figure 4-42.
- project these tangent points to the next view.
- add the runouts, as shown in Figure 4-43.

Take care to study in which direction the runouts must be drawn.

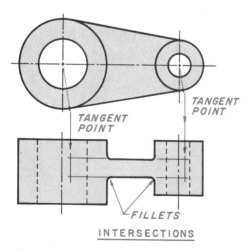

Figure 4-42 Locate the tangent points

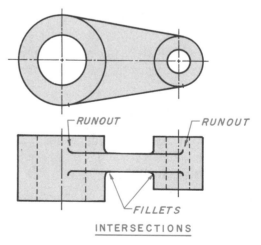

Figure 4-43 Complete runouts at tangent points

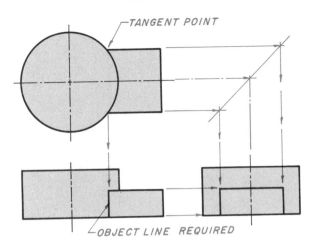

Figure 4-44 Intersecting surfaces

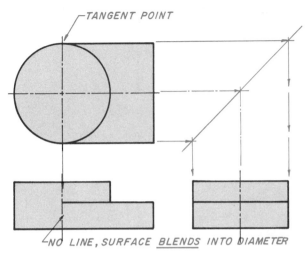

Figure 4-45 Intersecting surfaces

Treatment of Intersecting Surfaces

Figures 4-44 through 4-47 show how to illustrate various intersecting surfaces. In these examples, rounds, fillets, and runouts have been omitted to simplify the drawings. In actual practice, rounds, fillets, and/or runouts probably would have been added. In Figure 4-44, the flat surface meets the cylinder sharply (see top view), making a visible edge necessary in the front view. The tangent point in the top view is the exact location of the object line in the front view and right-side view.

In Figure 4-45, the flat surface blends into the cylinder, stopping at the point of tangency. Notice in the top view, the tangent point is located at the center line, and, in the front view, blends in at that exact point.

In Figure 4-46, the surface blends into the cylinder, stopping at the point of tangency, similar to Figure 4-45 except on an angle. Again, the tangent point is projected into the front view and right-side view.

In Figure 4-47, the radius blends into the base, stopping at the point of tangency. Notice that a line is drawn in the top view extending to the tangent point of the arc in the front view.

Curve Plotting

Some multiview drawings have a curved surface that must be projected into other views. The following steps are used to plot a curved surface.

Step 1. Lightly complete the basic shape of each view, Figure 4-48. Locate the center of the arc, point X.

Step 2. From this point, divide the arc into equal spaces (any spacing will do), Figure 4-49. In this example, 30° spaces have been used. The more spaces, the more accurate the plotted curve will be. Increments of 10° would be more accurate than 30° but would take more drawing time.

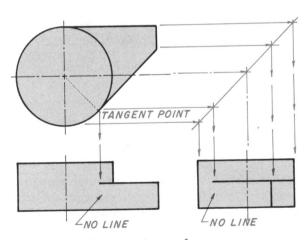

Figure 4-46 Intersecting surfaces

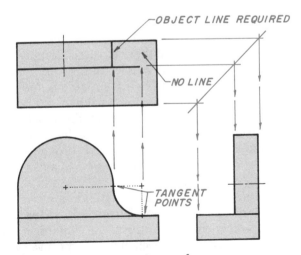

Figure 4-47 Intersecting surfaces

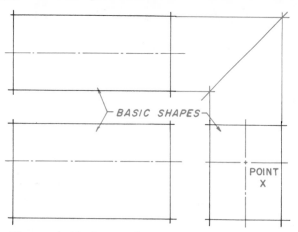

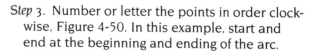

Figure 4-48 Curve plotting—Step 1

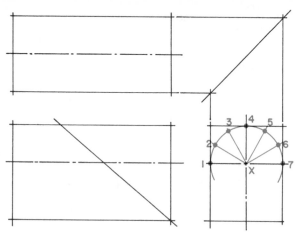

Figure 4-49 Curve plotting—Step 2

Step 3. Number or letter the points in order clockwise, Figure 4-50. In this example, start and end at the beginning and ending of the arc.

Step 4. Project these points up from the right-side view to the 45° axis line, and to the left to the top view, Figure 4-51.

Step 5. Project over to the slanted surface in the front view and locate each point on the slanted surface.

Step 6. Project these points up from the front view to the top view.

Step 7. Where the lines projected from the given points in the right-side view intersect with those from the front view is the exact location of points 1 through 7 in the top view (see Figure 4-51). Number all points at their intersection.

Step 8. Using an irregular curve, complete the drawing, as shown in Figure 4-52.

Cylindrical Intersections

A drafter can often indicate the intersection of cylinders, without actually showing the line of intersection between them. If there is a considerable difference in the diameters of each of the intersecting cylinders, then a straight line is capable of indicating the cylindrical intersection, as shown in Figures 4-53A and 4-53B. If there is little difference between the intersecting cylindrical diameters, then a simple arc is constructed or approximated through three principal points, as shown in Figures 4-54A and 4-54B.

On occasion, it is necessary to define the actual edge of intersection between two cylinders, especially in sheet metal construction. Figure 4-55 shows line elements on each of two intersecting cylindrical surfaces that are parallel to each of their respective cylindrical axes, and to a common (frontal) viewing plane.

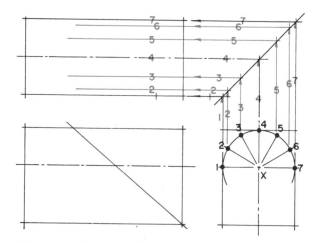

Figure 4-50 Curve plotting—Step 3

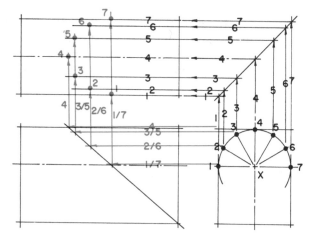

Figure 4-51 Curve plotting—Step 4

Note that each of the intersecting line elements, 1 through 4, are the same distance from the frontal viewing plane. Using this relationship to construct a line of intersection, line element locations are reflected on the perimeter of the smaller intersecting

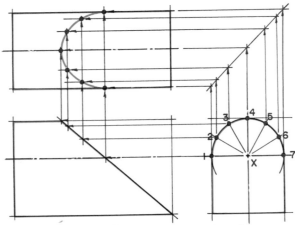

Figure 4-52 Curve plotting—completed

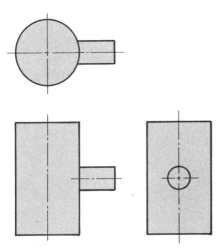

Figure 4-53A
Cylindrical intersections
(straight line)

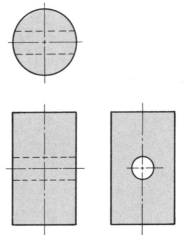

Figure 4-53B
Cylindrical intersections
(straight line)

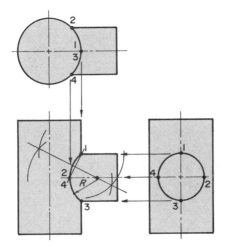

Figure 4-54A
Cylindrical intersections (arc)

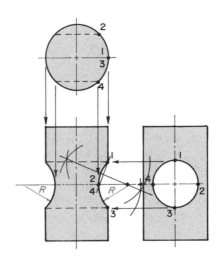

Figure 4-54B
Cylindrical intersections (arc)

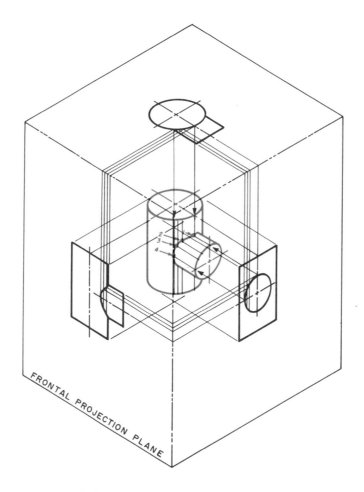

Figure 4-55 Two intersecting cylindrical surfaces

cylinder, Figure 4-56A, (Step 1). These line elements are projected in the front view. A fold line location is reflected between the front and left-side views in Figure 4-56B (Step 2) and its corresponding position is located between the front and top views by maintaining distance A at both locations. Distances 1 through 4 are transferred from the left-side view to the top view, locating the corresponding line elements 1 through 4 on the larger cylinder. As each corresponding line element from each cylinder is parallel to the front viewing plane and the same distance from it, they must intersect. A curve fitted through these points of intersection, 1 through 4, indicates the edge of intersection of the two cylinders. Additional points of intersection of corresponding line elements would be required to complete the edge of intersection in Figure 4-56B.

Incomplete Views

A drawing can be drawn absolutely *correctly*, but may sometimes still be confusing, as in Figure 4-57. Notice how the left and right-side views are difficult to understand because they overlap. Under these conditions, the drafter has the option of leaving out some features in order to make the views more clearly understood, Figure 4-58. Notice that the left and right-side views, while *not* drawn correctly, are much easier to understand. The right-side view illustrates only the right-side details, and the left-side view illustrates only the left-side details. This is actually incorrect, but is the "standard" used.

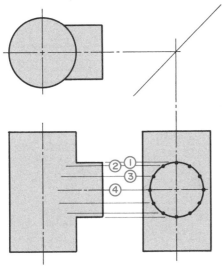

Figure 4-56A Step 1

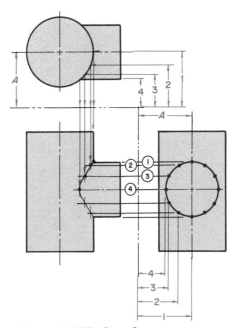

Figure 4-56B Step 2

Figure 4-57 Overlapping views—
poor practice

CORRECT BUT POOR PRACTICE

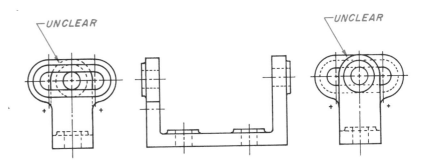

INCORRECT BUT BETTER PRACTICE

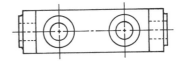

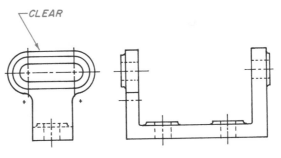

Figure 4-58 Omitting some details—
better practice

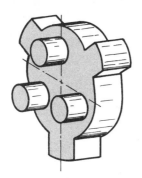

Figure 4-59A Isometric view of a symmetrical part

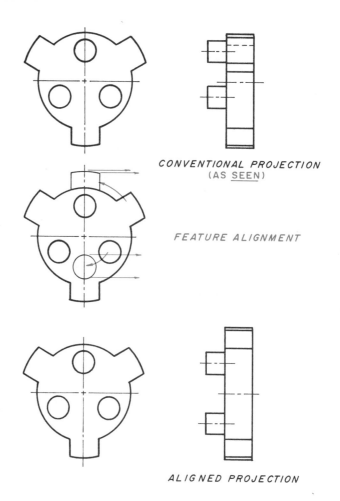

CONVENTIONAL PROJECTION
(AS <u>SEEN</u>)

FEATURE ALIGNMENT

ALIGNED PROJECTION

Figure 4-59B Aligned projection

Aligned Features

It is often of greater value to show features at a real distance from a line of symmetry, rather than in the projected or foreshortened position. Figure 4-59A is an isometric view of a symmetrical part. Figure 4-59B indicates how this symmetrical part appears confusing in conventional projection (top illustration) but gives a better indication of the part's symmetry when projected from the aligned position (lower illustration).

How to Represent Holes

Many objects contain various kinds of holes within their boundaries. The drafter must be able to identify and draw each kind of hole used.

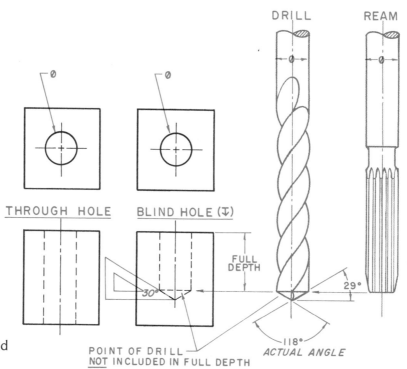

DRILL   REAM

THROUGH HOLE   BLIND HOLE (⊥)

FULL DEPTH

30°

POINT OF DRILL
<u>NOT</u> INCLUDED IN FULL DEPTH

29°

118°
ACTUAL ANGLE

Figure 4-60
Through hole and blind hole

Plain Holes

Two major kinds of plain holes are a through hole and a blind hole, Figure 4-60. A *through hole*, as its name implies, is a hole that goes completely through the object. A blind hole is a hole that is made to a specific depth. To the left in the figure is a top view and a side view of a through hole. Holes are usually produced with a drill. A drill cannot produce holes of an exact size; therefore, a ream must be used for a hole of an exact size. The hole is first drilled to a size approximately .015 smaller than the exact size required in order to prepare it for reaming. For example, a reamed hole of 1.0-inch diameter would first require a drilled hole of .985 diameter. The actual size difference between the size of the drill and ream is called the *allowance*. A standard ream can have a tolerance of +.0005.

Blind Holes

A *blind hole* is a hole that is drilled to a specific depth. Referring back to Figure 4-60, note that the tip of the drill is conical and has a taper. This taper is approximately 118° for general purpose work. In representing the tip of the drill on the drawing, a 30°-60° triangle is used. The *full depth* of the drilled hole includes only the cylindrical portion of the drill.

Tapered Holes

Another type of hole is the *tapered hole*, Figure 4-61. A tapered hole must first be drilled using a drill approximately .015 less in diameter than the smaller diameter of the tapered hole. The tapered portion is actually achieved by use of a tapered ream of the required taper.

Countersunk Holes

A *countersunk hole* is first drilled to a diameter slightly larger than the body of the fastener that is to be used. The upper portion is enlarged conically to a specified angle and to a depth that will produce a specified diameter, Figure 4-62. Most fasteners are manufactured with an 82° inclusive angle. In representing this angle, the 45° triangle is used, as illustrated. The taper is cut into the part until a specified diameter is achieved, usually slightly larger than that of the head of the fastener.

Counterbored Holes

A *counterbored hole* is first drilled to a diameter slightly larger than the body of the fastener that is to be used.

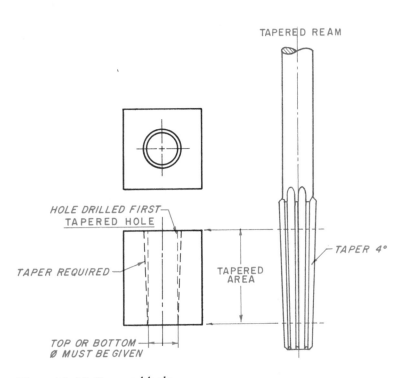

Figure 4-61 Tapered hole

Figure 4-62
Countersunk hole

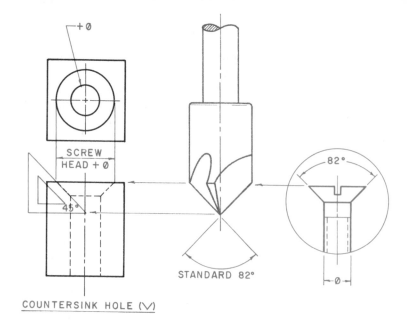

The upper portion is enlarged cylindrically to a specified diameter and depth, Figure 4-63. A counterbore is usually slightly larger than the head of the fastener, and slightly deeper than the height of the fastener. A pilot of the counterbore is used to help guide the cutter portion and to keep it centered.

Spotface

A spotface is used in conjunction with some holes. A *spotfaced hole* is used to form a flat surface for a head of a fastener, Figure 4-64. A spotface is very similar to a counterbore, except that the depth is not specified as this is left up to the craftsperson at the time of manufacture. To illustrate the depth of the spotface, show it at a depth of .06. Think of the spotface as a washer used to provide a flat surface for the head of a fastener.

Conventional Breaks

Long objects, such as a pipe, as illustrated in Part A of Figure 4-65, would appear very small if drawn 1/4 size scale in order to fit it on a sheet of paper. By using a *conventional break*, the same pipe can be drawn *full size* with a central portion "broken" or removed, as illustrated in Part B of Figure 4-65. This gives a much clearer understanding of the object. Note that the broken-out or removed section must be the same section throughout its entire length.

Kinds of Conventional Breaks

There are three major kinds of "breaks" used to remove center portions of an object: the "S" break, the "Z" break, and the freehand break. The "S" *break* is used for round objects. To draw the "S" break, refer

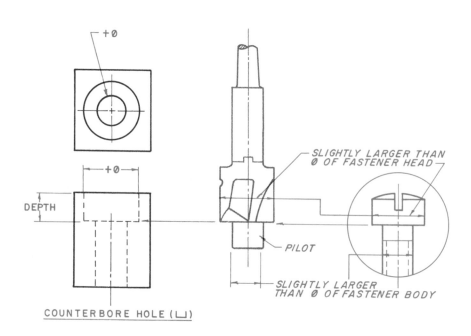

Figure 4-63
Counterbored hole

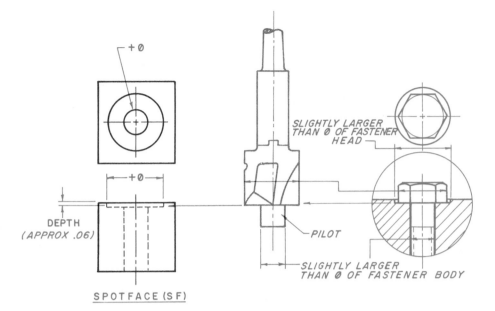

Figure 4-64 Spotfaced hole

DEPTH
(APPROX .06)

SPOTFACE (SF)

+ Ø

SLIGHTLY LARGER
THAN Ø OF FASTENER
HEAD

PILOT

SLIGHTLY LARGER
THAN Ø OF FASTENER BODY

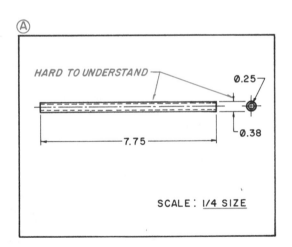

Ⓐ

HARD TO UNDERSTAND

Ø.25

Ø.38

7.75

SCALE : 1/4 SIZE

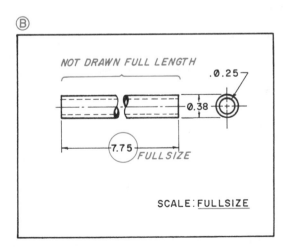

Ⓑ

NOT DRAWN FULL LENGTH

.Ø.25

Ø.38

7.75 FULL SIZE

SCALE : FULLSIZE

Figure 4-65 Conventional breaks

to Figure 4-66, Steps 1 through 5. A template can also be used to draw the "S" break. The "Z" *break* is usually used for thin, long or wide parts, Figure 4-67, Part A. The "Z" itself is drawn freehand, as illustrated. The *freehand break* is used to illustrate long, rectangular parts, Figure 4-67, Part B.

Visualization

Visualization is the process of recreating a three-dimensional image of an object in a person's mind, using the evidence and clues provided by orthographic drawings or other presentations. This process has been the focus of much investigation, often used for studying a person's ability to perceive (such as in intelligence tests). The goal of reading an orthographic drawing is to visualize accurately information about the relative positions of an object's surfaces and geometric features.

There is some evidence that an active "what if" imagination is a key ingredient in the visualization process. For example, in attempting to find the cause of a discovered broken window, a "what if" attitude seeks further clues, and a nearby baseball would provide a different probability of the event than would a scattering of bird feathers. Correspondingly, a solid-line circle in a view causes a "what if" visualizer to seek further evidence. If an adjoining view has a solid-line rectangle, bisected by a center line (axis) that

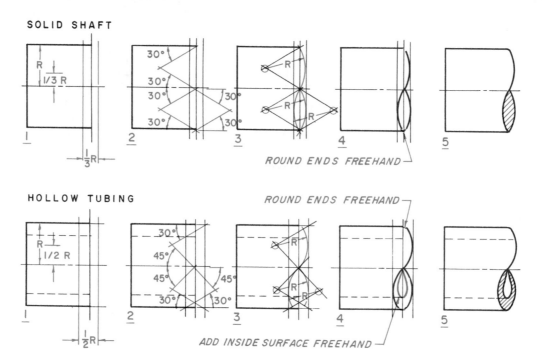

Figure 4-66 "S" break

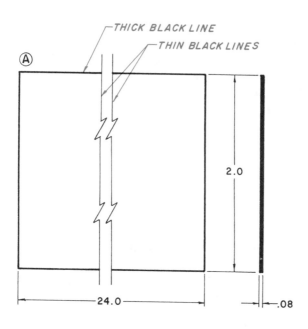

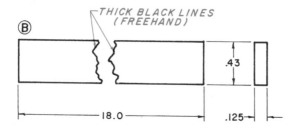

Figure 4-67 "Z" break and freehand break

aligns with the circle's center, a solid cylinder is indicated (see Figure 4-68). If the aligned rectangle has hidden lines, a hole is indicated (see Figure 4-69). A single orthographic view seldom is capable of providing three-dimensional evidence, and the reader must consider two or more views simultaneously in seeking an understanding of feature outlines.

Engineering drawings all follow the same procedures, and none is any more or less difficult to read than another. However, the quantity of geometric information varies considerably among drawings, and the time required to interpret the information varies accordingly. Visualization practice using simple drawings is required by most individuals to gain experience in mental three-dimensional image construction. Even experienced drafters often use modeling clay to form a three-dimensional object from orthographic views. In such cases, it is best to cut away the outline of each orthographic view, rather than piece together components to form the outline. Once a *brilliant* (or *gestalt*) image of the object has been perceived three-dimensionally, the possibility of error in its drawing or construction will be minimized, as it will be "real" to the visualizer.

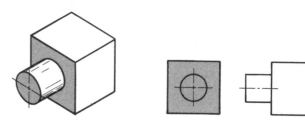

Figure 4-68 Visualization of an object

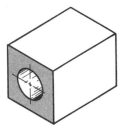

Figure 4-69 Visualization of an object

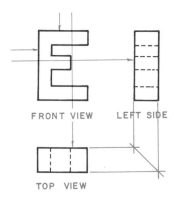

FRONT VIEW   LEFT SIDE

TOP VIEW

Figure 4-70
First-angle projection

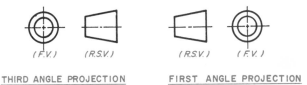

(F.V.)   (R.S.V.)   (R.S.V.)   (F.V.)

THIRD ANGLE PROJECTION   FIRST ANGLE PROJECTION

Figure 4-71A
Third-angle projection
symbol

Figure 4-71B
First-angle projection
symbol

symbol used to denote first-angle projection. Both symbols illustrate the front view (F.V.) and the right-side view (R.S.V.). The symbols are usually placed within the title block of the drawing.

First-Angle Projection

While the United States and Canada use the third-angle projection multiview system, most of the rest of the world uses first-angle projection. In *third-angle projection*, the top view is projected directly above the front view. The right side is projected directly to the right of the front view, as has been explained throughout this chapter.

First-angle projection also starts with the front view as the most important view and starting point, but *rotates* the views from the front view, Figure 4-70. Starting from the front view, rotating the object to the right, you will be viewing the left side; therefore, the left side appears and is drawn to the right of the front view. Again, starting from the front view, rotating the object down toward the viewer, you will be viewing the top view; therefore, the top view appears and is drawn below the front view. Each viewing plane displays an image from the object's far side, rather than the near-side image of third-angle projection.

Due to the increase in international exchange of parts and drawings, the projection method used should be indicated on the drawing. If it is not indicated, simply note the drawing's country of origin. Figure 4-71A illustrates the symbol used to denote third-angle projection. Figure 4-71B illustrates the

Review

1. Why is a sketch so important?

2. Explain the following terms (use illustrations if necessary): runout, rounds, and fillets.

3. Explain the difference between a perspective view and a multiview of an object.

4. What is considered the work area?

5. Where is the third-angle projection multiview system used? Where is the first-angle projection multiview system used?

6. What is the standard depth calloff for a spotface?

7. For what is the lightly constructed 45° angle projection line used?

8. In representing the tip of a drill in a blind hole, what triangle is used?

9. List three important considerations that must be used in positioning the front view.

10. How many views are used to describe an object?

11. To illustrate a countersunk hole, what triangle is used to draw the countersunk portion of the hole?

12. When and why should a drawing be lightly numbered?

13. List three major kinds of conventional breaks and note why each is used.

14. When is it permissible to use an incomplete view?

15. Explain in full the mathematical procedure used to center a typical three-view drawing with a 1" (25 mm) space between views within the work area. Make a sketch if necessary.

Chapter Four Problems

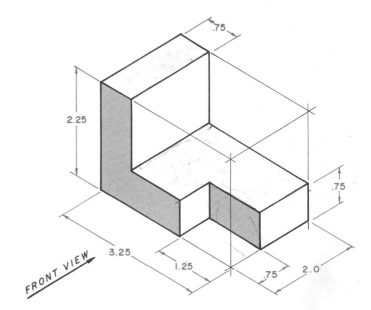

Problem 4-1

The following problems are intended to give the beginning drafter practice in visualizing the multiview system, practice in choosing and sketching the required view, practice is using drafting instruments and using correct line thickness. As these are beginning problems, no dimensions will be used at this time.

The steps to follow in laying out all drawings throughout this book are:

Step 1. Study the problem carefully.

Step 2. Choose the view with the most detail as the front view.

Step 3. Position the front view so that there will be the least number of hidden lines in the other views.

Step 4. Make a sketch of all required views.

Step 5. Center the required views within the work area with a 1-inch (25 mm) space between each view.

Step 6. Use light projection lines. Do not erase them.

Step 7. Lightly complete all views.

Step 8. Check to see that all views are centered within the work area.

Step 9. Check to see that there is a 1-inch (25 mm) space between all views.

Step 10. Carefully check all dimensions in all views.

Step 11. Darken in all views using correct line thickness.

Step 12. Recheck all work, and, if correct, neatly fill out the title block using light guidelines and neat lettering.

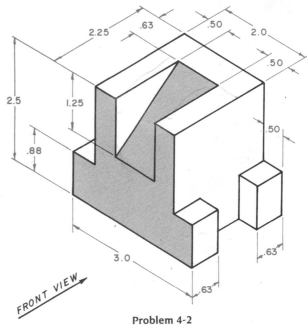

Problem 4-2

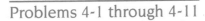

Problems 4-1 through 4-11

Construct a 3-view drawing of each object, using the listed steps.

Problem 4-3

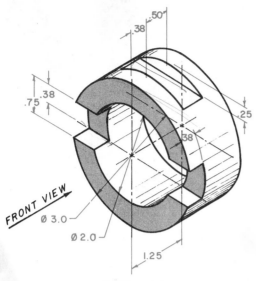

.50
.38
.75 .38
.25
.38

FRONT VIEW
Ø 3.0
Ø 2.0
1.25

Problem 4-4

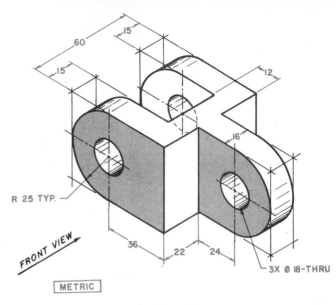

60
15
15
12
16
R 25 TYP.
FRONT VIEW
36
22
24
3X Ø 18-THRU

METRIC

Problem 4-7

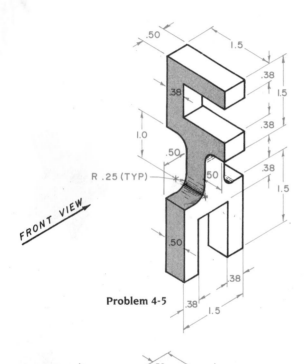

.50
1.5
.38
.38
1.5
.38
1.0
.50
.50
.38
1.5
R .25 (TYP)
FRONT VIEW
.50
.38
.38
1.5

Problem 4-5

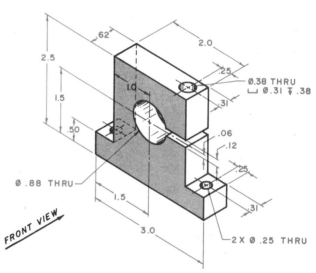

62
2.0
2.5
.25
0.38 THRU
⌴ Ø .31 ⌵ .38
1.0
.31
1.5
.50
.06
.12
.25
Ø .88 THRU
1.5
.31
3.0
2 X Ø .25 THRU
FRONT VIEW

Problem 4-8

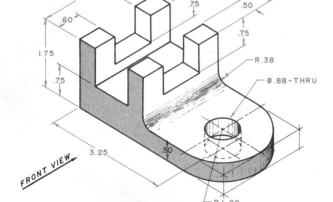

2.40
.60
.50
2.0
.60
.75
.50
1.75
.75
.75
R.38
Ø .88-THRU
3.25
.50
FRONT VIEW
R1.20

Problem 4-6

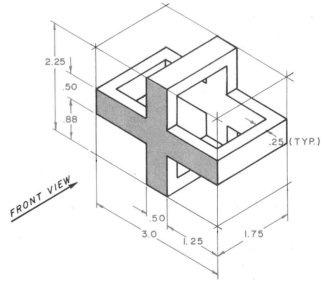

2.25
.50
.88
.25 (TYP.)
FRONT VIEW
.50
3.0
1.25
1.75

Problem 4-9

178   Section 2

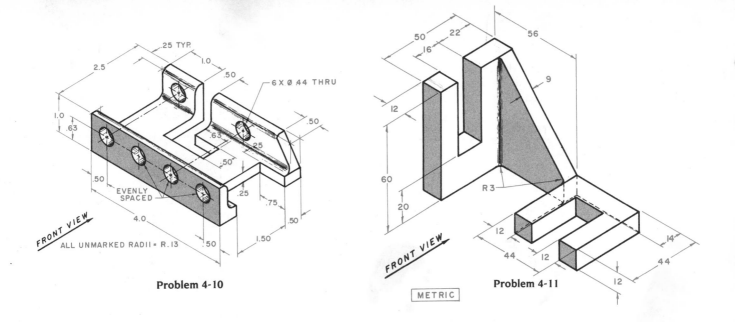

Problem 4-10

ALL UNMARKED RADII = R.13

FRONT VIEW

METRIC

Problem 4-11

Problem 4-12

Construct a 3-view drawing of this object. Project points 1, 2 and 3 into the right-side view and into the top view. Complete all views using the listed steps.

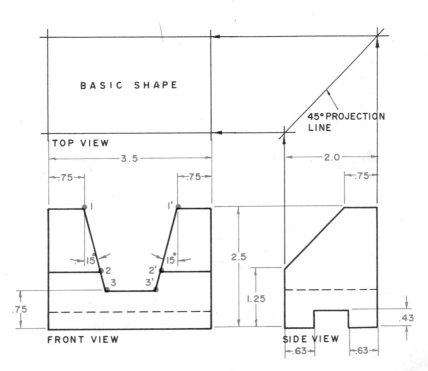

BASIC SHAPE

TOP VIEW

45° PROJECTION LINE

FRONT VIEW

SIDE VIEW

Problem 4-12

Problem 4-13

Construct a 3-view drawing of this object. Project points
1 through 8 into the top view and into the front view. Complete all views using the listed steps.

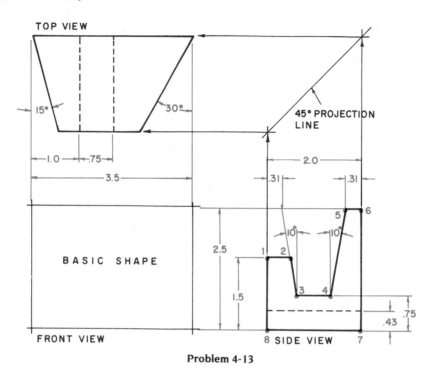

Problem 4-13

Problem 4-14

Construct a 3-view drawing of this object. Project points
1 through 6 into the front view and into the right-side view.
Complete all views using the listed steps.

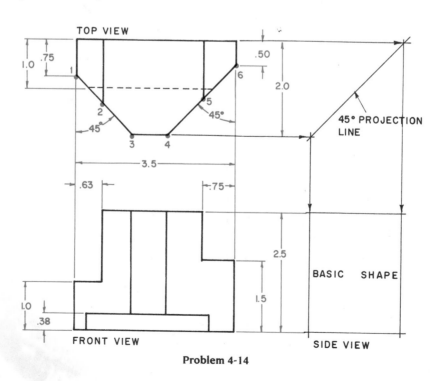

Problem 4-14

Problem 4-15

Construct a 3-view drawing of this object. Project points
1 through 7 into the right-side view and into the front view.
Complete all views using the listed steps.

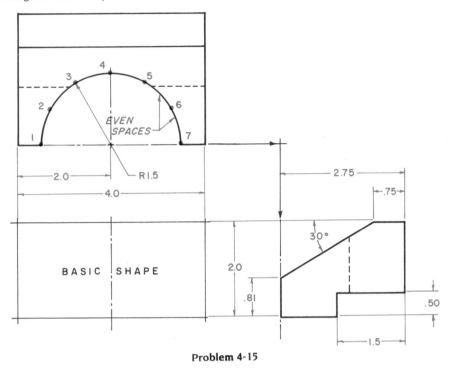

EVEN SPACES

R1.5

2.0

4.0

BASIC SHAPE

2.75

.75

30°

2.0

.81

.50

1.5

Problem 4-15

Problem 4-16

Construct a 3-view drawing of this object. Project points
1 through 7 into the right-side view and into the top view.
Complete all views using the listed steps.

Problem 4-17

Construct a 3-view drawing of this object. Locate and
number 12 points around the 1.5 diameter hole. Project
points 1 through 12 into the front view and into the right-
side view. Complete all views using the listed steps.

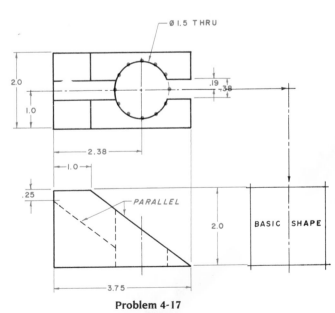

BASIC SHAPE

3.5

1.75

R 1.25

2.0

.88

30°

2.0

1.06

.63 .25

Problem 4-16

Ø 1.5 THRU

2.0

1.0

.19

.38

2.38

1.0

.25

PARALLEL

2.0

BASIC SHAPE

3.75

Problem 4-17

Problems 4-18 through 4-25

Construct a 3-view drawing of each object, using the listed steps.

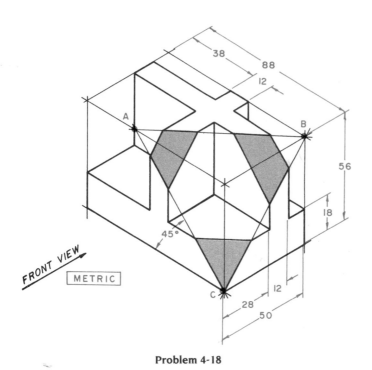

Problem 4-18

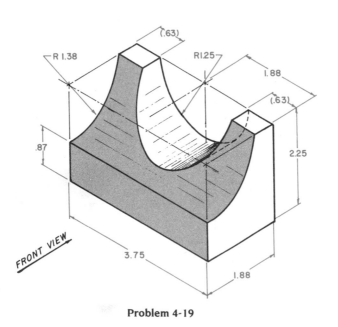

Problem 4-19

Problem 4-20

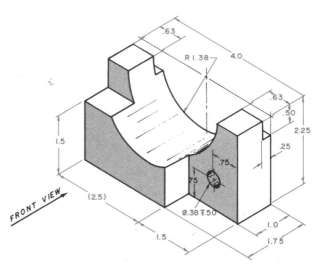

Problem 4-21

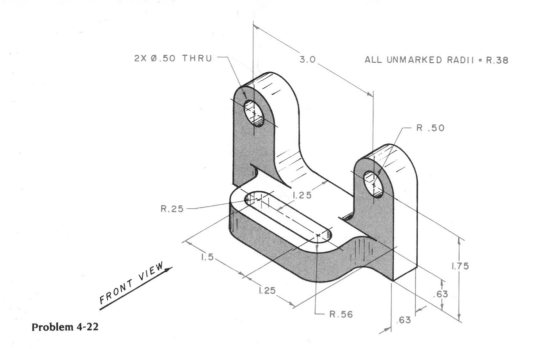

2X Ø.50 THRU

3.0

ALL UNMARKED RADII = R.38

R .50

1.25

R.25

1.5

1.25

FRONT VIEW

R.56

.63

.63

1.75

Problem 4-22

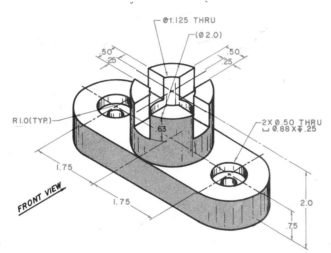

Ø1.125 THRU

(Ø2.0)

.50
.25

.50
.25

R1.0(TYP.)

.63

2X Ø.50 THRU
⌴ Ø.88 X ↧.25

1.75

1.75

2.0

.75

FRONT VIEW

Problem 4-23

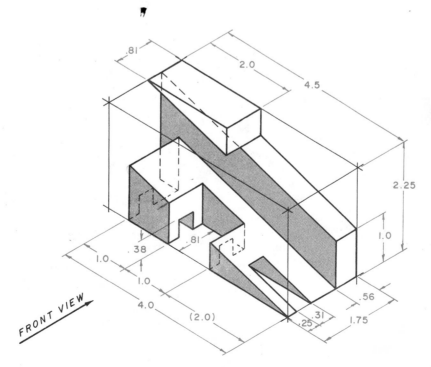

.81

2.0

4.5

2.25

1.0

38

.81

1.0

1.0

.56

4.0

(2.0)

.31

.25

1.75

FRONT VIEW

Problem 4-24

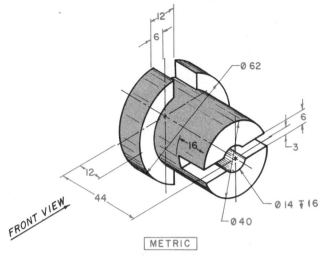

Problem 4-26

Construct a 4-view drawing of this object (front, top, right, and left-side views). Use the listed steps.

METRIC

Problem 4-25

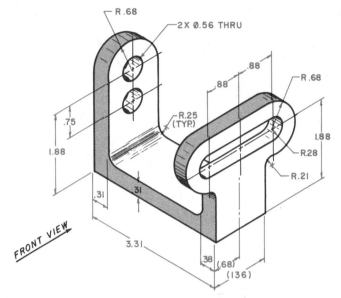

Problem 4-26

Problems 4-27 through 4-36

Construct a 3-view drawing of each object, using the listed steps.

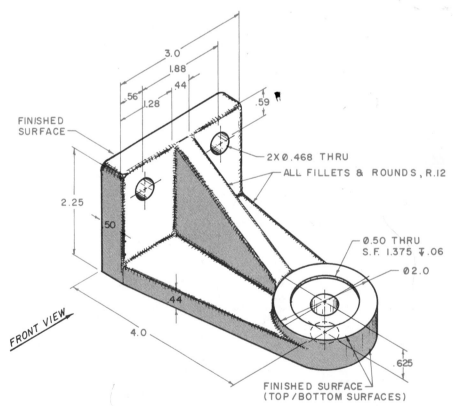

Problem 4-27

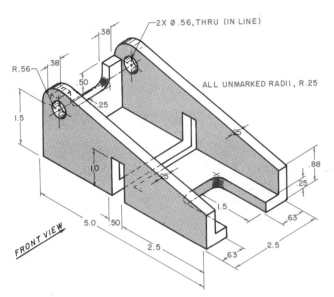

2X Ø.56, THRU (IN LINE)

R.56

ALL UNMARKED RADII, R.25

Problem 4-28

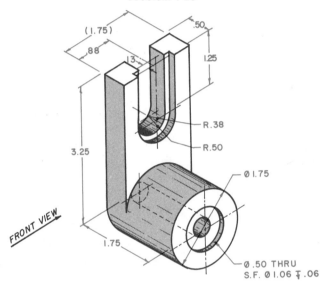

(1.75)

R.38

R.50

Ø 1.75

Ø .50 THRU
S.F. Ø 1.06 ⊤ .06

Problem 4-29

2X Ø 10, THRU

R 6 (TYP.)

METRIC

Problem 4-30

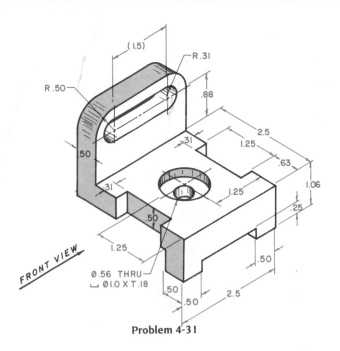

(1.5)

R.31

R.50

Ø.56 THRU
⌴ Ø1.0 X T.18

Problem 4-31

Ø 2.0

Ø.38 ⊤.88

Ø1.5 (TYP.)

Problem 4-32

2X Ø.25 THRU
⌴ Ø.50 ⊤.31

Ø .75

Ø .31 THRU

ALL UNMARKED RADII, R.12

R.50

FLAT SURFACE

Problem 4-33

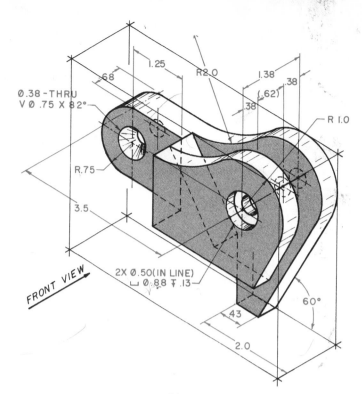

Problem 4-34

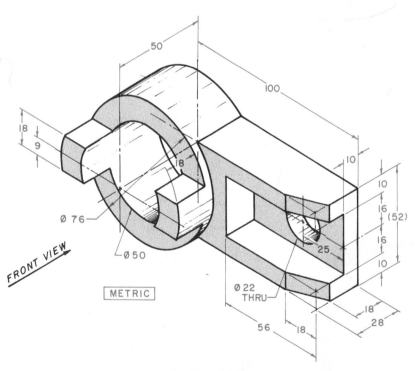

Problem 4-35

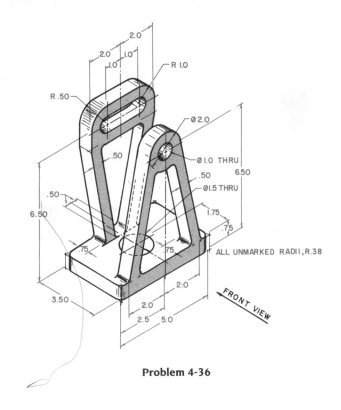

Problem 4-36

Problem 4-37

Choose the front view and construct a 3-view drawing of this object, using the listed steps.

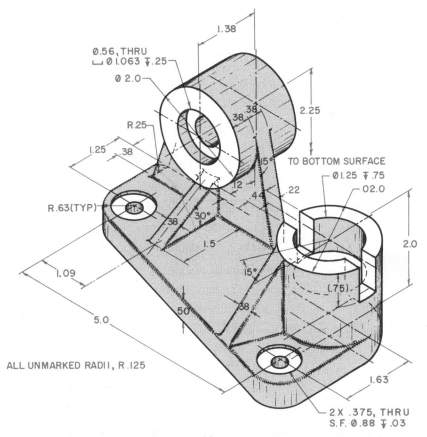

Problem 4-37

1. What is the name of the company that developed the drawing?
2. What is the name of the part?
3. Of what material is the part made?
4. What view of the part is "View I"?
5. What view of the part is "View II"?
6. How many counterbored holes does the part contain?
7. What type of finish is to be applied to the part?
8. How deep are the counterbores in the part?
9. What are the specifications for the chamfer applied to the part?
10. What are the thread specifications for the threaded hole in the part?

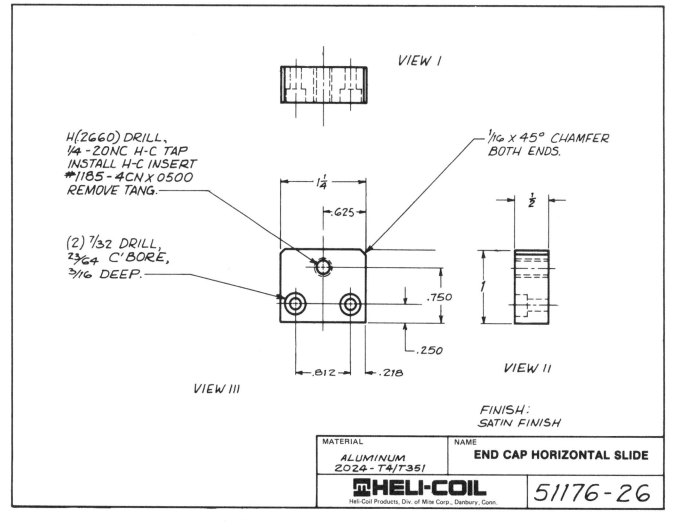

VIEW I

H.(2660) DRILL,
1/4 - 20NC H-C TAP
INSTALL H-C INSERT
#1185 - 4CN X 0500
REMOVE TANG.

(2) 7/32 DRILL,
23/64 C'BORE,
3/16 DEEP.

1/16 X 45° CHAMFER
BOTH ENDS.

1 1/4

.625

.750

.250

.812   .218

VIEW III

1/2

1

VIEW II

FINISH:
SATIN FINISH

MATERIAL	NAME
ALUMINUM 2024 - T4/T351	**END CAP HORIZONTAL SLIDE**

☰HELI-COIL Heli-Coil Products, Div. of Mite Corp., Danbury, Conn.   51176 - 26

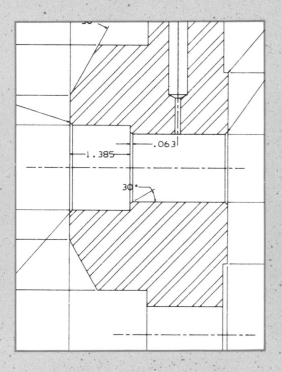

This chapter is an extension of Chapter 4, Multiview Drawings, with the addition of various drafting rules and practices associated with making one view into a sectional view. Covered in this chapter are full section, offset section, half section, broken-out section, revolved section, removed section, auxiliary view, thinwall section, and assembly section. The beginning drafter must fully understand each kind of sectional view, and know where to use each to best illustrate the object.

CHAPTER FIVE

Sectional Views

Sectional Views

The conventional method used to draw an object using the multiview or orthographic method of representation is discussed in Chapter 4. This system is excellent to illustrate various external features. Using this method, complicated interior features are illustrated with hidden lines. These interior features can be shown more clearly by the use of sectional views. A *sectional view* is a view of an imaginary surface, exposed by an imaginary slicing-open of an object, allowing interior details to become visible. A sectional view is sometimes referred to as a *cross section* or simply *section*, Figure 5-1.

Note that, as a rule, all hidden lines are omitted in the sectional view. However, they can be added to illustrate some important features *if absolutely necessary*.

Cutting-Plane Line

The *cutting-plane line* indicates the path that an imaginary cutting plane follows to slice through an object. Think of the cutting-plane line as a saw blade that is used to cut through the object. The cutting-plane line

is represented by a thick black dashed line, as shown in Figure 5-2. The line on top is the newer cutting-plane line; sizes are approximated and spaced by eye. If the cutting-plane line passes through the center of the object, and it is very obvious where the cutting plane is located, it can be omitted. This is the drafter's decision.

Direction of Sight

The drafter must indicate the direction in which the object is to be viewed after it is sliced or cut through. This is accomplished by adding a short leader and arrowhead to the ends of the cutting-plane line, Figure 5-3. These arrowheads indicate the direction of sight. Letters are usually added to the ends of the cutting-plane line to indicate exactly what cutting plane is used. The cutting plane extends past the object by approximately .50, as shown.

Section Lining

Section lining shows where the cutting plane is sliced or cut and the surface or surfaces touched by

189

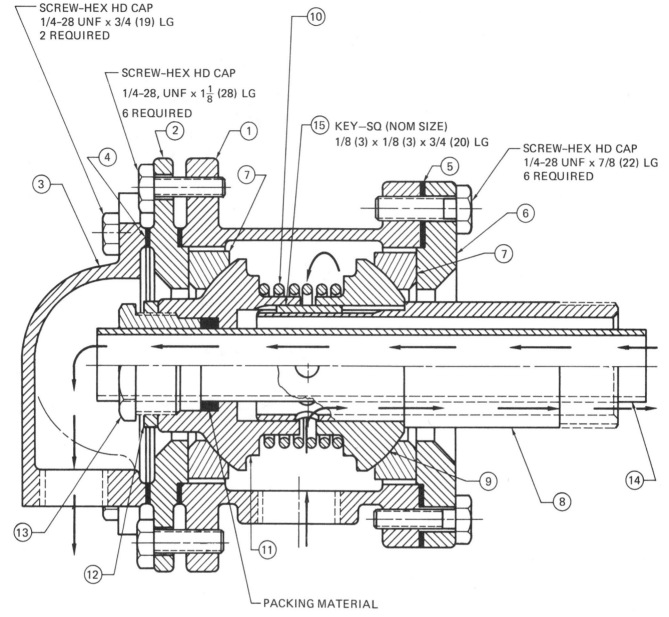

SCREW–HEX HD CAP
1/4–28 UNF x 3/4 (19) LG
2 REQUIRED

SCREW–HEX HD CAP
1/4–28, UNF x 1⅛ (28) LG
6 REQUIRED

KEY–SQ (NOM SIZE)
1/8 (3) x 1/8 (3) x 3/4 (20) LG

SCREW–HEX HD CAP
1/4–28 UNF x 7/8 (22) LG
6 REQUIRED

PACKING MATERIAL

Figure 5-1 Sectional view

the cutting-plane line. Section lining is represented by thin, black lines drawn at 45° to the horizon, unless there is some specific reason for using a different angle. Section lining is spaced by eye from 1/16" (1.5 mm) to 1/4" (6 mm) apart, depending upon the overall size of the object. The average spacing used for most drawings is .13" (3 mm), Figure 5-4. Section lines must be of uniform thickness (thin black) and evenly spaced by eye.

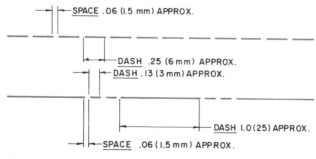

SPACE .06 (1.5 mm) APPROX.
DASH .25 (6 mm) APPROX.
DASH .13 (3 mm) APPROX.
DASH 1.0 (25) APPROX.
SPACE .06 (1.5 mm) APPROX.

Figure 5-2 Cutting-plane line

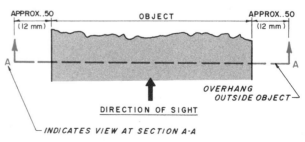

APPROX..50 (12 mm)
OBJECT
APPROX..50 (12 mm)
A
A
OVERHANG OUTSIDE OBJECT
DIRECTION OF SIGHT
INDICATES VIEW AT SECTION A-A

Figure 5-3 The direction of sight

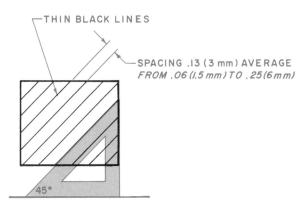

Figure 5-4 Section lining

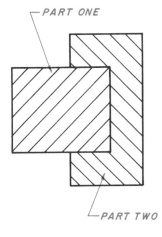

Figure 5-5
Two parts with
section lining

If a cutting plane passes through two parts, each part has section lines using a 45° angle or other principal angle. These section lines should not be aligned in the same direction, Figure 5-5. If the cutting plane passes through more than two parts, the section lining of each individual part must be drawn at different angles. When an angle other than 45° is used, the angle should be 30° or 60°. Section lining should *not* be parallel with the sides of the object to be section lined, Figure 5-6.

In past years, section lining used various symbols to indicate the type of material used to make the object. These symbols used only such general type identifications as cast iron, steel, brass, aluminum, and so forth. Today, because there are so many different kinds of material, section lining symbols have been eliminated, and a single, all-purpose section lining is used (as illustrated in Figure 5-4). Specific information as to the type of material is given in the title block under "material."

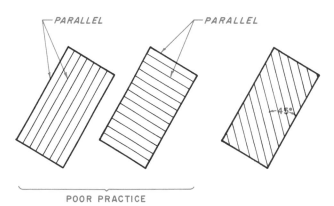

Figure 5-6 Section lining angle

Some sections are made up of a combination of the nine kinds of sections. Each is explained in full detail in the following paragraphs.

Multisection Views

When an object is very complicated and cut in more than one place, each cutting-plane line must be labeled starting with section A-A, followed by B-B, and so forth, Figure 5-7.

Kinds of Sections

Nine kinds of sections are used today in industry.

- Full section
- Offset section
- Half section
- Broken-out section
- Revolved section
  (rotated section)
- Removed section
- Auxiliary section
- Thinwall section
- Assembly section

Full Section

A *full section* is simply a section of one of the regular multiviews that is sliced or cut completely in two. See the given problem, Figure 5-8, a regular three-view drawing of an object.

Determine which view contains many hard-to-understand hidden lines. In this example, it is the front view. Add a cutting plane to either the top view or right-side view. In this example, the top view is chosen. Indicate how the front view is to be viewed or the direction of sight. After determining where the object is to be sliced or cut and viewed, change the front view into section A-A, Figure 5-9.

Think of the object as a pictorial drawing, Figure 5-10. An imaginary cutting-plane line is passed through the object, Figure 5-11. The front portion is

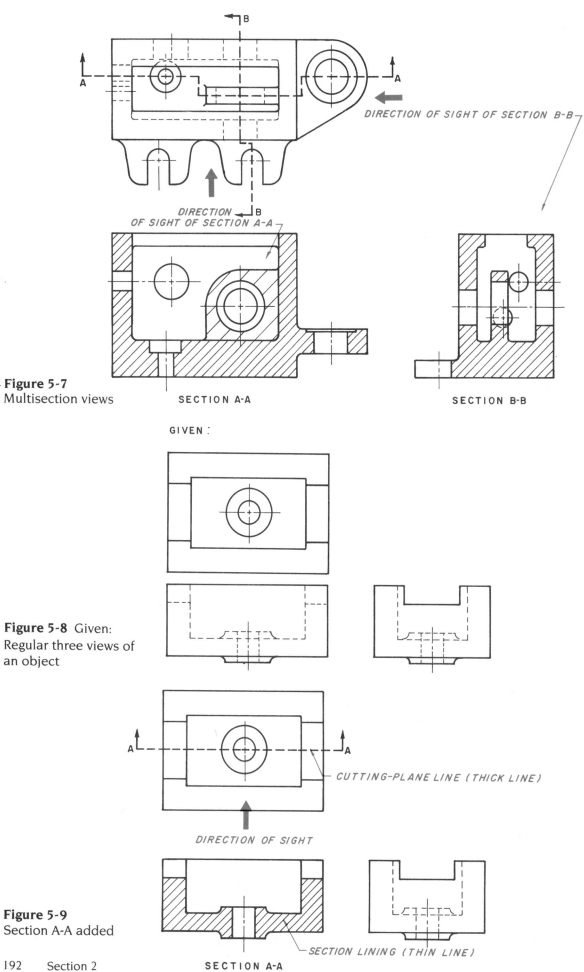

Figure 5-7
Multisection views

DIRECTION OF SIGHT OF SECTION B-B

DIRECTION OF SIGHT OF SECTION A-A

SECTION A-A

SECTION B-B

GIVEN :

Figure 5-8 Given:
Regular three views of
an object

CUTTING-PLANE LINE (THICK LINE)

DIRECTION OF SIGHT

Figure 5-9
Section A-A added

SECTION LINING (THIN LINE)

SECTION A-A

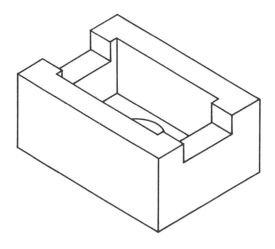

Figure 5-10 Pictorial view of the object

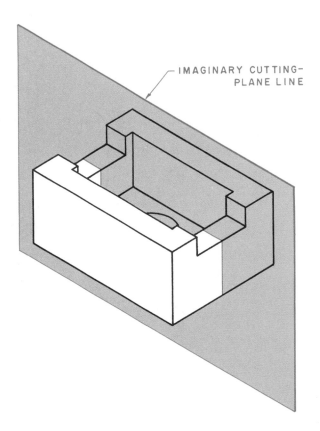

IMAGINARY CUTTING-
PLANE LINE

Figure 5-11 Imaginary cutting-plane line added

removed and the remaining section is viewed by the direction of sight, Figure 5-12.

Notice that section lining is applied only to the area the imaginary cutting plane passed through. The back side of the hole and the back sides of the notches are *not* section lined.

Offset Section

Many times, important features do not fall in a straight line as they do in a full section. These important features can be illustrated in an *offset section* by bending or offsetting the cutting-plane line. An offset section is very similar to a full section, except that the cutting-plane line is not straight, Figure 5-13.

Note that the features of the countersunk holes A, projection B with its counterbore, and groove C with a shoulder are not aligned with one another. The cutting-plane line is added, and changes of direction (*staggers*) are formed by right angles to pass

through these features. An offset cutting-plane line A-A is added to the top view and the material behind the cutting plane is viewed in section A-A, Figure 5-14. The front view is changed into an offset section, similar to a full-sectional view. The actual bends of the cutting-plane lines are omitted in the offset section, Figure 5-15. By using a sectional view, another view often may be omitted. In this example, the right-side view could have been omitted, as it adds nothing to the drawing and takes extra time to draw.

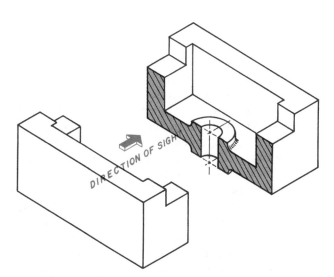

Figure 5-12 Pictorial view of the full section

GIVEN:

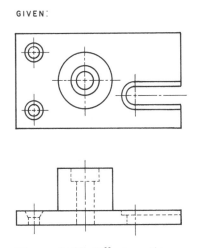
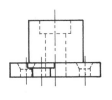

Figure 5-13 Offset section

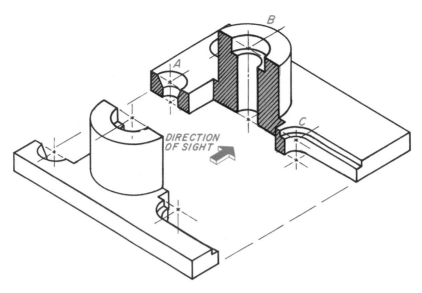

Figure 5-14 Pictorial view of the offset section

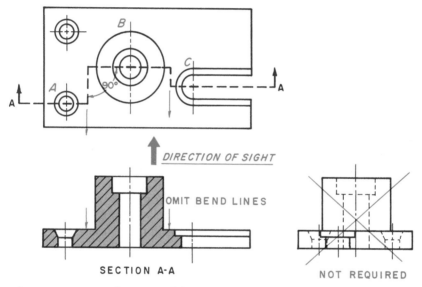

SECTION A-A

NOT REQUIRED

Figure 5-15 Bends omitted from section view

GIVEN:

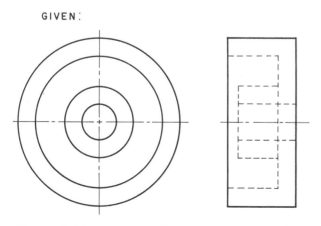

Figure 5-16 Given: Regular two views of an object

Half Section

In a *half section*, the object is cut only halfway through and a quarter section is removed, Figure 5-16. A cutting plane is added to the front view, with only one arrowhead to indicate the viewing direction. Also, a quarter section is removed and, in this example, the right side is sectioned accordingly, Figure 5-17. A pictorial view of this half section is illustrated in Figure 5-18. The visible half of the object that is not removed shows the *exterior* of the object, and the removed half shows the *interior* of the object. The half of the object not sectioned can be drawn as it would normally be drawn, with the appropriate hidden lines.

Half sections are best used when the object is *symmetrical*, that is, the exact same shape and size on both

Figure 5-17 Half section

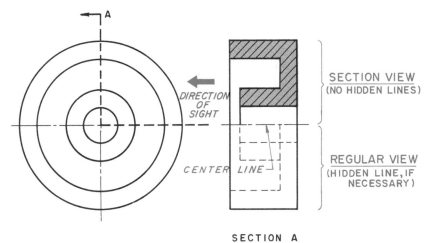

SECTION VIEW
(NO HIDDEN LINES)

DIRECTION
OF
SIGHT

CENTER LINE

REGULAR VIEW
(HIDDEN LINE, IF
NECESSARY)

SECTION A

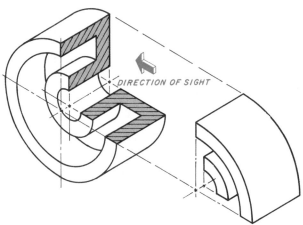

DIRECTION OF SIGHT

Figure 5-18 Pictorial view of the half section

GIVEN:

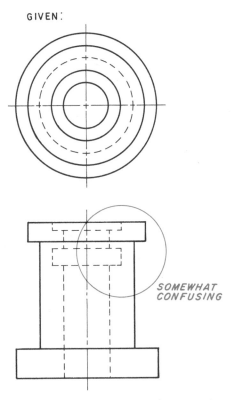

SOMEWHAT
CONFUSING

Figure 5-19 Given: Regular two views
of an object

sides of the cutting-plane line. A half-section view is capable of illustrating both the inside and the outside of an object in the same view. In this example, the top half of the right side illustrates the interior; the bottom half illustrates the exterior. A center line is used to separate the two halves of the half section (refer back to Figure 5-17). A solid line would indicate the presence of a real edge, which would be false information.

Broken-out Section

Sometimes, only a small area needs to be sectioned in order to make a particular feature or features easier to understand. In this case, a *broken-out section* is used. Given: Figure 5-19. As drawn, the top section is somewhat confusing and could create a question. To clarify this area, a portion is removed, Figure 5-20.

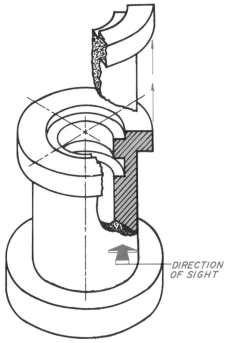

DIRECTION
OF SIGHT

Figure 5-20 Pictorial view of the
broken-out section

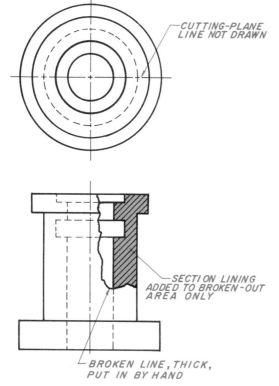

Figure 5-21 Broken-out section

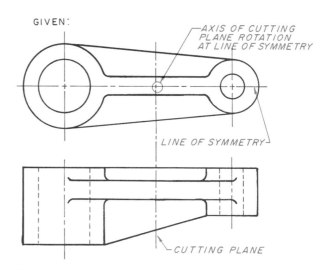

Figure 5-22 Given: Regular two views of an object

The finished drawing would be drawn as illustrated in Figure 5-21. The broken line is put in freehand, and is drawn as a visible thick line. The actual cutting-plane line is usually omitted.

Revolved Section (Rotated Section)

A *revolved section*, sometimes referred to as a *rotated section*, is used to illustrate the cross section of ribs, webs, bars, arms, spokes or other similar features of an object. Figure 5-22 is a two-view drawing of an arm. The cross-sectional shape of the center portion of the arm is not defined. In drafting, no feature should remain questionable, and a section through the center portion of the arm would provide the complete information.

A revolved section is made by assuming a cutting plane perpendicular to the axis of the feature of the object to be described, Figure 5-23. Note that the rotation point occurs at the cutting-plane location and, theoretically, will be rotated 90°. Rotate the imaginary cutting-plane line about the rotation point of the object, Figure 5-24. Notice that dimension X is transferred from the top view to the sectional view of the feature; in this example, the front view. Dimension Y in the top view is also transferred to the front view. The section is now drawn in place. The finished drawing is illustrated in Figure 5-25. Note that the break lines in the front view are on each side of the sectional view, and are put in freehand.

The revolved section is not used as much today as it was in past years. Revolved sections tend to be confusing, and often create problems for the people who must interpret the drawings. Today, it is recommended to use a removed section instead of a revolved or rotated section.

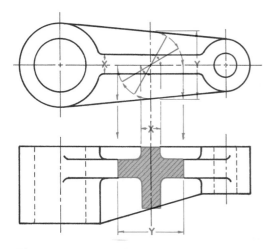

Figure 5-23 Revolved section

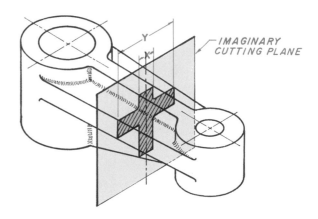

Figure 5-24 Pictorial view of a revolved section

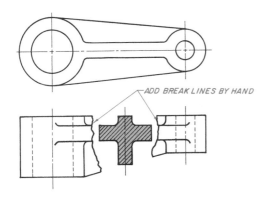

Figure 5-25 Revolved section view

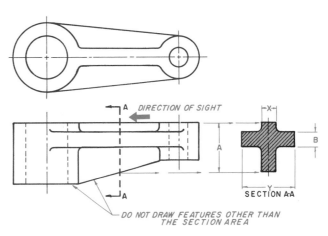

Figure 5-26 Removed section

Removed Section

A *removed section* is very similar to a rotated section except that, as the name implies, it is drawn removed or away from the regular views, Figure 5-26. The removed section, as with the revolved section, is also used to illustrate the cross section of ribs, webs, bars, arms, spokes or other similar features of an object.

A removed section is made by assuming that a cutting-plane, perpendicular to the axis of the feature of the object, is added through the area that is to be sectioned. (Refer back to Figures 5-23 and 5-24.) Transfer dimensions X and Y to the removed views, exactly as was done in the rotated section. Height features, such as dimensions A and B, are transferred from the front view in this example.

Note that a removed section must identify the cutting-plane line from which it was taken. In the sectional view, do not draw features other than the actual section.

Removed sections are labeled section A-A, section B-B, and so forth, corresponding to the letters at the ends of the cutting-plane line. The sections are usually placed on the drawing in alphabetical order from left to right or from top to bottom, away from the regular views.

Sometimes a removed section is simply drawn on a center line that is extended from the object, Figure 5-27. A removed section can be drawn to an enlarged scale if necessary to illustrate and/or dimension a small feature. The scale of the removed section must be indicated directly below the sectional view, Figure 5-28.

In the field of mechanical drafting, the removed section should be drawn on the same page as the regular views. If there is not room enough on the same page and the removed section is drawn on another page, a page number cross reference must be given as to where the removed section may be found. The page where the removed section is located must refer back to the page from which the section is taken. For example, section A-A on sheet 2 of 4.

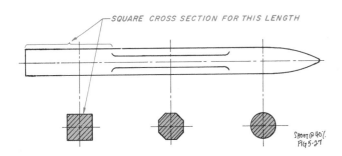

Figure 5-27 Removed section view

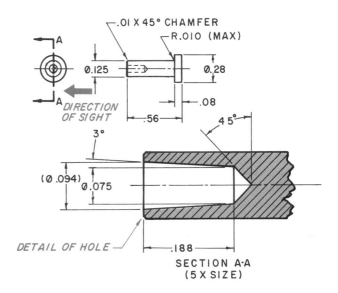

Figure 5-28 Enlarged removed section

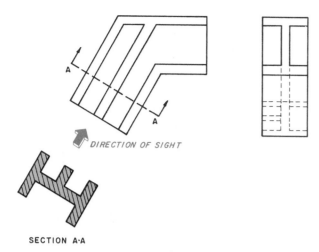

SECTION A-A

Figure 5-29 Auxiliary section

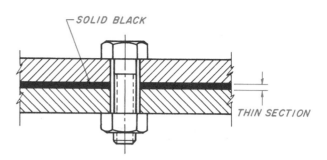

Figure 5-30 Thinwall section

Auxiliary Section

If a sectional view of an object is intended to illustrate the true size and shape of an object's boundary, the cutting-plane path must be perpendicular to the axis or surfaces of the object. An *auxiliary section* is projected in the same way as any normal auxiliary view, and it provides an option of orienting the cutting plane at any desired angle, Figure 5-29.

Thinwall Section

Any very thin object that is drawn in section, such as sheet metal, a gasket or a shim, should be filled-in solid black, as it is impossible to show the actual section lining. This is called a *thinwall section*, Figure 5-30. If several thin pieces that are filled-in solid black are touching one another, a small white space is left between the solid thinwall section, Figure 5-31.

Assembly Section

When a sectional drawing is made up of two or more parts it is called an *assembly section*, Figure 5-32. An assembly section can be a full section (as it is in

this example), an offset section, a half section or a combination of the various kinds of sectional views. The assembly section shows how the various parts go together.

Each part in the assembly must be labeled with a name, part or plan number, and the quantity required for one complete assembly. If the assembly section does not have many parts, this information is added by a note alongside each part. If the assembly has many parts, and there is not enough room to prevent the drawing from appearing cluttered, each individual part may be identified by a number within a circle called a *balloon*. The balloon callout system is used in this example. A table must be added to the drawing, listing the name, part or plan number, the quantity required for one complete assembly, and a cross reference to the corresponding balloon number. This is called a *parts list*. The exact form of the list varies from company to company. Figure 5-33 is an example of a parts list used with the balloon system of callouts. Notice that entries are sometimes listed in reverse (bottom to top) order, as illustrated.

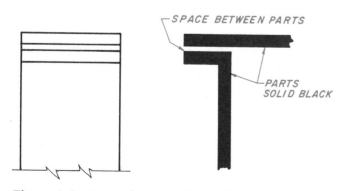

Figure 5-31 Space between thinwall sections

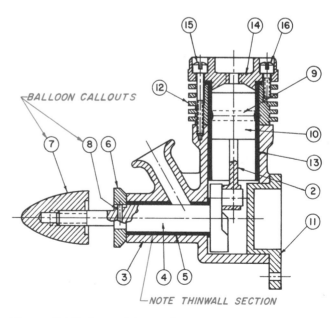

Figure 5-32 Assembly section

NO	TITLE	PART NO.	NO REQ'D
5	PIVOT PIN	A 9 2 0 0 1	6
4	ARM	B 1 1 6 2 7	1
3	CENTER SHAFT	A 1 1 6 2 6	2
2	BASE	C 1 1 6 2 5	1
1	MAIN FRAME	C 1 1 6 2 4	1
NO	TITLE	PART NO.	NO REQ'D

USUAL CALLOUT INFORMATION

Figure 5-33 Example of a parts list

Sections through Ribs or Webs

True projection of a sectioned view often produces incorrect impressions of the actual shape of the object. Figure 5-34 has a given front view and a right-side view. Its pictorial view would look like Figure 5-35. A full section A-A would appear as it does in Figure 5-36. This is a true projection of section A-A, as the cutting-plane line passes through the rib.

However, such a sectional view gives an incorrect impression of the object's actual shape, and is poor drawing practice. It misleads the viewer into thinking the object is actually shaped as it is in Figure 5-37. The conventional practice used to illustrate this section is to draw the section view as illustrated in Figure 5-38, which is not a true projection. Note that the web or rib is not section lined.

Some companies use another method to compensate for this problem. It is somewhat of a middle ground or a combination of true projection and correct representation, Figure 5-39. This is called *alternate section lining*. Section lining, over the rib or web section, is drawn using every *other* section line, and the actual shape is indicated by hidden lines. However, most companies do not use alternate section lining.

Another example of a cutting-plane passing through a rib or web is shown in Figure 5-40. This example is a true projection, but it is poor drafting practice, as it gives the impression that the center portion is a thick, solid mass. Figure 5-41 is drawn incorrectly, but does not give the false impression of the object's center portion. This is the *conventional practice* used.

Holes, Ribs and Webs, Spokes and Keyways

Holes located around a bolt circle are sometimes not aligned with the cutting-plane line, Figure 5-42. The cutting plane passes through only one hole. This is a true projection of the object, but poor drafting practice. In actual practice, the top hole is theoretically revolved to the cutting-plane line and projected

GIVEN:

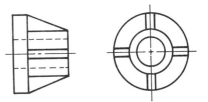

Figure 5-34 Given: Two-view drawing of an object

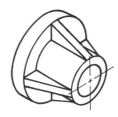

PICTORIAL VIEW

Figure 5-35 Pictorial view of an object

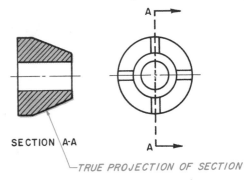

SECTION A-A

TRUE PROJECTION OF SECTION

Figure 5-36 True projection of an object

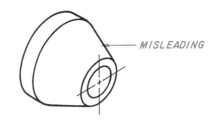

MISLEADING

PICTORIAL VIEW

Figure 5-37 True projection can be misleading

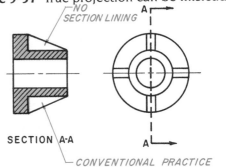

NO SECTION LINING

SECTION A-A

CONVENTIONAL PRACTICE

Figure 5-38 Conventional practice — web or rib *not* sectioned

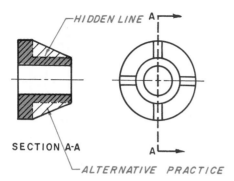

Figure 5-39 Alternate conventional practice

to the sectional view, Figure 5-43. This practice is called *aligning of features*.

Ribs and Webs

Ribs or webs sometimes do not align with the cutting-plane line, Figure 5-44. The cutting plane passes through only one web and only one hole. This is a true projection of the object, but poor drafting practice. In actual practice, one of the webs is theoretically revolved up to the cutting-plane line and projected to the sectional view, Figure 5-45. Notice that

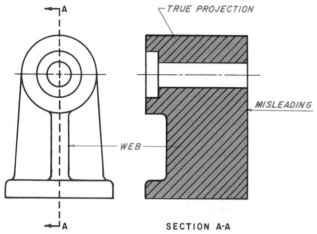

Figure 5-40 Example of true projection

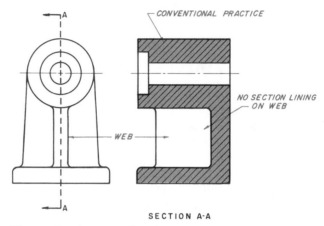

Figure 5-41 Example of conventional practice

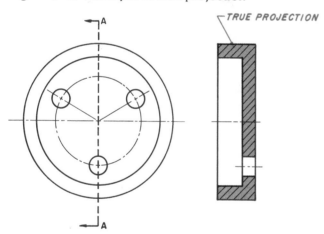

Figure 5-42 Holes using true projection

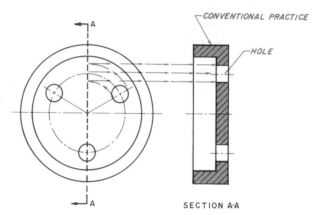

Figure 5-43 Holes using conventional practice (aligning of features)

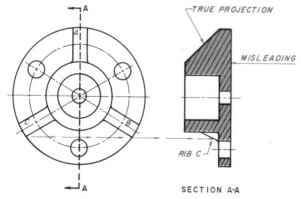

Figure 5-44 Rib or web using true projection

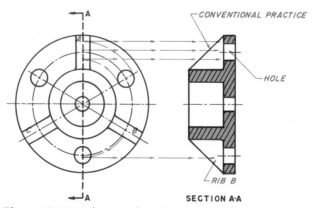

Figure 5-45 Rib or web using conventional practice

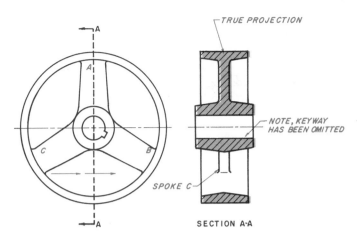

Figure 5-46 Spokes and keyway using true projection

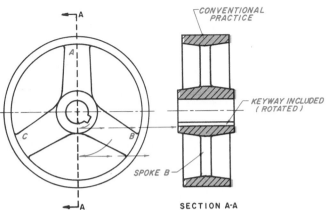

Figure 5-47 Spokes and keyway using conventional practice

the bottom hole is unaffected and is projected normally. This is another example of aligning of features.

Spokes and Keyways

Spokes and keyways and other important features sometimes do not align with the cutting-plane line, Figure 5-46. The cutting-plane line passes through only one spoke and misses the keyway completely. This also is a true projection of the object, but poor drafting practice. In conventional practice, spoke B is revolved to the cutting-plane line and projected to the sectional view, Figure 5-47. The keyway is also projected as illustrated. This is another example of aligning of features.

Aligned Sections

Arms and other similar features are revolved to alignment in the cutting plane, as were spokes in the preceding section. This procedure is used if the cutting-plane line cannot align completely with the object, as illustrated in Figure 5-48.

The arm or feature is now revolved to the imaginary cutting plane, and projected down to the sectional view, Figure 5-49. The actual cutting-plane line is bent and drawn through the arm or feature and then revolved to a straight, aligned vertical position. Notice that section lining is *not* applied to the arm, and is also omitted from the web area.

Fasteners and Shafts in Section

If a cutting plane passes *lengthwise* through any kind of fastener or shaft, the fastener or shaft is *not* sectioned. Section lining of a fastener or shaft would have no interior detail, thus it would serve no purpose and only add confusion to the drawing, Figure

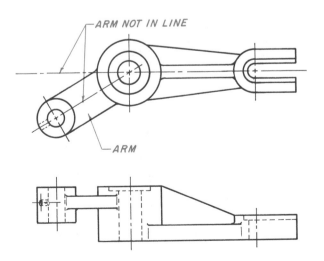

Figure 5-48 Two-view drawing

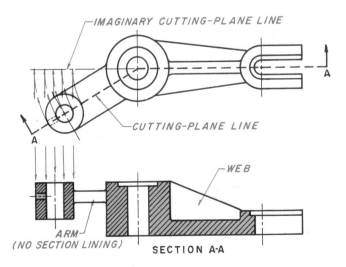

Figure 5-49 Aligned section

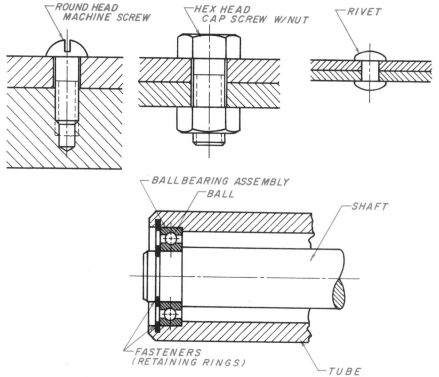

Figure 5-50 Parts *not* sectioned

5-50. The round head machine screw, the hex head cap screw w/nut, and the rivet are not sectioned. The other objects in the figure such as fasteners, ball bearing rollers, and so forth, are also not sectioned.

If a cutting-plane line passes *perpendicularly* through the axis of a fastener or shaft, section lining *is* added to the fastener or shaft, Figure 5-51. The end view has section lining added as shown.

Intersections in Section

Where an intersection of a small or relatively unimportant feature is cut by a cutting-plane line, it is not drawn as a true projection, Figure 5-52. Since a true projection takes drafting time, it is preferred that it be disregarded, and the feature drawn, using conventional practice, as shown in Figure 5-53. This procedure is much quicker and more easily understood.

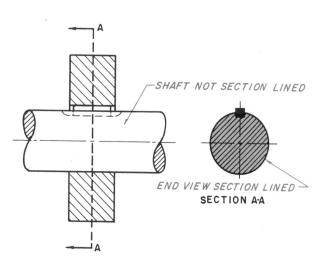

Figure 5-51 Shaft sectioned in end view only

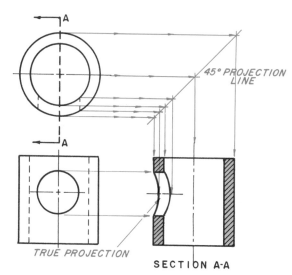

Figure 5-52 Intersection using true projection

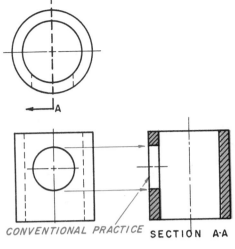

Figure 5-53 Intersection using conventional practice

Review

1. Explain the difference between true projection and conventional practice. Which is used in a sectional view and why?

2. Explain the difference between a revolved section and a removed section. Which is recommended today?

3. Are hidden lines used in a sectional view? Why?

4. Why is a removed section sometimes drawn at a larger scale?

5. List the nine kinds of sectional views and describe the various features of each.

6. What is alternate section lining? Where is it used?

7. List two major functions of an assembly drawing.

8. Explain the practice used for drawing intersections of small or unimportant features that are cut by a cutting-plane line.

9. What kind of sectional view illustrates both the exterior and interior of the object?

10. What must be done if a removed section is placed on another page other than the page on which the cutting-plane line is placed?

11. Explain the two methods used in regard to a cutting-plane line passing through the long dimension and perpendicularly to the center of a fastener or shaft.

12. What must be included for each part in an assembly section? Explain the two methods used to accomplish this.

Chapter Five Problems

The following problems are intended to give the beginning drafter practice in using the various kinds of sectional views used in industry. As these are beginning problems, no dimensions will be used at this time.

The steps to follow in laying out all problems in this chapter are:

Step 1. Study the problem carefully.

Step 2. Choose the view with the most detail as the front view.

Step 3. Position the front view so there will be the least amount of hidden lines in the other views.

Step 4. Make a sketch of all required views.

Step 5. Determine what should be drawn in section, what type of section should be used, and where to place the cutting-plane line.

Step 6. Center the required views within the work area with a 1-inch (25-mm) space between each view.

Step 7. Use light projection lines. Do not erase them.

Step 8. Lightly complete all views.

Step 9. Check to see that all views are centered within the work area.

Step 10. Check to see that there is a 1-inch (25-mm) space between all views.

Step 11. Carefully check all dimensions in all views.

Step 12. Darken in all views using correct line thickness.

Step 13. Add a cutting-plane line and section lining as required.

Step 14. Recheck all work, and, if correct, neatly fill out the title block using light guidelines and neat lettering.

Problem 5-1

Center three views within the work area, and make the front view a full section.

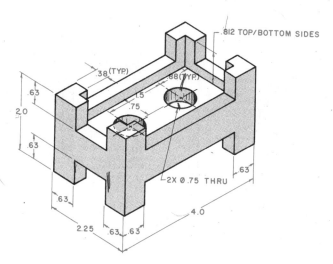

Problem 5-1

Problem 5-2

Center two views within the work area, and make one view a full section. Use correct drafting practices for the ribs.

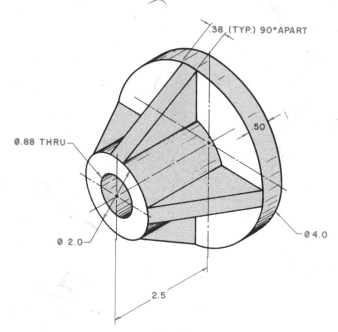

Problem 5-2

Problem 5-3

Center two views within the work area, and make one view a full section. Use correct drafting practices for the holes.

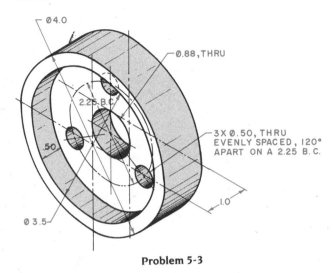

Ø4.0

Ø.88, THRU

2.25 B.C.

3X Ø.50, THRU
EVENLY SPACED, 120°
APART ON A 2.25 B.C.

.50

Ø 3.5

1.0

Problem 5-3

Problem 5-4

Center the front view and top view within the work area. Make one view a full section.

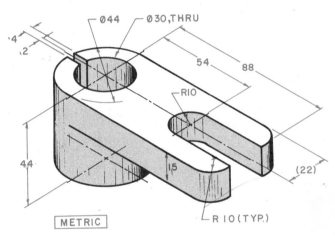

Ø44   Ø30, THRU

.4

.2

54   88

R10

4.4

1.5

(22)

METRIC

R 10 (TYP.)

Problem 5-4

Problem 5-5

Center two views within the work area, and make one view a full section. Use correct drafting practices for the arms, horizontal hole, and keyway.

Full section

P.646

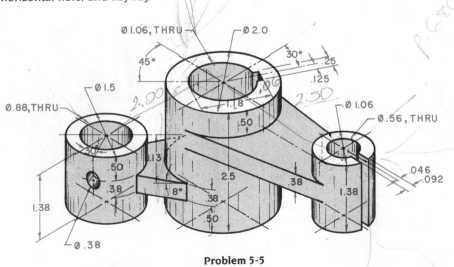

Ø 1.06, THRU   Ø 2.0

45°   30°   .25

Ø 1.5   .06   .125

Ø.88, THRU   2.00   1.18   2.50   Ø 1.06

.50   Ø.56, THRU

1.13   .50   2.5   .38   .046
.092

.50   8°   .38   1.38

1.38   .38

50

Ø.38

Problem 5-5

Problem 5-6

Center two views within the work area, and make one view a full section. Use correct drafting practices for the keyway, ribs, and holes.

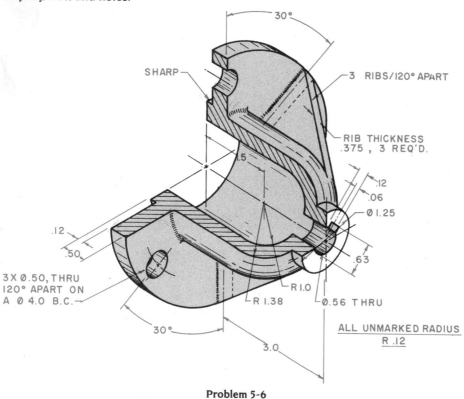

SHARP

30°

3 RIBS/120° APART

RIB THICKNESS
.375 , 3 REQ'D.

.12
.06
Ø 1.25
.5
.63
.12
.50
3 X Ø.50, THRU
120° APART ON
A Ø 4.0 B.C.
R 1.0
R 1.38
Ø.56 THRU
30°
3.0

ALL UNMARKED RADIUS
R .12

Problem 5-6

Problem 5-7

Center three views within the work area, and make one view an offset section. Be sure to include three major features.

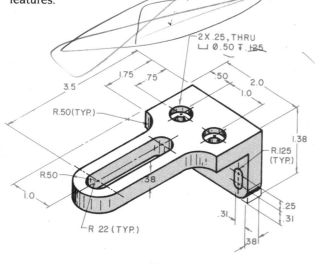

2 X .25, THRU
⌴ Ø .50 ⊺ .125

1.75 .75 .50 2.0
3.5 1.0
R.50 (TYP.)
R.50
1.0
R 22 (TYP.)
1.38
R.125
(TYP.)
.25
.31 .31
.38
.38

Problem 5-7

Problem 5-8

Center three views within the work area, and make one view an offset section. Be sure to include three major features.

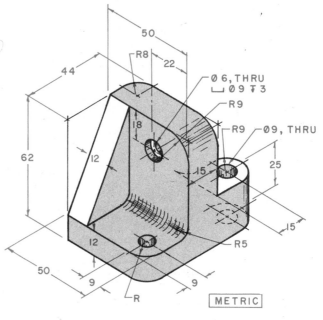

50
R8 22
44
Ø 6, THRU
⌴ Ø 9 ⊺ 3
R9
18
R9 Ø 9, THRU
62 12
15 25
12
50
9 9
R R5
15

METRIC

Problem 5-8

Problem 5-9

Center three views within the work area, and make one view an offset section. Be sure to include as many of the important features as possible.

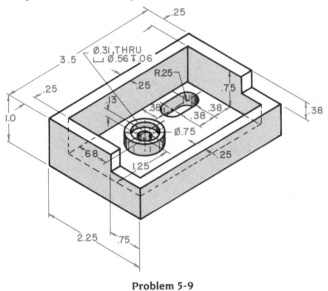

Problem 5-9

Problem 5-10

Center three views within the work area, and make one view an offset section. Be sure to include as many of the important features as possible.

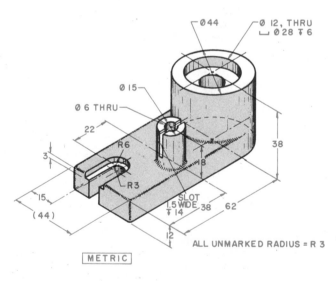

Problem 5-10

Problem 5-11

Center two views within the work area, and make one view an offset section. Be sure to include as many of the important features as possible.

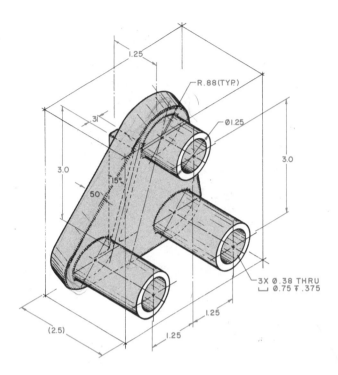

Problem 5-11

Problem 5-12

Center the front view and top view within the work area. Make one view a half section.

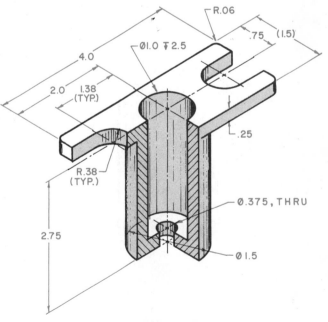

Problem 5-12

Problem 5-13

Center two views within the work area, and make one view a half section.

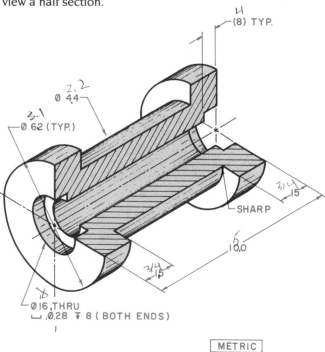

Problem 5-13

Problem 5-14

Center the two views within the work area, and make one view a half section.

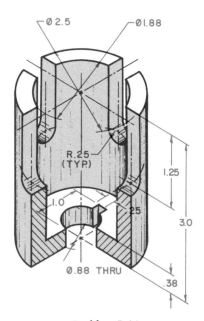

Problem 5-14

Problem 5-15

Center two views within the work area, and make one view a half section.

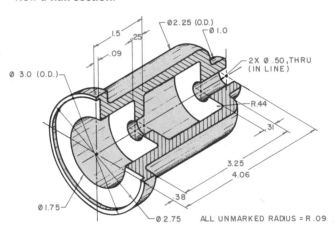

Problem 5-15

Problem 5-16

Center two views within the work area, and make one view a half section.

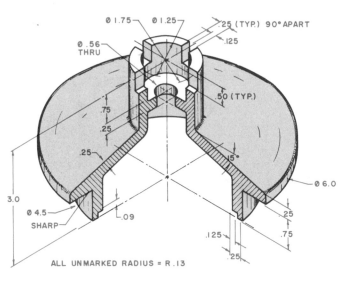

ALL UNMARKED RADIUS = R.13

Problem 5-16

Problem 5-17

Center the required views within the work area, and make one view a broken-out section to illustrate the complicated interior area.

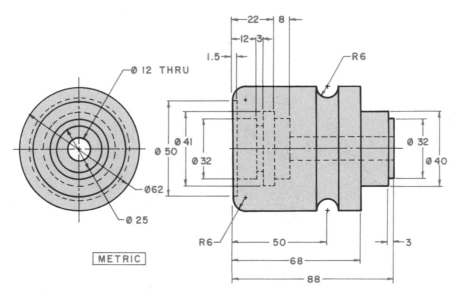

METRIC

Problem 5-17

Problem 5-18

Center the required views within the work area, and make one view a broken-out section as required.

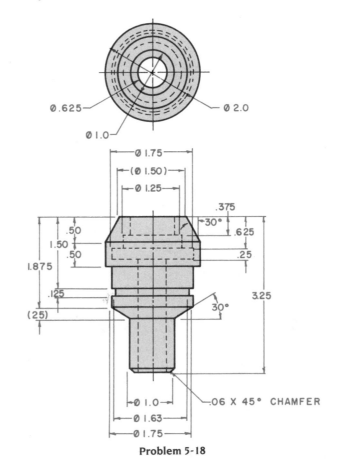

Problem 5-18

Problem 5-19

Center three views within the work area, and add removed sections A-A and B-B.

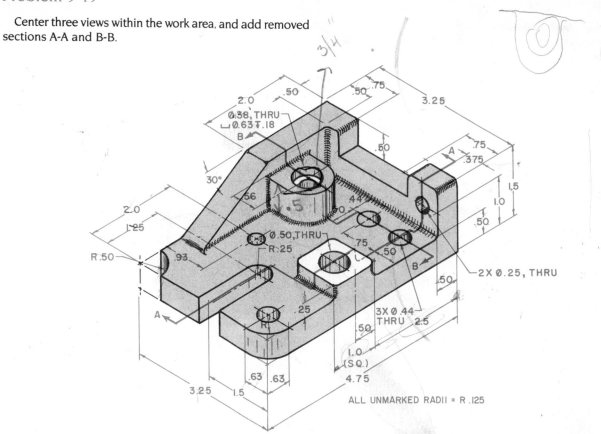

Problem 5-19

Problem 5-20

Center two views within the work area, and add removed sections A-A, B-B, and C-C.

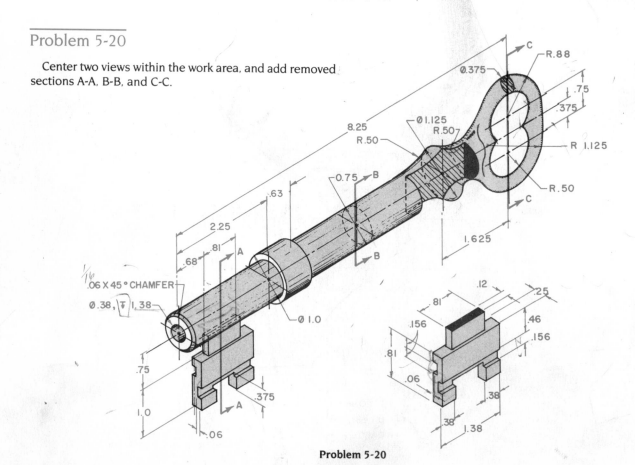

Problem 5-20

Problem 5-21

Center the required views within the work area, and add removed section as required.

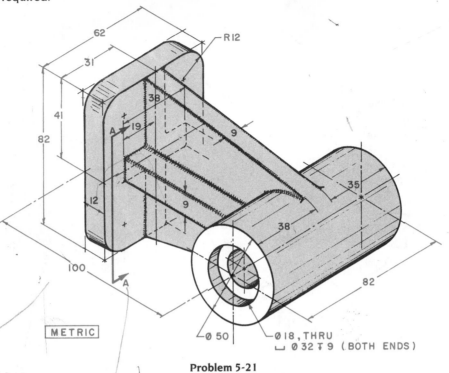

METRIC

Problem 5-21

Problem 5-22

Center the required views within the work area, and add removed section as required.

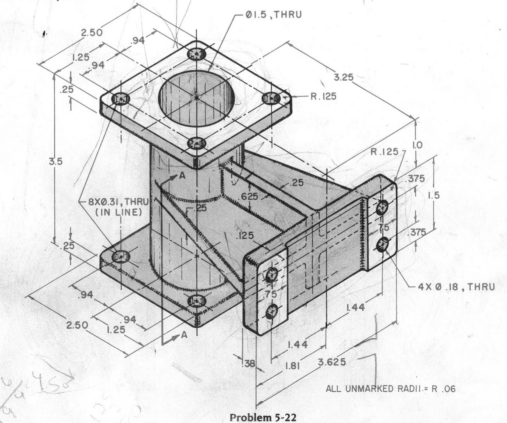

ALL UNMARKED RADII.= R .06

Problem 5-22

Problem 5-23

Center the required views within the work area, and add removed sections A-A and B-B.

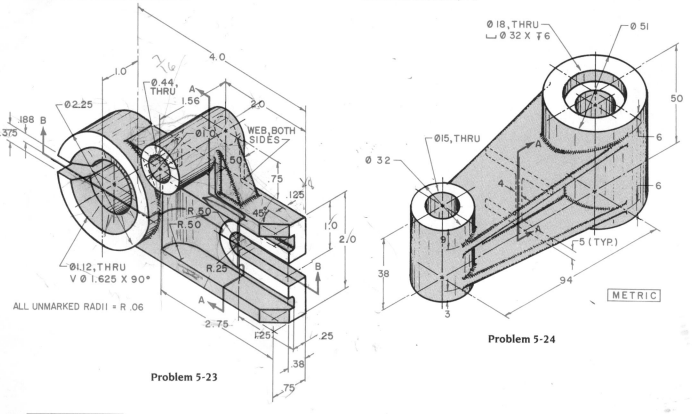

Problem 5-23

Problem 5-24

Center the required views within the work area, and add removed section A-A.

METRIC

Problem 5-24

Problem 5-25

Center the four views within the work area. Make the top view section A-A and the right-side view section B-B.

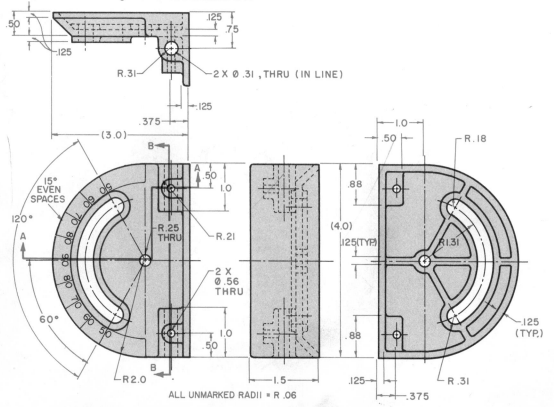

Problem 5-25

Problem 5-26

Center the required views within the work area, and add removed sections A-A and B-B.

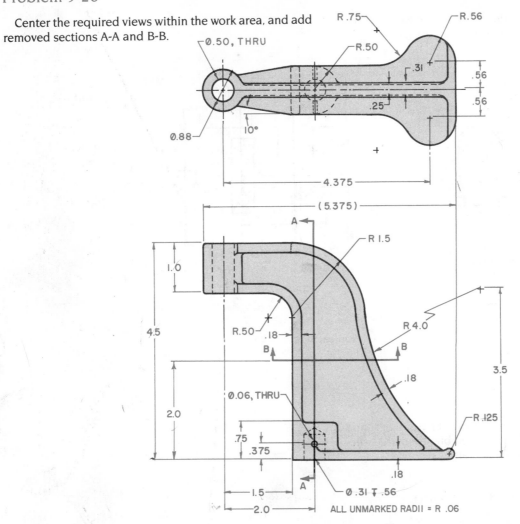

Problem 5-26

Problem 5-27

Center the required views within the work area, and add removed sections A-A and B-B.

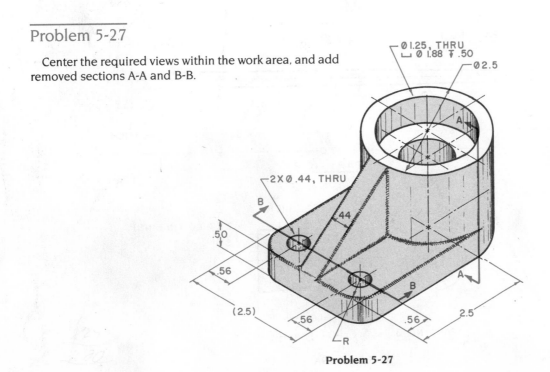

Problem 5-27

Problem 5-28

Center the front view, side view and removed sections
A-A, B-B, and C-C within the work area.

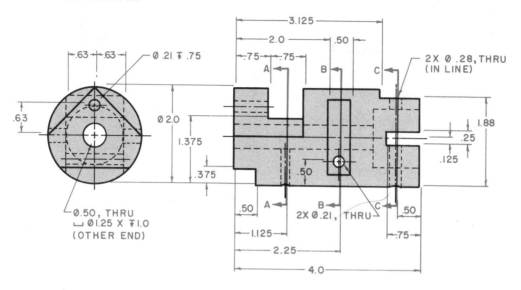

Problem 5-28

Problem 5-29

Make a two-view assembly drawing of parts 1, 2, 3, and
4. Make one view a full-section assembly. Use correct section lining, and all conventional drafting practices.

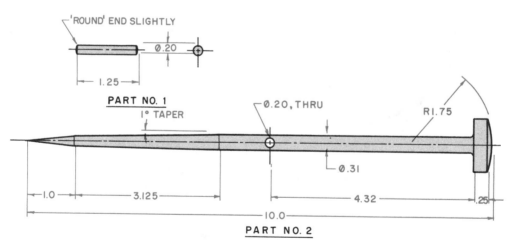

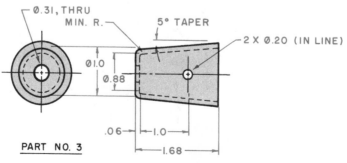

Problem 5-29A

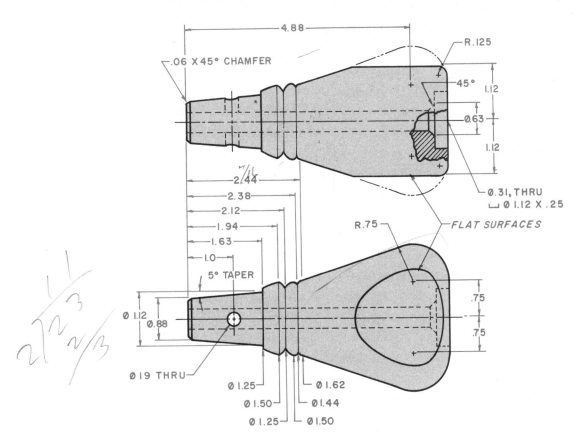

Problem 5-29B

Problem 5-30

Make a two-view assembly drawing of parts 1, 2, and 3. Make one view a full-section assembly. Use correct section lining and all conventional drafting practices.

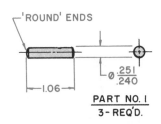

'ROUND' ENDS

$\varnothing \frac{.251}{.240}$

1.06

PART NO. 1
3- REQ'D.

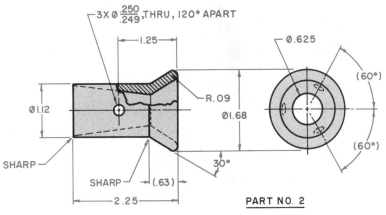

$3X \varnothing \frac{.250}{.249}$, THRU, 120° APART

1.25

R.09

$\varnothing$.625

$\varnothing$1.12

$\varnothing$1.68

(60°)

(60°)

SHARP

SHARP

30°

(.63)

2.25

PART NO. 2

Problem 5-30A

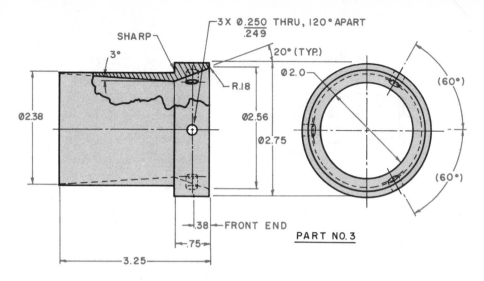

SHARP

3°

3X Ø.250 THRU, 120° APART
.249

20° (TYP.)

R.18

Ø2.0

Ø2.38

Ø2.56

Ø2.75

(60°)

(60°)

.38 — FRONT END

.75

3.25

PART NO. 3

Problem 5-30B

Problem 5-31

Make a two-view assembly drawing of parts 1, 2, and 3. Make one view a full-section assembly. Use correct section lining and all conventional drafting practices.

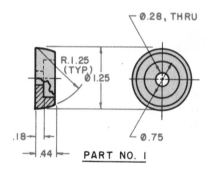

Ø.28, THRU

R.1.25
(TYP.)

Ø1.25

.18

.44

Ø.75

PART NO. 1

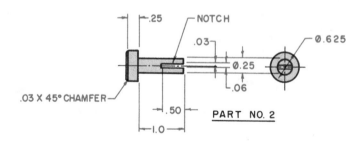

.25

NOTCH

.03

Ø.625

Ø.25

.06

.03 X 45° CHAMFER

.50

1.0

PART NO. 2

Problem 5-31A

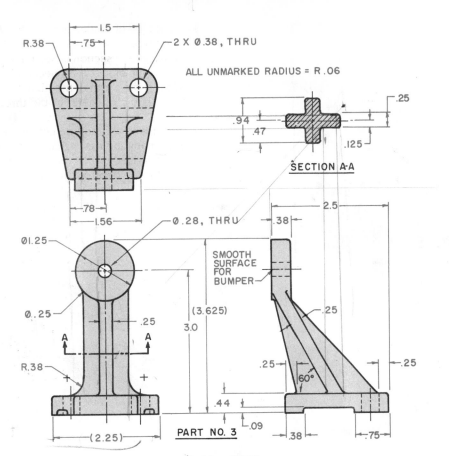

2 X Ø.38, THRU

R.38

ALL UNMARKED RADIUS = R.06

SECTION A·A

.94
.47
.125
.25

1.5
.75

.78
1.56

Ø1.25

Ø.28, THRU

Ø.25

R.38

SMOOTH
SURFACE
FOR
BUMPER

2.5
.38

(3.625)
3.0

.25
.25
.25
60°
.25

.44
.09

A   A

.25

(2.25)

PART NO. 3

.38
.75

Problem 5-31B

Chapter Five
Industry Print

1. What type of section is Section A-A?
2. What type of section is Section E?
3. What type of section is Section C-C?
4. What type of section is Section D-D?
5. What is the width of the rib at Section D-D?
6. What is the diameter of the through hole at Section C-C?
7. How many sections are cut on View II?
8. How many surfaces in Section A-A are to be machined?
9. What is the depth of the part in Section C-C?
10. What is the depth of the rib at Section B-B?

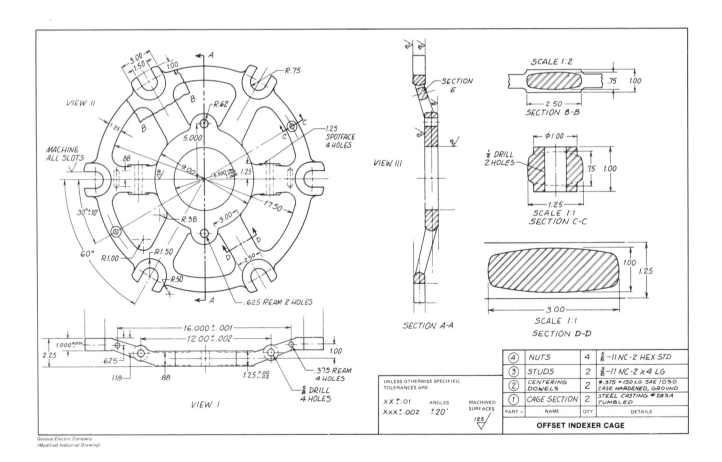

General Electric Company
(Modified Industrial Drawing)

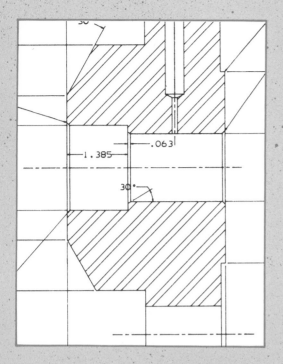

This chapter also is an extension of Chapter 4, Multiview Drawings, with the addition of an auxiliary view to show the true size and shape of a surface not on the usual planes of projection. The beginning drafter must fully understand how to lay out an auxiliary view and, if necessary, to add a secondary auxiliary view. The student must know the various standard drafting practices associated with auxiliary views.

CHAPTER SIX

Auxiliary Views

Auxiliary Views Defined

Many objects have inclined surfaces that are not always parallel to the regular planes of projection. For example, in Figure 6-1, the front view is correct as shown, but the top and right-side views do *not* correctly represent the inclined surface. To truly represent the inclined surface and to show its true shape, an auxiliary view must be drawn. An *auxiliary view* has a line of sight that is perpendicular to the inclined surface, as viewed looking directly at the inclined surface. Auxiliary views are always projected 90° from the inclined surface.

An auxiliary view serves three purposes:

- It illustrates the true size of a surface.
- It illustrates the true shape of a surface, including all true angles and/or arcs.
- It is used to project and complete other views.

An auxiliary view can be constructed from any of the regular views. An auxiliary view projected from the front view would appear as it does in Figure 6-1. This is referred to as a *front view auxiliary*. An auxiliary view projected from the top view would appear as it does in Figure 6-2. This is referred to as a *top view*

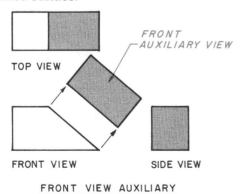

FRONT VIEW AUXILIARY

Figure 6-1 Front view auxiliary view

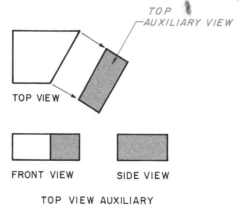

TOP VIEW AUXILIARY

Figure 6-2 Top view auxiliary view

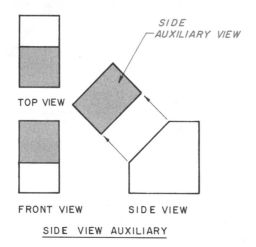

SIDE VIEW AUXILIARY

Figure 6-3 Side view auxiliary view

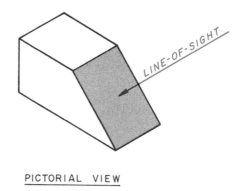

PICTORIAL VIEW

Figure 6-4 Drawing an auxiliary view

auxiliary. An auxiliary view projected from the right-side view would appear as it does in Figure 6-3. This is referred to as a *side view auxiliary*. Note in each case that the auxiliary view is projected 90° from the inclined or slanted surface, and is viewed from a line of sight 90° to the inclined or slanted surface, or as viewed looking directly down upon the inclined surface.

Hidden Lines in an Auxiliary View

Hidden lines should be omitted in an auxiliary view, unless they are needed for clarity. This is the drafter's prerogative or decision.

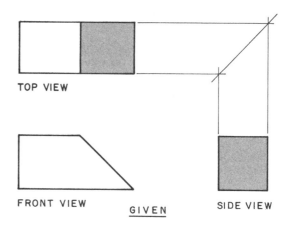

Figure 6-5 Given: Three views of an object

How To Draw an Auxiliary View

Given: The pictorial view of an object, Figure 6-4. Notice the inclined surface. As the inclined surface is on the front view, this will be a front view auxiliary. The usual three views of an object, front view, top view and right-side view, are shown in Figure 6-5.

Step 1. Label all important points of the auxiliary view, as illustrated in Figure 6-6A.

Step 2. Construct a *reference* line, which is also the edge view of a reference plane, in the right-side view. Always construct a reference line so that it is vertical and passes through as many points as possible. In this example, it passes through points a-d, Figure 6-6B.

Step 3. Draw light projection lines 90° from the inclined surface, and construct a reference line *parallel* to the inclined surface at any convenient distance, as shown in Figure 6-6C. Label all important points established thus far. Notice points a and d are *on* the reference line.

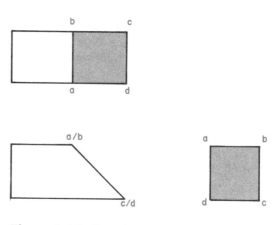

Figure 6-6A Step I

Step 4. In the right-side view, measure the distance each point is *from* the reference line. Project these distances back to the inclined surface of the front view and up 90° from the inclined surface to the reference line above. Transfer each distance, and lightly label each point as illustrated in Figure 6-6D.

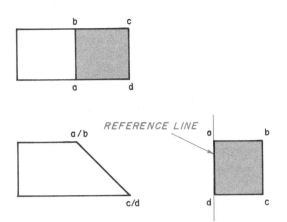

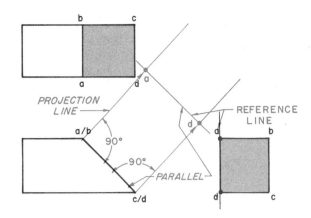

Figure 6-6B Step 2

Figure 6-6C Step 3

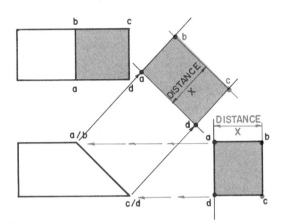

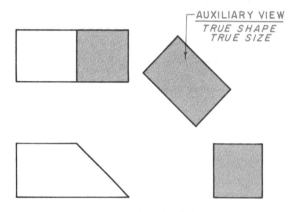

Figure 6-6D Step 4

Figure 6-6E Step 5 finished drawing

Step 5. Recheck all work. Be sure:

- The projection is 90° from the inclined surface, in this example the front view.
- The reference line is *parallel* to the inclined surface of the front view.
- All distances have been transferred accurately.

If correct, carefully darken in and complete all views. The final finished drawing will appear as it does in Figure 6-6E. Notice that *only* the inclined surface is projected into the auxiliary view. Anything else would be foreshortened and, thus, not of true size or shape, and therefore of no use. Good drafting practice is to project *only* the surface of the inclined line, Figure 6-7.

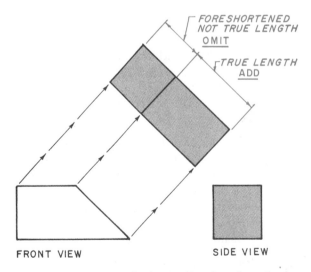

Figure 6-7 Draw only the inclined surface in the auxiliary view

How To Project a Round Surface from an Inclined or Slanted Surface

Given: The usual three views of an object, front view, right-side view and unfinished top view,

are shown in Figure 6-8. (The usual 45° projection angle line will be omitted, as a slightly newer projection method will be used to complete the top view.) Refer also to the pictorial drawing of this object, Figure 6-9.

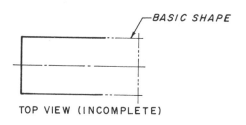

TOP VIEW (INCOMPLETE)

BASIC SHAPE

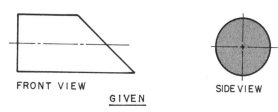

FRONT VIEW

GIVEN

SIDE VIEW

Figure 6-8 Projecting a round surface from an inclined surface

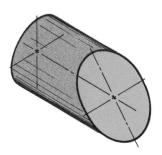

Figure 6-9 Pictorial view of an object

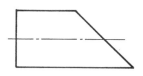

Figure 6-10A Step 1

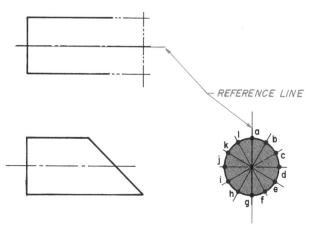

REFERENCE LINE

Figure 6-10B Step 2

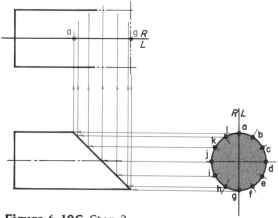

Figure 6-10C Step 3

Step 1. Divide the rounded view into equal spaces. In this example, using a 30° triangle, the right-side view is rounded and divided into 12 equal parts, Figure 6-10A. Letter each point clockwise, as shown in the figure.

Step 2. Construct a vertical reference line in the right-side view so it passes through the center; in this example, through points a and g. (Always place the reference line through the center of any symmetrical object.) Construct a reference line in the top view which runs through the center, as illustrated in Figure 6-10B.

Step 3. Draw light projection lines from the 12 points in the right-side view to the inclined edge in the front view. Project these same 12 points directly up to the top view from the inclined edge in the front view. Notice points a and g are *on* the reference line, Figure 6-10C.

Step 4. In the right-side view, measure the distance each point is *from* the reference line.

Transfer each of these distances from the right-side view reference line to the top view reference line. Label each point lightly as each is found, Figure 6-10D.

Step 5. Lightly connect all points and, if correct, darken in all views. This completes the top view, Figure 6-10E. Always darken in the compass and irregular curve layout work first. This completes the top view.

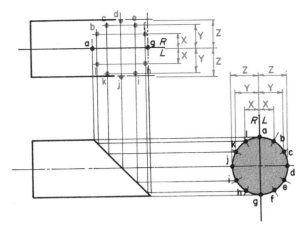

Figure 6-10D Step 4

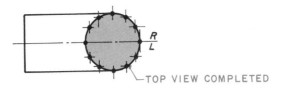
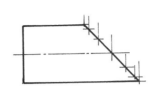

Figure 6-10E Step 5

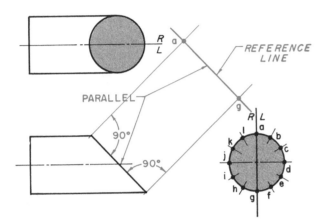

Figure 6-10F Step 6

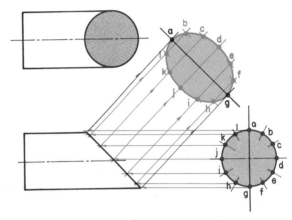

Figure 6-10G Step 7

How to Draw an Auxiliary View of a Round Surface

Step 6. Draw light projection lines 90° from the inclined surface of the front view. Construct the reference line *parallel* to the inclined surface at any convenient distance. Label the points that are on the reference line; in this example, a and g, Figure 6-10F.

Step 7. Draw light projection lines from the 12 points in the right-side view to the inclined edge in the front view (Step 3). Project these same 12 points directly up to the auxiliary view from the inclined edge in the front view. Again, notice points a and g are *on* the reference line, Figure 6-10G. In the right-side view, measure the distance each point is from the reference line. Transfer each of these distances from the right-side view reference line to the auxiliary view reference line. Label each point lightly as each is found.

Step 8. Lightly connect all points and, if correct, darken in the auxiliary view. This completes the

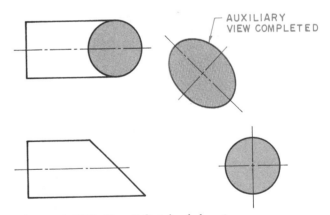

Figure 6-10H Step 8 finished drawing

auxiliary view, Figure 6-10H. Notice that only the inclined surface has been projected into the auxiliary view.

How to Plot an Irregular Curved Surface

Given: The usual three views; front view (incomplete), top view, and the right-side view, Figure 6-11. This is an example of a top view auxiliary.

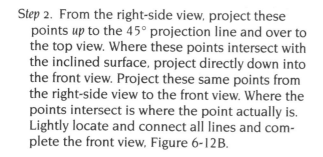

TOP VIEW

FRONT VIEW

SIDE VIEW

Figure 6-11 Plotting an irregular curved surface

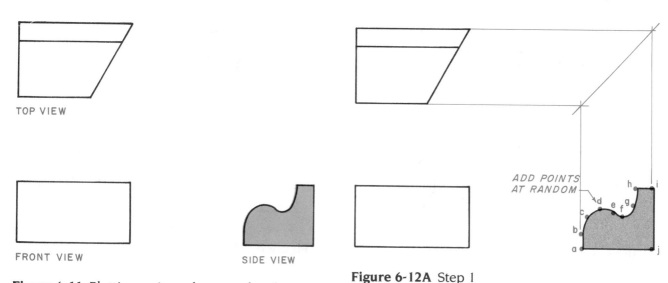

ADD POINTS AT RANDOM

Figure 6-12A Step 1

Step 1. Add various points at random along the curved surface. Even spaces are not necessary, but try to choose points that pick up high and low points along the line, Figure 6-12A. Always label the points in a clockwise direction.

Step 2. From the right-side view, project these points *up* to the 45° projection line and over to the top view. Where these points intersect with the inclined surface, project directly down into the front view. Project these same points from the right-side view to the front view. Where the points intersect is where the point actually is. Lightly locate and connect all lines and complete the front view, Figure 6-12B.

Step 3. Establish a reference line in the right-side view. Construct this reference line so it passes through as many points as possible; in this example, points a and j, Figure 6-12C.

Step 4. Add projection lines at 90° from the inclined surface. Add a reference line parallel to the inclined surface. Notice that points a and j again fall *on* the reference line, Figure 6-12D.

Step 5. Project all points from the right-side view up and over to the inclined surface of the top view. Where the points intersect with the inclined surface, project 90° from the inclined surface. Transfer all distances from the reference line in the right-side view to the reference line in the auxiliary view. Lightly connect all points and, if correct, darken in all views, Figure 6-12E.

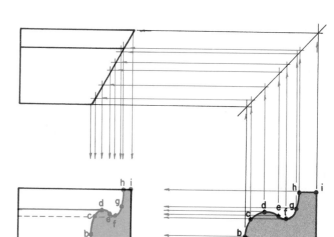

Figure 6-12B Step 2

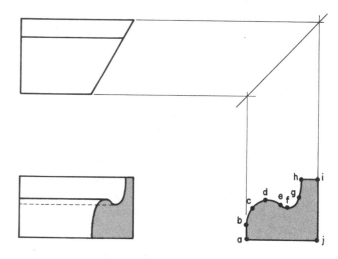

Figure 6-12C Step 3

224   Section 2

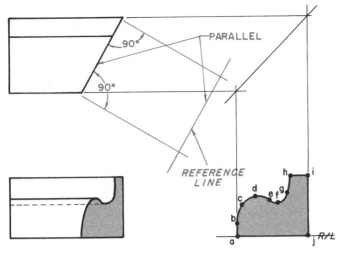

Figure 6-12D Step 4

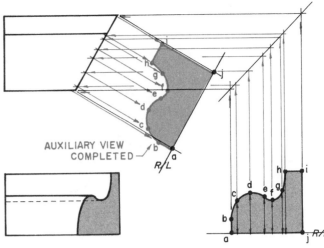

Figure 6-12E Step 5 finished drawing

Secondary Auxiliary Views

Up to this point, primary auxiliary views have been dealt with. A *primary auxiliary view* can be projected from any of the regular views, as has been illustrated thus far.

Sometimes, a primary auxiliary view is not enough to fully illustrate an object; a secondary auxiliary view is needed. A *secondary auxiliary view* is projected directly *from* the auxiliary view. Many times, the auxiliary view and/or some of the other views cannot be fully *completed* until the secondary view has been completed first, Figure 6-13.

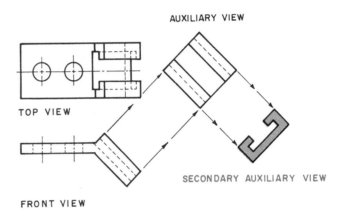

Figure 6-13 Secondary auxiliary view

Partial Views

The use of a *partial auxiliary view* makes it possible to eliminate one or more of the regular views which, in turn, saves drafting time and cost. Figure 6-14 is an example of a front view, top view, right-side view and auxiliary view. Note, the auxiliary view is always drawn partial; only the *inclined* surface is drawn on the auxiliary view. By using a partial top and right-side view in conjunction with the particular auxiliary view, the drawing can be simplified without detracting from its clarity; in fact, in most cases, these partial views make the drawing easier to read, Figure 6-15.

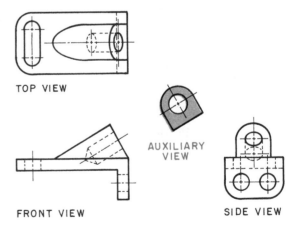

Figure 6-14 Given: Regular three views

Auxiliary Section

An *auxiliary section*, as its name implies, is an auxiliary view in section. An auxiliary section is drawn exactly as is any removed sectional view, and is pro-

jected in exactly the same way as any auxiliary view, Figure 6-16. All the usual auxiliary view rules apply, and generally only the surface cut by the cutting-plane line is drawn.

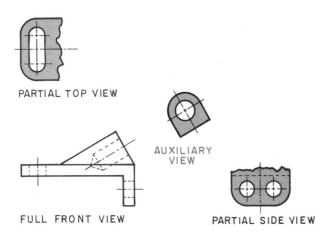

Figure 6-15 Partial auxiliary views

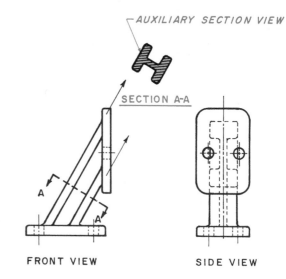

Figure 6-16 Auxiliary section

Half Auxiliary Views

If an auxiliary view is symmetrical, and space is limited, it is permitted to draw only half of the auxiliary view, Figure 6-17. Use of the half auxiliary view saves some time, but it should only be used as a last resort, as it could be confusing to those interpreting the drawing. Always draw the *nearest* half, as shown in the figure.

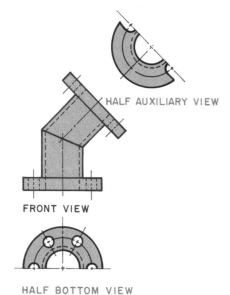

Figure 6-17 Half auxiliary view

Review

1. What three purposes does an auxiliary view serve?

2. Name the three major kinds of auxiliary views.

3. What must be done first if the projected surface is round or has a radius?

4. Explain the use of partial views as used in conjunction with an auxiliary view.

5. What is the practice for the use of hidden lines in an auxiliary view?

6. How should the regular views and the auxiliary view be placed within the work area?

7. Explain the use of a reference line. Where should it be drawn, and at what angle?

8. Projection lines *must* be drawn at what angle from the edge view?

9. When and why is a half auxiliary view used?

10. What is an auxiliary sectional view?

11. Explain the use of a secondary auxiliary view.

Chapter Six
Problems

The following problems are intended to give the beginning drafter practice in sketching and laying out multiviews with an auxiliary view.

The steps to follow in laying out any drawing with an auxiliary view are:

Step 1. Study the problem carefully.

Step 2. Choose the view with the most detail as the front view.

Step 3. Position the front view so there will be the least number of hidden lines in the other views.

Step 4. Determine which view from which to project the auxiliary view.

Step 5. Make a sketch of all views, including the auxiliary view.

Step 6. Center the required views within the work area with approximately 1-inch (25-mm) space between the views. Adjust the regular views to accommodate the auxiliary view.

Step 7. Use light projection lines. Do *not* erase them.

Step 8. Lightly complete all views.

Step 9. Check to see that all views are centered within the work area.

Step 10. Carefully check all dimensions in all views.

Step 11. Darken in all views using correct line thickness.

Step 12. Recheck all work, and, if correct, neatly fill out the title block using light guidelines and neat lettering.

Problems 6-1 through 6-4

Draw the front view, top view, right-side view and auxiliary view. Complete all views using the listed steps.

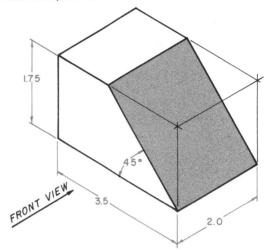

Problem 6-1

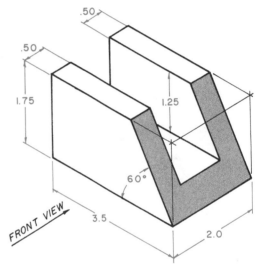

Problem 6-2

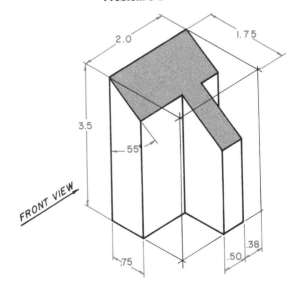

Problem 6-3

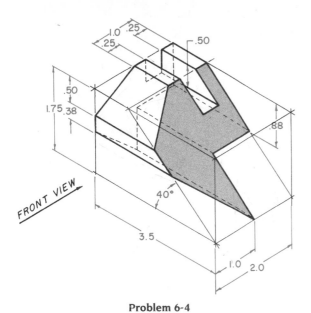

Problem 6-4

Problems 6-5 and 6-6

Draw the front view, top view, right-side view and *two* auxiliary views to illustrate the true size and shape of the slanted surfaces. Complete all views using the listed steps.

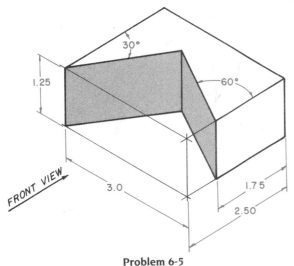

Problem 6-5

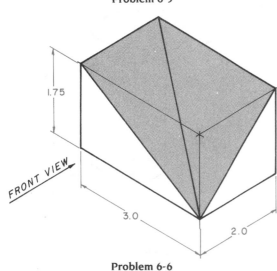

Problem 6-6

Problems 6-7 through 6-13

Draw the front view, top view, right-side view and auxiliary view. Complete all views using the listed steps.

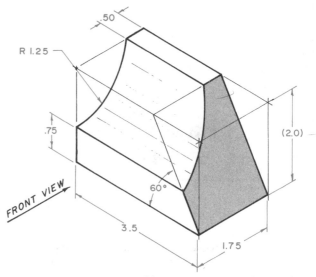

Problem 6-7

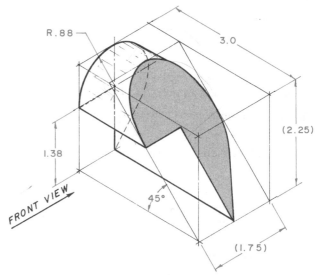

Problem 6-8

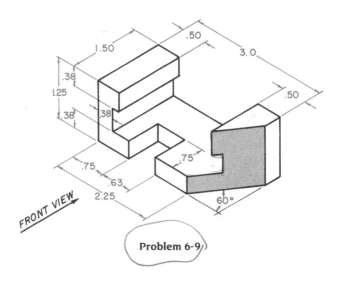

Problem 6-9

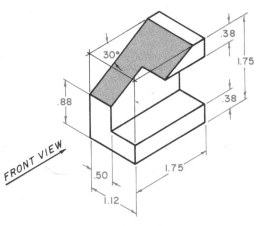

.38
1.75
30°
1.75
.88
.38
.50
1.12

Problem 6-10

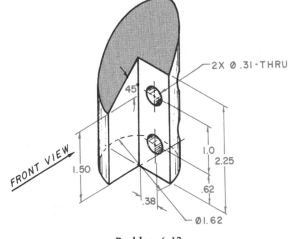

2X Ø .31·THRU
45
1.0
2.25
1.50
.62
.38
Ø1.62

Problem 6-13

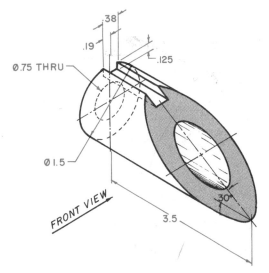

.38
.19
.125
Ø .75 THRU
Ø 1.5
30°
3.5

Problem 6-11

Problem 6-14

Draw the front view, top view, right-side view and *two* auxiliary views to illustrate the true size and shape of all surfaces. Complete all views using the listed steps.

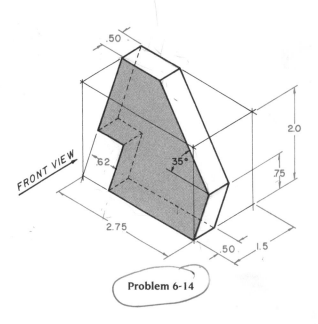

.50
2.0
.62
35°
.75
2.75
.50
1.5

Problem 6-14

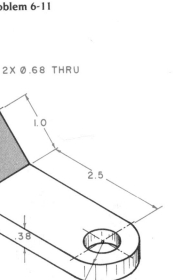

2X Ø.68 THRU
1.0
2.5
45°
.38
R.75 (TYP.)

Problem 6-12

Problems 6-15 through 6-22

Draw the required views to fully illustrate each object.
Complete all views using the listed steps.

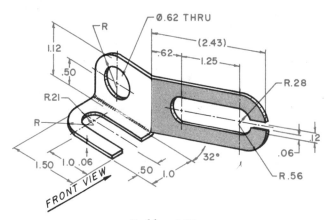

Problem 6-15

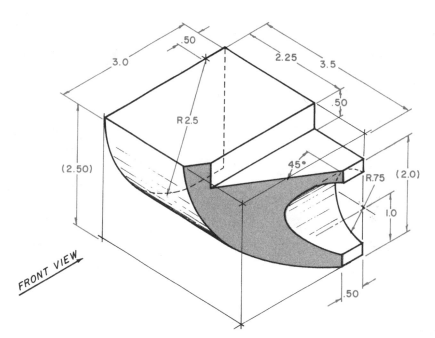

Problem 6-16

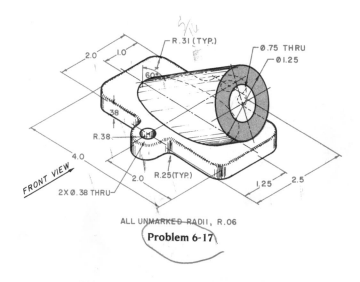

ALL UNMARKED RADII, R.06

Problem 6-17

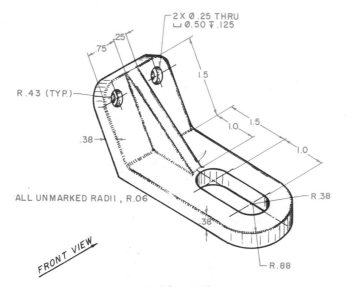

2X Ø.25 THRU
⌴ Ø.50 ▼.125

.25

.75

1.5

R.43 (TYP.)

1.0   1.5

.38   1.0

ALL UNMARKED RADII, R.06

R.38

.38

FRONT VIEW

R.88

Problem 6-18

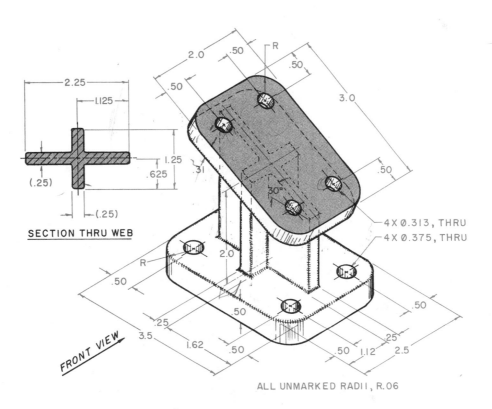

2.25

1.125

2.0   .50   R

.50   .50

.50

1.25

3.0

.625

.31

(.25)

30°   .50

(.25)

SECTION THRU WEB

4X Ø.313, THRU

4X Ø.375, THRU

R

2.0

.50

.50

.25

3.5   .50

FRONT VIEW

1.62   .50   .50   1.12   2.5

.25

ALL UNMARKED RADII, R.06

Problem 6-19

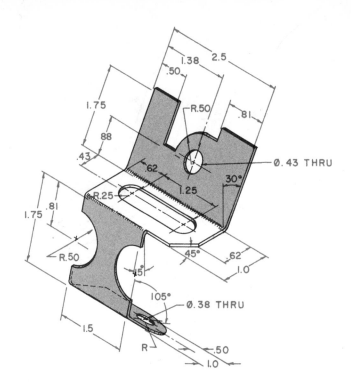

Problem 6-20

Dimensions visible: 2.5, 1.38, .50, 1.75, R.50, .81, 88, .43, .62, R.25, 1.25, Ø.43 THRU, 30°, 1.75, .81, R.50, 45°, 45°, .62, 1.0, 105°, Ø.38 THRU, R, 1.5, .50, 1.0

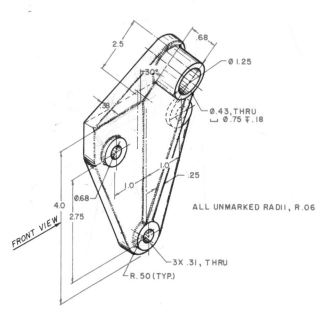

Problem 6-21

Dimensions visible: 2.5, .68, Ø 1.25, 30°, .38, Ø.43, THRU, ⌴ Ø.75 ⫫ .18, 1.0, .25, Ø.68, 1.0, ALL UNMARKED RADII, R.06, 4.0, 2.75, FRONT VIEW, 3X .31, THRU, R.50 (TYP.)

3 views & Auxilary

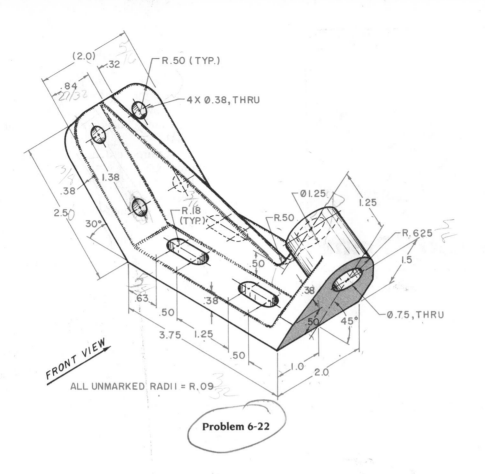

(2.0)
.32
.84
27/32
R.50 (TYP.)
4X Ø.38, THRU
1.38
3/8
.38
2.50
30°
R.18 (TYP.)
R.50
Ø1.25
1.25
R.625
1.5
.50
.38
Ø.75, THRU
.63
.50
.38
.50
45°
3.75 1.25
.50
1.0
2.0

FRONT VIEW

ALL UNMARKED RADII = R.09

Problem 6-22

Chapter Six
Industry Print

1. How many auxiliary views appear on this drawing?
2. Is View I an auxiliary view?
3. Is View II an auxiliary view?
4. Is View III an auxiliary view?
5. Identify the three orthographic views that appear on this drawing.
6. How many holes appear in the auxiliary view of the part?
7. What purpose does an auxiliary view(s) serve in this drawing?
8. In what view does the auxiliary surface appear as an edge?
9. What is the thickness of the web of the part?
10. What is the diameter of the ream through the part?

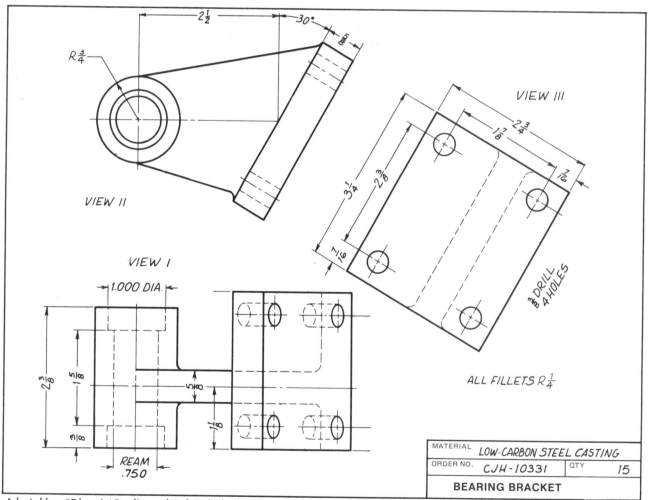

MATERIAL	*LOW-CARBON STEEL CASTING*		
ORDER NO. *CJH-10331*		QTY	*15*
BEARING BRACKET			

Adapted from "Blueprint Reading and Technical Sketching for Industry" and used by permission of Thomas P. Olivo and C. Thomas Olivo, Published by Delmar Publishers Inc.

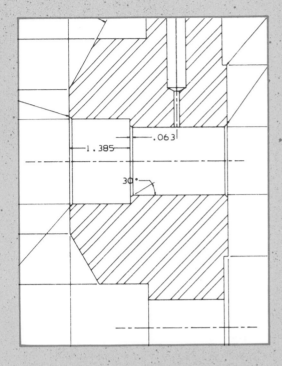

The basic techniques associated with descriptive geometry must be thoroughly mastered. Many procedures of this chapter are used in solving the more advanced technical problems that cannot be solved by any other method. It is important that sufficient time is given to this very important material, as it is incorporated into many other areas of drafting.

CHAPTER SEVEN

Descriptive Geometry

Descriptive Geometry Projection

Auxiliary views are views that have been projected 90° from an inclined surface and viewed looking directly down upon that surface. As studied in Chapter 6, the auxiliary view shows true size and true shape of an object, and is sometimes used to complete other views. If a surface is *skewed*, that is, slanting in more than one direction and not 90° from one of the other views, the auxiliary projection method cannot be used to find the true size and true shape. A completely different projection method must be incorporated. The true shape and true size of the skewed surface is a secondary auxiliary based on advanced orthographic theories called *descriptive geometry*.

Using the descriptive geometry method of projection not only gives true size and shape, but it also can be used to find intersections, true distances of lines in space, true angles between surfaces, and exact piercing points. Descriptive geometry graphically shows the solution to problems dealing with points, lines and planes, and their relationship in space. In order to be able to apply descriptive geometry to various drafting problems, the drafter must know and understand the various basic steps involved. This chapter explains these basic steps. The basic theories covered in this chapter are:

- Projecting a line into other views
- Projecting points into other views
- Determining the true length of a line
- Determining the point view of a line
- Finding the true distance between a line and a point in space
- Determining the true distance between two lines in space
- Projecting a plane surface in space
- Developing an edge view of a plane surface in space
- Determining the true distance between a plane surface and a point in space
- Determining the true angle between plane surfaces in space

Steps Used

All problems, regardless of their complexity, use the same procedures or steps. These steps ultimately must be done in the same order each time, although some intermediate steps may be selected or are optional. Like climbing a flight of stairs, each step must be executed, and only experience can provide the ability to perform multiple steps simultaneously. An example of the sequential steps needed in descriptive geometry problems is as follows. In order to find an edge view of a plane (flat surface), a true length line must be located in that plane (Step 1); a point view of the true length line must be projected, which yields an edge view of the plane (Step 2); see Figure 7-1. By projective means, it would be impossible to find the edge view of an oblique surface from normal views without performing Steps 1 and 2.

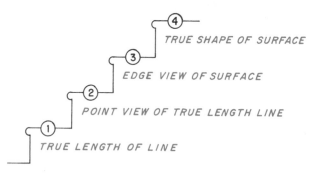

Figure 7-1 Steps used to solve descriptive geometric problems

Notations

In order to progress through sequential steps it is important to keep track of all views and points in space. (Recall that a straight line is composed of two spatially located end points.) To do this, a system of *notations* or labeling is advisable. Each view and each point should be labeled to provide an accurate identity of each at all times. For purposes of this text, the following notations are used in this chapter.

Each *point* in space is called out in *lowercase letters*.

Example:

a, b, c, d

Each *line* in space is identified by the *two lowercase end points*.

Example:

Line a-b

Each *line* in space is assumed to end at the indicated end points, unless it is specified as an *Extended Line* (continues without end).

Example:

Line p-q and Extended Line j-k

Each *plane* in space, such as a triangle, is called out in lowercase letters.

Example:

Triangle abc

Each *plane* in space is assumed to be limited (stops at the indicated boundaries) unless otherwise defined as an *Unlimited Plane* (limitless extension beyond the indicated boundaries).

Example:

Plane defg and Unlimited Plane mno

Each *viewing plane* is called out in *uppercase letters*.

Example:

F = Frontal viewing plane (a principal plane).

T = Top (horizontal) viewing plane (a principal plane).
R = Right side (profile) viewing plane (principal plane).
L = Left side (profile) viewing plane (principal plane).
A(digit) = an auxiliary viewing plane, with the number of successive projections from a principal viewing plane.
B(digit) = same as A(digit), but using a different series of successive projections.

A combination of lowercase and uppercase letters is used to fully describe each point in space, and the view in which it is located.

Example:

aF would be point *a* location as seen in the Frontal viewing plane.
aT-bT would be line *a-b* location as seen in the Top (horizontal) viewing plane.
aR-bR-cR would be plane *a-b-c* as seen in the Right side (profile) viewing plane.

Fold Lines

A fold line is represented by a thin black line, similar to a phantom line. It indicates a 90° intersection of two viewing planes. Each fold line must be labeled using uppercase letters, Figure 7-2. In this example, the fold line is placed between the front view and the top view. This placement eliminates the usual 45° projection line, Figure 7-3. It is important to add these notations in order to keep track of exactly which viewing plane is being executed.

Figure 7-2 Fold line

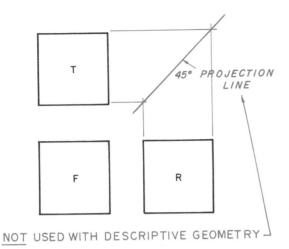

Figure 7-3 Regular multiview drawing using 45° projection line

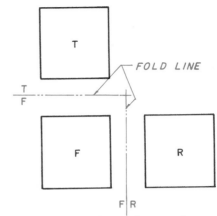

Figure 7-4 Regular multiview drawing using fold lines instead of 45° projection line

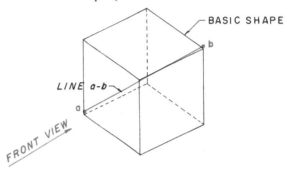

Figure 7-5 Spatial positioning of line a-b

How To Project a Line into Other Views

Fold lines are placed between successive orthographic views and labeled with uppercase letters, Figure 7-4. Any appropriate spacing is permissible, but the distance from the image to the frontal viewing plane fold line *must* be equal at both the top and right-side views. The spatial positioning of a line a-b (surrounded by an imaginary transparent box) is shown in Figure 7-5. Point a is closer to the frontal viewing plane than is point b, but is farther below the top (horizontal) viewing plane.

Represented on a normal layout drawing, it would be illustrated and labeled as shown in Figure 7-6.

The customary 45° projection line is *not* used in descriptive geometry projection. All points are projected along light projection lines drawn at 90° to the fold lines. All measurements are taken *along* these light projection lines and measured *from* the fold line *to* the point in space, Figure 7-7.

An important rule to remember is to *always skip-a-view* between all measurements. In Figure 7-7, notice in the right-side view that dimension X (distance from F/R fold line to point aR) is projected into and through the front view and *up* to the top view and transferred into the top view. (Distance from F/T fold line to point aT.) The front view was skipped.

Again, referring to the right-side view of Figure 7-7, dimension Y (distance from F/R fold line to point bR) is transferred into the top view (distance from F/T fold line to point bT) and the front view was skipped. In each of the instances, the distance is measured as the perpendicular distance from the fold line to the point.

Example:

To project a top view of a line from a given front and right-side view.

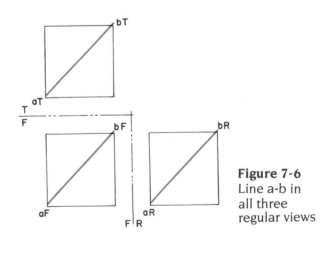

Figure 7-6 Line a-b in all three regular views

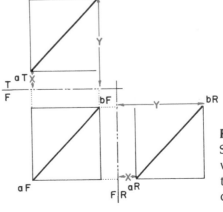

Figure 7-7 Skip-a-view when transferring distances

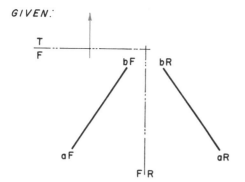

Figure 7-8 Locating line a-b in top view

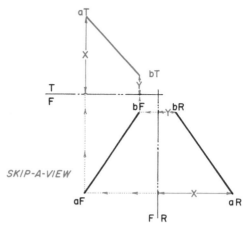

Figure 7-9 Line a-b in the top view using the fold line

Given: Line a-b in the front view and line a-b in the right-side view, Figure 7-8. Extend projection lines into the top view from the end points in the front view, aF and bF. Find the distances X and Y from the line end points in the right-side view aR and bR to the frontal viewing plane at fold line F/R and transfer them into the top view. Label all points and fold lines in all views. Be sure that all projections are made at 90° from the relevant fold lines, Figure 7-9.

How To Locate a Point in Space (Right View)

A point in space is projected and measured in exactly the same way as a line in space, except that the point is a line with a single end point, Figure 7-10A.

Example:

To project the right-side view of a point from a given top and front view.

Given: Point a in the front view and top view (Figure 7-10A).

Project point aF from the front view into the right-side view. Point aR must lie on this projection line. Find X, which is the distance from fold line F/T to aT in the top view and project it into and through the front view, Figure 7-10B. Project it over into the right-side view. Transfer distance X to find point a. Be sure to always label all points and fold lines in all views.

How To Find the True Length of a Line

Any line that is parallel to a fold line will appear in its true length in the next successive view adjoining that fold line.

To find the true length of any line:

Step 1. Draw a fold line parallel to the line that the true length is required to be. This can be done at any convenient distance, such as approximately one-half inch.

Step 2. Label the fold line "A" for auxiliary view.

Step 3. Extend projection lines from the end points of the line being projected into the auxiliary view. These must always be at 90° to the fold line.

Step 4. Transfer the end point distances from the fold line in the second preceding view from the one being drawn, to locate the corresponding end points in the view being drawn.

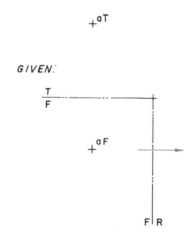

Figure 7-10A Locating a point in space (right view)

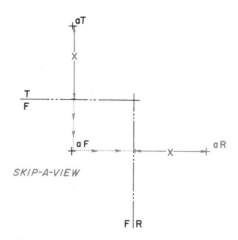

Figure 7-10B Step 1

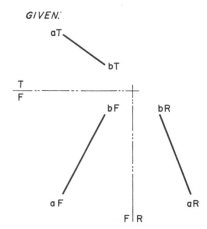

Figure 7-11A Finding the true
length of a line

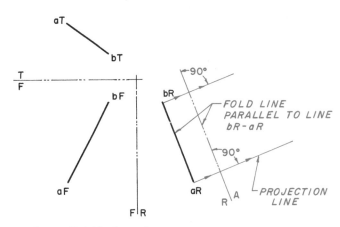

Figure 7-11B Step 1

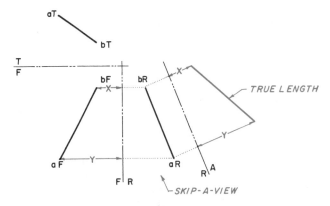

Figure 7-11C Step 2

Example:

Refer to Figure 7-11A.

Given: Line a-b in the front view, side view, and
top view. The problem is to find the true length
of line a-b. A true length can be projected from
any of the three principal views by placing a
fold line parallel to the line in any view. The
right-side view is selected in this example.

Step 1. Draw a fold line parallel to line a-b, and
label it as shown in Figure 7-11B. Extend light
projection lines at 90° to the fold line from
points a and b into the auxiliary view.

Step 2. Determine distance X and Y from the front
view to the near-fold line and transfer them
into the auxiliary view, as shown in Figure
7-11C. Label all points and fold lines. The result
will be the actual true length of line a-b.

Use these steps to find the true length of any line.

How To Construct a Point View of a Line

To construct a point view of a line:

Step 1. Find the true length of the line.

Step 2. Draw a fold line perpendicular to the true
length line at any convenient distance from
either end of the true length line.

Step 3. Label the fold line A-B (B indicates a secondary auxiliary view).

Step 4. Extend a light projection line from the
true length line into the secondary auxiliary
view.

Step 5. Transfer the distance of the line end points
into the secondary auxiliary view (B) from the
corresponding points in the second preceding
view.

Example:

Refer to Figure 7-12A.

Given: Line a-b in the front view, side view, and
auxiliary view. The true length is located in the
auxiliary viewing plane (A), which is projected
from the front view. (The true length could have
been projected from any of the given views.)

Step 1. At any convenient distance from either
end, draw a fold line A/B perpendicular to the
true length of line a-b. This will establish a
viewing plane that is perpendicular to the direction of the line's path. Extend a projection line
aligned with the true length line into this new

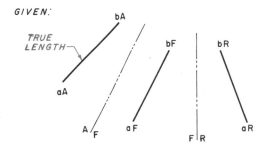

Figure 7-12A Constructing a point view of a line

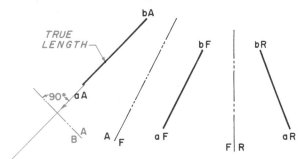

Figure 7-12B Step 1

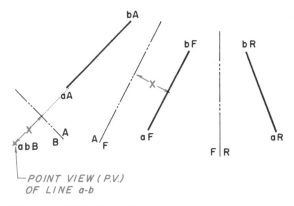

Figure 7-12C Step 2

secondary auxiliary view. The projection lines from both points a and b appear to align in this common projection line. Therefore, both points a and b must be located on this projection line in the secondary auxiliary view. In Figure 7-12B, the fold line A/B was added to the left of the true length line a-b.

Step 2. The front view is the second preceding view to the secondary auxiliary view being constructed. Therefore, distance X in the front view from the fold line T/A to line a/b is transferred to the secondary auxiliary view from the fold line A/B, Figure 7-12C. As points a and b in the front view are both at the same distance from the fold line T/A, and they both lie on the same projection line in the secondary auxiliary view, they will appear to coincide at the same location. This provides evidence that the end view of the line has been achieved.

How To Find the True Distance Between a Line and a Point in Space

The true distance between a line and a point in space will be evident in a view that shows the end point of the line and that point, simultaneously.

Step 1. Project an auxiliary view (viewing plane A) to find the true length of the line, and project the point into the same view.

Step 2. Project a secondary auxiliary view (B) to find the point view of the line, and project the point into the same view.

Step 3. The observable distance between the end view of the line and the point will be the actual distance between them. The location in the line that is nearest to the point is on the path that is perpendicular to the line from the point, but this is not discernable in the secondary auxiliary view.

Example:

Refer to Figure 7-13A.

Given: Line a-b and point c in the front view and right-side view.

Step 1. Project the true length of line a-b by placing a fold line parallel to a given view of the line, and project point c into the auxiliary view. Recall that point locations are transferred from the second preceding view. Label all fold lines and all points, Figure 7-13B.

Step 2. Draw a fold line perpendicular to the true length of line a-b, and label the fold line. Project line a-b into the secondary auxiliary view (B) to find the point view of line a-b. Project point c along into the secondary auxiliary view (B). The actual distance between the point view of line a-b/B and point cB is evident, Figure 7-13C.

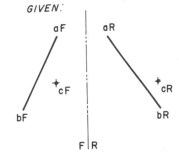

Figure 7-13A
Finding the true distance between a line and a point in space

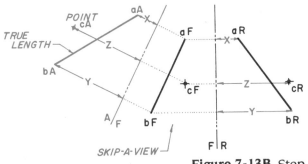

Figure 7-13B Step 1

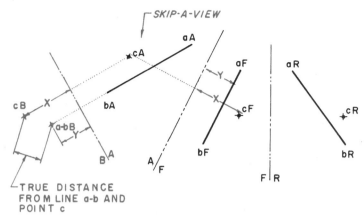

Figure 7-13C Step 2

The location in the line a-b that is nearest to point c lies on a path that is perpendicular to line ab and passes through c. Any path that is perpendicular to a line will appear perpendicular in a view where the line is true length. Therefore, the path can be drawn in the preceding view to locate point p on the line. Point p can be projected to its correct location on the original views of line a-b.

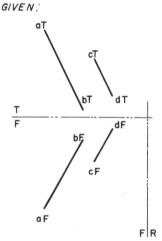

Figure 7-14A Finding the true distance between two parallel lines

How To Find the True Distance Between Two Parallel Lines

If two lines are actually parallel, they will appear parallel in all views. A view is needed to show the end view of both lines simultaneously, where the real distance between them will be apparent.

Step 1. Find the true length of each of the two lines. Projecting a view that will provide the true length of a line will automatically provide the true length of any other parallel line that is projected into the same view.

Step 2. Find the point view of each of the two lines. Projecting a view that will provide the end view of a line will automatically provide the end view of any other parallel line that is projected into the same view.

Step 3. The true distance between the parallel lines is the straight-line path between their end points. There is no single location within the length of the lines where this occurs, as each location in a line has a corresponding closest point location on the other line, each on the path connecting them and perpendicular to their respective lines.

Example:

Refer to Figure 7-14A.

Given: The parallel lines a-b and c-d, in a front view, right side view, and a top view.

Step 1. Find the true lengths of line a-b and line c-d in auxiliary view A. The two parallel lines will also be parallel in the auxiliary, if they are actually parallel. See Figure 7-14B.

Step 2. Draw a fold line perpendicular to the true length lines, a-b and c-d. Project the lines into the secondary auxiliary view (B) to find the point view of the two lines a-b and c-d. Measure the true distance between the point views of lines a-b/B and c-d/B. See Figure 7-14C.

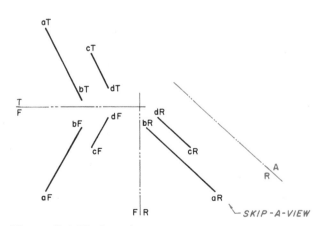

Figure 7-14B Step 1

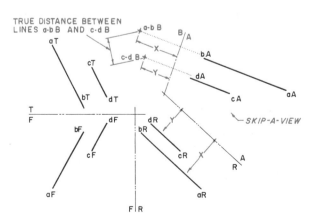

Figure 7-14C Step 2

How To Find the True Distance Between Two Nonparallel (or Skewed) Lines

If two lines are not parallel, they will appear nonparallel in at least one view. It is wise to first check two nonparallel (or skewed) lines to determine if they actually intersect, which would make the distance between them to be zero. This can be verified by projecting the apparent point of intersection of a view to the adjoining view. If the apparent points of intersection align, then the lines actually intersect.

To determine the true distance between two nonparallel lines:

Step 1. Project the true length of either one of the two lines, and project the other line into that auxiliary view (A).

Step 2. Find the point view of the true length line, and project the other line into that secondary auxiliary view (B).

Step 3. Measure the true distance between the point view line at a location that is perpendicular to the other line.

Example:

Refer to Figure 7-15A.

Given: Nonparallel lines a-b and c-d in a front view and right-side view.

Step 1. Project the true length of line a-b in an auxiliary view (A), and project line c-d along into that auxiliary view, Figure 7-15B. Notice that c-d is not the true length.

Step 2. Project the point view of line a-b into the secondary view (B) and project line c-d into this view. Measure the true distance between the point view of line a-b/B, perpendicular from line c-d/B, Figure 7-15C.

How To Project a Plane into Another View

A limited plane is located by points in space joined by straight lines. In order to transfer a plane, find at least three points in the plane and project each point into the next view.

Example:

Refer to Figure 7-16A.

Given: Triangular plane abc shown in the top view and front view.

Step 1. In the top view, find the distance from the fold line to points aT, bT, and cT, Figure 7-16B.

Step 2. Project these same points from the front view into the right-side view, and transfer the corresponding distances from the top view points to the top view fold line into the right-side view, Figure 7-16C. Construct straight lines between points a, b, and c. This transfers the plane surface.

Figure 7-15A Finding the true distance between two nonparallel (or skewed) lines

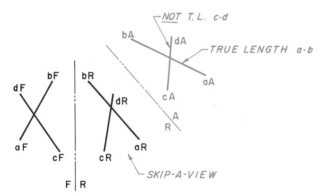

Figure 7-15B Step 1

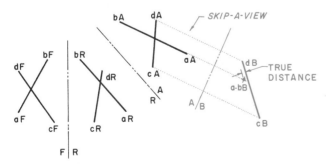

Figure 7-15C Step 2

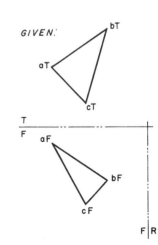

Figure 7-16A
Projecting a plane into another view

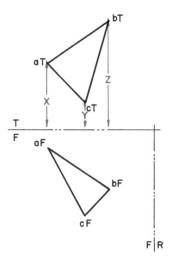

Figure 7-16B Step 1

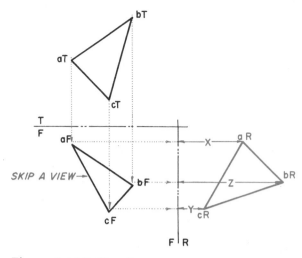

Figure 7-16C Step 2

How To Construct an Edge View of a Plane Surface

If the end view of a line that is in a plane is shown, then the edge view of that plane is shown also. Recall that to find the end view of a line, a true length must be found first. When the boundaries of a plane are provided, projecting a view to find the true length of a selected boundary is a simple procedure. It is necessary only to locate a fold line parallel to the boundary to ensure its true length in the next projected view.

A shorter method of securing a true length line in a plane is also used, if a true length line is not present already. A line may be added on the plane in any view, arranged to be parallel to an adjoining fold line, and to have its terminations at the plane boundaries. Projecting the added-line terminations to the corresponding boundaries in the adjoining view relocates the added line in that view. Moreover, it will be a true length line, as it was made to be parallel to the fold line in the preceding view.

Whichever method is used to find a true length line in the plane, all point locations within the plane should be projected to the view where the true length line exists, if not there already. A fold line perpendicular to the true length line will provide a direction for projecting the edge view of the plane. Projecting all the points in the plane will provide evidence of this, as they will all align into a single, straight path.

Example:

Refer to Figure 7-17A.

Given: Plane surface abc in the top view and front view.

Step 1. Construct a line parallel to fold line F/R and through one or more points if possible. In this example, the line passes through point c. Label this newly constructed line cF/xF, Figure 7-17B.

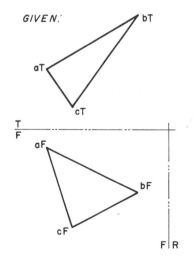

Figure 7-17A Constructing an edge view of a plane surface

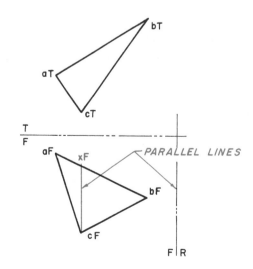

Figure 7-17B Step 1

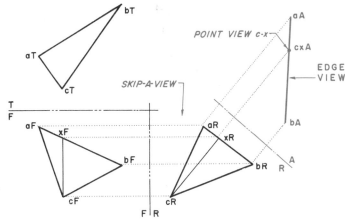

Figure 7-17C Step 2

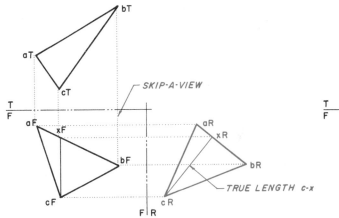

Figure 7-17D Step 3

Step 2. Construct the true length of line cF/xF in the next view, Figure 7-17C.

Step 3. Construct the point view of the true length line c-R/xR. Project the remaining points of the plane surface into this view. Also, label all points and fold lines. If done correctly, the result should be a straight line which passes through the point view. This straight line is the edge view of the plane, Figure 7-17D.

How To Find the True Distance Between a Plane Surface and a Point in Space

Finding the true distance between a given point and a plane in space requires a view that shows the edge view of the plane and the point in the same view. The distance from the point to the plane is the path that is perpendicular to the plane, and passing through the point.

Step 1. Add a line on the plane parallel to a fold line, if a true length line is not already available in the plane.

Step 2. Project the added line to an adjoining view, where it will be seen in true length.

Step 3. Project a point view of the added true length line.

Step 4 .Project the points of the plane into this view. The result should be a straight line which passes through the added line end view, which is the edge view. Project the point in space into this view. Measure the perpendicular distance from the plane's edge view to the point in space. This is the actual true distance between the plane surface and the point in space.

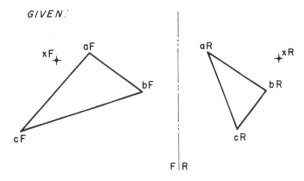

Figure 7-18A Finding the true distance between a plane surface and a point in space

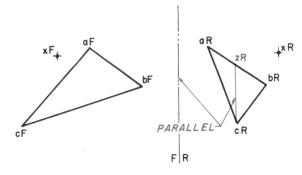

Figure 7-18B Step 1

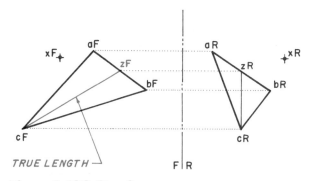

Figure 7-18C Step 2

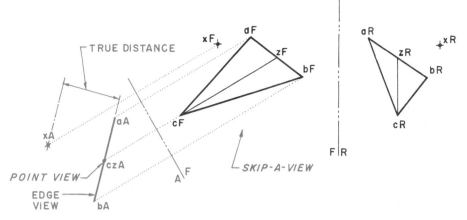

Figure 7-18D Step 3

Example:

Refer to Figure 7-18A.

Given: Triangular plane abc and point X in the front view and right-side view.

Step 1. Construct a line parallel to the fold line F/R and through one or more points. Point c is a convenient line termination. Label this newly constructed line cR/zR, Figure 7-18B.

Step 2. Construct the true length of line cR/zR in the front view, and label the line cF/zF, Figure 7-18C. Bring point X into this view, also. Label all points and fold lines.

Step 3. Project the end view of the true length line cF/zF in the auxiliary view, Figure 7-18D. Project the points of the plane surface into this view, and also point X, projecting 90° from the fold line. Label all points and fold lines. Measure the perpendicular distance from the edge view to the point in space. This is the true distance between the plane surface and the point X in space.

How To Find the True Angle Between Two Planes (Dihedral Angle)

The edge of intersection between two planes is a line that is common to both planes. The end view of that line will provide the edge view of both planes simultaneously, and the angle between them will be evident.

To find the true angle between two surfaces:

Step 1. Construct the true length of the intersection between the two surfaces, and project all other points of both planes into that auxiliary view.

Step 2. Project the end view of the true length edge of intersection line and project the points of both planes into this secondary auxiliary view. Label all points and fold lines. Measure the true angle between the two surfaces.

Example:

Refer to Figure 7-19A.

Given: Plane abc and plane abd that intersect one edge, a-b, in the front view, side view, and top view.

Step 1. Project the true length of the edge of intersection a-b, and project all other points into this auxiliary view (A), Figure 7-19B.

Step 2. Project the point view of the true length of the edge of intersection a-b and project all other points into this secondary auxiliary view (B), Figure 7-19C. Measure the true angle between two surfaces.

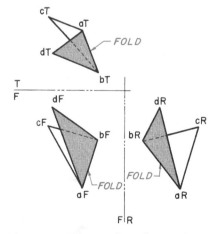

Figure 7-19A Finding the angle between two surfaces

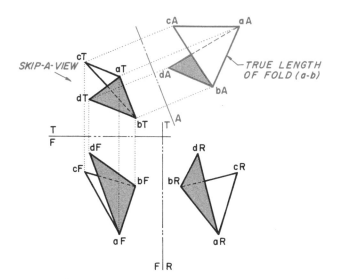

Figure 7-19B Step 1

How To Determine the Visibility of Lines

To determine which line of an apparent intersection of two lines is closer to the viewer, the following steps are used.

Step 1. From the exact crossover point of the lines, project to an adjoining view.

Step 2. In the adjoining view, determine which of the lines is closest to the fold line between the views, on the projection line from the first view (or the first line that the projection line encounters on its path from the first view).

Step 3. Whichever line is closest to the fold line at that point only is the line that is in front of the other line in the first view.

Example:

Refer to Figure 7-20A.

Given: Lines a-b and c-d in the front view and top view.

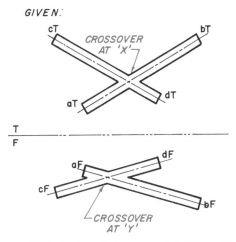

Figure 7-20A Determining the visibility of lines

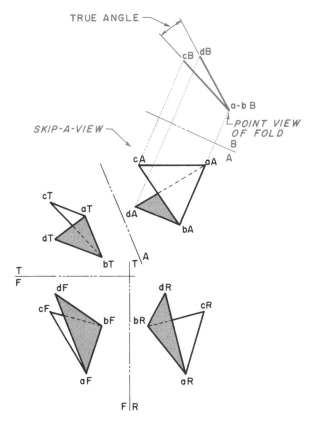

Figure 7-19C Step 2

Step 1. At the exact crossover of lines a-b and c-d in the top view, project down through the fold line to the corresponding lines in the front view. At this exact point along the lines, line c-d is closer to the fold line and line a-b is farther away from the fold line; thus, in the top view, line c-d is in front of line a-b, Figure 7-20B.

Step 2. See 7-20C. At the exact crossover of lines a-b and c-d in the front view, project up through the fold line to the corresponding lines in the top view. At this exact point along the lines, line a-b is closer to the fold line and line

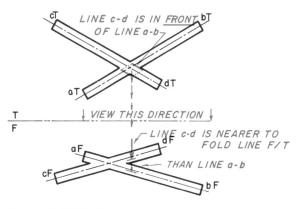

Figure 7-20B Step 1

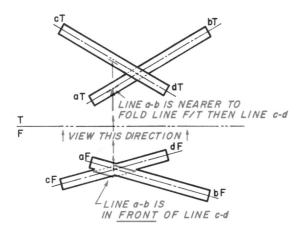

Figure 7-20C Step 2

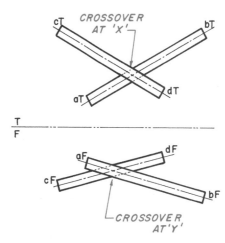

Figure 7-20D Step 3

c-d is farther away from the fold line; thus, in the front view, line a-b is in front of line c-d.

Step 3. The end result is drawn (in Figure 7-20D) to illustrate this crossover.

How To Determine the Piercing Point by Inspection

The *piercing point* is the exact location of the intersection of a surface and a line. The exact piercing point is determined from the view where the surface appears as an edge view. In example 7-21, Part A, the top view illustrates the edge view of the surface. The piercing point is established in the top view and projected down into the front view, Figure 7-21, Part B. In example 7-22, Part A, the front view illustrates the edge view of the surface. The piercing point is established in the front view and projected up into the top view, Figure 7-22, Part B.

How To Determine the Piercing Point by Construction

To determine the exact piercing point by construction:

Step 1. Construct a true length of a line on the plane surface.

Step 2. Project a point view of the true length line, and an edge view of the plane surface. The edge view of the plane lies in the path of the line in this view, indicating the piercing point location.

Step 3. Transfer the piercing point back into the other views.

Step 4. Determine the visibility of lines to find which part of the line is visible from the viewing direction. Use the fold line to determine visibility, if necessary.

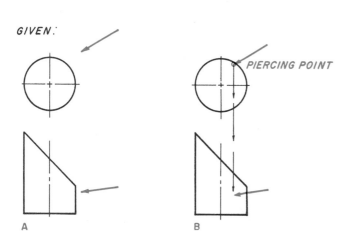

Figure 7-21 Determining the piercing point by inspection

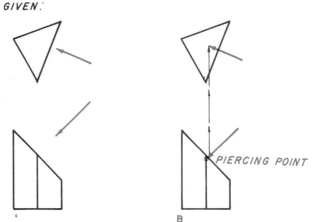

Figure 7-22 Determining the piercing point by inspection

Example:

Refer to Figure 7-23A.

Given: Plane surface abc and line x-y, in the top view and front view.

Step 1. In the front view of plane abc, construct a line parallel to a fold line (bF-dF) and find its true length in the top view, Figure 7-23B.

Step 2. Project an auxiliary view to find the point view of the true length line from the top view, and the edge view of the plane. The line inter-

section with the edge view of the plane indicates the piercing point location, Figure 7-23C.

Step 3. Project the piercing point back into other views, Figure 7-23D.

Step 4. In each view that the piercing point is being projected to, the portion of the line that is visible can be determined by checking its preceding view, Figure 7-23E.

Notice in the auxiliary view that line x-y is closer to the fold line from the piercing point to point yA than the edge view of the plane

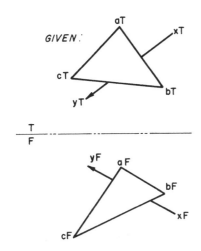

Figure 7-23A Determining the piercing point by construction

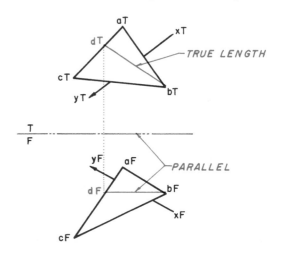

Figure 7-23B Step I

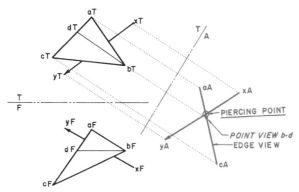

Figure 7-23C Step 2

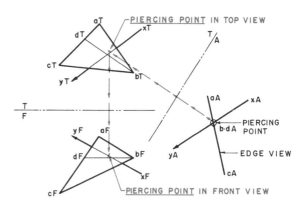

Figure 7-23D Step 3

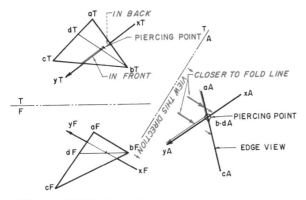

Figure 7-23E Step 4

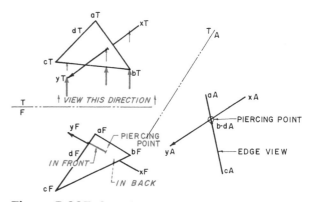

Figure 7-23F Step 5

GIVEN:

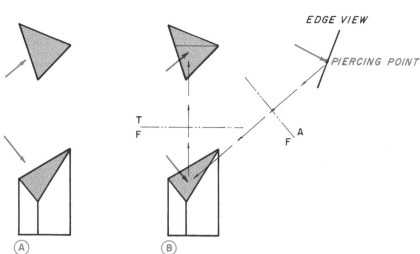

Figure 7-24
Determining the piercing point by construction

surface from the piercing point to point cA. Therefore, in the top view, line x-y is seen only from the piercing point to yT.

Step 5. By viewing back into the top view, determine the visibility of lines of the plane surface and line x-y in the front view, Figure 7-23F. Notice in the top view that line x-y is closer to the fold line from the piercing point to point yT than the edge view of the plane surface cT/bT; therefore, in the front view, line x-y is seen only from the piercing point to yF.

When the edge view does not appear in the regular views, it must be constructed using the preceding steps, Figure 7-24, Part A. Once the edge view is constructed, the piercing point can easily be seen. The piercing point is now projected down into the front view and up into the top view, Figure 7-24, Part B.

How To Determine the Piercing Point by Line Projection

An alternate method of determining the piercing point simply locates the path of a cutting plane that passes through the line, leaving a "scar" on the surface of the given plane. Since the line lies in the cutting plane, an intersection of the line and the scar on the given plane locates the piercing point. The end points of the scar are found by aligning the cutting plane with the given line in one view, where it crosses the boundary of the given plane in two places. Projecting the cutting plane intersection to the corresponding boundaries of the given plane in the next view, locates the scar end points in that view, and provides a view of the scar that the given line can cross at the piercing point.

Example:

See Figure 7-25, Part A.

GIVEN:

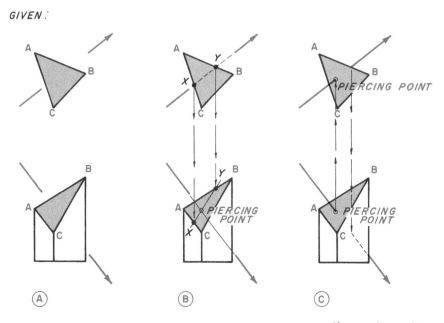

Figure 7-25
Determining piercing points by line projection (skewed surface)

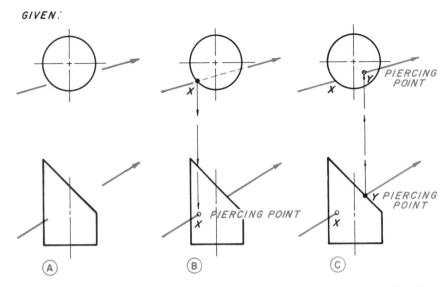

Figure 7-26 Determining piercing points by line projection (cylinder)

Given: Skewed surface abc, and a line.

Step 1. In the top view, lightly draw the piercing line and determine points X and Y where they cross the boundary of plane abc. The piercing point is located somewhere between these points.

Step 2. Project points X and Y down into the front view to the corresponding edges of the plane abc, Figure 7-25, Part B. Where line X-Y crosses the given line is the exact piercing point.

Step 3. Project the piercing point to the top view, Figure 7-25, Part C. Visibility of the piercing line is determined by inspection.

Figure 7-26, Part A, illustrates a cylinder pierced by a line. Lightly draw the piercing line in the top view and find point X. Project point X down into the front view, Figure 7-26, Part B, to find piercing point X. Because the piercing line exits the skewed surface, piercing point Y is found on the edge view of the

front view, Figure 7-26, Part C. Project piercing point Y up into the top view. Visibility of the piercing line is determined by inspection.

If the object is a cone, line segments must be located and drawn from the base of the cone up to the vertex of the cone.

Given: A cone with a piercing line, Figure 7-27, Part A.

Lightly draw the piercing line in the top view and find points X and Y, Figure 7-27, Part B. Project points X and Y down to the base of the cone. Project light line segments from points X and Y on the base up to the vertex of the cone. Where these lines cross the piercing line is the location of the two piercing points. Visibility of the piercing line is determined by inspection, Figure 7-27, Part C.

If the object is spherical, an imaginary flat surface on the sphere must be established along the piercing line.

GIVEN:

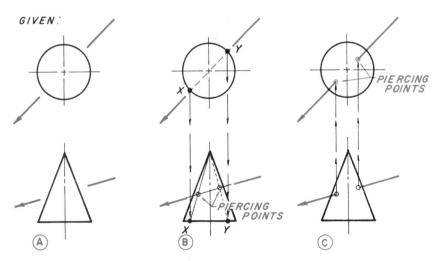

Figure 7-27 Determining piercing points by line projection (cone)

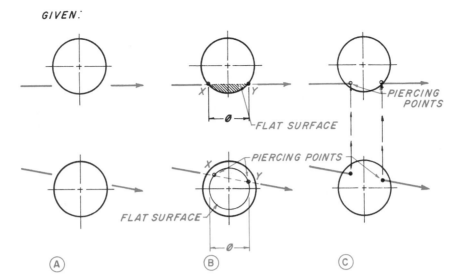

Figure 7-28 Determining piercing points by line projection (sphere)

Given: A sphere with a piercing line, Figure 7-28, Part A.

Lightly draw the piercing line in the top view. Where the piercing line intersects with the sphere is the location of the imaginary flat surface, Figure 7-28, Part B. This also establishes the diameter of the flat surface. Transfer the imaginary flat surface to the front view. Where the piercing line intersects the imaginary flat surface is the exact piercing point locations. Once the piercing point locations are found they are projected up into the top view, Figure 7-28, Part C. Visibility of the piercing line is determined by inspection.

How To Find the Intersection of Two Planes by Line Projection

The intersection of two planes can be determined by generating an edge view of one of the planes.

Another method, somewhat simpler, is the line-projection method.

Given: Two plane surfaces, Figure 7-29, Part A.

In the top view, locate points X and Y on edge ab of one of the planes. Project points X and Y down into the front view as illustrated in Figure 7-29, Part B, and draw a light line from X-Y in the front view. The piercing point is where line X-Y crosses the edge ab. Project points from edge ac of the same plane. Project these points down into the front view. Draw a line connecting these points; where they cross edge ac is the exact piercing point. Complete the views as illustrated in Figure 7-29, Part C.

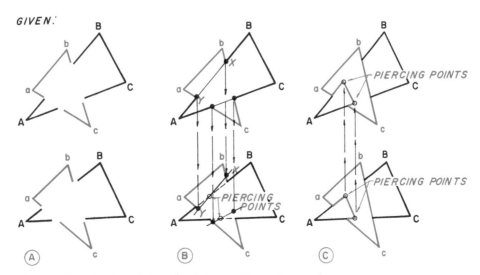

Figure 7-29 Determining the intersection of two planes

Figure 7-30 Determining the intersection of a cylinder and a plane surface

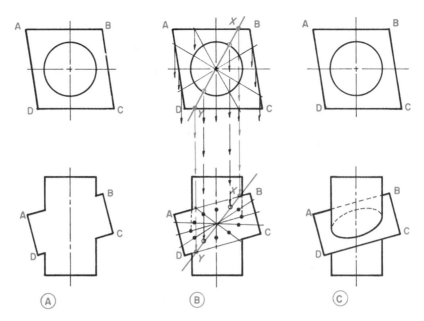

How To Find the Intersection of a Cylinder and a Plane Surface by Line Projection

The intersection of a cylinder and a plane surface can also be determined by generating an edge view of the plane surface. Another method is the line-projection method.

Given: A cylinder and a plane surface, Figure 7-30, Part A.

Divide the circle into equal parts; in this example, 12 equal parts. Extend these lines out to the edges of the plane surface, Figure 7-30, Part B. The example illustrated uses points 1 o'clock and 7 o'clock to

find points X and Y on the edge of the plane surface. Project points X and Y down into the front view. Draw a line from X-Y in the front view. Points 1 o'clock and 7 o'clock are located somewhere on this line. To find their exact locations, project points 1 and 7 from the top view to the line in the front view. Continue around the various points to locate each of the 12 points. Complete the drawing as shown in Figure 7-30, Part C.

How To Find the Intersection of a Sphere and a Plane Surface

Given: A sphere and a plane, Figure 7-31, Part A.

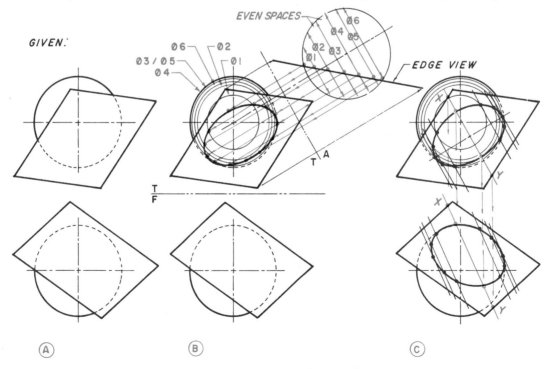

Figure 7-31 Determining the intersection of a sphere and a plane surface

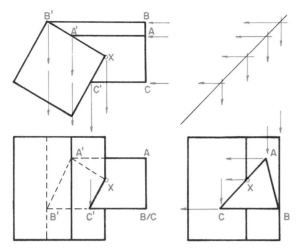

Figure 7-32 Finding the intersection of two prisms (method 1)

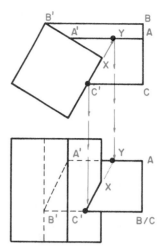

Figure 7-33 Finding the intersection of two prisms (method 2)

Construct an edge view of the plane, Figure 7-31, Part B. Divide the sphere into even spaces as illustrated. Think of each evenly spaced line as an imaginary sliced-off portion of the sphere. Draw each imaginary slice in the top view and where the projection from the corresponding point on the edge view cross is the piercing point. Complete the top view as illustrated. Once these points have been found in the top view, the line-projection method is used to transfer them to the front view, Figure 7-31, Part C.

How To Find the Intersection of Two Prisms (Method 1)

Given: The top, front, and right-side views of two intersecting prisms, Figure 7-32.

Label the various points as illustrated. Project each point to the 45° projection and into the next view and where they intersect is the location of each point, as illustrated. The only point in question is point X; using the three views, it can be easily found.

How To Find the Intersection of Two Prisms (Method 2)

Using only two views, point X creates a problem to locate point X in the front view. Refer to Figure 7-33. In the top view extend line C'-X to where it crosses line segment A-A'. Project point Y down into the front view to line A-A'. Draw a line from C' to Y to locate point X in the front view.

Review

1. Why is descriptive geometry used by a drafter?

2. Explain the basic steps involved in laying out the true shape of a plane.

3. What are notations and why are they so important?

4. What is a point view of a line?

5. Explain the basic steps involved in laying out the true distance between two parallel lines in space.

6. Explain the basic steps involved in laying out the true distance between two nonparallel lines.

7. What is an edge view?

8. How is a point called-out in a given view?

9. What is an important rule to remember in projecting from one view to another?

10. Explain the basic steps involved to find the true length of any line.

11. What is a fold line and what does it represent?

12. Explain the basic steps involved to find the point view of a line as projected from its true length?

Chapter Seven
Problems

The following problems are intended to give the beginning drafter practice in using the various principles of descriptive geometry. Problems 1 through 13 deal with each of the various principles used to solve actual design problems. Problems 14 through 32 apply these principles to develop required views of various objects.

The steps to follow in laying out problems 1 through 13 and 23 through 32 are:

1. On an 8½ x 11 sheet of paper with a sharp 4-H lead, locate all lines, points, and fold lines per the given dimensions.

2. Complete the problem per the given instructions. Project in the direction of the large arrow.

The steps to follow in laying out problems 14 through 27 are:

Step 1. Study the problem carefully.

Step 2. Using the given front view, make a sketch of all required views.

Step 3. Center the required views within the work area with a 1-inch (25-mm) space.

Step 4. Use light projection lines. Do *not* erase them.

Step 5. Lightly complete all views.

Step 6. Check to see that all views are centered within the work area.

Step 7. Check that there is a 1-inch (25-mm) space between all views.

Step 8. Carefully check all dimensions in all views.

Step 9. Darken in all views using correct line thickness.

Step 10. Recheck all work, and, if correct, neatly fill out the title block using light guidelines and neat lettering.

Problem 7-1

Locate line a-b and fold lines per given dimensions. Locate line a-b in the right-side view. Label all points in all views.

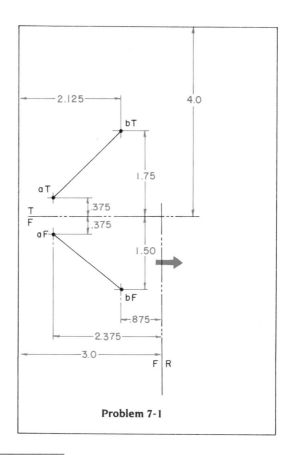

Problem 7-1

Problem 7-2

Locate line a-b and fold lines per given dimensions. Locate line a-b in the top view. Locate point c on line a-b in all views. Label all points in all views.

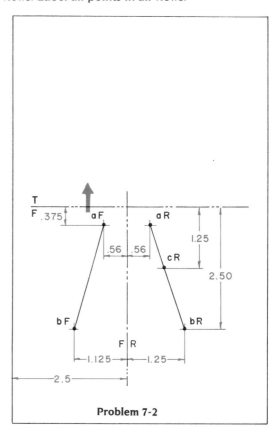

Problem 7-2

Problem 7-3

Locate line a-b and fold lines per given dimensions. Locate line a-b in the right view. Find the true length of line a-b, projecting from the front view. Label all points in all views.

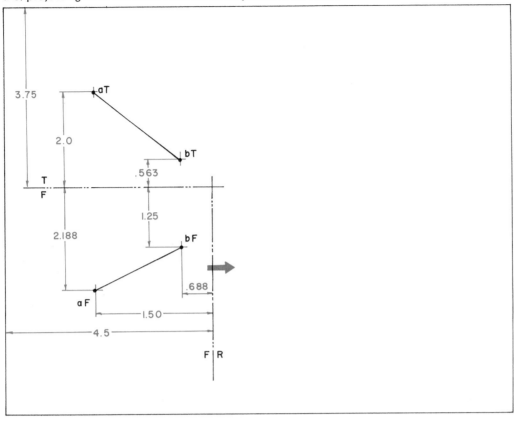

Problem 7-3

Problem 7-4

Locate the line a-b and fold lines per given instructions. Locate line a-b in the top view. Find the point view of line a-b, projecting from the top view. Label all points in all views.

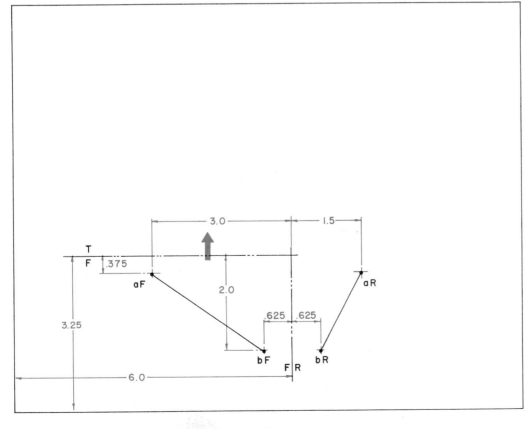

Problem 7-4

Problem 7-5

Locate line a-b, point c, and fold lines per given dimensions. Find the true distance from line a-b to point c. Project from the right view. Label all points in all views.

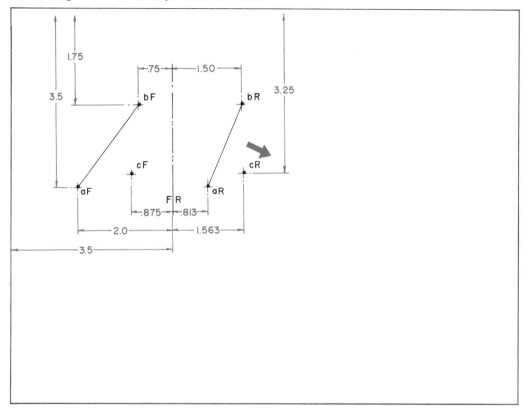

Problem 7-6

Locate parallel lines a-b and c-d, and fold lines per given dimensions. Find the true distance between lines a-b and c-d. Project from the front view. Label all points in all views.

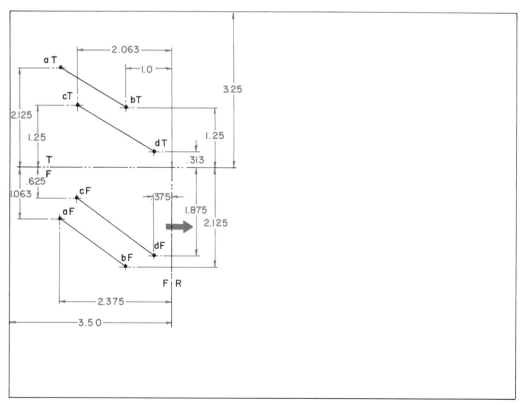

Problem 7-7

Locate lines a-b and c-d, and fold lines per given instructions. Find the true distance between lines a-b and c-d. Project from the top view. Label all points in all views.

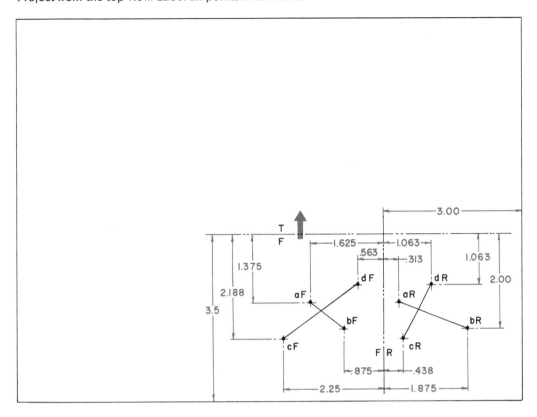

Problem 7-7

Problem 7-8

Locate points a, b, and c, and fold lines per given dimensions. Connect points a, b, and c to form a plane. Project plane abc into the top view. Label all points in all views.

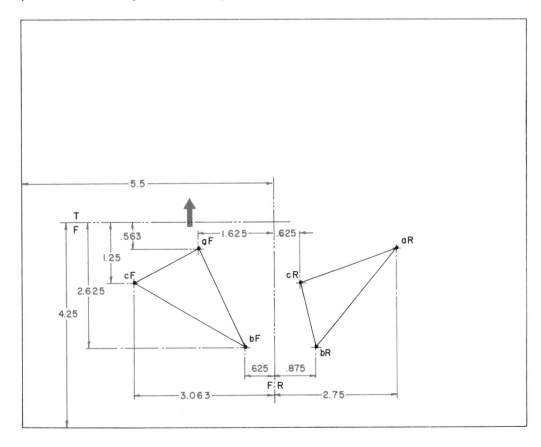

Problem 7-8

Problem 7-9

Locate points a, b, c, and d, point x, and fold lines per given dimensions. Connect points a, b, c, and d to form a plane. Project plane abcd into the right view; locate point x in all views. Label all points in all views.

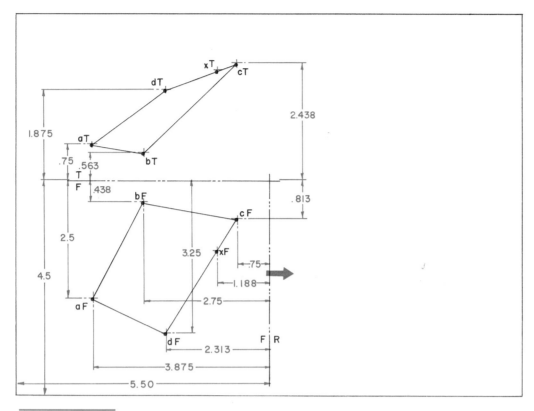

Problem 7-9

Problem 7-10

Locate points a, b, c, and fold lines per given dimensions. Connect points a, b, and c to form a plane. Draw a line from point cF, parallel to fold line F-R; label this line cF-xF. Find the edge view of plane surface abc. Project from the right view. Label all points in all views.

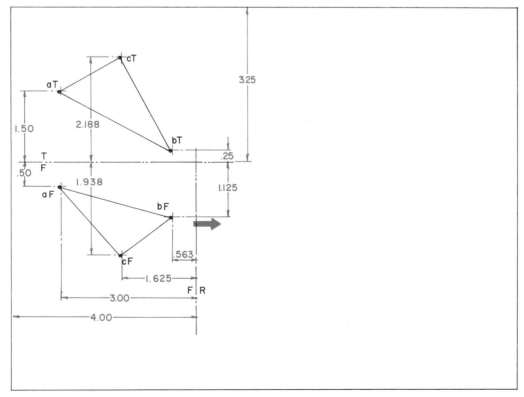

Problem 7-10

Problem 7-11

Locate points a, b, and c, and fold lines per given dimensions. Connect points a, b, and c to form a plane. Draw a line from point bF, parallel to fold line F-T; label this line bF-xF. Find the edge view of the plane abc and project from the front view. Label all points in all views.

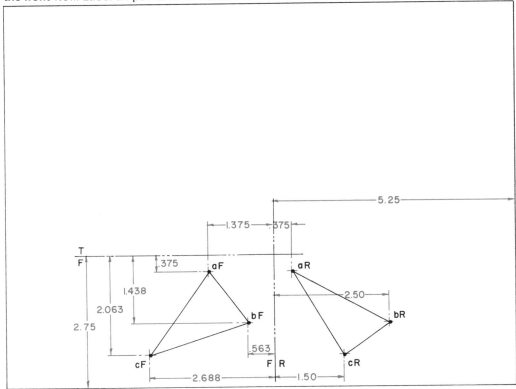

Problem 7-11

Problem 7-12

Locate points a, b, and c, point x, and fold lines per given dimensions. Connect points a, b, and c to form a plane. Find the true distance between plane surface abc, and point x. Project from the front view. Label all points in all views.

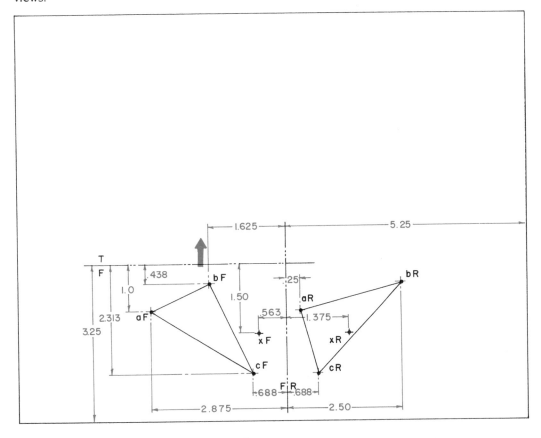

Problem 7-12

Problem 7-13

Locate points a, b, c, and d, and fold lines per given dimensions. Connect points a, b, c, and d to form a folded object. Find the true angle between the plane surfaces as viewed directly along the fold, a-b. Project from the front view. Label all points in all views.

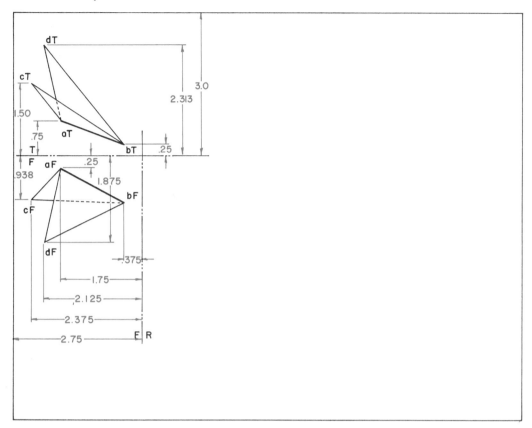

Problem 7-13

Problems 7-14 through 7-22

Draw the front view, top view, right view, and auxiliary view per the listed steps.

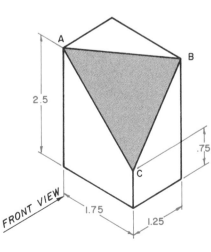

Problem 7-14

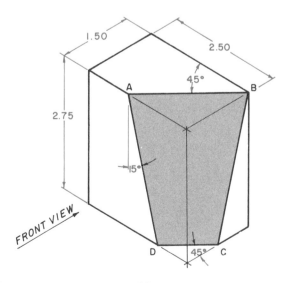

Problem 7-15

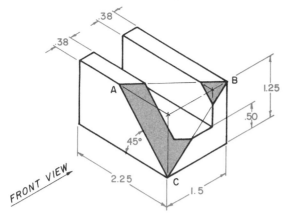

.38
.38
A
B
1.25
45°
.50
2.25
C
1.5
FRONT VIEW

Problem 7-16

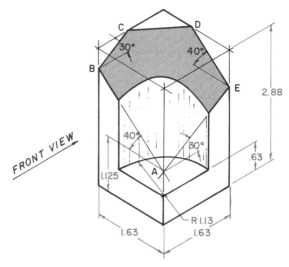

C
D
30°
40°
B
E
2.88
40°
30°
1.125
A
.63
FRONT VIEW
1.63
R1.13
1.63

Problem 7-19

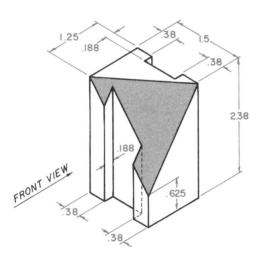

1.25
.188
.38
1.5
.38
2.38
.188
.625
FRONT VIEW
.38
.38

Problem 7-17

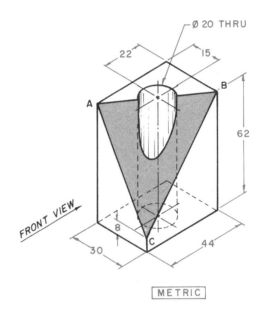

Ø 20 THRU
22
15
A
B
62
40
8
C
44
30
FRONT VIEW

METRIC

Problem 7-20

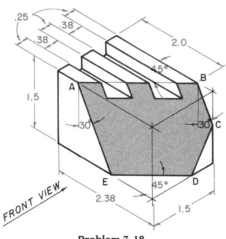

.25
.38
.38
2.0
45°
A
B
1.5
30
30 C
E
45°
D
2.38
1.5
FRONT VIEW

Problem 7-18

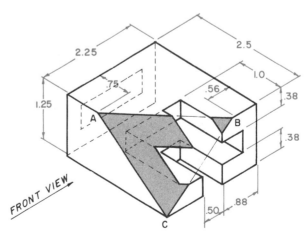

2.25
2.5
.75
1.0
1.25
.56
.38
A
B
.38
FRONT VIEW
.50
.88
C

Problem 7-21

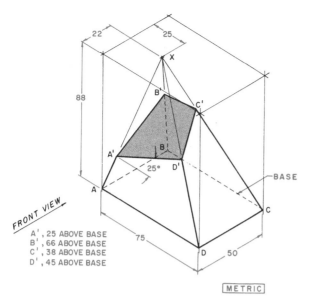

FRONT VIEW

A', 25 ABOVE BASE
B', 66 ABOVE BASE
C', 38 ABOVE BASE
D', 45 ABOVE BASE

METRIC

Problem 7-22

Problems 7-24 and 7-25

Draw the front view, top view, right view, and auxiliary view per the listed steps.

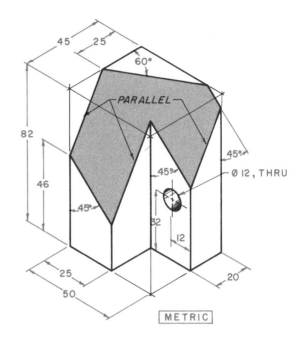

PARALLEL

Ø 12, THRU

METRIC

Problem 7-24

Problem 7-23

Draw the front view, top view, right view, and required auxiliary views per the listed steps. Find the true angle at abc, as viewed along the fold.

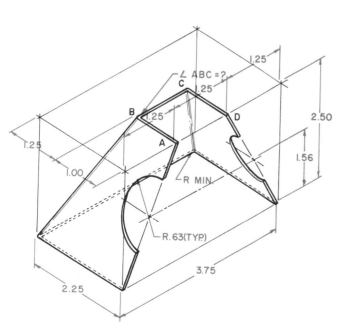

∠ ABC = ?

R MIN

R .63 (TYP.)

Problem 7-23

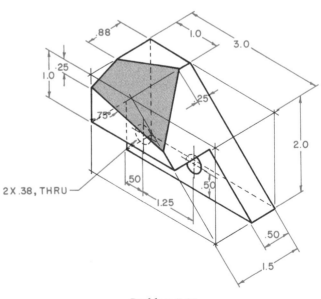

2 X .38, THRU

Problem 7-25

Problems 7-26 and 7-27

Draw the front view, top view, right view, and required auxiliary views per the listed steps. Find the true angle between the various surfaces as viewed along the fold.

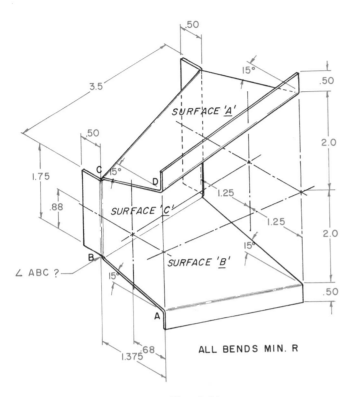

Problem 7-26

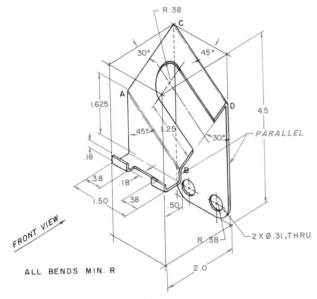

Problem 7-27

Problem 7-28

Complete the front view using the line-projection method. Draw the correct visibility of all lines in the front view.

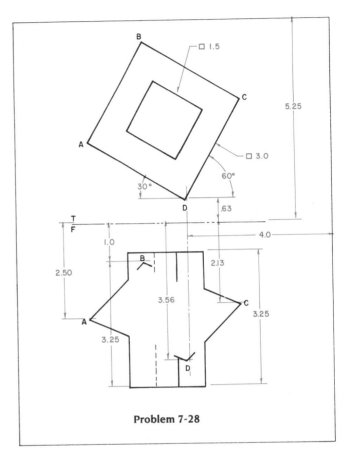

Problem 7-28

Problem 7-29

Complete the top and front views using the line-projection method. Draw the correct visibility of all lines by inspection.

Problem 7-30

Complete the front view by inspection and projection. Draw the correct visibility of all lines.

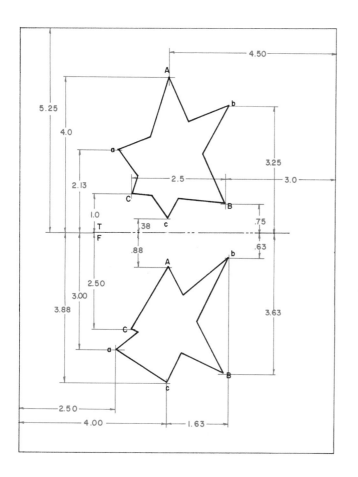

Problem 7-29

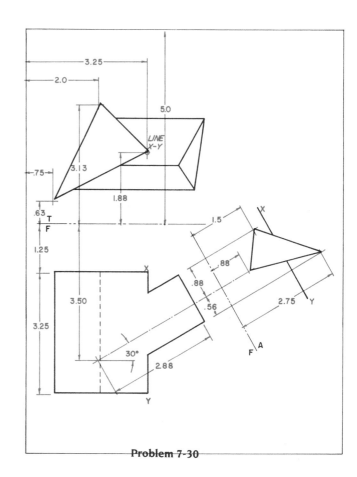

Problem 7-30

Problem 7-31

Complete the front view using the line-projection method. Draw the correct visibility of all lines.

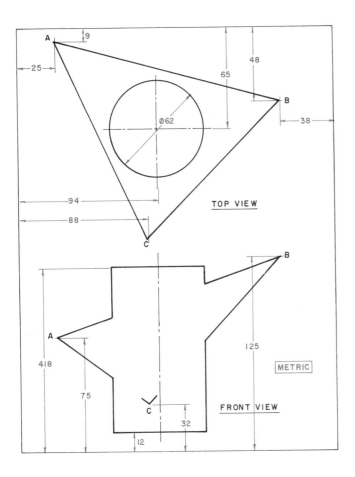

Problem 7-31

Problem 7-32

Complete the top and front views using the line-projection method. Do not use the right-side view for solving the problem. Draw the correct visibility of all lines.

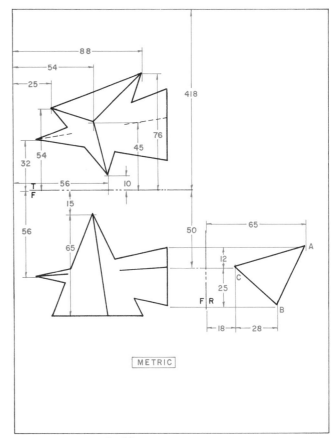

Problem 7-32

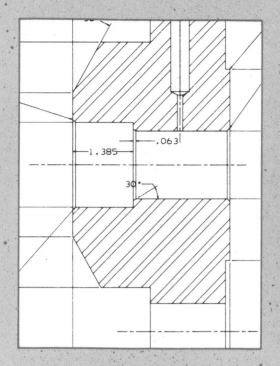

The methods and techniques learned in Chapters 3, 4, 6 and 7 are used in solving sheet metal development problems. This chapter gives the student an opportunity to learn the use of parallel line, radial line, and triangulation developments. Laps, seams, and tabs are studied also, to determine how many are required and the best location for each. Extensive study is done on bend allowances, using the bend allowance charts found in the Appendix.

CHAPTER EIGHT

Patterns and Developments

Developments

A *development* is the pattern or template of a shape that is laid out in a single flat plane in preparation for the bending or folding of a material to a required shape. Surface developments are used in many different industries. Some examples of objects requiring developments are cereal boxes, tool boxes, funnels, air-conditioning ducts, and simple mail boxes.

Three major kinds of surface developments are parallel line development, radial line development, and triangulation development. *Parallel line developments* are used for objects having parallel fold lines. *Radial line developments* are those whose fold lines radiate from one point. See Figure 8-1. *Triangulation developments* are the development process of breaking up an object into a series of triangular plane surfaces, Figure 8-2. Each kind of development is explained here in full.

Surfaces

A *development surface* is the exterior and/or interior of the sheet material used to form an object. The various kinds of surfaces include plane surface, single-curved surface, and double-curved surface.

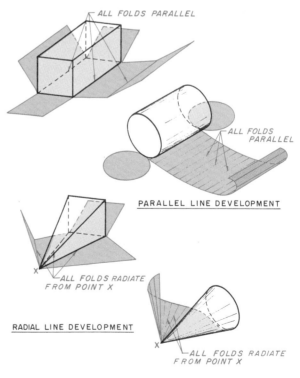

Figure 8-1 Parallel line development and radial line development

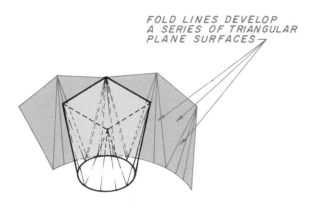

FOLD LINES DEVELOP
A SERIES OF TRIANGULAR
PLANE SURFACES

RADIAL LINE DEVELOPMENT

Figure 8-2 Triangular development

If any two points anywhere on a surface are connected to form a straight line, and that line rests upon the surface, it is a *plane surface*. If all points on a surface can be interconnected to form straight lines with-out exception, it is a *flat-plane surface*. The top of a drawing board is an example of a flat-plane surface. A flat-plane surface can have three or more straight edges. Such objects as a cube or a pyramid are bounded by plane surfaces. If a surface can be unrolled to form a plane, it is a *single-curved surface*. A cylinder or a cone is an example of a *single-curved surface*.

A surface that cannot be developed because it is neither a plane surface nor a curved surface is a *warped surface*. An automobile fender is an example of a warped surface. This kind of surface is usually stamped or pressed into shape. An object fully formed by curved lines with no straight lines is a *double-curved surface*. A sphere is an example of a double-curved surface. This type of surface cannot be developed exactly by using flat patterns; only an approximate development can be made.

Laps and Seams

Extra material must be provided for laps and seams. Many kinds of seams are available, Figure 8-3.

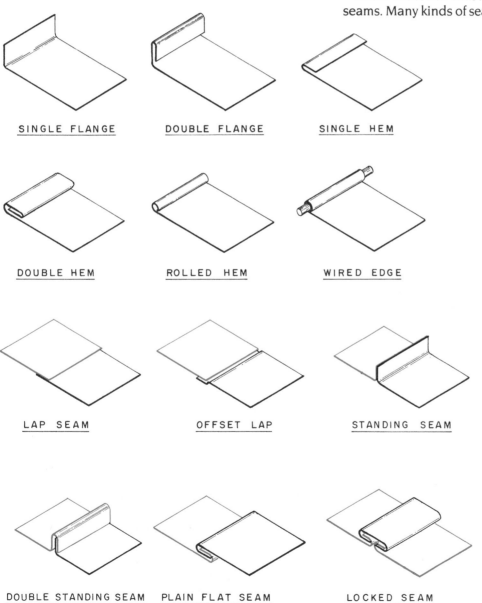

SINGLE FLANGE   DOUBLE FLANGE   SINGLE HEM

DOUBLE HEM   ROLLED HEM   WIRED EDGE

LAP SEAM   OFFSET LAP   STANDING SEAM

DOUBLE STANDING SEAM   PLAIN FLAT SEAM   LOCKED SEAM

Figure 8-3 Many types of seams are available

 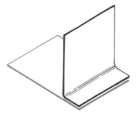 

SINGLE FLANGE CORNER   SINGLE FLANGE CORNER   OVERLAP FLANGE CORNER

DOUBLE FLANGE CORNER

Figure 8-3 (Continued)

When making a choice, the drafter must take into account the thickness of the metal so that crowding of the metal at the joints is not a problem. Allowance must be made for the gluing, soldering, riveting or welding processes for joints. The method of fastening joints together varies with the material and, accordingly, the elimination of rough or sharp edges. Tabs for the drawing problems at the end of this chapter should be designed to support the joining surface, Figure 8-4.

Design Practices

The drafter should lay out developments according to the dimensions of stock materials for economy and the best use of materials and labor. The stock material area required to cut a pattern should be kept to the smallest convenient size. It is good practice to put the seam at the shortest joint, and to attach tops and bottoms along the longest possible seam or bend to reduce the length requiring soldering, riveting or welding. It is assumed that the inside surface of the final object is the side that the pattern defines. (The important dimension sizes are usually the inside surfaces.) Fold or bend locations in the material are shown on the pattern with thin solid lines, and are locations that are also assumed to occur on the inside surface of the final object.

Thickness of Material

The actual thickness of sheet metal is specified by gage numbers. Each gage number indicates a partic-

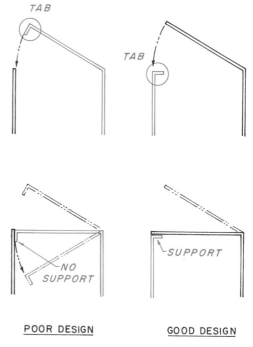

Figure 8-4 Tabs should support the
joining surfaces

ular thickness of material. Figure 8-5 lists the gage number and gage sizes for sheet and plate steel. These are nominal thicknesses, subject to permissible tolerances. Today, there is considerable confusion when using the gage numbering system.

In 1893, Congress established the United States Standard Gage, and it was primarily a *weight* gage rather than a *thickness* gage. It was derived from the weight of wrought iron. At that time, the weight of

GAGE	ALUMINUM		BRASS		STEEL	
	THICKNESS	WT./SQ. FT.	THICKNESS	WT./SQ. FT.	THICKNESS	WT./SQ. FT.
8	.1285	1.812	.1285	5.662	.1644	6.875
9	.1144	1.613	.1144	5.041	.1494	6.250
10	.1019	1.440	.1019	4.490	.1345	5.625
11	.0907	1.300	.0907	3.997	.1196	5.000
12	.0808	1.160	.0808	3.560	.1046	4.375
13	.0720	1.020	.0720	3.173	.0897	3.750
14	.0641	.907	.0641	2.825	.0747	3.125
15	.0571	.805	.0571	2.516	.0673	2.812
16	.0508	.720	.0508	2.238	.0598	2.500
17	.0453	.639	.0453	1.996	.0538	2.250
18	.0403	.580	.0403	1.776	.0478	2.000
19	.0359	.506	.0359	1.582	.0418	1.750
20	.0320	.461	.0320	1.410	.0359	1.500
21	.0285	.402	.0285	1.256	.0329	1.375
22	.0253	.364	.0253	1.119	.0299	1.250
23	.0226	.318	.0226	.996	.0269	1.125
24	.0201	.289	.0201	.886	.0239	1.000
25	.0179	.252	.0179	.789	.0209	.875
26	.0159	.224	.0159	.700	.0179	.750
27	.0142	.200	.0142	.626	.0164	.688
28	.0126	.178	.0126	.555	.0149	.625
29	.0113	.159	.0113	.498	.0135	.562
30	.0100	.141	.0100	.441	.0120	.500
31	.0089	.126	.0089	.392	.0105	.438
32	.0080	.112	.0080	.353	.0097	.406
33	.0071	.100	.0071	.313	.0090	.375
34	.0063	.089	.0063	.278	.0082	.344
35	.0056	.079	.0056	.247	.0075	.312
36	.0050	.071	.0050	.220	.0067	.281
37	.0045	.063	.0045	.198	.0064	.266
38	.0040	.056	.0040	.176	.0060	.250
39	.0035	.050	.0035	.154	-	-
40	.0031	.044	.0031	.137	-	-

SHEET METAL

Figure 8-5 Thickness of material

wrought iron was calculated at 480 pounds per cubic foot; thus, a plate 12 inches square and 1 inch thick weighs 40 pounds. A No. 3 U.S. gage represents a wrought iron plate weighing 10 pounds per square foot. Therefore, if a weight per square foot 1 inch thick is 40 pounds, the plate thickness for a No. 3 gage equals 10 ÷ 40 = 0.25 inch, which is the original thickness equivalent for a No. 3 U.S. gage. Since this and all other gage numbers were based on the weight of wrought iron, they are not correct for steel. To add to the confusion, there is considerable variation in the gage thickness for different kinds of

material. For example, a gage used for such nonferrous materials as brass and copper is sometimes also used to specify a thickness for steel or vice versa.

Today, to help eliminate the problems, the decimal or metric system of indicating gage size is now replacing the older gage size numbering system.

Parallel Line Development

In parallel line developments all fold lines are parallel. (Refer back to Figure 8-1.)

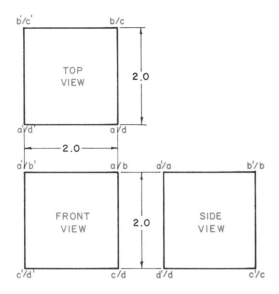

TYPICAL THREE-VIEW
DRAWING

Figure 8-6 Simple multiview drawing
of a cube

A three-view drawing of a simple 2-inch cube is
shown in Figure 8-6. A cube is a specific kind of prism.
Notice that all the fold lines that form the lateral
surfaces of a prism are parallel, Figure 8-7A. If the
cube were to be unfolded, it would appear as it does
in Figures 8-7B, 8-7C, and 8-7D. The finished development is shown in Figure 8-7E.

Notice that the cube was developed or laid out
from a baseline, sometimes referred to as the *stretchout
line*. Fold lines are always oriented 90° from the
baseline, as illustrated. If the pattern were to be cut
from a flat sheet of material, tabs would be needed
to fasten the cube together. Tabs must be positioned

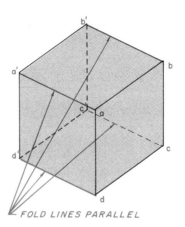

Figure 8-7A All fold lines are
parallel

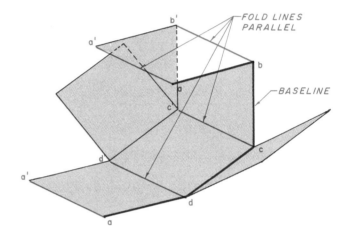

Figure 8-7B Cube as it unfolds

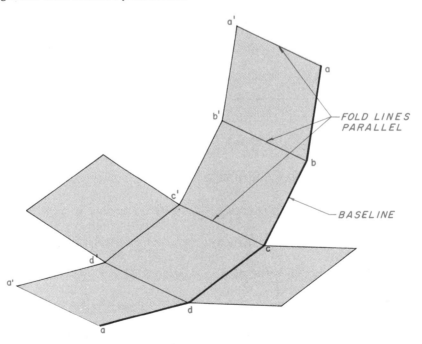

Figure 8-7C Cube as it unfolds further

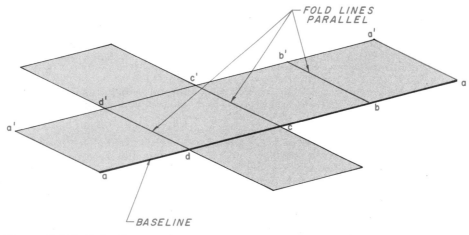

Figure 8-7D Cube flattened out

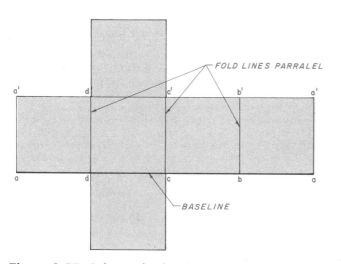

Figure 8-7E Cube in the finished development

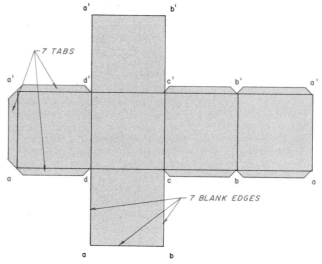

Figure 8-7F Cube flattened out with tabs added

so as to make each meet with an untabbed edge, Figure 8-7F. Notice that there are seven tabs and seven blank edges. Each tab folds to meet a particular blank edge.

How To Develop a Truncated Prism Using Parallel Line Development

Given: A prism with a front view, top view, and auxiliary view, Figure 8-8A.

Step 1. Locate the baseline, which must always be 90° from the parallel fold lines. Label each fold line clockwise in alphabetical or numerical order. Always start with the shortest fold line, Figure 8-8B. Either fold line a-a' or b-b' would be eligible, and a-a' is selected. Notice that the starting fold line is also the finishing fold line, and each is labeled at the same location. The finishing fold line is a-a'. All real (or true) lengths of the fold lines are seen in the front view and the real distances between their locations are seen in the top view.

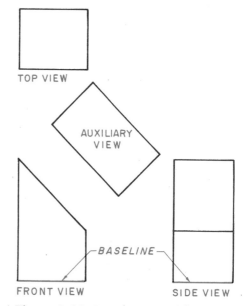

Figure 8-8A Development of a truncated prism using parallel line development

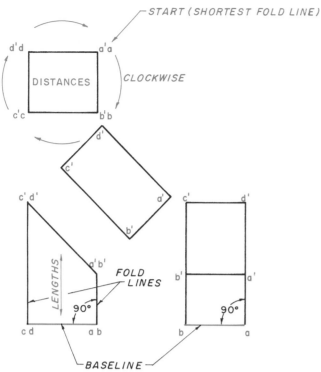

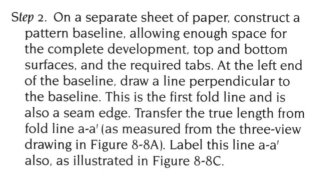

Step 2. On a separate sheet of paper, construct a pattern baseline, allowing enough space for the complete development, top and bottom surfaces, and the required tabs. At the left end of the baseline, draw a line perpendicular to the baseline. This is the first fold line and is also a seam edge. Transfer the true length from fold line a-a' (as measured from the three-view drawing in Figure 8-8A). Label this line a-a' also, as illustrated in Figure 8-8C.

Step 3. From the top view of Figure 8-8A, obtain the true distance from the first fold line location at a-a' to the second fold line location at b-b'. On the pattern baseline, transfer this distance, measuring from the first fold line a-a', and draw the next fold line parallel to the first fold line. Label it b-b', Figure 8-8D.

Step 4. Repeat the first three steps, alternating true fold line lengths and true distances between fold lines until all the true lengths and true distances are transferred. Each of the true

Figure 8-8B Step 1

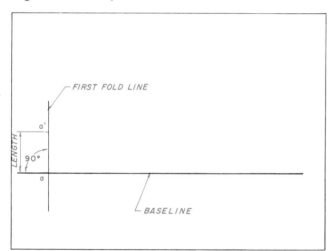

Figure 8-8C Step 2

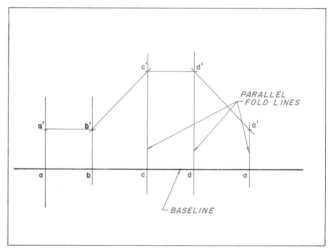

Figure 8-8E Step 4

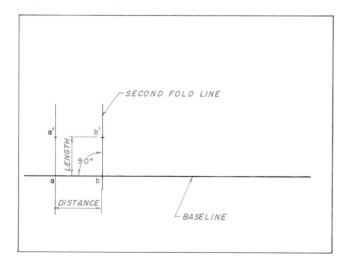

Figure 8-8D Step 3

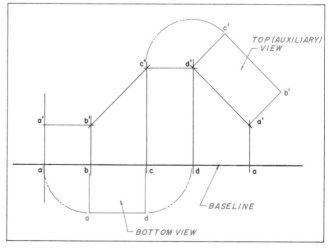

Figure 8-8F Step 5

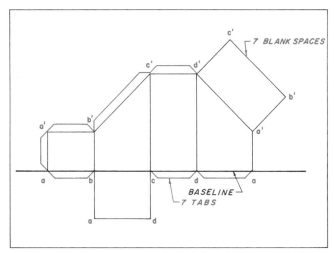

Figure 8-8G Step 6 completed development of the prism

lengths a-a' through e-e' are transferred from the front view, and the true distances between them are found from the top view of Figure 8-8A. connect the points as illustrated in Figure 8-8E.

Step 5. The edge selected for attachment of the top surface should be made to keep the rectangle size from which the whole pattern is cut as small as possible. Line a'b' would be the best, except an acute angle cutout c'b'a' is difficult to make. The fold line a'd' is next best, as shown. From Figure 8-8A, transfer the true size and shape of the top surface (auxiliary view) and the bottom surface, Figure 8-8F.

Step 6. Add tabs enough to fasten the edges. The locations should be selected so as not to enlarge the material area needed for the pattern cutout, Figure 8-8G. Check to verify that the number of tabs equals the number of blank edges in the same manner as is illustrated in Figure 8-7F. This completes the parallel line development of the object in Figure 8-8A.

How To Develop a Truncated Cylinder Using Parallel Line Development

Given: A cylinder with a front view, top view, and auxiliary view, Figure 8-9A. In developing a cylinder, the exact procedure is used as when developing a prism. Because the cylinder has a rounded surface, line segments called *station lines* must be assumed or chosen on the lateral surface. This is explained in Step 1.

Step 1. Locate a baseline which must always be 90° from the fold lines, sometimes referred to as *parallel station lines*. The bottom corner of the lateral surface is selected as a convenient location because it already exists. In the view where

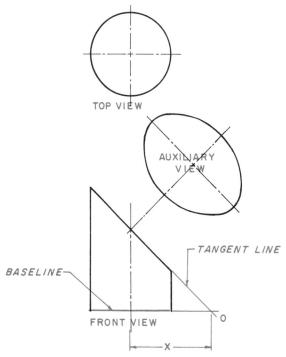

Figure 8-9A Development of a truncated cylinder using parallel line development

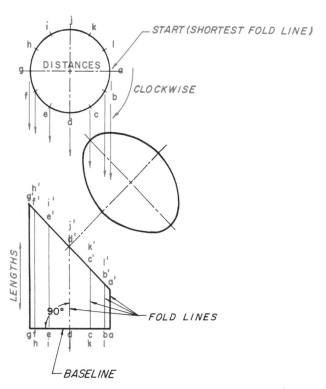

Figure 8-9B Step 1

the cylinder appears as a diameter (the top view in this example), divide the diameter into equal divisions, Figure 8-9B. This example uses 12 equal divisions, but any number of appropriate equal divisions would do. These divisions

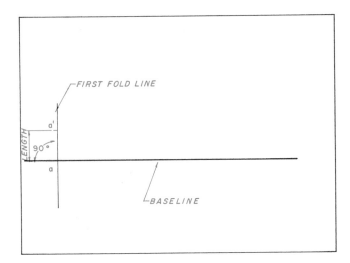

Figure 8-9C Step 2

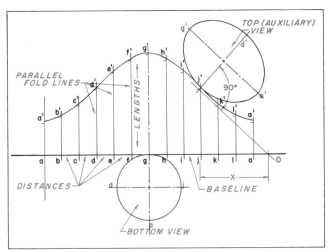

Figure 8-9D Steps 3 and 4

locate fold line positions, which are used only for construction and layout purposes. Starting with the shortest fold line, consecutively label each division; in this example, a through l. Note that a-a' marks the beginning and ending of the same line which meets from opposite directions. The true "distances" between station lines can be approximated in this top view. Project each point into the next (front) view in order to find the true "lengths" of the fold lines. It should be reemphasized that the fold lines must be 90° from the baseline.

Step 2. On a separate sheet of paper, construct a pattern baseline allowing enough space for the complete development, both ends, and the required tabs. At the left end of the pattern baseline draw a line perpendicular to the baseline. This is the first fold line location a-a'. Transfer the true length a-a' from the front view of Figure 8-9B. Label this line as illustrated in Figure 8-9C.

Step 3. Two methods are used for finding the distances between station lines. Method A is an approximate method that assumes the cylinder to be a prism, whereas Method B is mathematically correct.

Method A: From the top view of Figure 8-9B, obtain the direct chordal distance from the first station line a-a' to the second fold line b-b'. On the pattern baseline, draw the second fold line parallel to the first station line a-a' at the chordal distance found from the top view, and label it b-b', Figure 8-9D. Repeat these steps, alternating chordal lengths and true distances until all station lines have been located and drawn. Note that the last fold line length a-a' is a duplicate of the first.

Method B: The required length to form the cylinder's circumference can be calculated using the following formula: circumference = cylinder diameter x pi. (Pi = 3.1416, approximately.) The baseline is extended to this length, and divided into the same number of divisions as was done at Step 1 in the top view of the cylinder, Figure 8-9B. The division can best be performed as a construction exercise, rather than mathematically. Refer back to Chapter 3, "Geometric Construction." Each of the 13 division locations (from 12 spaces) is the correct location of the required station lines, and can be successively labeled.

Step 4. Add the true size and shape of the bottom surface (bottom view), and the top surface (auxiliary view). The attachment locations of these surfaces are particularly difficult, as the mathematical point of tangency is not easily isolated on the pattern. The top surface location is chosen to allow the widest access of cutting to this point. The bottom surface can be added to any of the fold lines. In order to locate the top surface, the distance X from center to point 0 is transferred from Figure 8-9B to the development, and projected 90° as illustrated in Figure 8-9D.

Step 5. Add the tabs necessary for edge attachments. In cylinders, a notched tab must be used in order to accommodate crowding by surface curvature. A tab must also be added at fold line a-a' in order to attach edge a-a' to the other edge a-a'; the top (auxiliary) and bottom fold to meet the notched tabs, Figure 8-9E.

Regardless of the shape, whether it be a cylinder or a prism, parallel line developments are used with

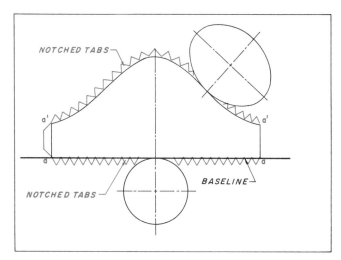

Figure 8-9E Step 5 completed development of the cylinder

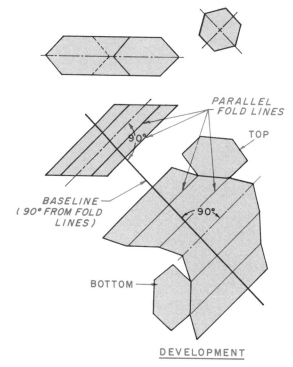

Figure 8-10 Creating a new baseline

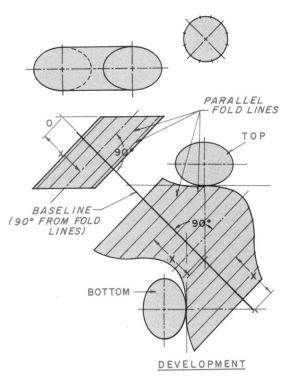

Figure 8-11 Using station lines instead of fold lines

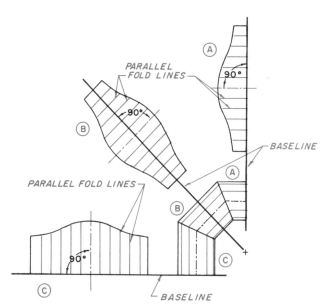

Figure 8-12 Series of patterns used to form an elbow

essentially the same procedure. Figure 8-10 shows a prism truncated at both ends, which necessitates creating a new baseline other than a part edge to ensure that all fold lines (instead of station lines) are at 90° to the baseline. Figure 8-11 is a cylindrical prism, truncated at each end, which requires the selection of station lines instead of using existing fold lines. Figure 8-12 shows a series of truncated cylinders, similar in construction to the preceding views, whose patterns are used to form an elbow.

Radial Line Development

Radial line development is different from parallel line development in that all fold lines or line segments radiate from one point. (Review Figure 8-1.) As in all patterns, true lengths and true distances must be used in laying out the developments.

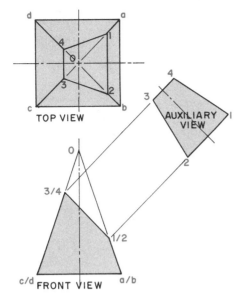

Figure 8-13A Development of a pyramid using radial line development

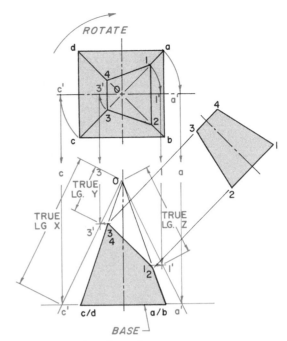

Figure 8-13B Step 1

How To Develop a Truncated Pyramid Using Radial Line Development

Given: A two-view drawing, and an auxiliary view of a pyramid with the top cut off at an angle (truncated), Figure 8-13A. Label points clockwise starting with the shortest fold line. In this example, either a-1 or b-2 could be used. Line a-1 is selected, but the fold line location a-1 is also the meeting corner of edges a-1 and a-1. True distances between the end points of the fold lines are evident in the top view. For example, a-b, b-c, c-d, and d-a are true length.

Step 1. To find the true lengths of the fold lines, they must be rotated to a position that is parallel to the frontal viewing plane, Figure 8-13B. In the top view, point a is rotated as shown, and projected into the front view. This revolved line from a' to 0 is its true length. Similarly, point c in the top view is rotated and projected into the front view, giving the true length of c'-0. The same procedure is used to find the true lengths 3'-0 and 1'-0. To construct a development of this object, continue with the following steps.

Step 2. Using the true length from 0 to a (a'-0), swing the arc as shown in Figure 8-13C. On this arc, mark off as chord lengths the true distances between the fold line end points a to b, b to c, c to d, and d to e, as transferred from the top view of Figure 8-13B. Label all points as shown. Connect the fold lines a to 0, b to 0, c to 0, d to 0, and a to 0.

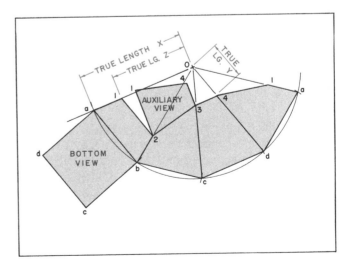

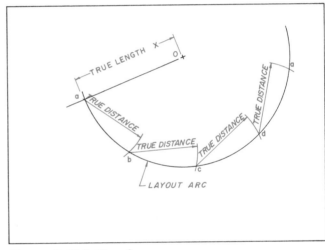

Figure 8-13C Step 2

Figure 8-13D Step 3

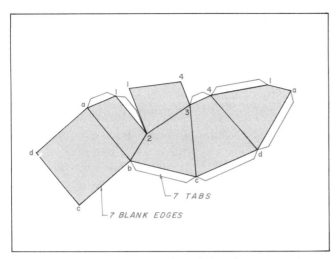

Figure 8-13E Step 4 completed development of a pyramid

Step 3. Using the true lengths found in Figure 8-13B, locate points 1, 2, 3, 4, and 1 on the respective fold lines 0-a, 0-b, 0-c, and 0-a. Note that 0-1 is the same length as both 0-2 and 0-1, and 0-3 is the same length as 0-4. Add the auxiliary and bottom views, transferring them directly from the original drawing 8-13A. See Figure 8-13D.

Step 4. Add the required tabs as illustrated in Figure 8-13E. Check to be sure the number of tabs equals the number of blank edges. In this example, there are seven tabs and seven blank spaces.

How To Develop a Truncated Cone Using Radial Line Development

As with cylindrical parallel line developments, station lines are line segments which must be positioned on the lateral surface in order to draw a development.

Given: A two-view drawing of a round cone, with the top cut off at an angle (truncated), and having no fold lines, Figure 8-14A. It is necessary to first complete the given views, and to draw a true view of the inclined flat surface. To project the flat surface to the top view, the front view is sliced at points A through G. These seven slices are projected into the top view and appear as circles, correspondingly labeled A through G. The points of intersection of the edge view with each slice is projected into the top view. To construct the auxiliary view, project from the seven point intersections perpendicular from the edge view of the flat surface to any convenient distance for the auxiliary view. Draw and use the center line as a reference line, and transfer all point-to-center line distances from the top view to the auxiliary view; refer to reference distances X.

Step 1. To construct a development of this object, station lines need to be established, and their true lengths and true distances must be determined. In the top view of Figure 8-14B, divide the base circle into 12 equal parts, and label each point. In this example, numbers 1 through

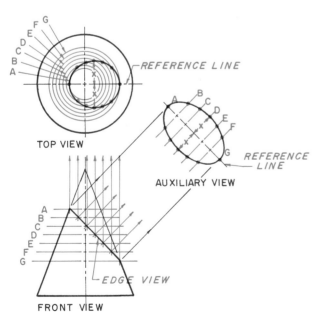

Figure 8-14A Development of a truncated cone using radial line development

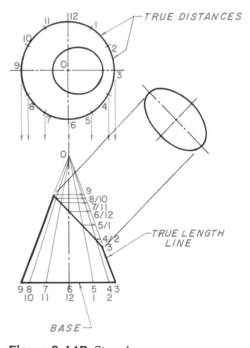

Figure 8-14B Step 1

12 are arbitrarily used, positioned as illustrated. The station line 0-3 must have two identities as two edges will meet at this point. The true distances between station lines are located between their end points on the cone base, seen in the top view. Project the 12 points down into the base of the front view. These are the end points of the station lines in the front view. Draw a line from each of these end points up to point 0.

Step 2. There are two methods of drawing the pattern outline. Each begins with swinging an arc whose radius is the true length of the lateral distance from the apex to the base, or length 0-3 or 0-9. The compass radius can be set to this distance on Figure 8-14B. This distance can also be derived mathematically if the diameter of the base of the cone is known, and the altitude (perpendicular distance from base to apex) is known. The formula is:

radius = the square root of [(½ cone base diameter)² + (altitude)²]

After swinging this arc, choose Method A, which approximates the cone as a pyramid, or Method B, which is mathematically correct.

Method A: In the top view of Figure 8-14B, the direct chordal distance between any two station line end points, say 5 and 6 (they are all equal), is an approximation of the arc length between them. On the 0-3 radius just drawn, strike off this chordal distance between station line end points repeatedly to equal the same number of spaces on the arc as the top view of the cone was divided into in Step 1, and ending at point 3, Figure 8-14C. Connect the station line end points from point 0 to each of these arc intersections, and proceed to Step 3.

Method B: The central angle of the pattern is determined by a pattern arc length needed to equal the circumference of the base of the cone. The ratio of the central angle (A°) to a full-circle 360° is the same as the circumference of the base of the cone is to the full circumference generated by the pattern's radius.

$$\frac{A°}{360°} = \frac{\text{circumference of base of cone}}{\text{circumference of pattern}}$$

$$\frac{A°}{360°} = \frac{\text{pi} \times \text{cone diameter at base}}{\text{pi} \times \text{pattern diameter}}$$

$$\frac{A°}{360°} = \frac{\text{cone base diameter}}{2 \times \text{pattern radius}}$$

$$A° = \frac{180 \times \text{cone base diameter}}{\text{pattern radius (a'−0)}}$$

After the central pattern radius A° is found, the angle between successive station lines (a°) is found by dividing the same number of conic divisions done in Step 1 into A°. The chordal lengths (D) of these divisions is found by using the formula:

$$D = 2 \times \text{pattern radius} \times [\sin(½ a°)]$$

This chordal distance can then be struck off on the pattern radius to form the same number of spaces as the conic divisions in Step 1, and labeled accordingly. See Figure 8-14C. Connect these station line end points from point 0 to each of the intersections, and proceed to Step 3.

Step 3. To draw the edge of intersection that the cone's lateral surfaces makes with the upper flat surface, true lengths from the apex to this edge at each station line must be determined. Only 3'-0 and 9'-0 are visible as true lengths in the front view of Figure 8-14B. To find the other true lengths, each segment must be rotated to a position parallel to the frontal viewing plane;

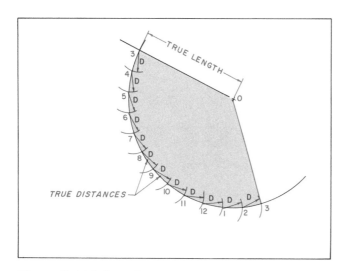

Figure 8-14C Step 2

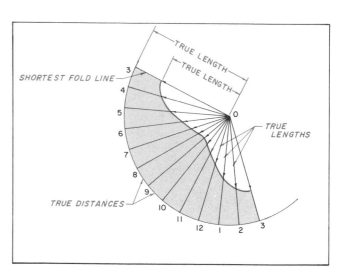

Figure 8-14D Step 3

in this example, on line 3-0. Project the inter-section of each station line at the truncated surface edge to line 3-0 on the cone profile in the front view (on line 3-0).

Starting with the shortest line segments of the cone, swing true lengths arc 0-3' on the corresponding pattern station lines, as illustrated in Figure 8-14D. Swing the apex-to-edge true length found for each station point, at the corresponding station line on the pattern. Repeat for all 13 stations, 3 through 12 and back to 3. Label each point as illustrated.

Draw light line segments between each of the station line intersection positions found on the corresponding pattern station lines. This completes the true contour of the top edge, Figure 8-14D.

Step 4. Add the top, bottom, and split tabs, as illustrated in Figure 8-14E.

Triangulation Development

Triangulation is the third major method used to lay out a surface development. *Triangulation* is a method of dividing a surface into a number of triangles and then transferring each triangle's true size and shape to the development. (Review Figure 8-2.) As with parallel line and radial line developments, *true lengths* and *true distances* must be used exclusively in pattern constructions.

True-Length Diagram

Before any layouts can be started, true lengths and true distances of the object's boundary edges must be determined. A true-length diagram is usually used

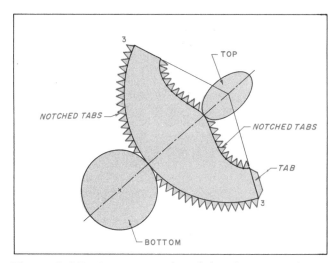

Figure 8-14E Step 4 completed development of a pyramid

to develop true lengths and true distances of these edges. A true-length diagram is often a more rapid method to obtain the needed projections than by descriptive geometry methods.

How To Develop a Transitional Piece Using Triangulation Development

Given: Figure 8-15A describes a transitional piece using a top view, front view, and isometric view. Each of the object's corners are labeled as illustrated. Note that the true distances A to B, B to C, C to D, D to A, 1 to 2, 2 to 3, 3 to 4, and 4 to 1 can be measured directly from the top view. As drawn, the lengths A-1, B-1, B-2, C-2, C-3, D-3, D-4, and A-4 are not shown in their true lengths. A true-length diagram must be used to find these.

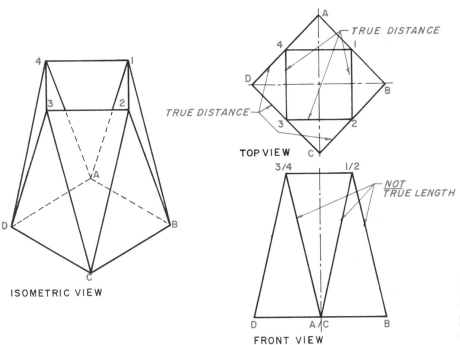

Figure 8-15A
Development of a transitional piece using triangulation development

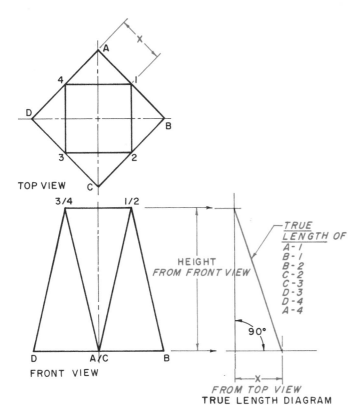

TOP VIEW

FRONT VIEW

TRUE LENGTH DIAGRAM

TRUE LENGTH OF
A-1
B-1
B-2
C-2
C-3
D-3
D-4
A-4

Figure 8-15B Step 1 true-length diagram

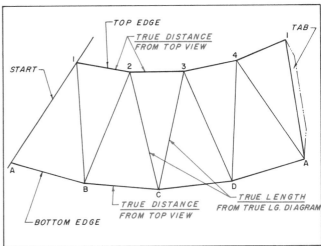

Figure 8-15C Step 2

Step 1. A true-length diagram is a combination of two views. In this example, it is a combination of the top view and the front view, Figure 8-15B. The height between points A and 1 is shown projected from the front view to the true-length diagram. The distance X between corresponding line end points A and 1 is found in the top view, and transferred to the true-length diagram. The illustrated diagonal line, drawn in the true-length diagram, is the true length of A-1. As all the heights and top-view distances between end points are the same for each of the lateral edges, then A-1 is also the same length as B-1, B-2, C-2, C-3, D-3, D-4, and A-4. These true lengths are needed to lay out the development.

Step 2. Starting from line A-1 and using true lengths and true distances, reconstruct each triangle representing a surface of the object, as illustrated in Figure 8-15C. (Review Chapter 2 for aid in transferring a triangle.) Notice in this example that all true distances are transferred from the top view, and all true distances are transferred from the true-length diagram. A tab

is added to join A-1 to A'-1'. This completes the transitional piece development as drawn in Figure 8-15A.

How To Develop a Transitional Piece with a Round End Using Triangular Development

A transitional piece that is square or rectangular at one end and round at the other is laid out using a procedure very.similar to the one outlined previously. The only difference is that station lines must be positioned on the lateral surface of the round-to-corner transitions. Figure 8-16A illustrates a two-view transitional piece with a round top end and a rectangular bottom end. In this example, the top end is divided into 12 equal spaces. Each point is numbered clockwise as illustrated. The bottom four corners making up the bottom rectangle are labeled A through D, Figure 8-16B. The true distances from A-B, B-C, C-D, and D-A' are found in the top view. Station lines W, X, Y, and Z are not true lengths, but the true lengths must be determined with a true-length diagram. Note that in the true-length diagram, the true length of line W, from A' to 0, is also the same lengths for B-6, C-6, and D-12. Lines X, Y, and Z are also used four times at corresponding locations. Segment the object into triangles and develop the pattern using true lengths and true distances to reconstruct each adjoining segment, Figure 8-16C.

How To Develop a Transitional Piece with No Fold Lines Using Triangular Development

Figure 8-17A shows a transitional piece with no evident fold lines or line segments. It is developed in the same manner as any other triangulation development.

Figure 8-16A Development of a transitional piece with a round end using triangulation

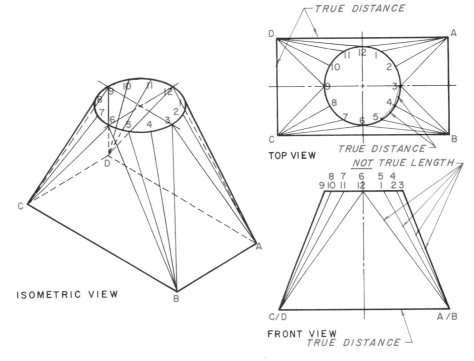

ISOMETRIC VIEW

TOP VIEW

NOT TRUE LENGTH

FRONT VIEW
TRUE DISTANCE

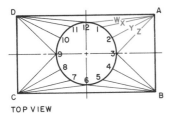

TOP VIEW

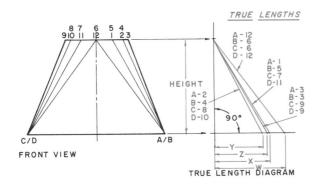

FRONT VIEW

TRUE LENGTHS

TRUE LENGTH DIAGRAM

Figure 8-16B True-length diagram

Figure 8-16C Completed development of a transitional piece with a round end

Figure 8-17A Development of a transitional piece with no fold lines

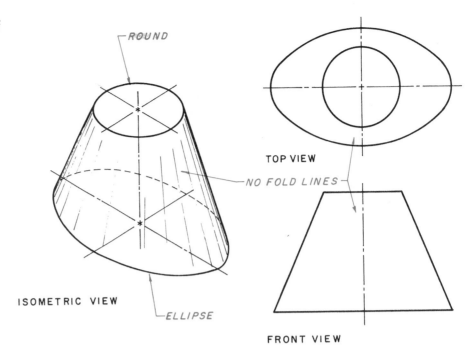

ROUND

TOP VIEW

NO FOLD LINES

ISOMETRIC VIEW

ELLIPSE

FRONT VIEW

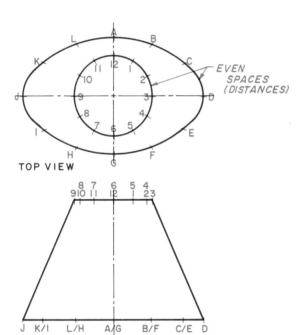

EVEN
SPACES
(DISTANCES)

TOP VIEW

FRONT VIEW

Figure 8-17B Step 1

Step 1. Divide the top and bottom surfaces into equal spaces as illustrated in Figure 8-17B. In this example, the top edge is divided into 12 equal spaces, 0 through 12. The bottom edge is divided into the exact same number of equal spaces, and lettered A through L to A.

Step 2. Connect points 12 to A, 1 to B, 2 to C, and consecutively to points L to 11 with solid lines, as illustrated in Figure 8-17C. Dash lines are added to segment the object into various triangles.

Step 3. True distances from 1 to 2, 2 to 3, 3 to 4, and so on through 11 to 12, and the true distances from A to B, B to C through to L to A' are found in the top view, Figure 8-17C. The other fold lines, connecting the top and bottom are not true lengths and, therefore, true-length diagrams must be made, one for the solid lines and the other for the dash lines, Figure 8-17D. The development is laid out by constructing connecting triangles using the lengths of each leg of each triangle, Figure 8-17E.

Notches

Some developments require notches. Two major types of notches are usually used: a sharp V or a rounded V, Figure 8-18. The sharp V is used only if a minimal force is tending to part the material. The sharp point of the V will tear or crack under stress. The rounded V is used where parts would be under greater stress. The radius of the vertex of the rounded V should be at least twice the thickness of the metal used, and larger if possible.

Bends

When bending sheet metal to form a corner, rib, or design, the outer surface stretches and the inner surface compresses. The length of material needed by each bend must be calculated and added to the straight unbent portions of the pattern. The material needed by each bend is the length of the neutral

Figure 8-17C Step 2

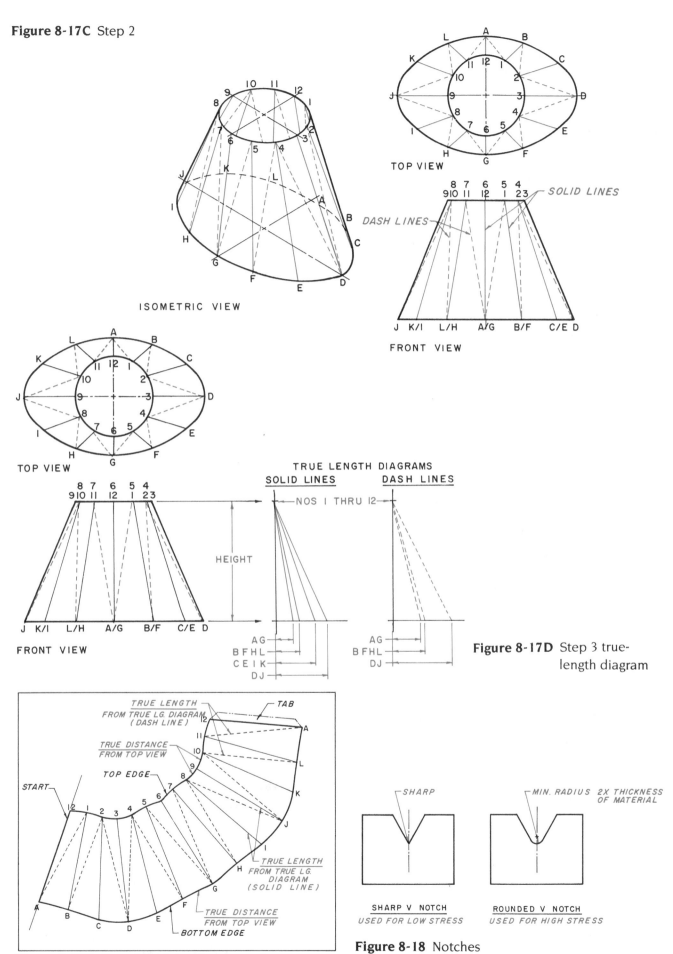

ISOMETRIC VIEW

TOP VIEW

SOLID LINES

DASH LINES

FRONT VIEW

TOP VIEW

TRUE LENGTH DIAGRAMS
SOLID LINES   DASH LINES

NOS I THRU 12

HEIGHT

FRONT VIEW

AG
BFHL
CEIK
DJ

AG
BFHL
DJ

Figure 8-17D Step 3 true-
length diagram

TRUE LENGTH
FROM TRUE LG. DIAGRAM
(DASH LINE)

TAB

TRUE DISTANCE
FROM TOP VIEW

TOP EDGE

START

TRUE LENGTH
FROM TRUE LG.
DIAGRAM
(SOLID LINE)

TRUE DISTANCE
FROM TOP VIEW

BOTTOM EDGE

SHARP

MIN. RADIUS 2X THICKNESS
OF MATERIAL

SHARP V NOTCH
USED FOR LOW STRESS

ROUNDED V NOTCH
USED FOR HIGH STRESS

Figure 8-17E Completed development of a
transitional piece

Figure 8-18 Notches

axis within the material, where neither compression nor stretching occurs. This is calculated either by formulas or by using various charts available for this purpose.

Bend allowance charts are included in the Appendix of this text, and are much faster to use than a formula. Two basic kinds of charts are used to calculate bend allowance, in both the English and metric systems. One type is used for 90° bends; the other for bends from 1° through 180°. The total length of a pattern is called the *developed length*. To determine the developed length, the stretched-out flat pattern must include all straight sides, plus the calculated bend allowances.

How To Find All Straight Sides of a 90° Bend

Figure 8-19A shows a simple 90° bend with an inside radius of .25, a sheet metal thickness of .125, and legs of 2.0 and 3.0 (the English system inch is used in this example). Bend radii are always measured from the surface closest to the bend radius center.

Step 1. Locate the tangent points at the ends of the straight sides.

Step 2. Add the thickness of the sheet metal to the bend radius .25 (.125 + .25 = .375).

Step 3. Subtract the sum of the sheet metal thickness .125 and the radius .25 from the 3.00 overall length of the object (3.00 − .375 = 2.625).

Step 4. Subtract the sum of the sheet metal thickness .125 and the bend radius .25 from the 2.00 overall height of the object (2.00 − .375 = 1.625).

Step 5. Add the two straight sides together (2.625 + 1.625 = 4.250). This is the total length of the straight sides of this object, Figure 8-19B.

How To Find the Bend Allowance of a 90° Bend

Review Figure 8-19A.

Step 1. Note the metal thickness, in this example .125, and the inside radius, in this example .25.

Step 2. Refer to the bend allowance chart in inches for 90° bends in the Appendix at the end of this text.

Step 3. The left-hand column gives various metal thicknesses. Go down the left-hand column until the required size or the closest size is found. In this example .125.

Step 4. Along the top of the chart is listed various inside radii; go across the top of the chart to

the required size or the closest radius, in this example .25.

Step 5. From the .125 number in the left-hand column project across to the right; from the .25 number along the top of the chart, project down to where the two columns intersect. Given is the bend allowance for material .125 thick with a .25 radius. In this example the bend allowance is .480, Figure 8-19C.

Step 6. The stretched-out dimension is found by adding the straight sides to the bend allowance; in this example, 4.250 + .480 = 4.730 (see Figure 8-19C).

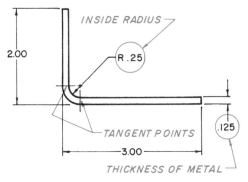

Figure 8-19A Bend allowance of 90° bend

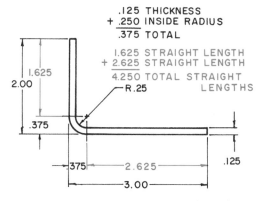

Figure 8-19B Total of straight lengths

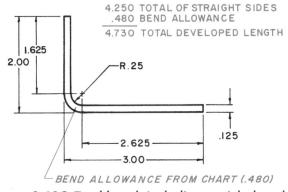

Figure 8-19C Total length including straight lengths and bend allowance

How To Find the Total Straight-Side Length Adjoining a Bend Other Than 90°

Figure 8-20A shows a simple 30° bend with an inside radius of 6.35 and a sheet metal thickness of 3.175 (the metric system is used in this example).

Step 1. Locate the tangent points at the ends of the straight sides.

Step 2. Determine the length of the straight sides and add them together (64.0 + 50.0 = 114.0). This is the total length of the straight sides of this object.

How To Find the Bend Allowance of Other Than a 90° Bend

Review Figure 8-20A.

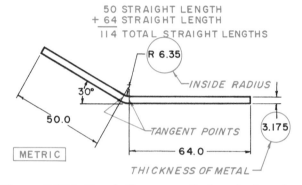

Figure 8-20A Bend allowance of a bend other than 90° — total of straight lengths

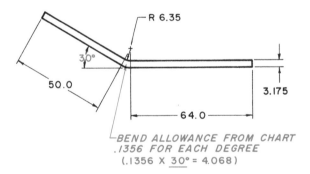

Figure 8-20B Total length of bend allowance

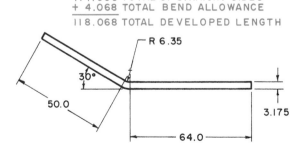

Figure 8-20C Total length including straight lengths and bend allowance

Step 1. Note the metal thickness, in this example 3.175, and the inside radius, in this case R6.35.

Step 2. Refer to the bend allowance chart in millimetres for 1° bends in the Appendix at the end of this text.

Step 3. The left-hand column gives various metal thicknesses. Go down the left-hand column until the required size or the closest size is found. In this example 3.175.

Step 4. Along the top of the chart is listed various inside radii. Go across the top of the chart to the required size or closest radius, in this example 6.35.

Step 5. From the 3.175 number of the left-hand column, project across to the right; from the 6.35 number along the top of the chart, project down to where the two columns intersect. Given is the factor used to calculate the bend allowance. In this example .1356.

Step 6. Multiply this factor times the actual degrees in the bend, from a straight 180° line, Figure 8-20B. This is the bend allowance (.1356 x 30° = 4.068).

Step 7. Add the straight-side lengths to the bend allowance to get the total developed length: 114.0 + 4.068 = 118.068, Figure 8-20C.

Note: A 30° bend dimension as illustrated in Figure 8-21A is actually 150° (180° − 30° = 150°). See Figure 8-21B. The total bend must be calculated from a straight piece that is actually 180° before bending.

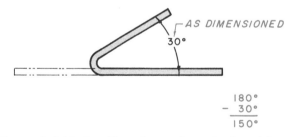

Figure 8-21A Total bend must be calculated from straight line dimension — given is 30°

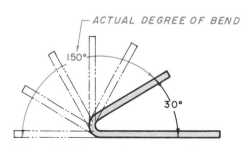

Figure 8-21B As dimensioned 30° is actually a 150° bend from a straight line

Review

1. What is a transitional piece and which kind of development would be used to develop a pattern?

2. What are the two major kinds of notches used in a development?

3. What does a gage number represent?

4. In parallel line development, at what angle to the fold lines must the baseline be projected?

5. Why is it important to develop the pattern or template with the inside surface up?

6. List the three major kinds of developments used to develop a pattern of an object.

7. Why are tabs used?

8. What is bend allowance, and why is it used?

9. Explain what must be done if a transitional piece does not have fold lines.

10. What is a true-length diagram and why is it used?

11. Why is the gage system of calling out the thickness of a material being phased out?

12. What two elements must be located or laid out before a development can be actually started?

Chapter Eight Problems

The following problems are intended to give the beginning drafter practice in sketching, laying out auxiliary views if required, and drawing the stretched-out development of various objects. These problems will use one or more of the three standard methods of development: *parallel line developments*, *radial line developments* and *triangulation developments*. The student will also practice calculating developed lengths using various charts to determine full length before bending.

The steps to follow in laying out any object that is to be developed are:

Step 1. Study the problem carefully.

Step 2. Determine which method or methods of development will be used (parallel line/radial line/triangulation).

Step 3. On a scrap sheet of paper, draw to scale enough views as required to lay out *all* true lengths and true distances. Add the end view and auxiliary views, in scale, if necessary. Draw with a *sharp* 4-H lead, and work as accurately as possible.

Step 4. Label each point, if necessary, in order to keep track of progress.

Step 5. Check all lengths, and auxiliary views if included.

Step 6. Starting from a baseline, lightly lay out the development. Use *true lengths* and *true distances* for *all* measurements. (Be sure to *start* the seam at the *shortest* fold line possible.)

Step 7. Label each point, if necessary, in order to keep track of progress.

Step 8. Use *phantom lines* to represent all *fold lines*.

Step 9. Add tabs as required, and check to see that there are enough but without duplications.

Step 10. Recheck all true lengths and true distances.

Step 11. Darken in all lines.

Problems 8-1 through 8-18

Using the parallel line development method, develop each object starting from the given *seam*. Label each point clockwise; add ends and/or auxiliary views as necessary to develop a complete object. Add tabs .125 (3) × 45° to suit. Use phantom lines for all fold lines.

Extra assignment(s): Cut out the development(s) and glue or tape it together to prove its accuracy.

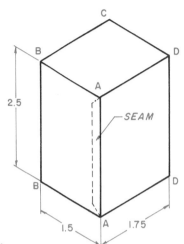

Problem 8-1

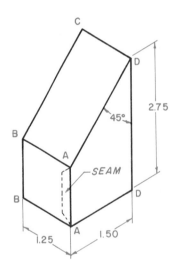

Problem 8-2

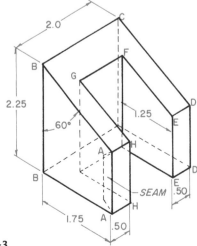

Problem 8-3

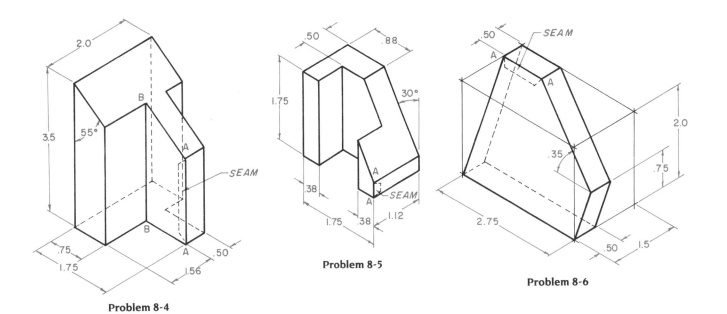

Problem 8-4

Problem 8-5

Problem 8-6

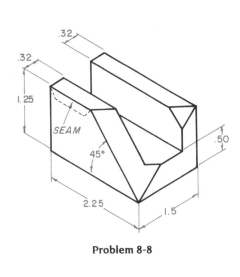

METRIC

Problem 8-7

Problem 8-8

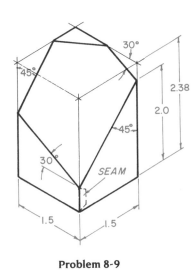

Problem 8-9

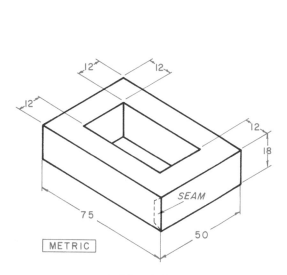

METRIC

Problem 8-10

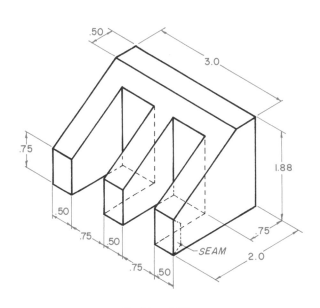

Problem 8-11

288   Section 2

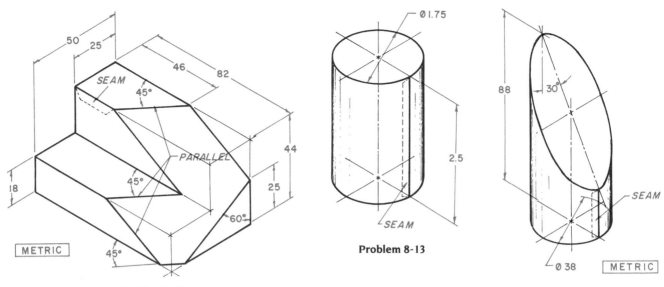

Problem 8-12

Problem 8-13

Problem 8-14

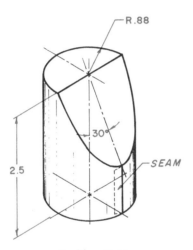

Problem 8-15

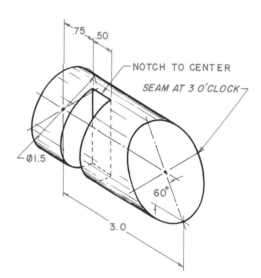

Problem 8-16

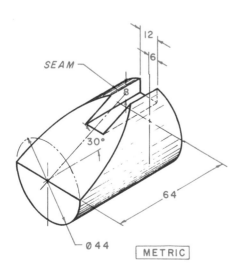

Problem 8-17

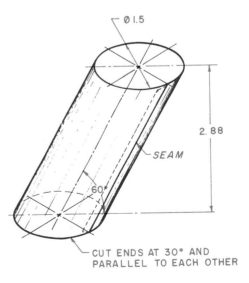

Problem 8-18

Problems 8-19 through 8-23

Using the parallel line development method, develop each object starting from the given *seams*. Add .125 (3) × 45° tabs as required to hold parts together. Design parts so there are as many identical parts as possible. Use phantom lines for all fold lines. Extra assignment(s): Cut out the development(s) and glue or tape it together to prove its accuracy.

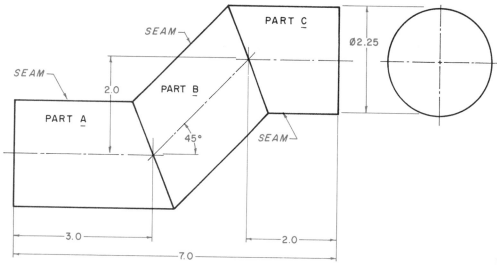

Problem 8-19

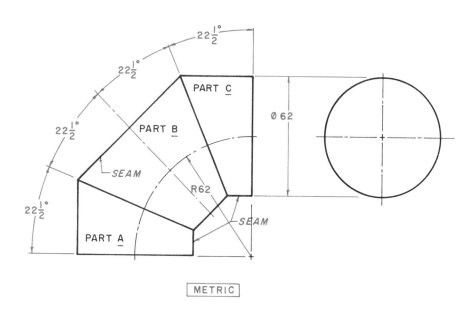

METRIC

Problem 8-20

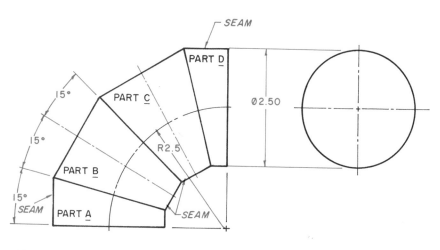

Problem 8-21

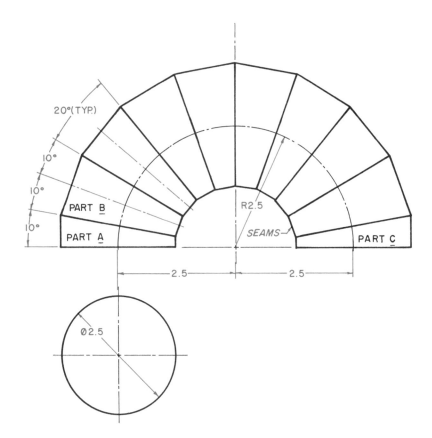

PART B

PART A

20°(TYP.)

10°

10°

10°

R2.5

SEAMS

PART C

2.5

2.5

Ø2.5

Problem 8-22

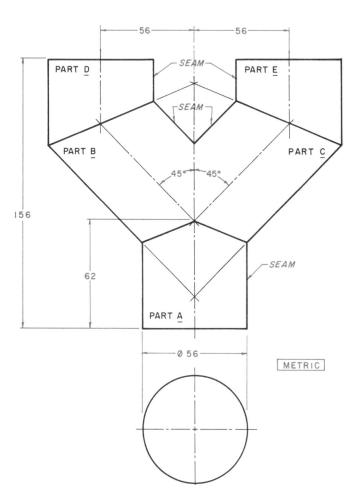

56

56

PART D

SEAM

PART E

SEAM

PART B

PART C

45° 45°

156

SEAM

62

PART A

Ø 56

METRIC

Problem 8-23

Problems 8-24 through 8-26

Develop the objects, using given dimensions. Using *bend allowance charts*, design layouts to include material for bends.

Extra assignment(s): Cut out the development(s) and glue or tape it together to prove its accuracy.

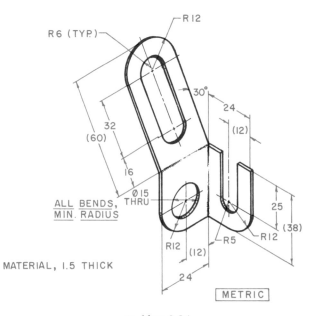

R 6 (TYP.)

R 12

30°

32

24

(60)

(12)

16

Ø 15
THRU

25

ALL BENDS,
MIN. RADIUS

(38)

R12

R5

(12)

R12

MATERIAL, 1.5 THICK

24

METRIC

Problem 8-24

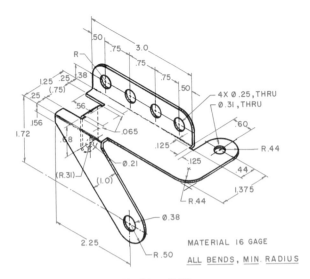

Problem 8-25

MATERIAL 16 GAGE

ALL BENDS, MIN. RADIUS

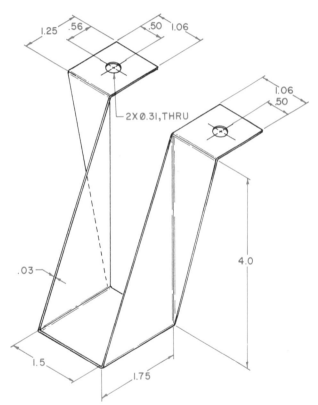

Problem 8-26

2X Ø.31, THRU

Problems 8-27 through 8-30

Carefully lay out a full-size multiview drawing of the object, include all intersection lines. Make a full-size development of the object. Add tabs .125 (3) × 45° as required to hold parts together.

Extra assignment(s): Cut out the development(s) and glue or tape it together to prove its accuracy.

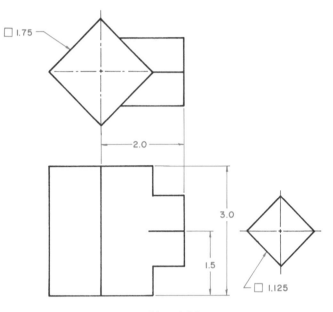

Problem 8-27

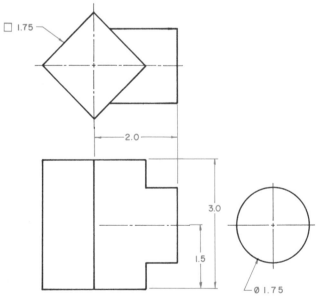

Problem 8-28

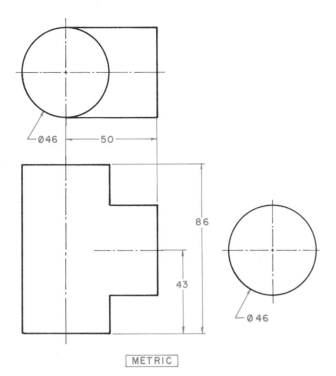

Ø46 ─ 50

86

43

Ø 46

⌷ METRIC ⌷

Problem 8-29

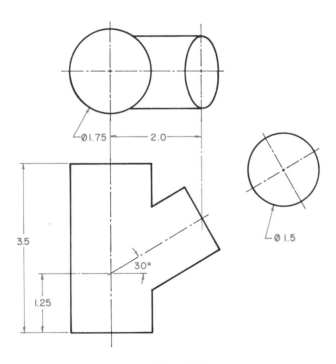

Ø1.75 ─ 2.0

Ø 1.5

3.5

30°

1.25

Problem 8-30

Problems 8-31 through 8-42

Using the radial line development method, develop each object starting from the given *seam*. Label each point clockwise; add ends and/or auxiliary views as necessary to develop a complete object. Add tabs .125 (3) × 45° to suit. Use phantom lines for all fold lines.

Extra assignment(s): Cut out the development(s) and glue or tape it together to prove its accuracy.

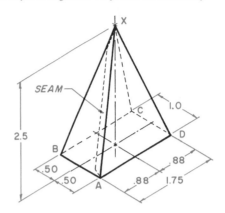

X

SEAM

2.5

C

1.0

B

D

.50

88

.50

A

.88

1.75

Problem 8-31

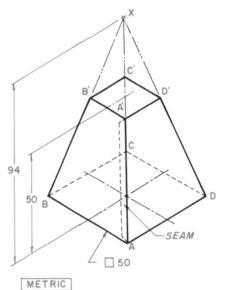

X

C'

B'

D'

A'

C

94

50

B

D

A

SEAM

⌷ 50

⌷ METRIC ⌷

Problem 8-32

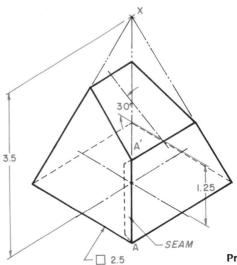

X

30°

3.5

A'

1.25

A

SEAM

⌷ 2.5

Problem 8-33

Chapter 8   293

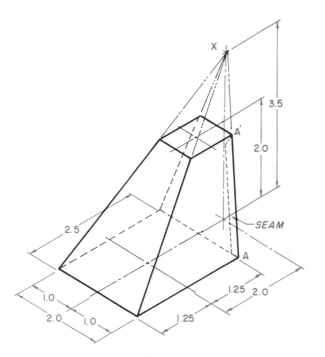

Problem 8-34

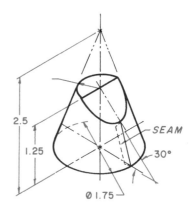

Problem 8-37

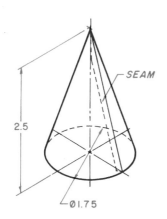

Problem 8-35

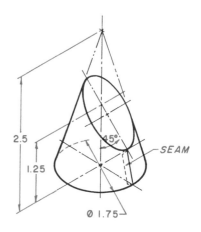

Problem 8-38

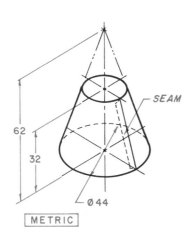

METRIC

Problem 8-36

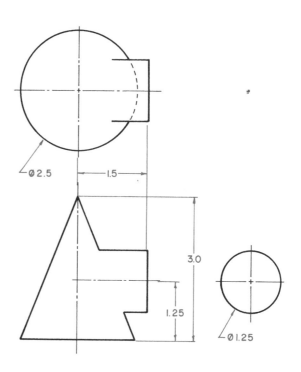

Problem 8-39

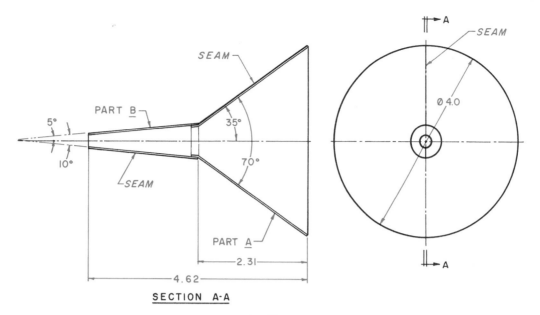

SECTION A-A

Problem 8-40

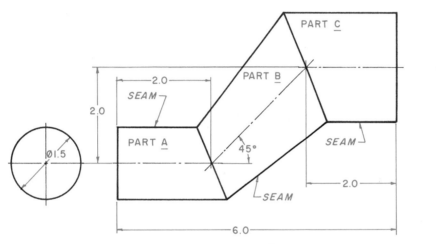

Problem 8-41

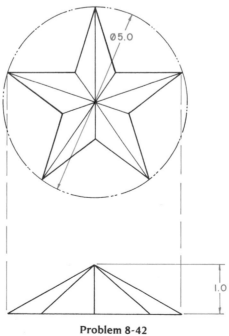

Problem 8-42

Problems 8-43 through 8-46

Using the triangulation development method, develop each object starting from the given *seam*. Label each point clockwise: add tabs .125 (3) × 45° to suit. Use phantom lines for all fold lines. Extra assignment(s): Cut out the development(s) and glue or tape it together to prove its accuracy.

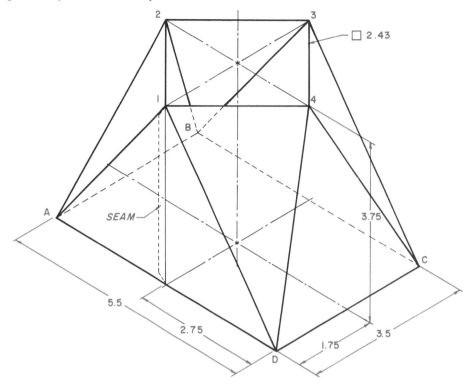

▢ 2.43

SEAM

3.75

5.5

2.75

1.75

3.5

Problem 8-43

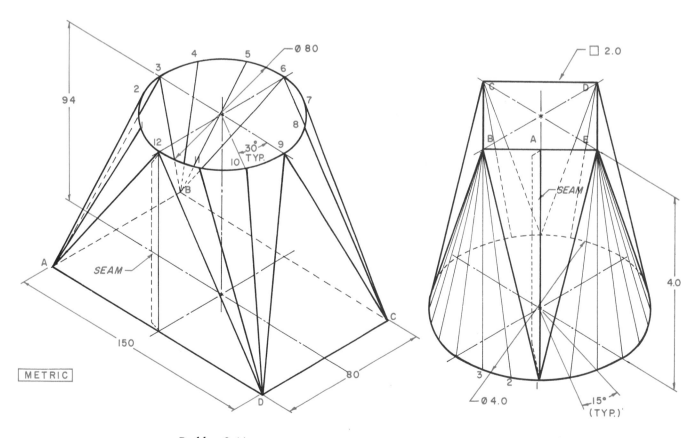

Ø 80

94

30° TYP.

SEAM

150

80

METRIC

Problem 8-44

▢ 2.0

SEAM

4.0

Ø 4.0

15°
(TYP.)

Problem 8-45

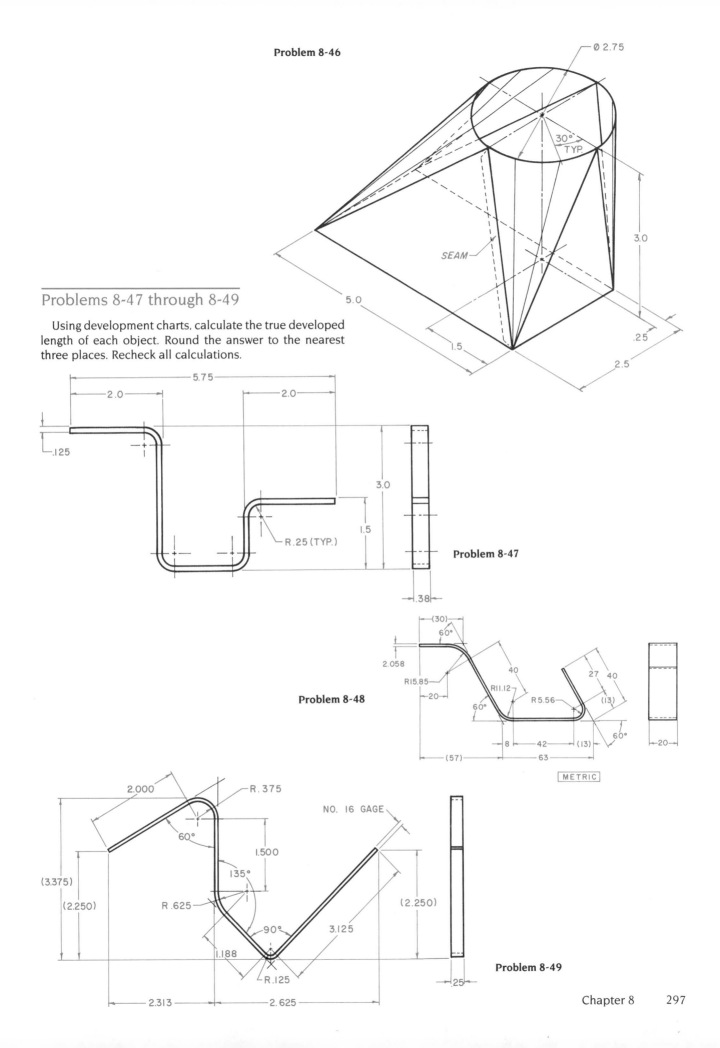

Problem 8-46

Ø 2.75

30°
TYP.

3.0

SEAM

5.0

.25

1.5

2.5

Problems 8-47 through 8-49

Using development charts, calculate the true developed length of each object. Round the answer to the nearest three places. Recheck all calculations.

5.75

2.0

2.0

.125

3.0

1.5

R.25 (TYP.)

Problem 8-47

.38

(30)

60°

2.058

40

R15.85

27   40

RII.12

(13)

60°   R5.56

8   42   (13)

60°

(57)

63

20

Problem 8-48

METRIC

2.000

R.375

NO. 16 GAGE

60°

1.500

135°

(3.375)

(2.250)

R.625

(2.250)

90°   3.125

1.188

R.125

.25

Problem 8-49

2.313   2.625

1. What is the scale of the drawing?
2. When was the drawing checked?
3. What product will be fabricated using this drawing?
4. How many edges must be formed by bending to make the finished product from the flat piece?
5. How thick is the material for the finished product?
6. What is the diameter of hole A?
7. How many times is hole A used?
8. What is the overall length of the flat piece?
9. What is the overall width of the flat piece?
10. What is the purpose of hole C?

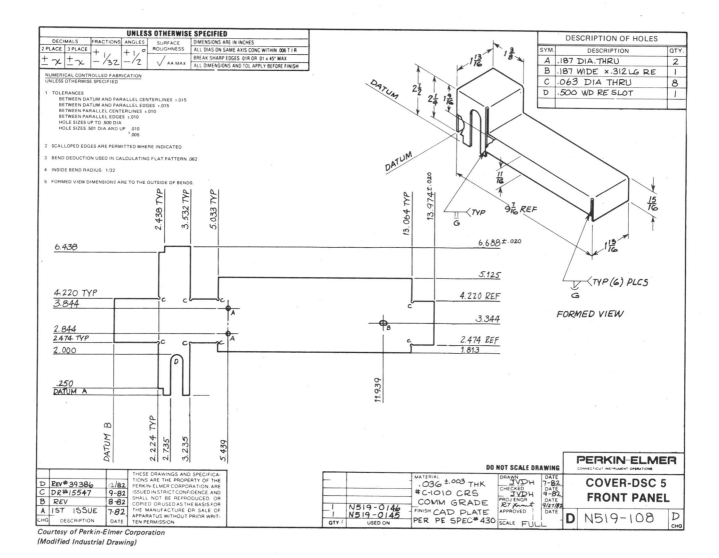

UNLESS OTHERWISE SPECIFIED

DECIMALS		FRACTIONS	ANGLES	SURFACE ROUGHNESS	DIMENSIONS ARE IN INCHES
2 PLACE	3 PLACE	$\pm \frac{1}{32}$	$\pm \frac{1}{2}°$	√ AA MAX	ALL DIAS ON SAME AXIS CONC WITHIN .006 T I R
$\pm x$	$\pm x$				BREAK SHARP EDGES .01R OR .01 x 45° MAX
					ALL DIMENSIONS AND TOL APPLY BEFORE FINISH

NUMERICAL CONTROLLED FABRICATION
UNLESS OTHERWISE SPECIFIED

1 TOLERANCES
    BETWEEN DATUM AND PARALLEL CENTERLINES ±.015
    BETWEEN DATUM AND PARALLEL EDGES ±.015
    BETWEEN PARALLEL CENTERLINES ±.010
    BETWEEN PARALLEL EDGES ±.010
    HOLE SIZES UP TO .500 DIA
    HOLE SIZES .501 DIA AND UP +.010 −.005

2 SCALLOPED EDGES ARE PERMITTED WHERE INDICATED

3 BEND DEDUCTION USED IN CALCULATING FLAT PATTERN .062

4 INSIDE BEND RADIUS: 1/32

5 FORMED VIEW DIMENSIONS ARE TO THE OUTSIDE OF BENDS.

DESCRIPTION OF HOLES

SYM.	DESCRIPTION	QTY.
A	.187 DIA. THRU	2
B	.187 WIDE x .312 LG RE	1
C	.063 DIA THRU	8
D	.500 WD RE SLOT	1

DATUM

DATUM

$2\frac{1}{2}$ $2\frac{1}{4}$ $1\frac{2}{16}$

$1\frac{13}{16}$ $1\frac{3}{8}$

$\frac{11}{16}$

$9\frac{7}{16}$ REF

13.064 TYP
$13.974 \pm .020$

√ G TYP

$\frac{15}{16}$

$1\frac{13}{16}$

√ G TYP (6) PLCS

FORMED VIEW

2.438 TYP
3.532 TYP
5.033 TYP

6.438

6.688 ±.020

5.125

4.220 TYP
3.844

4.220 REF

2.844
2.474 TYP
2.000

3.344

2.474 REF
1.813

.250
DATUM A

11.939

DATUM B

2.224 TYP
2.735
3.235

5.439

DO NOT SCALE DRAWING

PERKIN-ELMER
CONNECTICUT INSTRUMENT OPERATIONS

CHG	DESCRIPTION	DATE
D	REV # 39386	12/82
C	DR # 15547	9-82
B	REV	8-82
A	1ST ISSUE	7-82

QTY.	USED ON
1	N519-0146
1	N519-0145

MATERIAL .036 ±.003 THK # C-1010 CRS COMM GRADE
FINISH CAD PLATE PER PE SPEC # 430

		DATE
DRAWN	JVDH	7-82
CHECKED	JVDH	9-82
PROJ ENGR	RT Lunt	9/27/82
APPROVED		DATE

COVER-DSC 5 FRONT PANEL

SCALE FULL

D N519-108 D CHG

Courtesy of Perkin-Elmer Corporation
(Modified Industrial Drawing)

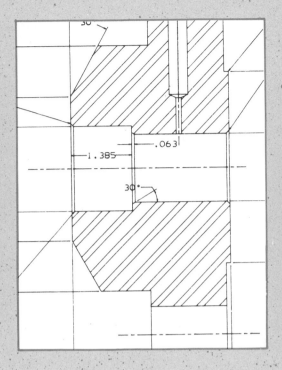

This chapter covers in depth the fundamentals of dimensioning and notation. All dimensioning techniques presented are in accordance with ANSI Y14.5M-1982, the latest edition of the dimensioning standard. Major topics covered include specifying the scale, dimensioning systems, general rules of dimensioning, specific dimensioning techniques, rules for applying notes on drawings, general notes, detail notes, writing notes, note specifications, and sample notes.

CHAPTER NINE

Dimensioning and Notation

One of the most fundamental drafting tasks is to meet the requirements of the engineering definition of the part, while providing for the most economical production process and the interchangeability considerations. All of this is accomplished by the use of proper dimensioning and notation on drawings. *Dimensioning* is the process wherein size and location data for the subject of a technical drawing are provided. *Notation* is the process wherein needed information not covered by dimensions is placed on a technical drawing.

It is critical that drafters, designers, and engineers be proficient in standard dimensioning practices. The most widely accepted dimensioning standard is American National Standards Institute document Y14.5M-1982 (ANSI Y14.5M-1982). Similar standards are produced by the International Standards Organization (ISO). However, unless otherwise specified, ANSI Y14.5M-1982 is the standard used for guiding dimensioning practices.

Modern dimensioning practices described in ANSI Y14.5M-1982 apply in most instances where interchangeability of parts is a major consideration. The concept dictates that parts produced from a drawing at one manufacturing site will be interchangea-

ble with those produced at another manufacturing site. Automotive parts are an excellent example of production for interchangeability. Some parts are manufactured in America, some in Europe, and some in Japan, but they must all fit together in one car during assembly. Although interchangeability is not a factor with all parts that are produced, the drafter should still use the basic dimensioning principles of ANSI Y14.5M-1982. This is particularly important when the parts will be produced by such ever increasing automated or semiautomated processes as numerical control or computer-aided manufacturing.

Specifying the Scale

As has already been learned, technical drawings are usually made to scale. When this is the case, the scale should be indicated in the appropriate space in the title block. On occasion, one or more dimensions on a technical drawing may be made NOT TO SCALE (NTS). Any NTS dimensions should be distinguished from scaled dimensions. This is accomplished by underlining NTS dimensions with a thick straight line. The older version, a wavy line, is no longer used.

Although a scale may have been indicated in the title block, it is frequently necessary to add an enlarged view of a characteristic. The scale for such enlarged views should be shown adjacent to it, even though a different overall scale applies to the entire drawing.

Dimensioning Systems

Three dimensioning systems are used on technical drawings in the United States: metric dimensioning, decimal-inch dimensioning, and fractional dimensioning. Certain rules of practice pertain to each of these dimensioning systems with which drafters should be familiar.

Metric Dimensioning

The standard metric unit of measurement for use on technical drawings is the millimetre (0.001 metre) or .039 inch. Figure 9-1 is a chart of the various metric units of measurement less than a metre showing where the millimetre fits in.

When using metric dimensioning, several general rules should be observed. When a dimension is less than one millimetre, a zero must be placed to the left of the decimal point, Figure 9-2, Part A. When a metric dimension is a whole number, neither the zero nor the decimal is required, Figure 9-2, Part B. When a metric dimension consists of a whole number and a decimal portion of another millimetre, it is written as follows: whole number first, decimal point next, and, finally, the decimal part of the number. The decimal part of the number is *not* followed by a zero in metric dimensioning, Figure 9-2, Part C. Individual digits in metric dimensions are not separated by commas or spaces, Figure 9-3. Drawings prepared with metric dimensions are identified with the word METRIC contained in a small rectangle below the part.

Decimal-inch Dimensioning

Decimal-inch dimensioning is frequently used in the dimensioning of technical drawings. It is a much less cumbersome system for mechanical drawings than is the fractional system, and it is still used more than the metric system.

When using the decimal-inch dimensioning system, several rules should be observed. If a dimension is less than one inch, only a decimal point and the numbers are required. A zero is not required to precede the decimal point, Figure 9-4, Part A. The number of places beyond the decimal that a decimal-inch dimension is carried is determined by the specified tolerance for the part in question, Figure 9-4, Part B. In this figure, a tolerance of .001 (three places to the right of the decimal) is specified. Consequently, the dimension 1.637 is carried out three places to the right of the decimal.

There are no specified sizes for decimal points, but they should be made dark enough and large enough to be seen, and to reproduce through any normal reproduction process (diazo, photocopy or microfilm).

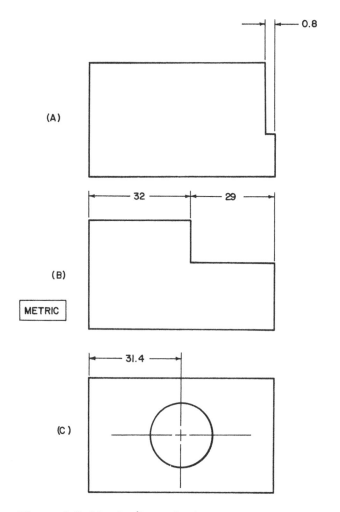

Figure 9-2 Metric dimensioning

UNIT	MULTIPLE OF A METRE
METRE	1
DECIMETRE	0.1
CENTIMETRE	0.01
MILLIMETRE	0.001

Figure 9-1 Metric linear measurements

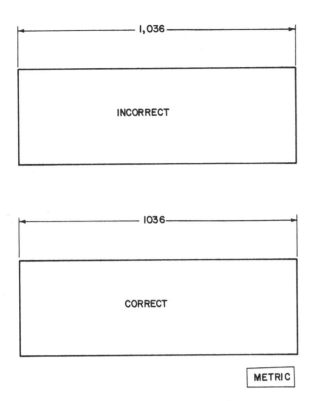

Figure 9-3 No commas in metric dimensions

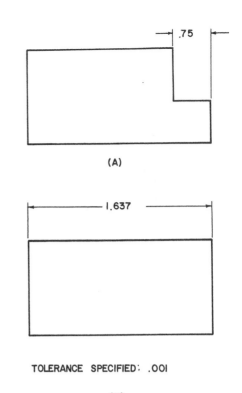

TOLERANCE SPECIFIED: .OOI

(B)

Figure 9-4 Decimal-inch dimensioning

Fractional Dimensioning

Fractional dimensioning is not frequently used on mechanical technical drawings. Its primary use is on architectural and structural engineering drawings. However, since it is occasionally still used on mechanical drawings, drafters should be familiar with this system.

When using fractional dimensions on mechanical drawings, several rules should be observed. The line separating the numerator and denominator of a fraction should be a horizontal line, not an inclined line, Figure 9-5. Full-inch dimensions should be a minimum of one-eighth inch in height. The combined height of the numerator, denominator, and horizontal line of a fraction should be one-quarter inch, Figure 9-6 for A-, B-, and C-size drawings, and a minimum of five-sixteenths inch for D, E, and F sizes.

Figure 9-5 Horizontal line is the correct method

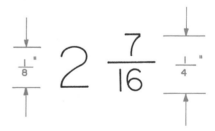

Figure 9-6 Proportions for fractions

Dimension Components

Several components are common to all dimensioning systems. These include extension lines, dimension lines, leader lines, arrowheads, and the actual numbers or dimensions. Drafters and engineers should be knowledgeable in the proper use of these components.

Extension Lines

An *extension line* is a thin, solid line that extends from the object in question or some feature of the object. Several rules should be observed when placing extension lines on technical drawings.

There should be a small but visible gap between the object or object feature and the beginning of an

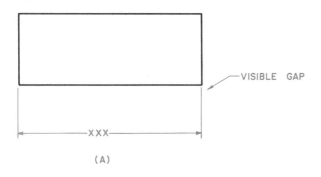

(A)

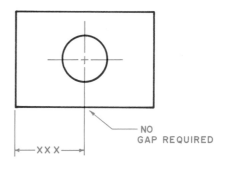

Figure 9-8 Center lines as extension lines

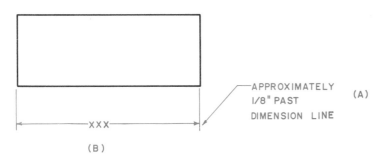

(B)

Figure 9-7 Drawing extension lines

extension line, Figure 9-7, Part A. Extension lines should extend uniformly beyond dimension lines a distance of approximately one-eighth inch, Figure 9-7, Part B. Extension lines that originate on the object, such as center lines, may cross visible lines with no gap required, Figure 9-8.

Dimension Lines

A *dimension line* is a thin, solid line used to indicate graphically the linear distance being dimensioned. Dimension lines are normally broken for placement of the dimension, Figure 9-9, Part A. If a horizontal dimension line is not broken, the dimension is placed above the dimension line with guidelines parallel to it, Figure 9-9, Part B.

When dimensioning multiple features of an object, dimensions should be aligned uniformly rather than staggered or randomly scattered about the object, Figure 9-10.

Dimension lines are drawn parallel to the direction of measurement. Sufficient distance between the object and the dimension lines and between successive dimension lines is important so that cramped and crowded dimensions do not result. The first dimension line should be at least three-eighths inch

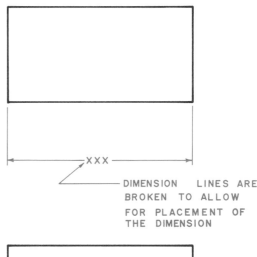

(A)

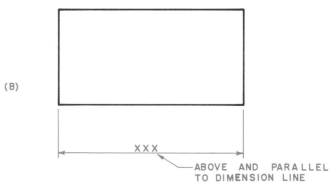

(B)

Figure 9-9 Placement of dimensions

away from the object, Figure 9-11, Part A. Successive dimension lines should be at least one-quarter inch apart. If using metric dimensions, the first dimension line should be at least 10 millimetres away from the object. Successive lines should have at least 6 millimetres between them, Figure 9-11, Part B.

When the shape of an object requires a series of parallel dimension lines, the breaks and the dimensions should be staggered to make it easier to read the dimensions, Figure 9-12.

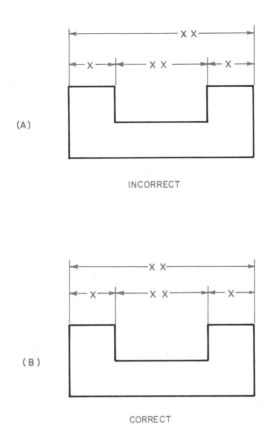

(A)

INCORRECT

(B)

CORRECT

Figure 9-10 Proper placement of dimensions

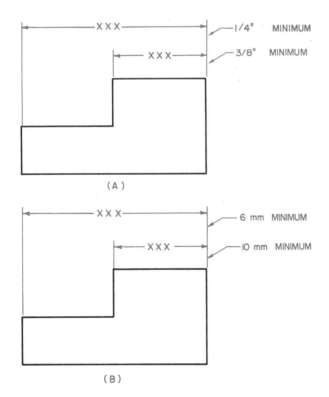

(A)

(B)

Figure 9-11 Successive dimension lines

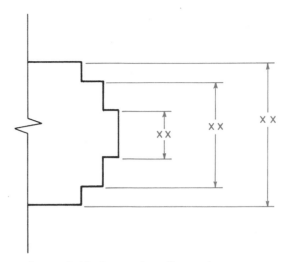

Figure 9-12 Staggering dimensions

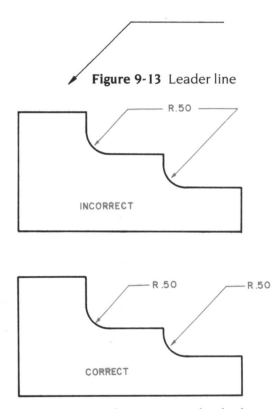

Figure 9-13 Leader line

R.50

INCORRECT

R.50   R.50

CORRECT

Figure 9-14 One dimension per leader line is preferred

Leader Lines

A *leader line* is a thin line that begins horizontally, breaks at an angle, and terminates in an arrowhead or, on occasion, a dot, Figure 9-13. Leader lines are used for tying dimensions, notes, symbols or other data to a specific point on a drawing.

When using a leader line to direct a dimension to its appropriate feature on a drawing, a dimension for each leader line is the preferred method, Figure 9-14. More than one leader line extending from the same dimension can create a confusing situation that is difficult to interpret. Leader lines pointing to the center of a circle should be directed toward but not extended into the circle center, Figure 9-15.

Arrowheads

An arrowhead is the most commonly used termination symbol for dimension and leader lines. Arrowheads should be approximately three times as long as they are wide. They should be large enough to be seen, but small enough that they do not detract from the appearance of the drawing. A commonly accepted length for arrowheads is one-eighth inch. Arrowheads may be slightly larger or smaller than this but, regardless of their size, they should be uniform throughout a drawing, Figure 9-16. Although the same standard applies to drawings prepared on a CAD system, arrowheads do not have to be filled in on drawings prepared using automated processes. Some CAD systems fill in arrowheads and some do not. Either open or filled in arrowheads have been considered acceptable since the advent of CAD.

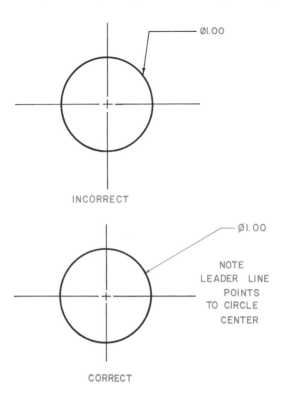

Figure 9-15 Leader lines point at the center of a circle

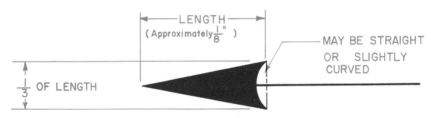

NOTES

1. ARROWHEADS SHOULD BE UNIFORM IN SIZE AND SHAPE THROUGHOUT A DRAWING.

2. ARROWHEADS DO NOT HAVE TO BE FILLED IN WHEN USING A CAD SYSTEM.

Figure 9-16 Proper size and shape of arrowheads

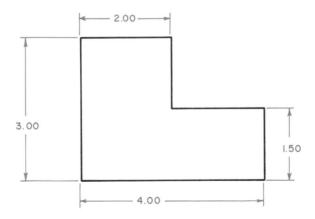

UNIDIRECTIONAL DIMENSIONING
ALL DIMENSIONS ARE READ FROM THE
BOTTOM OF THE DRAWING

Figure 9-17 Unidirectional dimensioning

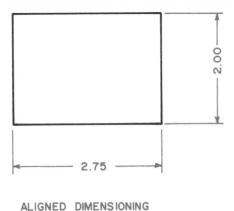

ALIGNED DIMENSIONING

Figure 9-18 Aligned dimensioning

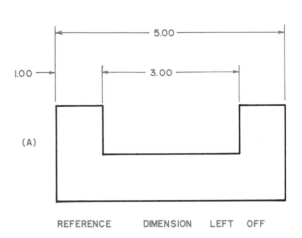

(A)

REFERENCE   DIMENSION   LEFT   OFF

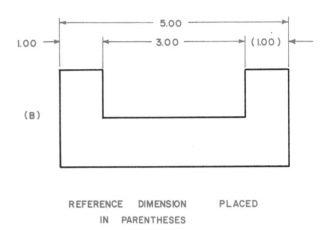

(B)

REFERENCE   DIMENSION   PLACED
IN   PARENTHESES

Figure 9-19 Reference dimensions

General Rules of Dimensioning

A number of specific rules of dimensioning are used for dealing with the various individual dimensioning situations that arise. The specific situations are covered in this chapter. However, before dealing with specific rules, the general rules which apply in all cases should be examined. Following are these general rules of dimensioning.

1. Unidirectional placement of dimensions on drawings is the preferred method. This means that all dimensions can be read from the bottom of the drawing, Figure 9-17.

2. Aligned dimensioning is the less preferred method for placing dimensions on drawings. When aligned dimensions are used, they should be placed so that they can be read from the bottom or right side of the drafting sheet, Figure 9-18.

3. When an overall dimension is given, one intermediate dimension may be either left off or placed in parentheses to indicate that it is a reference dimension, Figure 9-19.

4. Dimensions should be placed outside of the outline of an object, except in cases where the required extension and/or leader lines would be unusually long.

5. Unnecessary dimensioning should be avoided. Only those dimensions required to manufacture the part should be given, but given only once.

6. Enough dimensions should be provided so that the various tradespeople who will be using the drawing do not have to calculate, scale or estimate.

7. Features of an object should be dimensioned on the view or views where they appear true size and true shape.

8. Avoid dimensioning a feature where it appears as a hidden line, Figure 9-20.

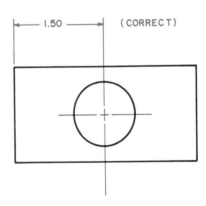

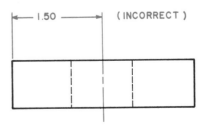

Figure 9-20 Avoid dimensioning
hidden lines

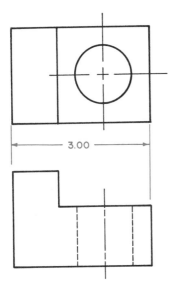

Figure 9-21 Placement of
common
dimensions
between views

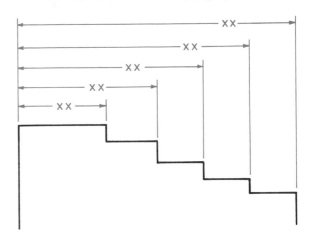

Figure 9-22 The topmost dimension in a series
should be the longest

9. Dimensions that are common to two views
   should be placed between the views, as long
   as they can be clearly read, Figure 9-21.

10. In a series of dimensions, begin with the
    shortest and work out and away from the
    object so that the last dimension is the
    longest, Figure 9-22.

11. No other type of line should cross a dimen-
    sion line unless it is absolutely unavoidable,
    Figure 9-23.

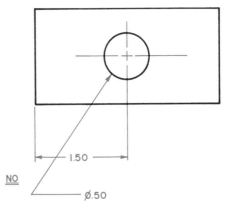

Figure 9-23 Avoid crossing dimension
lines

Specific Dimensioning Techniques

All of the information on dimensioning so far has
been of a general nature. The following sections deal
with the techniques used for applying dimensions to
specific situations that are recurrent in drafting. With
a thorough knowledge of the general information presented earlier, and the specific information presented
in these sections, drafters will be able to dimension
any situation confronted on technical drawings.

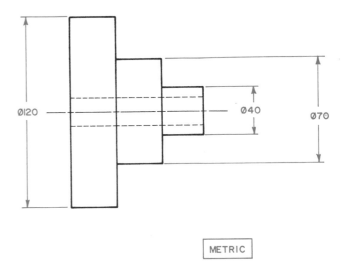

Figure 9-24 Dimensioning diameters in
longitudinal views

Dimensioning Diameters

Diameters may be dimensioned using extension
and dimension lines or using leader lines. In either
case, the diameter dimension is preceded by a diameter symbol ⌀. If the diameter is a spherical diameter, the diameter symbol is preceded by a capital

"S." Figure 9-24 illustrates how diameters may be
dimensioned in longitudinal views. Figure 9-25 illustrates how diameters are dimensioned in views where
they appear as circles. Figure 9-26 illustrates the
dimensioning of a spherical diameter.

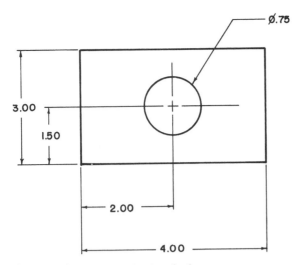

Figure 9-25 Dimensioning holes

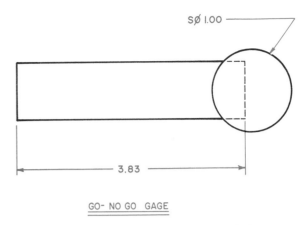

GO- NO GO GAGE

Figure 9-26 Dimensioning a spherical diameter

Dimensioning Radii

The symbol for radius is "R." The symbol for spherical radius is "SR." Each time a radius is called out, it must be preceded by either the R or SR symbol. Radii are dimensioned using leader lines which terminate in arrowheads. Figures 9-27 and 9-28 illustrate various ways in which radii are dimensioned.

Dimensioning Chords, Arcs, and Angles

Chords, arcs, and angles are dimensioned in a similar manner. When dimensioning a chord, the dimension line should be perpendicular to the extension lines and parallel to the chord, Figure 9-29.

When dimensioning an arc, the dimension line runs concurrent with the arc curve, but the extension lines are either vertical or horizontal, Figure 9-30. An arc symbol is placed above the dimension.

When dimensioning an angle, the extension lines extend from the sides forming the angle, and the dimension line forms an arc, Figure 9-31.

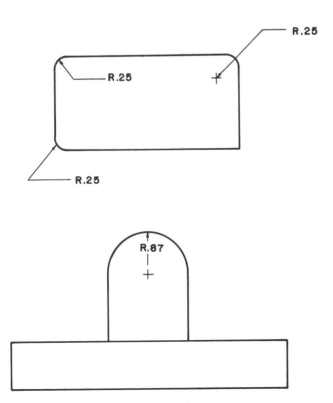

Figure 9-27 Dimensioning radii

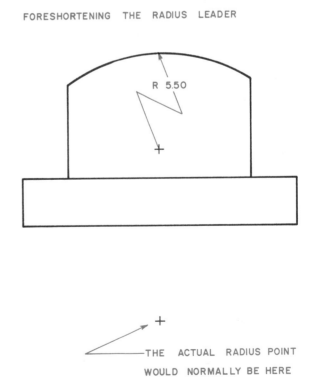

FORESHORTENING THE RADIUS LEADER

Figure 9-28 Dimensioning the radius

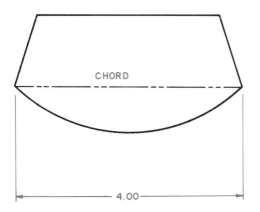

Figure 9-29 Dimensioning a chord

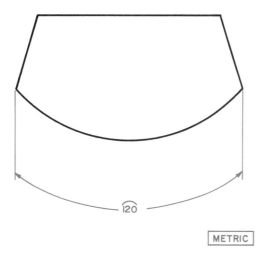

Figure 9-30 Dimensioning an arc

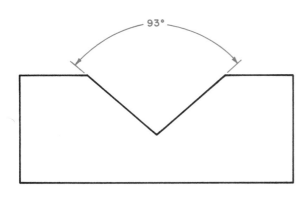

Figure 9-31 Dimensioning an angle

Dimensioning Rounded Ends

Parts with rounded ends are of two types: fully rounded and partially rounded. A part with a fully rounded end or ends is dimensioned using overall length and width dimensions. Radii are indicated, but not dimensioned, Figure 9-32.

Parts with only partially rounded ends require dimensioned radii. Overall dimensions for length and width should also be given, Figure 9-33.

Dimensioning Rounded Corners

Parts with rounded corners are common in manufacturing. The part is dimensioned as if the rounded corner is square. Then, a radius dimension is added for the rounded corner, Figure 9-34.

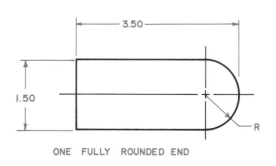

ONE FULLY ROUNDED END

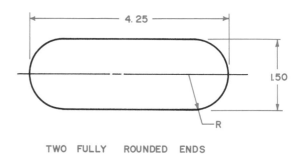

TWO FULLY ROUNDED ENDS

Figure 9-32 Dimensioning rounded ends

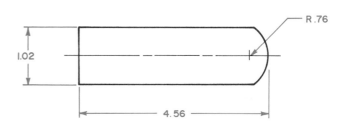

Figure 9-33 Dimensioning a partially rounded part

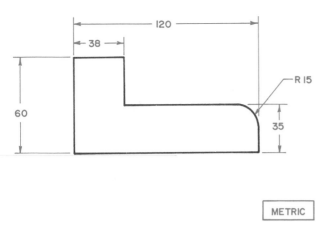

Figure 9-34 Dimensioning a rounded corner

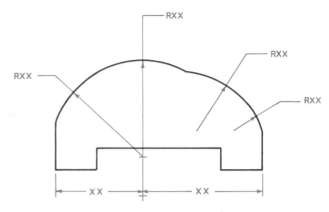

Figure 9-35 Dimensioning arc outlines

Dimensioning Arc Outlines

An object profile consisting of a series of arcs is common in manufacturing. Each arc in the series has a radius and each radius must be dimensioned. Radius points which fall outside of the object must be located using normal coordinate dimensions. Other radii are located according to their points of tangency.

Figure 9-35 contains an example of an object consisting of a series of arcs. Notice how the unusually long radii are dimensioned.

Dimensioning Round Holes

One of the most common dimensioning situations is the round hole. Holes drilled in objects may either pass through the object or penetrate it to a specified depth. Depending on the nature of the drawing, it may or may not be clear graphically which is the case.

When it is not clear as to whether a hole passes through an object or only penetrates it partially, the condition must be clarified by the dimension. Figures 9-36 through 9-39 illustrate the various methods used for dimensioning round holes.

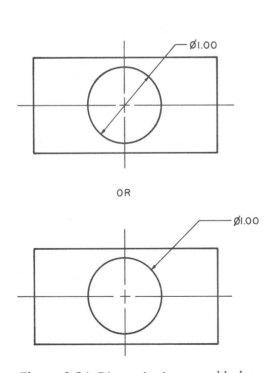

Figure 9-36 Dimensioning round holes

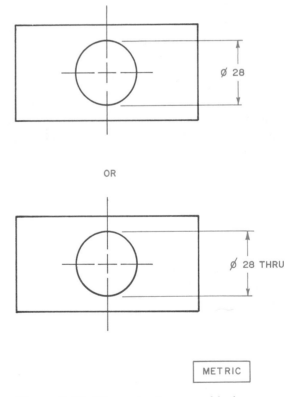

Figure 9-37 Dimensioning round holes

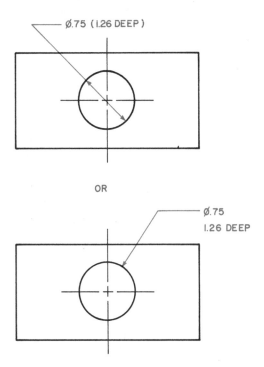

Figure 9-38 Dimensioning round holes

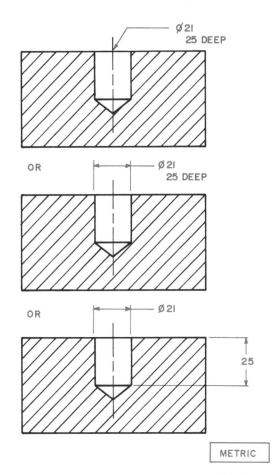

METRIC

Figure 9-39 Dimensioning round holes

Dimensioning Slotted Holes

Slotted holes are a common dimensioning situation. Several different methods are used in dimensioning slotted holes. The most commonly used method involves dimensioning the overall length and width of the slot and calling out, but not dimensioning, the radii, Figure 9-40.

Another method involves dimensioning the overall width of the slot, the distance between the radii centers, and calling out, but not dimensioning, the radii, Figure 9-41.

Still another method involves using leader lines to call out the overall length and width of the slot and calling out, but not dimensioning, the radii, Figure 9-42.

Dimensioning Counterbored Holes

Holes in manufactured parts are frequently *counterbored* to allow for a flush fitting of a fastener. Two methods are used for dimensioning counterbored holes.

The diameter of the hole, the diameter of the counterbore (CBORE), and the depth of the counterbore may be called out using a leader line, Figure 9-43. Another method involves calling out the diameter of the hole and the counterbore on a plan or frontal view and dimensioning the depth of the counterbore on a longitudinal or depth view, Figure 9-44.

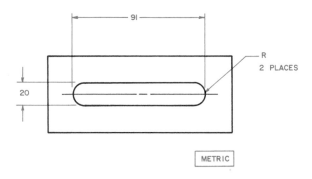

Figure 9-40 Dimensioning a slotted hole

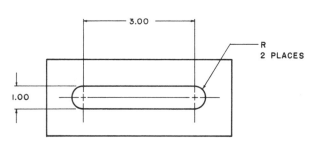

Figure 9-41 Dimensioning a slotted hole

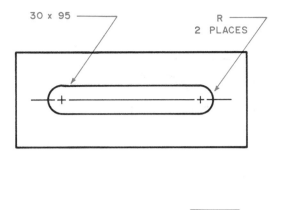

METRIC

Figure 9-42 Dimensioning a slotted hole

Dimensioning Countersunk Holes

Countersinking a hole in manufacturing a part is a process similar to counterboring. The major difference is that the countersink (CSK) is at an angle to accommodate a different head style on a fastener or for clearance. A countersunk hole may be dimensioned in one of two ways.

The first method involves calling out the diameter of the hole, the diameter of the countersink, and the angle of the countersink using a leader line, Figure 9-45. The second method involves calling out the diameter of the hole and the diameter of the countersink on a plan or frontal view and dimensioning the countersink angle on a longitudinal or depth view, Figure 9-46.

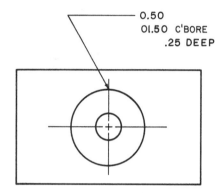

Figure 9-43 Dimensioning a
counterbored hole

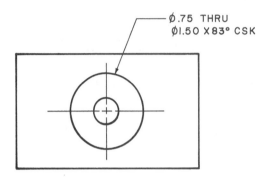

Figure 9-45 Dimensioning a
countersunk hole

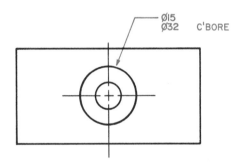

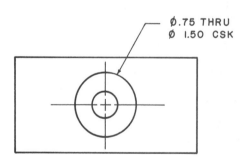

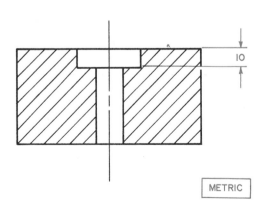

METRIC

Figure 9-44 Dimensioning a counterbored hole

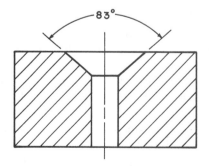

Figure 9-46 Dimensioning a countersunk hole

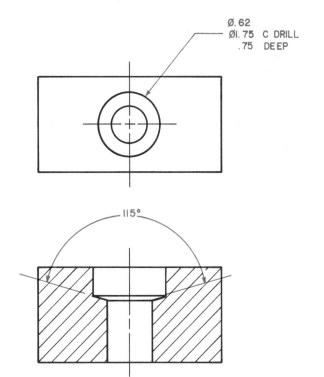

Figure 9-47 Dimensioning a counterdrilled hole

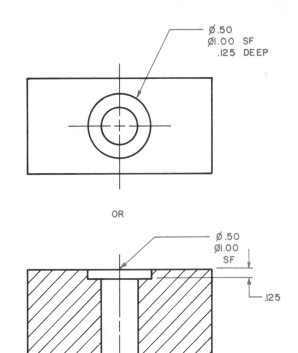

OR

Figure 9-48 Dimensioning a spotface

Dimensioning Counterdrilled Holes

Dimensioning a *counterdrilled* hole involves specifying the diameter of the hole, the diameter of the counterdrill (CDRILL), the depth of the counterdrill, and the angle of the included angle of the counterdrill. The first three items of information can be placed on a plan or frontal view. The angle, which is optional in terms of dimensioning, may be shown on a longitudinal or depth view, Figure 9-47.

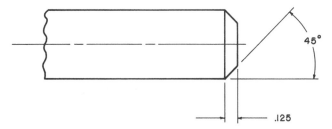

Figure 9-49 Dimensioning a chamfer

Dimensioning Spotfaces

A *spotface* (SF) is frequently used in manufacturing parts to provide an accurately machined surface seat for a fastener or a washer. A spotface involves dimensioning the diameter of the drilled hole, dimensioning the diameter of the spotface, and dimensioning either the depth of the spotface or the remaining depth of the part; usually a flange, Figure 9-48.

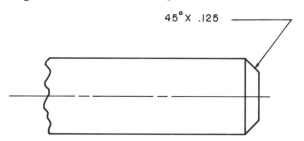

Figure 9-50 Dimensioning a chamfer

Dimensioning Chamfers

Chamfers are used to avoid sharp edges on machined parts and to provide essential clearances so as to avoid interferences. Several methods are used for dimensioning chamfers. In the first method, an angle and a dimension are used, Figure 9-49. In the second method, the chamfer angle and linear dimension are called out using a leader line, Figure 9-50.

Internal chamfers require a chamfer angle and a diameter dimension, as shown in Figure 9-51. Chamfers applied to edges of less than 90° require an angle

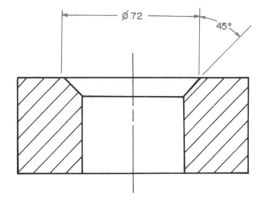

Figure 9-51 Dimensioning an internal chamfer

METRIC

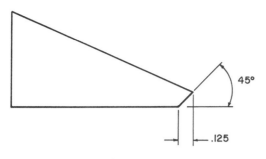

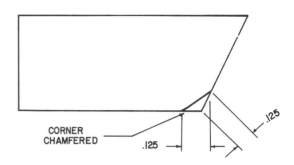

Figure 9-52 Dimensioning a chamfer on edges less than 90°

Figure 9-53 Dimensioning a chamfer on edges of more than 90°

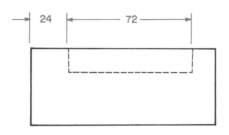

Figure 9-54 Dimensioning a keyseat

METRIC

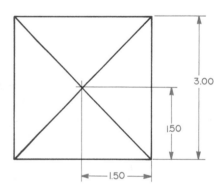

and a linear dimension, as shown in Figure 9-52. Chamfers applied to edges of more than 90° require linear dimensions with extension lines that are perpendicular to the subject edges, Figure 9-53.

Dimensioning Keyseats

A *keyseat* is a slot in a shaft to hold a key. Keyseats represent a fairly common dimensioning situation on manufactured parts. Dimensioning a keyseat involves dimensioning the width, depth, and length, as shown in Figure 9-54. Notice that the keyseat must be located if it does not extend the entire length of the part. It must also be located if it is allowed to run out at the shaft.

Dimensioning Geometric Shapes

Special techniques are called for when dimensioning such geometric shapes as pyramids, cones, spheres, and tapered objects. A rule of thumb to follow when dimensioning such shapes is to *provide only those dimensions necessary to define the shape and nothing more.*

A pyramid requires base and height dimensions. An intermediate dimension showing the distance from the apex of the pyramid to any corner in a top orthographic view can also be provided, but is not required, Figure 9-55.

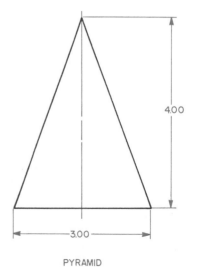

PYRAMID

Figure 9-55 Dimensioning a pyramid

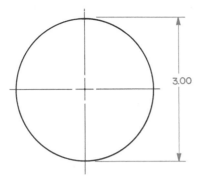

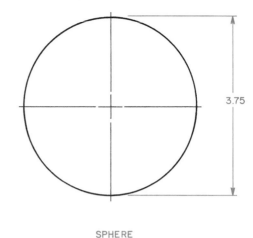

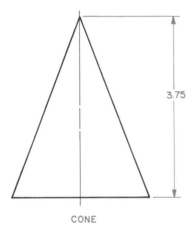

CONE

Figure 9-56 Dimensioning a cone

SPHERE

Figure 9-57 Dimensioning a sphere

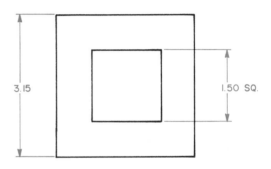

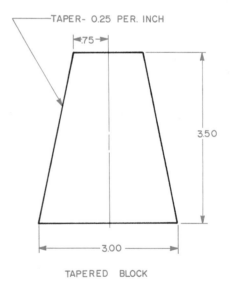

TAPERED BLOCK

Figure 9-58 Dimensioning a tapered object

A cone requires only two dimensions: a height dimension and a base diameter dimension, Figure 9-56. Dimensioning a sphere is even simpler. A sphere requires only a diameter dimension, Figure 9-57.

Tapered objects may be cones or pyramids in their overall shape. Tapered objects are dimensioned according to the rules of their basic shapes and, then, a note specifying the degree of taper is added using a leader line, Figure 9-58.

Locational Dimensioning Systems

Individual and multiple features on manufactured parts must be located from some *datum* (reference point or plane) and, on occasion, with respect to one another. Holes are the most frequently dimensioned features located with respect to some datum. Their location with regard to a specified datum is critical in some cases. In other cases, hole locations with regard to other holes is critical. Which case applies depends on the function and nature of the part and its related features in the final assembly process.

Two basic dimensioning systems are used for locating features on manufactured parts: rectangular coordinate dimensioning and polar coordinate dimen-

sioning. *Rectangular coordinate dimensioning* involves locating features using linear dimensions from specified X-Y or X-Y-Z axes. *Polar coordinate dimensioning* involves the use of both linear dimensions and angular dimensions for locating features.

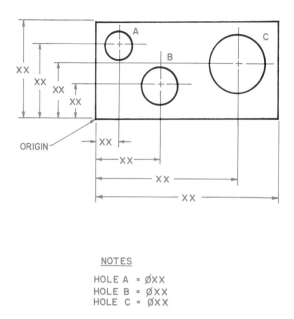

NOTES

HOLE A = ∅XX
HOLE B = ∅XX
HOLE C = ∅XX

Figure 9-59 Rectangular coordinate dimensions

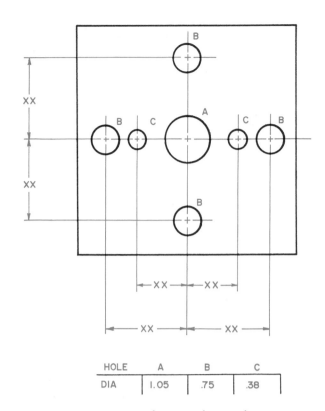

HOLE	A	B	C
DIA	1.05	.75	.38

Figure 9-60 Rectangular coordinate dimensions

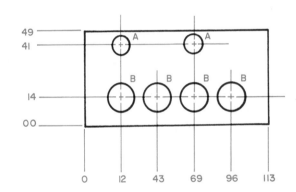

HOLE DATA			
HOLE	QUAN	DIA	DEPTH
A	2	7	10
B	4	12	10

METRIC

Figure 9-61 Rectangular coordinate dimensions

Rectangular Coordinate Dimensions

Rectangular coordinate dimensions can be applied to a part in several different ways: with linear dimensions, in table form, and in a modified dimensioning format in which dimension lines are left off.

The first method is illustrated in Figures 9-59 and 9-60. In Figure 9-59, the X-Y coordinate system of the rectangular part is used in locating the various holes from a hypothetical zero origin. Notice that the hole diameters are specified in a note rather than with leaders. A more complex part would require that the holes be shown in a table. This is done to avoid overcrowding the drawing, and is a method frequently used with rectangular coordinate dimensioning. Moreover, it lends itself to production by numerical control (NC), computer numerical control (CNC), or computer-aided manufacturing (CAM).

Figure 9-60 illustrates how rectangular coordinate dimensioning is applied to circular parts. The datum plane for referencing dimensions in this illustration is the center line of the part. The various other centers are located in relation to the center of the part. Notice that the hole diameters have been specified in tabular form. This is another method frequently used with rectangular coordinate dimensioning to avoid overly complicating the drawing.

Another method used in rectangular coordinate dimensioning involves giving dimensions from an X-Y or X-Y-Z origin, but leaving off the dimension lines. The dimensions are placed at the ends of the extension lines, Figure 9-61. The hole sizes are tabulated. Notice that each dimension refers to the distance from the origin to the hole center.

A final method used in rectangular coordinate dimensioning takes the method illustrated in Figure 9-61 one step farther. In this method, dimensions and hole sizes are tabulated, Figure 9-62. Dimensions are referenced to a three-dimensional X-Y-Z axis. The only dimensions placed on the part are overall length, width, and depth.

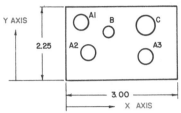

DIMENSIONING DATA			
HOLE	X	Y	Z
A I	.48	1.50	THRU
A2	.90	.60	
A3	2.38	.60	
B	1.60	1.00	
C	2.38	1.50	THRU

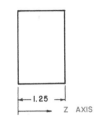

HOLE DATA		
HOLE	DIA.	QUANT.
A	.38	3
B	.25	1
C	.50	1

HOLE	A	B
DIA	26	12

Figure 9-62 Tabulating dimensions

Polar Coordinate Dimensioning

In *polar coordinate dimensioning*, linear dimensions and angular dimensions are used to locate features from a fixed point. This concept is illustrated in Figure 9-63. Hole diameters may still be tabulated for the purpose of simplicity.

Control of Surface Quality

The quality or texture of the surface of a metal part can vary according to the purpose of the part and its interaction with mating parts. A *finished surface* is one that has been machined through various processes to a specified texture. The actual texture is specified on drawings using finish symbols. Over the years, the symbols used by drafters and designers for specifying surface texture have changed. The evolution of these symbols is illustrated in Figure 9-64. Figure 9-65 shows the proper proportions for the finish symbol used today.

Surface texture is defined as the distance between the lowest and highest points of irregularity. Such measurements are made in microinches or micrometres, Figure 9-66. Figure 9-67 contains a chart of common finished surface textures.

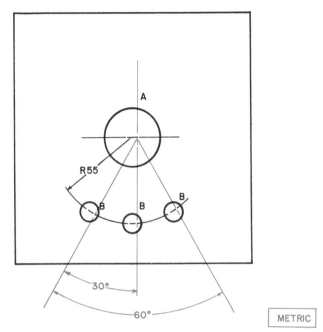

Figure 9-63 Polar coordinate dimensions

Notation

A good drawing may be defined as one that contains all of the information required by the various design and manufacturing people who will use it in producing the subject part. Most of this information can be conveyed graphically, using standard dimensioning practices. However, it is not uncommon to encounter a situation in which all of the needed information cannot be communicated graphically. In these cases, notes are used to communicate or clarify the designer's intent.

Notes are brief, carefully worded statements placed on drawings to convey information not covered or

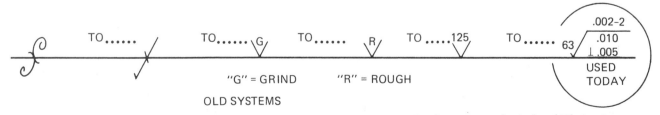

Figure 9-64 Evolution of the finish symbol (*From Drafting for Trades and Industry — Mechanical and Electronic, John Nelson. Delmar Publishers Inc.*)

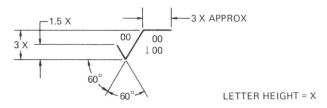

Figure 9-65 Proportions of the finish symbol (*From Drafting for Trades and Industry — Mechanical and Electronic, John Nelson. Delmar Publishers Inc.*)

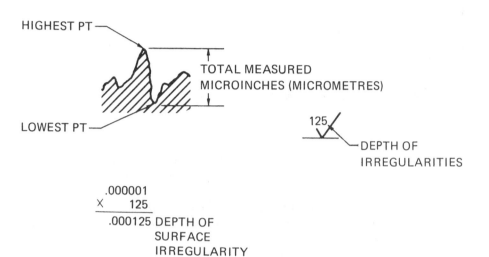

Figure 9-66 Measuring surface texture (*From Drafting for Trades and Industry — Mechanical and Electronic, John Nelson. Delmar Publishers Inc.*)

ROUGHNESS		KIND OF SURFACE	USAGE
μm	μin.		
12.5	500	Rough	Used where vibration or stress concentration are not critical and close tolerances are not required.
6.3	250	Medium	For general use where stress requirements and appearance are of minimal importance.
3.2	125	Average smooth	For mating surfaces of parts held together by bolts and rivets with no motion between them.
1.6	63	Smoother than average finish	For close fits or stressed parts except rotating shafts, axles, and parts subject to vibrations.
0.8	32	Fine finish	Used for such applications as bearings.
0.4	16	Very fine finish	Used where smoothness is of primary importance such as high-speed shaft bearings.
0.2	8	Extremely fine finish	Use for such parts as surfaces of cylinders (engines).
0.1	4	Super fine finish	Used on areas where surfaces slide and lubrication is not dependable.

Figure 9-67 Surface texture chart (*From Drafting for Trades and Industry — Mechanical and Electronic, John Nelson. Delmar Publishers Inc.*)

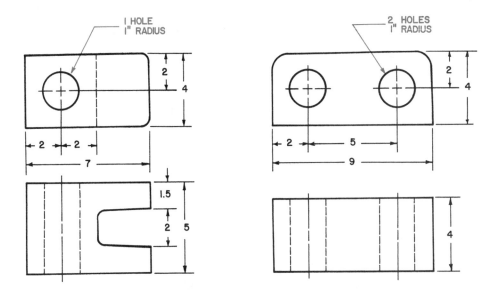

Figure 9-68 Notes on drawings improve communication

Rules for Applying Notes on Drawings

not adequately explained using graphics, Figure 9-68. Notes should be clearly worded so as to allow only one correct interpretation.

There are no ANSI standards specifically governing the use of notes on technical drawings. However, several rules of a general nature should be observed. These rules apply to both general notes and the more specific detail notes.

Notes may be lettered freehand, entered using a keyboard in a computer-aided drafting (CAD) system, or through any one of several mechanical lettering processes. Sample notes are included on the drawings in Figures 9-69 and 9-70. In any case, regardless of how they are put on the drawing, notes should be oriented horizontally on the drafting sheet, Figure 9-71. General notes should be located directly above the title block, Figure 9-72. When using manual processes to apply general notes such as those in Figure 9-72, the first note is placed directly above the title block, the second is placed on top of it and so on up the line. This allows notes to be added as needed without renumbering. However, when using a CAD system, this is not necessary since one of the advantages of CAD is that notes can be renumbered and rearranged automatically. Detail notes should

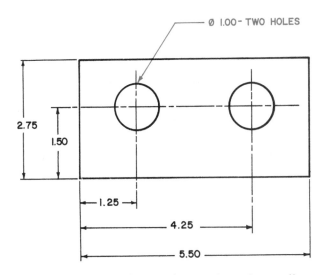

Figure 9-69 Sample note lettered mechanically

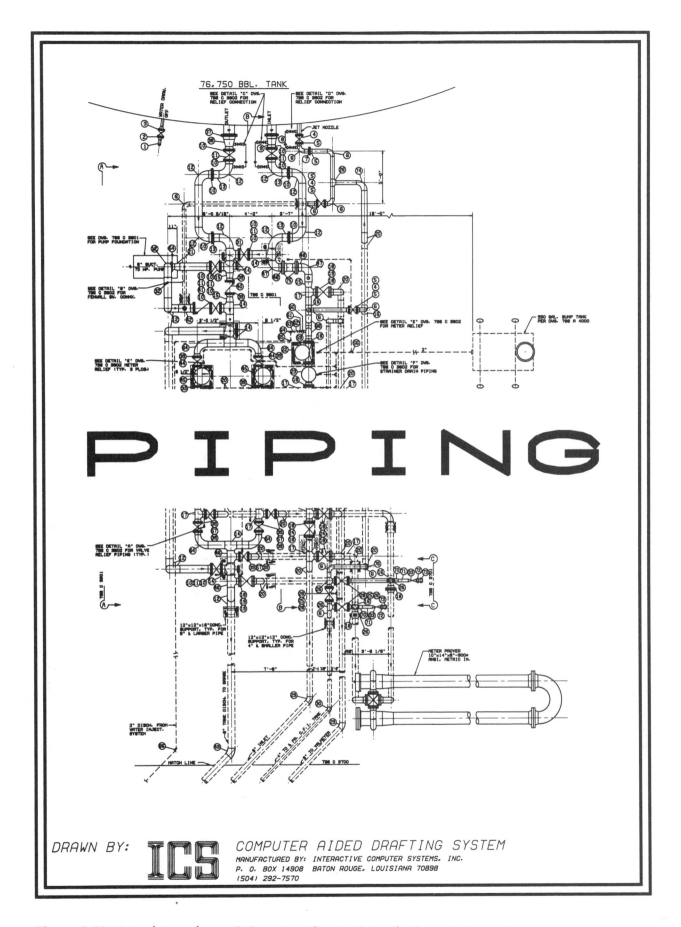

Figure 9-70 Notes lettered on a CAD system (*Courtesy Interactive Computer Systems, Inc.*)

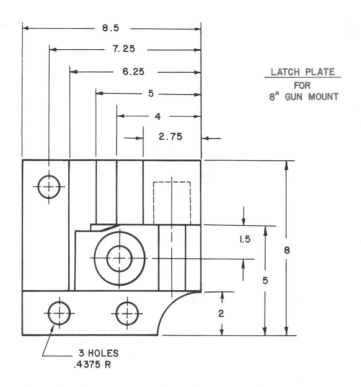

Figure 9-71 Notes are placed horizontally

be located as close as possible to the detail they are describing, and connected to it by a leader line, Figure 9-73.

Notes should be applied to drawings after all graphics have been completed. This prevents notes and graphics from overlapping, and minimizes other technique problems such as smudging the worksheet and frequent erasing of notes as changes are required on a drawing.

Two basic types of notes are used on technical drawings: general notes and detail notes. They serve different purposes and, therefore, must be examined separately.

General Notes

General notes are broad items of information which have a job- or project-wide application rather than relating to just one single element of a project or a part. They are usually placed immediately above the title block on the drawing sheet and numbered sequentially.

Information placed in general notes includes such characteristics of a product as finish specifications; standard sizes of fillets, rounds, and radii; heat treatment specifications; cleaning instructions; general tolerancing data; hardness testing instructions; and stamping specifications. Figures 9-74 and 9-75 are examples of drawings containing general notes.

Detail Notes

Detail notes are specific notes that pertain to one particular element or characteristic of a part. They are placed as near as possible to the characteristic to which they apply, and are connected using a leader line, Figure 9-76.

Detail notes should not be placed on views, Figure 9-77. The only exception to this rule is in cases where a great deal of open space exists on a view,

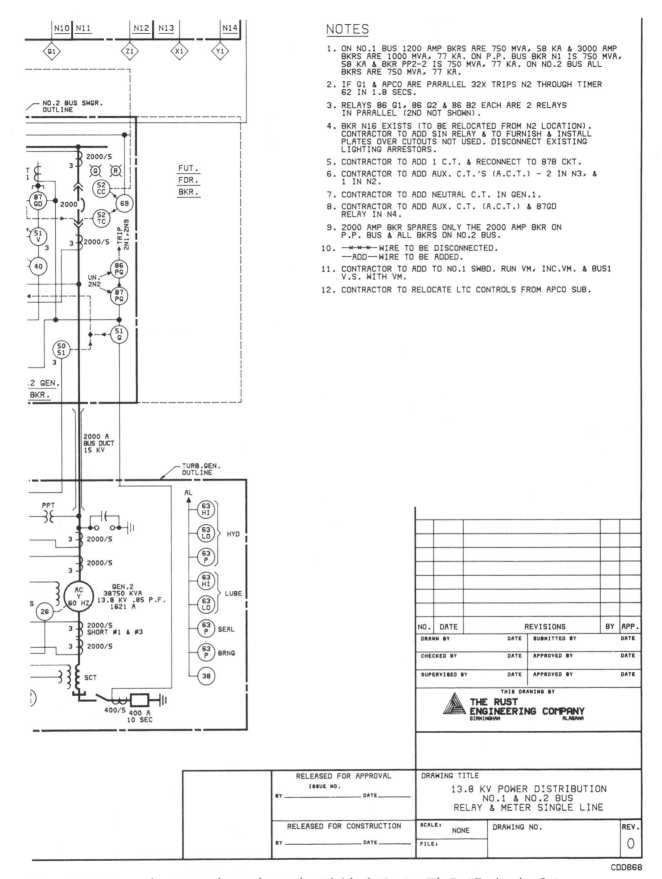

1. ON NO.1 BUS 1200 AMP BKRS ARE 750 MVA, 58 KA & 3000 AMP BKRS ARE 1000 MVA, 77 KA. ON P.P. BUS BKR N1 IS 750 MVA, 58 KA & BKR PP2-2 IS 750 MVA, 77 KA. ON NO.2 BUS ALL BKRS ARE 750 MVA, 77 KA.

2. IF G1 & APCO ARE PARALLEL 32X TRIPS N2 THROUGH TIMER 62 IN 1.8 SECS.

3. RELAYS 86 G1, 86 G2 & 86 B2 EACH ARE 2 RELAYS IN PARALLEL (2ND NOT SHOWN).

4. BKR N16 EXISTS (TO BE RELOCATED FROM N2 LOCATION). CONTRACTOR TO ADD SIN RELAY & TO FURNISH & INSTALL PLATES OVER CUTOUTS NOT USED. DISCONNECT EXISTING LIGHTING ARRESTORS.

5. CONTRACTOR TO ADD 1 C.T. & RECONNECT TO 87B CKT.

6. CONTRACTOR TO ADD AUX. C.T.'S (A.C.T.) - 2 IN N3, & 1 IN N2.

7. CONTRACTOR TO ADD NEUTRAL C.T. IN GEN.1.

8. CONTRACTOR TO ADD AUX. C.T. (A.C.T.) & 87GD RELAY IN N4.

9. 2000 AMP BKR SPARES ONLY THE 2000 AMP BKR ON P.P. BUS & ALL BKRS ON NO.2 BUS.

10. —✶—✶—✶— WIRE TO BE DISCONNECTED.
    —ADD— WIRE TO BE ADDED.

11. CONTRACTOR TO ADD TO NO.1 SWBD. RUN VM, INC.VM. & BUS1 V.S. WITH VM.

12. CONTRACTOR TO RELOCATE LTC CONTROLS FROM APCO SUB.

NO.	DATE		REVISIONS		BY	APP.

DRAWN BY	DATE	SUBMITTED BY	DATE
CHECKED BY	DATE	APPROVED BY	DATE
SUPERVISED BY	DATE	APPROVED BY	DATE

THIS DRAWING BY

THE RUST ENGINEERING COMPANY
BIRMINGHAM   ALABAMA

RELEASED FOR APPROVAL	DRAWING TITLE
ISSUE NO.	13.8 KV POWER DISTRIBUTION NO.1 & NO.2 BUS RELAY & METER SINGLE LINE
BY _____ DATE _____	

RELEASED FOR CONSTRUCTION	SCALE: NONE	DRAWING NO.	REV.
BY _____ DATE _____	FILE:		0

CDD868

Figure 9-72 General notes are located over the titleblock (*Courtesy The Rust Engineering Co.*)

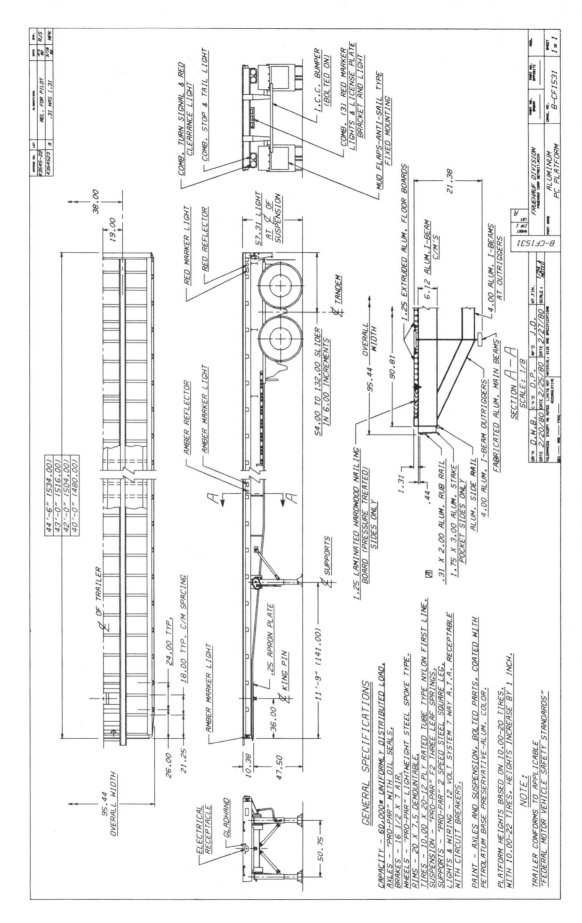

Figure 9-73 Detail notes using leader lines (*Courtesy Fruehauf Division, Fruehauf Corp.*)

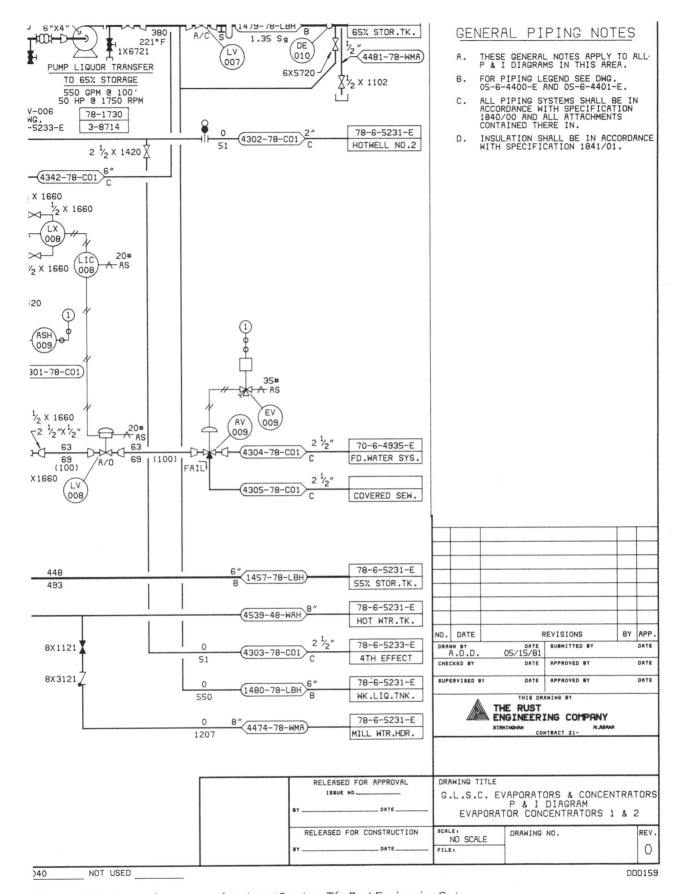

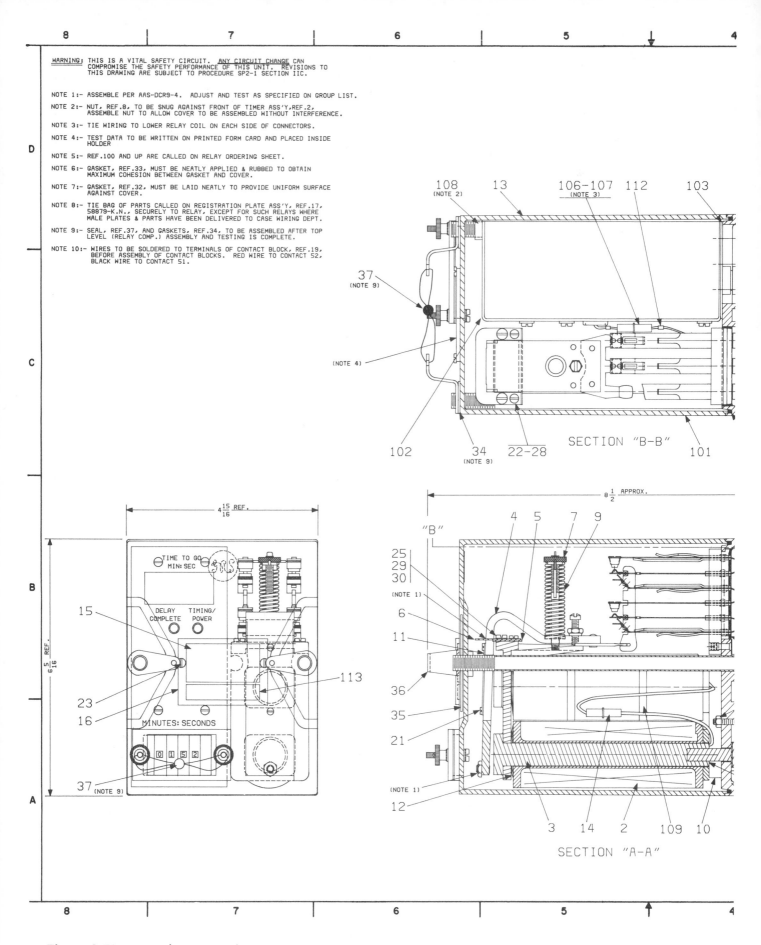

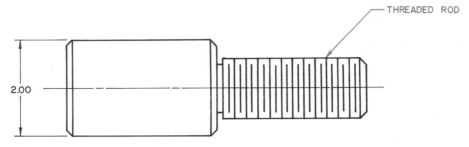

Figure 9-76 Detail note connected with a leader line

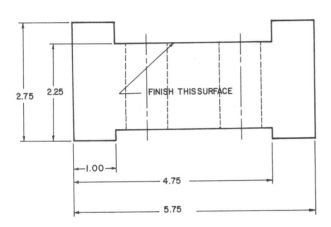

Figure 9-77 Poor placement of a detail note

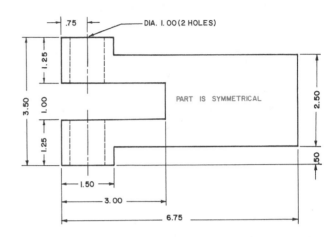

Figure 9-78 Acceptable placement of a detail note when there is space

but very little around it, Figure 9-78. Notes should never be superimposed over other data such as dimensions, lines or symbols, Figure 9-79.

Common sense is the best rule to follow in applying detail notes. Since detail notes are used to more completely communicate or to clarify intent, they should be placed as close as possible to the element to which they pertain and in such a way as to be easily read.

Writing Notes

The written word lends itself to interpretations that may vary. Thus, notes used on drawings must convey the exact intent of the designer. Consequently, drafters must be especially careful in the preparation of notes.

The first step is to place in a legend all abbreviations used in the notes, Figure 9-80. This ensures that all readers of the notes interpret the abbreviations consistently and correctly.

Another technique which will limit misinterpretation of notes is punctuation or indentation. Notes containing more than one sentence should be properly punctuated so that readers know where one sentence leaves off and the next one begins, Figure 9-81. Indentation is used when the length of a note requires

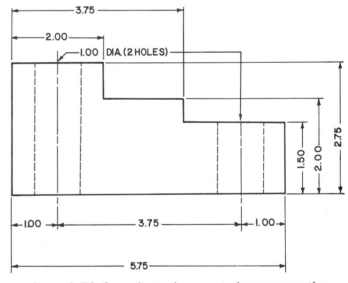

Figure 9-79 Superimposing a note is poor practice

more than one line. The second line is indented so that readers know it is part of the first line, Figure 9-82.

Another technique is particularly useful for beginning drafters. It is called *verification*. Notes to be placed on drawings are first jotted down, legibly, on a print of the drawing. Another drafter, preferably one with experience, is then asked to read the notes to ensure that they are not open to multiple interpretations.

LEGEND

1. ASSY = ASSEMBLY
2. FAO = FINISH ALL OVER
3. MS = MACHINE STEEL
4. FAB = FABRICATE
5. CARB = CARBURIZE
6. DVTL = DOVETAIL
7. HT TR = HEAT TREAT
8. KST = KEYSEAT
9. TPR = TAPER
10. SF = SPOTFACED

Figure 9-80 Abbreviations are explained in a legend

Note Specifications

A number of specifications should be observed when writing notes. The most widely accepted lettering style for notes is uppercase (all capital) block Gothic letters in a vertical format. Some companies accept uppercase block Gothic lettering in an inclined or slanted format. The proper lettering height is one-quarter inch for titles such as GENERAL NOTES, and one-eighth to three-sixteenths inch for actual notes, Figure 9-83.

Spacing between successive lines of the same note should be approximately one-sixteenth inch. Spacing between separate notes should be approximately one-eighth inch, Figure 9-84.

There are no hard and fast rules governing the length of a line of notation, but from four to six words per line is widely accepted, Figure 9-85. When a note string will contain more than six words, it should be divided into more than one line. The number of words in the multiple line note should be divided so as to balance the finished note. One long line followed by a drastically shorter line is bad form, Figure 9-86.

On occasion, the last word in a note string will have to be hyphenated. There are rules governing the dividing of words according to syllables. If the proper point of division is not obvious from the makeup of the word, consult a dictionary. Do not hyphenate abbreviated words.

Sample Notes

The following is a list of sample notes of the types frequently used on technical drawings:

- All fillets and rounds R1/8
- 45 degree chamfer all edges

NOTE

POWER BRUSH ALL SURFACES. APPLY THREE COATS PAINT. STAMP PART WITH NUMBER 01929612.

Figure 9-81 Proper punctuation of notes is important

NOTE

ALL DIMENSIONS AND CONDITIONS SHALL BE FIELD CHECKED AND VERIFIED BY PROPER TRADES

Figure 9-82 Indenting for clarity

GENERAL NOTES ← LETTERS .25 HIGH

ALL FILLETS AND ROUNDS .125 DIA.
ALL SURFACES MUST BE FREE OF BURRS.
LETTERS .125-.1875 HIGH

Figure 9-83 Proper lettering height is important

NOTES

1. SPACING BETWEEN LINES OF THE SAME NOTE SHOULD BE .0625.

2. SPACING BETWEEN SEPARATE NOTES SHOULD BE .1250.

Figure 9-84 Spacing of notes

NOTES
NOTES SHOULD BE LIMITED TO 4 TO 6 WORDS PER LINE

FOUR TO SIX WORDS

Figure 9-85 Number of words per line limitations are important

NOTES
AVOID LONG LINE IN A NOTE FOLLOWED BY ONE SHORT LINE

Figure 9-86 Lines in notes should be approximately equal

- Surfaces A & B parallel
- Stock .125 thick
- Heat treat
- All internal radii R.0625
- FAO
- Ream for 1/4 dowels
- All bends R5/32
- CBORE from bottom
- Identical bolts—both sides
- 3/32 × 1/16 oil groove
- 87 CSK 3/4 dia—4 holes
- Rounds .125 R unless otherwise specified
- 4 holes equally spaced
- 1/32 × 45 chamfer
- #7 Drill .75 deep
- Machine steel—4 reqd.
- Power brush all ext surfaces
- Sandblast before painting

Notice that these sample notes are very brief, concise, and to the point. Words such as "a," "an," "the," and "are" are used only sparingly. The notes are not always complete sentences in terms of proper grammar, but they are complete thoughts in terms of communication. Notice also the use of abbreviations such as FAO (finish all over), CBORE (counterbore), and CSK (countersink). Abbreviations can be used frequently to cut down on the length of notes and the time required to letter them. However, when used, abbreviations should always be placed in a legend to ensure consistency of interpretation.

Review

1. What is the most frequently used dimensioning standard?

2. How does a drafter indicate on a drawing that a dimension is NTS?

3. Of the three dimensioning systems used, which is most frequently used on mechanical technical drawings?

4. Illustrate how to label a metric dimension that is less than one millimetre.

5. How should the line separating the numerator and denominator of a fraction be drawn?

6. List the five components of all dimensioning systems.

7. How far should an extension line extend beyond the dimension line?

8. What is the recommended length/width ratio of arrowheads?

9. What is the keyword to remember in making arrowheads, regardless of the size?

10. What is unidirectional dimensioning?

11. Define the term *notes*.

12. Describe the proper location of general notes.

13. How should notes be oriented on the drafting sheet?

14. Define the term *general notes*.

15. What is the proper height for lettering note titles?

16. Define the term *verification* as it relates to notes.

17. Explain the significance of balance in terms of notation.

Chapter Nine Problems

The following problems are intended to give the beginning drafter practice in applying the various dimensioning techniques required on technical drawings in industry. Students should use the scale provided for determining dimensions, and apply the dimensions according to the specifications set forth in this chapter.

The steps to follow in laying out all problems in this chapter are:

Step 1. Study the problem carefully.

Step 2. Using the scale provided, determine all necessary dimensions.

Step 3. Using construction lines, lay out three complete views being careful to place them properly in the work area.

Step 4. Darken all lines using the proper line thicknesses.

Step 5. Using construction lines, lay out all required dimension, extension, and leader lines.

Step 6. Darken in all extension, dimension, and leader lines using the proper line thicknesses.

Step 7. Add dimensions.

Step 8. Check to make sure all dimensions have been properly placed on the drawing according to ANSI Y14.5M-1982.

Note: Some of these problems do not follow current drafting standards. You are to use the information here to develop properly drawn and dimensioned drawings.

Problems 9-1 through 9-35

In the art for Problems 9-1 through 9-35, undimensioned drawings of mechanical parts are provided. Draw three views and properly dimension each according to ANSI Y14.5M-1982.

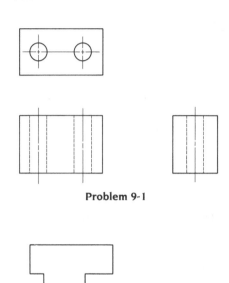

Problem 9-1

Problem 9-2

Problem 9-3

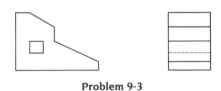

Problem 9-4

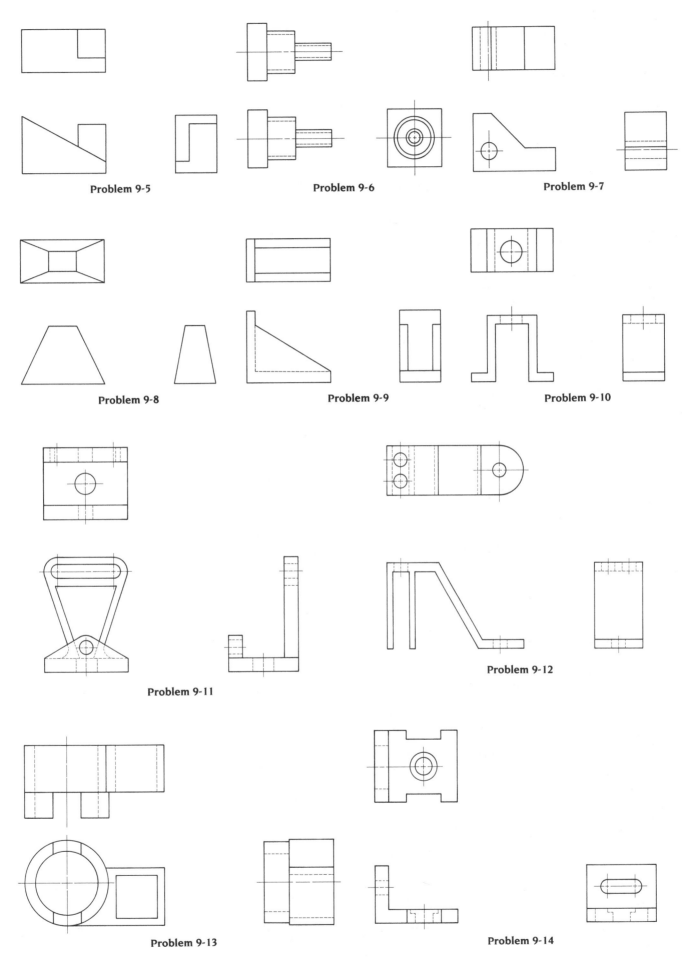

Problem 9-5

Problem 9-6

Problem 9-7

Problem 9-8

Problem 9-9

Problem 9-10

Problem 9-11

Problem 9-12

Problem 9-13

Problem 9-14

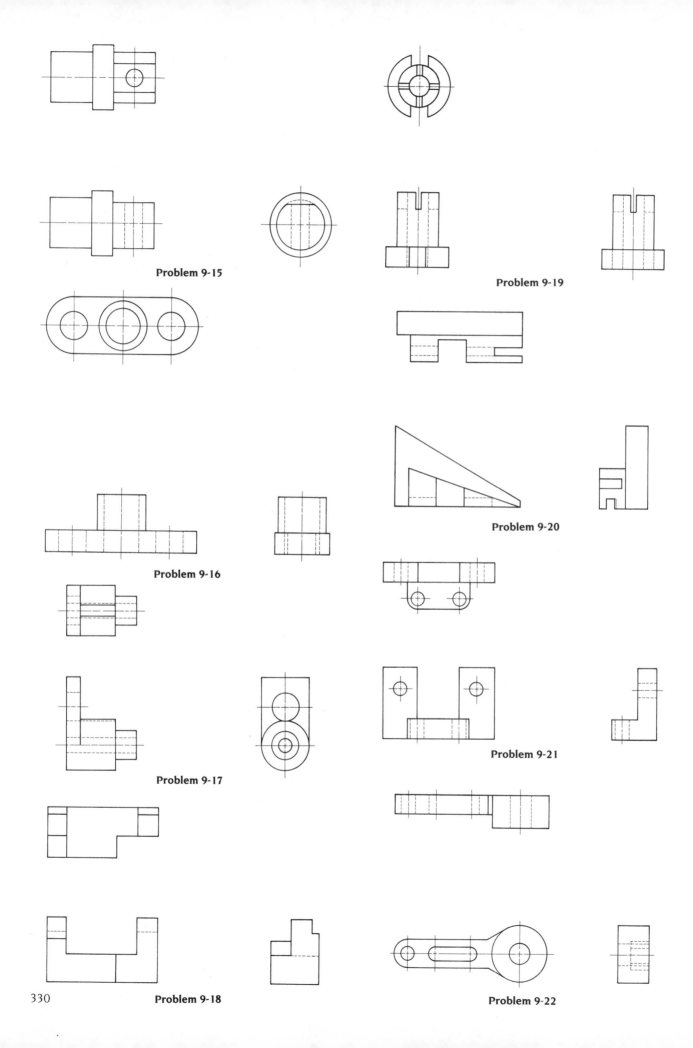

Problem 9-15

Problem 9-19

Problem 9-16

Problem 9-20

Problem 9-17

Problem 9-21

330 **Problem 9-18**

Problem 9-22

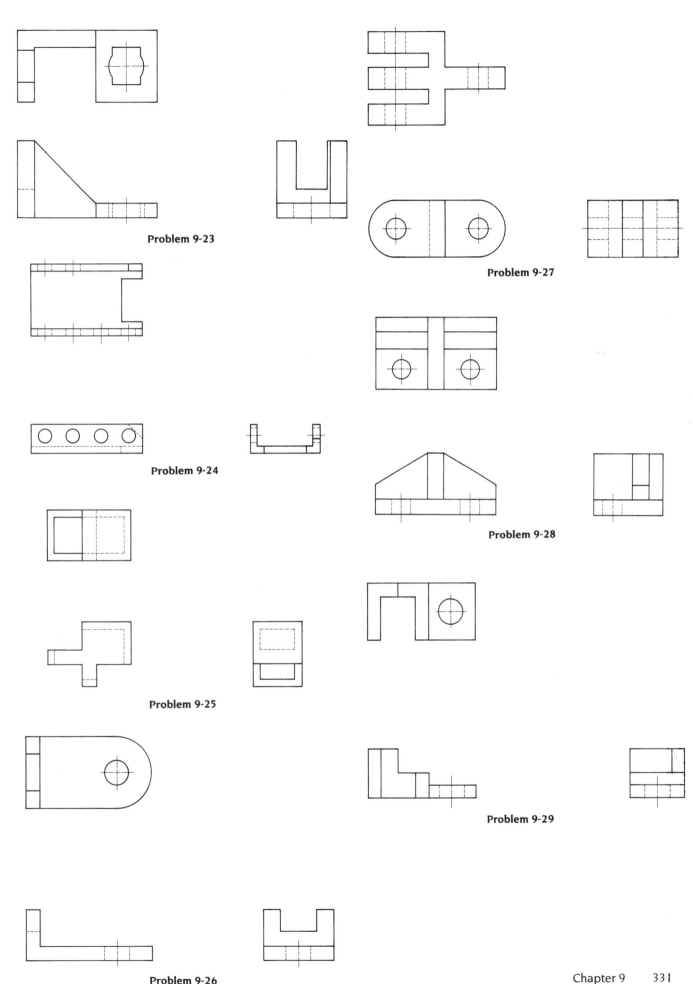

Problem 9-23

Problem 9-27

Problem 9-24

Problem 9-28

Problem 9-25

Problem 9-29

Problem 9-26

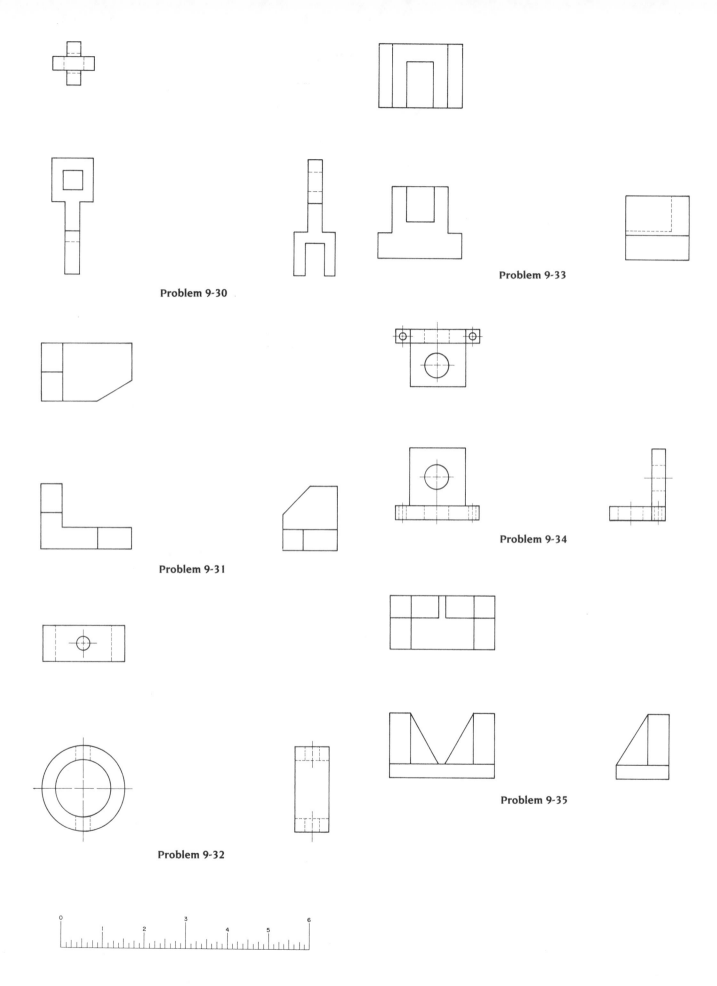

Problem 9-30

Problem 9-33

Problem 9-31

Problem 9-34

Problem 9-32

Problem 9-35

0 1 2 3 4 5 6

9-36 Given the top and front view, complete the drawing and add the following general note: SANDBLAST BEFORE PAINTING.

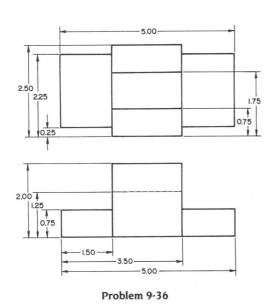

Problem 9-36

9-37 Given the top and front view, complete the drawing and add the following general note: ALL SURFACES MUST BE FLAT AND FREE OF BURRS.

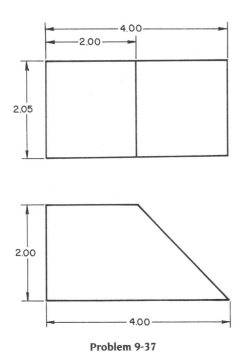

Problem 9-37

9-38 Given the top and front view, complete the drawing and add the following general notes: 1) REMOVE ALL SHARP CORNERS, 2) HONE HOLE, and 3) STAMP WITH PART NUMBER MDX 150SX.

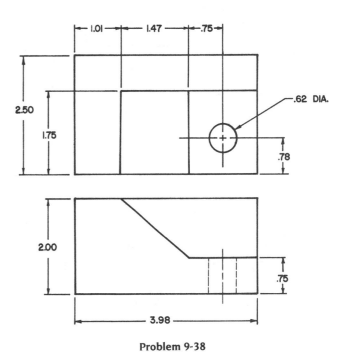

Problem 9-38

9-39 Given the top and front view, complete the drawing by adding the left-side view and add the following detail note with a leader line pointing to the inclined surface: GRIND THIS SURFACE.

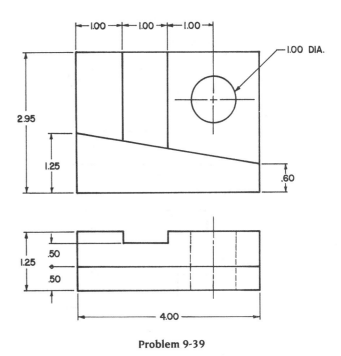

Problem 9-39

9-40 Given the top and front view, complete the drawing and add the following detail note: .50 DIA (2 HOLES).

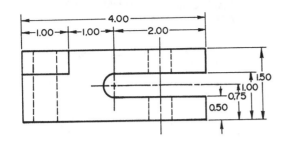

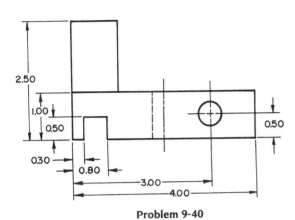

Problem 9-40

9-42 Given the top and front view, complete the drawing by adding the left-side view and add the following general notes: TOLERANCES ON ALL DIMENSIONS PLUS OR MINUS .001 UNLESS OTHERWISE SPECIFIED, and PART SHOULD BE SOLVENT-CLEANED BEFORE ASSEMBLY.

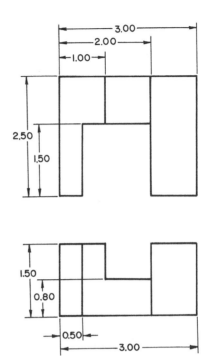

Problem 9-42

9-41 Given the top and front view, complete the drawing and add the following notes: .75 DIA HOLE, and FINISH TOP SURFACE OF PART.

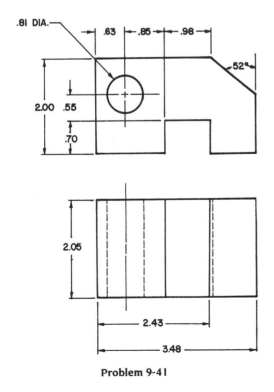

Problem 9-41

9-43 Given the top and front view, complete the drawing and add the following notes: .50 DIA HOLE, and FINISH ALL NORMAL SURFACES.

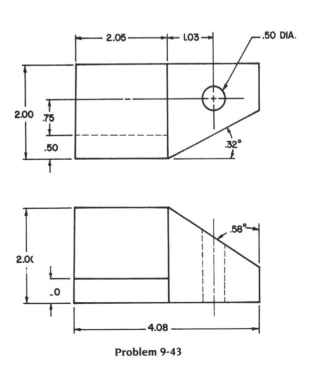

Problem 9-43

9-44 Given the top and front view, complete the drawing by adding the left-side view and add the following notes: 1.00 DIA HOLE, and ALL SURFACES MUST BE FLAT AND FREE OF BURRS.

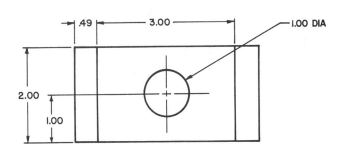

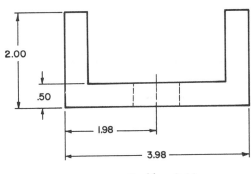

Problem 9-44

9-46 Given the top and front view, complete the drawing and add the following notes: .50 DIA (2 HOLES), and PACK CARBURIZE HARDEN.

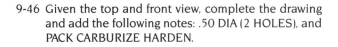
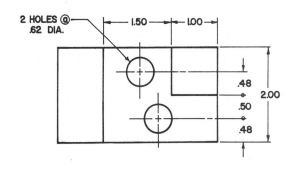

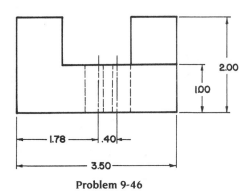

Problem 9-46

9-45 Given the top and front view, complete the drawing and add the following notes: STAMP PART NUMBER ZZ1456 ON BOTTOM OF PART, and .50 DIA HOLE.

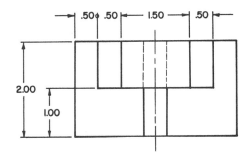

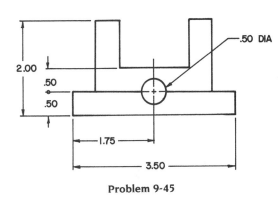

Problem 9-45

9-47 Given the top and front view, complete the drawing and add the following notes: 1.00 DIA HOLE, and POWER BRUSH AND PAINT ALL UNMACHINED SURFACES.

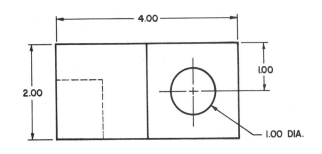

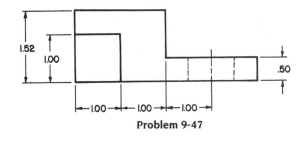

Problem 9-47

9-48 Given the top and front view, complete the drawing and add the following note: TOLERANCES ON ALL DIMENSIONS PLUS OR MINUS .005 UNLESS OTHERWISE SPECIFIED.

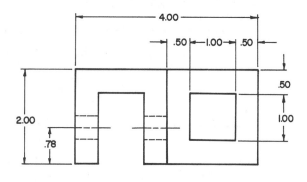

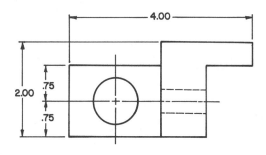

Problem 9-48

9-49 Given the top and front view, complete the drawing and add the following notes: 1.00 DIA HOLE, and .50 DIA HOLE.

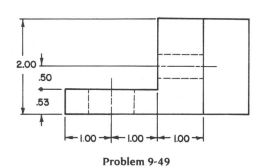

Problem 9-49

9-50 Given the top and front view, complete the drawing and add the following notes: 1.00 DIA HOLE, CASE HARDEN .257 DEEP, and STAMP NUMBER FFR 763-009 ON BOTTOM OF PART.

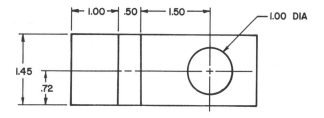

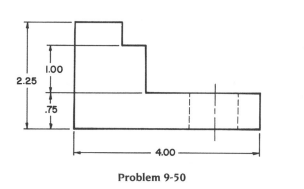

Problem 9-50

1. Unless otherwise specified, fractional dimensions are to be held to what tolerance?
2. Unless otherwise specified, angles are to be held to what tolerance?
3. What is the overall diameter of the part?
4. What is the overall depth of the part?
5. What size are fillets and rounds on the part?
6. Four holes are countersunk. What is the countersink

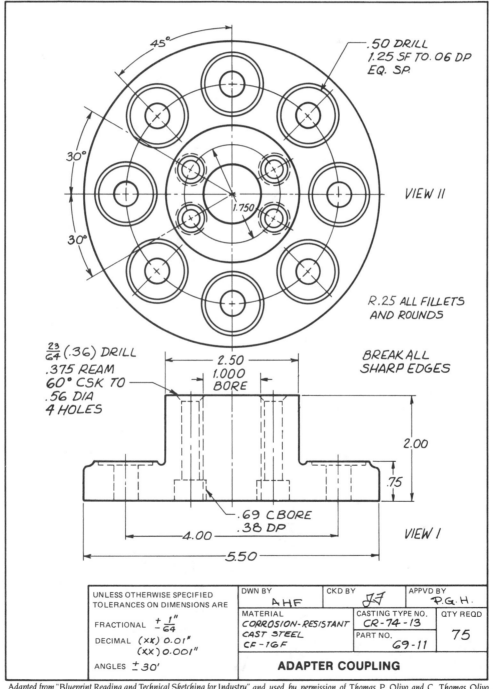

45°

.50 DRILL
1.25 SF TO .06 DP
EQ. SP.

30°

30°

1.750

VIEW II

R.25 ALL FILLETS
AND ROUNDS

BREAK ALL
SHARP EDGES

$\frac{23}{64}$ (.36) DRILL
.375 REAM
60° CSK TO
.56 DIA
4 HOLES

2.50
1.000
BORE

2.00

.75

.69 C BORE
.38 DP

4.00

5.50

VIEW I

UNLESS OTHERWISE SPECIFIED TOLERANCES ON DIMENSIONS ARE	DWN BY AHF	CKD BY	APPVD BY P.G.H.	
FRACTIONAL $\pm \frac{1}{64}''$	MATERIAL CORROSION-RESISTANT CAST STEEL CF-16F	CASTING TYPE NO. CR-74-13		QTY REQD
DECIMAL (XX) 0.01" (XX) 0.001"		PART NO. 69-11		75
ANGLES $\pm$ 30'	**ADAPTER COUPLING**			

Adapted from "Blueprint Reading and Technical Sketching for Industry" and used by permission of Thomas P. Olivo and C. Thomas Olivo, Published by Delmar Publishers Inc.

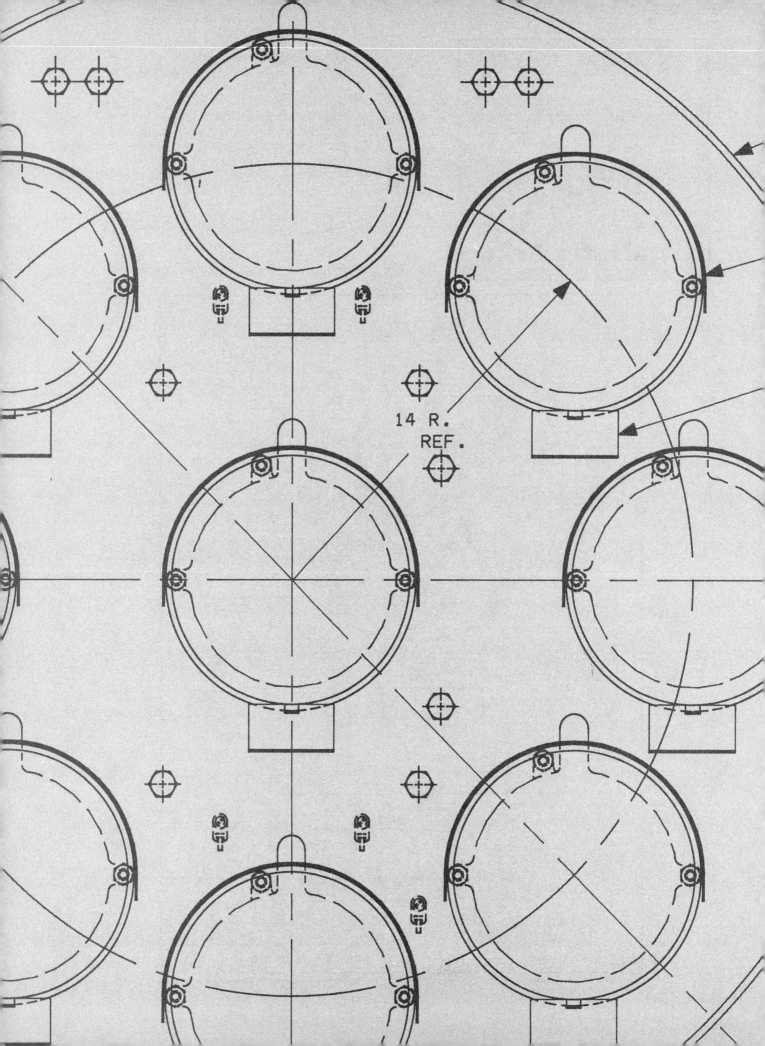

14 R.
REF.

SECTION THREE
COMPUTER-AIDED DRAFTING

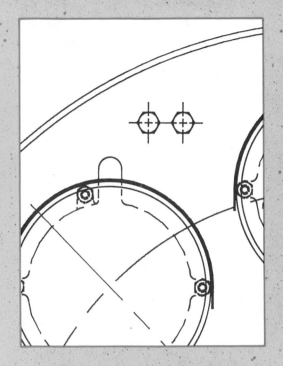

This chapter provides an in-depth treatment of modern computer-aided drafting (CAD) technology. The relationship of manual drafting and CAD is explained. The major topics covered include an overview of CAD, CAD systems, CAD software, CAD users, modern CAD system configurations, and the advantages of CAD.

CHAPTER TEN

Computer-Aided Drafting Technology

Overview of CAD

The dawning of the age of computers brought significant changes to the fields of design, drafting, engineering, and the various other fields related to these areas. You learned in the Introduction that the fundamental nucleus of these fields — the design process — was altered by computers in two ways: 1) the fourth step in the traditional design process, the making and testing of models or prototypes, was replaced by the developing and testing of three-dimensional computer models; and 2) the tools and techniques used in accomplishing each step of the process were significantly changed.

During the 1960s, shortly after the development of the integrated circuit (IC), the computer began to be used to save time, work, and money in hundreds of different fields. One of the more innovative uses of the computer was in the area of design and drafting, Figure 10-1.

Defining Terms

The rapid growth of the use of computers in the world of design and drafting brought forth a deluge of new, and often confusing, terms. Such terms as

CAD, CADD, CG, CAE, AD and hundreds of others spawned by the computer revolution began to be heard in drafting and design departments. As a result, drafting and design practitioners found themselves confused by it all.

The reader should be conversant in the language of computers as it applies to design and drafting. The first step is to develop an understanding of the term "computer-aided drafting" or CAD.

Before attempting to form a definition for CAD, the definition of "drafting" should be thoroughly understood. Most people think of drafting as simply the drawing of plans. True, drafters do draw plans. However, this is a very limited definition because drafters do much more than draw plans.

To understand drafting in its broadest sense, one must begin with the design process. The *design process* is an organized, systematic procedure used to accomplish a design that is needed to solve a problem or meet a need, Figure 10-2. Each step in the design process requires various types of documentation, such as sketches, preliminary drawings, working drawings, calculations, bills of material, parts lists, and schedules.

340

Figure 10-1 CAD system (*Courtesy Vector Automation*)

Producing or "drafting" this documentation is the job of the drafter. Therefore, *drafting* means producing the documentation required in support of the design process. With this understanding of drafting

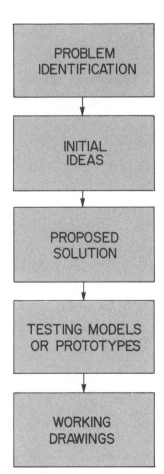

Figure 10-2 The modern design process

as a concept, one may now begin to develop an understanding of CAD. *Computer-aided drafting* or CAD means using the computer and peripheral devices in producing the documentation needed in support of the design process. or, put another way, CAD means using the computer and peripheral devices to do drafting.

A well-informed drafter should be able to distinguish CAD from among the various other related terms frequently heard in modern drafting settings, such as AD, CADD, CAE, CG, and CAM.

CAD is usually taken to mean computer-aided drafting, but it can also be used to mean *computer-aided design*. It is up to the person using the term to make the distinction. AD means *automated drafting*, and it is synonymous with CAD or computer-aided drafting. Originally, AD was used when referring to the use of the computer to do drafting, and CAD was used when referring to the use of the computer as a design tool. However, AD did not catch on because it is a less distinctive term than CAD. Consequently, CAD began to be used to mean both computer-aided drafting and computer-aided design.

CADD is the acronym for *computer-aided design and drafting*. This is a broad concept in which a computer system is used for both design and drafting.

CAE is the acronym for *computer-aided engineering*. This is a newer term that is replacing the term CAD when it is used to mean computer-aided design. The term CAE serves to eliminate the computer-aided design/computer-aided drafting confusion when using the term CAD.

CG is short for *computer graphics*. This is a term that is used and misused a great deal. It is frequently used in place of CAD (drafting). However, in reality, it is a much broader term. CG means using the computer and peripheral devices for producing any type of graphic image, technical or artistic.

Figure 10-3 Manual pencil and pointer (*Courtesy Hearlihy & Co.*)

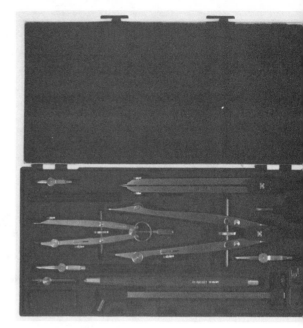

Figure 10-4 Manual drafting instrument set (*Courtesy Hearlihy & Co.*)

None of these terms has been formally standardized. However, there is a definite trend toward greater use of CAD when speaking of drafting, and CAE when speaking of design. CADD is the more general term used when speaking of both.

The manufacturing equivalent of CAD is *computer-aided manufacturing* or CAM. CAM refers to computer-controlled automation of the various manufacturing processes used in modern industry. CAD/CAM is the ultimate in productivity improvement because it links together the processes of design, drafting, and manufacturing. The CAD/CAM process involves designing a product on the computer, using the computer to produce necessary documentation, and using the data base stored in the computer from the design and drafting phases to issue the manufacturing instructions to automated machines and industrial robots.

Differences between Manual Drafting and CAD

The traditional methods used in producing the documentation needed in support of the design process are collectively known as *manual drafting*. The most obvious differences between manual drafting and CAD can be summarized in two words: tools and techniques. The tools of the manual drafter are such items

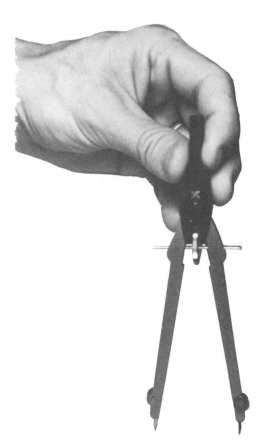

Figure 10-5 Manual instrument — large bow compass (*Courtesy Hearlihy & Co.*)

as triangles, scales, mechanical pencils, templates, erasing shields, erasers, tape, and lead pointers. Figures 10-3 through 10-11 contain examples of the tools of the manual drafter. Contrast these manual tools

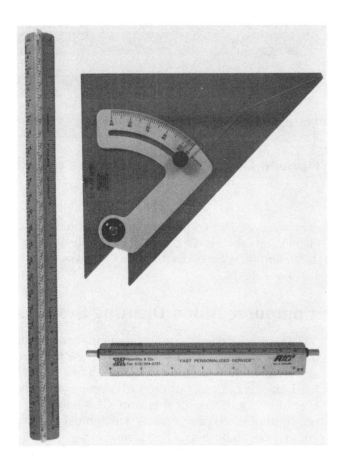

Figure 10-6 Manual drafting instruments (*Courtesy Hearlihy & Co.*)

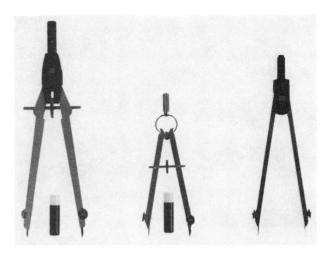

Figure 10-8 Compasses for manual drafting (*Courtesy Hearlihy & Co.*)

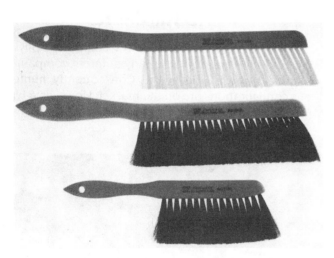

Figure 10-7 Brushes for manual drafting (*Courtesy Hearlihy & Co.*)

Figure 10-9 Lead holder and cleaner for manual drafting (*Courtesy Hearlihy & Co.*)

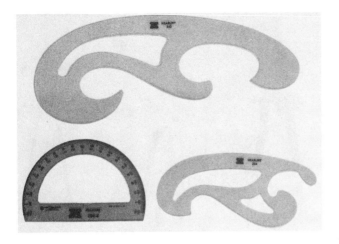

Figure 10-10 Protractor and French curves (*Courtesy Hearlihy & Co.*)

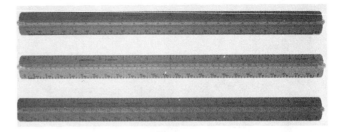

Figure 10-11 Scales for manual drafting (*Courtesy Hearlihy & Co.*)

of material, and schedules are normally stored in folders in filing cabinets. CAD technicians store all design documentation on magnetic tapes or disks, Figures 10-13 and 10-14.

Computer-Aided Drafting Systems

CAD *system* is one of the most frequently heard terms in modern drafting rooms. People who use this term are usually referring to a configuration of computer hardware or the tools of the CAD technician. However, this is a misuse of the term. CAD hardware is only one of three components which must be present in order to have a CAD system.

A CAD system actually consists of hardware, software, and users, Figure 10-15. A collection of CAD hardware is properly referred to as a *hardware configuration*. In order to turn a hardware configuration into a CAD system, software and well-trained users must be added.

CAD Hardware

Hardware is the term given to the computer's physical equipment or devices. Many different companies manufacture CAD hardware. Consequently, numerous configurations are on the market. However, a typical hardware configuration consists of a graphics

with such tools of the CAD technician as the graphics display, text display, keyboard, digitizer, light pen or puck, function menu, and plotter, Figure 10-12.

The techniques used by manual drafters in documenting a design include freehand lettering, manual scaling, and mechanical line work. CAD technicians accomplish these same tasks by interacting with a keyboard, graphics display, text display, digitizer, and function menu, Figure 10-12.

Another important difference between manual drafting and CAD is in the methods used for storing design documentation. Manual drafters store drawings in large flat files. Calculations, parts lists, bills

Figure 10-12 Tools of the CAD technician (*Courtesy Harvey Dean*)

Figure 10-13 Storage on magnetic tape (*Courtesy Clinton Corn Processing Co.*)

Figure 10-14 Magnetic disk storage (*Courtesy Deborah M. Goetsch*)

display, a keyboard, a text display, a digitizer equipped with either a light pen or a puck, function menus, a plotter, and a processor, Figure 10-16.

The Graphics Display

The *graphics display* is an output device resembling a television that displays the data on which the CAD technician is working. As a drawing or any other type of documentation is created, it is displayed on the screen of the graphics terminal, Figure 10-17.

CAD SYSTEM=

HARDWARE + SOFTWARE + USERS

Figure 10-15 Components of a CAD system

Figure 10-16 Typical hardware configuration (*Courtesy Bausch & Lomb*)

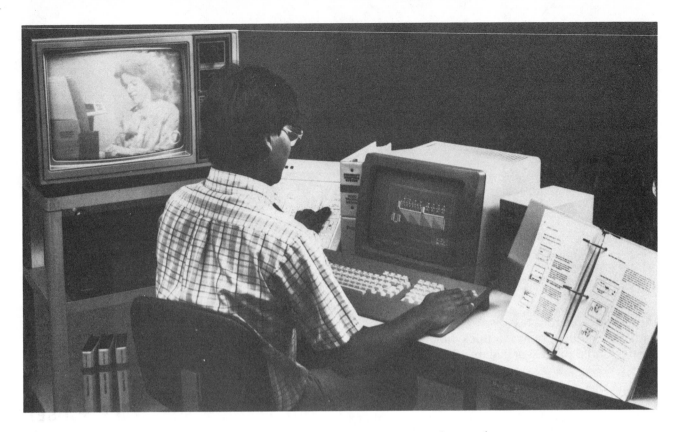

Figure 10-17 Viewing the display on a graphics terminal (*Courtesy Bausch & Lomb*)

Figure 10-18 Typical keyboard (*Courtesy The Rust Engineering Co.*)

There are three types of graphics displays: 1) refresh displays, 2) raster displays, and 3) storage tube displays. In a *refresh display*, the image is traced on the back of the screen by an electron beam. This image must be constantly retraced or "refreshed." *Raster displays* create images by illuminating *pixels* (picture elements) on the screen. The more pixels available on the screen for illumination, the higher the quality of the image. *High resolution* produces a high-quality image. *Storage tube displays* create images by tracing them on the back of the display screen with an electron beam like refresh displays. However, unlike refresh displays, storage tube images are stored and do not have to be refreshed. Graphics displays may be color, monochrome or amber devices.

The Keyboard

The keyboard is one of the most frequently used input devices in a CAD system. *Keyboards* are used for entering text, system commands, X-Y or X-Y-Z coordinates, and any other type of alphanumeric input.

Most CAD systems have keyboards which contain both the standard alphanumeric keys and special auxiliary keys, Figure 10-18. The special auxiliary keys may be part of a separate numeric pad or a group of display control keys.

The Text Display

The *text display* is an output device with a screen for displaying user prompts, instructions, and other alphanumeric data. Most text displays are noncolor cathode-ray tube (CRT) devices, Figure 10-19.

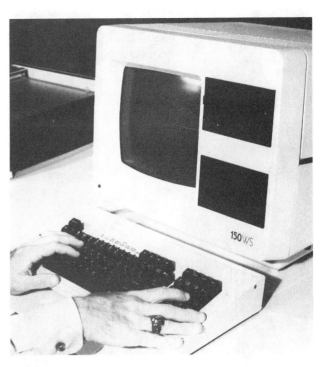

Figure 10-19 Text display (*Courtesy Deborah M. Goetsch*)

The Digitizer

The *digitizer* is a special electromechanical input device that resembles an electronic tablet. In fact, the digitizer is frequently referred to as the *tablet*. Some CAD systems have digitizers that are as small as tablets, and some have digitizers that are as large as conventional drafting tables, Figures 10-20 and 10-21.

A digitizer can be used for a number of different functions. It can be used in conjunction with a light

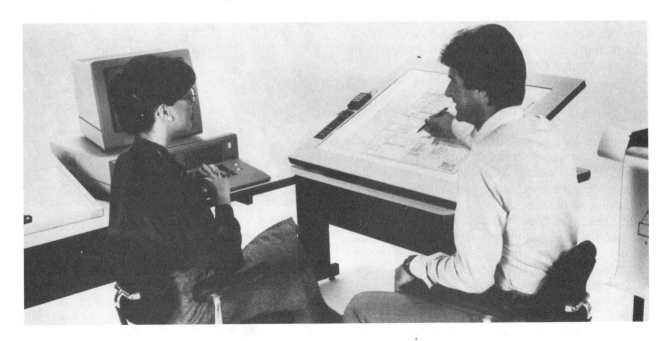

Figure 10-20 Digitizing tablet (left) and digitizing table (right) (*Courtesy Bausch & Lomb*)

Figure 10-21 Complete hardware configuration with large digitizing table (*Courtesy Auto-Trol Technology Corp.*)

Figure 10-22 Digitizer with a light pen (*Courtesy Deborah M. Goetsch*)

pen or a puck to control the screen cursor, Figures 10-22 and 10-23. A *cursor* is the small crosshairs symbol on the graphics display used in creating, locating, and manipulating graphic and alphanumeric data.

A digitizer can be used as a place to mount menus for giving system commands or calling up stored symbols from memory. It can also be used for electronically tracing (digitizing) graphic data so that they can be entered into a CAD system.

Function Menus

Menus are lists of system commands and stored data, or overlays that serve the same purpose, which correspond with storage locations in a CAD system's memory. They are used as a means of speeding and simplifying human interaction with the CAD system. One of the fastest, easiest ways to enter a system command is to have the command on a menu. Just as you need a menu when you sit down in a restaurant so that you know what options you have for ordering, you need a menu when you sit down at a CAD system so that you know what options you have for commanding the system.

The most frequently used types of menus are *screen-displayed menus* and *tablet-mounted menus*. A command on

Figure 10-23 Digitizer with a puck (*Courtesy Summagraphics Corp.*)

a screen-displayed menu may be activated by touching it with a light pen, moving the screen cursor to its location on the screen, or pressing a key that corresponds with the command. Tablet-mounted menu commands are activated by touching the appropri-

Figure 10-24 Activating menu commands (*Courtesy Cascade Development Corp.*)

ate position on the menu with a light pen or the sight of the puck. Figure 10-24.

In addition to system commands, menus may contain frequently used symbols or items of graphic data. If a certain symbol is used frequently, it may be created once, saved, and assigned to a specific position on a menu. Then, when needed, it can be called up from storage in a matter of milliseconds rather than having to be completely redrawn each time it is needed on a drawing.

The Plotter

The *plotter* is the output device which actually draws (plots) drawings and other types of documentation. Three types of plotters are used in CAD systems: pen plotters, electrostatic plotters, and photoplotters.

Pen plotters may be bed plotters, (Figure 10-25), or drum plotters, Figure 10-26. They create drawings by converting into lines X-Y coordinates received from the processor in the form of electrical impulses. Pen plotters are rated according to their accuracy, repeatability, resolution, and plotting speed. Most pen plotters can plot drawings accurately to plus or minus .001" or better.

Repeatability in plotting is the ability to redraw lines without creating a double image. This is important because on many plotters line thickness or density is increased by retracing lines until the desired density is achieved.

Resolution in plotting is judged by the shortest line segment the plotter is able to plot. This becomes a critical factor when plotting circles, arcs, and curves, all of which are formed by a series of short, straight-line segments. The shorter the line segments, the better the resolution. Resolution of .001" is common for pen plotters. A *plotting speed* of 40 inches per second is a common, and frequently exceeded, mean plotting speed with modern plotters.

Electrostatic plotters are much faster than pen plotters, but they are not as accurate and they require special electrostatic paper. This type of plotting paper is not dimensionally stable; a factor which limits the applications of electrostatic plotters.

Photoplotters are used in those situations requiring extremely close tolerances and dimensional stability, such as in the production of printed circuit board

Figure 10-25
B-size bed plotter
(*Courtesy Deborah M. Goetsch*)

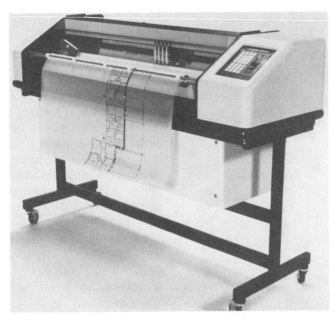

Figure 10-26 Roll-fed drum plotter (*Courtesy Bausch & Lomb*)

(PCB) artwork. Of the three types of plotters, the pen plotter is the one most frequently used.

The Processor

The *processor* is the computer in a computer-aided drafting system. As in any computer, processors have a memory component and a logic component. In addition, many CAD system processors have an added feature: a graphics processor, Figure 10-27.

A processor's memory is rated according to the amount of data it can hold. Individual memory location sizes are rated in *bits* or binary digits. This is why some processors are referred to as 8-, 16-, or 32-bit processors. An 8-bit processor is slower than a 16- or 32-bit processor. The 32-bit processor is the fastest.

Overall memory size is rated in *bytes* (the commonly accepted ratio is 1 byte = 8 bits). Because a byte represents a relatively small amount of data, memory sizes are usually given in kilobytes. A 256-kilo-

Figure 10-27 One type of graphics processor (*Courtesy Calma Company*)

byte memory would not be uncommon in a CAD system's processor.

The term *kilo* normally means 1000. However, when used in conjunction with computers, it means 1024. To determine the actual size of a 256-kilobyte processor, multiply $1024 \times 256 = 262,144$. To convert this to bits, multiply $262.144 \times 8 = 2,097,152$.

The processor console also houses a disk or tape drive unit for secondary storage devices such as disks, diskettes or magnetic tapes, Figure 10-28.

CAD Software

Software is the name given to the various programs that actually make the CAD system work, and any printed materials such as instruction sheets or operator's manuals, for instance. There are three basic types of software for CAD systems: 1) operational software; 2) application software; and 3) user-defined software.

Operational software consists of programs which make the system perform general operational tasks such as accepting commands, storing data, retrieving data, and so forth.

Application software consists of programs that command the CAD system to perform drafting functions in such specific application areas as architectural, mechanical, structural, electronic, and civil drafting.

User-defined software usually is restricted to such various user-specific needs as symbols that are added to a symbols library and a menu. Users do not normally write the code for CAD software.

Figure 10-28 Tape drive unit (*Courtesy Clinton Corn Processing Co.*)

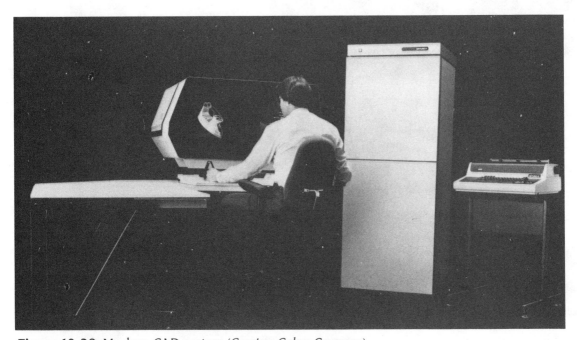

Figure 10-29 Modern CAD system (*Courtesy Calma Company*)

CAD Users

Users of CAD systems can be placed in two categories: 1) CAD system operators, and 2) CAD technicians. CAD *operators* are people who are able to operate a CAD system in terms of inputting, manipulating, and outputting data. However, they have little or no background in design, drafting or engineering.

CAD *technicians*, on the other hand, have both system operational skills and strong backgrounds in design, drafting or engineering. For CAD technicians, a CAD system is simply a tool to help them do their jobs faster and better.

Modern CAD System Configurations

There are more than 100 turnkey CAD systems on the market now, and the list will probably continue to increase. Some are general-purpose systems which can be used in a number of different drafting fields. Others are dedicated systems designed for one specific application.

Modern CAD systems range from systems based on large mainframe computers to small microcomputer-based configurations. Figures 10-29, 10-30, and 10-31 show examples of several modern CAD systems.

Advantages of CAD

The applications of computer-aided drafting grew rapidly because of its many advantages over manual drafting. CAD is being used in a variety of settings in both large and small companies. When compared with manual drafting, CAD is more accurate, faster, neater, more consistent, and better in terms of storing, correcting, and revising drawings, Figure 10-32.

CAD hardware is consistently accurate to better than .001 inch. Manual scaling and drawing cannot match this figure consistently, even under the most positive conditions. In fact, inaccuracy is a built-in human characteristic. Such typical drafting tasks as lettering and line work are slow and tedious for manual drafters; but, for CAD technicians, they are fast and easy because they are accomplished by simply pressing buttons and executing commands.

The computer actually does the work for the person and it does it very fast. Using the computer, a drafting technician can produce lettering as much as 25 times faster than by doing it manually. A drafting technician with typing skills can improve on this figure considerably.

Whereas a manual drafter might draw at an average rate of four to eight inches per minute, it is not uncommon for a CAD system's plotter to create drawings at a rate of 40 inches per *second*.

Figure 10-30 Modern CAD system (*Courtesy Vector Automation*)

Figure 10-31 Modern CAD system (*Courtesy Bausch & Lomb*)

Neatness and consistency in line work and lettering are the major weaknesses of manual drafting and the major strengths of CAD. Since documentation in CAD is produced electromechanically, it is not subject to the negative impact of such factors as fatigue, boredom or a lack of skills. Figures 10-33 and 10-34 illustrate the neatness and consistency that is characteristic of documentation prepared on a CAD system.

Some of the most important advantages of CAD can be found in the area of corrections and revisions. On the job, a great deal of time is devoted to correcting and revising drawings and other types of documentation. Time spent correcting and revising documentation is nonproductive time. Because of the drawing manipulation capabilities of CAD systems (for instance, such command functions as MOVE, COPY, DELETE, ROTATE, and ZOOM), correcting and revising time can be cut by as much as 75 percent in many cases. The time and work-saving potential of CAD in terms of corrections and revisions is even more important than that associated with the original creation of design documentation.

An additional advantage of CAD over manual drafting involves storage of documentation. In CAD, one 5.25-inch floppy disk might hold as much documentation as a large file drawer. With CAD, years of drafting work can be stored in a container the size of a shoe box. Backup disks or copies of file disks can be made quickly, easily, and inexpensively, and stored in protective containers. Thus, the deterioration problems normally associated with traditional drafting media are eliminated.

MORE ACCURATE

FASTER

NEATER

MORE CONSISTENT

BETTER STORAGE

EASIER CORRECTIONS

EASIER REVISIONS

Figure 10-32 Advantages of CAD

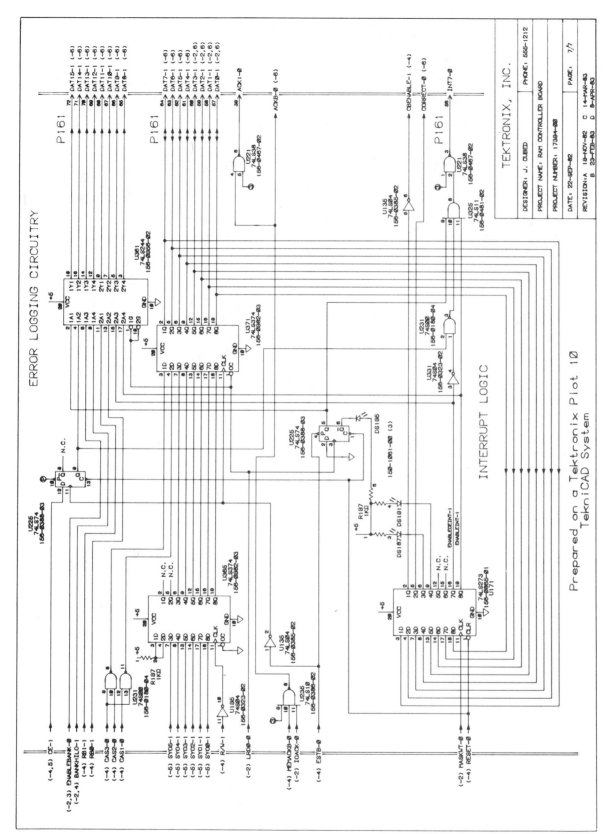

Figure 10-33 Drawing produced on a CAD system (*Courtesy Tektronix, Inc.*)

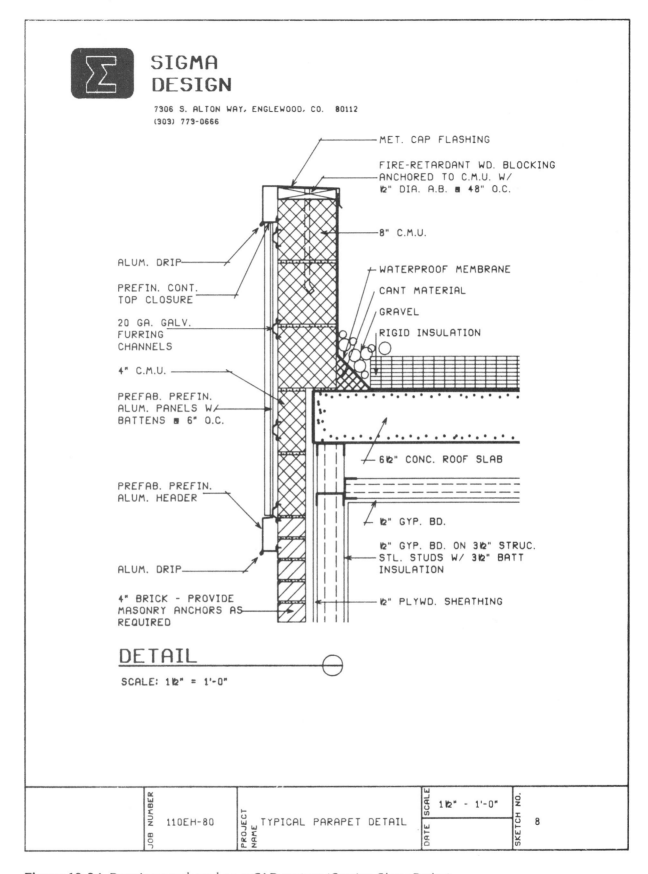

SIGMA
DESIGN

7306 S. ALTON WAY, ENGLEWOOD, CO. 80112
(303) 773-0666

MET. CAP FLASHING

FIRE-RETARDANT WD. BLOCKING
ANCHORED TO C.M.U. W/
½" DIA. A.B. ⌀ 48" O.C.

8" C.M.U.

WATERPROOF MEMBRANE

CANT MATERIAL

GRAVEL

RIGID INSULATION

ALUM. DRIP

PREFIN. CONT.
TOP CLOSURE

20 GA. GALV.
FURRING
CHANNELS

4" C.M.U.

PREFAB. PREFIN.
ALUM. PANELS W/
BATTENS ⌀ 6" O.C.

PREFAB. PREFIN.
ALUM. HEADER

6½" CONC. ROOF SLAB

½" GYP. BD.

½" GYP. BD. ON 3½" STRUC.
STL. STUDS W/ 3½" BATT
INSULATION

ALUM. DRIP

4" BRICK - PROVIDE
MASONRY ANCHORS AS
REQUIRED

½" PLYWD. SHEATHING

DETAIL

SCALE: 1½" = 1'-0"

| JOB NUMBER | 110EH-80 | PROJECT NAME | TYPICAL PARAPET DETAIL | DATE | SCALE | 1½" - 1'-0" | SKETCH NO. | 8 |

Figure 10-34 Drawing produced on a CAD system (*Courtesy Sigma Design*)

Review

1. Explain how CAD changed the design process.

2. Define the term *drafting*.

3. Define the term *computer-aided drafting*.

4. What do each of the following acronyms stand for? CAD, AD, CADD, CAE, CG, CAM.

5. What are the most obvious differences between CAD and manual drafting?

6. List the typical tools of the CAD technician.

7. How do CAD technicians store design documentation?

8. What three components must be present in order to have a CAD system?

9. What are the components in a typical CAD hardware configuration?

10. What are the three types of graphics displays?

11. Name two categories of keys found on CAD system keyboards.

12. What does CRT stand for?

13. Explain three uses of a digitizer.

14. What is a function menu?

15. Name two different types of function menus.

16. Name three different types of plotters used in CAD systems.

17. What is the processor? Name its two most important components.

18. Name three types of CAD software.

19. List five advantages of CAD over manual drafting.

20. How will CAD affect your future as a drafting technician?

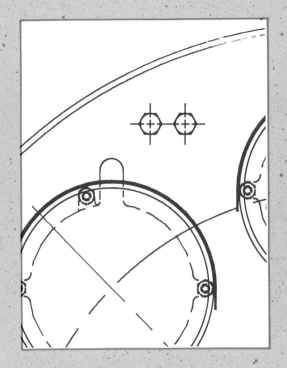

This chapter covers the typical operations used in producing technical drawings on a modern CAD system. The major topics covered are general system operation, input commands, manipulation commands, and output commands. Students will develop a basic understanding of how to operate the modern CAD systems available in the workplace. Problems are provided at the end of the chapter for those who have a CAD system available.

CHAPTER ELEVEN

Computer-Aided Drafting Operations

The tools of the CAD technician, as discussed in Chapter 10, consist of the keyboard, graphics display, text display, digitizer with puck or light pen, function menu, and plotter. The technician uses these tools in the three phases of producing design documentation in CAD: inputting data, manipulating data, and outputting data.

The basic operation of any computer system includes input, processing, and output. However, the processes involve actual human interaction to ensure that these drafting tasks are accomplished in the desired format. The computer or processor performs the processing tasks. To use a CAD system as a tool in producing the documentation for the design process, the drafter must first develop some basic operational skills, the subject of this chapter.

General System Operation

Just as one must learn how to manipulate pencils, pens, triangles, templates, and various other tools before using them to perform manual drafting tasks, one must also learn a number of general system operational skills before using a CAD system to perform drafting tasks. These general skills include: keying, digitizing, cursor control, entering system commands, and activating menu options.

Keying

Keying is simply typing, Figure 11-1. Although CAD technicians do not have to be accomplished typists, since only a minimum of keying is done on most CAD

Figure 11-1 Keying (*Courtesy Deborah M. Goetsch*)

357

systems, they do need to be familiar with the locations of the various keys on the keyboard.

Keying is the process used for entering text on drawings, entering certain system commands, entering dimensions not entered automatically, and for logging-on to the system.

Digitizing

Digitizing is a process through which graphic data, such as a sketch or a drawing, is converted to digital data as it is input. It involves placing the sketch or drawing on the digitizing tablet and tracing it electronically with a puck, Figure 11-2. A light pen is sometimes used instead of a puck.

When digitizing, it is not necessary to trace an entire line as it is in manual tracing. Rather, the cross hairs in the puck's sight are aligned with the end points of lines and the puck button is pushed. Each time the button is pushed, the computer calculates the X-Y coordinates for that point and stores them in the computer's memory. Consequently, to the computer a drawing is just a series of X-Y coordinates.

Cursor Control

The CAD equivalent of the point of the drafter's pencil is a small cross-hair symbol on the graphics display called a cursor, Figure 11-3. It is used for drawing graphic data, specifying the location of text and graphic data on the display, identifying data that are to be deleted or otherwise manipulated, and for identifying commands on screen menus that are to be activated.

The cursor can be controlled in a number of different ways: with horizontal and vertical thumbwheels; with a trackball or joystick, such as those found on many video games; with a light pen; or with a puck and a digitizing tablet. This last method, puck and digitizing tablet, is the most frequently used method on modern CAD systems.

Entering System Commands and Activating Menu Options

In manual drafting, drafters do the drawing themselves. However, in CAD, drafters only give commands; the computer actually does the work. Consequently, performing drafting tasks on a CAD system is similar to creating a drawing by giving commands to another drafter.

To simplify this process, all of the many commands needed to communicate with the CAD system are contained in the memory component of the processor. Consequently, CAD technicians need only activate the proper memory location in order to command the system to perform a certain task. Common commands include DRAW LINE, DRAW CIRCLE, DRAW RECTANGLE, ENTER TEXT, ADD DIMENSION, DELETE LINE, and so forth.

Figure 11-2 Digitizing with a puck (*Courtesy Harvey Dean*)

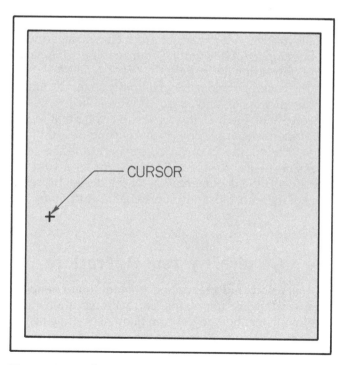

Figure 11-3 The cursor

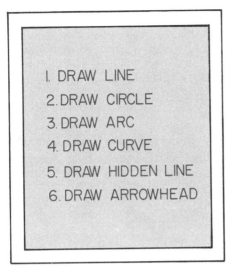

Figure 11-4 Sample menu

```
I. DRAW LINE
2. DRAW CIRCLE
3. DRAW ARC
4. DRAW CURVE
5. DRAW HIDDEN LINE
6. DRAW ARROWHEAD
```

Figure 11-5 Activating a menu command with a puck (*Courtesy Deborah M. Goetsch*)

Each command that is stored in memory is listed on a menu so that CAD technicians know what commands are available to them. The menu may be displayed on the screen of the text terminal or drawn on a paper or polyester film overlay and affixed to a digitizing tablet.

To activate a screen-displayed menu option, the CAD technician may touch the option on the screen with a light pen or key the option's code. For example, the menu in Figure 11-4 lists six options. To activate the DRAW CIRCLE option, the technician would touch the number 2 on the screen with a light pen or key the number 2 on the keyboard and, then, press either the RETURN or ENTER key. Pressing the RETURN or ENTER key is a necessary activation step in most cases when entering commands from a keyboard.

To activate a tablet-mounted menu command, the technician would align the sight of the puck with the menu position occupied by the option in question, and then press the puck button, Figure 11-5.

On some systems, commands that are normally entered from a screen or tablet-mounted menu can also be entered by keying. For example, in Figure 11-4 it would be possible on some systems to activate the number 1 option by typing DRAW LINE and pressing the ENTER or RETURN key. This, of course, would take longer than activating a menu option and, therefore, it is usually the least used method of entering commands on a CAD system.

Input Commands

Input commands are those commands given to the CAD system to make it create drawings, parts lists, bills of material, and all of the various other types of documentation associated with drafting. Drawings consist of two types of data: graphic and text.

```
              MASTER  MENU
 I. SOLID LINES        13. SCALE
 2. HIDDEN LINES       14. ROTATE
 3. ARCS               15. MIRROR
 4. CURVES             16. MARK DELETE
 5. AUTO DIMENSION     17. UNDELETE
 6. TEXT SIZE          18. BOX DELETE
 7. TEXT               19. PLOT
 8. RECTANGLES         20. PLOT SPECS
 9. ELLIPSES           21. ARROWHEADS
10. POLYGONS           22. FILLETS
11. COPY               23. ROUNDS
12. MOVE               24. SAVE
```

Figure 11-6 Screen-mounted menu

All graphic data are created using standard geometric shapes in the proper combinations. These shapes include straight lines of different types (i.e. solid, hidden, stitched, and so forth) arcs, circles, curves, rectangles, ellipses and polygons, among others. All text data are created using letters, numbers, and other keyboard characters in combination or separately.

Input commands which cause the computer to create all graphic and text data can be stored in a CAD system's memory and assigned to a menu. Figure 11-6 is an example of a screen-mounted menu. The input commands on this menu are number coded 1, 2, 3, 4, 5, 7, 8, 9, 10, 21, 22, 23, and 24. These are the commands a CAD technician would use in creating documentation.

With a menu displayed on the text terminal (Figure 11-6), creating documentation is simply a matter of activating menu commands and following the directions provided in the *prompts* which correspond with each command. However, before getting this far, the CAD technician must first complete such preliminaries as logging-on to the system, calling-up stored work files, and, when necessary, creating new work files.

Logging-on, Calling-up Files, and Creating New Files

Every CAD system has a "log-on" sequence that must be completed in order to gain access to the system. Logging-on prevents unauthorized use of the system. The *log-on sequence* for most CAD systems involves keying answers to questions that are displayed on the text terminal as each step in the sequence is satisfied. For example, after turning on the system, a CAD technician might go through the following procedure:

The following message appears on the text display once the power has been turned on:

WELCOME TO THE MEGADRAFT CAD SYSTEM. WHAT IS YOUR USER CODE?

Each operator of this particular system has been assigned a user code that only the operator, and perhaps the drafting manager, knows. The code must be entered in order to proceed. Let us say the operator's social security number is used as the code. The operator keys:

276-98-9790

The operator then presses the RETURN key. At this point a new prompt appears on the text display:

OLD OR NEW FILE?

Documentation in CAD is stored in files. Each file has a code or designation that distinguishes it. After deciding to work on an old file, the operator types OLD, and then presses the RETURN key. At this point, a new prompt appears on the text display:

FILE NAME?

Let us say the code for the desired work file is JOB NUMBER 1701. The operator types this and presses the RETURN. A new prompt appears on the text display:

FILE ACCESSED. DO YOU NEED A LISTING OF FILE'S CONTENTS?

This file might contain several drawings, parts lists, and various other types of documentation, each with its own access code. The user may not remember the code of the drawing needed. When this is the case, a listing of the contents of the file can be brought to the text display by pressing the Y key for "yes." When this is done, a complete list of the contents of the subject file will be displayed on the graphics terminal, Figure 11-7.

To call-up one of the items within the file, the user types in its code (say number 1 to see the ASSEMBLY DRAWING) and presses the RETURN key. The subject data will immediately appear on the graphics display, Figure 11-8.

JOB NUMBER 1701
1. ASSEMBLY DRAWING
2. SUB-ASSEMBLIES
3. SECTIONAL VIEWS
4. PARTS DETAILS
5. PARTS LIST

Figure 11-7 File contents displayed on the graphics terminal

Figure 11-8 Dual graphics display in background (*Courtesy Bausch & Lomb*)

When the code for the desired data is known, the user indicates that a listing is not required by pressing the N key for "no." When this is done, another prompt appears:

CODE PLEASE

At this point, the user enters the appropriate code and then desired data appears on the display screen.

In summary, the following is a typical log-on sequence for a CAD system:

WELCOME TO THE MEGADRAFT CAD SYSTEM.

WHAT IS YOUR USER CODE?

Drafter keys response.

OLD OR NEW FILE?

Drafter keys response.

FILE NAME?

Drafter keys response.

FILE ACCESSED. DO YOU NEED A LISTING OF FILE'S CONTENTS?

Drafter keys response.

To create a new file, the operator would press the N key for "new" when confronted by the OLD OR NEW FILE prompt. The operator will then receive a blank graphics display screen on which to begin creating the desired documentation. Now the job is one of activating menu commands and responding to prompts.

Giving Commands and Responding to Prompts

Refer again to Figure 11-6 for an example of a screen-mounted menu. It contains a list of input, manipulation, and output commands. The input commands on this menu are: SOLID LINES, HIDDEN LINES, ARCS, CURVES, AUTO DIMENSION, TEXT, RECTANGLES, ELLIPSES, POLYGONS, ARROWHEADS, FILLETS, ROUNDS, and SAVE. By activating these input commands and responding to their corresponding prompts, CAD technicians can create all of the various types of documentation required to support the design process. Examples of how several of these commands and their prompts are used are presented in the following paragraphs.

Solid Lines. The menu in Figure 11-6 provides for three different types of lines: solid lines, hidden lines, and dimension lines (through the AUTO DIMENSION function). This means that the SOLID LINES function is used to create object lines, leader lines, center lines, and other nonhidden lines.

Figure 11-9 Making a solid line

To create solid lines, activate the SOLID LINES option on the master menu by touching the number 1 position with the light pen (the master menu is constantly displayed on the screen of the text display once the operator has logged-on). A prompt will appear on the text display below the master menu:

INDICATE POINT OF BEGINNING

Using the puck and digitizer, the CAD technician now moves the cursor on the graphics display to the point for the beginning of the line and presses the puck button. A new prompt will appear on the text display:

INDICATE END POINT

The cursor is now moved to the point on the graphics display which is to be the end point of the line and the puck button is pressed. This establishes the line. The process may be repeated as many times as necessary. The system will stay in the SOLID LINES mode until the CAD technician activates a different menu option. The SOLID LINES function is illustrated in Figure 11-9. Examine the solid lines in Figure 11-10.

Arcs and Circles. The ARCS command is used for creating arcs and circles. To use this function, the ARCS command is activated by touching the number 3 position with the light pen. A prompt will appear on the text display:

ARC OR CIRCLE?

To make an arc, type in the word "arc" and press the RETURN key. A new prompt will appear on the text display:

INDICATE ARC CENTER

Move the cursor to the point on the graphics display that will be the center of the arc and press the puck button. A new prompt will appear on the text display:

INDICATE ARC BEGINNING

Move the cursor to the beginning point of the arc and press the puck button. A new prompt will appear on the text display:

INDICATE ARC ENDING

Move the cursor to the end point of the arc and press the puck button. The arc will be created in either

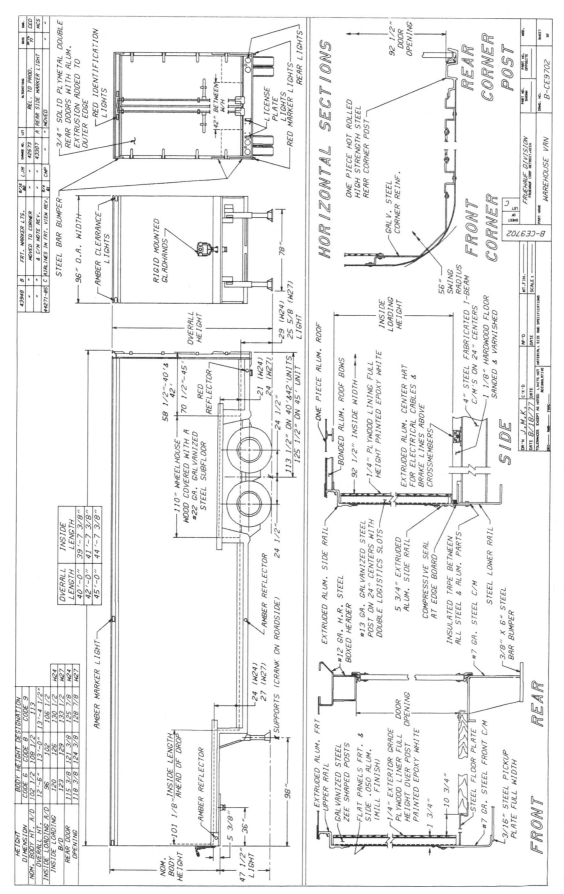

Figure 11-10 Sample CAD drawing (*Courtesy Freuhauf Division, Fruehauf Corp.*)

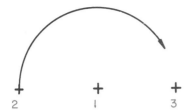

Figure 11-11 Making an arc

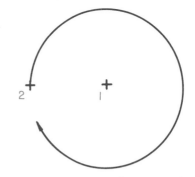

Figure 11-12 Making a circle

Figure 11-13 Making a rectangle

a clockwise or a counterclockwise direction, depending on the system's software. This procedure is illustrated in Figure 11-11.

To make a circle, activate the ARC command by touching the number 3 position with the light pen. A prompt will appear on the text display:

ARC OR CIRCLE?

Type the word "circle" and press the RETURN key. A new prompt will appear on the text display:

INDICATE CIRCLE CENTER

Move the cursor to the point on the graphics display which will be the center of the circle and press the puck button. A new prompt will appear on the text display:

INDICATE RADIUS

Move the cursor away from the center point a distance equal to the radius and press the puck button. The circle will be created in either a clockwise or a counterclockwise direction, depending on the system's software. The procedure for creating circles is illustrated in Figure 11-12. The truck wheels in Figure 11-10 were created using a similar procedure.

Rectangles. Any rectangular object can be created using the SOLID LINES function, but a faster, easier way is to use the RECTANGLES function. To create rectangles using this function, activate the RECTANGLES command by touching the number 8 position with the light pen. A prompt will appear on the text display:

INDICATE CORNER

Move the cursor to a point on the graphics display which corresponds with any corner of the rectangle being created and press the puck button. A new prompt will appear on the text display:

INDICATE OPPOSITE CORNER

Move the cursor diagonally to a point that represents the opposite corner of the rectangle being cre-

ated and press the puck button. Using these two opposite points, the computer will calculate and draw all four sides of the rectangle. The procedure used in creating rectangles is illustrated in Figure 11-13. Examine the rectangles in Figure 11-14.

Automatic Dimensioning. One of the advantages of CAD is that the computer is able to calculate dimensions for the user. This saves time and is very accurate. To automatically dimension objects, activate the AUTO DIMENSION command by touching the number 5 position with the light pen. A prompt will appear on the text display:

INDICATE DIMENSION BEGINNING

Move the cursor to the point on the graphics display where the dimension will begin and press the puck button. A new prompt will appear on the graphics display:

INDICATE DIMENSION ENDING

Move the cursor to the point on the graphics display where the dimension will end and press the puck button. A new prompt will appear on the text display:

INDICATE DISTANCE FROM OBJECT

Move the cursor to a point on the graphics display to indicate how far away from the object the dimension line should be and press the puck button. At this point, the extension lines, the dimension line, arrowheads, and the dimension will appear on the graphics display as specified. The procedure used in automatically dimensioning an object is illustrated

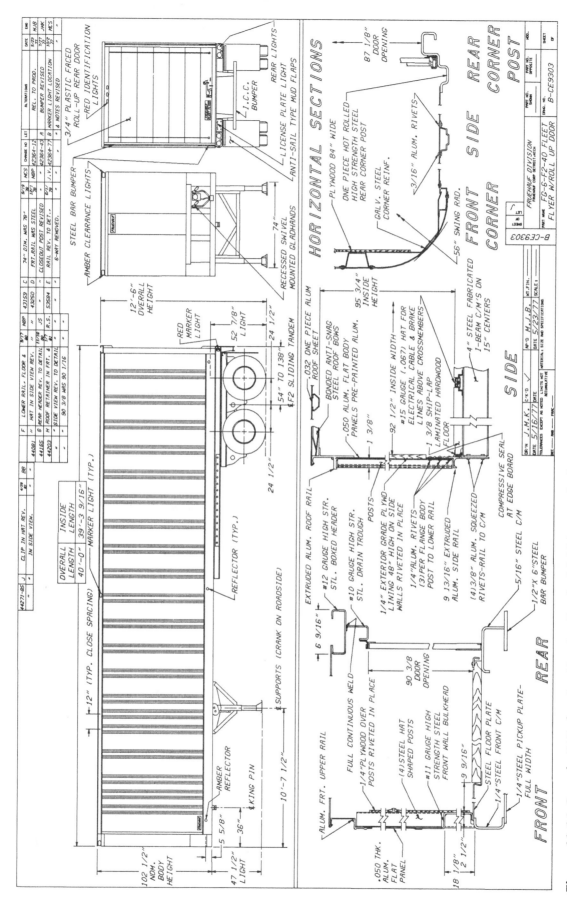

Figure 11-14 Sample CAD drawing (*Courtesy Fruehauf Division, Fruehauf Corp.*)

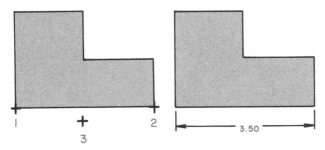

Figure 11-15 Automatic dimensioning

in Figure 11-15. Examine the dimensions in Figure 11-16.

Text. The TEXT command is used for entering words, notes, dimensions, and any other type of annotation. To use this function, the TEXT command is activated by touching the number 7 position with the light pen. A prompt will appear on the text display:

ENTER TEXT (UP TO 80 CHARACTERS)

At this point, type in the desired text. As you type, the text data will appear on the text display. Notice that the text is limited to 80 characters per string. When the typing is completed, the technician should check the spelling, wording, and spacing. If there are mistakes, the BACKSPACE key is used to make the necessary corrections. If the data are correct, press the RETURN key. A new prompt will appear on the text display:

INDICATE BEGINNING OF TEXT STRING

At this point, move the cursor to a point which marks the lower left-hand corner of the first letter in that string and press the puck button. The text data will appear in the specified location.

Manipulation Commands

Manipulation commands are those commands which allow drafters to alter data that have been entered into the system. The manipulation functions on the sample menu in Figure 11-6 are numbers 6, 11, 12, 13, 14, 15, 16, 17, and 18. These functions allow the CAD technician to perform such tasks as altering the size of text data, and copying, moving, scaling, rotating, mirroring, and deleting both alphanumeric and graphic data. Two other commonly used manipulation functions do not appear on the menu in Figure 11-6. They are ZOOM-IN and ZOOM-OUT. Each of these commands is activated by pressing a special control key located on the CAD system keyboard rather than using the menu.

Activating Manipulation Commands

By activating the desired manipulation commands and responding to their corresponding prompts, CAD technicians can perform all of the various manipulations of data. Examples of how several of the manipulation commands in Figure 11-6 and their prompts are actually used are presented in the following paragraphs. The ZOOM-IN AND ZOOM-OUT commands are also discussed.

Move. The MOVE command is used for rearranging data that have been incorrectly positioned on the graphics display (i.e., a front view that was placed in the position that should be occupied by the top view). To use this function, activate the MOVE command by touching the number 12 position on the menu with the light pen. A prompt will appear on the text display:

BOX DATA TO BE MOVED

Create a box around the data to be moved by 1) moving the cursor to a point on the screen that is above and to the left of the data to be moved and pressing the puck button, and 2) moving the cursor diagonally to a point that is below and to the right of the data to be moved and pressing the puck button. Make sure that the box encompasses all of the data to be moved, but only the data to be moved, Figure 11-17. At this point, a new prompt will appear on the text display:

INDICATE MOVE FROM POINT

The CAD technician must select a convenient point to use as a marker point when moving data. The point is used for aligning the data in its new position and is known as the "move from point." See point 1 in Figure 11-17. To identify the "move from point," move the cursor to the desired point, and press the puck button. A new prompt will appear on the text display:

INDICATE MOVE TO POINT

The "move to point" is the point with which you will align the cursor in specifying the new location of the data. To identify this point, move the cursor to the desired position and press the puck button. The data will appear in the position specified, but it will blink on and off at this point. This is represented by the phantom lines in Figure 11-17. A new prompt will appear on the text display:

NEW POSITION CORRECT? Y OR N

If the new position is the correct position, type Y for "yes." The blinking lines of the moved data will solidify and the data in the old position will disappear. If the new position is incorrect, type N for "no," and the blinking will stop, the lines will disappear, and the process may be repeated. This last prompt is simply a built-in double check.

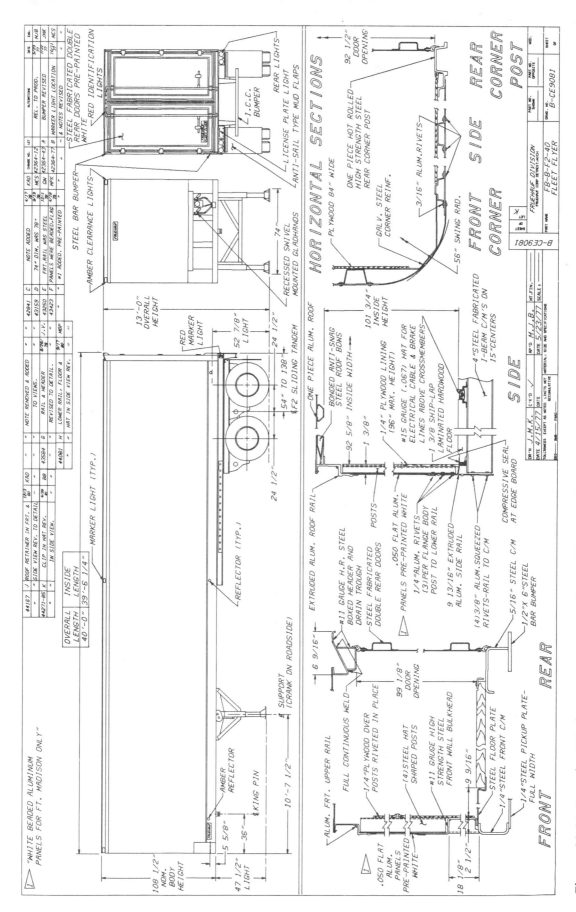

Figure 11-16 Sample CAD drawing (Courtesy Fruehauf Division. Fruehauf Corp.)

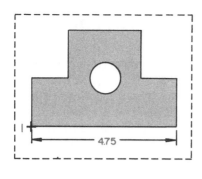

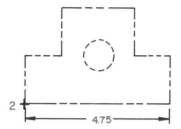

Figure 11-17 The MOVE command

Copy. The COPY command saves CAD technicians time when working on a drawing that contains the same object or symbol repeatedly when that symbol is not contained in a symbols library. Once the object has been created the first time, it can simply be copied as many times as necessary.

To use this function, activate the copy command by touching the number 11 position on the menu with the light pen. A prompt will appear on the text display:

BOX DATA TO BE COPIED

The box is created exactly as it was in the MOVE command. With the box completed, a new prompt will appear on the text display:

INDICATE COPY FROM POINT

This point (point 1 in Figure 11-18) serves the same purpose as the "move from point." It is selected by

aligning the cursor with the desired point and pressing the puck botton. A new prompt will appear on the text display:

INDICATE COPY TO POINT

This point (point 2 in Figure 11-18) serves the same purpose as the "move to point." It is specified by aligning the cursor with the desired position and pressing the puck button. At this point, the copy of the original data will appear in the specified location on the graphics display in a blinking mode. This is represented by phantom lines in Figure 11-18. At this point, a new prompt will appear on the text display:

COPIED POSITION CORRECT? Y OR N

If the new position is correct, type Y for "yes." The blinking lines of the copied data will solidify and the old view will remain. The process may be repeated as many times as necessary. If the new copy, at any time, is not properly positioned, type N for "no," and it will disappear.

Rotate. The ROTATE command allows CAD technicians to revolve data into different angles. To use this function, activate the ROTATE command by touching the number 14 position on the menu with the light pen. A prompt will appear on the text display:

BOX DATA TO BE ROTATED

Create a box around the data to be rotated in the normal manner. Once this step has been accomplished, a new prompt will appear on the text display:

INDICATE ROTATION POINT

Move the cursor to the desired point of rotation, Figure 11-19, and press the puck button. At this point a new prompt will appear on the text display:

INDICATE NEW AXIS

Move the cursor from the point of rotation to create a line at the desired angle of the rotated object and press the puck button, Figure 11-19. The data will now revolve around the rotation point into the new angle, as shown in the figure.

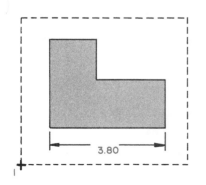

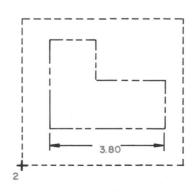

Figure 11-18 The COPY command

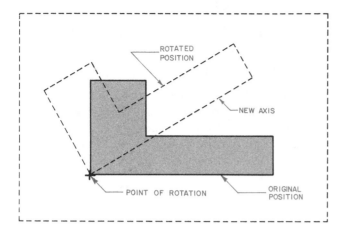

Figure 11-19 The ROTATE command

ZOOM-IN *and* **ZOOM-OUT.** The ZOOM functions allow CAD technicians to move in close to a drawing for intricate detail work and move away from a drawing for pan views. Each ZOOM command in either direction changes the frame in multiples of two (either doubling or halving the frame). To use this function, simply press the ZOOM-IN or the ZOOM-OUT button on the keyboard.

Figure 11-20 shows an object in its normal frame (center), the same object after one ZOOM-OUT command (left), and the same object after one ZOOM-IN command (right).

If the CAD technician desires to return to the normal or beginning frame while zoomed in or out, the BASE key on the keyboard is pressed and the normal frame is returned. It should be noted that zooming in or out does not alter the size of the object being worked on. It simply moves the eye closer to or farther away from the object, giving the appearance of scaling it up or down.

Output Commands

Once data have been input and manipulated, the CAD technician may desire to output them in hardcopy form. The most commonly used type of hard copy in CAD is the pen-plotted drawing. The sample menu in Figure 11-6 contains two output commands for plotting data: PLOT and PLOT SPECS (numbers 19 and 20).

Using the PLOT and PLOT SPECS Commands

A CAD system's pen plotter (the most frequently used type of plotter) is set up to output hard copies of data according to certain specifications which may be adjusted. These specifications include the *window parameters* (the size of the area in which data will be plotted) and the plotting speed. A pen plotter will automatically do the necessary computations to plot data within a window of virtually any size. Plotting speeds for pen plotters may be adjusted from approximately 16 inches per second to as much as 60 inches per second, depending on the type of pen and the type of media being used.

Window and speed specifications are normally set and left. When this is the case, only the PLOT command is needed to accomplish the plotting task. Should a CAD technician find it necessary to alter the specifications, the PLOT SPECS command is used.

PLOT. The PLOT command is used to make penplotted hard copies of data that are in the system. Before using the PLOT command, the SAVE command must be given. This ensures that the data to be plotted are first stored in a workfile so that they are not lost after plotting. The SAVE command is activated by touching the number 24 position on the menu with the light pen, (Figure 11-6). At this point, a prompt will appear on the text display:

DATA SAVED. WILL YOU PLOT? Y OR N

If you desire to plot the data, type Y for "yes." If not, type N for "no." If you type Y a new prompt will appear on the text display:

GIVE PLOT COMMAND

Activate the PLOT command by touching the number 19 position on the menu with the light pen. A prompt will appear on the text display:

PEN(S) INSTALLED? Y CONTINUES.

Type Y if you have remembered to place the pen or pens in their holders. A new prompt will appear on the text display:

MEDIA LOADED? Y CONTINUES.

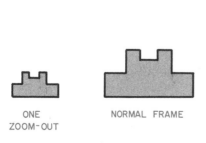

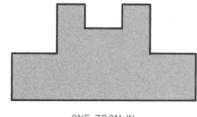

Figure 11-20 The ZOOM command

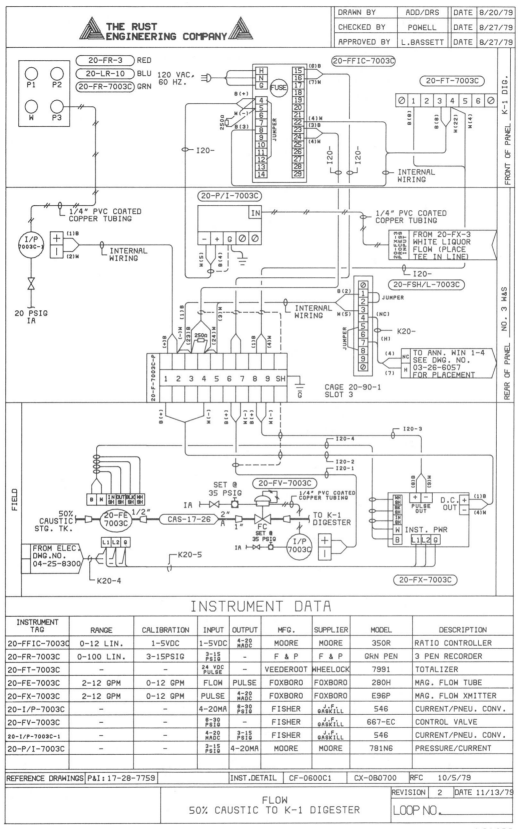

Figure 11-21 Sample CAD drawing (*Courtesy The Rust Engineering Company*)

LA1468

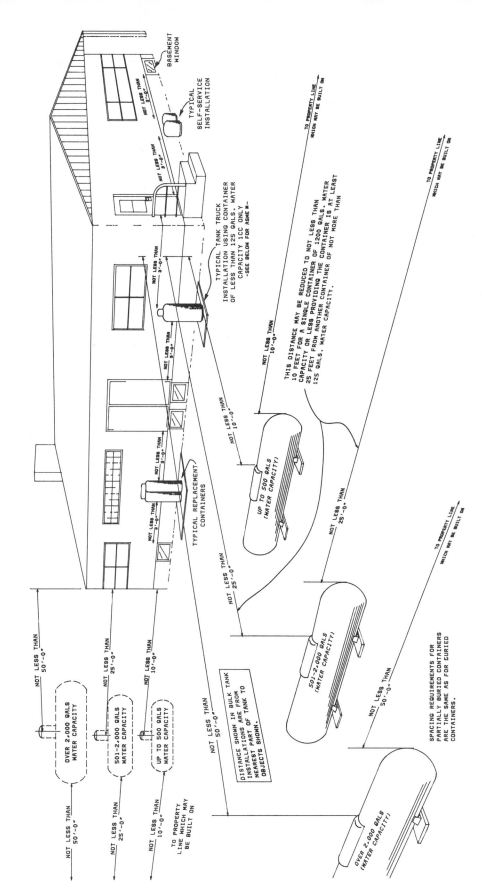

Figure 11-22 Sample CAD drawing (*Courtesy Phillips 66*)

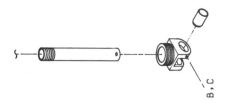

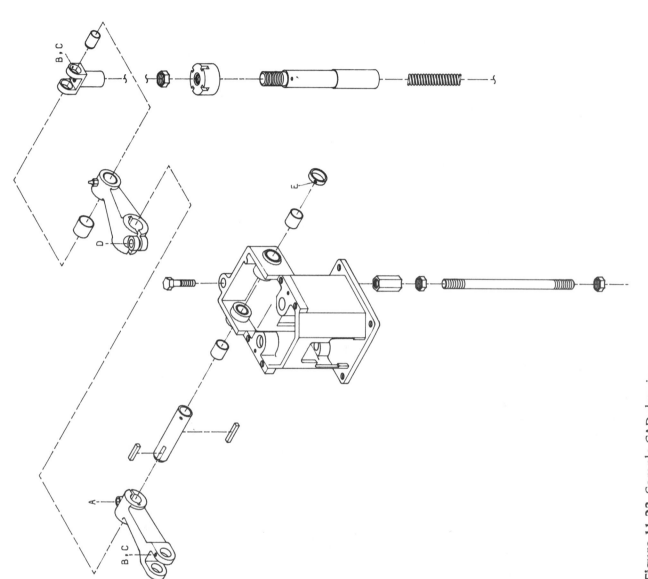

Figure 11-23 Sample CAD drawing

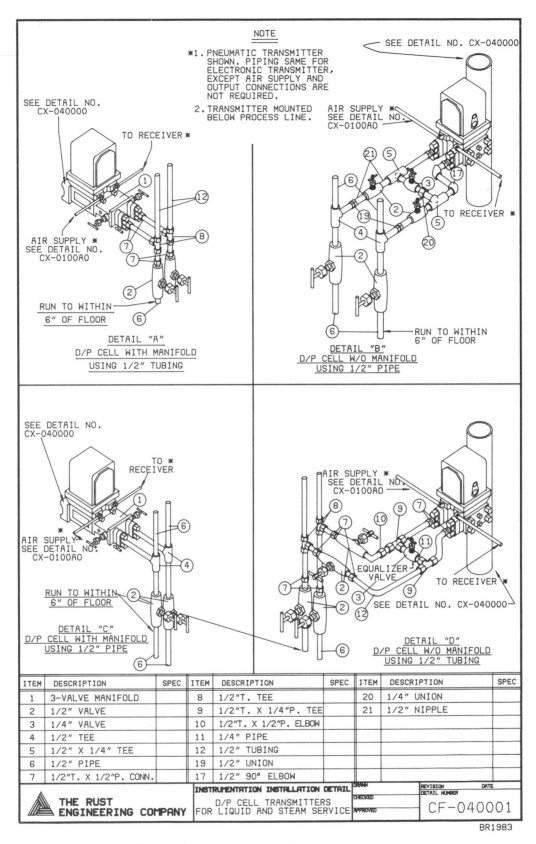

Figure 11-24 Sample CAD drawing (*Courtesy The Rust Engineering Company*)

Type Y if the paper, vellum or polyester film is loaded. Once you have pressed the Y key, the data will be plotted at the rate set. Examine the plotted drawings in Figures 11-21 through 11-24.

PLOT SPECS. Should you ever wish to alter the size of the plotting window or change the plotting speed, use the PLOT SPECS command. To activate the PLOT SPECS command, touch the number 20 position on the menu (Figure 11-6) with the light pen. A prompt will appear on the text display:

(A) WINDOW SIZE
(B) PLOTTING SPEED
(C) WINDOW SIZE AND PLOTTING SPEED

Make the desired selection by pressing the appropriate key. If you press the C key, a new prompt will appear on the text display:

NEW WINDOW SIZE IN INCHES (HORIZONTAL THEN VERTICAL)

Type in the appropriate numbers, giving the horizontal distance for the window first and the vertical second. Once this has been done, a new prompt will appear on the text display:

NEW PLOTTING SPEED (INCHES PER SECOND)

Type in the appropriate numbers. When this has been done, a new prompt will appear on the display verifying the changes that have been made. For example, had you wanted to specify a window size of 10 inches by 8 inches and a plotting speed of 40 inches per second, the verification prompt would read:

WINDOW SIZE: 10 × 8
PLOTTING SPEED: 40

If these figures reflect the desired changes, press the RETURN key once to get out of the PLOT SPECS mode. If you have entered incorrect data, type the word "ERROR" and the A, B, C selection prompt will again appear on the text display so that the necessary changes can be made.

Review

1. What are the three phases of producing design documentation in CAD?

2. List five general operational skills needed to operate any CAD system.

3. Name four ways in which keying is used in CAD.

4. What is the cursor?

5. Name five different ways of controlling the cursor.

6. List five hypothetical commands you might give a computer to have it create a drawing.

7. Name three different types of function menus.

8. Drawings consist of two types of data. What are they?

9. List at least seven standard geometric shapes used in creating graphic data.

10. Explain the term log-on sequence.

Chapter Eleven Problems

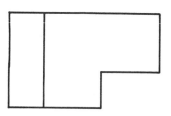

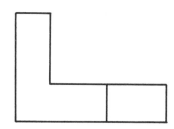

Problems 11-1 and 11-7

The following problems are included to give drafting students practice in inputting, manipulating, and outputting data on a CAD system. A scale has been provided so that students can determine dimensions. Those students who have access to a CAD system should use that system to complete the problems. Students who do not have access to a CAD system should develop detailed, step-by-step explanations of how each problem would be accomplished on a CAD system. The material in this chapter should be used as a guide in preparing such explanations.

Problems 11-1 through 11-6
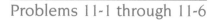

Using the scale provided, determine the dimensions for the object in Problems 11-1 through 11-6, and then sketch it full scale on grid paper. Tape your sketch to the digitizer and complete the following tasks:

a. Input the sketch by digitizing it.

b. Add a right-side view.

c. Dimension the object using the automatic dimensioning (AUTO DIMENSION) function.

d. Plot the completed drawing.

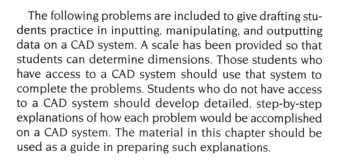

Problems 11-7 through 11-12

Using the scale provided, determine the dimensions for the object in Problems 11-7 through 11-12, then complete the following tasks:

a. Develop the drawing without digitizing it, using the appropriate hardware and commands.

b. Add a right-side view.

c. Enter all dimensions.

d. Plot the completed drawing.

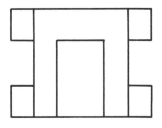

Problems 11-2 and 11-8

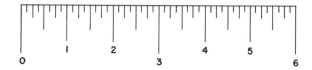

Scale for Problems 11-1 through 11-12

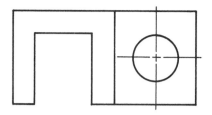

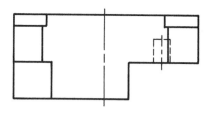

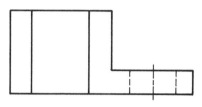

Problems 11-3 and 11-9

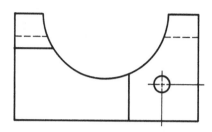

Problems 11-5 and 11-11

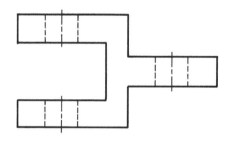

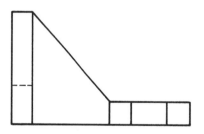

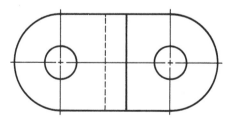

Problems 11-4 and 11-10

Problems 11-6 and 11-12

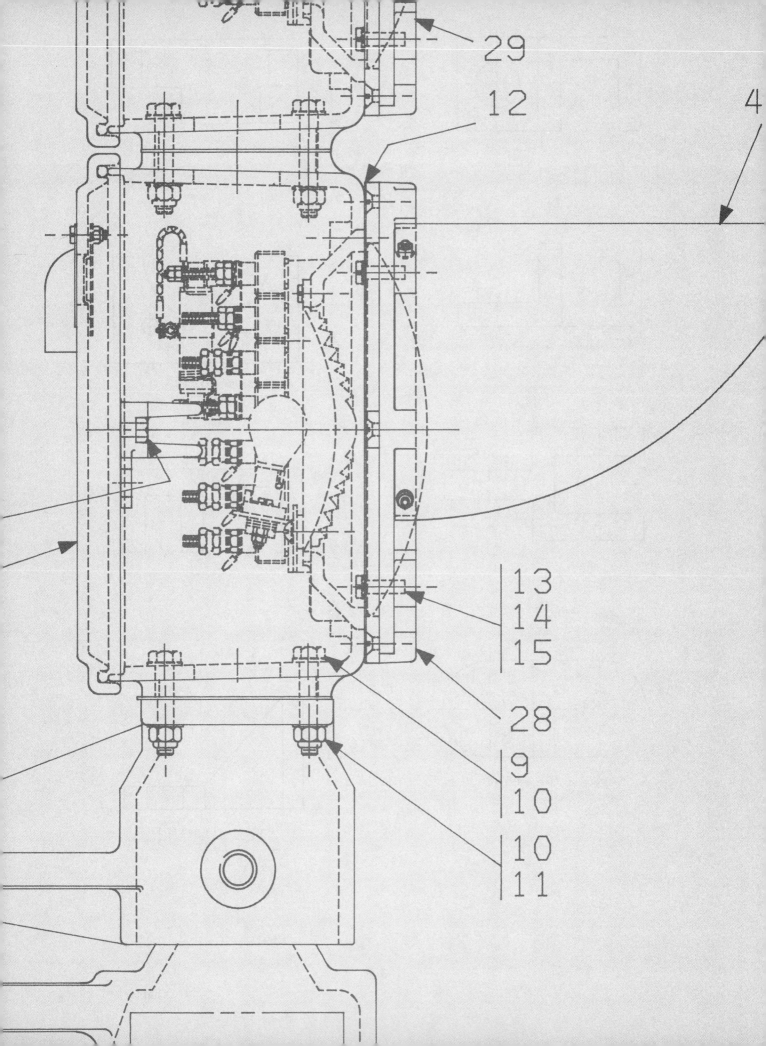

29

12

4

13
14
15

28

9
10

10
11

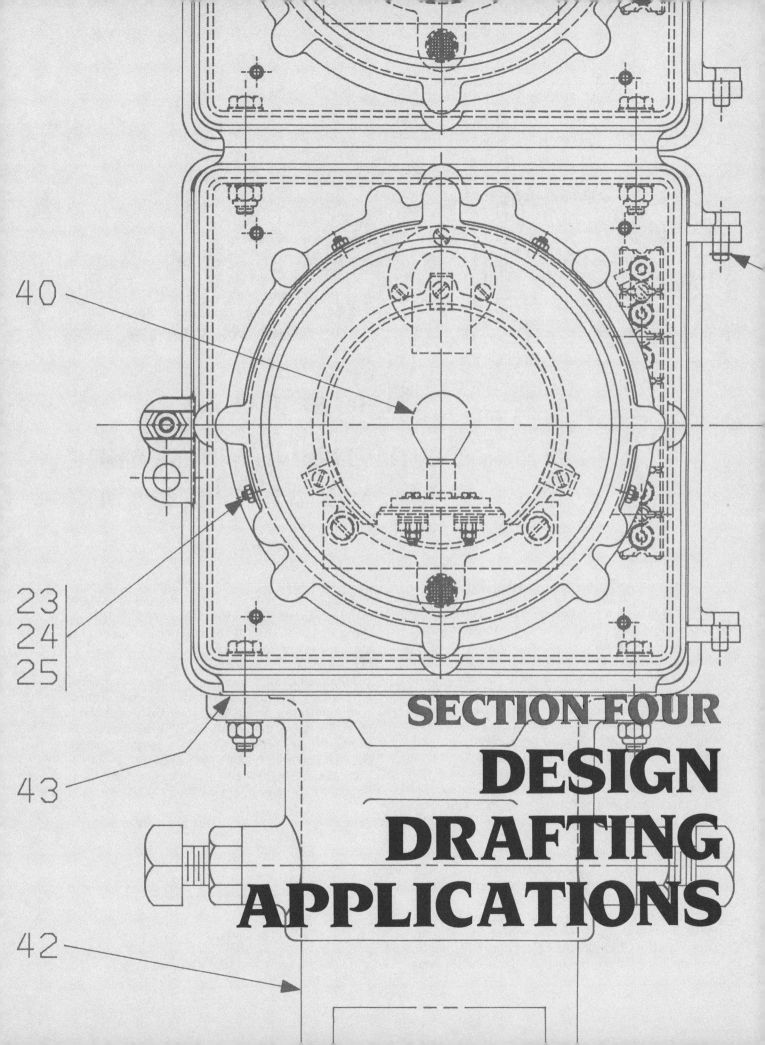

40

23
24
25

43

42

SECTION FOUR

DESIGN
DRAFTING
APPLICATIONS

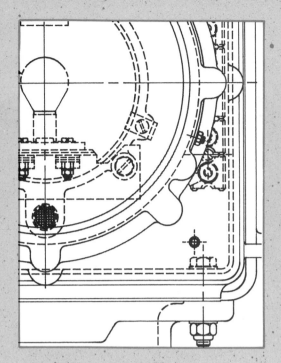

This chapter covers the advanced dimensioning techniques of geometric dimensioning and tolerancing according to ANSI Y14.5M-1982. Major topics covered include general tolerancing, datums, feature control symbols, true position, flatness, straightness, circularity, cylindricity, angularity, parallelism, perpendicularity, profile, and runout.

CHAPTER TWELVE

Geometric Dimensioning and Tolerancing

General Tolerancing

The industrial revolution created a need for mass production; assembling interchangeable parts on an assembly line to turn out great quantities of a given finished product. Interchangeability of parts was the key. If a particular product was composed of 100 parts, each individual part could be produced in quantity, checked for accuracy, stored, and used as necessary.

Since it was humanly and technologically impossible to have every individual part produced exactly alike (it still is), the concept of geometric and positional tolerancing was introduced. *Tolerancing* means setting acceptable limits of deviation. For example, if a mass produced part is to be 4″ in length under ideal conditions, but is acceptable as long as it is not less than 3.99″ and not longer than 4.01″, there is a tolerance of plus or minus .01″, Figure 12-1. This type of tolerance is called a *size tolerance*.

There are three different types of size tolerances: unilateral and bilateral, shown in Figure 12-2, and limit dimensioning. When a *unilateral tolerance* is applied to a dimension, the tolerance applies in one direction only (for example, the object may be larger but not smaller, or it may be smaller but not larger). When

a *bilateral tolerance* is applied to a dimension, the tolerance applies in both directions, but not necessarily evenly distributed. In *limit dimensioning*, the high limit is placed above the low value. When placed in a single line, the low limit precedes the high limit and the two are separated by a dash.

Tolerancing size dimensions offers a number of advantages. It allows for acceptable error without compromises in design, cuts down on unacceptable parts, decreases manufacturing time, and makes the product less expensive to produce. However, it soon became apparent that in spite of advantages gained from size tolerances, tolerancing only the size of an object was not enough. Other characteristics of

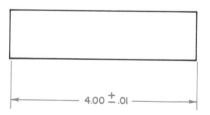

Figure 12-1 Size tolerance

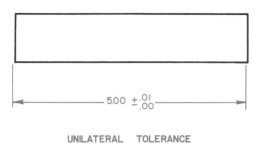

UNILATERAL TOLERANCE

5.00 $\pm$ .01 / .00

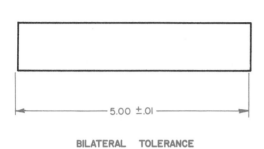

5.00 $\pm$ .01

BILATERAL TOLERANCE

Figure 12-2 Two types of tolerances

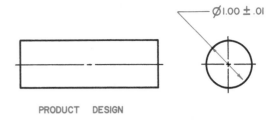

$\emptyset$ 1.00 $\pm$ .01

PRODUCT DESIGN

1.01 (MMC ENVELOPE)

$\emptyset$ 1.00 $\pm$ .01

.99 $\emptyset$ (AS PRODUCED)

MANUFACTURED PRODUCT

Figure 12-3 Tolerance of form

objects also needed to be toleranced, such as location of features, orientation, form, runout, and profile.

In order for parts to be acceptable, depending on their use, they need to be straight, round, cylindrical, flat, angular and so forth. This concept is illustrated in Figure 12-3. The object depicted is a shaft that is to be manufactured to within plus or minus .01 of 1.00 inch in diameter. The finished product meets the size specifications but, since it is not straight, the part might be rejected.

The need to tolerance more than just the size of objects led to the development of a more precise system of tolerancing called geometric dimensioning and positional tolerancing. This new practice improved on conventional tolerancing significantly by allowing designers to tolerance size, form, orientation, profile, location, and runout. In turn, these are the characteristics that make it possible to achieve a high degree of interchangeability.

Geometric Dimensioning and Positional Tolerancing Defined

Geometric dimensioning and positional tolerancing is a dimensioning practice which allows designers to set tolerance limits not just for the size of an object, but for all of the various critical characteristics of a part.

In applying geometric dimensioning and tolerancing to a part, the designer must examine it in terms of its function and its relationship to mating parts.

Figure 12-4 is an example of a drawing of an object that has been geometrically dimensioned and toleranced. It is taken from ANSI Standard Y14.5M-1982, the dimensioning standards manual produced by the American National Standards Institute. This manual is a necessary reference for drafters and designers involved in geometric dimensioning and positional tolerancing.

The key to learning geometric dimensioning and positional tolerancing is to learn the various building blocks which make up the system, as well as how to properly apply them. Figure 12-5 contains a chart of the building blocks of the geometric dimensioning and tolerancing system. In addition to the standard building blocks shown in the figure, several modifying symbols are used when applying geometric tolerancing, as discussed in detail in upcoming paragraphs.

Another concept that must be understood in order to effectively apply geometric tolerancing is the concept of datums. For skilled, experienced designers, the geometric building blocks, modifiers, and datums blend together as a single concept. However, for the purpose of learning, they are dealt with separately, and undertaken step-by-step as individual concepts. They are presented now in the following order: modifiers, datums, and geometric building blocks.

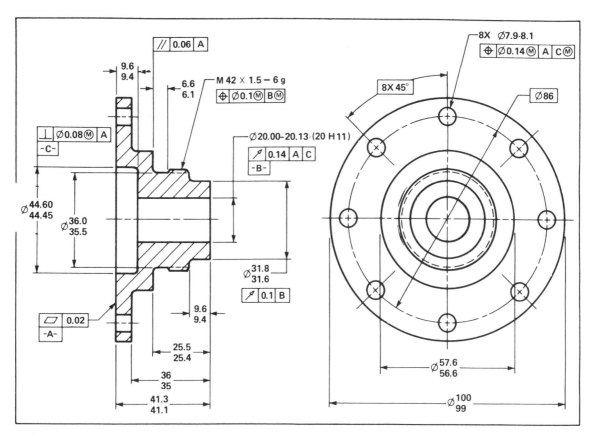

Figure 12-4 Geometrically dimensioned and toleranced drawing (*From* ANSI Y14.5M-1982)

Modifiers

Modifiers are symbols that can be attached to the standard geometric building blocks to alter their application or interpretation. The proper use of modifiers is fundamental to effective geometric tolerancing. Four modifiers are frequently used: maximum material condition, regardless of feature size, least material condition, and projected tolerance zone, Figure 12-6.

Maximum Material Condition

Maximum material condition, abbreviated MMC, is the condition of a characteristic when the most material exists. For example, the MMC of the external feature in Figure 12-7 is .77 inch. This is the MMC because it represents the condition where the most material exists on the part being manufactured. The MMC of the internal feature in the figure is .73 inch. This is the MMC because the most material exists when the hole is produced at the smallest allowable size.

In using this concept, the designer must remember that the MMC of an internal feature is the smallest allowable size. The MMC of an external feature is the largest allowable size within specified tolerance limits inclusive. A rule of thumb to remember is that MMC means *most material*.

Regardless of Feature Size

Regardless of feature size, abbreviated RFS, is a modifier which tells machinists that a tolerance of form or position or any characteristic must be maintained, regardless of the actual produced size of the object. This concept is illustrated in Figure 12-8. In the RFS example, the object is acceptable if produced in sizes from 1.01 inches to .99 inch inclusive. The form control is axis straightness to a tolerance of .02 inch regardless of feature size. This means that the .02 inch axis straightness tolerance must be adhered to, regardless of the produced size of the part.

Contrast this with the MMC example. In this case, the produced sizes are still 1.01 inches, 1.00 inch, and .99 inch. However, because of the MMC modifier, the .02 inch axis straightness tolerance applies only at MMC or 1.01 inches.

If the produced size is smaller, the straightness tolerance can be increased proportionally. Of course, this makes the MMC modifier more popular with machinists for several reasons: 1) it allows them greater room for error without actually increasing the tolerance, 2) it decreases the number of parts rejected, 3) it cuts down on unacceptable parts, 4) it decreases the number of inspections required, and 5) it allows the use of functional gaging. All of these advantages

	TYPE OF TOLERANCE	CHARACTERISTIC	SYMBOL	SEE:
FOR INDIVIDUAL FEATURES	FORM	STRAIGHTNESS	—	6.4.1
		FLATNESS	▱	6.4.2
		CIRCULARITY (ROUNDNESS)	○	6.4.3
		CYLINDRICITY	⌭	6.4.4
FOR INDIVIDUAL OR RELATED FEATURES	PROFILE	PROFILE OF A LINE	⌒	6.5.2 (b)
		PROFILE OF A SURFACE	⌓	6.5.2 (a)
FOR RELATED FEATURES	ORIENTATION	ANGULARITY	∠	6.6.2
		PERPENDICULARITY	⊥	6.6.4
		PARALLELISM	//	6.6.3
	LOCATION	POSITION	⊕	5.2
		CONCENTRICITY	◎	5.11.3
	RUNOUT	CIRCULAR RUNOUT	↗ *	6.7.2.1
		TOTAL RUNOUT	↗↗ *	6.7.2.2

*Arrowhead(s) may be filled in.

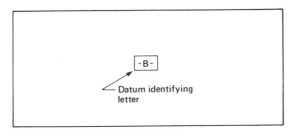

Figure 12-5 Building blocks of the geometric dimensioning and tolerancing system (*From* ANSI Y14.5M-1982)

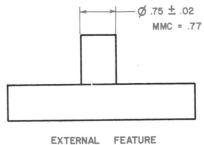

EXTERNAL FEATURE

MODIFIER	SYMBOL
MAXIMUM MATERIAL CONDITION	Ⓜ
REGARDLESS OF FEATURE SIZE	Ⓢ
LEAST MATERIAL CONDITION	Ⓛ
PROJECTED TOLERANCE ZONE	Ⓟ

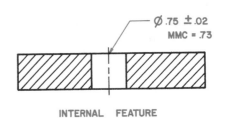

INTERNAL FEATURE

Figure 12-6 Modifiers used when applying geometric tolerancing

Figure 12-7 MMC of an external and an internal feature

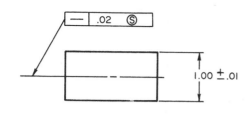

PRODUCED SIZE	TOLERANCE (FORM)
1.01	.02
1.00	.02
.99	.02

RFS

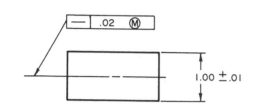

PRODUCED SIZE	TOLERANCE (FORM)
1.01	.02
1.00	.03
.99	.04

MMC

Figure 12-8 Regardless of feature size (RFS)

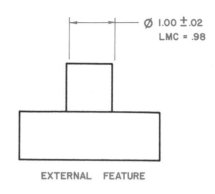

EXTERNAL FEATURE

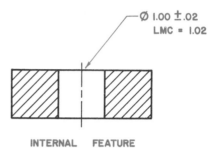

INTERNAL FEATURE

Figure 12-9 Least material condition (LMC)

be extended a specified distance beyond a given surface. This concept is discussed further later in this chapter under the heading "True Position."

Datums

Datums are points, lines, axes, surfaces or planes used for referencing features of an object. Datums are identified on drawings by datum "flags" or symbols, Figure 12-10. This figure shows several ways in which datum feature symbols are placed on drawings.

In Part A of the figure, the datum symbol is attached to an extension line. In this example, DATUM A would be the plane upon which the part is resting. In Parts B, C, and D, the datum symbol is associated with a dimension by a dimension line, a leader line or by appearing under a dimension. In these cases, the symbol applies to the entire feature rather than just one surface as in object A.

Establishing Datums

In establishing datums, designers must consider the function of the part, the manufacturing processes that will be used in producing the part, how the part will be inspected, and the part's relationship to other parts after assembly. Designers and drafters must also understand the difference between a datum feature and a datum surface.

A *datum plane* is a theoretically perfect plane from which measurements are made. A *datum surface* is the

translate into substantial financial savings while, at the same time, making it possible to produce interchangeable parts at minimum expense.

Least Material Condition

Least material condition, abbreviated LMC, is the opposite of MMC. It refers to the condition in which the least material exists. This concept is illustrated in Figure 12-9.

In the top example, the external feature of the part is acceptable if produced in sizes ranging from .98 inch to 1.02 inches inclusive. The least material exists at .98 inch. Consequently, .98 inch is the LMC.

In the bottom example, the internal feature (hole) is acceptable if produced in sizes ranging from .98 inch to 1.02 inches inclusive. The least material exists at 1.02 inches. Consequently, 1.02 represents the LMC.

Projected Tolerance Zone

Projected tolerance zone is a modifier that allows a tolerance zone established by a locational tolerance to

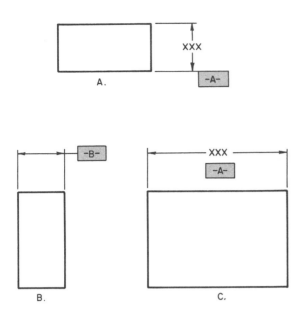

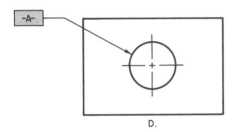

Figure 12-10 Datum flags

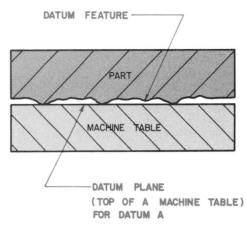

Figure 12-11 Establishing datums

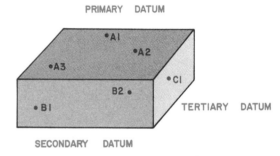

Figure 12-12 Specified point contact for establishing datums

inexact corresponding surface of the object. This concept is illustrated in Figure 12-11.

Notice the irregularities on the datum surface. The high points on the datum surface actually establish the datum plane, which, in this case, is the top of a machine table. All measurements referenced to DATUM A are measured from the theoretically perfect datum plane. High point contact is used for establishing datums when the entire surface in question will be a machined surface. On rougher more irregular surfaces, such as those associated with castings, specified point contacts are used for establishing datums. In these cases, only specified points on surfaces are employed to establish datum features.

When specified point contact is used for establishing datums, a minimum of three points is required for the primary datum, a minimum of two for the secondary, and a minimum of one for the tertiary, Figure 12-12. These points are normally located with basic dimensions and identified with datum target symbols, Figure 12-13. A *basic dimension* is a theoretically exact dimension enclosed in a rectangular box.

In Figure 12-13, DATUM A is the top of the object and it is established by points A1, A2, and A3. The characters in each datum target symbol identify the

datum by letter and target numbers. DATUM B is the front of the object and DATUM C is the right side.

Notice that a datum is called out on a drawing in the view where the surface in question appears as an edge. Notice also that the secondary datum must be perpendicular to the first, and the tertiary datum must be perpendicular to both the primary and secondary datums. These three datums establish what is called the datum frame. The *datum frame* is a hypothetical three-dimensional box into which the object being produced fits and from which measurements can be made.

Round and cylindrically shaped objects require a datum axis. The center line of a cylindrical object can be a datum, as can the end of the object. This concept is illustrated in Figure 12-14. DATUM A is the left-hand end of the cylinder. DATUM B is the diameter.

Feature Control Symbol

The *feature control symbol* is a rectangular box in which all data referring to the subject feature control are placed, including: the symbol, datum references, the

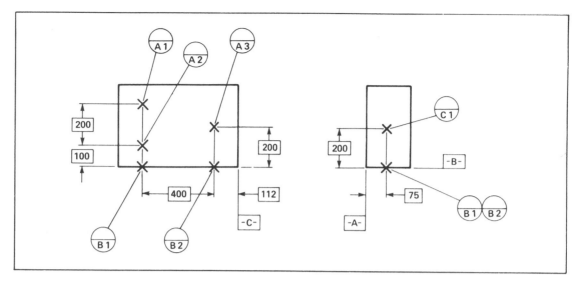

Figure 12-13 Datum target symbols with basic dimensions (*From* ANSI Y14.5M-1982)

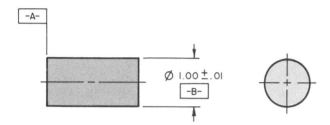

Figure 12-14 Datum axis

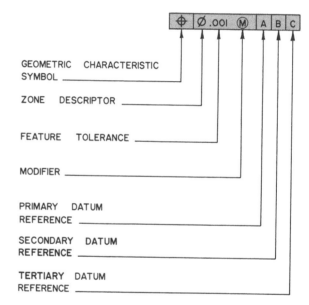

Figure 12-15 Order of elements in a feature control symbol

feature control tolerance, and modifiers. These various feature control elements are separated by vertical lines. (Figure 12-4 contains a drawing showing how feature control symbols are actually composed.)

The order of the data contained in a feature control frame is important. The first element is the feature control symbol. Next is the zone descriptor, such as a diameter symbol where applicable. Then, there is the feature control tolerance, modifiers when used, and datum references listed in order from left to right, Figure 12-15.

True Position

True position is the theoretically exact location of the center line of a product feature such as a hole. The tolerance zone created by a position tolerance is an imaginary cylinder, the diameter of which is equal to the stated position tolerance. The dimensions used to locate a feature, that is to have a position tolerance, must be basic dimensions.

Figure 12-16 contains an example of a part with two holes drilled through it. The holes have a position tolerance relative to three datums: A, B, and C.

The holes are located by basic dimensions. The feature control frame states that the positions of the center lines of the holes must fall within cylindrical tolerance zones having diameters of .030 inch at MMC relative to DATUMS A, B, and C. The modifier indicates that the .030 inch tolerance applies only at MMC. As the holes are produced larger than MMC, the diameter of the tolerance zones can be increased correspondingly.

Figure 12-17 illustrates the concept of the cylindrical tolerance zone from Figure 12-16. The feature control frame is repeated showing a .030 inch diameter tolerance zone. The broken-out section of the object from Figure 12-16 provides the interpretation. The cylindrical tolerance zone is shown in phantom

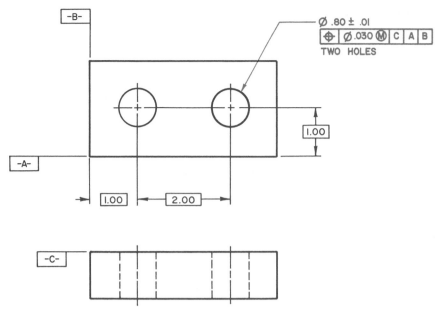

Figure 12-16 True position

lines. The center line of the hole is acceptable as long as it falls anywhere within the hypothetical cylinder.

Using the Projected Tolerance Zone Modifier

As stated previously, a *projected tolerance zone* is a tolerance zone that extends beyond the surface of a part. When a projected tolerance zone modifier is used, the surface in question must be identified as a datum and the length of the projected tolerance zone specified.

The tolerance zone illustrated in Figure 12-18 has a depth equal to that of the part through which it extends. On occasion, there is a need to extend a tolerance zone beyond the surface of a part. ANSI

Y14.5M-1982 recommends the use of the projected tolerance zone concept when the variation in perpendiculars of threaded or press-fit holes could cause fasteners, such as screws, studs, or pins, to interfere with mating parts. In these cases, the projected tolerance zone concept is used.

Flatness

Flatness is a feature control of a surface which requires all elements of the surface to lie within two hypothetical parallel planes. When flatness is the feature control, a datum reference is neither required nor proper.

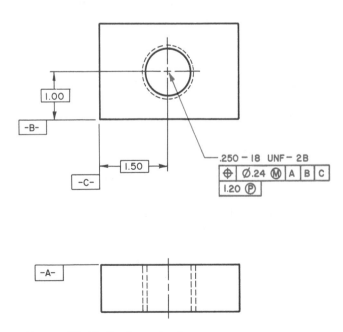

Figure 12-17 Cylindrical tolerance zone

Figure 12-18 Projected tolerance zone

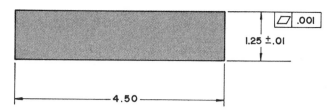

Figure 12-19 Flatness

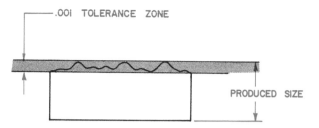

Figure 12-20 Flatness interpreted

Figure 12-21 Straightness

Figure 12-19 shows how flatness is called-out on a drawing. A part that is 4.50 inches long and 1.25 inches thick, plus or minus .01 inch, is required to be flat on its top surface. The feature control frame is placed on that surface and is interpreted as follows: this surface must be flat within .001 inch.

Figure 12-20 interprets the meaning of Figure 12-19. The tolerance zone is established by the highest points on the surface in question within the limits of the size tolerance, regardless of the produced size, and extending down .001 inch. All points on the top surface of the object must lie between the two parallel planes that are .001 inch apart.

Straightness

Straightness is a feature control of one single element of a surface that must be a straight line to within the stated tolerance. It differs from flatness in that flatness covers an entire surface rather than just one element on a surface. A straightness tolerance yields a tolerance zone of specified width between which all points on the line in question must lie. Straightness is generally applied to longitudinal elements.

Figure 12-21 shows how a straightness tolerance is applied on a drawing. The feature control frame states that any longitudinal element of the surface in the direction indicated must lie between two parallel straight lines that are .001 inch apart. Figure 12-22 illustrates this interpretation.

Straightness, like flatness, does not require a datum reference. Straightness is not additive to the size tolerance and must be contained within the limits of the size tolerance.

Circularity (Roundness)

Circularity, sometimes referred to as *roundness,* is a feature control for a surface of revolution (cylinder, sphere, cone, and so forth). It specifies that all points of a surface must be equidistant from the center line or axis of the object in question. The tolerance zone for circularity is formed by two concentric and coplanar circles between which all points on the surface of revolution must lie.

Figure 12-23 illustrates how circularity is called-out on a drawing. Figure 12-24 provides an interpretation of what the circularity tolerance in Figure 12-23

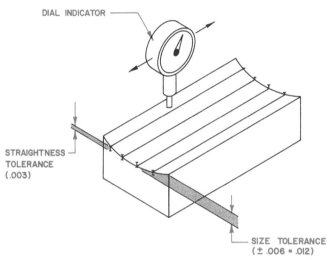

Figure 12-22 Straightness interpreted

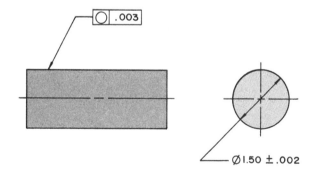

Figure 12-23 Circularity

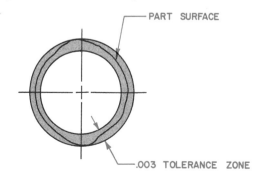

Figure 12-24 Circularity interpreted

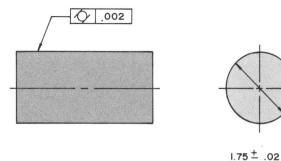

Figure 12-25 Cylindricity

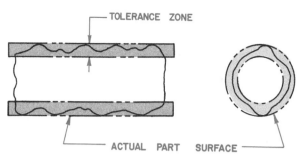

Figure 12-26 Cylindricity interpreted

actually means. At any selected cross section of the part, all points on the surface must fall within the zone created by the two concentric circles. At any point where circularity is measured, it must fall within the size tolerance. Notice that a circularity tolerance cannot specify a datum reference.

Cylindricity

Cylindricity is a feature control in which all elements of a surface of revolution form a cylinder. It gives the effect of circularity extended the entire length of the object, rather than just as a specified cross section. The tolerance zone is formed by two hypothetical concentric cylinders.

Figure 12-25 illustrates how cylindricity is called-out on a drawing. Notice that a cylindricity tolerance does not require a datum reference.

Figure 12-26 provides an illustration of what the cylindricity tolerance actually means. Two hypothetical concentric cylinders form the tolerance zone. The outside cylinder is established by the outer limits of the object at its produced size within specified size limits. The inner cylinder is smaller (on radius) by a distance equal to the cylindricity tolerance.

Cylindricity requires that all elements on the surface fall within the size tolerance and the tolerance established by the feature control.

Angularity

Angularity is a feature control in which a given surface, axis, or center plane must form a specified angle with a datum other than 90°. Consequently, an angularity tolerance requires a datum reference. The tolerance zone formed by an angularity callout consists of two hypothetical parallel planes which form the specified angle with the datum. All points on the angular surface or along the angular axis must lie between these parallel planes.

Figure 12-27 illustrates how an angularity tolerance is called out on a drawing. Notice that the specified angle is enclosed in a BASIC box. This is required when applying an angularity tolerance. Figure 12-28 provides an interpretation of what the angularity tolerance in Figure 12-27 actually means. Notice that

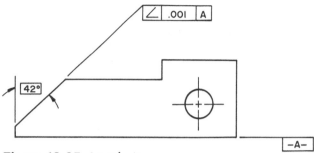

Figure 12-27 Angularity

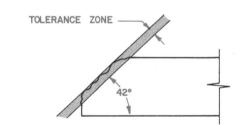

Figure 12-28 Angularity interpreted

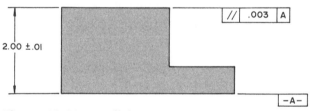

Figure 12-29 Parallelism

the outside plane of the tolerance zone is established by the outermost extremities of the angular surface. The inner plane is measured in a distance equal to the angularity tolerance. All elements of the toleranced surface must lie within the size tolerance limits.

Parallelism

Parallelism is a feature control that specifies that all points on a surface, center plane, axis or line must be equidistant from a datum. Consequently, a parallelism tolerance must have a datum reference. A parallelism tolerance zone is formed by two hypothetical planes that are parallel to a specified datum. They are spaced apart at a distance equal to the parallelism tolerance.

Figure 12-29 illustrates how a parallelism is called-out on a drawing. Figure 12-30 provides an interpre-

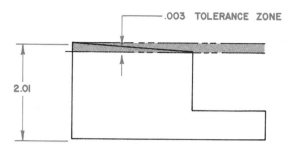

Figure 12-30 Parallelism interpreted

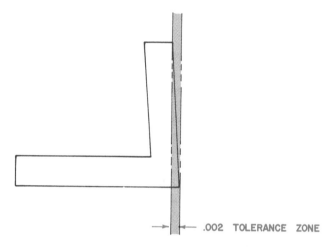

Figure 12-31 Perpendicularity

tation of what the parallelism tolerance in Figure 12-29 actually means. Notice that all elements of the toleranced surface must fall within the size limits.

Perpendicularity

Perpendicularity is a feature control which specifies that all elements of a surface, axis, center plane, or line form a 90° angle with a datum. Consequently, a perpendicularity tolerance requires a datum reference. A perpendicularity tolerance is formed by two hypothetical parallel planes.

Figure 12-31 illustrates how a perpendicularity tolerance is called-out on a drawing. Figure 12-32 provides an interpretation of what the perpendicularity tolerance actually means. The elements of the toleranced surface must fall within the size limits and between two hypothetical parallel planes that are a distance apart equal to the perpendicularity tolerance.

Figure 12-32 Perpendicularity interpreted

Profile

Profile is a feature control that specifies the amount of allowance variance of a surface or line elements on a surface. There are three different variations of the profile tolerance: bilateral, unilateral up, and unilateral down, Figures 12-33, 12-34, and 12-35. A profile tolerance is normally used for controlling arcs,

curves, and other unusual profiles not covered by the other feature controls. It is a valuable feature control for use on objects that are so irregular that other feature controls do not easily apply.

When applying a profile tolerance, the symbol used indicates whether the designer intends profile of a line or profile of a surface, Figure 12-36. *Profile of a line* establishes a tolerance for a given single element of a surface. *Profile of a surface* applies to the entire

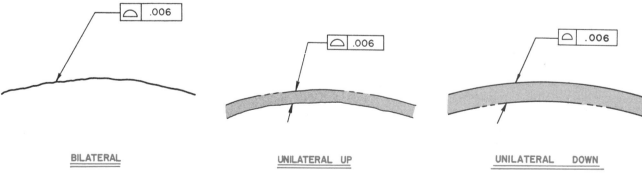

Figure 12-33 Bilateral profile tolerance

Figure 12-34 Unilateral up profile

Figure 12-35 Unilateral down profile

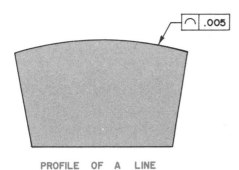

PROFILE OF A LINE

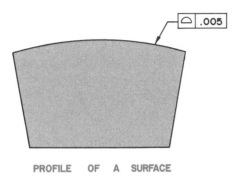

PROFILE OF A SURFACE

Figure 12-36 Profile callouts

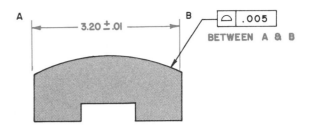

BETWEEN A & B

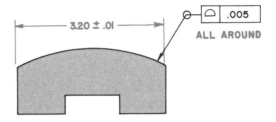

ALL AROUND

Figure 12-37 Profile "ALL AROUND"

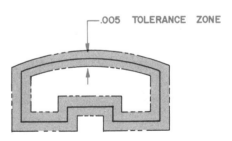

Figure 12-38 Interpretation of "BETWEEN A & B"

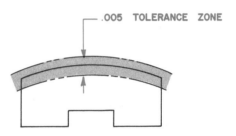

Figure 12-39 Interpretation of "ALL AROUND"

surface. The difference between profile of a line and profile of a surface is similar to the difference between circularity and cylindricity.

When using a profile tolerance, drafters and designers should remember to use phantom lines to indicate whether the tolerance is applied unilaterally up or unilaterally down. A bilateral profile tolerance requires no phantom lines. An ALL AROUND symbol should also be placed on the leader line of the feature control frame to specify whether the tolerance applies ALL AROUND or between specific points on the object, Figure 12-37.

Figure 12-38 provides an interpretation of what the BETWEEN A & B profile tolerance in Figure 12-37 actually means. The rounded top surface, and only the top surface, of the object must fall within the specified tolerance zone. Figure 12-39 provides an interpretation of what the ALL AROUND profile tolerance in Figure 12-37 actually means. The entire surface of the object, all around the object, must fall within the specified tolerance zone.

Runout

Runout is a feature control that limits the amount of deviation from perfect form allowed on surfaces or rotation through one full rotation of the object about its axis. Revolution of the object is around a datum axis. Consequently, a runout tolerance does require a datum reference.

Runout is most frequently used on objects consisting of a series of concentric cylinders and other shapes of revolution that have circular cross sections; usually, the types of objects manufactured on lathes, Figure 12-40.

Notice in Figure 12-40 that DATUM A is the center point of the left-hand end surface and DATUM B is the center point of the right-hand surface. The axis of revolution is the axis running from DATUM A to DATUM B. In the feature control frame, this is indicated by an A-B datum reference. Notice also that each successive part of the object may have a different runout tolerance, depending on its function and relationship to other parts in the final assembly.

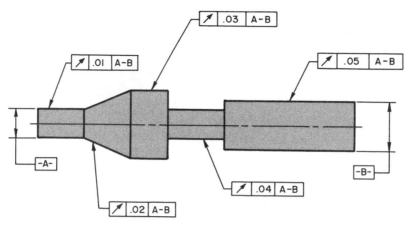

Figure 12-40 Runout

Review

1. Define the term *tolerancing*.

2. What are the two types of size tolerances?

3. What led to the development of geometric dimensioning and tolerancing?

4. Define the term *geometric dimensioning and tolerancing*.

5. Identify the ANSI standard that pertains to geometric dimensioning and tolerancing.

6. Sketch the symbols for the following:
   a. Flatness   e. Perpendicularity
   b. Circularity   f. Parallelism
   c. Straightness   g. Angularity
   d. True position

7. Explain the term *maximum material condition*.

8. Explain the term *regardless of feature size*.

9. Explain the term *least material condition*.

10. What is a datum?

11. How is a datum established on a machined surface?

12. How is a datum established on a cast surface?

13. Sketch a sample feature control symbol that illustrates the proper order of elements.

14. Which feature controls *do not* require a datum reference?

15. Which feature controls *must* have a datum reference?

Chapter Twelve Problems

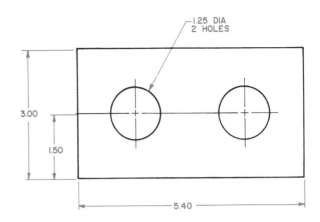

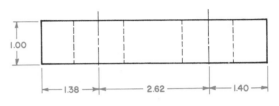

Problem 12-1

The following problems are intended to give beginning drafters practice in applying the principles of geometric dimensioning and tolerancing.

The steps to follow in completing the problems are:

Step 1. Study the problem carefully.

Step 2. Make a checklist of tasks you will need to complete.

Step 3. Center the required view or views in the work area.

Step 4. Include all dimensions according to ANSI Y14.5M-1982.

Step 5. Re-check all work. If it's correct, neatly fill out the title block using light guidelines and freehand lettering.

Note: These problems do not follow current drafting standards. You are to use the information shown here to develop properly drawn, dimensioned, and toleranced drawings.

Problems 12-1 through 12-10

12-1. Using the scale provided, compute all dimensions for the object in Problem 12-1, and draw it full scale. Establish Datums A, B, and C. Use basic dimensions and apply a true position tolerance of 0.001 to the holes.

12-2. Using the scale provided, compute all dimensions for the object in Problem 12-2, and draw it full scale. Establish Datums A, B, and C. Use basic dimensions and apply a true position tolerance of 0.001 to the holes. Also apply a 0.003 FLATNESS tolerance to the top surface.

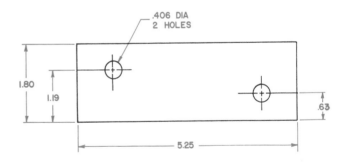

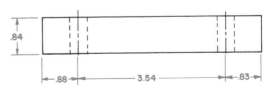

Problem 12-2

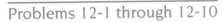

Scale for Problems 12-1 through 12-10

12-3. Using the scale provided, compute all dimensions for the object in Problem 12-3, and draw it full scale. Apply a STRAIGHTNESS tolerance of 0.002 to the object.

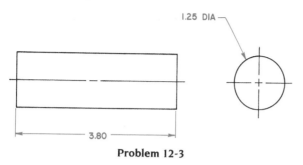

1.25 DIA

3.80

Problem 12-3

12-4. Using the scale provided, compute all dimensions for the object in Problem 12-4, and draw it full scale. Establish Datums A and B. Apply a TRUE POSITION tolerance of 0.001 to the hole through the cylinder. Apply a FLATNESS tolerance of 0.002 to the flat surface. Apply a CIRCULARITY tolerance of 0.001 to the element indicated.

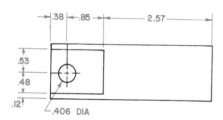

.38 .85 2.57

.53 .48 .12

.406 DIA

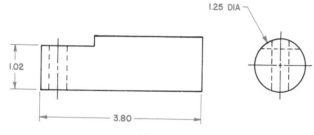

1.25 DIA

1.02

3.80

Problem 12-4

12-5. Using the scale provided, compute all dimensions for the object in Problem 12-5, and draw it full scale. Apply a CYLINDRICITY tolerance of 0.001 to the entire object.

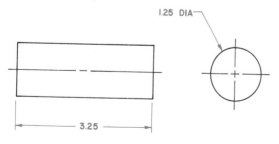

1.25 DIA

3.25

Problem 12-5

12-6. Using the scale provided, compute all dimensions for the object in Problem 12-6, and draw it full scale. Establish Datums A, B, and C. Apply a TRUE POSITION tolerance of 0.003 to the hole. Apply an ANGULARITY tolerance of 0.001 to the sloped surface.

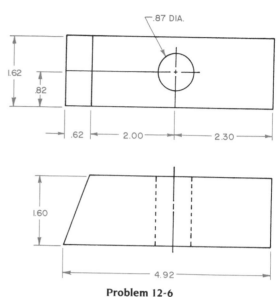

.87 DIA.

1.62

.82

.62 2.00 2.30

1.60

4.92

Problem 12-6

12-7. Using the scale provided, compute all dimensions for the object in Problem 12-7, and draw it full scale. Establish Datums A, B, and C. Apply a TRUE POSITION tolerance of 0.001 to the hole. Apply a PARALLELISM tolerance of 0.002 to Surface A.

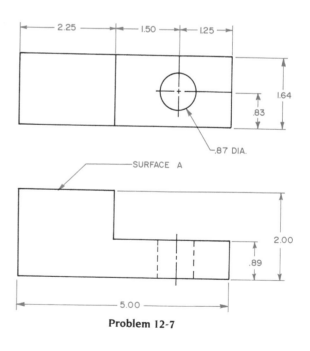

2.25 1.50 1.25

1.64

.83

.87 DIA.

SURFACE A

2.00

.89

5.00

Problem 12-7

0 1 2 3 4 5 6

12-8. Using the scale provided, compute all dimensions for the object in Problem 12-8, and draw it full scale. Establish Datums A, B, and C. Apply a TRUE POSITION tolerance of 0.002 to the hole. Apply a PERPENDICULARITY tolerance of 0.001 to Surface A.

12-9. Using the scale provided, compute all dimensions for the object in Problem 12-9, and draw it full scale. Apply a SURFACE PROFILE tolerance of 0.001 to the top surface of the object between Points A and B.

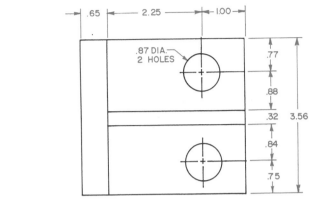

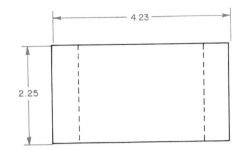

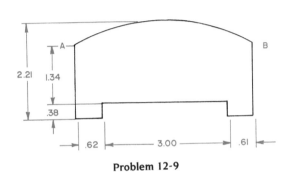

Problem 12-9

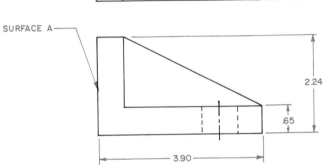

Problem 12-8

12-10. Using the scale provided, compute all dimensions for the object in Problem 12-10, and draw it full scale. Establish Datum A-B axis. Apply a RUNOUT tolerance of 0.001 to each of the five parts of the object.

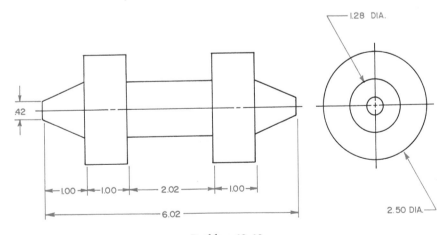

Problem 12-10

Drawings of mechanical parts are shown in Problems 12-11 through 12-34. Each part is dimensioned. Examine each part closely and apply the appropriate feature controls. Use a standard tolerance of .003.

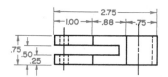

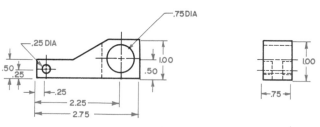

Problem 12-11

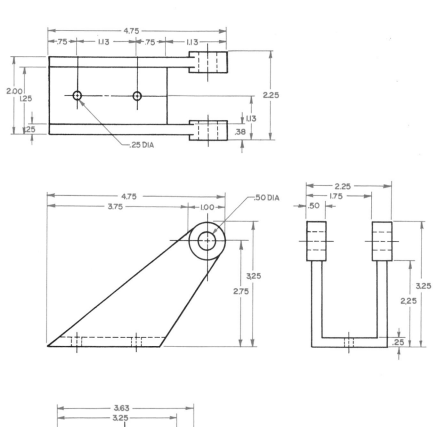

Problem 12-12

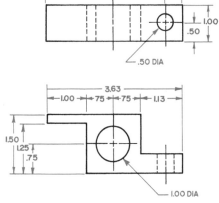

Problem 12-13

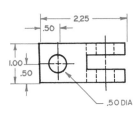

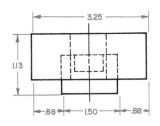

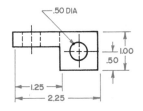

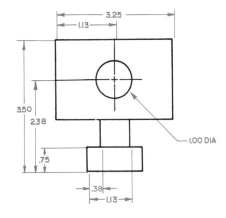

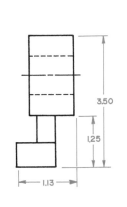

Problem 12-14

Problem 12-17

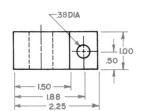

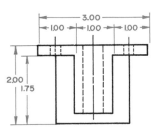

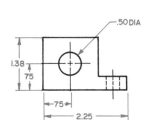

Problem 12-15

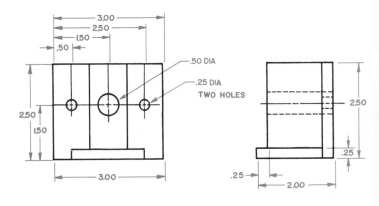

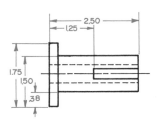

Problem 12-18

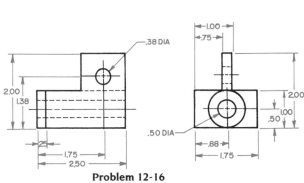

Problem 12-16

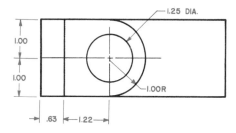

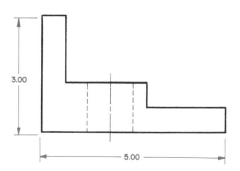

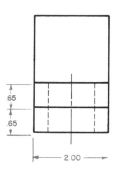

Problem 12-19

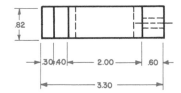

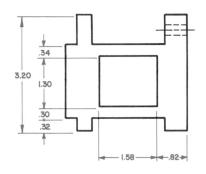

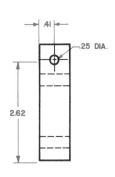

Problem 12-20

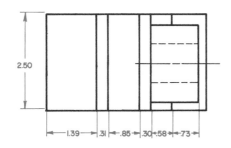

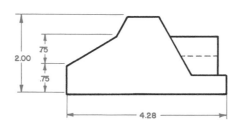

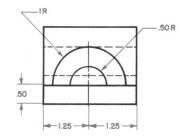

Problem 12-21

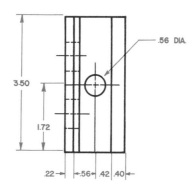

.56 DIA.

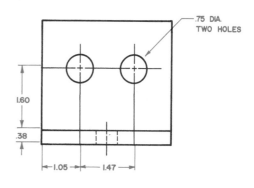

.75 DIA.
TWO HOLES

Problem 12-22

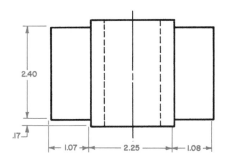

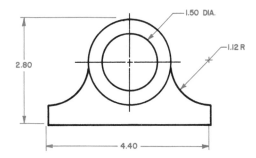

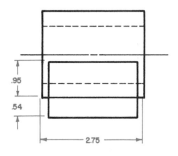

Problem 12-23

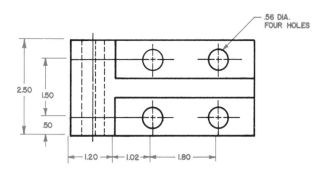

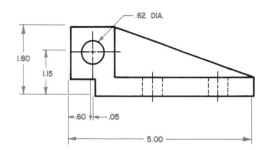

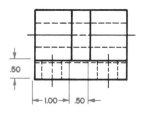

Problem 12-24

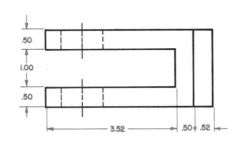

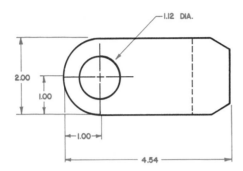

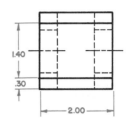

Problem 12-25

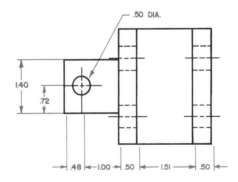

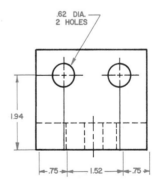

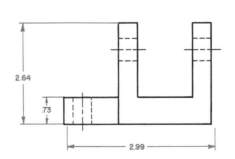

Problem 12-26

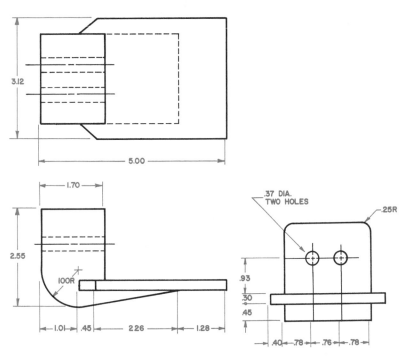

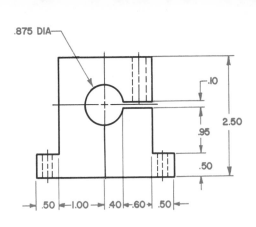

.875 DIA

.10

2.50

.95

.50

.50 |←1.00→| .40 |←.60→| .50 |←

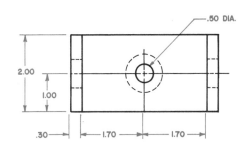

.50 DIA.

2.00

1.00

.30 |← 1.70 → |← 1.70 →

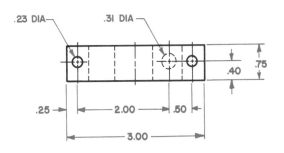

.23 DIA   .31 DIA

.75

.40

.25 →   2.00   .50 |←

3.00

Problem 12-27

Problem 12-29

.37 DIA.
TWO HOLES

.25R

.93

.30

.45

.40 .78 .76 .78

2.10

.75 DIA.

1.43

.25

1.50   1.00

4.00

.50 .50 .50

Problem 12-28

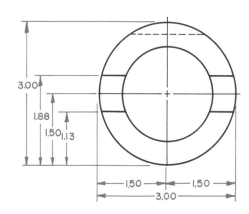

1.25

.88

.38

.63 |← 1.63 →

3.00

3.00

1.88

1.50 1.13

1.50   1.50

3.00

Problem 12-30

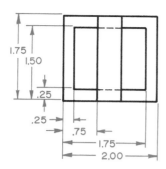

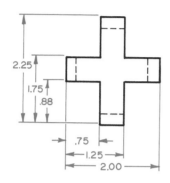

Problem 12-31

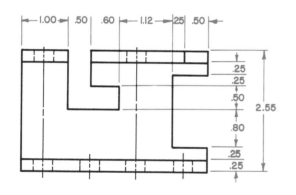

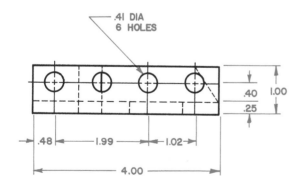

.41 DIA
6 HOLES

Problem 12-32

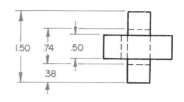

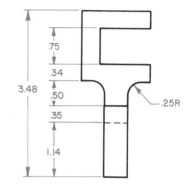

Problem 12-33

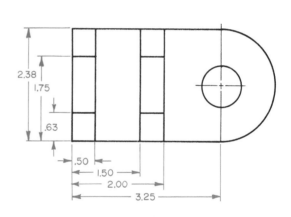

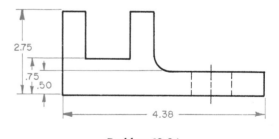

Problem 12-34

1. Three holes in the part are reamed to .078. What is the positional tolerance of these holes?
2. Describe the shape of the tolerance zone for the positional tolerance in Question 1.
3. Three holes in the part are tapped. What is the positional tolerance for these holes?
4. How many different types of geometric dimensioning and tolerancing controls are applied to the part?
5. What tolerancing controls are applied in regard to Question 4?
6. What is the controlling datum on the part?
7. Unless otherwise specified, two-place decimals should be held to what tolerance?
8. Unless otherwise specified, three-place decimals should be held to what tolerance?
9. Is Datum A called out on the radius or diameter?
10. What parallelism tolerance is applied to the part?

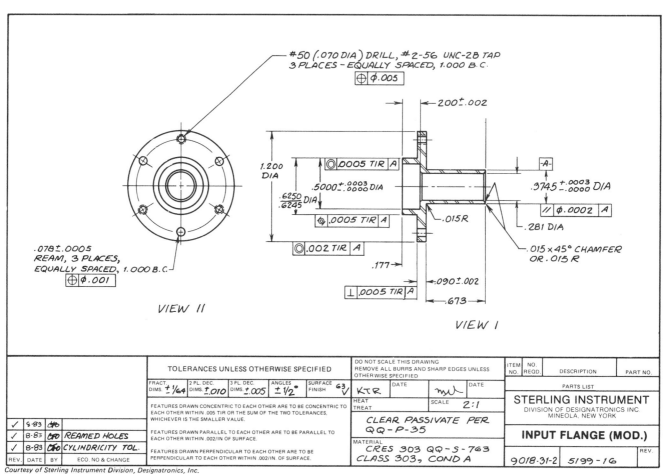

Courtesy of Sterling Instrument Division, Designatronics, Inc.
(Modified Industry Blueprint)

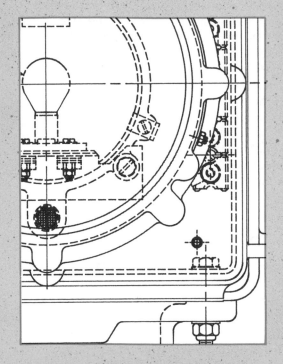

This chapter covers all terminology associated with the major kinds of fasteners, and illustrates the fasteners used in industry today. How to interpret and draw tabulated fastener standard-size drawings is covered. In-depth study is devoted to where to use groove pins and retaining rings, and how to design them into existing assemblies.

CHAPTER THIRTEEN

Fasteners

Classifications of Fasteners

As a new product is developed, determining how to fasten it together is a major consideration. The product must be assembled quickly, using standard, easily available, low-cost fasteners. Some products are designed to be taken apart easily—others are designed to be permanently assembled. Many considerations are required as to what kind, type, and material of fastener to be used. Sometimes the stress load upon a joint must be considered. There are two major classifications of fasteners: permanent and temporary. *Permanent* fasteners are used when parts will not be disassembled. *Temporary* fasteners are used when the parts will be disassembled at some future time.

Permanent fastening methods include welding, brazing, stapling, nailing, gluing, and riveting. Temporary fasteners include screws, bolts, keys, and pins.

Many temporary fasteners include threads in their design. In early days, there was no such thing as standardization. Nuts and bolts from one company would not fit nuts and bolts from another company. In 1841, Sir Joseph Whitworth worked toward some kind of standardization through England. His efforts were

finally accepted, and England came up with a standard thread form called Whitworth Threads.

In 1864, the United States tried to develop a standardization of its own but, because it would not interchange with the English Whitworth Threads, it was not adopted at that time. It was not until 1935 that the United States adopted the American Standard Thread. It was actually the same 60° V-thread form proposed back in 1864. Still, there was no standardization between countries. This created many problems, but nothing was done until World War II, which changeability of parts that, in 1948, the United States, Canada, and the United Kingdom developed the Unified Screw Thread. It was a compromise between the newer American Standard Thread and the old Whitworth Threads.

Today, with the changeover to the metric system, new standards are being developed. The International Organization for Standardization (ISO) was formed to develop a single international system using metric screw threads. This new ISO standard will be united with the American National Standards Institute (ANSI) standards. At the present time, we are in a transitional period and a combination of both systems is still being used.

Threads

Threads are used for four basic applications:

1. to fasten parts together, such as a nut and a bolt.

2. for fine adjustment between parts in relation to each other, such as the fine adjusting screw on a surveyor's transit.

3. for fine measurement, such as a micrometer.

4. to transmit motion or power, such as an automatic screw threading attachment on a lathe or a house jack.

There are many types and sizes of fasteners, each designed for a particular function. Permanently fastening parts together by welding or brazing is discussed in Chapter 19. Although screw threads have other important uses, such as adjusting parts and measuring and transmitting power, only their use as a fastener and only the most used kinds of fasteners are discussed in this chapter.

Thread Terms

Refer to Figure 13-1 for the following terms.

- *External thread*—Threads located on the outside of a part, such as those on a bolt.
- *Internal thread*—Threads located on the inside of a part, such as those on a nut.
- *Axis*—A longitudinal center line of the thread.
- *Major diameter*—The largest diameter of a screw thread, both external and internal.
- *Minor diameter*—The smallest diameter of a screw thread, both external and internal.
- *Pitch diameter*—The diameter of an imaginary diameter centrally located between the major diameter and the minor diameter.

- *Pitch*—The distance from a point on a screw thread to a corresponding point on the next thread, as measured parallel to the axis.
- *Root*—The bottom point joining the sides of a thread.
- *Crest*—The top point joining the sides of a thread.
- *Depth of thread*—The distance between the crest and the root of the thread, as measured at a right angle to the axis.
- *Angle of thread*—The included angle between the sides of the thread.
- *Series of thread*—A standard number of threads per inch (TPI) for each standard diameter.

Screw Thread Forms

The *form* of a screw thread is actually its profile shape. There are many kinds of screw thread forms. Seven major kinds are discussed next.

Unified National Thread Form

The Unified National thread form has been the standard thread used in the United States, Canada, and the United Kingdom since 1948, Figure 13-2A. This thread form is used mostly for fasteners and adjustments.

ISO Metric Thread Form

The ISO metric thread form is the new standard to be used throughout the world. Its form or profile is very similar to that of the Unified National thread, except that the thread depth is slightly less, Figure 13-2B. This thread form is used mostly for fasteners and adjustments.

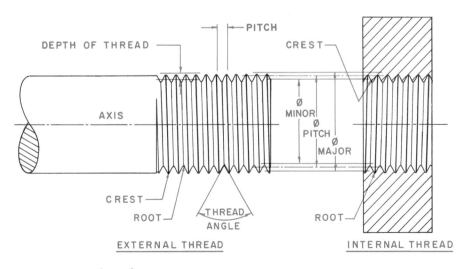

Figure 13-1 Thread terms

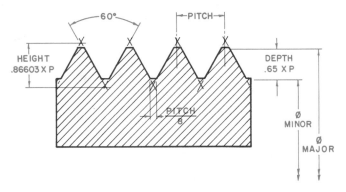

Figure 13-2A Unified national thread form (UN)

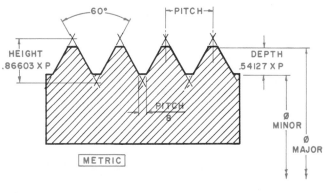

Figure 13-2B ISO metric thread form

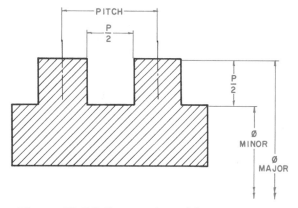

Figure 13-2C Square thread form

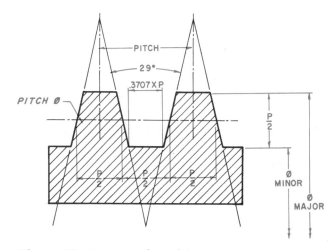

Figure 13-2D Acme thread form

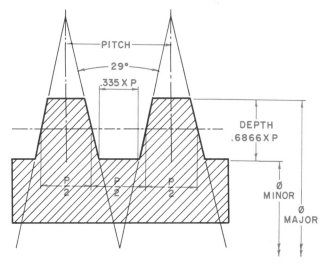

Figure 13-2E Worm thread form

Square Thread Form

The square thread's profile is exactly as its name implies; that is, square. The faces of the teeth are at right angles to the axis and, theoretically, this is the best thread to transmit power, Figure 13-2C. Because this thread is difficult to manufacture, it is being replaced by the Acme thread.

Acme Thread Form

The Acme thread is a slight modification of the square thread. It is easier to manufacture and is actually stronger than the square thread, Figure 13-2D. It, too, is used to transmit power.

Worm Thread Form

The worm thread is similar to the Acme thread, and is used primarily to transmit power, Figure 13-2E.

Knuckle Thread Form

The knuckle thread is usually rolled from sheet metal and is used, slightly modified, in electric light

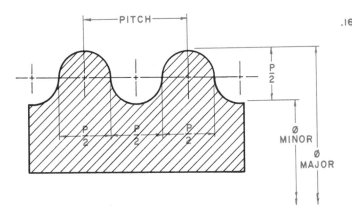

Figure 13-2F Knuckle thread form

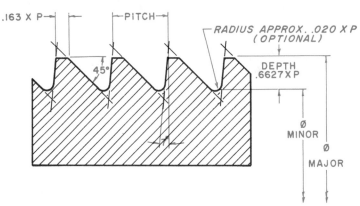

Figure 13-2G Buttress thread form

bulbs, electric light sockets, and sometimes for bottle tops. The knuckle thread is sometimes cast, Figure 13-2F.

Buttress Thread Form

The buttress thread has certain advantages in applications involving exceptionally high stress along its axis in *one* direction only. Examples of applications are the breech assemblies of large guns, airplane propeller hubs, and columns for hydraulic presses, Figure 13-2G.

Tap and Die

Various methods are used to produce inside and outside threads. The simplest method uses thread-cutting tools called taps and dies. The *tap* cuts internal threads; the *die* cuts external threads, Figure 13-3. In making an internal threaded hole, a tap-drilled hole must be drilled first. This hole is approximately the same diameter as the minor diameter of the threads.

Notice how the tap is tapered at the end; this taper allows the tap to start into the tap-drilled hole. This tapered area contains only partial threads.

Threads per Inch (TPI)

One method of measuring threads per inch (TPI) is to place a standard scale on the crests of the threads, parallel to the axis, and count the number of full threads within one inch of the scale, Figure 13-4. If only part of an inch of stock is threaded, count the number of full threads in one half inch and multiply by two to determine TPI.

A simple, more accurate method of determining threads per inch is to use a screw thread gage, Figure 13-5. By trial and error, the various fingers or leaves of the gage are placed over the threads until one is found that fits exactly into all the threads. Threads per inch are then read directly on each leaf of the gage, Figure 13-6.

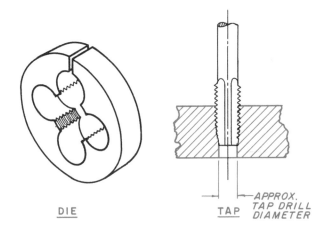

Figure 13-3 Tap and die

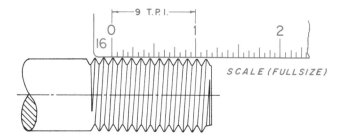

Figure 13-4 Use of a scale to calculate threads per inch (TPI)

Pitch

The *pitch* of any thread, regardless of its thread form or profile, is the distance from one point on a thread to the corresponding point on the adjacent thread as measured parallel to its axis, Figure 13-7. Pitch is found by dividing the TPI into one inch.

In this example, a coarse thread pitch, there are 10 threads in one measured inch; 10 TPI divided into one inch equals a pitch of 10. In a fine thread of the same diameter there are 20 threads in one measured inch; 20 TPI divided into one inch equals a pitch of 20, Figure 13-8.

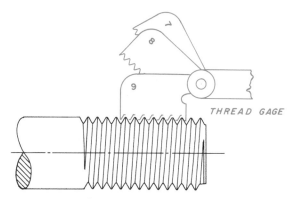

Figure 13-5 Use of a screw thread gage

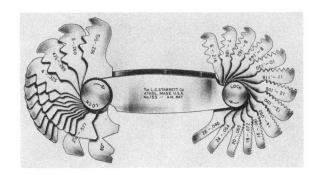

Figure 13-6 Reading screw thread gage (*Courtesy The L.S. Starrett Co.*)

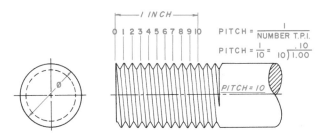

Figure 13-7 Coarse thread pitch

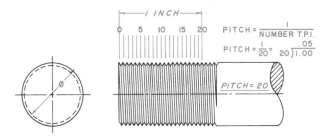

Figure 13-8 Fine thread pitch

For metric threads, the pitch is specified in milli-metres. Pitch for a metric thread is included in its call-off designation. For example: M10 × 1.5. The 1.5 indicates the pitch; therefore, it does not as a rule have to be calculated.

Single and Multiple Threads

A *single thread* is composed of one continuous ridge. The lead of a single thread is equal to the pitch. Lead is the distance a screw thread advances axially in one full turn. Most threads are single threads.

Multiple threads are made up of two or more continuous ridges following side-by-side. The lead of a double thread is equal to twice the pitch. The lead of a triple thread is equal to three times the pitch, Figure 13-9.

Multiple threads are used when speed or travel distance is an important design factor. A good example of a double or triple thread is found in an inexpensive ball-point pen. Take a ball-point pen apart and study the end of the external threads. There will probably be two or three ridges starting at the end of the threads. Notice how fast the parts screw together. This speed, not power, is the characteristic of multiple threads.

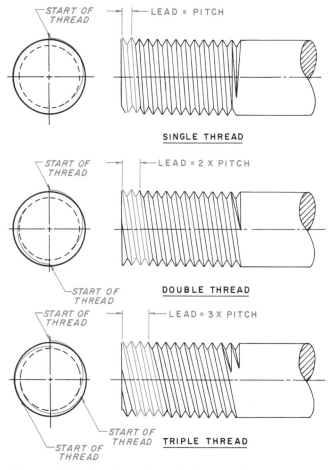

Figure 13-9 Single and multiple threads

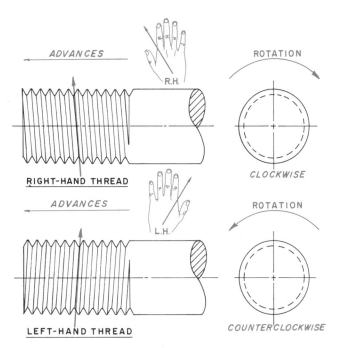

Figure 13-10 Right-hand and left-hand threads

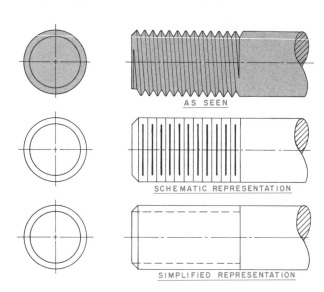

Figure 13-11 Thread representation

Right-Hand and Left-Hand Threads

Threads can be either right-handed or left-handed. To distinguish between a right-hand and a left-hand thread, use this simple trick. A right-hand thread winding tends to lean toward the left. If the thread leans toward the left, the right-hand thumb points in the same direction. If the thread leans to the right, Figure 13-10, the left-hand thumb leans in that direction indicating that it is a left-hand thread.

Thread Representation

The top illustration of Figure 13-11 shows a normal view of an external thread. To draw a thread exactly as it will actually look takes too much drafting time. To help speed up the drawing of threads, one of two basic systems is used and each is described and illustrated. The schematic system of representing threads was developed approximately in 1940, and is still used somewhat today. The simplified system of representing threads was developed 15 years later, and is actually quicker and in greater use today.

How To Draw Threads Using the Schematic System

Step 1. Refer to Figure 13-12. Lightly draw the major diameter, and locate the approximate length of full threads.

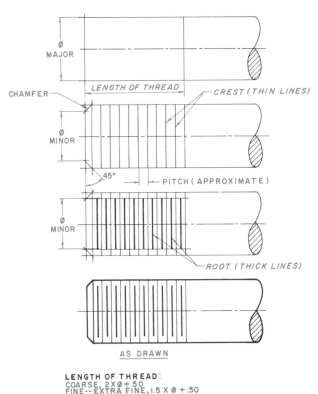

Figure 13-12 How to draw threads using the schematic system

Step 2. Lightly locate the minor diameter and draw the 45° chamfered ends as illustrated. Draw lines to represent the crest of the threads spaced approximately equal to the pitch.

Step 3. Draw slightly thicker lines centered between the crest lines to the minor diameter. These lines represent the root of the threads.

Step 4. Check all work and darken in. Notice the crest lines are *thin* black lines and the root lines are *thick* black lines.

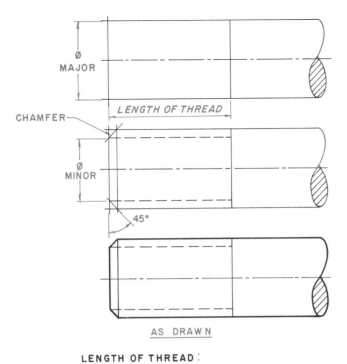

LENGTH OF THREAD:
COARSE, 2 X Ø + .50
FINE--EXTRA FINE, 1.5 X Ø + .50

Figure 13-13 How to draw threads using the simplified system

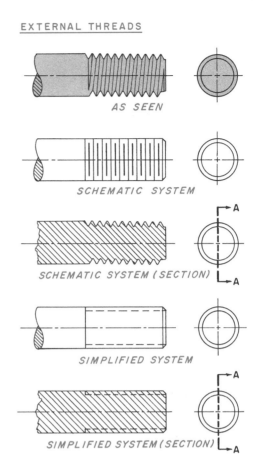

AS SEEN

SCHEMATIC SYSTEM

SCHEMATIC SYSTEM (SECTION)

SIMPLIFIED SYSTEM

SIMPLIFIED SYSTEM (SECTION)

Figure 13-14 Standard external thread representation

How To Draw Threads Using the Simplified System

Step 1. Refer to Figure 13-13. Lightly draw the major diameter and locate the approximate length of full threads.

Step 2. Lightly locate the minor diameter and draw the 45° chamfered ends as illustrated. Draw dash lines along the minor diameter. This represents the root of the threads.

Step 3. Check all work and darken in. The dash lines are *thin* black lines.

Standard External Thread Representation

The most recent standard to illustrate external threads using either the schematic or simplified system is illustrated in Figure 13-14. Note how section views are illustrated using schematic and simplified systems.

Standard Internal Thread Representation

There are two major kinds of interior holes: through holes and blind holes. A *through hole*, as its name implies, goes completely through an object. A *blind hole* is a hole that does *not* go completely through an object. In the manufacture of a blind hole, a tap drill must be drilled into the part first, Figure 13-15. To illustrate a tap drill, use the 30°-60° triangle. This is *not* the actual angle of a drill point but is close enough for illustration. The tap is now turned into the tap drill hole. Because of the taper on the tap, *full* threads do not extend to the bottom of the hole (refer back to Figure 13-3). The drafter illustrates the tap drill and the full threaded section as shown to the right in Figure 13-15.

Using the schematic system to represent a through hole is illustrated in Figure 13-16. A blind hole and a section view are drawn as illustrated in Figure 13-17.

Using the simplified system to represent a through hole is illustrated in Figure 13-18. A blind hole and a section view are drawn as illustrated in Figure 13-19.

Thread Relief (Undercut)

On exterior threads, it is impossible to make perfectly uniform threads up to a shoulder; thus, the

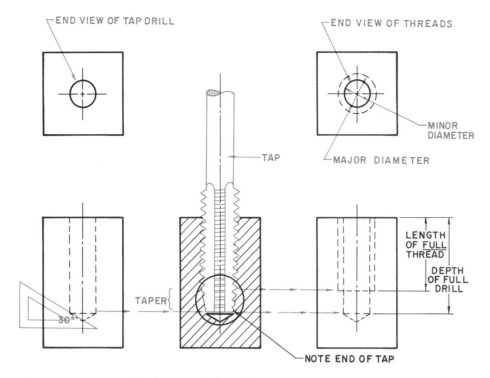

Figure 13-15 Standard internal thread representation

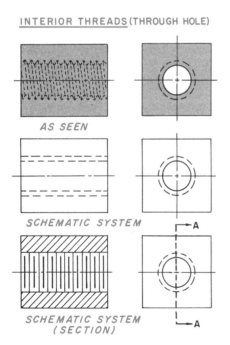

Figure 13-16 Standard internal thread representation for a through hole (schematic system)

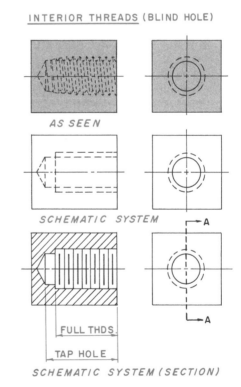

Figure 13-17 Standard internal thread representation for a blind hold (schematic system)

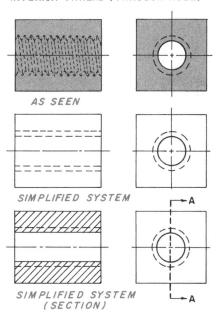

INTERIOR THREAD (THROUGH HOLE)

AS SEEN

SIMPLIFIED SYSTEM

SIMPLIFIED SYSTEM (SECTION)

Figure 13-18 Standard internal thread representation for a through hole (simplified system)

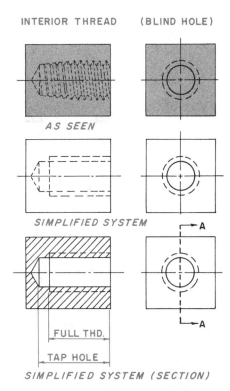

INTERIOR THREAD (BLIND HOLE)

AS SEEN

SIMPLIFIED SYSTEM

FULL THD.

TAP HOLE

SIMPLIFIED SYSTEM (SECTION)

Figure 13-19 Standard internal thread representation for a blind hold (simplified system)

threads tend to run out, as illustrated in Figure 13-20. Where mating parts must be held tightly against the shoulder, the last one or two threads must be removed or *relieved*. This is usually done no farther than to the depth of the threads so as not to weaken the fastener. The simplified system of thread representation is illustrated at the bottom of Figure 13-20.

Full interior threads cannot be manufactured to the end of a blind hole. One way to eliminate this problem is to call-off a thread relief or undercut, as illustrated in Figure 13-21. The bottom illustration is as it would be drawn by the drafter.

Screw, Bolt, and Stud

Figure 13-22 illustrates and describes a screw, a bolt, and a stud. A *screw* is a fastener that does not use a nut and is screwed directly into a part.

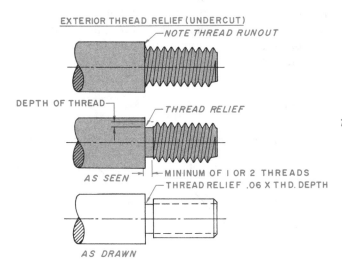

EXTERIOR THREAD RELIEF (UNDERCUT)
NOTE THREAD RUNOUT

DEPTH OF THREAD
THREAD RELIEF

AS SEEN
MININUM OF I OR 2 THREADS
THREAD RELIEF .06 X THD. DEPTH

AS DRAWN

Figure 13-20 External thread relief (undercut)

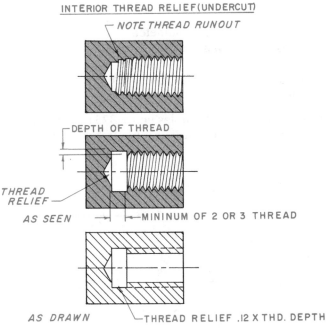

INTERIOR THREAD RELIEF (UNDERCUT)
NOTE THREAD RUNOUT

DEPTH OF THREAD

THREAD RELIEF

AS SEEN
MININUM OF 2 OR 3 THREAD

AS DRAWN
THREAD RELIEF .12 X THD. DEPTH

Figure 13-21 Internal thread relief (undercut)

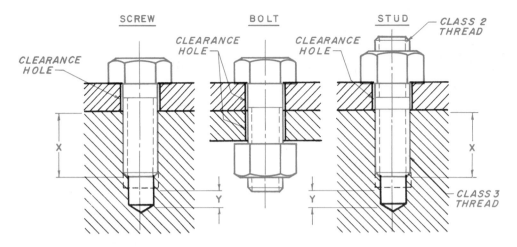

X = MINIMUM THREADS REQUIRED:
  STEEL, X = OUTSIDE DIAMETER
  CAST IRON / BRASS / BRONZE , X = 1.5 X OUTSIDE DIAMETER
  ALUMINUM / ZINC / PLASTIC, X = 2 X OUTSIDE DIAMETER

Y = MINIMUM SPACE = 2 X PITCH LENGTH

CLEARANCE HOLE: 0 TO .375 (9) = .03 (1) LARGER THAN OUTSIDE DIAMETER
  .375 (9) UP   = .06 (2) LARGER THAN OUTSIDE DIAMETER

Figure 13-22 Screw, bolt and stud

A *bolt* is a fastener that passes directly through parts to hold them together, and uses a nut to tighten or hold the parts together.

A *stud* is a fastener that is a steel rod with threads at both ends. It is screwed into a blind hole and holds other parts together by a nut on its free end. In general practice, a stud has either fine threads at one end and coarse threads at the other, or Class 3-fit threads at one end and Class 2-fit threads at the other end. Class of fit is fully explained later in this chapter under "Classes of Fit."

The minimum full thread length for a screw or a stud is:

In steel: equal to the diameter.
In cast iron, brass, bronze: equal to 1.5 times the diameter.
In aluminum, zinc, plastic: equal to 2 times the diameter.

The clearance hole for holes up to .375 (9) diameter is approximately .03 oversize; for larger holes, .06 oversize.

Machine Screws

Machine screw sizes run from .021 (.3) to .750 (20) in diameter. There are eight standard head forms. Four major kinds are illustrated in Figure 13-23. *Machine screws* are used for screwing into thin materials. Most machine screws are threaded within a thread or two to the head. Although these are screws, machine screws sometimes incorporate a hex-head nut to fasten parts together.

The length of a machine screw is measured from the top surface to the part to be held together to the end of the screw (refer again to Figure 13-23).

Cap Screws

Cap screw sizes run from .250 (6) and up. There are five standard head forms, Figure 13-24. A *cap screw* is usually used as a true screw, and it passes through a clearance hole in one part and screws into another part.

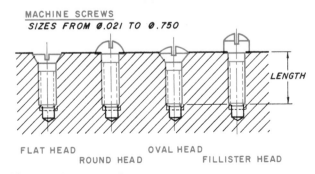

Figure 13-23 Machine screws

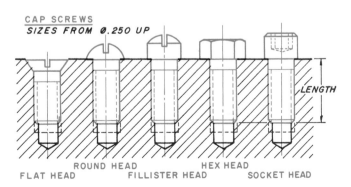

Figure 13-24 Cap screws

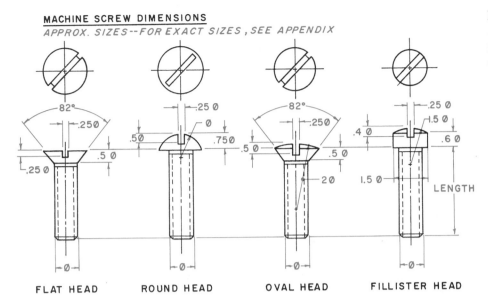

Figure 13-25
Approximate sizes for
machine screws
or cap screws

MACHINE SCREW DIMENSIONS
APPROX. SIZES--FOR EXACT SIZES , SEE APPENDIX

FLAT HEAD   ROUND HEAD   OVAL HEAD   FILLISTER HEAD

How To Draw a Machine Screw or Cap Screw

The exact dimensions of machine screws and cap screws are given in the Appendix of the text but, in actual practice, they are seldom used for drawing purposes. See Figures 13-25 and 13-26, which show the various sizes as they are proportioned in regard to the diameter of the fastener. Various fastener templates are now available to further speed up drafting time.

Set Screws

A *set screw* is used to prevent motion between mating parts, such as the hub of a pulley on a shaft. The set screw is screwed into and through one part so that it applies pressure against another part, thus preventing motion. Set screws are usually manufactured of steel, and are hardened to make them stronger than the average fastener.

Set screws have various kinds of heads and many kinds of points. Figure 13-27 illustrates a few of the more common kinds of set screws. Set screws are manufactured in many standard lengths of very small increments, so almost any required length is probably "standard." Exact sizes and lengths can be found in the Appendix. As with machine screws and cap screws, in actual practice, the actual drawing of set screws is done using their proportions in relationship to their diameters.

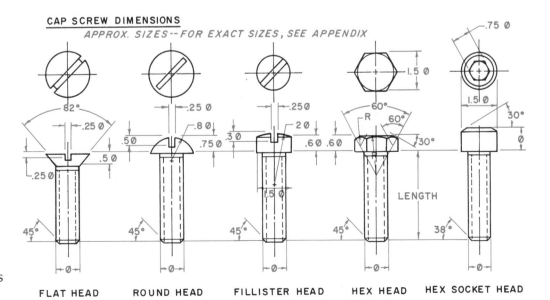

CAP SCREW DIMENSIONS
APPROX. SIZES--FOR EXACT SIZES , SEE APPENDIX

Figure 13-26
Approximate sizes
for machine screws
or cap screws

FLAT HEAD   ROUND HEAD   FILLISTER HEAD   HEX HEAD   HEX SOCKET HEAD

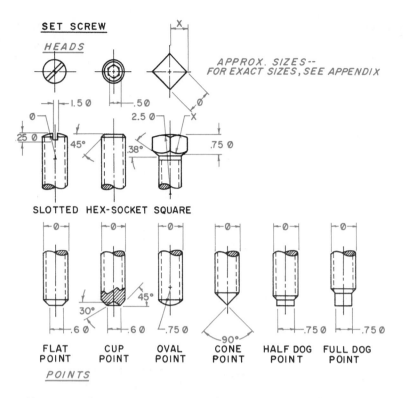

Figure 13-27 Approximate sizes for set screws and set screw points

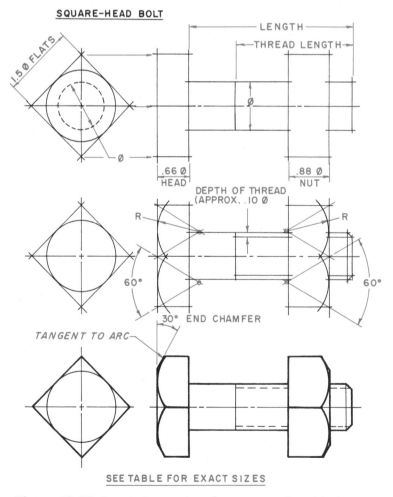

Figure 13-28 Approximate sizes for a square-head bolt

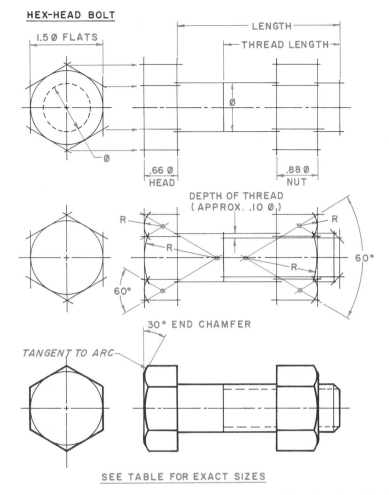

Figure 13-29 Approximate size for a hex-head bolt

How To Draw Square- and Hex-Head Bolts

Exact dimensions for square- and hex-head bolts are given in the Appendix of the text, but, in actual practice, they are drawn using the proportions as given in Figures 13-28 and 13-29. Notice that the heads are shown in the profile so three surfaces are seen in the front view. In the event a square- or hex-head bolt must be illustrated 90°, the proportions as illustrated in Figure 13-30 are used.

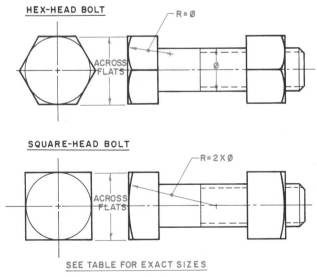

Figure 13-30 Side view of hex- and square-head bolts

Nuts, Bolts, and Other Fasteners in Section

If the cutting plane passes through the axis of any fastener, the fastener is *not* sectioned. It is treated exactly as a shaft and drawn exactly as it is viewed. Refer to Figure 13-31. The illustration at the left is drawn correctly. The figure at the right is drawn incorrectly (notice how difficult it is to understand, especially the nut).

Thread Call-offs

Although not all companies use the exact same call-offs for various fasteners, it is important that all drafters within one company use the same method. One method used to call-off fasteners is illustrated

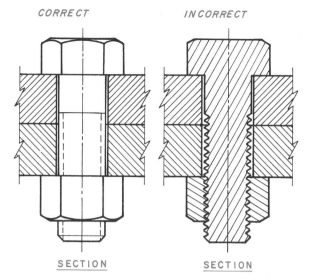

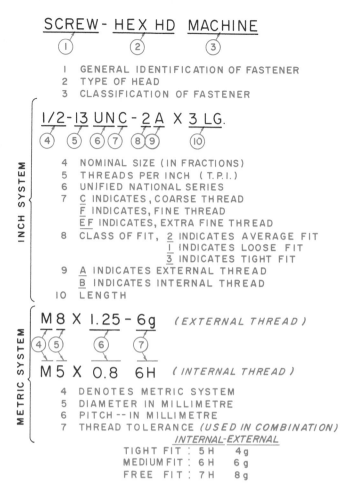

Figure 13-31 Fasteners in section

SCREW - HEX HD MACHINE
1   2   3

1 GENERAL IDENTIFICATION OF FASTENER
2 TYPE OF HEAD
3 CLASSIFICATION OF FASTENER

INCH SYSTEM

1/2-13 UNC - 2A X 3 LG.
4 5   6 7 8 9   10

4 NOMINAL SIZE (IN FRACTIONS)
5 THREADS PER INCH (T.P.I.)
6 UNIFIED NATIONAL SERIES
7 C INDICATES, COARSE THREAD
  F INDICATES, FINE THREAD
  EF INDICATES, EXTRA FINE THREAD
8 CLASS OF FIT, 2 INDICATES AVERAGE FIT
                1 INDICATES LOOSE FIT
                3 INDICATES TIGHT FIT
9 A INDICATES EXTERNAL THREAD
  B INDICATES INTERNAL THREAD
10 LENGTH

METRIC SYSTEM

M 8 X 1.25 - 6g (EXTERNAL THREAD)
4 5   6   7

M 5 X 0.8 6H (INTERNAL THREAD)

4 DENOTES METRIC SYSTEM
5 DIAMETER IN MILLIMETRE
6 PITCH -- IN MILLIMETRE
7 THREAD TOLERANCE (USED IN COMBINATION)
                    INTERNAL-EXTERNAL
        TIGHT FIT : 5 H   4 g
        MEDIUM FIT : 6 H   6 g
        FREE FIT : 7 H   8 g

Figure 13-32 Thread call-off

in Figure 13-32. Regardless of which system is used, the first line contains the fastener's general identification, type of head, and classification. The second line contains all exact detailed information. All threads are assumed to be right hand (R.H.), unless otherwise noted. If a thread is to be left hand (L.H.), it is noted at the end of the second line.

Various Kinds of Heads

Many different kinds of screw heads are used today. Figure 13-33 illustrates a few of the standard heads.

Rivets

Rivets are permanent fasteners, usually used to hold sheet metal together. Most rivets are made of wrought iron or soft steel and, for aircraft and space missiles, copper, aluminum, alloy or other exotic metals.

Riveted joints are classified by applications, such as pressure vessels, structural and machine members. For data concerning joints for pressure vessels refer to such sources as ASME boiler codes. For data

concerning larger field structural rivets, such as bridges, buildings and ships, see ANSI standards or a *Machinery's Handbook*. This chapter covers information for small-size rivets for machine-member riveted joints used for lighter mass produced applications.

Two kinds of basic rivet joints are the lap joint, and the butt joint, Figure 13-34. In the *lap joint*, the parts overlap each other, and are held together by one or more rows of rivets. In the butt joint, the parts are butted, and are held together by a cover plate or butt strap which is riveted to both parts.

Slotted   PHILLIPS*   Hex Cap   TORX*   Clutch Type G   SCRULOX*   Multi-Spline

Triple Square   TRI-WING*   Clutch Type A   TORQ-SET*   Slab Head   POZIDRIV*   Reed & Prince (Frearson)

Figure 13-33 Kinds of screw heads

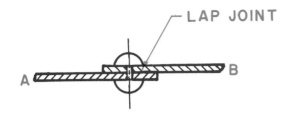

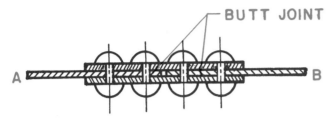

Figure 13-34 Two basic rivet joints (lap joint and butt joint)

Factors to consider are: type of joint, pitch of rivets, type and diameter of rivet, rivet material, and size of clearance holes, Figure 13-35. The diameter of a rivet is calculated from the thickness of metal, and commonly ranges between

$$d = 1.2 \sqrt{t} \text{ and } d = 1.4 \sqrt{t}$$

where d is the diameter, and t is the thickness of the plate.

Size and Type of Hole

Rivet holes must be punched, punched and then reamed, or drilled. As a general rule, holes are usu-

ally made .06 (1.5 mm) larger in diameter than the nominal rivet diameter.

Rivet Symbols

Rivets applied in mass produced applications are represented in Figure 13-36, illustrating the kind of rivet, to which side it is applied, and if it is to be countersunk, and so forth.

Kinds of Rivets

There are many different kinds of small-size rivets. The five major kinds are truss head, button head, pan head, countersunk head, and flat head. The countersunk-head rivet is not as strong as the other kinds of rivets; therefore, more rivets must be used to gain strength equal to the other types.

Drawing of Rivets

American standard small solid rivets are shown in their approximate standard proportions in Figure 13-36. These sizes are close enough to be used for drawing the rivet if necessary. For exact sizes, data must be obtained from other sources.

End Points

The end of a standard rivet is usually cut off straight. If a point is required, a standard size point is illustrated at the lower end of a countersunk head (refer back to Figure 13-36).

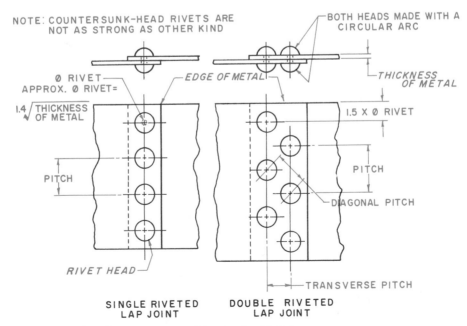

Figure 13-35 Factors to consider in rivet joints

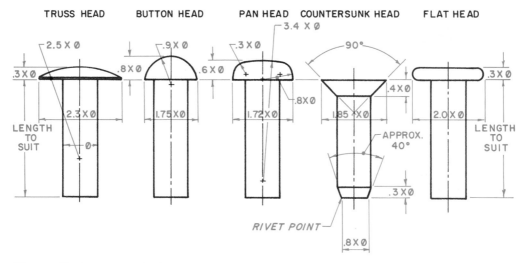

Figure 13-36 Approximate sizes for standard small rivets

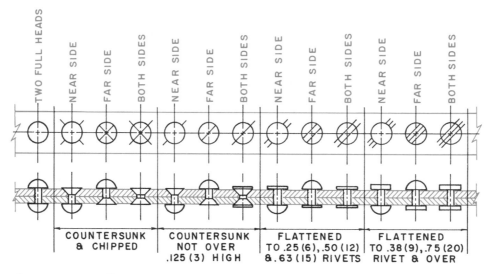

Figure 13-36A Illustrated rivet code

The drafter must indicate what kind of rivet is to be used, and on which side the head is to be positioned. To illustrate this, the drafter uses a code such as that illustrated in Figure 13-36A.

Keys and Keyseats

Key

A *key* is a demountable part that provides a positive means of transferring torque between a shaft and a hub.

Keys are used to prevent slippage and to transmit torque between a shaft and a hub. There are many kinds of keys. The five major kinds used in industry today are illustrated in Figure 13-37: square key, flat key, gib head key, Pratt & Whitney key and Woodruff key. Where a lot of torque is present, a double key and keyseat is often used. In extreme conditions, a spline is machined into the shaft and into the hub. For spline information, refer to a *Machinery's Handbook*.

Keyseat

A *keyseat* is an auxiliary-located rectangular groove machined into the shaft and/or hub to receive the key.

Classes of Fit

There are three classifications of fit:

Class 1. A side surface clearance fit obtained by using bar stock key and keyseat tolerances. This is a relatively free fit.

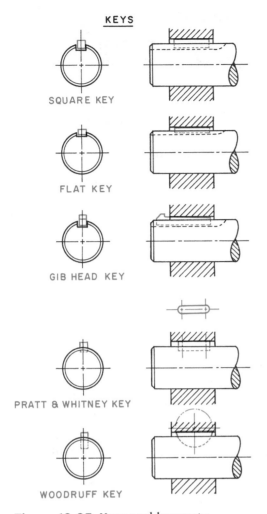

KEYS

SQUARE KEY

FLAT KEY

GIB HEAD KEY

PRATT & WHITNEY KEY

WOODRUFF KEY

Figure 13-37 Keys and keyseats

Class 2. A possible side surface interference or side surface clearance fit obtained by using bar stock key and keyseat tolerances. This is a relatively tight fit.

Class 3. A side surface interference fit obtained by interference fit tolerances. This is a very tight fit and has not been generally standardized.

Key Sizes

For a general rule, the key width is about one-fourth the nominal diameter of the shaft. For exact recommended key sizes, refer to the Appendix in the text or a *Machinery's Handbook*.

Dimensioning Keyseats

Methods of dimensioning a stock key are shown in Figure 13-38.

For dimensioning a Woodruff keyseat, the key number must be included, Figure 13-39. Refer to the key size in the Appendix to obtain the exact sizes. To dimension a Pratt & Whitney keyseat, see Figure 13-40.

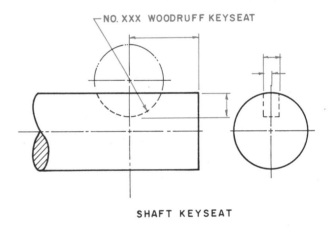

NO. XXX WOODRUFF KEYSEAT

SHAFT KEYSEAT

Figure 13-39 Woodruff keyseat

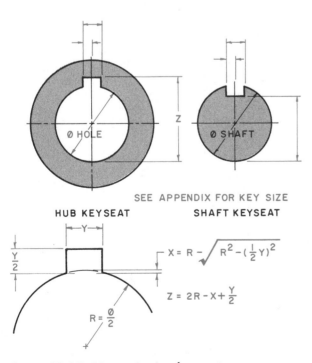

Ø HOLE

Ø SHAFT

SEE APPENDIX FOR KEY SIZE

HUB KEYSEAT   SHAFT KEYSEAT

$$X = R - \sqrt{R^2 - \left(\tfrac{1}{2}Y\right)^2}$$

$$Z = 2R - X + \frac{Y}{2}$$

$$R = \frac{\emptyset}{2}$$

Figure 13-38 Dimensioning keyseats

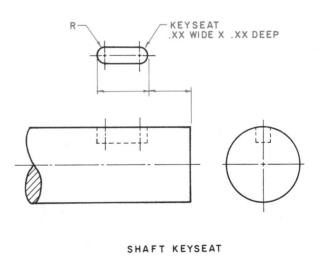

R

KEYSEAT
.XX WIDE X .XX DEEP

SHAFT KEYSEAT

Figure 13-40 Pratt & Whitney keyseat

Figure 13-41 Grooved fastener

Grooved Fasteners

Many types of fasteners are used in industry, each with its own application. Threaded fasteners, such as nuts and bolts, are used to hold parts in tension. *Grooved fasteners* are used to solve metal-to-metal pinning needs with shear application, Figure 13-41.

Grooved fasteners have great holding power and are resistant to shock, vibration, and fatigue. They are available in a wide range of types, sizes, and materials. A grooved fastener often has a better appearance than most other methods of fastening. This can be important to the overall design, if the fastener is visible.

Grooved fasteners have three parallel grooves, equally spaced, impressed longitudinally on their exterior surface. To make these grooves, a grooving tool is pressed *below* the surface to displace a carefully determined amount of material. *Nothing* is removed. The metal is displaced to each side, forming a raised portion or flute extending along each side of the groove, Figure 13-42. The crest of the flute constitutes the *expanded diameter* (Dx). The expanded diameter (Dx) is a few thousandths larger than the nominal diameter (D) of the stock.

Grooved fasteners can be custom manufactured in order to meet most any application. For example, a custom grooved pin can be made with a cross-drilled hole, a groove for a snap ring, or a threaded hole in one end. The groove length can also be varied and placed anywhere along the length of the pin.

Installation

The grooved fastener is forced into a drilled hole slightly larger than the nominal or specified diameter of the pin, Figure 13-43. The crest or flutes are forced back into the grooves when the fastener is driven into the hole. The resiliency of the metal forced back into the grooves creates powerful radial forces against the hole wall.

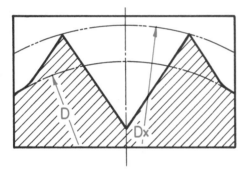

Figure 13-42 Enlarged sectional view of one of the grooves *before* inserting fastener

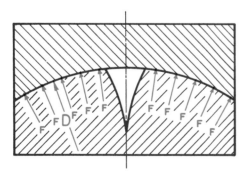

Figure 13-43 The same view *after* insertion. A powerful radial thrust is obtained.

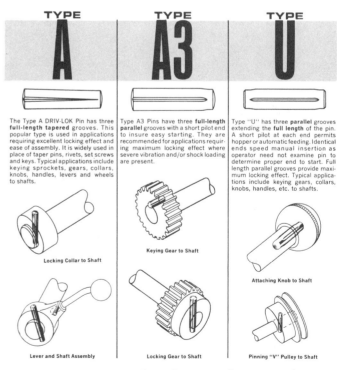

Figure 13-44 Typical applications for grooved pin types A, A3 and U

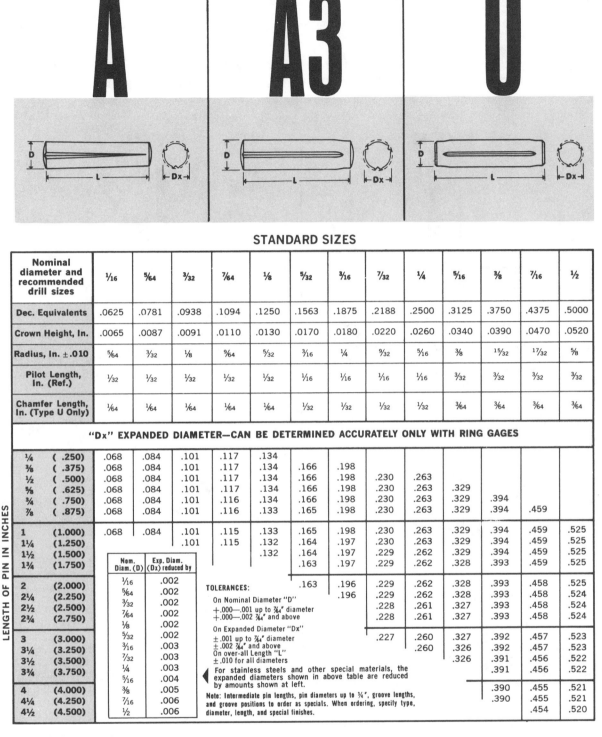

TYPE A | TYPE A3 | TYPE U

STANDARD SIZES

Nominal diameter and recommended drill sizes	1/16	5/64	3/32	7/64	1/8	5/32	3/16	7/32	1/4	5/16	3/8	7/16	1/2
Dec. Equivalents	.0625	.0781	.0938	.1094	.1250	.1563	.1875	.2188	.2500	.3125	.3750	.4375	.5000
Crown Height, In.	.0065	.0087	.0091	.0110	.0130	.0170	.0180	.0220	.0260	.0340	.0390	.0470	.0520
Radius, In. ±.010	5/64	3/32	1/8	9/64	5/32	3/16	1/4	9/32	5/16	3/8	15/32	17/32	5/8
Pilot Length, In. (Ref.)	1/32	1/32	1/32	1/32	1/32	1/16	1/16	1/16	1/16	3/32	3/32	3/32	3/32
Chamfer Length, In. (Type U Only)	1/64	1/64	1/64	1/64	1/64	1/32	1/32	1/32	1/32	3/64	3/64	3/64	3/64

"Dx" EXPANDED DIAMETER—CAN BE DETERMINED ACCURATELY ONLY WITH RING GAGES

LENGTH OF PIN IN INCHES

Length of pin	1/16	5/64	3/32	7/64	1/8	5/32	3/16	7/32	1/4	5/16	3/8	7/16	1/2
1/4 (.250)	.068	.084	.101	.117	.134								
3/8 (.375)	.068	.084	.101	.117	.134	.166	.198						
1/2 (.500)	.068	.084	.101	.117	.134	.166	.198	.230	.263				
5/8 (.625)	.068	.084	.101	.117	.134	.166	.198	.230	.263	.329			
3/4 (.750)	.068	.084	.101	.116	.134	.166	.198	.230	.263	.329	.394		
7/8 (.875)	.068	.084	.101	.116	.133	.165	.198	.230	.263	.329	.394	.459	
1 (1.000)	.068	.084	.101	.115	.133	.165	.198	.230	.263	.329	.394	.459	.525
1¼ (1.250)			.101	.115	.132	.164	.197	.230	.263	.329	.394	.459	.525
1½ (1.500)					.132	.164	.197	.229	.262	.329	.394	.459	.525
1¾ (1.750)						.163	.197	.229	.262	.328	.393	.459	.525
2 (2.000)						.163	.196	.229	.262	.328	.393	.458	.525
2¼ (2.250)							.196	.229	.262	.328	.393	.458	.524
2½ (2.500)								.228	.261	.327	.393	.458	.524
2¾ (2.750)								.228	.261	.327	.393	.458	.524
3 (3.000)								.227	.260	.327	.392	.457	.523
3¼ (3.250)									.260	.326	.392	.457	.523
3½ (3.500)										.326	.391	.456	.522
3¾ (3.750)											.391	.456	.522
4 (4.000)											.390	.455	.521
4¼ (4.250)											.390	.455	.521
4½ (4.500)												.454	.520

Nom. Diam. (D)	Exp. Diam. (Dx) reduced by
1/16	.002
5/64	.002
3/32	.002
7/64	.002
1/8	.002
5/32	.002
3/16	.003
7/32	.003
1/4	.003
5/16	.004
3/8	.005
7/16	.006
1/2	.006

TOLERANCES:

On Nominal Diameter "D"
+.000—.001 up to 7/64" diameter
+.000—.002 7/64" and above

On Expanded Diameter "Dx"
±.001 up to 7/64" diameter
±.002 7/64" and above

On over-all Length "L"
±.010 for all diameters

◄ For stainless steels and other special materials, the expanded diameters shown in above table are reduced by amounts shown at left.

Note: Intermediate pin lengths, pin diameters up to ¾", groove lengths, and groove positions to order as specials. When ordering, specify type, diameter, length, and special finishes.

Figure 13-45 Standard size chart of grooved pin types A, A3 and U

In many cases, grooved fasteners are lower in cost than knurled pins, taper pins, pins with cotter pins, rivets, set screws, keys or other methods used to fasten metal-to-metal parts together. The installation cost of grooved pins is invariably lower because of the required hole tolerances, and no special guides are required at assembly.

Material

Standard grooved fasteners are made of cold-drawn, low-carbon steel. The physical properties of this material are more than enough for ordinary applications. Alloy steel, hard brass, silicon bronze, stainless steel, and other exotic metals may also be specially ordered. These special materials are usually heat treated for optimum physical qualities.

Finish

Standard grooved fasteners have a finish of zinc electroplated, deposited approximately .00015 inch

deep. Chromate, brass, cadmium, and black oxide can also be specially ordered.

Standard Types

Study the various types of grooved fasteners and the related technical data in Figures 13-44 (page 420),

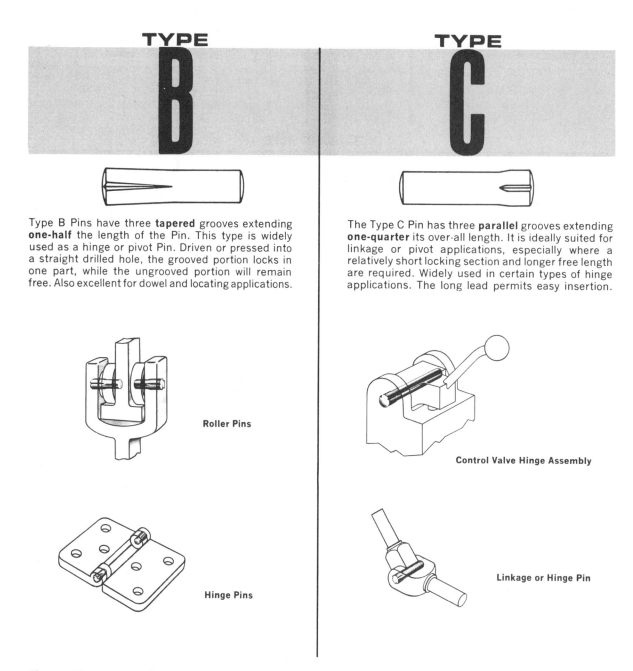

TYPE B

Type B Pins have three **tapered** grooves extending **one-half** the length of the Pin. This type is widely used as a hinge or pivot Pin. Driven or pressed into a straight drilled hole, the grooved portion locks in one part, while the ungrooved portion will remain free. Also excellent for dowel and locating applications.

Roller Pins

Hinge Pins

TYPE C

The Type C Pin has three **parallel** grooves extending **one-quarter** its over-all length. It is ideally suited for linkage or pivot applications, especially where a relatively short locking section and longer free length are required. Widely used in certain types of hinge applications. The long lead permits easy insertion.

Control Valve Hinge Assembly

Linkage or Hinge Pin

Figure 13-46 Typical applications for grooved pin types B and C

13-46, 13-48, 13-50, and 13-53. Note the various types of fasteners, how each functions, and what each replaces. For example, Type A is used in place of taper pins, rivets, set screws, and keys.

Standard Sizes

Refer to the standard size charts in Figures 13-45 (page 421), 13-47, 13-49, and 13-51. Across the top of each chart are the nominal sizes from 1/16-inch diameter to 1/2-inch diameter. At the left side of each

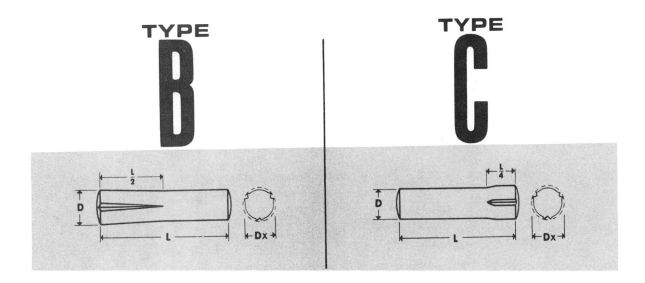

STANDARD SIZES

Nominal diameter and recommended drill sizes	1/16	5/64	3/32	7/64	1/8	5/32	3/16	7/32	1/4	5/16	3/8	7/16	1/2
Dec. Equivalents	.0625	.0781	.0938	.1094	.1250	.1563	.1875	.2188	.2500	.3125	.3750	.4375	.5000
Crown Height, In.	.0065	.0087	.0091	.0110	.0130	.0170	.0180	.0220	.0260	.0340	.0390	.0470	.0520
Radius, In. ±.010	5/64	3/32	1/8	9/64	5/32	3/16	1/4	9/32	5/16	3/8	15/32	17/32	5/8

"Dx" EXPANDED DIAMETER—CAN BE DETERMINED ACCURATELY ONLY WITH RING GAGES

LENGTH OF PIN IN INCHES	1/16	5/64	3/32	7/64	1/8	5/32	3/16	7/32	1/4	5/16	3/8	7/16	1/2
1/4 (.250)	.068	.084	.101	.117	.134								
3/8 (.375)	.068	.084	.101	.117	.134	.166	.198						
1/2 (.500)	.068	.084	.101	.117	.134	.166	.198		.230	.263	.329		
5/8 (.625)	.068	.084	.101	.117	.134	.166	.198		.230	.263	.329	.394	
3/4 (.750)	.068	.084	.101	.117	.134	.166	.198		.230	.263	.329	.394	
7/8 (.875)	.068	.084	.101	.117	.134	.166	.198		.230	.263	.329	.394	.459
1 (1.000)	.068	.084	.101	.117	.134	.166	.198		.230	.263	.329	.394	.459	.525
1 1/4 (1.250)			.101	.117	.134	.166	.198		.230	.263	.329	.394	.459	.525
1 1/2 (1.500)					.134	.166	.198		.230	.263	.329	.394	.459	.525
1 3/4 (1.750)						.165	.198		.230	.263	.329	.394	.459	.525
2 (2.000)						.165	.198		.230	.263	.329	.394	.459	.525
2 1/4 (2.250)							.197		.230	.263	.329	.394	.459	.525
2 1/2 (2.500)									.230	.263	.329	.394	.459	.525
2 3/4 (2.750)									.229	.262	.329	.394	.459	.525
3 (3.000)									.229	.262	.329	.394	.459	.525
3 1/4 (3.250)										.262	.329	.393	.459	.525
3 1/2 (3.500)											.328	.393	.459	.525
3 3/4 (3.750)												.393	.458	.525
4 (4.000)												.393	.458	.525
4 1/4 (4.250)												.393	.458	.524
4 1/2 (4.500)													.458	.524

Nom. Diam. (D)	Exp. Diam. (Dx) reduced by
1/16	.002
5/64	.002
3/32	.002
7/64	.002
1/8	.002
5/32	.002
3/16	.003
7/32	.003
1/4	.003
5/16	.004
3/8	.005
7/16	.006
1/2	.006

TOLERANCES:
On Nominal Diameter "D"
+.000—.001 up to 7/64" diameter
+.000—.002 7/64" and above
On Expanded Diameter "Dx"
±.001 up to 7/64" diameter
±.002 7/64" and above--
On over-all Length "L"
±.010 for all diameters

◄ For stainless steels and other special materials, the expanded diameters shown in above table are reduced by amounts shown at left.

Note: Intermediate pin lengths, pin diameters up to 3/4", groove lengths, and groove positions to order as specials. When ordering, specify type, diameter, length, and special finishes.

Figure 13-47 Standard size chart of grooved pin types B and C

TYPE D

TYPE E

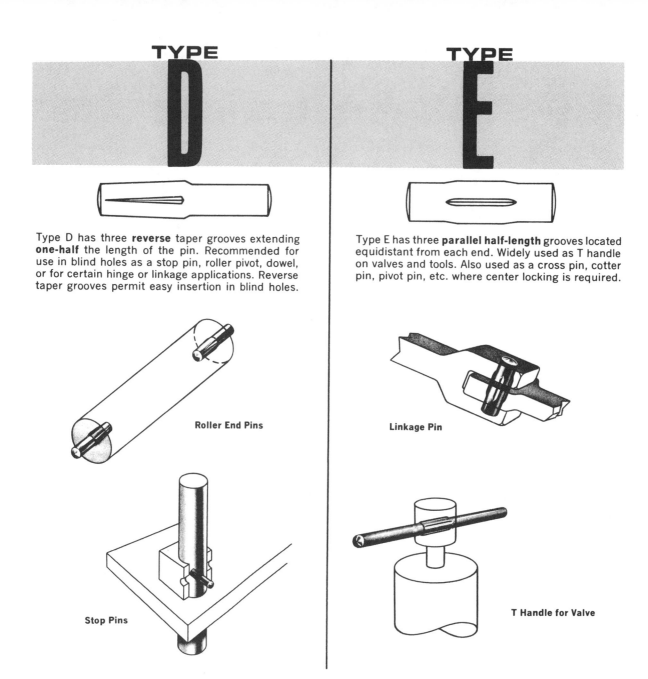

Type D has three **reverse** taper grooves extending **one-half** the length of the pin. Recommended for use in blind holes as a stop pin, roller pivot, dowel, or for certain hinge or linkage applications. Reverse taper grooves permit easy insertion in blind holes.

Type E has three **parallel half-length** grooves located equidistant from each end. Widely used as T handle on valves and tools. Also used as a cross pin, cotter pin, pivot pin, etc. where center locking is required.

Roller End Pins

Linkage Pin

Stop Pins

T Handle for Valve

Figure 13-48 Typical applications for grooved pin types D and E

TYPE D   # TYPE E

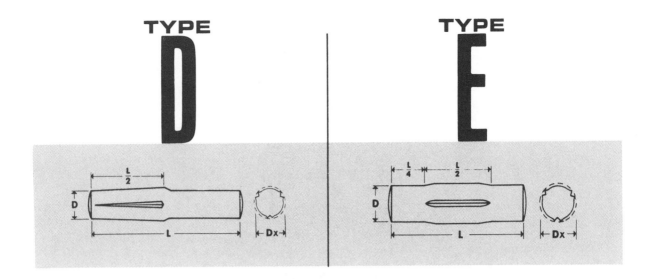

STANDARD SIZES

Nominal diameter and recommended drill sizes	1/16	5/64	3/32	7/64	1/8	5/32	3/16	7/32	1/4	5/16	3/8	7/16	1/2
Dec. Equivalents	.0625	.0781	.0938	.1094	.1250	.1563	.1875	.2188	.2500	.3125	.3750	.4375	.5000
Crown Height, In.	.0065	.0087	.0091	.0110	.0130	.0170	.0180	.0220	.0260	.0340	.0390	.0470	.0520
Radius, In. ±.010	5/64	3/32	1/8	9/64	5/32	3/16	1/4	9/32	5/16	3/8	15/32	17/32	5/8

"Dx" EXPANDED DIAMETER—CAN BE DETERMINED ACCURATELY ONLY WITH RING GAGES

LENGTH OF PIN IN INCHES

Length of pin	1/16	5/64	3/32	7/64	1/8	5/32	3/16	7/32	1/4	5/16	3/8	7/16	1/2
1/4 (.250)	.068	.084	.101	.117	.134								
3/8 (.375)	.068	.084	.101	.117	.134	.166	.198						
1/2 (.500)	.068	.084	.101	.117	.134	.166	.198	.230	.263				
5/8 (.625)	.068	.084	.101	.117	.134	.166	.198	.230	.263	.329			
3/4 (.750)	.068	.084	.101	.117	.134	.166	.198	.230	.263	.329	.394		
7/8 (.875)	.068	.084	.101	.117	.134	.166	.198	.230	.263	.329	.394	.459	
1 (1.000)	.068	.084	.101	.117	.134	.166	.198	.230	.263	.329	.394	.459	.525
1 1/4 (1.250)			.101	.117	.134	.166	.198	.230	.263	.329	.394	.459	.525
1 1/2 (1.500)					.134	.166	.198	.230	.263	.329	.394	.459	.525
1 3/4 (1.750)						.165	.198	.230	.263	.329	.394	.459	.525
2 (2.000)						.165	.198	.230	.263	.329	.394	.459	.525
2 1/4 (2.250)							.197	.230	.263	.329	.394	.459	.525
2 1/2 (2.500)								.230	.263	.329	.394	.459	.525
2 3/4 (2.750)								.229	.262	.329	.394	.459	.525
3 (3.000)								.229	.262	.329	.394	.459	.525
3 1/4 (3.250)									.262	.329	.393	.459	.525
3 1/2 (3.500)										.328	.393	.459	.525
3 3/4 (3.750)										.328	.393	.458	.525
4 (4.000)											.393	.458	.525
4 1/4 (4.250)											.393	.458	.524
4 1/2 (4.500)												.458	.524

Nom. Diam. (D)	Exp. Diam. (Dx) reduced by
1/16	.002
5/64	.002
3/32	.002
7/64	.002
1/8	.002
5/32	.002
3/16	.003
7/32	.003
1/4	.003
5/16	.004
3/8	.005
7/16	.006
1/2	.006

TOLERANCES:
On Nominal Diameter "D"
+.000—.001 up to 7/64" diameter
+.000—.002 7/64" and above
On Expanded Diameter "Dx"
±.001 up to 7/64" diameter
±.002 7/64" and above
On over-all Length "L"
±.010 for all diameters

◄ For stainless steels and other special materials, the expanded diameters shown in above table are reduced by amounts shown at left.

Note: Intermediate pin lengths, pin diameters up to 1/4", groove lengths, and groove positions to order as specials. When ordering, specify type, diameter, length, and special finishes.

Figure 13-49 Standard size chart of grooved pin types D and E

TYPE G

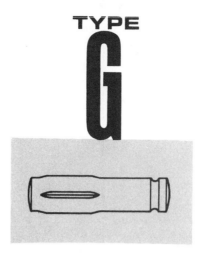

Typical applications

Type G has three **parallel half-length** grooves including pilot. It is a very versatile pin, suitable for use in both blind and through holes as a spring anchor pin. The annular groove opposite the locking end is used to anchor the end loop of a tension spring. If snap or retainer rings are to be used, special section annular grooves can be machined to order.

Spring Anchor Pin Used in Through Hole

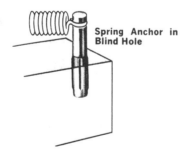

Spring Anchor in Blind Hole

Figure 13-50 Typical application for grooved pin type G

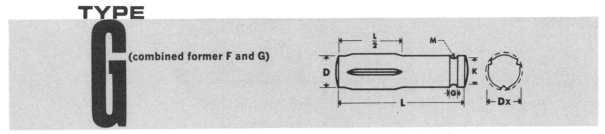

STANDARD SIZES

Nominal diameter and recommended drill sizes	3/32	7/64	1/8	5/32	3/16	7/32	1/4	5/16	3/8	7/16	1/2
Dec. Equivalents	.0938	.1094	.1250	.1563	.1875	.2188	.2500	.3125	.3750	.4375	.5000
Crown Height, In.	.0091	.0110	.0130	.0170	.0180	.0220	.0260	.0340	.0390	.0470	.0520
Radius, In., ±0.010	1/8	9/64	5/32	3/16	1/4	9/32	5/16	3/8	15/32	17/32	5/8
Pilot Length, In. (Ref.)	1/32	1/32	1/32	1/16	1/16	1/16	1/16	3/32	3/32	3/32	3/32
M Neck Radius (Ref.)	1/64	1/64	1/32	1/32	1/32	3/64	3/64	1/16	1/16	3/32	3/32
G Neck Width (Ref.)	1/32	1/32	1/16	1/16	1/16	3/32	3/32	1/8	1/8	3/16	3/16
Shoulder Width +.010−.000	1/32	1/32	1/32	3/64	3/64	1/16	1/16	3/32	1/8	1/8	1/8
K Neck Diameter ±.005	.062	.078	.083	.104	.125	.146	.167	.209	.250	.293	.312

"Dx" EXPANDED DIAMETER—CAN BE DETERMINED ACCURATELY ONLY WITH RING GAGES

LENGTH OF PIN IN INCHES		3/32	7/64	1/8	5/32	3/16	7/32	1/4	5/16	3/8	7/16	1/2
3/8	(.375)	.101	.117	.134	.166	.198						
1/2	(.500)	.101	.117	.134	.166	.198	.230	.263				
5/8	(.625)	.101	.117	.134	.166	.198	.230	.263	.329			
3/4	(.750)	.101	.117	.134	.166	.198	.230	.263	.329	.394		
7/8	(.875)	.101	.117	.134	.166	.198	.230	.263	.329	.394	.459	
1	(1.000)	.101	.117	.134	.166	.198	.230	.263	.329	.394	.459	.525
1¼	(1.250)	.101	.117	.134	.166	.198	.230	.263	.329	.394	.459	.525
1½	(1.500)			.134	.166	.198	.230	.263	.329	.394	.459	.525
1¾	(1.750)				.165	.198	.230	.263	.329	.394	.459	.525
2	(2.000)					.165	.198	.230	.263	.329	.394	.459	.525
2¼	(2.250)						.197	.230	.263	.329	.394	.459	.525
2½	(2.500)							.230	.263	.329	.394	.459	.525
2¾	(2.750)							.229	.262	.329	.394	.459	.525
3	(3.000)							.229	.262	.329	.394	.459	.525
3¼	(3.250)								.262	.328	.393	.459	.525
3½	(3.500)									.328	.393	.459	.525
3¾	(3.750)										.393	.458	.525
4	(4.000)										.393	.458	.525
4¼	(4.250)										.393	.458	.524
4½	(4.500)											.458	.524

Nom. Diam. (D)	Exp. Diam. (Dx) reduced by
3/32	.002
7/64	.002
1/8	.002
5/32	.002
3/16	.003
7/32	.003
1/4	.003
5/16	.004
3/8	.005
7/16	.006
1/2	.006

TOLERANCES:
On Nominal Diameter "D"
+.000—.001 up to 7/64" diameter
+.000—.002 7/64" and above
On Expanded Diameter "Dx"
± .001 up to 7/64" diameter
± .002 7/64" and above
On over-all Length "L"
± .010 for all diameters
◄ For stainless steels and other special materials, the expanded diameters shown in above table are reduced by amounts shown at left.
Note: Intermediate pin lengths, pin diameters up to ¼", groove lengths, and groove positions to order as specials. When ordering, specify type, diameter, length, and special finishes.

Figure 13-51 Standard size chart of groove pin type G

chart are listed the standard lengths from 1/4-inch long to 4 1/2-inches long. Various other technical information can be derived from the charts. For drilling procedure, hole tolerances, application data, and high-alloy pin applications, see Figures 13-52, 13-54, 13-55.

Grooved Studs

Grooved studs are widely used for fastening light metal or plastic parts to heavier members or assemblies, Figures 13-56 and 13-57. They replace screws, rivets, peened pins, and many other types of fasteners. The grooves function as any grooved fastener.

Pin Diameter	Decimal Equivalent	Recommended Drill Size	Hole Tolerances ADD To Nominal Diameter
1/16"	.0625	1/16"	.002"
5/64	.0781	5/64	.002"
3/32	.0938	3/32	.003"
7/64	.1094	7/64	.003"
1/8	.1250	1/8	.003"
5/32	.1563	5/32	.003"
3/16	.1875	3/16	.004"
7/32	.2188	7/32	.004"
1/4	.2500	1/4	.004"
5/16	.3125	5/16	.005"
3/8	.3750	3/8	.005"
7/16	.4375	7/16	.006"
1/2	.5000	1/2	.006"

Tolerances for drilled holes shown in table are based on depth to diameter ratio of approximately 5 to 1. Higher ratios may cause these figures to be exceeded, but in no case should they be exceeded by more than 10%. Specifications for holes having a depth to diameter ratio of 1 to 1 or less should be held extremely close; 60% of figures shown in table is recommended.

Undersized drills should never be used to produce holes for Driv-Lok Pins. This malpractice results from the false assumption that the pins will "hold better." Instead, they bend, damage the hole wall, crack castings, and peel their expanded flutes, thus reducing their retaining characteristics and preventing their reuse.

Holes made in hardened steel or cast iron are recommended to have a slight chamfer at the entrance. This eliminates shearing of the flutes as the pins are forced in.

Care should be exercised in all drilling for Driv-Lok Pins. Drills used should be new or properly ground with the aid of an approved grinding fixture. The drilling machine spindle must be in good condition and operated at correct speeds and feeds for the metal being drilled, and suitable coolant is always recommended. Drill jigs with accurate bushings always facilitate good drilling practice.

Figure 13-52 Drilling procedure and hole tolerance

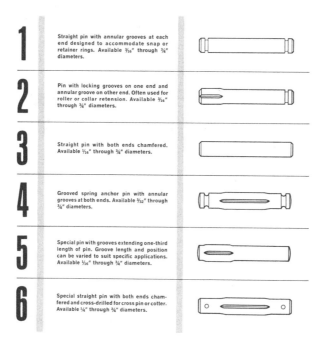

1 Straight pin with annular grooves at each end designed to accommodate snap or retainer rings. Available 3/16" through 3/4" diameters.

2 Pin with locking grooves on one end and annular groove on other end. Often used for roller or collar retension. Available 3/16" through 3/4" diameters.

3 Straight pin with both ends chamfered. Available 1/16" through 3/4" diameters.

4 Grooved spring anchor pin with annular grooves at both ends. Available 3/32" through 3/4" diameters.

5 Special pin with grooves extending one-third length of pin. Groove length and position can be varied to suit specific applications. Available 1/16" through 3/4" diameters.

6 Special straight pin with both ends chamfered and cross-drilled for cross pin or cotter. Available 1/4" through 3/4" diameters.

Figure 13-53 Special pins

RECOMMENDED PIN DIAMETER FOR VARIOUS SHAFT SIZES AND TORQUE TRANSMITTED BY PIN IN DOUBLE SHEAR							
Shaft Size	Pin Diameter	Torque Inch Lbs.	H.P. at 100 R.P.M.	Shaft Size	Pin Diameter	Torque Inch Lbs.	H.P. at 100 R.P.M.
3/16	1/16	4.6	.007	7/8	1/4	347	.555
7/32	5/64	8.4	.013	15/16	5/16	580	.927
1/4	3/32	13.7	.022	1	5/16	618	.990
5/16	7/64	23.6	.038	1 1/16	5/16	657	1.05
3/8	1/8	37.2	.060	1 1/8	3/8	1010	1.61
7/16	5/32	67.6	.108	1 3/16	3/8	1065	1.70
1/2	5/32	77.2	.124	1 1/4	3/8	1120	1.79
9/16	3/16	125.0	.200	1 5/16	7/16	1590	2.55
5/8	3/16	139.0	.222	1 3/8	7/16	1670	2.67
11/16	7/32	207.0	.332	1 7/16	7/16	1740	2.79
3/4	1/4	297.0	.476	1 1/2	1/2	2380	3.81
13/16	1/4	322	.516				

This table is a guide in selecting the proper size Driv-Lok Grooved Pin to use in keying machine members to shafts of given sizes and for specific load requirements. Torque and horsepower ratings are based on pins made of cold finished, low carbon steel and a safety factor of 8 is assumed.

MINIMUM SINGLE SHEAR VALUES (LBS.) OF DRIV-LOK PINS OF VARIOUS MATERIALS					
	MATERIAL				
DRIV-LOK PIN DIAM.	Cold Finished 1213 Steel	Shear-Proof® ALLOY STEEL R.C. 40 - 48	Brass	Silicon Bronze	Heat Treated Stainless Steel
1/16	200	363	124	186	308
5/64	312	562	192	288	478
3/32	442	798	272	408	680
7/64	605	1091	372	558	933
1/8	800	1443	492	738	1230
5/32	1240	2236	764	1145	1910
3/16	1790	3220	1100	1650	2750
7/32	2430	4386	1495	2240	3740
1/4	3190	5753	1960	2940	4910
5/16	4970	8974	3060	4580	7650
3/8	5810	12960	4420	6630	11050
7/16	7910	17580	6010	9010	15000
1/2	10300	23020	7850	11800	19640

Figure 13-54 Grooved pin application/engineering data

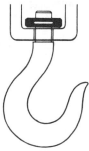

MATERIAL HANDLING EQUIPMENT—The Type E pin provides positive locking with a half-length groove in the center of the Pin. Extreme shear is exerted in this application, yet the Shear-Proof Pin is used with complete safety for both men and materials. Type E is a special Pin.

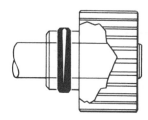

HEAVY-DUTY GEAR AND SHAFT ASSEMBLY—Type A Shear-Proof Pin as specified for this application to give maximum locking power over the entire pin and gear hub area. The Type A Pin, with grooves the full length of the Pin, is the standard stock Pin which meets most applications.

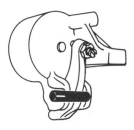

AUTOMATIC TRANSMISSION IN AUTOMOBILES—Special Type C was selected as a shaft in this transmission servo to replace a cross drilled shaft with a cross pin for holding shaft in position. This eliminated a costly drilling operation and the cross pin.

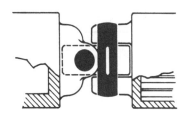

UNIVERSAL JOINTS IN HAND TOOLS—Special Type E Pin with center groove eliminates costly staking and grinding operations and improves product appearance. This Pin is easily installed, fits flush and permits plating before assembly.

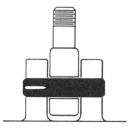

EYE BOLT HINGE PIN — Type C Pin, with quarter-length grooves, provides maximum ease of assembly. There is no interference until three-fourths of the pin is in position. The high safety factor inherent in Shear-Proof Pins makes them practical and efficient for such constant shear applications. Type C is a special Pin.

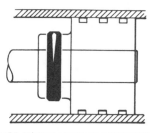

HIGH-PRESSURE PISTON AND ROD ASSEMBLY—Type B Pin was used here because the half-length grooves simplified the job of starting the pin into the hole. Ease of assembly was matched with sufficient locking power even when subjected to continuous, strong reciprocating forces. Type B is a special Pin.

Figure 13-55 High alloy shear-proof pins

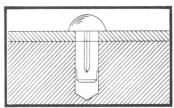

Stud Number	Nominal Shank Diameter	Recommended Drill Size	Head Dia. Max.	Head Dia. Min.	Head Height Max.	Head Height Min.
0	.067	51	.130	.120	.050	.040
2	.086	44	.162	.146	.070	.059
4	.104	37	.211	.193	.086	.075
6	.120	31	.260	.240	.103	.091
7	.136	29	.309	.287	.119	.107
8	.144	27	.309	.287	.119	.107
10	.161	20	.359	.334	.136	.124
12	.196	9	.408	.382	.152	.140
14	.221	2	.457	.429	.169	.156
16	.250	¼"	.472	.443	.174	.161

Maximum Expansion—Standard Lengths

Stud No.	⅛"	³⁄₁₆"	¼"	⁵⁄₁₆"	⅜"	½"
0	.074	.074	.074			
2	.096	.096	.095			
4		.115	.113	.113		
6			.132	.130	.130	
7				.147	.147	.144
8					.155	.153
10					.173	.171
12						.206
14						.234
16						.263

TOLERANCES:
On length ± .010
On Exp. Diameter ± .002
On Nominal Diameter + .000
—.002

Note: The expanded diameter can be determined accurately only with ring gages.

Figure 13-56 Standard studs

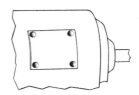

Fastening knobs
handles, etc.

Attaching nameplates,
instruction panels

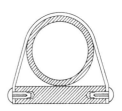

Widely used for
fastening brackets

Fastening spring assemblies
or control arms

1.	Flat head special stud with one-third length groove at lead end. Groove length can be varied.	
2.	Flat head special grooved stud with shoulder. Often hardened to provide wear surface in shoulder area.	
3.	Flat head grooved stud.	
4.	Round head reverse taper groove stud.	
5.	Stud with conical head and parallel grooves.	
6.	Round head stud with parallel grooves of special length.	
7.	Countersunk head grooved stud.	
8.	"T" head cotter used extensively in chain industry in place of cotter pins.	

SPECIAL STUD APPLICATIONS

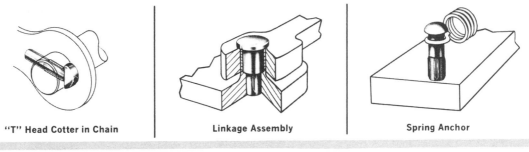

"T" Head Cotter in Chain   Linkage Assembly   Spring Anchor

*Available as specials through ½" diameter.

Figure 13-57 Special studs

Springs Pins

Another type of fastener is the *spring pin*, Figure 13-58. Spring pins are manufactured by cold forming strip metal in a progressive roll forming operation. After forming, the pins are broken off and deburred to eliminate any sharp edges. They are then heat treated to a R/C 46/53. This develops spring qualities or resiliency in the metal.

In the free state, the pins are larger in diameter than the hole into which they are to be inserted. The pins compress themselves as they are driven into the hole, thus exerting radial forces around the entire circumference of the hole.

Figure 13-58 Spring pin

Spring pins are made from a high carbon steel— 1074, stainless steel, either 420 heat treated or 300 series cold worked. Some spring pins are also made from brass or beryllium copper. Spring pins meet the demand of many industrial applications. Figures 13-59 and 13-60 list some suggested applications.

Figure 13-59
Spring pin application

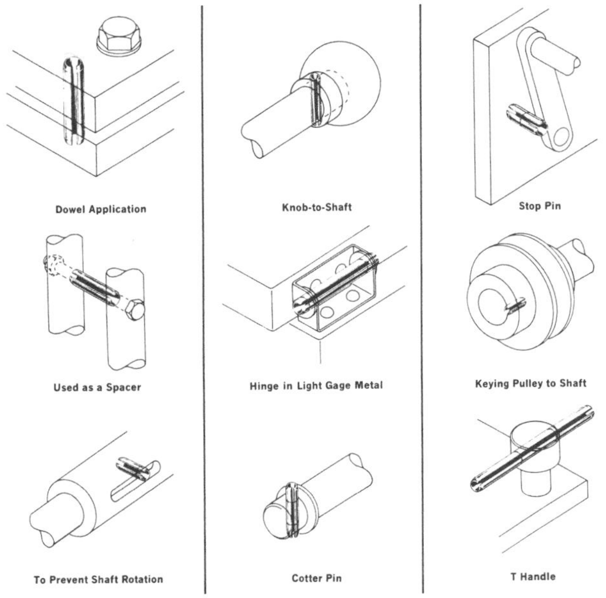

Dowel Application • Knob-to-Shaft • Stop Pin
Used as a Spacer • Hinge in Light Gage Metal • Keying Pulley to Shaft
To Prevent Shaft Rotation • Cotter Pin • T Handle

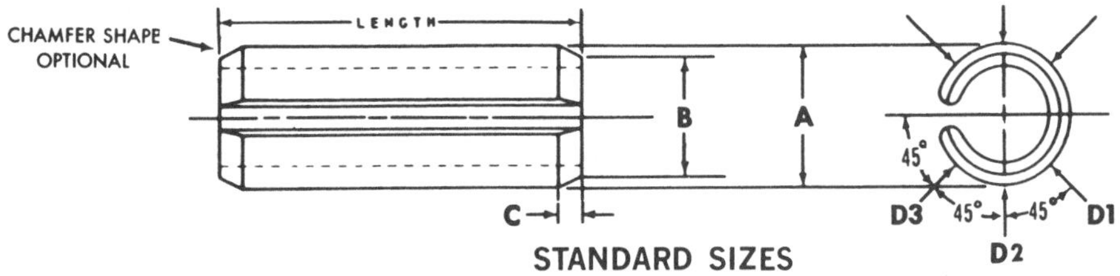

STANDARD SIZES

Nominal	A Minimum $\frac{1}{3}(D_1+D_2+D_3)$	A Maximum (Go Ring Gage)	B Max.	C Min.	C Max.	WALL THICKNESS	RECOMMENDED HOLE SIZE Min.	Max.	MINIMUM DOUBLE SHEAR STRENGTH POUNDS Carbon Steel and Stainless Steel
.062	.066	.069	.059	.007	.028	.012	.062	.065	425
.078	.083	.086	.075	.008	.032	.018	.078	.081	650
.094	.099	.103	.091	.008	.038	.022	.094	.097	1,000
.125	.131	.135	.122	.008	.044	.028	.125	.129	2,100
.156	.162	.167	.151	.010	.048	.032	.156	.160	3,000
.187	.194	.199	.182	.011	.055	.040	.187	.192	4,400
.219	.226	.232	.214	.011	.065	.048	.219	.224	5,700
.250	.258	.264	.245	.012	.065	.048	.250	.256	7,700
.312	.321	.328	.306	.014	.080	.062	.312	.318	11,500
.375	.385	.392	.368	.016	.095	.077	.375	.382	17,600
.437	.448	.456	.430	.017	.095	.077	.437	.445	20,000
.500	.513	.521	.485	.025	.110	.094	.500	.510	25,800

All dimensions listed on this page are in accordance with National Standards. Wall thicknesses within the Spring Pin industry are standard.

SPRING PIN WEIGHT PER 1000 PIECES
Material—Steel • Nominal Diameter

LENGTH	.062	.078	.094	.125	.156	.187
0.187	.10	0.18				
0.250	.15	0.22	0.33			
0.312	.19	0.28	0.41			
0.375	.23	0.34	0.50	0.89		
0.437	.27	0.40	0.58	1.00	1.50	
0.500	.30	0.46	0.66	1.20	1.70	2.50
0.562	.34	0.51	0.75	1.30	1.90	2.90
0.625	.38	0.57	0.83	1.50	2.10	3.20
0.687	.42	0.63	0.92	1.60	2.40	3.50
0.750	.46	0.69	1.00	1.80	2.60	3.80
0.812	.49	0.74	1.10	1.90	2.80	4.10
0.875	.53	0.80	1.20	2.10	3.00	4.50
0.937	.57	0.86	1.30	2.20	3.20	4.80
1.000	.61	0.92	1.40	2.40	3.40	5.10
1.125		1.00	1.50	2.70	3.80	5.70
1.250		1.20	1.70	3.00	4.30	6.40
1.375		1.30	1.80	3.30	4.70	7.00
1.500		1.40	2.00	3.60	5.10	7.60
1.625				3.90	5.50	8.30
1.750				4.20	6.00	8.90
1.875				4.50	6.40	9.60
2.000				4.70	6.80	10.0
2.250					7.80	12.0
2.500					8.60	13.0

LENGTH	.219	.250	.312	.375	.437	.500
0.562	4.00					
0.625	4.50	5.30				
0.687	4.90	5.90				
0.750	5.30	6.30	9.90	15.0		
0.812	5.80	6.90	11.0			
0.875	6.20	7.40	12.0			
0.937	6.70	8.00	12.0			
1.000	7.00	8.50	13.0	20.0	24.0	
1.125	7.90	9.50	14.0			
1.250	8.80	11.0	16.0	24.0	30.0	41.0
1.375	9.70	12.0	18.0			
1.500	11.0	13.0	19.0	29.0	36.0	49.0
1.625	12.0	14.0	21.0			
1.750	12.0	15.0	22.0	34.0	42.0	57.0
1.875	13.0	16.0	24.0			
2.000	14.0	17.0	25.0	38.0	48.0	65.0
2.250	16.0	19.0	29.0	43.0	54.0	73.0
2.500	18.0	21.0	32.0	48.0	60.0	81.0
2.750	19.0	23.0	35.0	53.0	66.0	89.0
3.000	21.0	25.0	38.0	58.0	72.0	97.0
3.250		28.0	41.0	62.0	77.0	105.0
3.500		31.0	44.0	67.0	83.0	114.0
3.750			48.0	72.0	89.0	122.0
4.000			51.0	77.0	95.0	130.0

Figure 13-60 Spring pin data

Fastening Systems

Fastening systems play a critical role in most product design. They often do more than position and secure components. In many cases, fastening systems have a direct effect upon the product's durability, reliability, size, and weight. They affect the speed with which the product may be assembled and disassembled, both during manufacturing and later in field service. Fastening systems also affect cost; not only for the fasteners, but for the machining and assembly operations they require.

Unless a drafter has had a great deal of experience in using different fastening devices and techniques, it is often difficult to choose a fastener that combines optimum function with maximum economy. A fastening system that is best for one product may not be desirable for another.

Retaining Rings

Retaining rings are precision-engineered fasteners. They provide removable shoulders for positioning or limiting the movement of parts in an assembly, Figure 13-61. Applications range from miniaturized electronic systems to massive earth-moving equipment. Retaining rings are used in automobiles, business machines, and complex components for guided missiles. They are found in such commonplace items as doorknobs to sophisticated underwater seismic cable connectors.

Typical ring applications are shown in Figures 13-62 through 13-66. From these figures, the drafter may examine the design of various types of rings, determine their purpose and function, and decide how they may be used to the best advantage.

Most retaining rings are made of materials that have good spring properties. This permits the rings to be deformed elastically to a substantial degree, yet still spring back to their original shape during assembly and disassembly. This allows most rings to function in one of two ways: 1) they may be sprung into a groove or other recess in a part, or 2) they may be seated on a part in a deformed condition so that they grip the part by frictional means. In either case, the rings form a fixed shoulder against which other components may be abutted and prevented from moving.

Unlike wire-formed rings which have a uniform section height, stamped rings have a tapered radial width. The width decreases symmetrically from the center section to the free ends. The tapered section per-

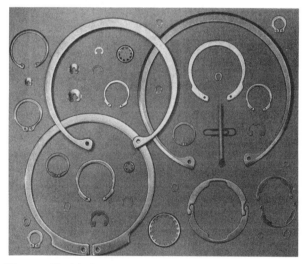

Figure 13-61 Retaining rings (*Courtesy Waldes Kohinoor, Inc.*)

Figure 13-62 Ring application in a precision differential (*Courtesy Waldes Kohinoor, Inc.*)

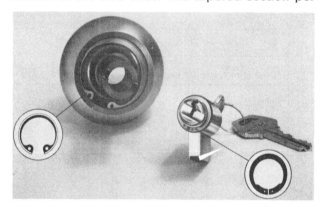

Figure 13-63 Ring application in a cylindrical lockset (*Courtesy Waldes Kohinoor, Inc.*)

Figure 13-64 Ring application in an electromagnetic clutch brake (*Courtesy Waldes Kohinoor, Inc.*)

Figure 13-65 Ring application in a road grader
(Courtesy Waldes Kohinoor, Inc.)

mits the rings to remain circular after they have been compressed for insertion into a bore or expanded for assembly over a shaft. Most rings are designed to be seated in grooves. This constant circularity assures maximum contact surface with the bottom of the groove. It is also an important factor in achieving high static and dynamic thrust load capacities.

Types of Retaining Rings

A great number of fastening requirements are involved in product design. This factor has led to the development of many different types of retaining rings. Standard rings are shown in Figure 13-67.

Limited space prevents describing in detail all the different ring types available. In general, however, retaining rings can be grouped into two major categories.

Internal rings for axial assembly are compressed for insertion into a bore or housing. They generally have a large gap and holes in the lugs, located at the free ends, for pliers which are used to grasp the rings securely during installation and removal, Figure 13-68.

External rings for axial assembly are expanded with pliers, Figure 13-69, so they may be slipped over the end of a shaft, stud, or similar part. They have a small gap and the lug position is reversed from that of the internal ring. Radially assembled external rings do not have holes for pliers. Instead, the rings have a large gap and are pushed into the shaft directly in the plane of the groove with a special application tool, Figure 13-70.

In addition to the tools shown here, retaining rings can be installed with automatic equipment. This equipment can be designed for specific, high-speed,

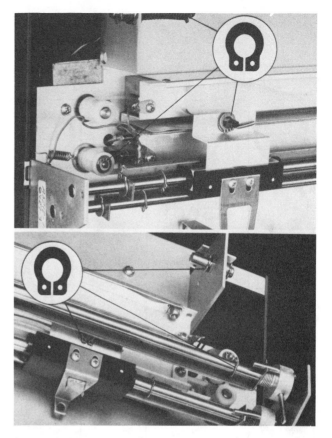

Figure 13-66 Ring application in strip chart recorder
(Courtesy Waldes Kohinoor, Inc.)

automatic assembly lines. Retaining ring grooves serve two purposes: 1) they assure precise seating of the ring in the assembly, and 2) they permit the ring to withstand heavy thrust loads. The grooves must be located accurately and precut in the housing or shaft before the rings are assembled. Shaft grooves often can be made at no additional cost during the cut-off and chamfering operations.

Self-locking rings do not require any grooves because they exert a frictional hold against axial displacement. They are used mainly as positioning and locking devices where the ring will be subjected to only moderate or light loading.

Bowed rings differ from conventional types in that they are bowed around an axis perpendicular to the diameter bisecting the gap. The bowed construction permits the rings to function as springs as well as fasteners. This provides resilient end play take-up in the assembly.

Beveled rings have a 15-degree bevel on the groove-engaging edge. They are installed in grooves having a comparable bevel on the load-bearing wall. When the ring is seated in the groove, it acts as a wedge against the retained part. Sometimes play develops between the ring and retained part because of accumulated tolerances or wear in the assembly. If play develops, the spring action of the ring causes the fastener to seat more deeply in its groove and move in an axial direction, automatically taking up the end

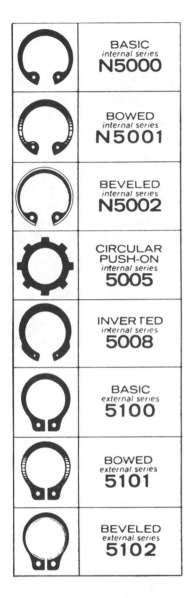

	BASIC *internal series* **N5000**
	BOWED *internal series* **N5001**
	BEVELED *internal series* **N5002**
	CIRCULAR PUSH-ON *internal series* **5005**
	INVERTED *internal series* **5008**
	BASIC *external series* **5100**
	BOWED *external series* **5101**
	BEVELED *external series* **5102**

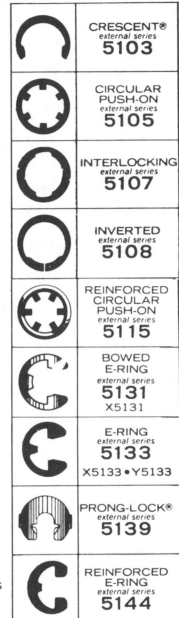

	CRESCENT® *external series* **5103**
	CIRCULAR PUSH-ON *external series* **5105**
	INTERLOCKING *external series* **5107**
	INVERTED *external series* **5108**
	REINFORCED CIRCULAR PUSH-ON *external series* **5115**
	BOWED E-RING *external series* **5131** X5131
	E-RING *external series* **5133** X5133 • Y5133
	PRONG-LOCK® *external series* **5139**
	REINFORCED E-RING *external series* **5144**

	HEAVY DUTY *external series* **5160**
	TRIANGULAR NUT *external series* **5300**
	KLIPRING® *external series* **5304 T-5304**
	TRIANGULAR PUSH-ON *external series* **5305**
	GRIPRING® *external series* **5555** D5555 • G5555
	MINIATURE HIGH-STRENGTH *external series* **5560**
	PERMANENT SHOULDER *external series* **5590**
	PRECISION SUPPORT WASHER **5900**

Figure 13-67 Standard retaining rings series (*Courtesy Waldes Kohinoor, Inc.*)

Figure 13-68 Internal ring pliers (*Courtesy Waldes Kohinoor, Inc.*)

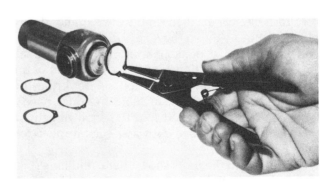

Figure 13-69 External ring pliers (*Courtesy Waldes Kohinoor, Inc.*)

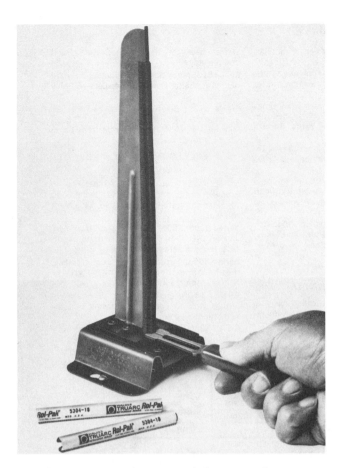

Figure 13-70 Applicator and dispenser for retaining rings (*Courtesy Waldes Kohinoor, Inc.*)

play. (Because self-locking rings can be seated at any point on a shaft or in a bore, they too can be used to compensate for tolerances and eliminate end play.)

Materials and Finishes

As indicated previously, retaining rings are made of materials having good spring properties. Some also have high tensile and yield strengths. They must also have an adequate ratio of ultimate tensile strength to elasticity. This permits the required deformation without too much permanent set. A ratio of 1:100 is satisfactory for most rings having the tapered-section design.

Standard material for most rings is carbon spring steel (SAE 1060-1090). For special applications, rings are also available in stainless steel (PH 15-7 Mo), beryllium copper (Alloy #25), and aluminum (Alclad 7075-T6). Rings are normally phosphate coated. Cadmium, zinc, and other platings and finishes are used for assemblies where extra corrosion resistance is needed or if the rings must withstand other unusual environmental conditions. Selection of the ring material and finish for a specific product design should be based upon the operating conditions under which

the ring must function. These may include temperature, the presence of corrosive elements, thrust loads, and other factors.

Selection Considerations

Selecting the best ring for a product load capacity is a critical factor in some product designs. There are other factors the drafter should consider, however, before selecting specific ring types for a given product.

- Will there be adequate clearance to assemble the ring with pliers or other tools?
- Must the ring take up accumulated tolerances, either resiliently or rigidly?
- Is it possible to machine a ring groove on the shaft or inside a bore?
- Should the ring be adjustable to several positions on a shaft?

Making the Choice

After considering all these conditions, the drafter may find that more than one ring type is suitable. How, then, can the drafter make the best choice?

The most important design criterion is the ability of the ring to do the fastening job required. Before a final selection of ring type is made, the drafter should consider savings which may be possible in various parts of the assembly. These include:

- The cost of installing the ring.
- Whether or not a groove is required.
- If the ring can be installed *permanently* or if it may have to be removed for field service.

A self-locking ring, for example, eliminates the need for the ring grooves. If the ring will be subjected to only moderate loading, this may be ideal for the assembly. But most self-locking rings must be destroyed for removal. If field service is anticipated, another style of ring should be adopted.

The ideal ring is the one which will function adequately and provide the most economical means of fastening. Retaining rings are designed primarily as shoulders for positioning and retaining machine components on shafts and in housings and bores. Different rings have been developed and manufactured to meet specific fastening needs and problems.

To ensure correct selection of the proper type for any individual application, rings have been grouped according to their basic function. The selector guides, Figures 13-71 and 13-72, provide a visual index to all standard types.

Figures 13-73 and 13-74 are from the latest *Truarc Technical Manual*. Figure 13-73 is a sample of an internal series (N5000); Figure 13-74 is from an external series (5100).

DESIGN FEATURES

RING TYPES FOR AXIAL ASSEMBLY

Series N5000, 5100: Tapered section assures constant circularity and groove pressure. Secure against heavy thrust loads and high rotational speeds.

Series 5008, 5108: Lugs inverted to abut groove bottom. Rings form high circular shoulder, concentric with bore or shaft. Good for parts having large corner radii or chamfers.

Series 5160: Heavy-duty ring resists high thrust, impact loads. Eliminates spacer washers in bearing assemblies.

Series 5560: New miniature, high-strength ring. Forms tamper-proof shoulder on small diameter shafts subject to heavy thrust loads.

Series 5590: Permanent-shoulder ring for small diameter shafts. When compressed into groove, notches deform to close gaps, reducing both I.D. and O.D.

RING TYPES FOR RADIAL ASSEMBLY

Series 5103: Forms narrow, uniformly concentric shoulder. Excellent for assemblies where clearance is limited.

Series 5133: Provides large shoulder on small diameter shafts. Installed in deep groove for added thrust capacity.

Series 5144: Reinforced to provide five times greater gripping strength, 50% higher rpm limits than conventional E-rings. Secure against rotation.

Series 5107: Two-part ring balanced to withstand high rpm's, heavy thrust loads, relative rotation between parts.

Series 5304: New high-strength ring for large bearing surface. Can be installed quickly with pliers or mallet, removed with ordinary screw driver.

Series T5304: Thinner model of 5304. Can be seated in same width grooves as E-rings, has more gripping power. Good for cast or molded grooves.

RING TYPES FOR TAKING UP END-PLAY

Series N5001, 5101: Bowed cylindrically to accommodate large tolerances, provide resilient end-play take-up.

Series N5002, 5102: Rings beveled 15° on groove-engaging edge for use in groove with similar bevel. Wedge action provides rigid end-play take-up.

Series 5131: Provides large shoulder on small diameter shafts. Bowed for resilient end-play take-up.

Series 5139: Bowed ring designed for use as shoulder against rotating parts. Prongs lock against shaft, prevent ring from being forced from groove.

SELF-LOCKING TYPE RINGS (No groove required)

Series 5115: Push-on type fastener for ungrooved shafts and studs. Has arched rim for extra strength, long prongs for wide shaft tolerances.

Series 5105, 5005: Flat rim, shorter prongs, smaller O.D. than 5115. For flat contact surface, better clearance.

Series 5555: Secure against axial displacement from either direction. No groove needed. Adjustable, reusable.

Series 5305: Dished body, three heavy prongs lock on shaft under spring tension. Withstands heavy thrust loads.

Series 5300: Free-spinning nut. Dished body flattens under torque, eliminating need for separate lock washers.

INTERNAL	BASIC **N5000** For housings and bores / Size Range .250—10.0 in. / 6.4—254.0 mm.	EXTERNAL	BOWED **5101** For shafts and pins / Size Range .188—1.750 in. / 4.8—44.4 mm.	EXTERNAL	REINFORCED **5115** For shafts and pins / Size Range .094—1.0 in. / •
EXTERNAL	TRIANGULAR NUT **5300** For threaded parts / Size Range 6-32 and 8-32 10-24 and 10-32 1/4-20 and 1/4-28				
INTERNAL	BOWED **N5001** For housings and bores / Size Range .250—1.750 in. / 6.4—44.4 mm.	EXTERNAL	BEVELED **5102** For shafts and pins / Size Range 1.0—10.0 in. / 25.4—254.0 mm.	EXTERNAL	BOWED E-RING **5131** For shafts and pins / Size Range .110—1.375 in. / 2.8—34.9 mm.
EXTERNAL	KLIPRING **5304** T-5304 For shafts and pins / Size Range .156—1.000 in. / 4.0—25.4 mm.				
INTERNAL	BEVELED **N5002** For housings and bores / Size Range 1.0—10.0 in. / 25.4—254.0 mm.	EXTERNAL	CRESCENT® **5103** For shafts and pins / Size Range .125—2.0 in. / 3.2—50.8 mm.	EXTERNAL	E-RING **5133** For shafts and pins / Size Range .040—1.375 in. / 1.0—34.9 mm.
EXTERNAL	TRIANGULAR **5305** For shafts and pins / Size Range .062—.438 in. / •				
INTERNAL	CIRCULAR **5005** For housings and bores / Size Range .312—2.0 in. / •	EXTERNAL	CIRCULAR **5105** For shafts and pins / Size Range .094—1.0 in. / •	EXTERNAL	PRONG-LOCK® **5139** For shafts and pins / Size Range .092—.438 in. / •
EXTERNAL	GRIPRING® **5555** For shafts and pins / Size Range 079—.750 in. / 2.0—19.0 mm.				
INTERNAL	INVERTED **5008** For housings and bores / Size Range .750—4.0 in. / 19.0—101.6 mm.	EXTERNAL	INTERLOCKING **5107** For shafts and pins / Size Range .469—3.375 in. / 11.9—85.7 mm.	EXTERNAL	REINFORCED E-RING **5144** For shafts and pins / Size Range .094—.562 in. / 2.4—14.3 mm.
EXTERNAL	HIGH-STRENGTH **5560** For shafts and pins / Size Range .101—.328 in. / •				
EXTERNAL	BASIC **5100** For shafts and pins / Size Range .125—10.0 in. / 3.2—254.0 mm.	EXTERNAL	INVERTED **5108** For shafts and pins / Size Range .500—4.0 in. / 12.7—101.6 mm.	EXTERNAL	HEAVY-DUTY **5160** For shafts and pins / Size Range .394—2.0 in. / 10.0—50.8 mm.
EXTERNAL	PERMANENT SHOULDER **5590** For shafts and pins / Size Range .250—.750 / 6.4—19.0 mm.				

Figure 13-71 Selector guide: standard ring series (*Courtesy Waldes Kohinoor, Inc.*)

The symbols listed below are used in the data charts for various ring types. Ring, groove and retained part dimensions are in inches; allowable thrust loads are in pounds.

SYMBOL		DEFINITION	RING SERIES WHERE APPLICABLE	SYMBOL		DEFINITION	RING SERIES WHERE APPLICABLE
A		Minimum gap width: internal ring installed in groove	N5000, N5001, N5002, 5008	$Ch_{max.}$		Maximum allowable chamfer height of retained part	N5000, N5001, N5002, 5008, 5100, 5101, 5102, 5103, 5107, 5108, 5131, 5133, 5144, 5160, 5555, 5560
B	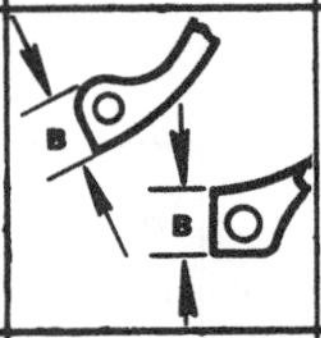	Lug height	N5000, N5001, N5002, 5100, 5101, 5102, 5160, 5555	D		Free diameter	N5000, N5001, N5002, 5008, 5100, 5101, 5102, 5103, 5107, 5108, 5131, 5133, 5144, 5160, 5304, 5555, 5560, 5590
b	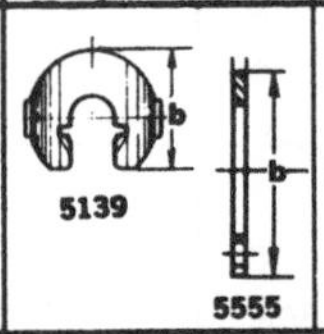 5139 5555	Ring height	5139, 5555, 5305, 5300, 5304	d		Nominal groove depth	ALL RINGS USED IN GROOVES
C	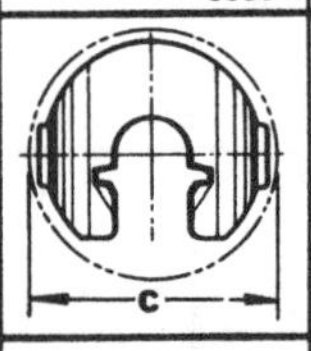	Clearance diameter	5139, 5590	E		Large section height	N5000, N5001, N5002, 5008, 5100, 5101, 5102, 5103, 5107, 5108, 5160, 5560
C_1	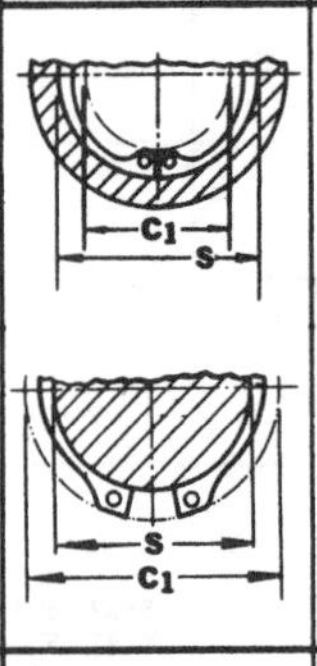	Clearance diameter: ring sprung into housing or over shaft, prior to installation in groove	N5000, N5001, N5002, 5008, 5100, 5101, 5102, 5108, 5160, 5555, 5560, 5590	e REF.		Distance from center of ring to outer edge (Reference)	5139
				G		Groove diameter	ALL RINGS USED IN GROOVES
C_2		Clearance diameter: ring installed in groove	N5000, N5001, N5002, 5008, 5100, 5101, 5102, 5103, 5107, 5108, 5131, 5133, 5144, 5160, 5560	H		Bow height	5139
				h	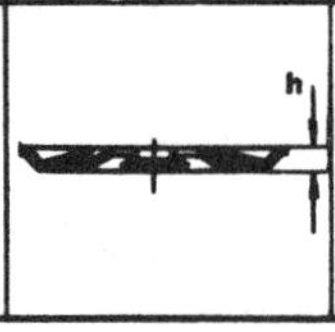	Ring height	5115

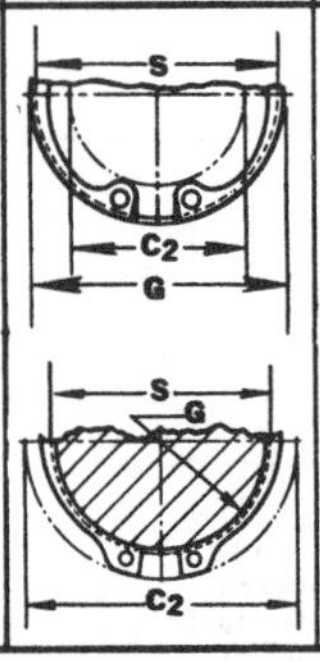

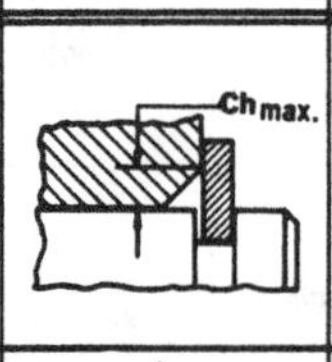

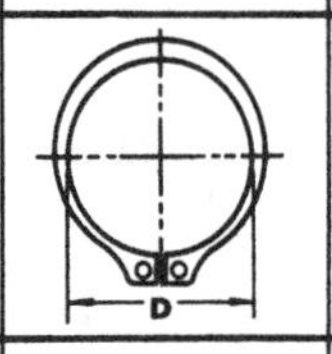

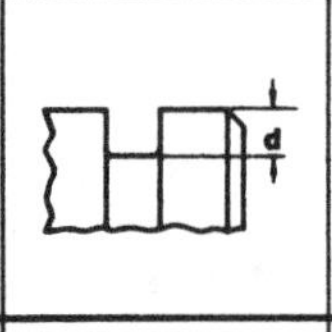

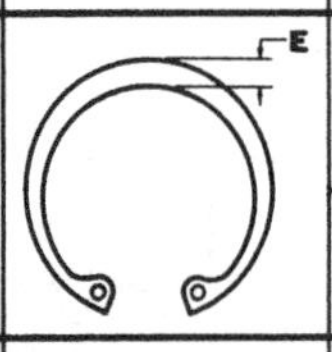

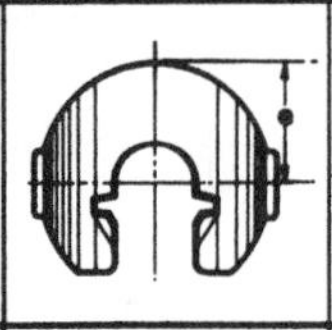

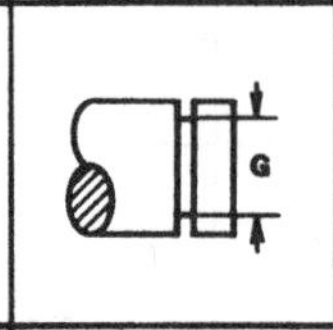

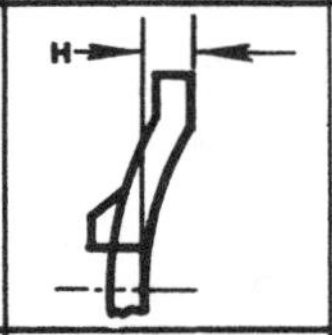

Figure 13-72 Definition symbols (*Courtesy Waldes Kohinoor, Inc.*)

SYMBOL		DEFINITION	RING SERIES WHERE APPLICABLE	SYMBOL		DEFINITION	RING SERIES WHERE APPLICABLE
J		Small section height	N5000, N5001, N5002, 5008, 5100, 5101, 5102, 5108, 5160, 5560	S		Shaft or housing diameter	ALL RINGS
K		Maximum gaging diameter: ring installed in groove	5100, 5101, 5102, 5108, 5160, 5560	t		Ring thickness	ALL RINGS
k		Overall ring width	5305, 5300	U		Ring thickness at beveled edge	N5002, 5102
L, M		L: Location of outer groove wall from plane of reference. M: Width of retained part	N5001, N5002, 5101, 5102, 5131, 5139	V		Overall bow height	N5001, 5101, 5131
P		Pliers hole diameter	N5000, N5001, N5002, 5008, 5100, 5101, 5102, 5108, 5160, 5555	W		Groove width	ALL RINGS USED IN GROOVES
p		Gap width	5139	X		Distance from outer groove wall to face of retained part	N5001, 5101, 5131, 5139
P_r, P'_r, P_g	P_r Allowable thrust load for ring (lbs.). P'_r Allowable assembly load with maximum corner radius or chamfer. P_g Allowable thrust load for groove (lbs.)		ALL RINGS USED IN GROOVES	Y		Free outside diameter	5103, 5133, 5144, 5131, 5304
R		Radius of groove bottom	ALL RINGS USED IN GROOVES	Z		Edge margin	ALL RINGS USED IN GROOVES
$R_{max.}$		Maximum allowable corner radius of retained part	N5000, N5001, N5002, 5008, 5100, 5101, 5102, 5103, 5107, 5108, 5131, 5133, 5144, 5160, 5555, 5560	Z_1		Minimum distance from face of retained part to end of shaft or housing	5115, 5005, 5105, 5305

Figure 13-72 (continued)

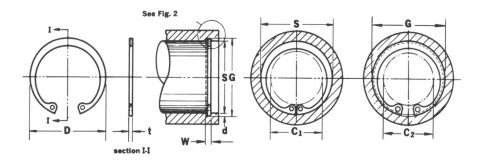

HOUSING DIA.			MIL-R-21248 MS 16625 INTERNAL SERIES N5000	TRUARC RING DIMENSIONS					GROOVE DIMENSIONS					APPLICATION DATA			
				FREE DIA.		THICKNESS		Approx. weight per 1000 pieces	DIAMETER		WIDTH		Nominal groove depth	CLEARANCE DIAMETER		ALLOW. THRUST LOAD (lbs.) Sharp corner abutment	
														When sprung into housing	When sprung into groove	RINGS (Standard material) Safety factor = 4	GROOVES (Cold rolled steel bores and housings) Safety factor = 2
Dec. equiv. inch	Approx. fract. equiv.	Approx. mm															
S	S	S	size—no.	D	tol.	t	tol.	lbs.	G	tol.	W	tol.	d	C₁	C₂	Pr	Pg
.250	¼	6.4	N5000-25	.280		.015		.08	.268	±.001 .0015 T.I.R.	.018	+.002 −.000	.009	.115	.133	420	190
.312	5/16	7.9	N5000-31	.346		.015		.11	.330		.018		.009	.173	.191	530	240
.375	3/8	9.5	N5000-37	.415		.025		.25	.397	±.002 .002 T.I.R.	.029		.011	.204	.226	1050	350
.438	7/16	11.1	N5000-43	.482		.025		.37	.461		.029		.012	.23	.254	1220	440
.453	29/64	11.5	N5000-45	.498		.025		.43	.477		.029		.012	.25	.274	1280	460
.500	½	12.7	N5000-50	.548	+.010 −.005	.035		.70	.530	±.002 .004 T.I.R.	.039		.015	.26	.29	1980	510
.512	—	13.0	N5000-51	.560		.035		.77	.542		.039		.015	.27	.30	2030	520
.562	9/16	14.3	N5000-56	.620		.035		.86	.596		.039		.017	.275	.305	2220	710
.625	5/8	15.9	N5000-62	.694		.035		1.0	.665		.039	+.003 −.000	.020	.34	.38	2470	1050
.688	11/16	17.5	N5000-68	.763		.035		1.2	.732		.039		.022	.40	.44	2700	1280
.750	¾	19.0	N5000-75	.831		.035		1.3	.796		.039		.023	.45	.49	3000	1460
.777	—	19.7	N5000-77	.859		.042		1.7	.825		.046		.024	.475	.52	4550	1580
.812	13/16	20.6	N5000-81	.901		.042		1.9	.862		.046		.025	.49	.54	4800	1710
.866	—	22.0	N5000-86	.961		.042		2.0	.920		.046		.027	.54	.59	5100	1980
.875	7/8	22.2	N5000-87	.971	+.015 −.010	.042		2.1	.931	±.003 .004 T.I.R.	.046		.028	.545	.60	5150	2080
.901	—	22.9	N5000-90	1.000		.042	±.002	2.2	.959		.046		.029	.565	.62	5350	2200
.938	15/16	23.8	N5000-93	1.041		.042		2.4	1.000		.046		.031	.61	.67	5600	2450
1.000	1	25.4	N5000-100	1.111		.042		2.7	1.066		.046		.033	.665	.73	5950	2800
1.023	—	26.0	N5000-102	1.136		.042		2.8	1.091		.046		.034	.69	.755	6050	3000
1.062	1 1/16	27.0	N5000-106	1.180		.050		3.7	1.130		.056		.034	.685	.75	7450	3050
1.125	1⅛	28.6	N5000-112	1.249		.050		4.0	1.197		.056		.036	.745	.815	7900	3400
1.181	—	30.0	N5000-118	1.319		.050		4.3	1.255		.056		.037	.79	.86	8400	3700
1.188	1 3/16	30.2	N5000-118	1.319		.050		4.3	1.262		.056		.037	.80	.87	8400	3700
1.250	1¼	31.7	N5000-125	1.388		.050		4.8	1.330		.056		.040	.875	.955	8800	4250
1.259	—	32.0	N5000-125	1.388	+.025 −.020	.050		4.8	1.339		.056		.040	.885	.965	8800	4250
1.312	1 5/16	33.3	N5000-131	1.456		.050		5.0	1.396	±.004 .005 T.I.R.	.056		.042	.93	1.01	9300	4700
1.375	1⅜	34.9	N5000-137	1.526		.050		5.1	1.461		.056		.043	.99	1.07	9700	5050
1.378	—	35.0	N5000-137	1.526		.050		5.1	1.464		.056		.043	.99	1.07	9700	5050
1.438	1 7/16	36.5	N5000-143	1.596		.050		5.8	1.528		.056	+.004 −.000	.045	1.06	1.15	10200	5500
1.456	—	37.0	N5000-145	1.616		.050		6.4	1.548		.056		.046	1.08	1.17	10300	5700
1.500	1½	38.1	N5000-150	1.660		.050		6.5	1.594		.056		.047	1.12	1.21	10550	6000
1.562	1 9/16	39.7	N5000-156	1.734		.062		8.9	1.658		.068		.048	1.14	1.23	13700	6350
1.575	—	40.0	N5000-156	1.734	+.035 −.025	.062	±.003	8.9	1.671	±.005 .005 T.I.R.	.068		.048	1.15	1.24	13700	6350
1.625	1⅝	41.3	N5000-162	1.804		.062		10.0	1.725		.068		.050	1.15	1.25	14200	6900
1.653	—	42.0	N5000-165	1.835		.062		10.4	1.755		.068		.051	1.17	1.27	14500	7200
1.688	1 11/16	42.9	N5000-168	1.874		.062		10.8	1.792		.068		.052	1.21	1.31	14800	7450

Figure 13-73 Internal series N5000 (*Courtesy Waldes Kohinoor, Inc.*)

Example (Refer to Figure 13-73):

A shaft of 1.000 (25.4) diameter would use a No. N5000-100.

- Free diameter = $1.111 \pm \genfrac{}{}{0pt}{}{.015}{.010}$
- Thickness = .042 ± .002
- Lug size = .155 ± .005
- Plier hole diameter = .062

The required groove must be:

- Diameter = 1.066 ± .003 (.004 TIR)
- Width = $.046 \pm \genfrac{}{}{0pt}{}{.003}{.000}$

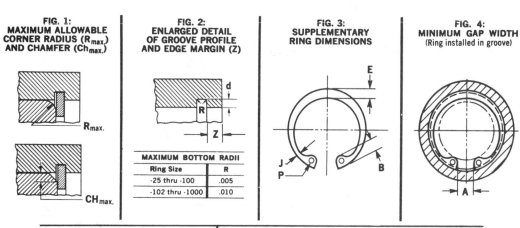

FIG. 1: MAXIMUM ALLOWABLE CORNER RADIUS (R_max.) AND CHAMFER (Ch_max.)

FIG. 2: ENLARGED DETAIL OF GROOVE PROFILE AND EDGE MARGIN (Z)

FIG. 3: SUPPLEMENTARY RING DIMENSIONS

FIG. 4: MINIMUM GAP WIDTH (Ring installed in groove)

MAXIMUM BOTTOM RADII	
Ring Size	**R**
-25 thru -100	.005
-102 thru -1000	.010

INTERNAL SERIES N5000 size—no.	SUPPLEMENTARY APPLICATION DATA				SUPPLEMENTARY RING DIMENSIONS								
	Maximum allowable corner radii and chamfers of retained parts (Fig. 1)		Allow. assembly load with R max. or Ch max.	Edge margin (Fig. 2)	LUG		LARGE SECTION		SMALL SECTION		HOLE DIAMETER		MIN. GAP WIDTH (Fig. 4) Ring installed in groove
	$R_{max.}$	$Ch_{max.}$	P'_r(lbs.)	Z	B	tol.	E	tol.	J	tol.	P	tol.	A
N5000-25	.011	.0085	190	.027	.065		.025	±.002	.015	±.002	.031		.047
N5000-31	.016	.013	190	.027	.066		.033		.018		.031		.055
N5000-37	.023	.018	530	.033	.082		.040		.028		.041		.063
N5000-43	.027	.021	530	.036	.098	±.003	.049	±.003	.029	±.003	.041		.063
N5000-45	.027	.021	530	.036	.098		.050		.030		.047		.071
N5000-50	.027	.021	1100	.045	.114		.053		.035		.047		.090
N5000-51	.027	.021	1100	.045	.114		.053		.035		.047		.092
N5000-56	.027	.021	1100	.051	.132		.053		.035		.047		.095
N5000-62	.027	.021	1100	.060	.132		.060	±.004	.035	±.004	.062		.104
N5000-68	.027	.021	1100	.066	.132		.063		.036		.062	+.010 −.002	.118
N5000-75	.032	.025	1100	.069	.142		.070		.040		.062		.143
N5000-77	.035	.028	1650	.072	.146		.074		.044		.062		.145
N5000-81	.035	.028	1650	.075	.155		.077		.044		.062		.153
N5000-86	.035	.028	1650	.081	.155		.081		.045		.062		.172
N5000-87	.035	.028	1650	.084	.155		.084	±.005	.045	±.005	.062		.179
N5000-90	.038	.030	1650	.087	.155		.087		.047		.062		.188
N5000-93	.038	.030	1650	.093	.155		.091		.050		.062		.200
N5000-100	.042	.034	1650	.099	.155		.104		.052		.062		.212
N5000-102	.042	.034	1650	.102	.155		.106		.054		.062		.220
N5000-106	.044	.035	2400	.102	.180		.110		.055		.078		.213
N5000-112	.047	.036	2400	.108	.180		.116		.057		.078		.232
(S=1.181) N5000-118	.047	.036	2400	.111	.180	±.005	.120		.058		.078		.226
(S=1.188) N5000-118	.047	.036	2400	.111	.180		.120		.058		.078		.245
(S=1.250) N5000-125	.048	.038	2400	.120	.180		.124		.062		.078		.265
(S=1.259) N5000-125	.048	.038	2400	.120	.180		.124	±.006	.062	±.006	.078		.290
N5000-131	.048	.038	2400	.126	.180		.130		.062		.078		.284
(S=1.375) N5000-137	.048	.038	2400	.129	.180		.130		.063		.078	+.015 −.002	.297
(S=1.378) N5000-137	.048	.038	2400	.129	.180		.130		.063		.078		.305
N5000-143	.048	.038	2400	.135	.180		.133		.065		.078		.313
N5000-145	.048	.038	2400	.138	.180		.133		.065		.078		.320
N5000-150	.048	.038	2400	.141	.180		.133		.066		.078		.340
(S=1.562) N5000-156	.064	.050	3900	.144	.202		.157		.078		.078		.338
(S=1.575) N5000-156	.064	.050	3900	.144	.202		.157		.078		.078		.374
N5000-162	.064	.050	3900	.150	.227		.164	±.007	.082	±.007	.078		.339
N5000-165	.064	.050	3900	.153	.227		.167		.083		.078		.348
N5000-168	.064	.050	3900	.156	.227		.170		.085		.078		.357

Figure 13-73 (continued)

Example (Refer to Figure 13-74):

A shaft of 4.000 (101.6) diameter would use a No. 5100-400.

- Free diameter = $3.700 \pm {.020 \atop .030}$
- Thickness = $.109 \pm .003$
- Lug size = $.352 \pm .005$
- Plier hole size = $.125 \pm {.015 \atop .002}$

The required groove must be:

- Diameter = $3.792 \pm .006$ (.006 TIR)
- Width = $.120 \pm {.005 \atop .000}$

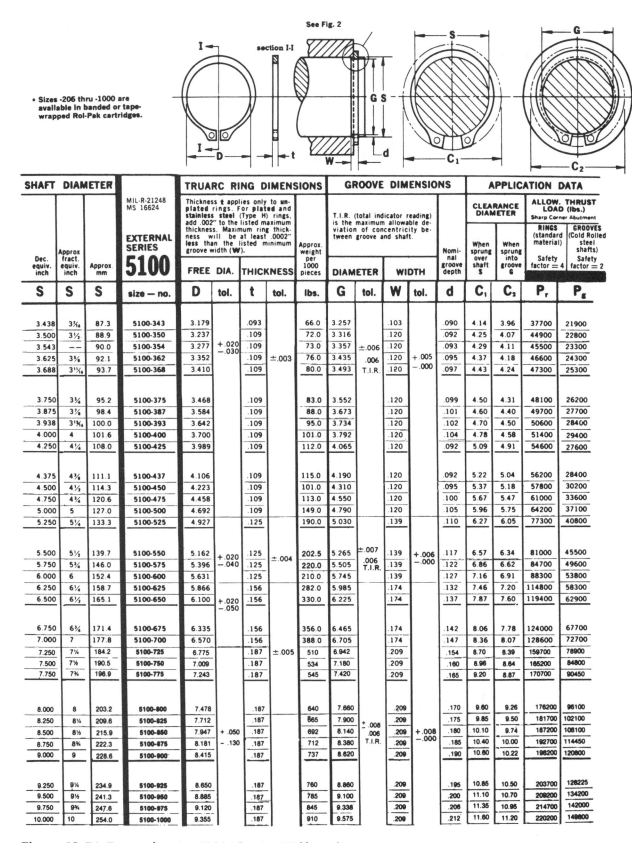

- Sizes -206 thru -1000 are available in banded or tape-wrapped Rol-Pak cartridges.

See Fig. 2

section I-I

SHAFT DIAMETER			MIL-R-21248 MS 16624 EXTERNAL SERIES **5100**	TRUARC RING DIMENSIONS					GROOVE DIMENSIONS					APPLICATION DATA			
				Thickness **t** applies only to un-plated rings. For plated and stainless steel (Type H) rings, add .002" to the listed maximum thickness. Maximum ring thickness will be at least .0002" less than the listed minimum groove width (**W**).				Approx. weight per 1000 pieces	T.I.R. (total indicator reading) is the maximum allowable deviation of concentricity between groove and shaft.				Nominal groove depth	CLEARANCE DIAMETER		ALLOW. THRUST LOAD (lbs.) Sharp Corner Abutment	
															When sprung over shaft	When sprung into groove	RINGS (standard material)	GROOVES (Cold Rolled steel shafts)
Dec. equiv. inch	Approx fract. equiv. inch	Approx mm	size — no.	FREE DIA.		THICKNESS			DIAMETER		WIDTH					Safety factor = 4	Safety factor = 2
S	S	S	size — no.	D	tol.	t	tol.	lbs.	G	tol.	W	tol.	d	C₁	C₂	Pᵣ	Pₘ
3.438	3⁷⁄₁₆	87.3	5100-343	3.179		.093		66.0	3.257		.103		.090	4.14	3.96	37700	21900
3.500	3½	88.9	5100-350	3.237	+.020	.109		72.0	3.316		.120		.092	4.25	4.07	44900	22800
3.543	––	90.0	5100-354	3.277	−.030	.109		73.0	3.357	±.006	.120	+.005	.093	4.29	4.11	45500	23300
3.625	3⅝	92.1	5100-362	3.352		.109	±.003	76.0	3.435	.006	.120	−.000	.095	4.37	4.18	46600	24300
3.688	3¹¹⁄₁₆	93.7	5100-368	3.410		.109		80.0	3.493	T.I.R.	.120		.097	4.43	4.24	47300	25300
3.750	3¾	95.2	5100-375	3.468		.109		83.0	3.552		.120		.099	4.50	4.31	48100	26200
3.875	3⅞	98.4	5100-387	3.584		.109		88.0	3.673		.120		.101	4.60	4.40	49700	27700
3.938	3¹⁵⁄₁₆	100.0	5100-393	3.642		.109		95.0	3.734		.120		.102	4.70	4.50	50600	28400
4.000	4	101.6	5100-400	3.700		.109		101.0	3.792		.120		.104	4.78	4.58	51400	29400
4.250	4¼	108.0	5100-425	3.989		.109		112.0	4.065		.120		.092	5.09	4.91	54600	27600
4.375	4⅜	111.1	5100-437	4.106		.109		115.0	4.190		.120		.092	5.22	5.04	56200	28400
4.500	4½	114.3	5100-450	4.223		.109		101.0	4.310		.120		.095	5.37	5.18	57800	30200
4.750	4¾	120.6	5100-475	4.458		.109		113.0	4.550		.120		.100	5.67	5.47	61000	33600
5.000	5	127.0	5100-500	4.692		.109		149.0	4.790		.120		.105	5.96	5.75	64200	37100
5.250	5¼	133.3	5100-525	4.927		.125		190.0	5.030		.139		.110	6.27	6.05	77300	40800
5.500	5½	139.7	5100-550	5.162	+.020	.125		202.5	5.265	±.007	.139	+.006	.117	6.57	6.34	81000	45500
5.750	5¾	146.0	5100-575	5.396	−.040	.125	±.004	220.0	5.505	.006	.139	−.000	.122	6.86	6.62	84700	49600
6.000	6	152.4	5100-600	5.631		.125		210.0	5.745	T.I.R.	.139		.127	7.16	6.91	88300	53800
6.250	6¼	158.7	5100-625	5.866		.156		282.0	5.985		.174		.132	7.46	7.20	114800	58300
6.500	6½	165.1	5100-650	6.100	+.020	.156		330.0	6.225		.174		.137	7.87	7.60	119400	62900
					−.050												
6.750	6¾	171.4	5100-675	6.335		.156		356.0	6.465		.174		.142	8.06	7.78	124000	67700
7.000	7	177.8	5100-700	6.570		.156		388.0	6.705		.174		.147	8.36	8.07	128600	72700
7.250	7¼	184.2	5100-725	6.775		.187	±.005	510	6.942		.209		.154	8.70	8.39	159700	78900
7.500	7½	190.5	5100-750	7.009		.187		534	7.180		.209		.160	8.96	8.64	165200	84800
7.750	7¾	196.9	5100-775	7.243		.187		545	7.420		.209		.165	9.20	8.87	170700	90450
8.000	8	203.2	5100-800	7.478		.187		640	7.660		.209		.170	9.60	9.26	176200	96100
8.250	8¼	209.6	5100-825	7.712		.187		565	7.900		.209		.175	9.85	9.50	181700	102100
8.500	8½	215.9	5100-850	7.947	+.050	.187		692	8.140	±.008	.209	+.008	.180	10.10	9.74	187200	108100
8.750	8¾	222.3	5100-875	8.181	−.130	.187		712	8.380	.006	.209	−.000	.185	10.40	10.00	192700	114450
9.000	9	228.6	5100-900	8.415		.187		737	8.620	T.I.R.	.209		.190	10.60	10.22	198200	120800
9.250	9¼	234.9	5100-925	8.650		.187		760	8.860		.209		.195	10.85	10.50	203700	126225
9.500	9½	241.3	5100-950	8.885		.187		785	9.100		.209		.200	11.10	10.70	209200	134200
9.750	9¾	247.6	5100-975	9.120		.187		845	9.338		.209		.205	11.35	10.95	214700	142000
10.000	10	254.0	5100-1000	9.355		.187		910	9.575		.209		.212	11.60	11.20	220200	149800

Figure 13-74 External series 5100 (*Courtesy Waldes Kohinoor, Inc.*)

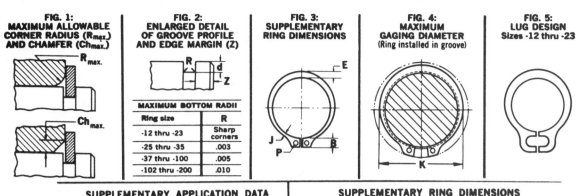

| | FIG. 1: MAXIMUM ALLOWABLE CORNER RADIUS (R_max.) AND CHAMFER (Ch_max.) | FIG. 2: ENLARGED DETAIL OF GROOVE PROFILE AND EDGE MARGIN (Z) | FIG. 3: SUPPLEMENTARY RING DIMENSIONS | FIG. 4: MAXIMUM GAGING DIAMETER (Ring installed in groove) | FIG. 5: LUG DESIGN Sizes -12 thru -23 |

MAXIMUM BOTTOM RADII

Ring size	R
-12 thru -23	Sharp corners
-25 thru -35	.003
-37 thru -100	.005
-102 thru -200	.010

	SUPPLEMENTARY APPLICATION DATA					SUPPLEMENTARY RING DIMENSIONS								
EXTERNAL SERIES 5100	Maximum allowable corner radii and chamfers of retained parts (Fig. 1)		Allow. assembly load with R max. or Ch_max.	Edge margin (Fig. 2)	Calculated RPM limits (Std. ring mat'l.) Apply req'd. safety factor	**(Fig. 3)**								**MAX. GAGING DIA.** Ring installed in groove (Fig. 4)
						LUG		**LARGE SECTION**		**SMALL SECTION**		**HOLE DIAMETER**		
size — no.	R_max.	Ch_max.	P'_r (lbs.)	Z		B	tol.	E	tol.	J	tol.	P	tol.	K
5100-343	.129	.077	7350	.270	5900	.308	±.005	.292	±.008	.148	±.008	.125		3.712
5100-350	.122	.073	10500	.276	5900	.328		.285		.148		.125		3.764
5100-354	.123	.074	10500	.279	5800	.328		.288		.149		.125		3.809
5100-362	.127	.076	10500	.285	5700	.328		.296		.153		.125		3.898
5100-368	.1295	.078	10500	.291	5600	.330		.302		.156		.125	+.015 -.002	3.966
5100-375	.133	.080	10500	.297	5500	.332		.310		.160		.125		4.037
5100-387	.137	.082	10500	.303	5100	.330		.318		.163		.125		4.169
5100-393	.137	.082	10500	.306	5200	.342		.318		.163		.125		4.230
5100-400	.135	.081	10500	.312	5000	.352		.318		.163		.125		4.288
5100-425	.146	.088	10500	.276	4800	.395		.318		.176		.125		4.558
5100-437	.146	.088	10500	.276	4700	.395		.318		.181		.125		4.683
5100-450	.102	.061	10500	.285	4500	.404		.285		.128		.125		4.730
5100-475	.115	.069	10500	.300	4200	.429		.303		.136		.125		4.996
5100-500	.165	.099	10500	.315	4000	.450	±.008	.360	±.010	.194	±.010	.156		5.346
5100-525	.169	.101	13500	.330	3900	.472		.372		.211		.156		5.605
5100-550	.175	.105	13500	.351	3700	.497		.390		.209		.156		5.867
5100-575	.184	.110	13500	.366	3500	.518		.408		.220		.156		6.134
5100-600	.143	.086	13500	.381	3400	.540		.381		.171		.156		6.302
5100-625	.148	.089	21000	.396	3100	.561		.396		.176		.156		6.568
5100-650	.191	.114	21000	.411	3000	.586		.438		.236		.156		6.905
5100-675	.200	.120	21000	.426	3000	.608		.456		.246		.187		7.172
5100-700	.208	.125	21000	.441	2900	.629		.474		.256		.187		7.439
5100-725	.214	.128	30000	.460	2800	.660		.490		.267		.187		7.700
5100-750	.220	.132	30000	.480	2700	.676		.507		.277		.187		7.963
5100-775	.227	.136	30000	.495	2600	.660		.523		.285		.187		8.228
5100-800	.235	.141	30000	.510	2500	.735	±.012	.540	±.015	.294	±.015	.187	+.020 -.005	8.493
5100-825	.242	.146	30000	.525	2400	.735		.556		.304		.187		8.758
5100-850	.250	.150	30000	.540	2300	.735		.573		.314		.187		9.023
5100-875	.258	.155	30000	.555	2200	.735		.591		.322		.187		9.280
5100-900	.267	.160	30000	.570	2200	.735		.609		.333		.187		9.557
5100-925	.274	.164	30000	.585	2100	.735		.625		.341		.187		9.830
5100-950	.281	.168	30000	.600	2100	.735		.642		.350		.187		10.086
5100-975	.287	.172	30000	.618	2000	.735		.658		.358		.187		10.340
5100-1000	.294	.176	30000	.636	2000	.735		.675		.367		.187		10.610

Figure 13-74 (continued)

Review

1. In addition to fastening parts together, what are two other uses for threads?

2. What number 5100 series retaining ring would be recommended for a shaft diameter of 203.2 mm?

3. List four factors to consider before selecting any type of retaining ring, other than load capacity.

4. Explain in full what ⅝ - 11 UNC - 2A X 2½ LG means.

5. What is the recommended hole size for a .312-inch (8) diameter spring pin?

6. If you have a ¼ - 20 UNC threaded screw and rotate it 10 full turns, how far will the end travel?

7. What series retaining ring would be recommended to provide rigid end play take-up?

8. How is a screw called-out that is 3 inches long, ⅜-inch in diameter, coarse, right-hand threads, average fit, and round-head style?

9. What is the free diameter of an external 5100-775 retaining ring? What is the tolerance of the dimension?

10. A 2-inch (50.8) diameter shaft uses what size square key?

11. What is the pliers hole diameter and tolerance for an internal N5000-131 retaining ring?

12. How is the depth of thread figured?

The following problems are intended to give the beginning drafter practice using various size charts of fastener and, by using these charts, practice in laying out and drawing the many fasteners used.

The steps to follow in laying out fasteners are:

Step 1. Study all required specifications.

Step 2. Using the appropriate size chart, for size dimension, *lightly* lay out fastener in place.

Step 3. Check each dimension.

Step 4. Darken in the fastener, starting with diameters and arcs first to correct line thickness. Use simplified thread representation.

Step 5. Neatly add all required call-off specifications, using the latest drafting standard.

Problem 13-1

On an A-size, 8½ x 11 sheet of paper, lay out the center lines per given dimensions.

Calculate standard thread lengths.

Draw the following fasteners on the center lines as illustrated.

A. Flat-head cap screw — ⅜ — 16 UNC — 2A x 2.0 lg.
B. Round-head cap screw — ½ — 13 UNC — 2A x 2.25 lg.
C. Fillister-head cap screw — ⅝ — 11 UNC — 2A x 2.5 lg.
D. Oval-head cap screw — ⅜ — 24 UNF — 2A x 1.75 lg.
E. Hex-head cap screw with hex nut — 1 — 8 UNC — 2A x 2.5 lg.
F. Socket-head cap screw — ½ — 20 UNF — 2A x 2.0 lg.
G. Cotter pin — ⅜ dia. x 3.5 lg.

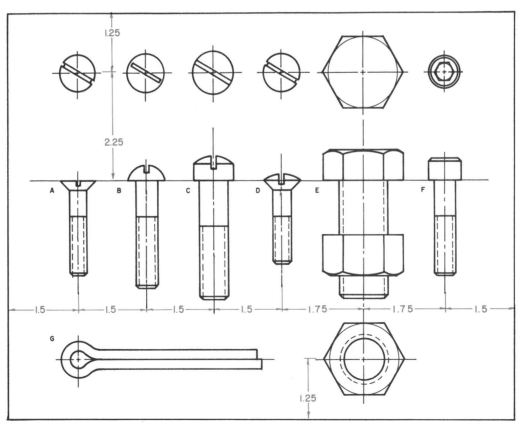

Problem 13-1

Problem 13-2

On an A-size, 8½ x 11 sheet of paper, lay out the center lines per given dimensions.

Calculate standard thread lengths.

Draw the following fasteners on the center lines as illustrated.

A. Hex-head bolt – 1⅜ – UNC x 3.0 lg.
B. Lock washer (For a ⅞ dia. cap screw).
C. Square nut – 1 – 12 UNF 2B.
D. Square-head cap screw – ¾ – 10 UNC x 3.0 lg.

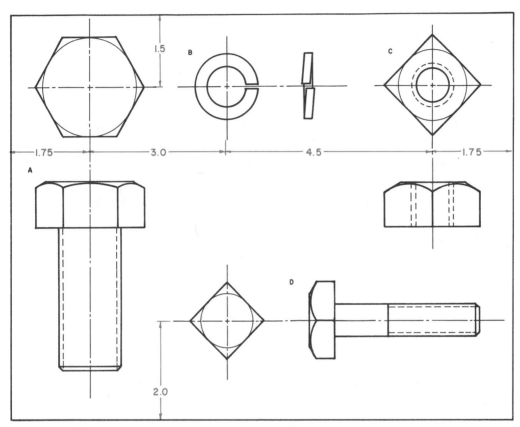

Problem 13-2

Problem 13-3

On an A-size, 8½ x 11 sheet of paper, lay out the center lines per given dimensions.

A. Calculate the required size of square key needed for a 2.25 ID collar. Draw collar and keyway, and dimension per drafting standards.
B. Calculate the required size of square key needed for the matching shaft of the collar. Draw the end view of the shaft and dimension per drafting standards.

C. Calculate the *minimum* length hex-head cap screw, ¾ —10 UNC — 2A required to safely secure part X to part Y (material: aluminum). Illustrate and call-off the required pilot drill size and recommended depth into part Y. Illustrate and call-off the required size and depth of *full* threads for the hex-head cap screw.

Illustrate and call-off the diameter for clearance hole in part X.

D. Draw a square-head set screw with a full dog point, ⅝ — 18 UNF x 3.0 lg.

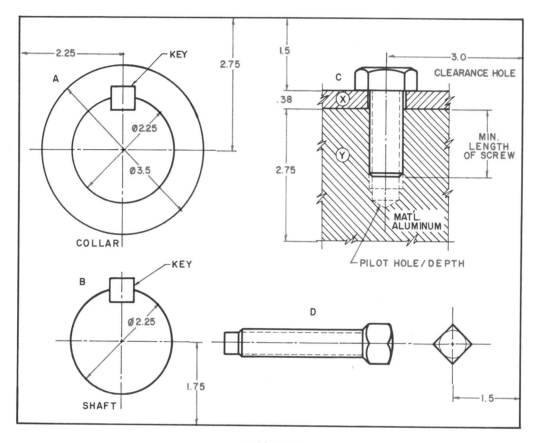

Problem 13-3

Problem 13-4

On an A-size, 8½ × 11 sheet of paper, lay out the center lines, part X, and part Y per given dimensions.

Calculate the standard thread lengths.

Draw the following fasteners on the center lines as illustrated.

Give *all* specific hole call-off information for each fastener in the space above each as illustrated (include clearance hole sizes, counterbore specifications, countersink specifications, spotface specifications, required tap drill size and depth for 75% thread; and tap size and required depth per latest drafting standards.

A. Socket-head cap screw — 7/3 — 9 UNC × 1.75 lg (calculate c'bore information).
B. Oval-head cap screw — ¾ — 10 UNC × 2.5 lg (calculate c'sink information).
C. Hex-head cap screw 1 — 12 UNF × 2.5 lg (calculate S.F. information).
D. Fillister-head cap screw with plain washer and hex-head nut — ¾ — 16 UNF × length to suit. (Calculate clearance hole required.)

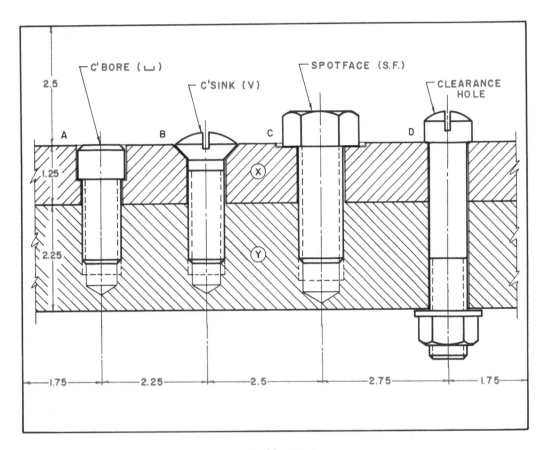

Problem 13-4

Problem 13-5

Given, an illustration of an old design (Figure 13-5A) that uses a washer and cotter pin to hold the .3543 dia. shaft in place. Using this design, it was necessary to locate and drill a hole for the cotter pin. The new design incorporated the use of one retaining ring to achieve the exact same function at a much lower cost. Choose the correct type and size retaining ring and fill in the required dimensions at A, B and C (see Figure 13-6A). Neatly letter all retaining ring data below the dimensioned shaft as illustrated.

Problem 13-6

Given, an illustration of an old design (Figure 13-5B) that uses a cover plate with 4 holes for 4 washers and 4 round-head machine screws to hold the shaft in place. Using this design required drilling 4 blind tap-drill holes which had to be tapped for the 4 round-head screws. The new design incorporated the use of one retaining ring to achieve the exact same function at a much lower cost. Choose the correct type and size retaining ring and fill in the required dimensions at A, B and C (see Figure 13-6B). Neatly print all retaining ring data below the dimensioned collar as illustrated.

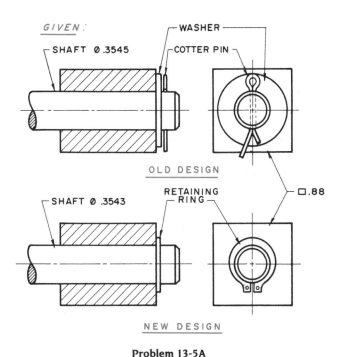

Problem 13-5A

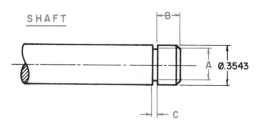

DATA:
RETAINING RING NUMBER _____
FREE DIAMETER _____
THICKNESS _____
LUG SIZE _____
LARGE SECTION _____
SMALL SECTION _____
PLIER HOLE SIZE _____
MAX. GAGING DIA. WHEN INSTALLED

_____

Problem 13-6A

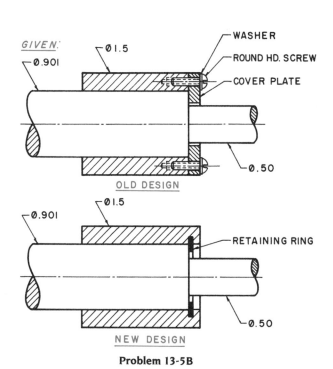

Problem 13-5B

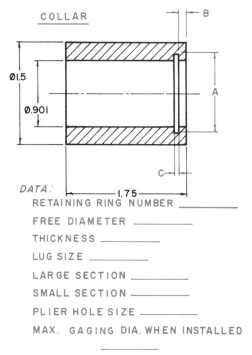

DATA:
RETAINING RING NUMBER _____
FREE DIAMETER _____
THICKNESS _____
LUG SIZE _____
LARGE SECTION _____
SMALL SECTION _____
PLIER HOLE SIZE _____
MAX. GAGING DIA. WHEN INSTALLED

_____

Problem 13-6B

Chapter Thirteen Industry Print

1. Threads are to be applied to how many parts of the shaft?
2. What are the thread specifications for the 1.00 part of the shaft?
3. What are the thread specifications for the .75 part of the shaft?
4. What are the thread specifications for the 1.07 part of the shaft?
5. What are the drill specifications for the centerdrilled holes?
6. What are the countersink specifications for the center-drilled holes?
7. An Acme thread is called-out for the 1.07 part of the shaft. What is the nominal outside diameter of the Acme thread?
8. What is the pitch of the Acme thread?
9. What is the lead of the Acme thread?
10. Is the Acme thread a right- or left-hand thread?

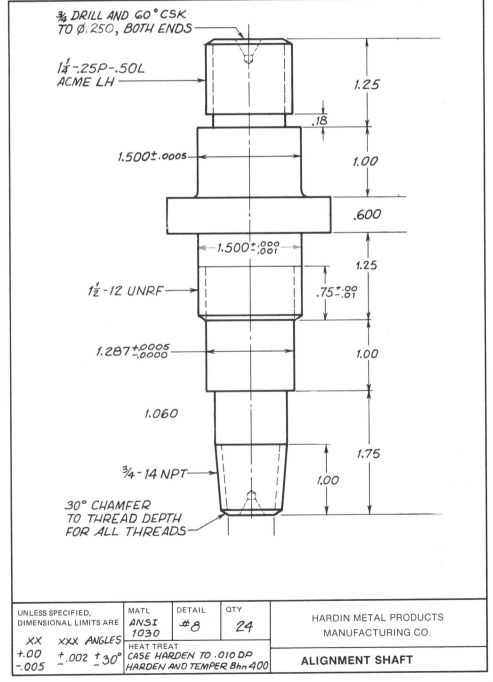

¾ DRILL AND 60° CSK TO Ø.250, BOTH ENDS

1¼-.25P-.50L ACME LH

1.25

.18

1.500±.0005

1.00

.600

1.500 +.000 -.001

1.25

1½-12 UNRF

.75 +.00 -.01

1.287 +.0005 -.0000

1.00

1.060

1.75

¾-14 NPT

1.00

30° CHAMFER TO THREAD DEPTH FOR ALL THREADS

UNLESS SPECIFIED, DIMENSIONAL LIMITS ARE	MATL ANSI 1030	DETAIL #8	QTY 24	HARDIN METAL PRODUCTS MANUFACTURING CO.
XX +.00 -.005	XXX ANGLES +.002 ±30° -.002 -30°	HEAT TREAT CASE HARDEN TO .010 DP HARDEN AND TEMPER Bhn 400		ALIGNMENT SHAFT

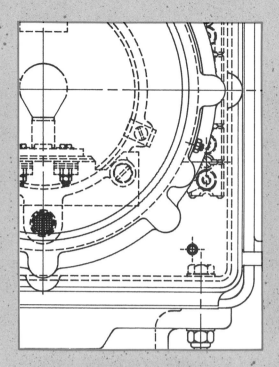

The most commonly used kinds of springs and the terminology associated with them are covered in this chapter. Extensive study of compression, extension, torsion and flat springs is fully covered, along with step-by-step study as to how to design and lay out a spring. Accuracy, line work, neatness, speed, correct dimensioning and required spring data should be stressed in all drawing problems at the end of this chapter.

CHAPTER FOURTEEN

Springs

A *spring* is a mechanical device that is used to store and apply mechanical energy. A spring can be designed and manufactured to apply a pushing action, a pulling action, a torque or twisting action, or a simple power action.

Product manufacturers usually use standard-size springs that are purchased from companies that specialize in making springs. These springs are usually mass produced and are, therefore, relatively inexpensive. Occasionally, however, a special spring must be designed to perform a special function. To be able to design special-function springs, the drafter or designer must know the many terms associated with springs, and how to design and construct special springs.

Spring Classification

A spring is generally classified as either a helical spring or a flat spring. *Helical springs* are usually cylindrical or conical; the *flat springs*, as their name implies, are usually flat.

Helical Springs

Three types of helical springs are compression springs, extension springs, and torsion springs, Figures 14-1A, 14-1B, and 14-1C.

Compression Spring

A *compression spring* offers resistance to a compressive force or applies a pushing action. In its free state, the coils of the compression spring do not touch. The types of ends of a compression spring are illustrated and described as follows.

Plain open end — Figure 14-2A. The plain open end spring is very unstable, has only point contact, and the ends are *not* perpendicular to the axis of the spring.

Plain closed end — Figure 14-2B. The plain closed end spring is a little more stable, and provides a round parallel surface contact perpendicular to the axis of the spring.

Ground open end — Figure 14-2C. The ground open end spring is much more stable, and provides a flat surface contact perpendicular to the axis of the springs.

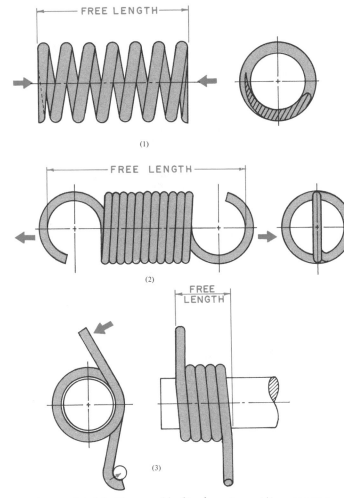

Figure 14-1A Types of helical springs: (1) compression spring, (2) extension spring, and (3) torsion spring

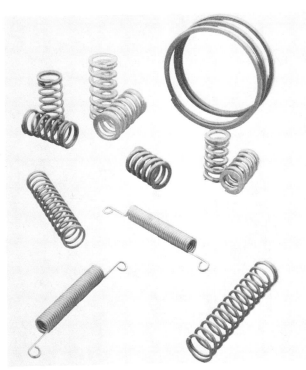

Figure 14-1B Varieties of compression and extension springs (*Courtesy* AMETEK, *Inc.*, *Hunter Spring Division*)

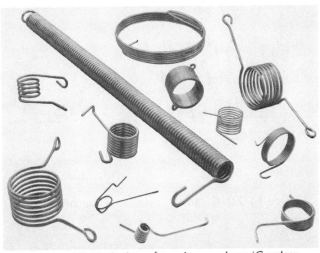

Figure 14-1C Varieties of torsion springs (*Courtesy* AMETEK, *Inc.*, *Hunter Spring Division*)

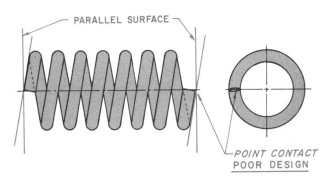

Figure 14-2A Plain open end compression spring

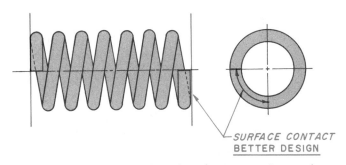

Figure 14-2B Plain closed end compression spring

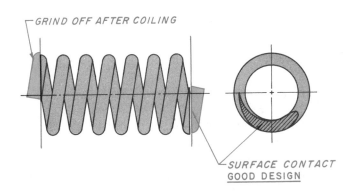

Figure 14-2C Ground open end compression spring

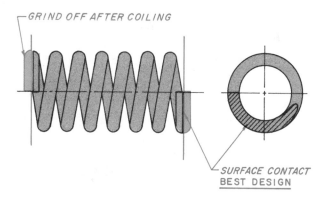

Figure 14-2D Ground closed end compression spring

Ground closed end — Figure 14-2D. The ground closed end spring is the most stable, and the best design. This design provides a large flat surface contact perpendicular to the axis of the spring.

A good example of a compression spring is the kind that is used in the front end of an automobile.

Extension Spring

An *extension spring* offers resistance to a pulling force. In its free state, the coils of the extension spring are usually either touching or very close. The ends of an extension spring are usually a hook or loop, Figure 14-3. Illustrated are a few of the many kinds of ends that are used. The loop or hook can be over the center or at the side. Each end is designed for a specific application or assembly. Regardless of the shape of the loop or hook, the overall length of an extension spring is measured from the *inside* of the loops or hook. A good example of an extension spring is the kind that is used to counterbalance a garage door.

Torsion Spring

A *torsion spring* offers resistance to a torque or twisting action. In its free state, the coils of a torsion spring are usually either touching or very close. The ends of a torsion spring are usually specially designed to fit a particular mechanical device, Figure 14-4. A good example of a torsion spring is the kind that is used to return a doorknob back to its original position.

Flat Springs

The other classification or type of spring is the flat spring. The *flat spring* is made of spring steel and is designed to perform a special function. Flat springs are not standard, and must be designed and manufactured to fit each particular need. The flat spring is considered a power spring, and it resists a pressure. A good example of a flat spring is the kind that is used for a door latch on a cabinet, Figure 14-5, or a

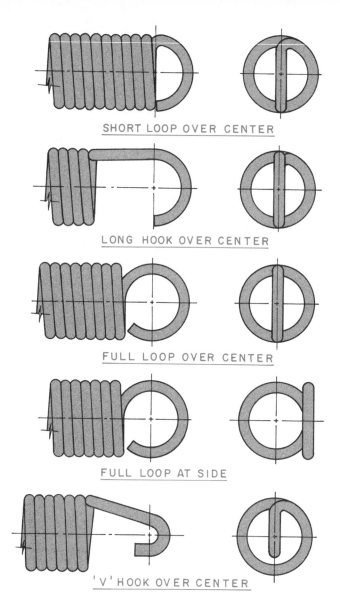

SHORT LOOP OVER CENTER

LONG HOOK OVER CENTER

FULL LOOP OVER CENTER

FULL LOOP AT SIDE

'V' HOOK OVER CENTER

Figure 14-3 Ends of extension spring

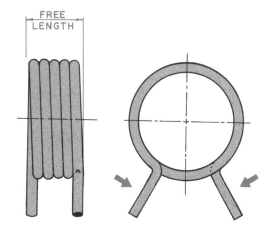

Figure 14-4 Torsion spring

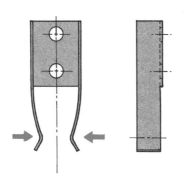

Figure 14-5 Flat spring

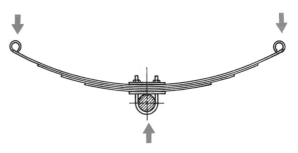

Figure 14-6 Leaf spring

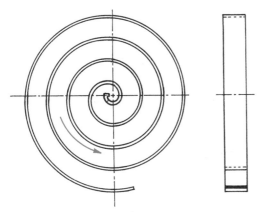

Figure 14-7 Special spring

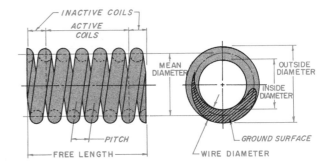

Figure 14-8 Spring terminology

leaf spring as used in a truck, Figure 14-6. A flat spiral spring used to power a windup clock or toy is another example, Figure 14-7.

Terminology of Springs

It is important to know and fully understand the various terms associated with springs and their design, Figure 14-8.

Free Length

Free length is the overall length of the spring when it is in its free state of unloaded condition. Free length of a compression spring is measured from the extreme ends of the spring. In an extension spring, the free length is measured inside the loops or hooks.

Solid Length

Solid length of a compression spring is the overall length of the spring when *all* coils are compressed together so they touch. This can be mathematically calculated. If the wire diameter is .500, the overall solid length is 3.25.

Outside Diameter

Outside diameter (O.D.) is the outside diameter of the spring (2 x wire size *plus* inside diameter).

Wire Size

Wire size is the diameter of the wire used to make up the spring.

Inside Diameter

Inside diameter (I.D.) is the inside diameter of the spring (2 x wire size *minus* outside diameter).

Active Coils

Active coils in a compression spring are usually the total coils minus the two end coils. In a compression spring, the coil at each end is considered nonfunctional. In an extension spring, active coils include *all* coils.

Loaded Length

Loaded length is the overall length of the spring with a special given or designed load applied to it.

Mean Diameter

Mean diameter is the theoretical diameter of the spring measured to the center of the wire diameter. This theoretical diameter is used by the drafter to lay out and draw a spring. To mathematically calculate the mean diameter, subtract the wire diameter from the outside diameter.

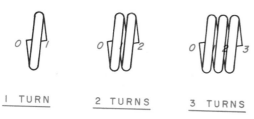

Figure 14-9 Coils or turns

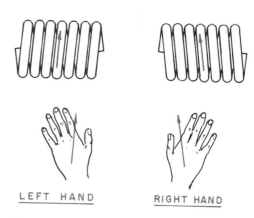

Figure 14-10 Direction of winding

Coil

Coil or *turn* is one full turn or 360° of the wire about the center axis, Figure 14-9.

Direction of Winding

Direction of winding is the direction in which the spring is wound. It can be wound right hand or left hand, Figure 14-10. The thumbs of your hands point toward the direction in which the coil windings are leaning. If the coil windings slant to the right, as the example to the left illustrates, it agrees with the direction of your left-hand thumb; thus, a left-hand winding. If the coil windings slant to the left as the example to the right illustrates, it agrees with the direction of your right-hand thumb; thus, a right-hand winding. If the coil winding direction is not called-off on the drawing, it will be manufactured with right-hand winding.

Required Spring Data

Each drawing must include complete dimensions and specific data. Besides the regular dimensions, i.e., outside diameter (O.D.) and/or inside diameter (I.D.) wire size; free length; and solid length (compression spring), the following data should be called-off at the lower right side of the work area:

- Material
- Number of coils (including active and inactive coils)
- Direction of winding (if not noted, it is assumed to be right hand)
- Torque data (if torsion spring)
- Finish
- Heat treatment specification
- Any other required data

How To Draw a Compression Spring

Given: Plain open ends, 1.50 outside diameter, .25 diameter wire size, 4.00 free length, oil tempered spring steel wire, 8 total coils (6 active coils), left-hand winding, heat treatment: heat to relieve coiling stresses, finish: black

paint. Make a rough sketch of the spring using the given specifications, Figure 14-11A. A rough sketch is extremely important in developing a new spring design.

Step 1. Measure and construct two vertical lines the required overall length (front view). Divide the space between the two vertical lines into 17 equal spaces (front view) 2 x total coils plus 1; in this example, $2 \times 8 + 1 = 17$, Figure 14-11B. Measure and construct the mean diameter. To calculate the mean diameter, subtract the wire diameter from the outside diameter (end view).

Step 2. Determine the direction of windings; refer to sketch and lightly construct the wire diameter accordingly, (front view). The wire diameters are constructed on the mean diameter. Lightly draw the outside diameter and inside diameter (end view), Figure 14-11C.

Step 3. Lightly construct the near side of the spring. Add short, light tangent points, Figure 14-11D.

Step 4. Lightly construct the rear side of the spring. Add ends of spring; in this example, plain open ends. Check all work against given requirements, Figure 14-11E.

Step 5. Recheck all work. Darken in spring using the latest drafting standards. Add end of spring, right end, dimensions and add all required spring data, Figure 14-11F.
Notice: outside diameter, wire size, free length, and solid length are given as regular dimensions. The material, number of coils, direction of winding, heat treatment specifications and finish requirements are listed below and to the right as a note. In this example, the outside diameter is the controlling dimension; that is, the outside dimension is more important than the inside dimension. Both inside and outside dimensions should not be added in the same drawing.

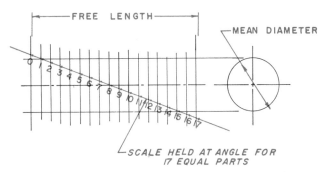

Figure 14-11A How to draw a compression spring with plain open ends

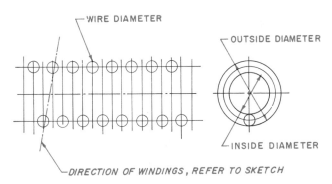

Figure 14-11B Step 1

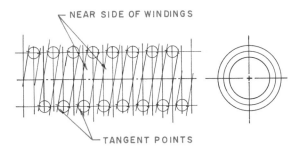

Figure 14-11C Step 2

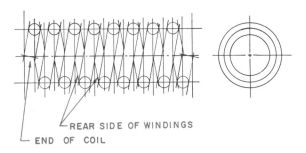

Figure 14-11D Step 3

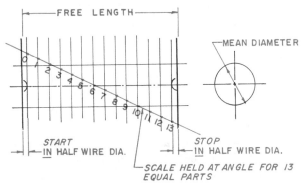

Figure 14-11E Step 4

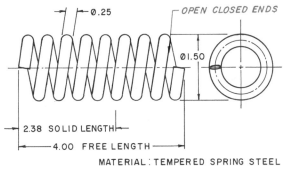

O.25

OPEN CLOSED ENDS

Ø1.50

2.38 SOLID LENGTH

4.00 FREE LENGTH

MATERIAL: TEMPERED SPRING STEEL
8 TOTAL COILS (6 ACTIVE)
LEFT-HAND COILS
HEAT TO RELIEVE COILING STRESSES
FINISH: BLACK

Figure 14-11F Step 5

How To Draw a Compression Spring (Plain Closed Ends)

Given: Plain closed ends, 20 mm inside diameter, 5 mm diameter wire size, 100 mm free length, material: hard drawn steel spring wire, 5 active coils (total 7 coils), left-hand winding, heat treatment: heat to relieve coiling stresses, finish: zinc plate. Make a rough sketch of the spring using the given specifications, Figure 14-12A.

Step 1. Measure and construct two vertical lines the required overall length (front view). Draw a line *inside* the two vertical lines equal to half the wire diameter from each end. Divide the space between the two vertical lines into 13 equal spaces (front view) 2 x total coils minus 1; in this example, $2 \times 7 - 1 = 13$, Figure 14-12B. Measure and construct the mean diam-

Figure 14-12A How to draw a compression spring with plain closed ends

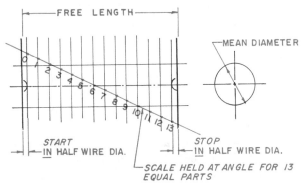

Figure 14-12B Step 1

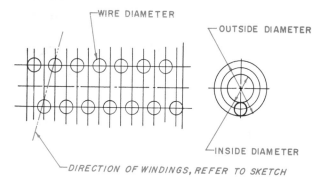

Figure 14-12C Step 2

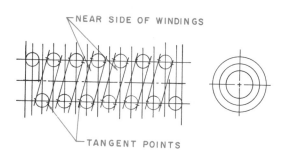

Figure 14-12D Step 3

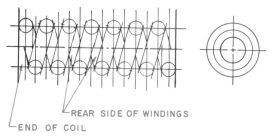

Figure 14-12E Step 4

eter. To calculate the mean diameter, add the wire diameter to the inside diameter (end view).

Step 2. Determine the direction of winding; refer to the sketch and lightly construct the wire diameter accordingly (front view). Lightly draw the outside diameter and inside diameter (end view), Figure 14-12C.

Step 3. Lightly construct the near side of the spring. Add short tangent points, Figure 14-12D.

Step 4. Lightly construct the rear side of the spring. Add ends of springs. In this example, plan closed ends. Check all work against given requirements, Figure 14-12E.

Step 5. Recheck all work. Darken in spring using the latest drafting standards. Add all required dimensions and spring data, Figure 14-12F.

How To Draw a Compression Spring (Ground Closed Ends)

(The following illustrations are for a ground closed end compression spring. The exact same steps can be used for a ground open end compression spring.)

Given: Ground closed ends, 1.125 outside diameter, .188 diameter wire size, 3.00 free length, material: hard drawn steel spring wire, 7 total coils (5 active coils), right-hand winding, heat treatment: heat to relieve coiling stresses, finish: zinc plate. Make a rough sketch of the spring using the given specifications, Figure 14-13A.

Step 1. Measure and construct two vertical lines the required free length (front view). Divide the space between the two vertical lines into 13 equal spaces (front view), 2 × total coils minus 1; in this example, 2 × 7 − 1 = 13, Figure 14-13B. Measure and construct the mean diameter. To calculate the mean diameter, subtract the wire diameter from the outside diameter (end view).

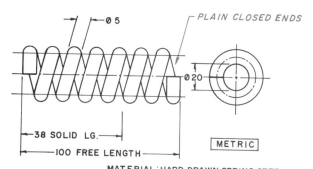

PLAIN CLOSED ENDS

Ø 5

Ø 20

38 SOLID LG.

100 FREE LENGTH

METRIC

MATERIAL: HARD DRAWN SPRING STEEL WIRE
5 ACTIVE COILS (7 TOTAL COILS)
LEFT-HAND COILS
HEAT TO RELIEVE COILING STRESSES
FINISH: ZINC PLATE

Figure 14-12F Step 5

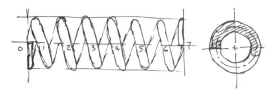

Figure 14-13A How to draw a compression spring with ground closed ends

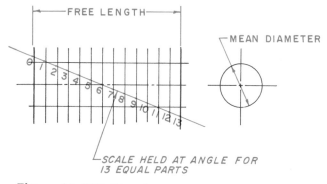

FREE LENGTH

MEAN DIAMETER

SCALE HELD AT ANGLE FOR 13 EQUAL PARTS

Figure 14-13B Step I

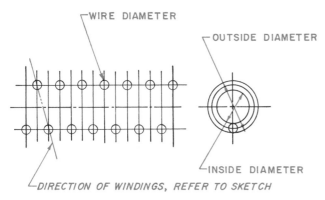

Figure 14-13C Step 2

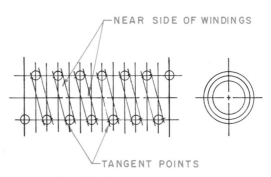

Figure 14-13D Step 3

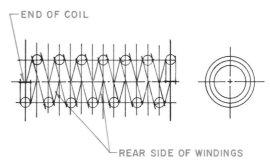

Figure 14-13E Step 4

Step 2. Determine the direction of winding; refer to the sketch and lightly construct the wire diameter accordingly (front view). Lightly draw the outside diameter and inside diameter (end view), Figure 14-13C.

Step 3. Lightly construct the rear side of the spring. Add short tangent points, Figure 14-13D.

Step 4. Lightly construct the rear side of the spring. Add ends of spring; in this example, ground closed end. Check all work against given requirements, Figure 14-13E.

Step 5. Recheck all work. Darken in spring using the latest drafting standards. Add all required dimensions and spring data, Figure 14-13F.

How To Draw an Extension Spring

Given: Full loop, over center each end 1.625 outside diameter, .188 diameter wire size, 4.750 approximate free length, material: hard drawn spring steel wire, heat treatment: heat to relieve coiling stresses, finish: black paint, windings as tight as possible. Make a rough sketch of the spring using the given specifications, Figure 14-14A.

Step 1. Draw the outside diameter, inside diameter and mean diameter (end view). Lightly draw the required free length. In the design of an extension spring, it is almost impossible to arrive at the *exact* free length. In designing an extension spring, it is best to design it slightly *shorter* than actually required. Construct the end loops with the *inside* diameter within the specified free length, Figure 14-14B. Note in this example the loops are the exact diameter as the spring itself.

Step 2. Starting at the intersection of the mean diameter of the right end loop, and the mean diameter of the spring, lower side, locate and

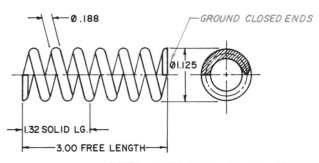

MATERIAL: HARD DRAWN SPRING STEEL WIRE
7 TOTAL COILS (5 ACTIVE COILS)
HEAT TO RELIEVE COILING STRESSES
FINISH: ZINC PLATE

Figure 14-13F Step 5

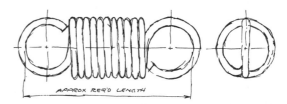

Figure 14-14A How to draw an extension spring with full loop ends

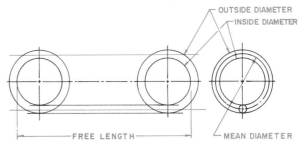

Figure 14-14B Step 1

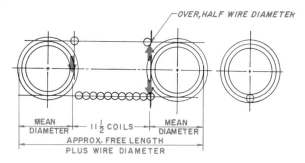

Figure 14-14C Step 2

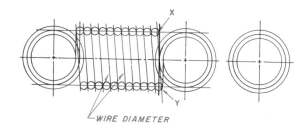

Figure 14-14D Step 3

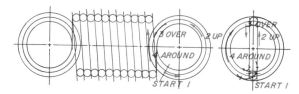

Figure 14-14E Step 4

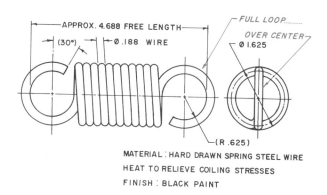

Figure 14-14F Step 5

draw the first lower wire diameter, Figure 14-14C. Project directly up to the mean diameter of the spring, upper side, and draw the first upper wire diameter *over* to the left, half a wire diameter as illustrated. Lay out, side-by-side, wire diameters touching, as illustrated; in this example, 10½ coils. Notice the coils must end on the mean diameter, left end. This may mean adjusting slightly the location of the left end loop. It is best to bring it *in* slightly, if necessary.

Step 3. Locate points X and Y, as illustrated in Figure 14-14D, and adjust the drafting machine at this angle and lock on this angle. Draw the wire loops.

Step 4. Lightly draw the loop in the right-side view. To calculate just what the loop will actually look like, number various points along the loop in the front view; in this example, point 1, starting point; point 2, up; point 3, over; and point 4, around. Project these points to the end view, as illustrated in Figure 14-14E.

Step 5. Recheck all work. Darken in spring using the latest drafting standards. Add all required dimensions and spring data, Figure 14-14F.

Other Spring Design Layout

When drawing any specially designed spring for a particular function, it is recommended to make a rough sketch of it, including all required specifications. A torsion spring is developed in a manner very similar to an extension spring with its tight coils. Usually, the required torque pressure is included in the given spring data. The actual designed deflection of the torsion spring is illustrated by phantom lines, and the angle is noted.

Standard Drafting Practices

Some company standards specify that the drafter not draw the complete spring because of the time and cost involved. A shortcut method to draw a spring

is shown in Figure 14-15. At top is the conventional method of representing a compression spring; below it is the same spring drawn by the schematic drawing system.

Another method of drawing a spring, especially a long spring, is illustrated in Figure 14-16. The incompleted coils are illustrated by phantom lines. If the company does not have a standard method of drawing springs it is the drafter's decision as to which method of representation is used. In most cases, it is best to draw the object, in this case the spring, in such a way that there is no question whatsoever as to exactly what is required. Of course, regardless of which system is used, full dimensions and specification data must be included.

Section View of a Spring

If a cutting plane line passes through the axis of a spring, the spring is drawn in one of two ways. A small spring is drawn with the back coils showing, as illustrated in Figure 14-17, Part A, with the section lining of the coils filled in solid. A larger spring is drawn

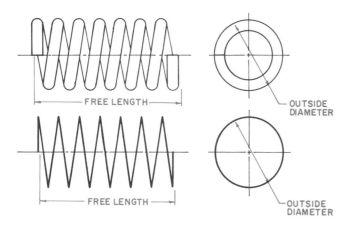

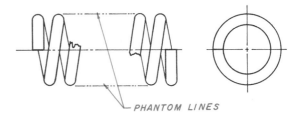

Figure 14-16 Incompleted coils with phantom lines

Figure 14-15 Schematic system of representing a compression spring

the same way, but uses standard section lining, Figure 14-17, Part B. Notice that both illustrations are right-hand springs, but, because the front half has been removed, only the rear half is seen which appears as left hand.

Isometric Views

A template, such as the one shown in Figure 14-18, makes it quick and simple to draw isometric views of light, medium, or heavyweight open springs. Lightweight springs are slightly smaller than the dimension printed on the template because they are drawn with the two smallest cutouts in any set. The thick lower edge and pinpoint centers ensure precise positioning to repeat each loop for any length spring.

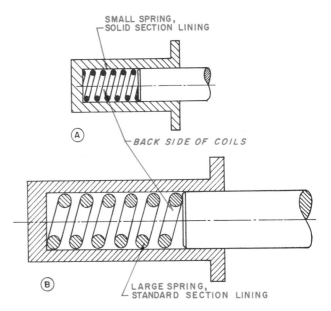

Figure 14-17 Section view of a spring

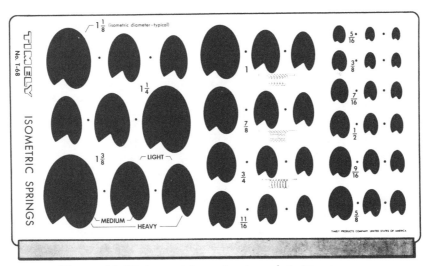

Figure 14-18 Isometric spring template (*Courtesy Timely Products Co.*)

Review

1. List two reasons why the schematic method of illustrating coils is sometimes used in industry.

2. List four major kinds of ends used on a compression spring. Which is the most stable?

3. Why is it incorrect to dimension both the inside diameter and the outside diameter of a spring on the same drawing? Which dimension should be given?

4. Other than the required dimensions, outside and/or inside diameter, wire size, free length, solid length (compression spring), what spring data must be noted below and to the right?

5. List two general classifications of springs.

6. What is the solid length of a compression spring with 8 active coils, a wire size of .375 diameter, mean diameter of 2.00, closed ground ends?

7. List three types of helical springs. Make a rough sketch of each illustrating how pressure is applied to each.

8. How is the section lining illustrated in a small spring where the cutting plane line passes through the center axis of the spring?

9. What is meant by the free length of a spring?

10. List three common uses for a torsion spring.

Chapter Fourteen Problems

The following problems are intended to give the beginning drafter practice in drawing various kinds of springs with special required specifications.

The steps to follow in laying out springs are:

Step 1. Study all required specifications.

Step 2. Make a rough sketch of the spring, incorporating *all* required specifications and required dimensions.

Step 3. Calculate the size of the basic shape of the front and end views.

Step 4. Center the two views within the work area with correct spacing for dimensions between views.

Step 5. Starting with the end view, or the view that contains the diameter, lightly draw the spring adhering to *all* specific required specifications. Take care to show correct direction of winding, correct type of ends, and total count of windings.

Step 6. Check to see that both views are centered within the work area.

Step 7. Check all dimensions in both views.

Step 8. Darken in both views, starting first with all diameters and arcs.

Step 9. Neatly add all dimensioning as required to fully describe the object using the latest drafting standards.

Step 10. Neatly add all required specifications under the two views. The drawing must include the following:
  Free length
  Wire size
  Total number of coils
  Type of ends
  Solid length
  Outside diameter
  Inside diameter
  Direction of windings
  Material (usually, hard drawn steel, spring wire)
  Required heat treating process (usually, heat to relieve coiling stresses)
  Any other special requirements

Step 11. For torsion-type springs, illustrate *working location* and travel with phantom lines and dimension.

Problems 14-1 through 14-7

Construct a 2-view drawing of each spring using the listed steps. Center the views on an A-size, 8½ × 11 sheet of paper with a 1″ (25 mm) space between views. Use correct line thicknesses and all drafting standards. Add all required dimensions and specifications per the latest drafting standards.

Compression-type spring
Free length 7.00
Wire size .25
12 active coils
 (14 *total* coils)
Plain open ends
O.D. 2.00/I.D. 1.50
R.H. winding
Calculate solid length

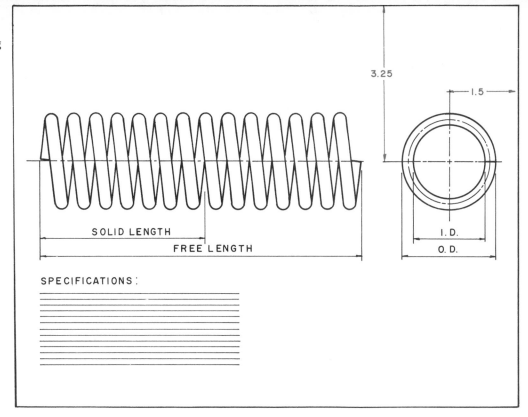

SOLID LENGTH

FREE LENGTH

I. D.

O. D.

3.25

1.5

SPECIFICATIONS:

Problem 14-1

Compression-type spring
Free length 5.25
Wire size .375
4 active coils (6 *total* coils)
Ground open ends
O.D. 2.75/I.D. 2.00
L.H. winding
Calculate solid length

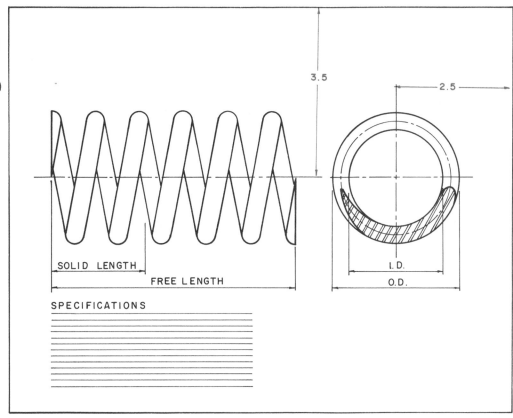

SOLID LENGTH

FREE LENGTH

I. D.

O.D.

3.5

2.5

SPECIFICATIONS

Problem 14-2

Compression-type spring
Free length 6.25
Wire size .375
9 total coils (7 *active* coils)
Plain closed ends
O.D. 2.00/I.D. 1.25
L.H. winding
Calculate solid length

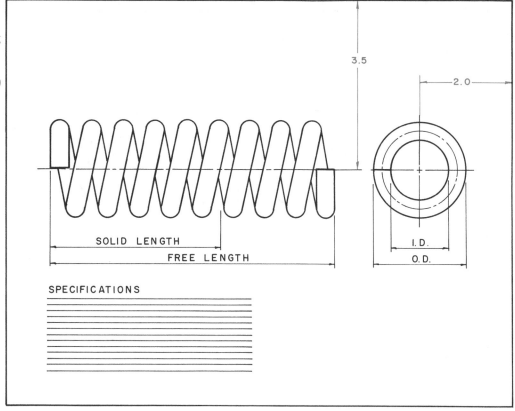

SOLID LENGTH

FREE LENGTH

I.D.

O.D.

3.5

2.0

SPECIFICATIONS

Problem 14-3

(Metric)
Compression-type spring
Free length 90
Wire size 12
4 total coils
 (2 active coils)
Ground closed ends
O.D. 96/I.D. 72
R.H. winding
Calculate solid length

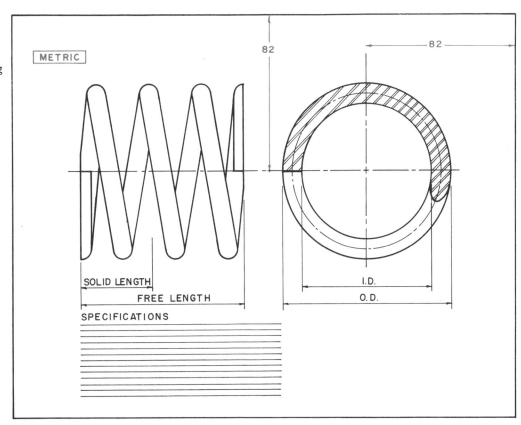

METRIC

82

82

SOLID LENGTH

FREE LENGTH

I.D.

O.D.

SPECIFICATIONS

Problem 14-4

Extension-type spring
Full loop over center,
  both ends
Approx. free length 7.0
Wire size .25
O.D. 2.0/I.D. 1.50
R.H. winding (This is
  standard for all
  extension springs)

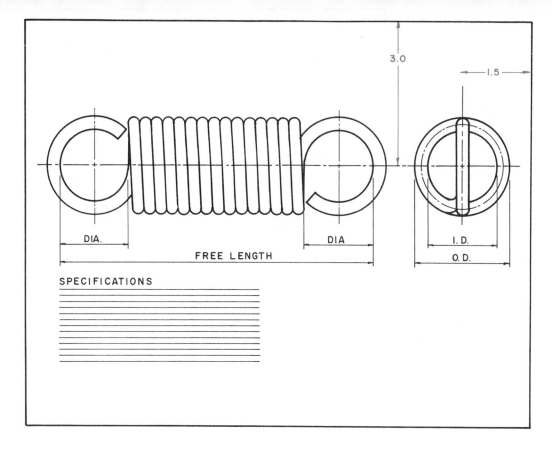

Problem 14-5

(Metric)
Extension-type spring
Full loop over center
  (right end)
Long hook over center
  (left end)
Free length 150
Wire size 5
12 total coils
65 coil length
O.D. 42/I.D. 32
Standard R.H. winding

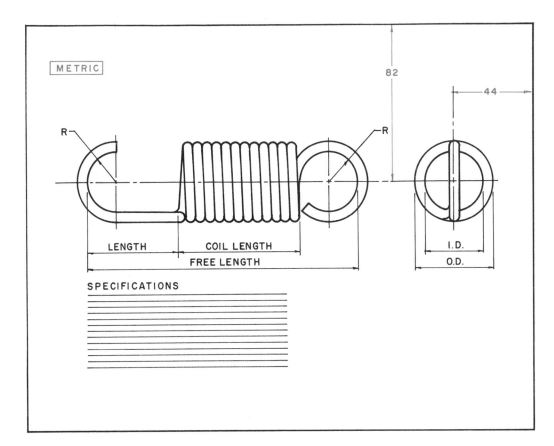

Problem 14-6

Torsion-type spring
Free length 3.625
Wire size .25
13 total coils
O.D. 2.0/I.D. 1.50
R.H. standard winding
90° maximum working
  flex (counterclockwise)
Arm lengths 2.0 (as
  shown)

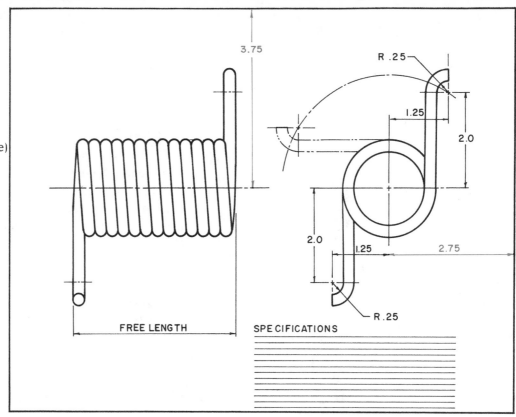

3.75

R .25

1.25

2.0

2.0

1.25

2.75

R .25

FREE LENGTH

SPECIFICATIONS

Problem 14-7

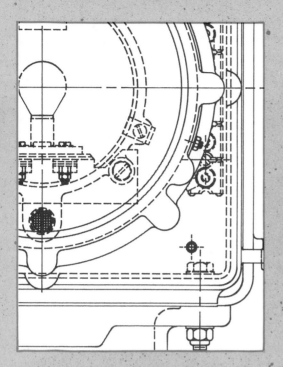

Various kinds of standard cam motions are studied in this chapter, including uniform velocity, modified uniform velocity, harmonic motion, and uniform acceleration. The terminology associated with cams, and how to dimension them are also covered. Cam designing with the aid of a displacement diagram is discussed. This material will provide the beginning designer with the required information to achieve any required cam motion.

CHAPTER FIFTEEN

Cams

Cam Principle

A *cam* produces a simple means to obtain irregular or specified predictable designed motion. These motions would be very difficult to obtain in any other way.

Figure 15-1 illustrates the basic principle and terms of a cam. In this example, a rotating shaft has an irregularly shaped disc attached to it. This disc is the cam. The follower, with a small roller attached to it, pushes against the cam. As the shaft is rotated the roller follows the irregular surface of the cam, rising or falling according to the profile of the cam. The roller is held tightly against the cam either by gravity or a spring.

Figure 15-2 illustrates a simple cam in action. Notice how the rotation of the shaft converts into an up and down motion of the follower. A cam using a flat-faced follower is shown in Figure 15-3. This type of cam, for example, is used to raise and lower the valve in an automobile engine. A modified cam follower is used to change the rotary motion of shaft A

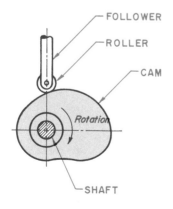

Figure 15-1 Basic cam principle and terms

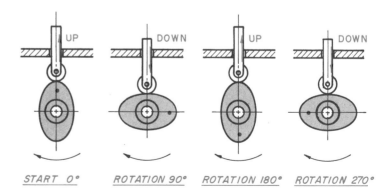

Figure 15-2 Simple cam in action

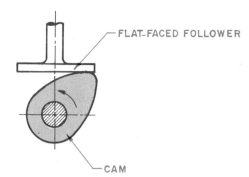

Figure 15-3 Flat-faced follower

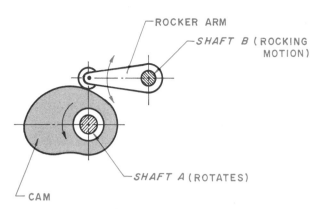

Figure 15-4 Rocker arm follower with wheel

into an up and down of the rocker arm, and then into a rocking motion of shaft B, Figure 15-4.

Basic Types of Followers

Speed of rotation and the actual load applied upon the lifter determines the type of follower to be used. There are various basic types of followers: roller, pointed, flat faced and spherical faced, Figure 15-5.

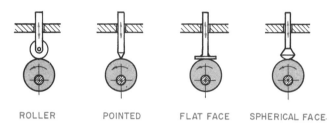

Figure 15-5 Basic types of followers

Cam Mechanism

Two major kinds of cams are used in industry: radial arm design and cylindrical design. The *radial arm* design changes a rotary motion into either an up and down motion or a rocking action as discussed previously. In the *cylindrical* design, a shaft rotates exactly as it does in the radial design, but the action or direction of the follower differs greatly. In the radial arm design, the follower operates *perpendicular* to the cam shaft. In the cylindrical design, the follower operates *parallel* to the cam shaft. See Figure 15-6 for a simplified example of a cylindrical design cam. Shaft A has an irregular groove cut into it. As the cam rotates, the follower traces along the groove. The follower is directly attached to a shaft that moves to the left and right. Notice, this shaft does *not* rotate; the motion is *parallel* to the axis of the rotating cam shaft.

Cams produce one of three kinds of motion: uniform velocity, harmonic, or uniform acceleration. Each is discussed in full later.

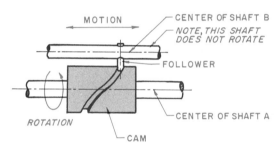

Figure 15-6 Cylindrical cam mechanism

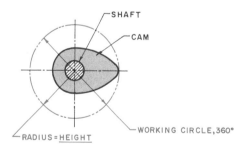

Figure 15-7 Cam terms and working circle

Cam Terms

Working Circle

The *working circle* is considered a distance equal to the distance from the center of the cam shaft to the highest point on the cam, Figure 15-7.

Displacement Diagram

The *displacement diagram* is a designed layout of the required motion of the cam. It is laid out on a grid, and its length represents one complete revolution of the cam. The length of the displacement diagram is usually drawn equal in length to the circumference of the working circle. This is not absolutely neces-

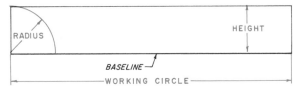

WORKING CIRCLE = <u>CIRCUMFERENCE OF CIRCLE</u> = 360°
BASELINE = CENTER OF WORKING CIRCLE

Figure 15-8 Displacement diagram — basic layout

sary, but it will give an in-scale idea of the cam's profile. The height of the displacement diagram is equal to the radius of a working circle, Figure 15-8. The bottom line of the displacement diagram is the *baseline*. All dimensions should be measured upward from the baseline. On the cam itself, think of the center of the cam as the baseline.

The length of the displacement diagram is divided into equal lines or grid, each of which represents degrees around the cam. These divisions can be 30°, 15° or even 10°. The finer the divisions, the more accurate is the final cam profile, Figure 15-9.

Dwell

Dwell is the period of time during which the follower does not move. This is shown on the displacement diagram by a straight horizontal line throughout the dwell angle. On the cam, the dwell is drawn by a radius.

Time Interval

The *time interval* is the time it takes the cam to move the follower to the designed height.

Rotation

Rotation of the cam is either clockwise or counter-clockwise. The actual cam profile is laid out *opposite* of the rotation.

Base Circle

The *base circle* is used to lay out an offset follower which is illustrated in detail later. The base circle is a circle with a radius equal to the distance from the center of the shaft to the center of the follower wheel

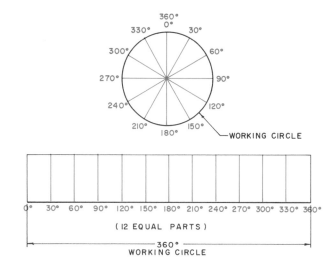

Figure 15-9 Displacement diagram with 30° increments

at its lowest position. On an offset follower, the base circle replaces the working circle.

Cam Motion

Four major types of curves are usually employed. Special irregular curves other than the four major types can be designed to meet a specific movement, if necessary.

The four types are: uniform velocity, modified uniform velocity, harmonic motion, and uniform acceleration. For comparison, the following four examples use the same working circle, same shaft, and same rise.

Uniform Velocity

In this type of motion, the cam follower moves with a *uniform velocity*; that is, it rises and falls at a constant speed, but the start and stop is very abrupt and rough, Figure 15-10.

Modified Uniform Velocity

Because of the abrupt and rough start and stop of the uniform velocity, it is modified slightly. *Modified uniform velocity* smoothes out the roughness slightly by adding a radius at the ends of the high and low points. This radius is equal to 1/3 the rise or fall, Figure 15-11. This radius smoothes out the start and stop somewhat, and is good for slow speed.

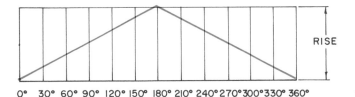

0° 30° 60° 90° 120° 150° 180° 210° 240°270°300°330° 360°

Figure 15-10 Uniform velocity cam

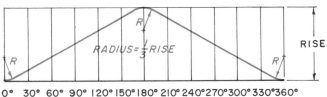

0° 30° 60° 90° 120° 150°180° 210° 240°270°300° 330°360°

Figure 15-11 Modified uniform velocity cam

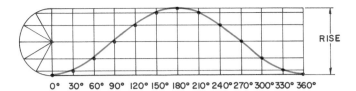

Figure 15-12 Harmonic motion cam

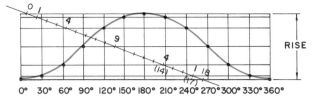

Figure 15-13 Uniform acceleration cam

Harmonic Motion

Harmonic motion is very smooth, but the speed is not uniform. Harmonic motion has a smooth start and stop, and is good for fast speed.

To lay out a harmonic motion on the displacement diagram, draw a semicircle whose diameter is equal to the designed rise or fall. Divide the semicircle into equal divisions; in this example, 30°. Divide the overall horizontal distance of the rise or fall equally into increments of the same number as the semicircle; in this example, 6 equal divisions. Projection points on the semicircle are projected horizontally to the corresponding vertical lines. For example, the first point on the semicircle is projected to the first vertical line, the second point on the semicircle is projected to the second vertical line, and so forth. Connect the curve with an irregular curve. This completes the harmonic curve, Figure 15-12.

Uniform Acceleration

Uniform acceleration is the smoothest motion of all cams, and its speed is constant throughout the cam travel. The uniform acceleration curve is actually a parabolic curve; the first half of the curve is exactly the reverse of the second half. This form is best for high speed.

To lay out a uniform acceleration motion on the displacement diagram, divide the designed rise or fall into 18 equal parts. To do this, place the edge of a scale with its "0" on the starting level of the rise or fall, and equally space 18 units of measure on the high or low elevation of the rise or fall, Figure 15-13.

Mark off points 1, 4, 9, 4 (14), and 1 (17), and draw horizontal lines. Note that 14 is actually 4, and 17 is actually 1. Divide the overall horizontal distance of the rise or fall into 12 equal increments. From points 1, 4, 9, 4 (14), and 1 (17), project points to the corresponding vertical lines. For example, the first point is from point 1 on the division line to the first vertical line, and then from point 4 on the division line to the second vertical, and so forth. Connect the curve with an irregular curve. This completes the uniform acceleration curve.

Combination of Motions

The illustrations thus far have covered the four major types of cams: uniform velocity, modified uniform velocity, harmonic motion, and uniform acceleration. Any combinations of these can be designed for a particular function or requirement. Figure 15-14 illustrates a displacement diagram with a 90° fall using a harmonic motion, followed by a 30° dwell. The cam then falls again 90° using a modified uniform velocity followed by a 30° dwell. The cam then rises using uniform acceleration motion followed by another 30° dwell.

Laying Out the Cam from the Displacement Diagram

Regardless of which type of motion is used, the layout process is exactly the same. The working circle is drawn equal to the height of the displacement

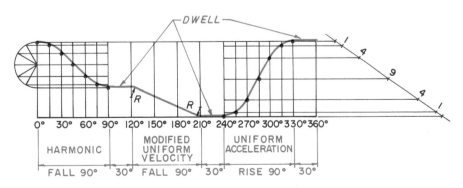

Figure 15-14 Combination of motions

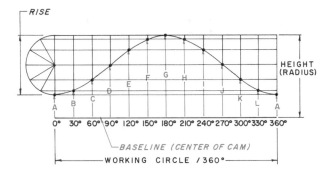

Figure 15-15A Displacement diagram example

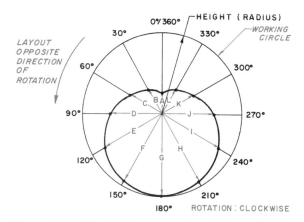

Figure 15-15B Heights transferred from displacement diagram

diagram. The circle of the cam must be divided into the exact same number of equal divisions as the displacement diagram, Figure 15-15A. Label the increments on both the displacement diagram and the working circle of the cam; in this example, harmonic motion is illustrated.

Notice that the even increment on the cam layout is labeled *opposite* of the rotation of the cam. In this example, rotation is clockwise; therefore, the labeling of the radial lines should be counterclockwise. Transfer each distance from the displacement diagram to the corresponding radial line, Figure 15-15B. Using an irregular curve, complete the cam layout.

Offset Follower

The cam follower is usually in line with the same center line as the cam (see Figure 15-1). Occasionally, however, when position or space is a problem, the follower can be designed to be on another center line, or offset as illustrated in Figure 15-16. In order to lay out an offset follower, it is necessary to use slightly different steps than those used for a regular cam.

How To Draw a Cam with an Offset Follower

Given: Cam data: harmonic motion cam, follower offset .625, base circle .88 radius, rise of 1.50, shaft diameter .375, direction of rotation: counterclockwise.
Follower data: Follower width .250, roller .375 diameter.

Step 1. Design and lay out a displacement diagram exactly as with any cam, Figure 15-17A.

Step 2. From the center of the shaft of the cam, construct the offset circle. The offset circle has a radius equal to the required follower offset. From the same swing point, lay out the base circle. The radius of the base circle is equal to the distance from the center of the shaft to the center of the follower *wheel* at its *lowest* position, Figure 15-17B.

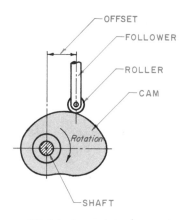

Figure 15-16 Associated cam terms

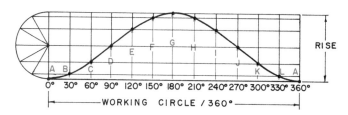

Figure 15-17A Drawing a cam with an offset follower design layout — Step 1

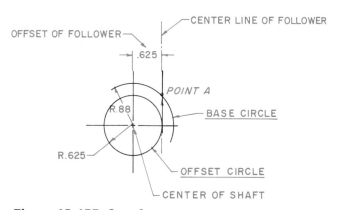

Figure 15-17B Step 2

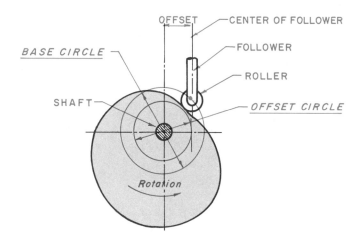

Figure 15-17C Step 3

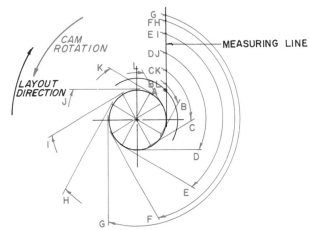

Figure 15-17D Step 4

Step 3. The point where it crosses a constructed vertical line from the center line of the follower is point A on the displacement diagram, Figure 15-17C. This represents the *lowest* position of the follower wheel. This vertical line is the measuring line and will be used to lay out the other points around the cam.

Step 4. Divide the offset circle into the same equal divisions as used on the displacement diagram; in this example, 12 equal parts. Lightly, consecutively, letter or number the divisions on the offset circle. Start at the lowest point of the cam; in this example, point A. Letter or number in a direction *opposite* of the cam rotation. Note that point A is the exact location where the center line of the follower is tangent with the offset circle. Construct a layout projection line 90° from the tangent point of the circle, Figure 15-17D.

Step 5. Transfer the distances from the displacement diagram to the measuring line, lightly letter or number each point. With the compass point on the center of the shaft of the cam, swing these distances to the corresponding radial lines (refer back to Figure 15-17D). Lightly letter or number each point. Do not forget to *lay out the cam opposite to the direction of rotation.* These points represent the path of the center of the follower wheel. Lightly connect these points together, Figure 15-17E.

Step 6. With the compass set at the radius for the roller, lightly draw the roller around the path of the follower. The inside, high points of these arcs represent the actual profile of the cam, Figure 15-17F.

Step 7. Check all construction work. If correct, darken and complete the cam, Figure 15-17G.

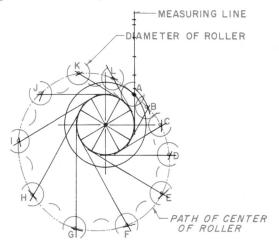

Figure 15-17E Step 5

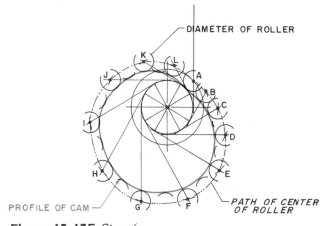

Figure 15-17F Step 6

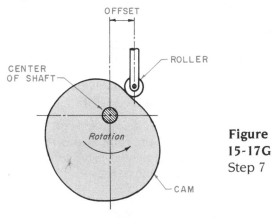

Figure 15-17G Step 7

473

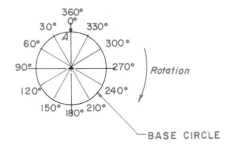

Figure 15-18A Drawing a cam with
a flat-faced follower
— Step 1

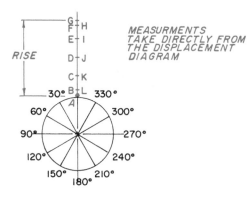

MEASURMENTS
TAKE DIRECTLY FROM
THE DISPLACEMENT
DIAGRAM

Figure 15-18B Step 2

How To Draw a Cam with a Flat-faced Follower

Given: Cam data: harmonic motion cam, base circle .75 radius, rise of 1.50, shaft diameter .375.,
direction of rotation: clockwise.
Follower data: flat-faced follower, 1.25 diameter distance across the flat surface.
Design and lay out a displacement diagram
exactly as with any cam. In this example, the
displacement diagram in Figure 15-17A will be
used.

Step 1. Construct the base circle. The base circle
is drawn with a radius equal to the distance
from the center of the shaft to the face of the
follower in its *lowest* position. Divide the base
circle into the same equal divisions as used on
the displacement diagram. Lightly consecutively letter or number the divisions on the
base circle, Figure 15-18A.

Step 2. Locate point A which represents the location of the flat-faced follower at its lowest point.
Construct a measuring line starting from the
base circle and transfer the distances from the
displacement diagram to the measuring line.
Letter or number each point, Figure 15-18B. Do
not forget to lay out the *cam opposite to the direction
of rotation*.

Step 3. Swing these distances to the corresponding radial lines, Figure 15-18C.

Step 4. From these distances, construct lines perpendicular from these radial lines at the point
of intersection, Figure 15-18D.

Step 5. Lightly construct the cam within these
perpendicular lines, Figure 15-18E.

Step 6. Check all construction work. If correct,
darken and complete the cam, Figure 15-18F.
Notice the position of the flat-faced follower at
the 0°, 90°, and 150° positions.

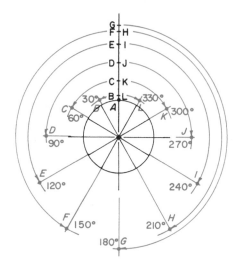

Figure 15-18C Step 3

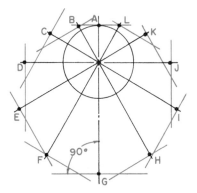

Figure 15-18D Step 4

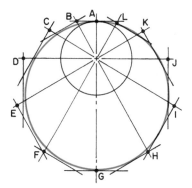

Figure 15-18E Step 5

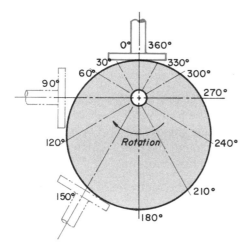

Figure 15-18F Step 6

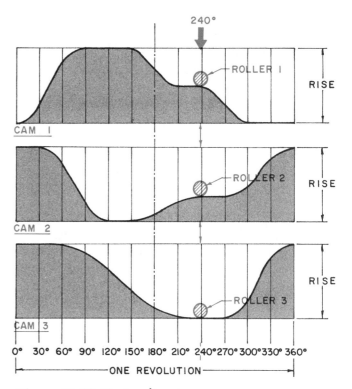

Figure 15-19 Timing diagram

Timing Diagram

Many times, more than one cam is attached to the same shaft. In this case, each cam works independently but must function in relation to the other cams. In order to be able to visualize graphically and study the interaction of the cams on the same shaft, place the various displacement diagrams in a column with the starting points and ending points in line, as illustrated in Figure 15-19. This represents one full revolution and is referred to as a *timing diagram*. For example, at 240° it is easy to visualize the exact position of each cam roller.

Dimensioning a Cam

A cam must be fully dimensioned. The cam size dimensions are constructed radiating from the center of the shaft by both radii length and degrees from the starting point, as illustrated in Figure 15-20. Other important dimensions pertain to the shaft hole size

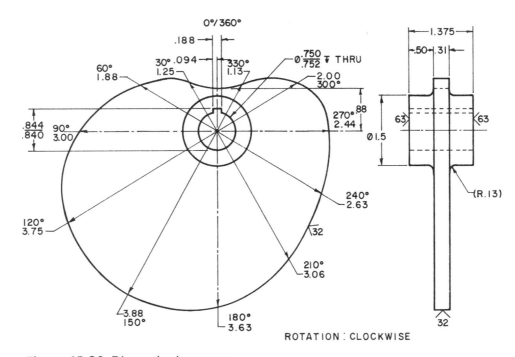

Figure 15-20 Dimensioning a cam

and keyway size, and are dimensioned according to most recent drafting standards.

Try not to place dimensions on the cam itself; place them around the perimeter, if possible.

Review

1. List three examples of uses of a simple cam.

2. What determines the type of follower to be used?

3. List the two major kinds of cams used in industry. Explain the function of each.

4. What is the working circle?

5. Explain what the displacement diagram is and why it is used.

6. What do the length and height of the displacement diagram represent on the actual cam?

7. List the four major types of curves usually used in the design and layout of the cam displacement diagram. Which one gives the smoothest motion?

8. What is meant by an offset follower?

9. Explain why a timing diagram is used.

10. Why is the cam laid out opposite of the direction of rotation?

Chapter Fifteen
Problems

The following problems are intended to give the beginning drafter practice in designing and laying out cam displacement diagrams according to the required specifications to do a particular function. The cam displacement diagram will then be transferred to an actual cam profile layout. The drafter will practice laying out uniform velocity, harmonic motion, and uniform acceleration motions.

The steps to follow in laying out cams are:

Step 1. Study all required specifications.

Step 2. Make a rough sketch of the cam displacement diagram incorporating *all* required specifications and required dimensions.

Step 3. Lightly lay out the displacement diagram. Space out horizontally as far as possible.

Step 4. Break down the area into equal spaces that correspond to degrees around the cam. *Label* each line indicating the degrees.

Step 5. Following the required specifications, lightly develop the cam's travel on the displacement diagram. (Show all light layout work — do not erase.)

Step 6. Check all work.

Step 7. Darken in displacement diagram.

Step 8. To lay out the cam profile, locate the center of the cam.

Step 9. Draw the shaft, hub, working circle, and key, if required, in place.

Step 10. Divide the working circle into equal degrees as required.

Step 11. Note *direction of travel*. Lightly letter the degrees around the working circle *opposite* of direction of travel.

Step 12. Locate the starting elevation and draw the cam follower *from* the starting point.

Step 13. Lightly draw in the rest of the cam, following details as required.

Step 14. Transfer all distances *from* the cam displacement diagram to the corresponding degree line *around* the cam layout.

Step 15. Lightly connect all points using an irregular curve.

Step 16. Check all work.

Step 17. Darken in all work.

Step 18. Neatly add all required specifications and dimensions to the drawing.

The drawing must include the following:
Shaft diameter
Hub size
Key or set screw size
Direction of rotation
Working circle size
All dimensions according to the most recent drafting standards. (For these problems, dimension only the front view.)

Problems 15-1A and 15-1B

Construct a cam displacement diagram on a A-size, 8½ × 11 sheet of paper per dimensions given.

Lay out a cam with the following specifications:

1. Working circle 5.0 dia.
2. Shaft dia. .75
3. Hub dia. 1.25
4. Square key .18
5. Rotation: counterclockwise
6. Follower dia. − 1.0
7. Start 1.0 from center

Required motion:

1. Rise 90°, modified uniform velocity 1.5
2. Dwell 60°
3. Fall 60°, modified uniform velocity .75
4. Dwell 30°
5. Fall 60°, modified uniform velocity .75
6. Dwell 60°

Construct the cam and follower on an A-size, 8½ × 11 sheet of paper per dimensions given, and transfer all dimensions from the displacement diagram to the cam layout.

Check all work and add all required dimensions and specifications according to the most recent drafting standards.

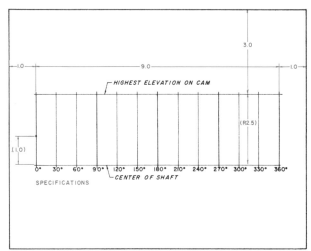

Problem 15-1A

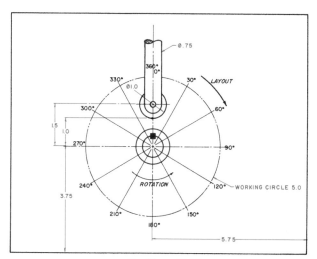

Problem 15-1B

Problems 15-2A and 15-2B

Construct a cam displacement diagram on an A-size, 8½ × 11 sheet of paper per dimensions given.

Lay out a cam with the following specifications:

1. Working circle 5.0 dia.
2. Shaft dia. .75
3. Hub dia. 1.25
4. Square key .18
5. Rotation: clockwise
6. Follower dia. 1.0
7. Start 1.0 from center

Required motion:

1. Rise 90°, harmonic motion 1.00
2. Dwell 30°
3. Rise 90°, harmonic motion .50
4. Dwell 30°
5. Fall 90°, harmonic motion 1.50
6. Dwell 30°

Construct the cam and follower on an A-size, 8½ × 11 sheet of paper per dimensions given, and transfer all dimensions from the displacement diagram to the cam layout.

Check all work and add all required dimensions and specifications according to the most recent drafting standards.

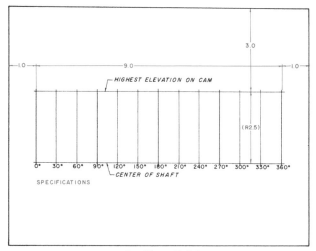

Problem 15-2A

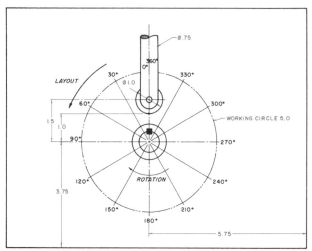

Problem 15-2B

Problems 15-3A and 15-3B

Construct a cam displacement diagram on an A-size, 8½ × 11 sheet of paper per dimensions given.

Lay out a cam with the following specifications:
1. Working circle 5.0
2. Shaft dia. .63
3. Hub dia. 1.0
4. Square key to suit
5. Rotation: clockwise
6. Follower dia. .75
7. Start 2.5 from center

Required motion:
1. Fall 180°, uniform acceleration 1.50
2. Dwell 45°
3. Rise 90°, uniform acceleration 1.50
4. Dwell 45°

Construct the cam and follower on an A-size, 8½ × 11 sheet of paper per dimensions given, and transfer all dimensions from the displacement diagram to the cam layout.

Check all work and add all required dimensions and specifications according to the most recent drafting standards.

Problems 15-4A and 15-4B

Construct a cam displacement diagram on an A-size, 8½ × 11 sheet of paper per dimensions given.

Lay out a cam with the following specifications (metric):
1. Working circle 245
2. Shaft dia. 28
3. Hub dia. 44
4. Square key to suit
5. Rotation: clockwise
6. Follower dia. 22
7. Start 32 from center

Required motion:
1. Rise 60°, harmonic motion 25
2. Dwell 15°
3. Rise 90°, harmonic motion 38
4. Dwell 15°
5. Fall 45°, harmonic motion 44
6. Dwell 15°
7. Rise 30°, harmonic motion 18
8. Dwell 15°
9. Fall to starting level 75°, harmonic motion 38

Construct the cam and follower on an A-size, 8½ × 11 sheet of paper per dimensions given, and transfer all dimensions from the displacement diagram to the cam layout.

Check all work and add all required dimensions and specifications according to the most recent drafting standards.

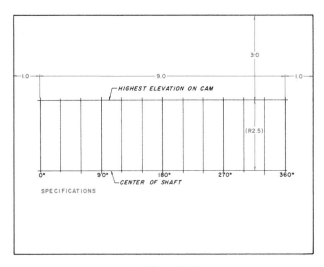

Problem 15-3A

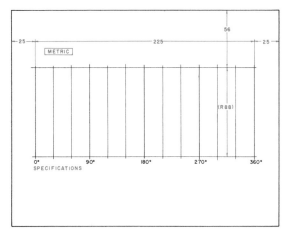

Problem 15-4A

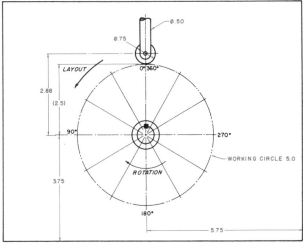

Problem 15-3B

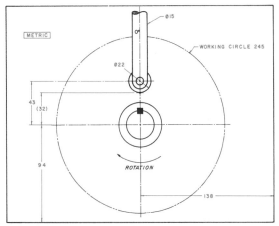

Problem 15-4B

Problems 15-5A and 15-5B

Construct a cam displacement diagram on an A-size, 8½ × 11 sheet of paper per dimensions given.
Lay out a cam with the following specifications:
1. Working circle 7.0
2. Shaft dia. .625
3. Hub dia. 1.0
4. Square key to suit
5. Rotation: counterclockwise
6. Follower dia. 1.0
7. Start 1.0 from center

Required motion:
1. Rise 120°, modified uniform velocity 2.5
2. Dwell 60°
3. Fall 30°, modified uniform velocity .75
4. Dwell 30°
5. Fall to starting level 90°, modified uniform velocity 1.75
6. Dwell 30°

Construct the cam and follower on an A-size, 8½ × 11 sheet of paper per dimensions given, and transfer all dimensions from the displacement diagram to the cam layout.
Check all work and add all required dimensions and specifications according to the most recent drafting standards.

Problems 15-6A and 15-6B

Construct a cam displacement diagram on an A-size, 8½ × 11 sheet of paper per dimensions given.
Lay out a cam with the following specifications:
1. Working circle 7.75
2. Shaft dia. .56
3. Hub dia. .88
4. Square key to suit
5. Rotation: clockwise
6. Follower dia. .375
7. Start 1.625 from center

Required motion:
1. Fall 90°, uniform acceleration 3.0
2. Dwell 15°
3. Rise 60°, uniform acceleration 2.0
4. Dwell 105°
5. Rise 90°, uniform acceleration to starting level (1.0)

Construct the cam and follower on an A-size, 8½ × 11 sheet of paper per dimensions given, and transfer all dimensions from the displacement diagram to the cam layout.
Check all work and add all required dimensions and specifications according to the most recent drafting standards.

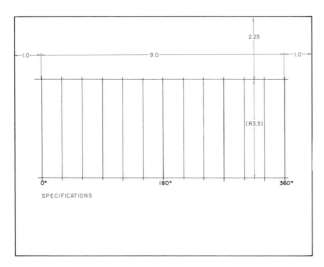

Problem 15-5A

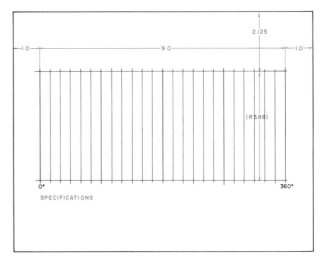

Problem 15-6A

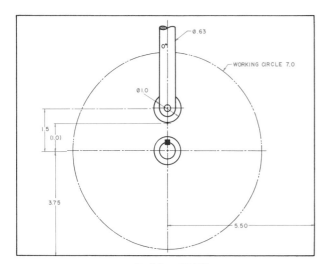

Problem 15-5B

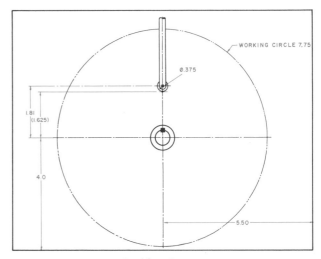

Problem 15-6B

Problems 15-7A and 15-7B

Construct a metric cam displacement diagram per dimensions given:

Lay out a cam with the following specifications:
1. Rise/fall 56
2. Shaft dia. 16
3. Hub dia. 32
4. Square key to suit
5. Rotation: clockwise
6. Follower: *flat-faced (34 dia.)
7. Base circle 26 radius
8. Harmonic motion/one full turn

Construct the cam and flat-faced follower per dimensions given.

Check all work and add required dimensions and specifications according to the most recent drafting standards, and using the metric system.

*Design the flat-faced follower to suit.

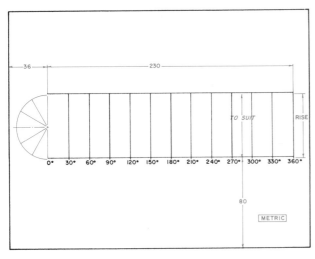

Problem 15-7A

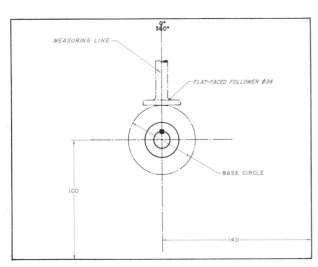

Problem 15-7B

Problems 15-8A and 15-8B

Construct a cam displacement diagram per dimensions given.

Lay out an offset cam with the following specifications:
1. Offset circle dia. 2.00
2. Shaft dia. .75
3. Hub dia. 1.25
4. Square key to suit
5. Rotation: clockwise
6. Follower dia. .75
7. Follower offset 1.00
8. Base circle 1.375 radius

Required motion:
1. Fall 75°, uniform acceleration 1.75
2. Dwell 10°
3. Rise 35°, harmonic motion .75
4. Dwell 5°
5. Fall 70°, modified uniform velocity 1.25
6. Dwell 15°
7. Rise 45°, harmonic motion .75
8. Dwell 15°
9. Rise 60°, uniform acceleration 1.50
10. Dwell 30°

Construct the cam and offset follower per dimensions given, and transfer all dimensions from the displacement diagram to the cam layout.

Check all work and add all required dimensions and specifications according to the most recent drafting standards.

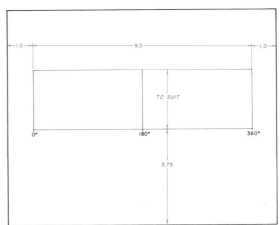

Problem 15-8A

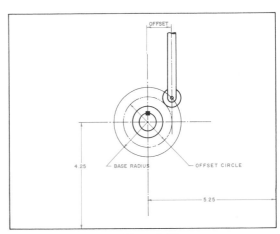

Problem 15-8B

1. What is the surface specification for the groove walls of the cam?
2. What is the surface specification for the bored holes that will fit on the shaft?
3. Of what material is the cam made?
4. What are the hardness specifications for the material?
5. How deep are the grooves in the cam?
6. What is the angle of the groove walls?
7. What are the corner specifications at the root of the grooves?
8. What is the "rise" dimension for the cam?
9. What is the "fall" dimension for the cam?
10. What form tolerance is specified on the two bored holes in the cam?

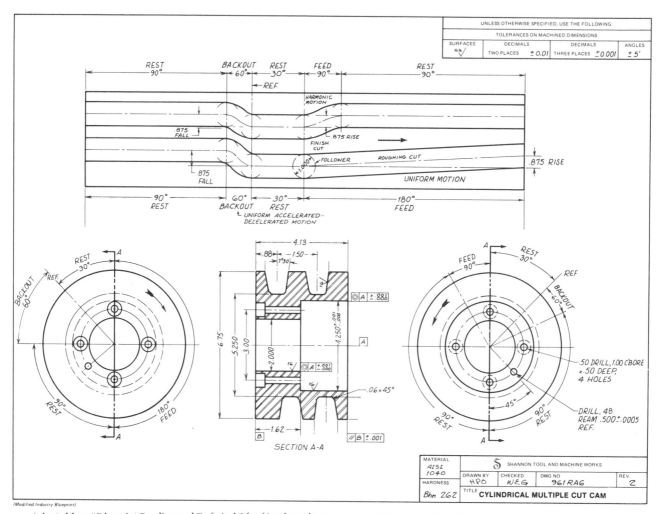

Adapted from "Blueprint Reading and Technical Sketching for Industry and used by permission of Thomas P. Olivo and C. Thomas Olivo, Published by Delmar Publishers Inc.

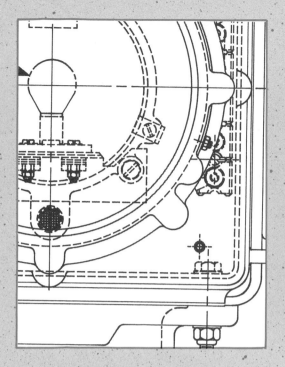

The major kinds of standard gears are explained in this chapter, along with the terms and formulas associated with gears. The major gear types included are: spur gears, pinion gears, helical gears, ring gears, bevel gears, worm gears, and rack and pinion gears. The study of gear ratios and gear chains is covered in order to give the beginning designer the required information to achieve the required ratios.

CHAPTER SIXTEEN

Gears

Gears transmit or transfer rotary motion from one shaft to another shaft. Gears can change the direction of rotation, speed up or slow down rotation, increase or reduce power, and change rotary motion into a reciprocating motion. There are various kinds of gears, each with their own function, Figure 16-1.

The drafter must be able to identify each kind of gear, know the various functions of each and be able to design and draw the various gears using correct terminology associated with gears.

Kinds of Gears

Gears are usually classified by the position or location of the shafts they connect. A spur gear, pinion gear or helical gear is usually used to connect shafts that are parallel to each other. Intersecting shafts at 90° are usually connected by a beveled gear or angle gear. Shafts that are not parallel to each other and that do not intersect use a worm and worm gear. In order to connect rotary motion into a reciprocating or back and forth motion, a rack gear and pinion would be used.

Spur Gear

The *spur gear* is the most commonly used gear, Figure 16-2. It is cylindrical in form, with teeth that are cut straight across the face of the gear. All teeth are parallel to the axis of the shaft. The spur gear is usually considered the *driven gear.*

Pinion Gear

The *pinion gear* is exactly like a spur gear but it is usually smaller, and has fewer teeth, Figure 16-2. The pinion gear is normally considered the *drive gear.*

Rack Gear

The *rack gear* is a type of spur gear, but its teeth are in a straight line or flat instead of in a cylindrical form, Figure 16-3. The rack gear is used to transfer circular motion into straight-line motion.

Ring Gear

The *ring gear* is similar to the spur, pinion, and rack gears, except that the teeth are internal, Figure 16-4.

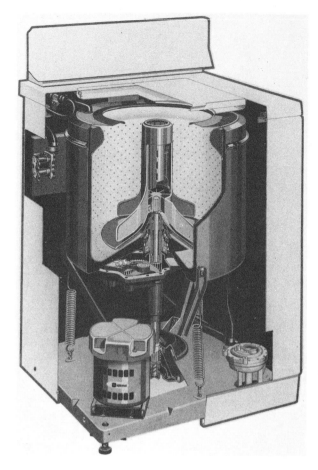

Figure 16-1 Example of gears in use (*Courtesy The Maytag Company*)

Figure 16-2 Spur gear/pinion gear (*Courtesy Boston Gear*)

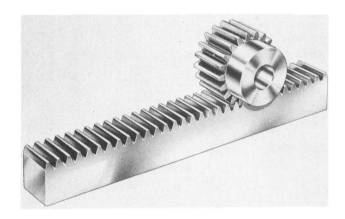

Figure 16-3 Rack gear (*Courtesy Boston Gear*)

Bevel Gear

A bevel gear is another gear commonly used, Figure 16-5. A *bevel gear* is cone shaped in form with straight teeth that are on an angle to the axis of the shaft. Bevel gears are used to transmit power and motion between intersecting shafts that are at 90° to each other.

Angle Gear

The *angle gear* is similar to a bevel gear, except that the angles are at other than 90° to each other.

Miter Gear

The *miter gear* is exactly the same as a bevel gear, except that both mating gears have the same number of teeth. The shafts are at 90° to each other, Figure 16-6.

Spiral Bevel or Miter Gears

Any bevel gears with curved teeth are called *spiral bevel gears*, Figure 16-7.

Worm Gear

Worm gears are used to transmit power and motion at a 90° angle between nonintersecting shafts, Figure 16-8. They are normally used as a speed reducer.

Figure 16-4 Ring gear (*Courtesy Boston Gear*)

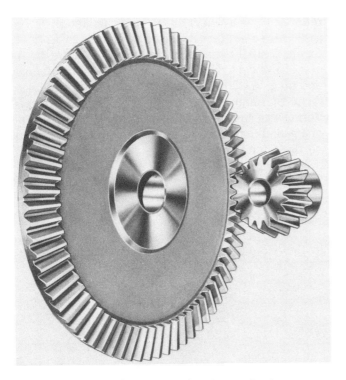

Figure 16-5 Bevel gear (*Courtesy Boston Gear*)

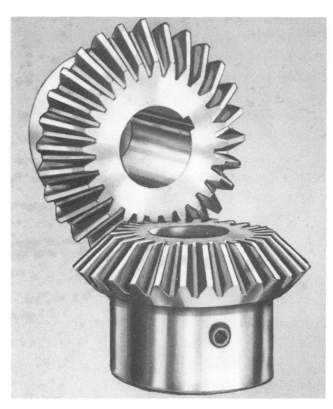

Figure 16-6 Miter gear (*Courtesy Boston Gear*)

Figure 16-7 Spiral bevel gear (*Courtesy Boston Gear*)

Figure 16-8 Worm gear (*Courtesy Boston Gear*)

Figure 16-9 Chain and sprockets (*Courtesy Boston Gear*)

The worm gear is round like a wheel. The worm is shaped like a screw, with threads (or teeth) wound around it. Because one full turn of the worm is required to advance the worm gear one tooth, a high-ratio speed reduction is achieved. The worm drives the larger worm gear.

Chain and Sprockets

A *chain and sprockets* are used to transmit motion and power to shafts that are parallel to each other, Figure 16-9. Sprockets are similar to spur and pinion gears, as the teeth are cut straight across the face of the sprocket and are parallel to the shaft. There are many types of chains and sprockets, but all use the same terms, formulas, ratios, and so forth.

Velocity of Feet Per Minute (F.P.M.)

A gear is very similar to a simple wheel. In the field of horology (the science of measuring time), a gear is

referred to as a wheel. In making calculations for a gear, it is sometimes easier to think of it as a wheel. Figure 16-10 illustrates the relationship of the diameter of a wheel to given revolutions per minute (R.P.M.) and time. Velocity is calculated by recording the distance that a given point on a wheel or gear travels during a certain period of time. To calculate velocity of a gear, the point is usually assumed to be located on the pitch circle. The pitch circle is discussed in detail later.

The formula is:

$$\pi \text{Dia.} \times \text{R.P.M.} \times \text{Time} = \text{Distance}$$

In this example, point a on the wheel will travel 785 feet.

$$\pi 2.0 \text{ Dia.} \times 500 \text{ R.P.M.} \times 3 \text{ Min.} = \text{Distance}$$
$$6.28 \times 500 \times 3 = 9420 \text{ inches}$$
$$9420 \text{ inches} = 785 \text{ feet}$$

If two wheels of the exact same diameter are placed together, as illustrated in Figure 16-11, and assuming there is no slippage, point a on wheel A will travel the exact same distance as point b on wheel B. Notice, wheel a is rotating clockwise, and wheel B is rotating counterclockwise. If wheel A is rotated at 500 R.P.M. for 3 minutes point a will travel 785 feet.

Gear Ratio

If two friction wheels of different diameters are placed together, Figure 16-12, and assuming there is no slippage, point a on wheel A will travel the same distance as point b on wheel B. In this example, wheel A is the driver.

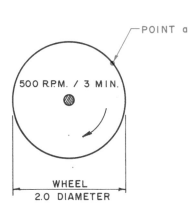

Figure 16-10
Relationship of diameter of a wheel to given revolutions per minute and time

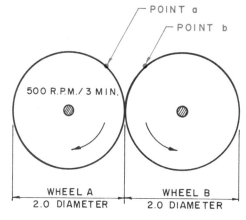

Figure 16-11
Relationship of two wheels of the exact same diameters

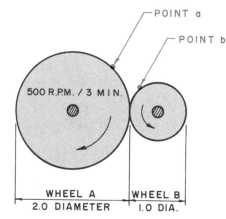

Figure 16-12
Relationship of two wheels of different diameters

π Diameter × revolutions per minute
× time = distance

WHEEL A
πDia. × R.P.M. × time =
π2.0 Dia. × 500 R.P.M.
    × 3 min. =
6.28 × 500 × 3 = 9420 =

WHEEL B
πDia. × R.P.M. × time
π1.0 Dia. × ? R.P.M.
    × 3 min.
3.14 × 3 = 9.42

9420 ÷ 9.42 = 1000 R.P.M. of Wheel B

Wheel A has a 2.0 diameter and turns at 500 R.P.M. Wheel B has a diameter of half of wheel A, but turns twice as fast, thus point a and point b travel the exact same distance. Note that both wheel A and wheel B travel at the same velocity.

The *ratio* between gears is:

$$\frac{\text{Diameter, wheel A}}{\text{Diameter, wheel B}} = \frac{2}{1} = 2{:}1$$

The ratio of 2:1 means that wheel B rotates two times each time wheel A rotates once.

The ratio of one gear to another can be calculated by using any of three different methods: the number of teeth of corresponding gears, pitch diameters of corresponding gears, or the R.P.M. between the gears.

Method 1
To find the ratio, divide the number of teeth on the spur gear by the number of teeth on the pinion gear:

$$\frac{\text{Number of teeth (spur gear)}}{\text{Number of teeth (pinion gear)}} = \text{Ratio}$$

Example:

If a spur gear has 60 teeth and a pinion gear has 30 teeth, the ratio would be:

$$\frac{\text{Number of teeth (spur gear)}}{\text{Number of teeth (pinion gear)}} = \frac{60}{30} = 2{:}1$$

The pinion gear will rotate two times for each full revolution of the spur gear.

Method 2
To find the ratio, divide the pitch diameter (D) of the spur gear by the pitch diameter (D) of the pinion gear:

$$\frac{\text{Pitch Diameter (D)(spur gear)}}{\text{Pitch Diameter (D)(pinion gear)}} = \text{Ratio}$$

Example:

If a spur gear has a pitch diameter (P.D.) of 4.000 and a pinion gear has a pitch diameter (P.D.) of 1.000, the ratio would be:

$$\frac{\text{P.D. (spur gear)}}{\text{P.D. (pinion gear)}} = \frac{4.000}{1.000} = 4{:}1$$

The pinion gear will rotate four times for each full revolution of the spur gear.

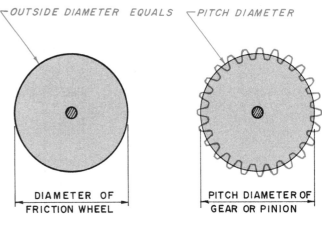

OUTSIDE DIAMETER EQUALS   PITCH DIAMETER

DIAMETER OF
FRICTION WHEEL

PITCH DIAMETER OF
GEAR OR PINION

Figure 16-13 Pitch diameter of a gear

Method 3
To find the ratio, divide the revolutions per minute (R.P.M.) of the pinion gear by the R.P.M. of the spur gear:

$$\frac{\text{R.P.M. (pinion gear)}}{\text{R.P.M. (spur gear)}} = \text{Ratio}$$

Example:

If a pinion gear rotates at 1000 R.P.M. and a spur gear rotates at 200 R.P.M., the ratio would be:

$$\frac{\text{R.P.M. (pinion gear)}}{\text{R.P.M. (spur gear)}} = \frac{1000}{200} = 5{:}1$$

The pinion gear will rotate five times for each full revolution of the spur gear.

Pitch Diameter

Think of the pitch diamter of a gear as the outside diameter of the friction wheel, Figure 6-13. All ratios as discussed so far with the friction wheel apply also to gears. To apply formulas, substitute the outside diameter of the friction wheel for the pitch diameter of the gear.

Gear Blank

Gears are usually cut from a gear blank. The gear blank must allow sufficient space for the gears and a method to attach the gear to the shaft. Illustrated in Figure 16-14 is a common gear blank with an outside diameter and face width for the gears to be cut into, a hub, and a hole for the set screw to secure the gear blank to the shaft. A half-section is used to illustrate the gear blank in this example. Two complete sets of dimensions must be applied to the gear drawing; those relating to the gear blank, as illustrated, and those relating to the teeth. Gear dimensions do not hold tight tolerancing, but all gear tooth dimensions must hold very tight tolerancing.

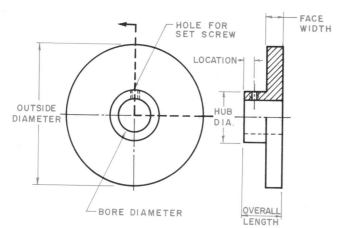

Figure 16-14 Gear blank

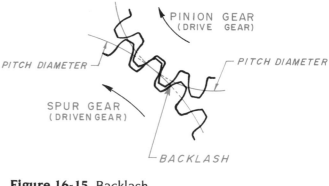

Figure 16-15 Backlash

Backlash

Backlash is the amount by which the gear tooth is less than the tooth space. It is measured by locking one gear and rocking the other back and forth, and measuring that rock at a known radius, usually the pitch diameter. The amount of backlash is the shortest distance between mating teeth as measured between the nondriving surfaces of adjacent teeth, Figure 16-15.

Basic Terminology

The drafter must know and understand all major terms associated with various kinds of gears used today in industry. Illustrated in Figure 16-16 are the basic terms associated with most gears, regardless of type.

Pitch Diameter (D)

The *pitch diameter* is the theoretical circle on which the teeth of mating gear mesh. Think of the pitch diameter on the gear as being the same as the outer diameter on the friction wheel. Note: The pitch diameter is considered the *nominal* size of the gear.

Root Diameter (DR)

The *root diameter* is the measurement over the extreme inner edges of teeth. It is equal to the pitch diameter minus 2 times the dedendum (b).

Outside Diameter (OD)

The *outside diameter* is the measurement over the extreme outer edge of teeth. It is equal to the pitch diameter, plus 2 times the addendum (a).

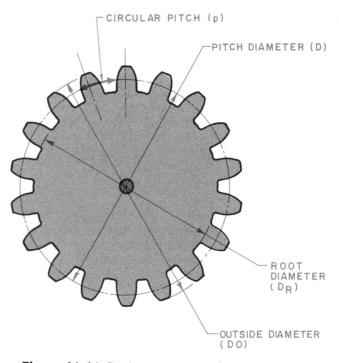

Figure 16-16 Basic gear terminology

Circular Pitch (p)

Circular pitch is the distance between corresponding points of adjacent teeth measured on the circumference of the pitch diameter.

Figure 16-17 illustrates the major terms associated with individual gears. There are many more technical terms associated with the actual form and cutting of gear teeth, but the following terms are the ones the drafter should be familiar with.

Working Depth (hk)

The *working depth* is the distance that a tooth projects into the mating space. It is equal to gear addendum (a) plus pinion addendum (a).

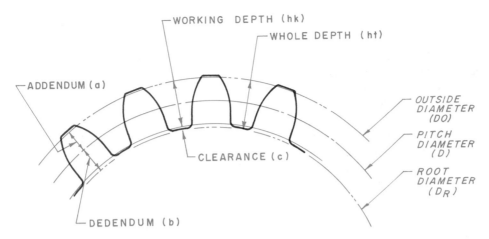

Figure 16-17 Major gear terms

Addendum (a)

The *addendum* is the radial distance from the pitch diameter to the top of the tooth. It is equal to $\frac{1}{P}$.

Dedendum (b)

The *dedendum* is the radial distance from the pitch diameter to the bottom of the tooth. It is equal to $\frac{1.157}{P}$.

Clearance (c)

Clearance is the space between the working depth and the whole depth. It is equal to dedendum (b) minus addendum (a).

Whole Depth (ht)

The *whole depth* is the total depth of a tooth space. It is equal to the addendum (a), plus the dedendum (b), plus the clearance (c).

Note: All notations in parentheses () correspond to those used in the most recent edition of *Machinery's Handbook* so that they agree if a more detailed formula is required by the drafter.

Diametral Pitch (P)

Diametral pitch (P) is a ratio equal to the number of teeth on the gear per inch of pitch diameter (D).

$$\text{Diametral pitch (P)} = \frac{\text{Number of teeth (N)}}{\text{Pitch diameter (D)}}$$

Example:

A gear with 48 teeth on a 3.0 pitch diameter (D) would have a diametral pitch (P) of 16.

$$P = \frac{N}{D} = \frac{48}{3.0} = 16$$

Gear Template

A quick and efficient method of illustrating teeth is by using a gear template, Figure 16-18. When using a gear template, the drafter must first calculate the diametral pitch (P). The numbers next to each tooth indicate the diametral pitch. The *lower* part of the template opening is used to draw spur gear *teeth*. The *top* portion of the opening is used to draw teeth of a *rack* or *worm*.

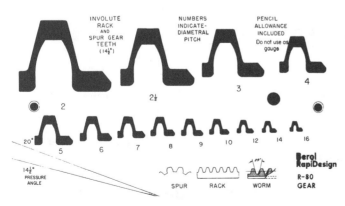

Figure 16-18 Gear template — spur/rack/worm (*Courtesy Modern School Supply*)

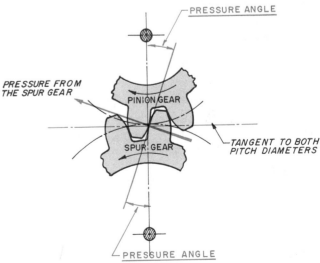

Figure 16-19 Pressure angle

Pressure Angle

The pressure angle determines the tooth form, Figure 16-19. The *pressure angle* is the angle at which pressure from the driver gear tooth passes to the tooth of the driven gear. The standard pressure angle is either 14½° or 20°. The 14½° pressure angle is still popular, especially when replacing old machinery, and because it is quiet. The 20° pressure angle is stronger, due to the heavier tooth section at the base of the tooth form.

Center-to-Center Distances

Calculating the center-to-center distance between two friction wheels is a simple math problem, Figure 16-20.

$$\text{Center-to-center distance} = \frac{\text{Outside diameter of wheel A} + \text{Outside diameter of wheel B}}{2}$$

In this example, $\frac{2.0 + 1.0}{2} = 1.5$ center-to-center distance

The exact same formula applies to a gear and pinion, except that the pitch diameter (D) is used in place of the outside diameter (OD). Refer to Figure 16-21.

$$\text{Center-to-center distance} = \frac{\text{Pitch diameter of gear} + \text{Pitch diameter of pinion}}{2}$$

Measurements Required to Use a Gear Tooth Caliper

A *gear tooth caliper* is used to check and measure an individual gear tooth. In order to use a gear tooth caliper, two parts of the individual tooth are used: the chordal thickness and the chordal addendum, Figure 16-22.

Chordal Thickness

The *chordal thickness* is the length of the chord measured straight across at the pitch diameter. It is measured straight across tooth, not along the pitch diameter as is the circular thickness.

$$\text{Chordal thickness} = \text{Sin}\left(\frac{90°}{N}\right) \times \text{Pitch dia.}$$

Chordal Addendum

The *chordal addendum* is the distance from the top of the tooth to the point at which the chordal thickness is measured.

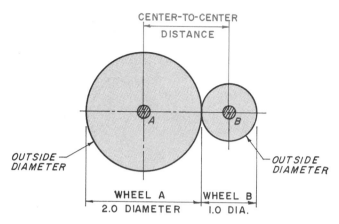

Figure 16-20 Center-to-center distances of two wheels

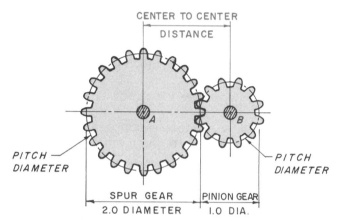

Figure 16-21 Center-to-center distances of two gears

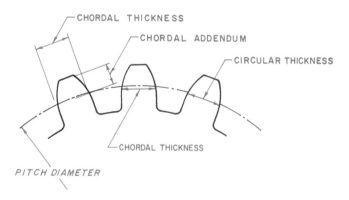

Figure 16-22 Measurements required to use a gear tooth caliper

$$\text{Chordal addendum} = \left[1 - \text{Cos}\left(\frac{90°}{N}\right)\right] \times \frac{\text{Pitch dia.}}{2} + \text{addendum}$$

Figures 16-23 and 16-24 show a gear tooth vernier caliper and how it is used.

Required Tooth-Cutting Data

Overall gear blank dimensions are shown on the detail drawing (as illustrated in Figure 16-15) and can

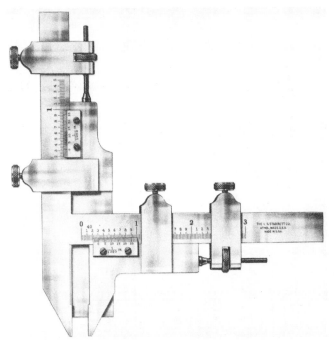

Figure 16-23 Gear tooth caliper (*Courtesy* L. S. Starrett Co.)

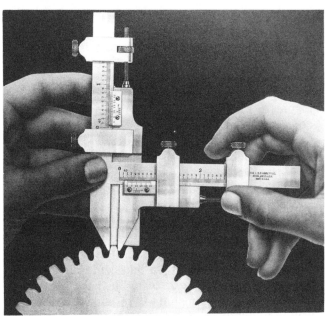

Figure 16-24 Using a gear tooth caliper (*Courtesy* L. S. *Starrett Co.*)

REQUIRED CUTTING DATA			
ITEM	TO FIND:	HAVING	FORMULA
1	Number of teeth (N)	D & P	D x P
		DO & P	(DO x P) - 2
2	Diametral pitch (P)	p	$\dfrac{3.1416}{P}$
		D & N	$\dfrac{N}{D}$
		DO & N	$\dfrac{N+2}{DO}$
3	Pressure angle (∅)	—	20° STANDARD 14°-30' OLD STANDARD
4	Pitch diameter (D)	N & P	$\dfrac{N}{P}$
		N & DO	$\dfrac{N \times DO}{N+2}$
		DO & P	$DO - \dfrac{2}{P}$
5	Whole depth (ht)	a & b	a + b
		P	$\dfrac{2.157}{P}$
6	Outside diameter (DO)	N & P	$\dfrac{N+2}{P}$
		D & P	$D + \dfrac{2}{P}$
		D & N	$\dfrac{(N+2) \times D}{N}$
7	Addendum (a)	P	$\dfrac{1}{P}$
8	Working depth (hk)	a	2 x a
		P	$\dfrac{2}{P}$
9	Circular thickness (t)	P	$\dfrac{3.1416}{2 \times P}$
10	Chordal thickness	N, D & a	$\sin\left(\dfrac{90°}{N}\right) \times D$
11	Chordal addendum	N, D & a	$\left[1 - \cos\left(\dfrac{90°}{N}\right)\right] \times \dfrac{D}{2} + a$
12	Dedendum (b)	P	$\dfrac{1.157}{P}$

Figure 16-25 Required tooth-cutting data for spur and pinion gears

hold loose tolerancing. The actual gear dimensions must hold tight tolerancing.

Twelve essential items of information must be calculated and listed on the gear detail drawing. This list is usually located at the lower right-hand side of the drawing, and generally consists of the data in Figure 16-25, but these items may vary slightly from company to company. The required material, heat-treating process, and other important information must also be included, usually in the title block.

In actual practice, the gear teeth are *not* drawn as it takes too much drawing time. A simplified method is used to illustrate the actual gear tooth, similar to that used to illustrate threads of a fastener, Figure 16-26. The outside diameter and the root diameter are illustrated by a thin phantom line. The pitch diameter is illustrated by a thin center line. Actual teeth are not drawn, except to illustrate some special feature in relationship to a tooth, a spline, a keyway or a locating point. In this example, all dimensions in color are related to the gear blank only. All gear dimensions are called-off in the cutting data box underneath.

Rack

A *rack* is simply a gear with teeth formed on a flat surface, Figure 16-27. A rack changes rotary motion into reciprocating motion. All terms and formulas associated with spur and pinion gears apply to the rack. The sides of the teeth are *straight*, not involute as on a spur gear. The teeth are inclined at the same angle as the pressure angle of the mating pinion gear. Note that the circular pitch of the pinion gear is the same as the linear pitch of the rack. All dimensions for heights of depth, pitch, and addendum lines are calculated from a datum or reference line. The rack is usually manufactured out of rectangular stock but, occasionally, is manufactured out of round stock to meet a specific design.

Two methods can be used to obtain full meshing of the pinion and the rack:

1. The rack is cut down so that it has less backlash.
2. The outside diameter of the pinion is increased slightly so that the pinion is *not* a standard size. If a spur gear is meshed with the over-sized pinion gear, its outside diameter is manufactured slightly undersize.

Both methods maintain the standard shaft center-to-center distances between the gear and the pinion. Standard tabulated charts are available to calculate these amounts of increased or decreased outside diameters.

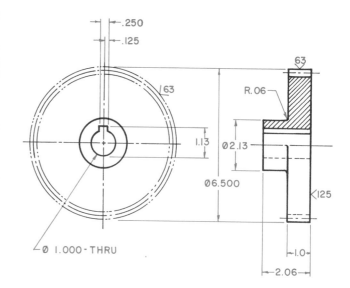

CUTTING DATA	
NUMBER OF TEETH	24
DIAMETRAL PITCH	4
PRESSURE ANGLE	20°
PITCH DIAMETER	6.000
WHOLE DEPTH	0.5393
OUTSIDE DIAMETER	6.500
ADDENDUM	.250
WORKING DEPTH	.500
CIRCULAR THICKNESS	.3925
CHORDAL THICKNESS	.2566
CHORDAL ADDENDUM	.3924
DEDENDUM	.289

Figure 16-26 Example of a spur gear detail drawing

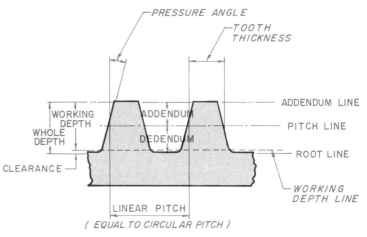

Figure 16-27 Gear terminology for a rack

The required tooth-cutting data for a rack are:

- Pressure angle (same as mating pinion gear)
- Tooth thickness at pitch line (same as mating pinion gear)
- Whole depth (same as mating pinion gear)
- Maximum allowance pitch variation
- Accumulated pitch error max. (over total length)

492   Section 4

MUST BE HELD VERY CLOSE TO ENSURE PROPER TOOTH BEARING

MOUNTING DISTANCE

PITCH APEX TO CROWN / CROWN TO BACK

C WHOLE DEPTH

ADDENDUM CROWN

DEDENDUM

BACK ANGLE

CONE DISTANCE

PITCH ANGLE

ROOT ANGLE

FACE

ADDENDUM ANGLE

PITCH ANGLE

ROOT ANGLE

FACE ANGLE

DEDENDUM ANGLE

PINION GEAR

CLEARANCE

BACK CONE

BACK CONE DISTANCE

BEVEL GEAR

PITCH DIAMETER

OUTSIDE DIAMETER

Bevel Gear

Bevel gears transmit power and motion between intersecting shafts at right angles to each other. Figure 16-28 illustrates the various terms associated with bevel gears. The required tooth-cutting data for bevel gears are similar to spur/pinion gears. These must be calculated and listed on the detail drawing. Figure 16-29 shows the order in which the data and formulas should be listed on the drawing. Bevel gears must be designed and drawn in pairs to ensure a perfect fit.

Worm and Worm Gear

The *worm* and *worm gear* is used to transmit power between nonintersecting shafts at right angles to each other. When using the worm and worm gear, a large speed ratio is possible as one revolution of a single-thread worm turns the worm gear only one tooth and one space.

Figure 16-28 Gear terminology for bevel gears

REQUIRED CUTTING DATA

ITEM	TO FIND:	HAVING	FORMULA SPUR	PINION
1	Number of teeth (N)	–	AS REQ'D.	
2	Diametral pitch (P)	p	$\frac{3.1416}{p}$	
3	Pressure angle (Ø)	–	20° STANDARD 14°-30' OLD STANDARD	
4	Cone distance (A)	D & d	$\sin d \,)\overline{\frac{D}{2}}$	
5	Pitch distance (D)	p	$\frac{N}{p}$	
6	Circular thickness (t)	p	$\frac{1.5708}{p}$	
7	Pitch angle (d)	N & d (of pinion)	90°–d(pinion)	$\tan d \frac{N \text{ pinion}}{N \text{ gear}}$
8	Root angle (ɣR)	d & δ	d – δ	
9	Addendum (a)	p	$\frac{1}{p}$	
10	Whole depth (ht)	p	$\frac{2.188}{p} + .002$	
11	Chordal thickness (C)	D & d	$\frac{1}{2}\left(\frac{D}{\cos d}\right)$	$1-\cos\left(\frac{\frac{90°}{N}}{\cos d}\right)+a$
12	Chordal addendum (aC)	d	$\sin\left(\frac{\frac{90°}{N}}{\cos d}\right)$	
13	Dedendum (bC)	P	$\frac{2.188}{P}$ –a(pinion)	$\frac{2.188}{P}$ –a (gear)
14	Outside diameter (DO)	D, a & d	D + (2 x a) x cos. d	
15	Face	A	1/3 A (max.)	
16	Circular pitch (p)	p & N	$\frac{3.1416 \times p}{N}$	
17	Ratio	N gear & N pinion	$\frac{N \text{ gear}}{N \text{ pinion}}$	
18	Back angle (ɣO)	–	SAME AS PITCH ANGLE	
19	Angle of shafts	–	90°	
20	Part number of mating gear	–	AS REQ'D.	
21	Dedendum angle δ	A & b	$\frac{b}{A} = \tan \delta$	

Figure 16-29 Required tooth-cutting data for bevel gears

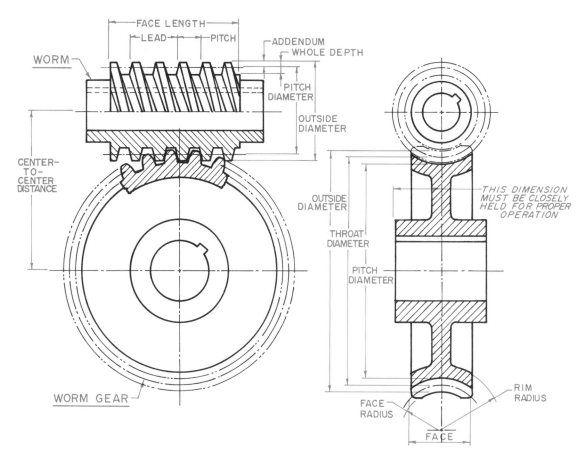

Figure 16-30 Example of worm and worm gear data

The worm's thread is similar in form to a rack tooth. The worm gear is similar in form to a spur gear, except that the teeth are twisted slightly and curved to fit the curvature of the worm.

When drawing the worm and worm gear, an approximate representation is used, Figure 16-30. Cutting data for the worm and worm gear must be listed on the drawing in the lower right side in the order illustrated in Figure 16-31 for the worm gear, and in Figure 16-32 for the worm.

Note: It is important that the mounting of a worm and gear set ensures that the central plane of the gear passes essentially through the axis of the worm. This may be accomplished by adjusting the gear axially at assembly by means of shims. When properly mounted and lubricated, worm gear sets will become more efficient after the initial breaking-in period.

Other information that must be included on a worm/worm gear detail drawing includes:

- Gear blank information
- Tooth-cutting data
- Reference to mating part

Gear Train

A *gear train* is two or more gears used to achieve a designed R.P.M. The ratio of the R.P.M. of the first gear to the R.P.M. of the final gear is called the *value* of the gear train.

Example:

$$\frac{\text{R.P.M. of shaft, first shaft}}{\text{R.P.M. of shaft, final shaft}} = \frac{\text{Value of}}{\text{gear train}}$$

Study the example on gear trains in Figure 16-33:

- Shafts 1 and 2 are connected by gear A and pinion B
- Shafts 2 and 3 are connected by gear C and pinion D
- Shafts 3 and 4 are connected by gear E and pinion F
- Gear A has 60 teeth
- Pinion B has 15 teeth
- Gear C has 40 teeth
- Pinion D has 20 teeth
- Gear E has 45 teeth
- Pinion F has 9 teeth

Problem:

The R.P.M. of shaft 1 is 200 R.P.M. What is the R.P.M. of shaft 4? What is the gear train value?
Each shaft taken individually:

$$\text{Ratio} = \frac{\text{N spur gear (driver gear)}}{\text{N pinion gear (drive gear)}}$$

R.P.M. of driven pinion = Ratio × R.P.M. driver

Figure 16-31 Required tooth-cutting data for a worm gear

REQUIRED CUTTING DATA			
ITEM	TO FIND:	HAVING	FORMULA
1	Number of teeth (N)	-	AS REQ'D.
2	Pitch diameter (D)	N & p	$\frac{N \times p}{3.1416}$
3	Addendum (a)	p	p x .3181
		P	$\frac{1}{P}$
4	Whole depth (ht)	p	p x .6866
		P	$\frac{2.157}{P}$
5	Lead (L) Right-Left	p & N	p x N
6	Worm part no.	-	AS REQ'D.
7	Pressure angle Ø	-	20° STANDARD 14°-30' OLD STANDARD
8	Outside diameter (DO)	Dt & Pa	Dt + .4775 x Pa
9	*Circular pitch (p)	P	$\frac{3.1416}{P}$
		L & N	$\frac{L}{N}$
10	Diametral pitch (P)	p	$\frac{3.1416}{p}$
11	Throat diameter (Dt)	D & Pa	D + .636 x Pa
12	Ratio of worm/worm gear	N worm & N worm gear	$\frac{N \text{ worm gear}}{N \text{ gear}}$
13	Center to center distance between worm & worm gear.	D worm & D worm gear	$\frac{D\text{worm} + D \text{ worm gear}}{2}$

Circular pitch (p) must be same as worm Axial pitch (Pa)

REQUIRED CUTTING DATA			
ITEM	TO FIND	HAVING:	FORMULA
1	Number of teeth (N)	P	$\frac{3.1416}{P}$
2	Pitch diameter (D)	Pa	(2.4 x Pa)+ 1.1
		DO & a	DO - (2 x a)
3	Axial pitch (Pa)	-	Distance from a point on one tooth to same point on next tooth
4	Lead (L) Right or Left	p & N	p x N
5	Lead angle (La)	L & D	$\frac{L}{3.1416 \times D}$ = tan La
6	Pressure angle (Ø)	—	20° STANDARD 14°-30' OLD STANDARD
7	Addendum (a)	p	p x .3183
		P	$\frac{1}{P}$
8	Whole depth (ht)	Pa	.686 x Pa
9	Chordal thickness	N, D & a	$\left[1 - \cos\left(\frac{90°}{N}\right)\right] \times \frac{D}{2} + a$
10	Chordal addendum	N, D & a	$\sin\left(\frac{90°}{N}\right) \times D$
11	Outside diameter (DO)	D & a	D + (2 x a)
12	Worm gear part no.	-	AS REQ'D.

Axial pitch (Pa) must be same as worm gear circular pitch (p)

Figure 16-32 Required tooth-cutting data for a worm

Step 1.
   a. Shaft 1 rotates at 200 R.P.M.
   b. Gear A has 60 teeth; pinion B has 15 teeth.
   c. This ratio is:
   $$\frac{\text{N spur gear}}{\text{N pinion gear}} = \frac{60}{15} = \frac{4}{1} \text{ or 4:1 ratio}$$
   d. R.P.M. of driven gear (shaft 2) = Ratio × .R.P.M. driver = 4 × 200 = 800 R.P.M.

Step 2.
   a. Shaft 2 rotates at 800 R.P.M.
   b. Gear C has 40 teeth; pinion D has 20 teeth.
   c. This ratio is:
   $$\frac{\text{N spur gear}}{\text{N pinion gear}} = \frac{40}{20} = \frac{2}{1} \text{ or 2:1 ratio}$$
   d. R.P.M. of driven gear (shaft 3) = Ratio × R.P.M. driver = 2 × 800 = 1600 R.P.M.

Step 3.
   a. Shaft 3 rotates at 1600 R.P.M.
   b. Gear E has 45 teeth; pinion F has 9 teeth.
   c. This ratio is:
   $$\frac{\text{N spur gear}}{\text{N pinion gear}} = \frac{45}{9} = \frac{5}{1} \text{ or 5:1 ratio}$$
   d. R.P.M. of driven gear (shaft 4) = Ratio × R.P.M. driver = 5 × 1600 = 800 R.P.M.

Step 4.
$$\frac{\text{Value of}}{\text{gear train}} = \frac{\text{R.P.M. last shaft}}{\text{R.P.M. first shaft}} = \frac{8000}{200} = 40 \text{ value}$$

To calculate the value of a *complete* gear train use the following formula:

The product of the number of teeth on all the drivers (spur gears) divided by the product of the number of teeth on all followers (pinion gears) equals the value of a complete gear train.

Example:

$$\frac{\text{N gear A} \times \text{N gear C} \times \text{N gear E}}{\text{N pinion B} \times \text{N pinion D} \times \text{N pinion F}}$$

$$\frac{60 \times 40 \times 45}{15 \times 20 \times 9} = \frac{108,000}{2700} = 40$$

In Figure 16-33, the gear train has been sketched with the gears in a neat row. This is done for the purpose of showing the sketch or basic design. In the final assembly, gears are actually clustered closer together to save space, and they are almost never in a row as shown.

Materials

Gears are made of many materials, such as brass, cast iron, steel, and plastic to mention but a few. Many bevel gears are forged, some are stamped from thin material, some are cast as a blank and machined, and others are die-cast to the exact size and shape.

Each application must be carefully analyzed. Metal gears have been used for years, but, if the load is not excessive, plastic gears have many advantages. Plastic runs quieter, has a self-lubricating effect, weighs much less, and costs much less. In addition, complicated multiple gears can be molded into a single piece which further reduces cost.

Design and Layout of Gears

The initial design of gears starts with the nominal size of the pitch diameters. The required speed ratio, loading, space limitations, and center-to-center distances are also important factors to consider. Mating gears and pinions must have equal diametral pitch in order to correctly mesh; therefore, the diametral pitch should be one of the first considerations. A complete analysis and design of a gear or a complete gear chain are very complex, and far beyond the scope of this text. Most designers try to use standard gears from a company that specializes in gears, gear chains, and gear design. Most gear manufacturing companies can be of assistance in designing gears for special applications. These companies can usually manufacture gears at a lower price than if they were made in-house.

Further analysis, in-depth study, and design data can be found in the most recent edition of *Machinery's Handbook.*

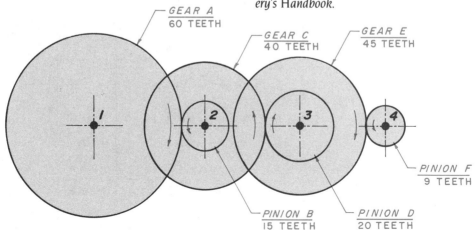

Figure 16-33 Gear train

Review

Calculate the following math problems; include all math work to illustrate how the answers were derived.

1. What is the outside diameter of a spur gear having a pitch diameter of 1.500 and 48 teeth?

2. What is the addendum and dedendum of a spur gear having a pitch diameter of 3.000 and 48 teeth?

3. With the addendum, dedendum, and pitch diameter of Problem 2, what is the root diameter and outside diameter?

4. How many teeth are on a spur gear having a pitch diameter of 1.750, diametral pitch of 20, and an outside diameter of 1.850?

5. What is the pitch diameter of a spur gear having a diametral pitch of 24, 78 teeth, and an outside diameter of 3.333?

6. What is the diametral pitch of a spur gear having 80 teeth and a 2.500 pitch diameter?

7. What is the root diameter of a spur gear having a pitch diameter of 2.000 and 32 teeth?

8. What is the pitch diameter of a spur gear having 40 teeth and a diametral pitch of 20?

9. What is the diametral pitch of a pinion gear having 75 teeth and 3.208 outside diameter? Use the formula $P = \dfrac{N}{D}$

10. How many teeth are required for a pinion gear with a mating spur gear having a P of 48, a 2.000 pitch diameter, and a 4:1 ratio?

11. What is the whole depth of a spur gear with the following specifications: 70 teeth, .7291 pitch diameter, .750 outside diameter, and 96 diametral pitch?

12. What is the circular pitch of a bevel pinion having a 16 diametral pitch, and 20 teeth?

13. A spur gear has a pitch diameter of 3.000 and a pinion gear has a pitch diameter of 1.500. If an R.P.M. 250 is needed at the shaft of the pinion gear, how fast should the spur gear be rotated?

14. What is the ratio between a spur gear with an R.P.M. of 175 and a pinion gear with an R.P.M. of 1050?

15. The ratio between a pair of bevel gears at a 90° angle and 28 P is:
    Gear: D = .9375, 45 teeth
    Pinion: D = .6250, 30 teeth

16. What is the outside diameter of a bevel gear with the following specifications: a 2.500 pitch diameter, a 48 diametral pitch, a 4:1 ratio, and 120 teeth?

17. What is the shaft center distance between a worm and a worm gear with the following specifications:
    Worm: Pitch diameter .500
                Diametral pitch 24
                Single thread
    Worm Gear: Pitch diameter 4.000
                Diametral pitch 24
                (N) Teeth 100

18. What is the gear ratio of a spur gear and pinion gear with the following specifications? (Calculate the answer using three different methods.)

Spur gear:	Pinion gear:
75 teeth	15 teeth
3.208 DO	.6250 D
3.125 D	.708 DO
725 R.P.M.	24 P
24 P	3/16 face
3/16 face	3625 R.P.M.

Chapter Sixteen Problems

The following problems are intended to give the beginning drafter practice in using the many formulas associated with gears, and practice in designing and laying out finished professional detail drawings of many major kinds of gears.

The steps to follow in laying out gears are:

Step 1. Study the problem carefully.

Step 2. Make a sketch if necessary.

Step 3. Do all math required for each problem. Keep all math work for rechecking.

Step 4. Center the required views within the work area.

Step 5. Include all dimensioning according to the most recent drafting standards.

Step 6. Add all required gear cutting specifications in the lower, right side of the paper.

Step 7. Recheck all work, and, if correct, neatly fill out the title block using light guidelines and neat lettering.

Problem 16-1

Use the following specifications to lay out a single-view drawing of a 2:1 ratio spur gear and pinion gear *in mesh*.

Spur gear:	Pinion gear:
88 teeth	hub diameter 1.625
pressure angle 20°	bore of 1.000 diameter
pitch diameter 5.500	pressure angle 20°
hub diameter 2.625	
bore of 2.000 diameter	

Use all standard drafting methods to illustrate the outside diameter, pitch diameter, and root diameter. Calculate and add the center-to-center distance between the shafts.

Problem 16-2

Draw a spur gear having the following specifications: 56 teeth, pressure angle 20°, pitch diameter 3.500, hub diameter 1.00 gear blank with overall size (width) 1.00, face size .50, bore .5625; use a ⅛'' set screw.

Complete two views using the half-section method of representing gears. Do not show the gear teeth; use the conventional method of illustrating teeth. Dimension per all standard practices. Calculate and add all standard required cutting data to the lower right side of the work area. Material S.A.E. #3120 cast steel. Heat treatment: carburize .015-.020 deep.

Problem 16-3

Complete a two-view detail drawing of a pinion gear having the following specifications: calculate and add all standard cutting data: 30 teeth, pressure angle 20°, pitch diameter 6.000, hub diameter 2.000, heat treatment: carburize .050 deep, gear blank with overall size (width) 3.500, face size 1.250, bore 1.000, material S.A.E. #4620 steel.

Problem 16-4

Make a detail drawing, completely dimensioned, of a spur gear with the following specifications: diametral pitch 16, pressure angle 20°, pitch diameter 5.500, whole depth .1348, outside diameter 5.625, addendum .0625, working depth .125, circular thickness .098, chordal thickness .0633, chordal addendum .0985, whole depth .1348, and dedendum .0723, material cast brass.

Design a simple gear blank with a set screw to fasten gear to a .75 dia. shaft.

Problem 16-5

Design and draw a detail drawing, completely dimensioned, of a pair of bevel gears having teeth of a 4.0 diametral pitch; the gear with 25 teeth, the pinion gear with 13 teeth, face width of 1.00, gear shaft 1.25 diameter, and pinion shaft .875 diameter. (Calculate for FN-2 fits with the shaft.)

Design the hub diameter to be approximately twice the diameter of the shaft diameter. Backing for the gear 1.375 and for the pinion .75. Design and dimension the remaining portion to suit, using standard drafting practices. Add all cutting data to the lower right side of the work area. *Material:* S.A.E. #3120 cast steel, heat treatment: carburize .015/.020 deep.

Problem 16-6

Make a design layout of a worm and worm gear with the following specifications: shaft diameter 1.25, single-thread worm, lead .75, worm gear with 28 teeth. Add all dimensions and cutting data as required.

Problem 16-7

Sketch a clock gear chain with the following specifications:

First gear (main wheel): 84 teeth
Second gear: pinion gear 8 teeth
           spur gear 60 teeth
Third gear (center shaft): pinion gear 12 teeth
                    spur gear 68 teeth
Fourth gear: pinion gear 7 teeth
           spur gear 66 teeth
Fifth gear (escapement gear): pinion gear 7 teeth
                    spur gear 33 teeth

Note: In actual practice, a clock's first gear (main wheel) is usually driven by a spring. The minute hand is connected to the shaft of the third gear (center shaft) and the fifth gear is the escapement gear.

Problem 16-8

Using the sketch of a clock gear chain in Problem 16-7, answer the following questions:

1. How many times will the second gear rotate for one full revolution of the first gear?

2. How many turns will the third gear rotate for one full revolution of the second gear?

3. How many teeth of the main wheel are required to turn the third gear shaft one complete revolution? (On a clock, one revolution of the third gear equals one hour. This wheel turns 24 times for one day.)

4. What is the required revolution fo the first gear (main wheel) for one day's running time (24 turns of the center wheel).

5. What is the value of the complete gear train?

6. As this is a clock gear chain in revolutions, how many will the fifth gear (escapement) make in twelve hours? (The third gear will turn 12 times.)

Problem 16-9

Design and develop a design layout two-view drawing of a gear train with the following specifications. Design a compact clustered arrangement of the gears within an 8.0 x 10.0 supporting plates area.

First shaft: 42 teeth/4.0 outside diameter-gear
Second shaft: 6 teeth/1.0 outside diameter-pinion
Third shaft: 6 teeth/.75 outside diameter-pinion
                30 teeth/2.5 outside diameter-gear
Fourth shaft: 6 teeth/.75 outside diameter-pinion
                 30 teeth/2.0 outside diameter-gear
Shaft sizes = .375 dia/supporting plates 1.5 apart

Problem 16-10

Locate a windup clock or some other such small gear chain device and make a design layout similar to that in Problem 16-9.

1. What type of gear does the drawing represent?
2. Of what material is the gear made?
3. What is the surface texture specification for the gear?
4. What is the upper-limit dimension of the keyway?
5. What is the lower-limit dimension of the keyway?
6. How many teeth does the gear have?
7. What is the width of the hub of the gear?
8. What is the maximum overall height of the keyway?
9. What is the maximum bore diameter of the gear?
10. What is the pitch cone angle for the gear teeth?

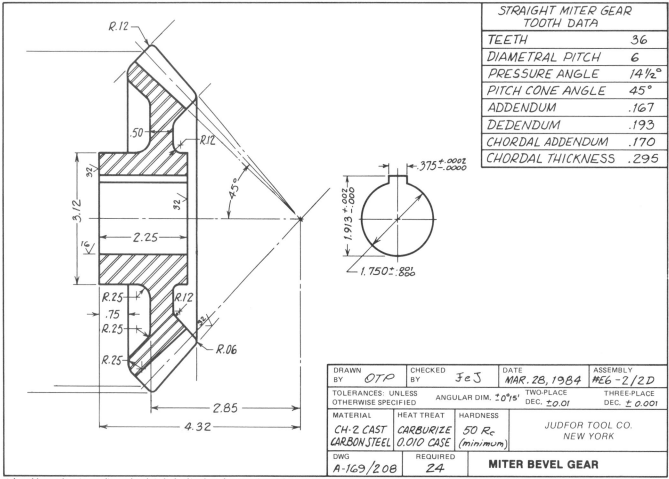

STRAIGHT MITER GEAR TOOTH DATA	
TEETH	36
DIAMETRAL PITCH	6
PRESSURE ANGLE	14½°
PITCH CONE ANGLE	45°
ADDENDUM	.167
DEDENDUM	.193
CHORDAL ADDENDUM	.170
CHORDAL THICKNESS	.295

DRAWN BY OTP	CHECKED BY JeJ	DATE MAR. 28, 1984	ASSEMBLY #E6 -2/2D
TOLERANCES: UNLESS OTHERWISE SPECIFIED	ANGULAR DIM. ±0°15'	TWO-PLACE DEC. ±0.01	THREE-PLACE DEC. ± 0.001
MATERIAL	HEAT TREAT	HARDNESS	JUDFOR TOOL CO. NEW YORK
CH-2 CAST CARBON STEEL	CARBURIZE 0.010 CASE	50 R꜀ (minimum)	
DWG A-169/208	REQUIRED 24	**MITER BEVEL GEAR**	

Adapted from "Blueprint Reading and Technical Sketching for Industry" and used by permission of Thomas P. Olivo and C. Thomas Olivo. Published by Delmar Publishers Inc.

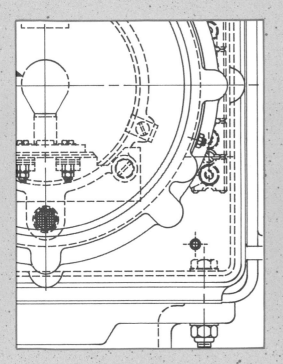

A hypothetical engineering department and the chain of command for this department are given in this chapter. Related information, such as engineering change order procedures, master lists, and the kinds of drawings used in industry, is also covered. The beginning drafter is given the opportunity to do basic designing and develop various tolerances for mating parts. The most recent dimensioning standards and the correct tolerances for mating parts should be stressed in all drawing problems at the end of this chapter.

CHAPTER SEVENTEEN
Assembly and Detail Drawings

The Engineering Department

Engineering departments vary from company to company, but most of them have various things in common. An example of a relatively small engineering department is illustrated in Figure 17-1.

An engineering department consists of various departments, usually headed by a chief engineer. An example of the various departments associated with a small engineering division would include the design section, the application section, and the quotation department. The *design section* is responsible for new design products and design improvement of existing products. The *application section* handles special order variations of existing products. For example, a scale company customizes a scale platform size to meet a particular customer need. The *quotation department* gives estimated costs or quotations to customers needing special modifications of standard products manufactured by the company. In some companies or engineering departments, these various departments may be broken down into actual fields of engineering, such as the mechanical section, the electronic section, the illustration section, and so forth.

Directly under the Chief Engineer are the various department heads, usually engineers, designers or supervisors. Regardless of their expertise, these department heads have drafters of many levels under them. Most engineering departments have a clear-cut departmental structure. Others move drafters back and forth among the various departments as needed.

New personnel should fully understand the engineering department structure of the organization. An engineering department must work as efficiently as possible in order to provide an orderly flow of drawings, parts lists, and drawing changes.

Drawing Revisions

Anyone having a suggestion to improve a part or assembly within the product, or a need to correct an error on a drawing, must bring it up at an *engineering change request* meeting. This is sometimes referred to as an ECR meeting, Figure 17-2. After a drawing has been released to the factory *no* change can be made to it without an *engineering change order* (ECO), even by the drafter who originally drew it. The change cannot be made unless all departments concerned agree to it, (refer back to Figure 17-2). The ECR and ECO explain *what* change is to be made, what it *was* before, *why* the change is to be made, whether it *affects other*

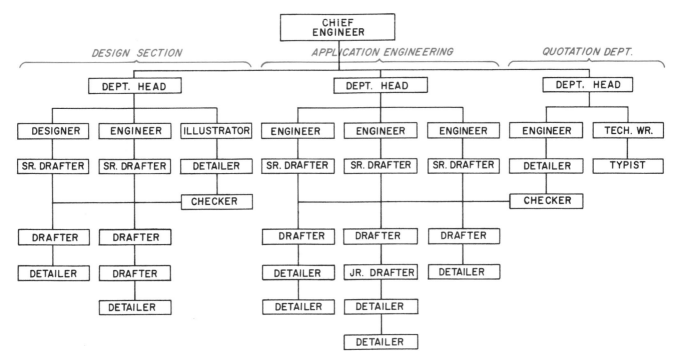

Figure 17-1 Chain of command in an engineering department

parts, and *who* suggested the change. The ECR and ECO sometimes list the departments throughout the company that will be directly affected by the change. When the engineering change request has been approved, the committee issues an engineering change order. The ECO is usually given to a beginning drafter or detailer to make the actual change.

Changes are made by erasures directly on the original drawing. Additions are simply drawn in on the original drawing. If a dimension is not noticeably affected by the change, the dimension is simply corrected and underlined with a heavy black line. This indicates that the dimension is correct, but slightly out-of-scale, Figure 17-3.

If the change is extensive, the drawing must be redrawn. The original drawing is *not* destroyed; it is stamped OBSOLETE and includes the note, "SUPERSEDED BY (the number of the corrected drawing)." The new drawing must have a note referring back to the superseded drawing. The change must be indicated, both as to *where* it has been made, and basically as to *what* was changed. Each company has its own standard method to indicate this information. The most common method is to place a letter or number in a small balloon near the dimension or dimensions where the change was made. Note in Figure 17-3 that the A within the balloon indicates which dimension was changed. The same letter or number

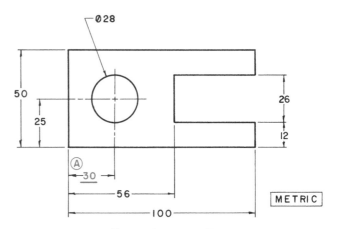

Figure 17-2 Engineering change request order procedure

Figure 17-3 Balloon change call-off

must be listed in the revision block, Figure 17-4, which can be located either to the left of the title block or on the top right corner of the drawing sheet. The revision block lists the letter or number of the change, what was changed, the date of the change, and who made the change. If the change is extensive, only the ECO number is listed. Anyone needing to know why the change was made can refer back to the ECR or ECO.

Some companies add the *last* changed letter to the drawing number which, in effect, actually changed the part number. Part number C-65498 would become C-65498-A. If changed again at a later date, it would become C-65498-B, and so on. In certain cases, appropriate administrative changes may be made without an engineering change order. Such simple changes as misspelled words or incorrect references to improve clarity are nonengineering changes that would not affect any other department.

Invention Agreement

Anyone who works in a job that is creative, such as a drafter, a designer or an engineer, must sign an agreement form giving the company the right to any new invention designed while working for the company. The form puts in writing that the employee will not reveal any of the company's discoveries or projects. The invention agreement is in effect from six months to two years after an employee leaves the company, depending upon the company. This is so employees will not invent something, quit, and patent it themselves.

A company is not obligated to, and does not usually, give extra pay for an invention an employee develops. This is what the employee is paid to do. However, a company usually recognizes talent, and will reward an employee with a promotion, stocks, bonds, or, possibly, a raise. Signing an employer/employee invention agreement is a normal request in any company. A prospective employee should read the contract carefully and understand it fully before signing it. An applicant who does not sign the agreement will probably not be offered the job.

Title Block

Title blocks vary from company to company. The latest ANSI standard title block is illustrated in Figure 17-5. The function of a title block is to indicate information that is not given directly on the drawing itself. The following information is usually found in the title block:

- Name and address of the company
- Drawing title
- Drawing or part number
- Scale of the drawing

Figure 17-4 Revision block (*Courtesy Bishop Graphics Accupress*)

- Name of the drafter and the date the drawing was completed
- Name of the checker and the date the drawing was checked
- Name of the chief official approving the drawing and the date it was approved
- Material the part is to be made of
- Tolerance/limits of dimensions
- Heat treatment requirements, if necessary
- Finish requirements, if necessary

Size of Lettering within Title Block

- General information, .125 (3) freehand letters/ .120 mechanical letters
- Drawing title, .250 (6) freehand letters/.240 mechanical letters
- Drawing number, .312 (8) freehand letters/.350 mechanical letters

Checking Procedure

The importance of absolute accuracy in engineering drawings cannot be stressed enough. The slightest error could cause tremendous and unnecessary expenses. In some fields of engineering, such as in aircraft or space products, an error could cost lives. Therefore, it is important to take every precaution against errors. In most engineering departments, checkers are used to verify all dimensions and all stress analysis, and to see that standard material and tools are used wherever possible. It is a checker's responsibility to check the drawing before it is released.

Drafter's Checklist

The drafter should check the drawing before giving it to the checker to further ensure accuracy. Some of the things to look for are:

- What is the drawing's general appearance (legibility, neatness, and so forth)?
- Does it follow all drawing and company standards?
- Are the dimensions and instructions clear and understandable?
- Is the drawing easy to understand?

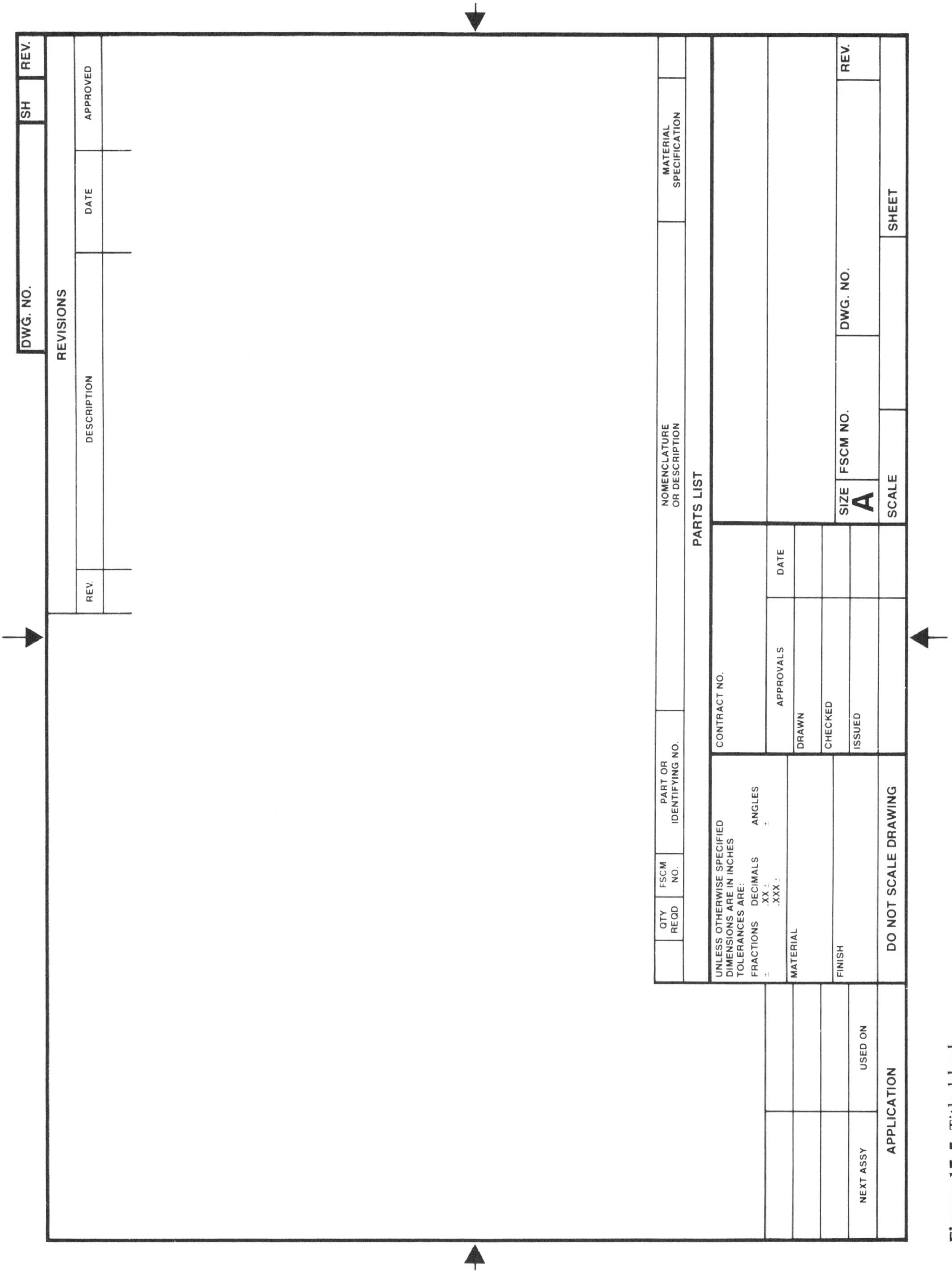

Figure 17-5 Title block

REV.

SH

DWG. NO.

REVISIONS

APPROVED

DATE

DESCRIPTION

REV.

MATERIAL
SPECIFICATION

NOMENCLATURE
OR DESCRIPTION

PARTS LIST

PART OR
IDENTIFYING NO.

FSCM
NO.

QTY
REQD

CONTRACT NO.

UNLESS OTHERWISE SPECIFIED
DIMENSIONS ARE IN INCHES
TOLERANCES ARE:

FRACTIONS   DECIMALS   ANGLES
.XX -
.XXX -

MATERIAL

FINISH

APPROVALS

DATE

DRAWN

CHECKED

ISSUED

SIZE
A

FSCM NO.

DWG. NO.

REV.

SCALE

SHEET

DO NOT SCALE DRAWING

NEXT ASSY   USED ON

APPLICATION

- Are all dimensions included? A machinist must neither have to calculate to find a size or location nor assume anything, and should not have any question whatsoever as to what is required.
- Are there unnecessary dimensions?
- Is the drawing prepared so the part may be manufactured in the most economical way?
- Will the part assemble with mating parts?
- Have all limits, tolerances, and allowances been properly analyzed for all moving parts?
- Have undesirable accumulations of tolerances been adequately analyzed?
- Are all notes added?
- Are finish texture symbols added?
- Are the material and treatment of each part adequate for the design?
- Is the title block complete? Does it include the title, number of the part, drafter's name, and any other required information?

Numbering System

The numbering system varies from company to company, but all companies have some kind of a system of identifying and recording drawings. There is no standard system used by all companies. Most companies assign a sequential number for each drawing as it is finished. A prefix letter is added to this number to indicate the drawing size. A drawing, drawn on a C-size sheet with an assigned number of 114937 would be indicated on the drawing by C-114937. The drawing number should be freehand lettered .312 (8) high or mechanically lettered .350 (9) high.

Parts List

A *parts list* is actually a bill of material that itemizes the parts needed to assemble *one* complete full assembly of a product. The parts list is usually on a separate A-size page or pages other than the assembly drawing. The parts list is usually assigned the same number as the corresponding assembly drawing. For example, an assembly drawing with a drawing number D-114937 would have a corresponding parts list number A-114937.

The parts list contains all drawings, all purchased parts, all drawing numbers, material of each part, and the quantity of individual parts needed for one complete assembly.

The drafter usually makes up the parts list. Companies use various methods to list the parts. Some list the parts in the order of size, some in the order of importance, and others list the parts in order of assembly. The latter method is preferred. Figure 17-6 shows an example of a parts list for a machine vise. The parts are listed in the order the drafter suggests for assembling the machine vise.

Note that purchased parts do not have a plan number; therefore, they are listed under "drawing number" as PURCH, indicating that these are standard parts to be purchased, not manufactured. Purchased parts must be listed giving complete specification so that there is no question whatsoever as to what must be purchased. In this example, items numbered 7, 8, 17, 23, and 24 are purchased parts.

Some companies use a system of indents to list the parts (refer back to Figure 17-6). There are four indents, indicating the various kinds of drawings. The first indent indicates the main assembly drawing. The second indent indicates all subassembly drawings used to make up the assembly. The third indent indicates all detail drawings and modified purchased parts with drawing numbers used to make up the assembly. The fourth indent indicates all purchased parts used to make up the assembly.

Not all companies use this sytem, but using the indent system gives a quick indication of which parts are used to make up the various subassemblies. Note on the example list that item number 3 is indented one place, indicating a subassembly. Items numbers 4, 5, and 6 are detail drawings used to make up the subassembly item number 3. Items number 7 and 8 are purchased parts used to complete the subassembly item number 3.

INDENT 1 = ASSEMBLY DRAWING
INDENT 2 = SUB-ASSEMBLY DRAWINGS
INDENT 3 = DETAIL DRAWINGS
INDENT 4 = PURCHASED PARTS

MASTER PARTS LIST

NO	DRAWING NO.				DESCRIPTION	MATERIAL	QUAN
1	D77942	VISE ASSEMBLY - MACHINE				AS NOTED	1
2							
3	C77947		BASE-VISE			AS NOTED	1
4	B77952			BASE-LOWER		C.I.	1
5	A77951			BASE-UPPER		C.I.	1
6	A77946			SPACER-BASE		STEEL	1
7	PURCH.				BOLT 1/2-13 UNC -2.0 LG.	STEEL	1
8	PURCH.				NUT 1/2-13 UNC	STEEL	1
9							
10	C77955		JAW-SLIDING			STEEL	1
11							
12	A77954		SCREW-VISE			STEEL	1
13	A77953		ROD-HANDLE			STEEL	1
14	A77956		BALL-HANDLE			STEEL	2
15							
16	A77961	PLATE-JAW				STEEL	2
17	PURCH.			SCREW 1/4-20 UNC-1.0LG.		STEEL	4
18							
19	A77962	COLLAR				STEEL	1
20							
21	A77841	KEY-SPECIAL VISE				STEEL	2
22							
23	PURCH.				BOLT 1/2-13 UNC-4.0LG.	STEEL	4
24	PURCH.				NUT 1/2-13 UNC	STEEL	4

JAN ENGINEERING PETERBOROUGH, N.H	Model No. 160	Parts Lister NELSON	Date 6 AUG 86
Title VISE ASSEMBLY-MACHINE		PAGE 1 OF 1 PAGES	Drawing No. A1198891

FIG. 17-6

Figure 17-6 Parts list

Personal Technical File

It is important to be able to locate technical information quickly. Usually, a company manufactures a certain kind of product or performs a particular kind of engineering, thereby using certain related technical information. A conscientious drafter should develop and keep updated a personal technical file. Any information, charts or reference materials used many times in the course of daily work should be copied and put into a 3-ring binder with index tabs for quick retrieval. The following is an example of what a technical file should contain:

- All company product data — especially those associated with daily assignments.
- Notes, copies or clippings from various technical magazines, literature, and so forth, associated with the company product.
- Information on standard materials commonly used in daily assignments.
- Miscellaneous related information that would make your job more efficient.
- Various page numbers in printed material that are helpful and often used for reference.
- Records of various supervisors, pay levels you have achieved with dates or promotions and job levels. Various personal information that may be needed at a future date.

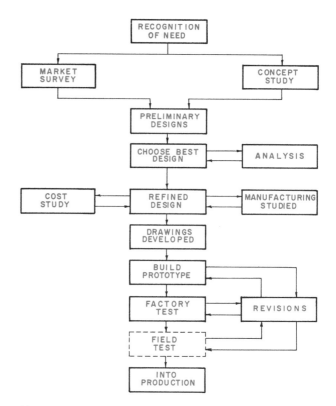

Figure 17-7 Design procedure

The Design Procedure

There are two types of design: conceptual design and scientific design. In *conceptual design*, much use is made of known technical information from such sources as reference books, technical manuals, and handbooks. A competent designer should be aware of the latest technical information available. In *scientific design*, use is made of established principles of math, physics, mechanics, and, sometimes, chemistry.

The designer must be able to communicate ideas both graphically and symbolically, and have the ability to analyze and adapt these ideas to new products. The designer must combine conceptual design and scientific design principles in order to solve a problem or to meet a particular need.

The actual process of developing a new product, from its established need and conception to its final production, consumes many hours and involves many highly trained, skilled personnel. The flow chart in Figure 17-7 gives an example of the process involved.

The activity begins with the *recognition of a need*. A *market survey* is conducted to ensure that the need is great enough to indeed design and manufacture a particular product. At the same time, a *concept study* is conducted to get an idea of the best design to meet the particular need. At this stage, many ideas are presented — usually in sketch form. Ideas are very broad and without restraint in order to arrive at a

totally new, fresh, and unique solution. The more ideas proposed, the greater are the chances of finding the best solution to the problem. Some companies put a group of designers together in what is referred to as a "think tank" to come up with many varied ideas. At this stage, no attempt is made to evaluate any of the ideas presented.

Design ideas and sketches are usually followed by a study of suitable materials and strength of materials, and, if needed, basic calculations related to velocity and/or acceleration.

Preliminary designs are developed, usually full size if possible. All drawing is done to exact size or scale using a very sharp 4-H or 6-H lead. No attempt is made to draw using the usual "thick-thin" alphabet of lines. At this time, the emphasis is on the strength of material, the clearance of moving parts, the use of standard materials, and the manufacturing processes. Costs are an important factor in this stage of the design process. The preliminary designs are studied and the overall design that best meets the required need, at the lowest cost, is selected.

The *best design* is further analyzed and a *cost study* is conducted. The *refined design* is now given to the drafters and detailers to develop the *working drawings* required for *manufacturing* the product.

The *prototype* is built from the working drawings to test both the overall design and to recheck all working drawings. The prototype is thoroughly *factory tested*.

Any improvements in the overall design or better manufacturing methods are studied and, if necessary, *revisions* are made to the working drawings. A product must be as foolproof as possible before it goes into final production. If time permits, the prototype is field tested. The working drawings are now released to the production division of the factory and put *into production*.

A product, such as an automobile, requires the previously described design process for each and every subassembly, Figure 17-8. Components in the console system in Figure 17-9, such as the laser key entry system, the automatic level attitude, spoiler control system, and the map navigation display system followed a similar design process.

The engineering department is responsible for maintaining and filing the final working drawings, and keeping up-to-date with all ECOs and records associated with changes. If the product is a new design, or if part of the product incorporates a new design feature, a *patent drawing* is developed and filed with the patent office.

Figure 17-8 Innovative automobile evolved from design process (*Courtesy Buick Public Relations*)

Figure 17-9 Touch control electronic console system was created using the design process steps (*Courtesy Buick Public Relations*)

Working Drawings

Many different kinds of drawings are associated with the development and production of a product. The drafter must be familiar with each kind and able to recognize each, know the company standards used for each kind, and be able to draw each kind of drawing.

The major kinds of working drawings are:

- Design layout
- Assembly drawing
- Subassembly drawing
- Detail drawing
- Purchased parts
- Modified purchased parts drawing

Design Layout

The design layout is usually done by the designer or engineer using a sharp 4-H or 6-H lead. This is either an exact size or scaled as large as space permits. The *design layout* is drawn from various sketches made in the first steps of the design process. The designer tries to use standard materials, standard manufacturing processes and, if needed, standard size fasteners. A design layout is not usually dimensioned, except for general overall dimensions. Many designers pencil in nominal sizes, special notes or requirements and important required tolerances or limits.

After the preliminary design layout has been approved, the detailer or drafter uses it to draw the necessary subassemblies and detail parts. As there are very few or no dimensions on the design layout, the detailer or drafter must carefully transfer each size dimension from the design layout to the working drawing. In transferring these size dimensions, the detailer or drafter must also try to use standard materials and standard manufacturing processes, as well as calculate all dimensions. Before starting a project, regardless of its complexity, the detailer or drafter must fully understand the function of the design. All tolerances should be kept as large as possible, except in the clearance of moving or mating parts. Tight or close tolerances add extra cost to the product.

Assembly Drawing

Any product that has more than one part must have an assembly drawing. The *assembly drawing* illustrates how a product is assembled when completed. The assembly drawing can have one, two, three or more views that are placed in the usual positions. Because the assembly drawing is used to show how the parts are assembled, a full section view is usually used. There is only one assembly drawing for any given product.

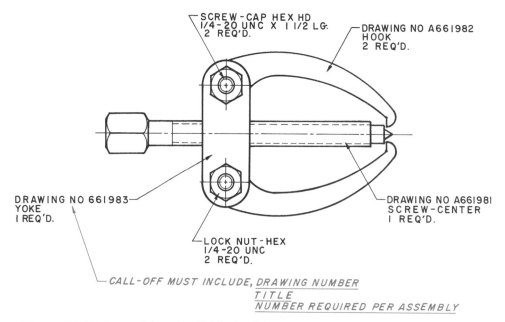

SCREW-CAP HEX HD
1/4-20 UNC X 1 1/2 LG.
2 REQ'D.

DRAWING NO A661982
HOOK
2 REQ'D.

DRAWING NO 661983
YOKE
1 REQ'D.

DRAWING NO A661981
SCREW-CENTER
1 REQ'D.

LOCK NUT-HEX
1/4-20 UNC
2 REQ'D.

CALL-OFF MUST INCLUDE, DRAWING NUMBER
TITLE
NUMBER REQUIRED PER ASSEMBLY

Figure 17-10 Assembly call-off (Method 1)

Assembly drawings are usually not dimensioned, except for general overall dimensions or to indicate the capabilities of the assembly. For example, a clamp assembly might add a dimension to illustrate how wide it opens. Assembly drawings, as a rule, do not contain hidden lines, unless they are absolutely necessary to illustrate some important feature that otherwise may be missed.

An assembly drawing must call-off each part that is used to make up the assembly. Each part must be called off by its part number, and title, and the total number required to make up one complete assembly. Companies usually use one of two methods to call-off the parts.

Method one: Each part has a leader line with the part number, title, and number required, Figure 17-10.

Method two: Each part has a leader line with a balloon call-off letter inside, Figure 17-11. This method uses a parts tabulation chart to also list the part number, title, and number required. Because the assembly drawing is made up of many different parts, each of which could be manufactured of a different material, the title block material box indicates "AS NOTED."

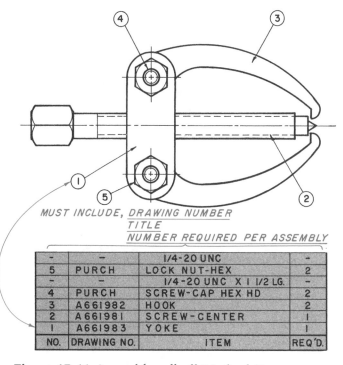

MUST INCLUDE, DRAWING NUMBER
TITLE
NUMBER REQUIRED PER ASSEMBLY

NO.	DRAWING NO.	ITEM	REQ'D.
-	-	1/4-20 UNC	-
5	PURCH	LOCK NUT-HEX	2
-	-	1/4-20 UNC X 1 1/2 LG.	-
4	PURCH	SCREW-CAP HEX HD	2
3	A661982	HOOK	2
2	A661981	SCREW-CENTER	1
1	A661983	YOKE	1

Figure 17-11 Assembly call-off (Method 2)

Subassembly Drawing

A subassembly is composed of two or more parts permanently fastened together. A *subassembly drawing* is very similar to an assembly drawing. All standard practices associated with an assembly drawing apply to the subassembly drawing. That is, no or few over-

all dimensions, no hidden lines unless absolutely necessary, material listing "AS NOTED," and call-offs using either method one or two.

Subassemblies sometimes require machining operations *after* assembly, Figure 17-12. For instance, a hole is drilled through the parts *after* assembly. This would have been much more difficult and costly if it had been done to each individual part. Thus, all the

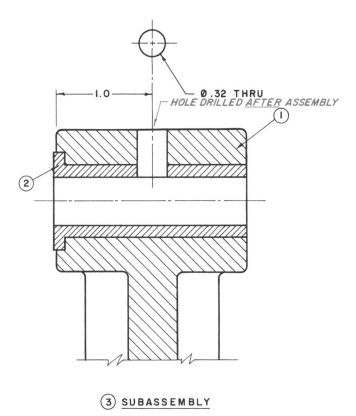

③ SUBASSEMBLY

Figure 17-12 Subassembly drawing

required dimensions to drill the hole must be added to the subassembly drawing.

The subassembly is used whenever two or more small assemblies are permanently made up; and, alone with other parts, makes up the final assembly. There can be any number of subassemblies in the final assembly. Think of a subassembly as the major components needed to put together an automobile. The engine is a subassembly, the water pump is a subassembly, and the transmission is a subassembly, and so forth. These subassemblies, together with various single parts, make up the final assembly of the automobile.

Note that subassemblies are usually stocked and purchased as a unit by themselves.

Detail Drawings

Every part must have its own fully dimensioned detail drawing, with its own drawing number and title block. All information needed to manufacture the part is contained in the *detailed drawing*, including as many views as are needed to fully illustrate and dimension the parts.

The general practice in mechanical drafting is to draw only one part to one sheet of paper, regardless of how small the part is.

Most companies use general detail drawings as just described. However, some companies break the detail

drawings into the various manufacturing processes, such as:

- Pattern detail drawings for castings
- Cast detail drawings for cast parts
- Machine detail drawings for machining processes
- Forging detail drawings for forged parts
- Welding detail drawings for welded parts
- Stamping detail drawings for stamped parts

Purchased Parts

A manufacturing company cannot make such standard items as nuts, bolts, screws, and washers as cheaply as companies that specialize in making these fasteners. All design, therefore, should specify standard parts whenever possible. Purchased parts are not drawn but are simply called-out on the assembly drawing. It is important that all specific information be given for the particular purchased part. A full description must be given as to diameter, length, material, and so forth. Odd purchased parts, such as a 12-volt D.C. motor would list its size and a note such as: "12-VOLT, D.C. MOTOR, G.E. NO. 776113 OR EQUAL."

Modified Purchased Parts

A purchased part that needs to be modified, however slightly, must be fully detailed and dimensioned as with any detail part, except that under "material" the purchased part size and description is given, Figure 17-13. In this example, a standard 1/4-20 UNC x 1.5 long hex-head machine screw has a special tip and a .125 dia. hole in it. The standard fastener is drawn with the modifications added and dimensioned. The standard fastener is called-out in the title block under "material." This indicates that this particular part must be manufactured from a standard fastener.

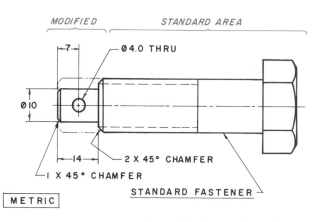

Figure 17-13 Drawing of modified purchased part

Pattern Drawings

All patent applications for a new idea or invention must include a pattern drawing to illustrate and explain its use. Pattern drawings must be mechanically correct, and must adhere to strict patent drawing regulations, Figure 17-14.

Patent drawings are more pictorial and explanatory in nature than regular detail or assembly drawings. Center lines, dimensions, and notes are omitted from patent drawings. All features and parts are identified by numbers which refer to the written explanation of description of the new idea or invention. Line shading is used to improve readability.

All patent drawings must be done in ink on heavy white paper, exactly 10.0 x 15.0 in size, with a 1.0 border on all sides. A space of not less than 1.25 from the shorter border is left blank for data to be added by the patent office.

Because of the many patent drawing rules and requirements, special drafters are employed to make only patent drawings. A complete list of these standards can be obtained from the Superintendent of Documents, U.S. Government Printing Office, Washington, DC 20402. Write for the publication, A *guide for patent drawings.*

Computer Drawings

Refer to Chapters 10 and 11 for information about *computer drawings.*

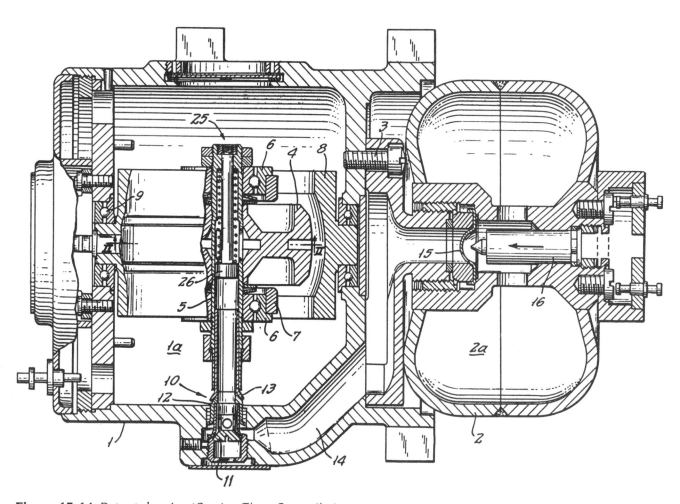

Figure 17-14 Patent drawing (*Courtesy Timex Corporation*)

Review

1. List the information that must be included in a title block according to the most recent ANSI standard.

2. What three general methods are used in listing parts of an assembly on a parts list?

3. Explain the design process.

4. What is a modified purchased part, and does it need a drawing? Explain.

5. Explain the drawing revision process.

6. What is the difference between an assembly drawing and a subassembly drawing?

7. What is the invention agreement, and who must sign it?

8. What is a design layout drawing and who usually draws it?

9. List the important parts of the drawing that should be checked by the drafter before submitting the final drawing to the checker.

10. What should be included in a personal technical file?

11. Explain in full what an assembly drawing is.

12. Explain what must be done if a drawing revision is extensive and the drawing must be completely redrawn.

The following problems are intended to give the beginning drafter knowledge of the many drafting procedures, and practice in developing detail drawings, subassembly drawings, and assembly drawings. Students should practice calculating tolerances for various fits of mating parts, and filling out parts lists.

The steps to follow in developing these problems are:

Step 1. Using a sharp 4-H lead, lay out a full scale design layout drawing of the problem, using the basic overall dimensions provided. Lay out and design any missing dimensions, using standard sizes and materials. (Use as many views as necessary to fully illustrate the assembly.)

Step 2. Calculate all required fits and tolerances as necessary on a separate sheet of paper.

Step 3. Center a fully dimensioned detail drawing of each part using the latest drafting standards. Apply tolerances and limits to mating parts. Assign the drawing number per the numbering provided on the design layout drawing, and include the material to be used.

Step 4. Develop *subassembly drawings* wherever two or more parts are assembled, *before* the final assembly, as needed. List each part and part number, using standard drafting practice. (Material is to be listed "AS NOTED.")

Step 5. Develop a final assembly drawing and list each part and part number, using standard drafting practice. (Material is to be listed "AS NOTED.") Use standard fasteners wherever possible.

Step 6. Develop a parts list of the assembly. List parts per standard drafting practices, in the recommended order of assembly.

Step 7. Recheck all drawings and related calculations for accuracy.

Step 8. Make a list of recommendations as to how the design could be improved.

Problems 17-1 through 17-8

Follow the eight listed steps to develop Problems 17-1 through 17-8.

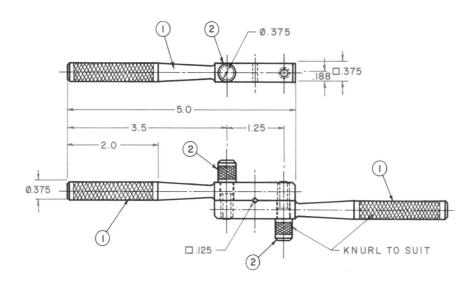

Problem 17-1

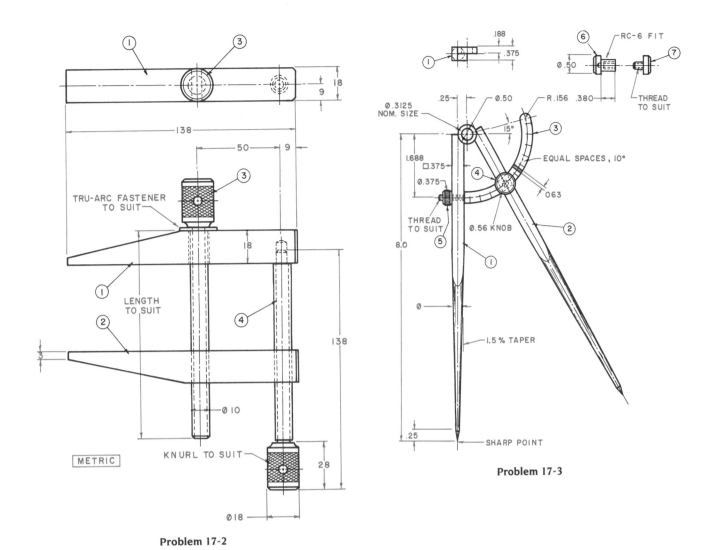

① ③

⑤ RC-6 FIT

① ⑥ ⑦

.188
.375
Ø.50
THREAD
TO SUIT

18
9
138
50
9

TRU-ARC FASTENER
TO SUIT

③

18

LENGTH
TO SUIT

①

②

3

Ø10

138

KNURL TO SUIT

METRIC

28

Ø18

④

Problem 17-2

Ø.3125
NOM. SIZE
.25
Ø.50
R.156
.380

⑥

15°
③

EQUAL SPACES, 10°

1.688
□.375
0.375

④

063

THREAD
TO SUIT

⑤

Ø.56 KNOB

②

8.0

①

0

1.5 % TAPER

.25

SHARP POINT

Problem 17-3

PAD SIZE
.75 X 1.0

BALL JOINT
Ø.25

④ ⑥

FASTENER TO SUIT

⑦

⑤

①

PAD SIZE
Ø1.0

PEEN OVER AT ASSEMBLY

Ø.50
THREAD TO SUIT

HANDLE Ø.88 X 3.0 LG

2.88

②

4.25

FASTENER TO SUIT

SLIP FIT (RC-6

.75

.88

.1875

.88

③

18.0

Problem 17-4

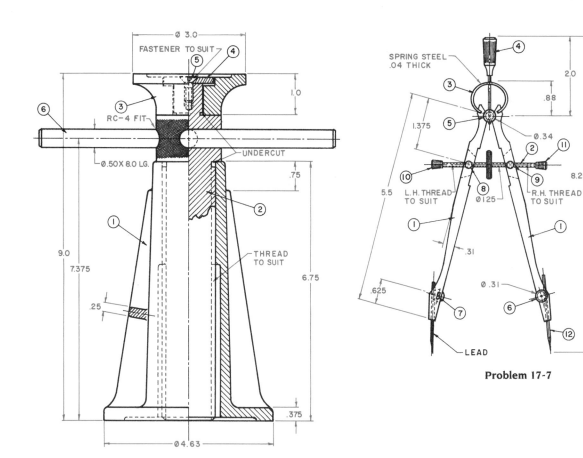

Ø 3.0
FASTENER TO SUIT ④
⑤
③
⑥
RC-4 FIT
UNDERCUT
Ø.50 X 8.0 LG.
②
①
THREAD TO SUIT
1.0
.75
6.75
9.0
7.375
.25
.375
Ø4.63

Problem 17-5

SPRING STEEL .04 THICK
④
③
⑤
②
⑪
⑩
⑧
⑨
L.H. THREAD TO SUIT
R.H. THREAD TO SUIT
Ø.34
Ø.125
1.375
5.5
.31
.625
⑦
Ø.31
⑥
LEAD
⑫
2.0
.88
④
③
⑤
②
0.75
8.25 (NOM.)
①
.25
⑥
⑦
⑫

Problem 17-7

③
Ø.125 X .188 LONG ④
RC-6 ①
Ø.125 ∓ 1.0
SPRING TO SUIT ⑤
Ø.625
③
RC-6
KNURL TO SUIT
10° TAPER
①
Ø.75
Ø.625
Ø1.063
①
.078
12° TAPER
②
Ø.344 ∓ 1.0 (AT REST)
.31
3.0
1.375

Problem 17-6

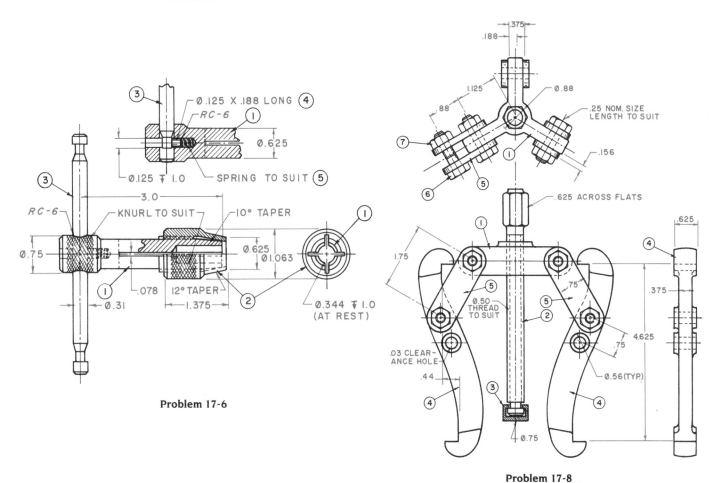

.375
.188
1.125
.88
Ø.88
.25 NOM. SIZE LENGTH TO SUIT
⑦
.156
⑥
⑤
.625 ACROSS FLATS
①
.625
④
1.75
⑤
Ø.50 THREAD TO SUIT
⑤
②
.75
4.625
.75
Ø.56 (TYP)
.03 CLEAR- ANCE HOLE
.44
④
③
④
Ø.75
.375

Problem 17-8

Problem 17-9

Follow the eight listed steps to develop Problem 17-9. Redesign the layout to include standard ball bearings in place of the bushings, part No. 4.

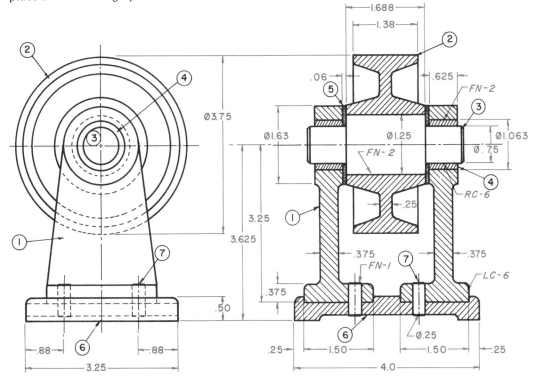

Problem 17-9

Problems 17-10 through 17-12

Follow the eight listed steps to develop Problems 17-10 through 17-12.

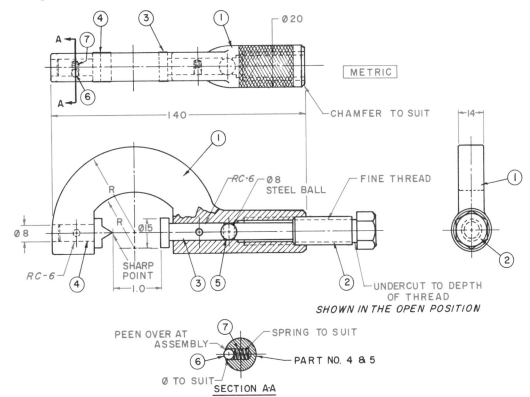

Problem 17-10

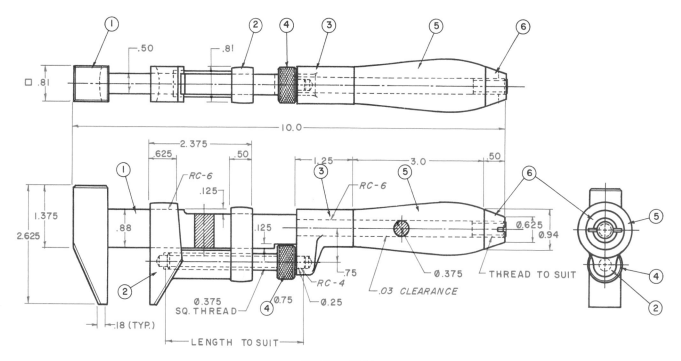

Problem 17-11

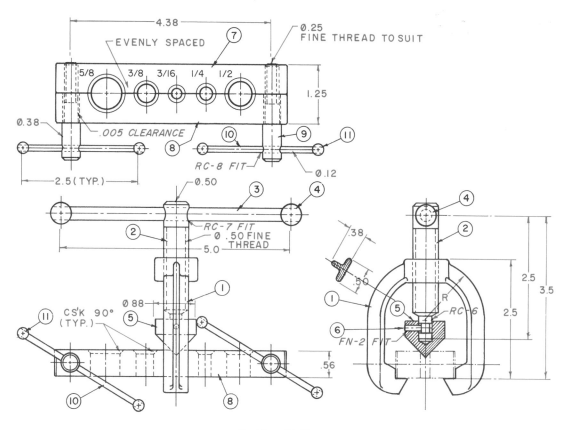

Problem 17-12

Problem 17-13

Follow the eight listed steps to develop Problem 17-13.
Redesign the layout to secure the round head locking screw,
part No. 10, from turning.

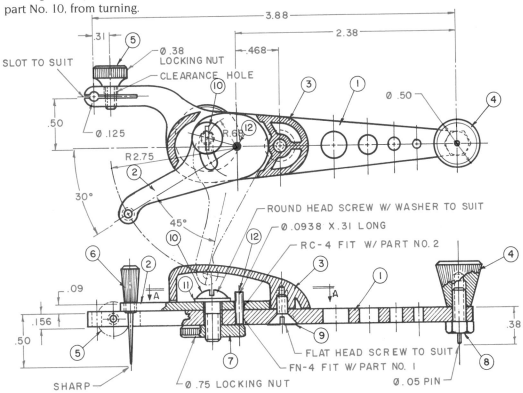

Problem 17-13

Problem 17-14

Follow the eight listed steps to develop Problem 17-14.

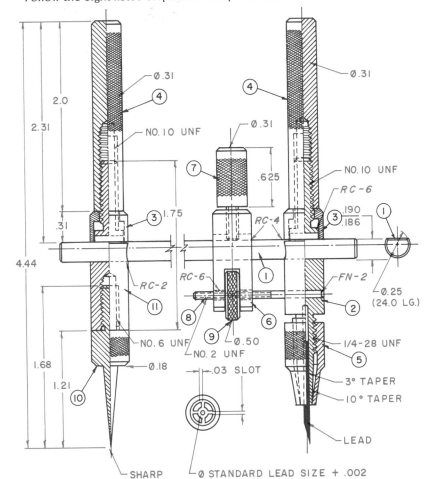

Problem 17-14

Using detail drawings 17-15-1 through 17-15-15, lay out a three-view assembly drawing. Call off parts per standard drafting practices. Calculate the following fits using each of the 15 detail drawings. Show all calculations.

Part numbers 1-12, FN-1/1.062
Part numbers 1-2, FN-2/.688
Part numbers 1-11, LC-4/1.50
Part numbers 3-4, RC-5/.562
Part numbers 4-5, RC-5/.375
Part numbers 4-8, LC-2/.125
Part numbers 9-14, FN-2/.250
Part numbers 13-14-15, RC-6/.938

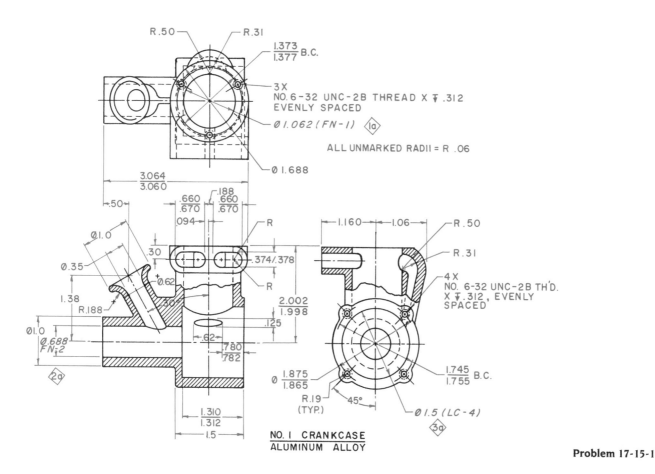

NO. 1 CRANKCASE
ALUMINUM ALLOY

Problem 17-15-1

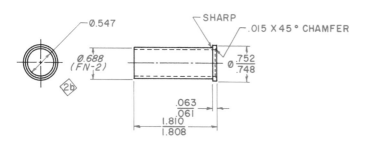

NO.2 BUSHING CRANKCASE
LEADED BRONZE

Problem 17-15-2

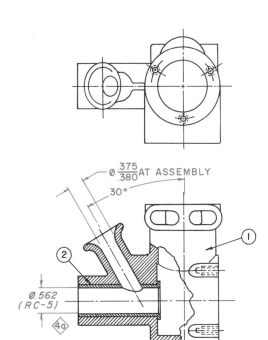

$\emptyset \frac{.375}{.380}$ AT ASSEMBLY

30°

② ① ①

$\emptyset 0.562$
(RC-5)

④ₐ

NO. 3 CRANKCASE SUBASSEMBLY
MAT'L. AS NOTED

Problem 17-15-3

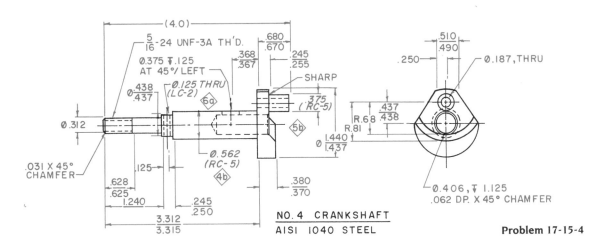

(4.0)

$\frac{5}{16}$-24 UNF-3A TH'D.

.680
.670

$\emptyset 0.375 \; \overline{\;} .125$
AT 45° LEFT

.368
.367

.245
.255

$\emptyset \frac{.438}{.437}$

$\emptyset 0.125$ THRU
(LC-2)

SHARP

⑤ₐ

$\frac{.375}{}$
(RC-5)

⑤ᵦ

$\emptyset 0.312$

R.68
R.81

.437
.438

.510
.490

.250

$\emptyset 0.187$, THRU

$\emptyset \frac{1.440}{1.437}$

.031 X 45°
CHAMFER

$\emptyset 0.562$
(RC-5)

④ᵦ

.125

.628
.625
1.240

.245
.250

.380
.370

$\emptyset 0.406, \; \overline{\;} \; 1.125$
.062 DP. X 45° CHAMFER

3.312
3.315

NO. 4 CRANKSHAFT
AISI 1040 STEEL

Problem 17-15-4

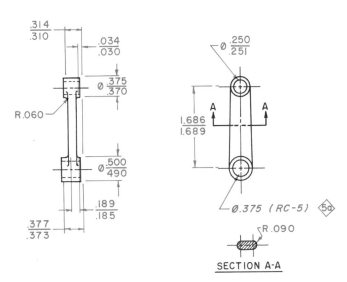

.314
.310

.034
.030

$\emptyset \frac{.375}{.370}$

R.060

$\emptyset \frac{.500}{.490}$

.189
.185

.377
.373

$\emptyset \frac{.250}{.251}$

A   A

1.686
1.689

$\emptyset 0.375$ (RC-5) ⑤𝒹

R.090

SECTION A-A

NO. 5 ROD-CONNECTING
ALUMINUM ALLOY 7075-T6

Problem 17-15-5

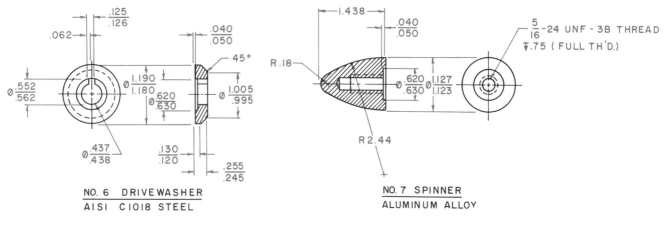

.125
.126

.062

.040
.050

45°

Ø .552
.562

Ø 1.190
1.180

Ø 1.005
.995

Ø .620
.630

Ø .437
.438

.130
.120

.255
.245

NO. 6 DRIVEWASHER
AISI C1018 STEEL

Problem 17-15-6

1.438

.040
.050

5/16 -24 UNF - 3B THREAD
↧.75 (FULL TH'D.)

R .18

Ø .620
.630

Ø 1.127
1.123

R 2.44

NO. 7 SPINNER
ALUMINUM ALLOY

Problem 17-15-7

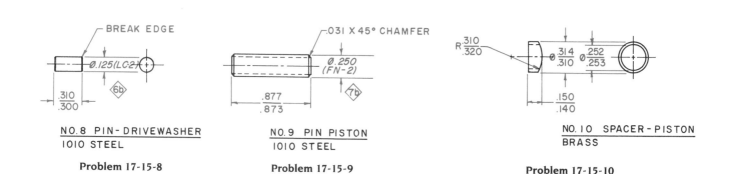

BREAK EDGE

Ø.125(LC2)

⬦6b

.310
.300

NO. 8 PIN-DRIVEWASHER
1010 STEEL

Problem 17-15-8

.031 X 45° CHAMFER

Ø .250
(FN-2)

.877
.873

⬦7b

NO. 9 PIN PISTON
1010 STEEL

Problem 17-15-9

R .310
.320

Ø .314
.310

Ø .252
.253

.150
.140

NO. 10 SPACER-PISTON
BRASS

Problem 17-15-10

.750
.740

.255
.245

2.625
2.615

1.750
B.C.

1.312

3 X Ø.166 EQUALLY SPACED
ON A 2.44 B.C.

R .25
(TYP.)

.938

30°

Ø 1.500
(LC-4)

1.240
1.250

⬦3b

.620
.630

2.44 B.C.

4 X Ø .166 THRU
⌴ Ø 0.250 ↧.156
EQUALLY SPACED
ON A 1.750 B.C

2.375
2.370

SHARP

45°

.875

1.750
1.740

NO. 11 BACKPLATE
CAST ALUMINUM

Problem 17-15-11

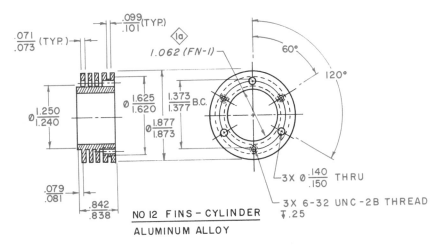

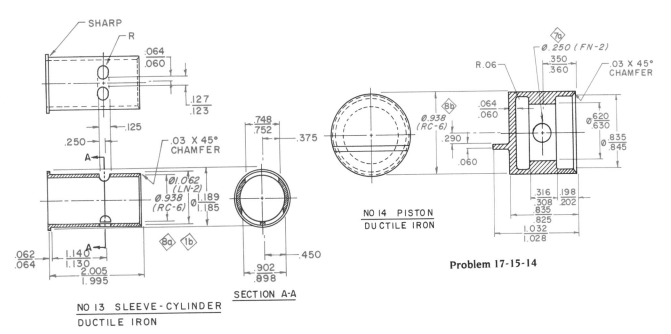

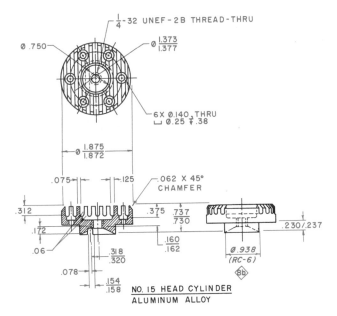

.099/.101 (TYP.)

.071/.073 (TYP.)

⬦1a

1.062 (FN-1)

60°

120°

Ø 1.625/1.620

1.373/1.377 B.C.

Ø 1.250/1.240

Ø 1.877/1.873

.079/.081

.842/.838

3X Ø .140/.150 THRU

3X 6-32 UNC-2B THREAD ⊽.25

NO 12 FINS-CYLINDER
ALUMINUM ALLOY

Problem 17-15-12

SHARP

R

.064/.060

.127/.123

.125

.250

A

.03 X 45° CHAMFER

.748/.752

.375

Ø1.062 (LN-2)

Ø .938 (RC-6)

.189/1.185

⬦8a ⬦1b

A

.062/.064

1.140/1.130

2.005/1.995

.902/.898

.450

SECTION A-A

NO 13 SLEEVE-CYLINDER
DUCTILE IRON

Problem 17-15-13

⬦7a

Ø.250 (FN-2)

R.06

.350/.360

.03 X 45° CHAMFER

Ø.938 (RC-6)

⬦8b

.064/.060

.290

.060

Ø .620/.630

Ø .835/.845

.316/.308

.835/.825

1.032/1.028

.198/.202

NO 14 PISTON
DUCTILE IRON

Problem 17-15-14

¼-32 UNEF-2B THREAD-THRU

Ø .750

Ø 1.373/1.377

6X Ø.140 THRU
⌴ Ø 0.25 ⊽.38

Ø 1.875/1.872

.075

.125

.062 X 45° CHAMFER

.312

.375

.737/.730

.230/.237

.172

.160/.162

.06

.318/.320

Ø .938 (RC-6)

⬦8b

.078

.154/.158

NO.15 HEAD CYLINDER
ALUMINUM ALLOY

Problem 17-15-15

Problem 17-16

Using detail drawings 17-16-1 through 17-16-7, lay out a one-view assembly drawing. Call off parts per standard drafting practices. Design special extra-fine threads where required.

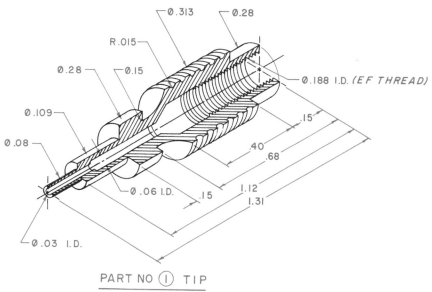

PART NO ① TIP

Problem 17-16-1

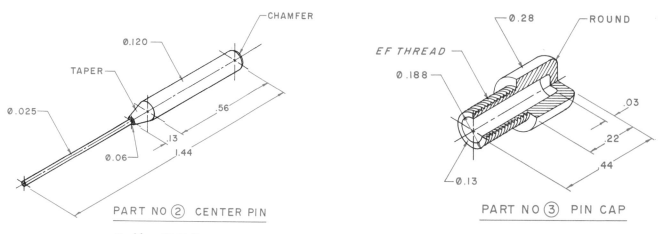

PART NO ② CENTER PIN

Problem 17-16-2

PART NO ③ PIN CAP

Problem 17-16-3

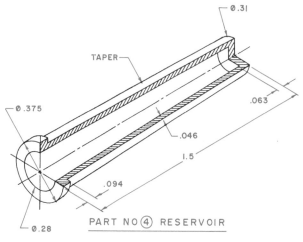

PART NO ④ RESERVOIR

Problem 17-16-4

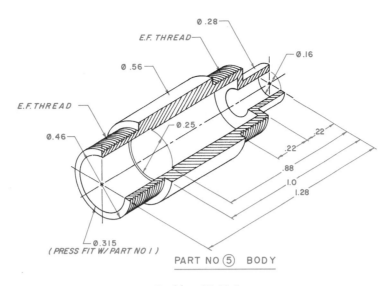

Ø.28
E.F. THREAD
Ø.16
Ø.56
E.F. THREAD
.22
Ø.46
.22
Ø.25
.88
1.0
1.28
Ø.315
(PRESS FIT W/ PART NO I)

PART NO ⑤ BODY

Problem 17-16-5

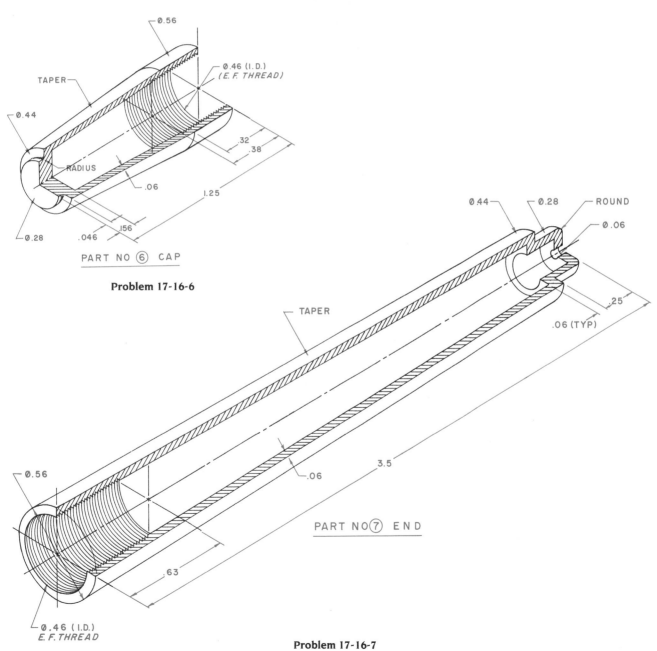

Ø.56
Ø.46 (I.D.)
(E. F. THREAD)
TAPER
Ø.44
.32
.38
RADIUS
.06
1.25
Ø.28
.046
.156

PART NO ⑥ CAP

Problem 17-16-6

Ø.44
Ø.28
ROUND
Ø.06

TAPER
.25
.06 (TYP)
.06
3.5

Ø.56

PART NO ⑦ END

.63

Ø.46 (I.D.)
E.F. THREAD

Problem 17-16-7

Follow the eight listed steps to develop Problem 17-17.

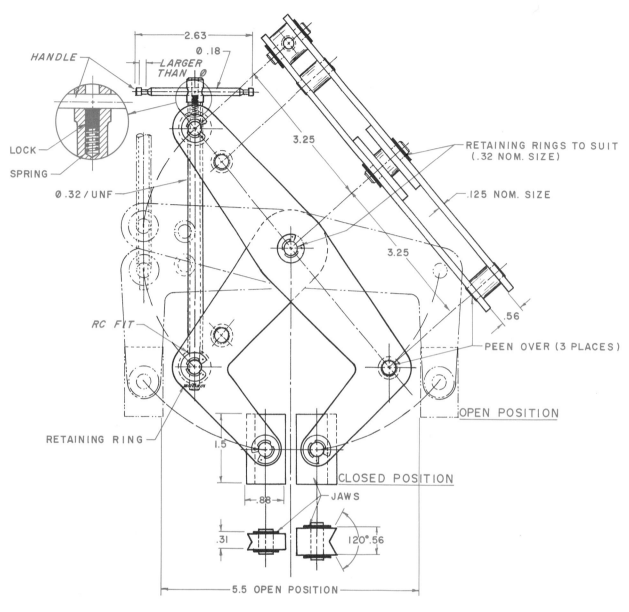

HANDLE

2.63

Ø .18

LARGER THAN Ø

LOCK

SPRING

Ø .32 / UNF

RC FIT

RETAINING RING

3.25

3.25

RETAINING RINGS TO SUIT (.32 NOM. SIZE)

.125 NOM. SIZE

.56

PEEN OVER (3 PLACES)

OPEN POSITION

CLOSED POSITION

1.5

.88

JAWS

.31

120° .56

5.5 OPEN POSITION

Problem 17-17

Follow the eight listed steps to develop Problem 17-18.

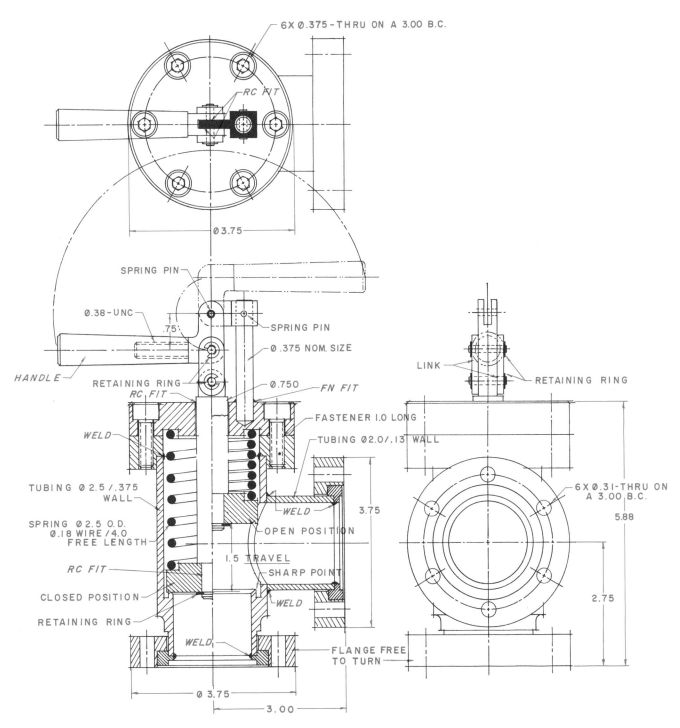

Problem 17-18

Problem 17-19

Follow the eight listed steps to develop Problem 17-19. Draw the clamping device in the position shown, and draw arm A at 30° and 60° to the left as shown. Design a complete pneumatic cylinder with a length to suit based on .75 extra interior space at each end of the cycle as shown. Improve the basic designs where necessary.

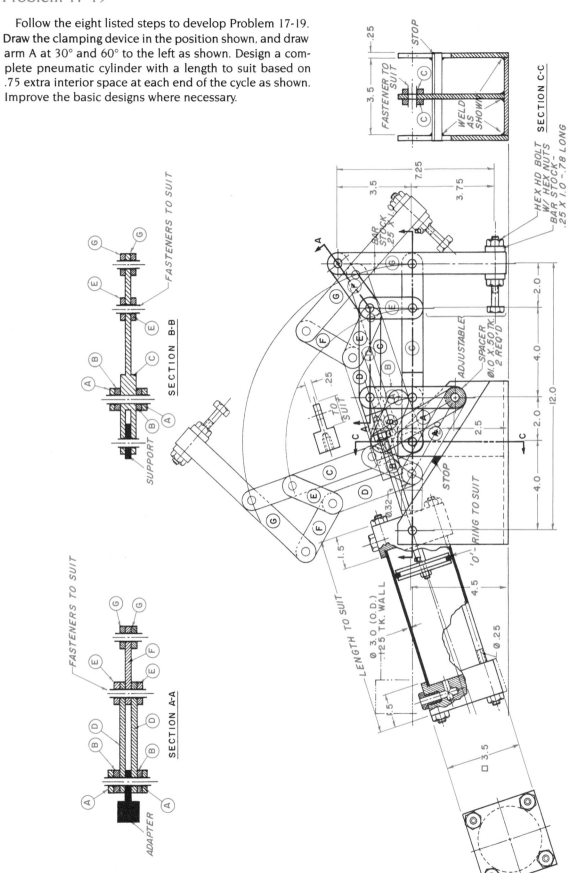

Problem 17-19

Problem 17-20

Choose an existing mechanical device of your choice and develop a design layout drawing of it. Improve the overall design for its function, or for better use of standard materials in place of special materials. Follow the eight listed steps.

1. How many parts make up the assembly?
2. How many screws are used in the assembly?
3. What are the part numbers of the screws used in the assembly?
4. Of what material is the base of the assembly made?
5. How many springs does the assembly contain?
6. How many revisions have been made to the assembly drawing?
7. When were the revisions made?
8. Who made the revisions?
9. What is the upper-limit thickness of the feeler gage?
10. Describe the dowel pins used in the assembly.

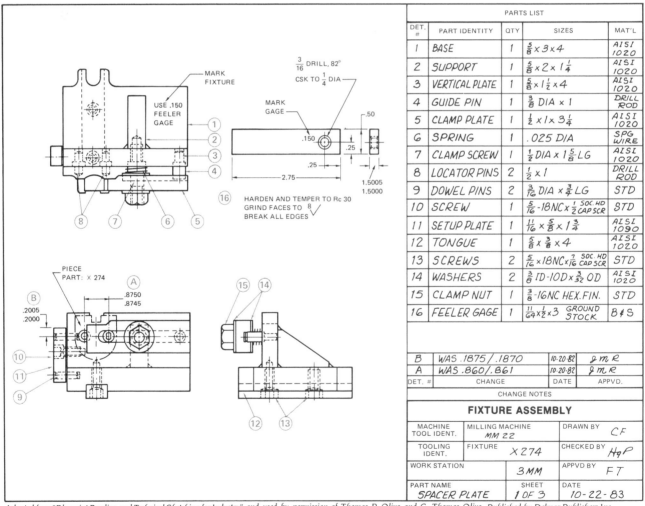

PARTS LIST				
DET. #	PART IDENTITY	QTY	SIZES	MAT'L
1	BASE	1	$\frac{5}{8} \times 3 \times 4$	AISI 1020
2	SUPPORT	1	$\frac{5}{8} \times 2 \times 1\frac{1}{4}$	AISI 1020
3	VERTICAL PLATE	1	$\frac{5}{8} \times 1\frac{1}{2} \times 4$	AISI 1020
4	GUIDE PIN	1	$\frac{3}{8}$ DIA $\times 1$	DRILL ROD
5	CLAMP PLATE	1	$\frac{1}{2} \times 1 \times 3\frac{1}{4}$	AISI 1020
6	SPRING	1	.025 DIA	SPG WIRE
7	CLAMP SCREW	1	$\frac{1}{2}$ DIA $\times 1\frac{5}{8}$ LG	AISI 1020
8	LOCATOR PINS	2	$\frac{1}{2} \times 1$	DRILL ROD
9	DOWEL PINS	2	$\frac{3}{16}$ DIA $\times \frac{3}{4}$ LG	STD
10	SCREW	1	$\frac{5}{16}$-18NC $\times \frac{1}{2}$ SOC. HD CAP SCR	STD
11	SETUP PLATE	1	$\frac{11}{16} \times \frac{5}{8} \times 1\frac{3}{4}$	AISI 1090
12	TONGUE	1	$\frac{5}{8} \times \frac{3}{8} \times 4$	AISI 1020
13	SCREWS	2	$\frac{5}{16} \times$18NC$\times \frac{7}{16}$ SOC. HD CAP SCR	STD
14	WASHERS	2	$\frac{3}{8}$ ID-10D$\times \frac{3}{32}$ OD	AISI 1020
15	CLAMP NUT	1	$\frac{3}{8}$-16NC HEX.FIN.	STD
16	FEELER GAGE	1	$\frac{11}{64} \times \frac{1}{2} \times 3$ GROUND STOCK	B & S

B	WAS .1875/.1870	10-20-82	J M R
A	WAS .860/.861	10-20-82	J M R
DET. #	CHANGE	DATE	APPVD.
CHANGE NOTES			

FIXTURE ASSEMBLY

MACHINE TOOL IDENT.	MILLING MACHINE MM 22	DRAWN BY CF
TOOLING IDENT.	FIXTURE X274	CHECKED BY HgP
WORK STATION	3MM	APPVD BY FT
PART NAME SPACER PLATE	SHEET 1 OF 3	DATE 10-22-83

MARK FIXTURE

USE .150 FEELER GAGE

$\frac{3}{16}$ DRILL, 82°
CSK TO $\frac{1}{4}$ DIA

MARK GAGE

.150

.50
.25
.25
2.75

1.5005
1.5000

16   HARDEN AND TEMPER TO Rc 30
GRIND FACES TO 8
BREAK ALL EDGES

PIECE PART: X 274

.8750
.8745

.2005
.2000

Adapted from "Blueprint Reading and Technical Sketching for Industry" and used by permission of Thomas P. Olivo and C. Thomas Olivo. Published by Delmar Publishers Inc.

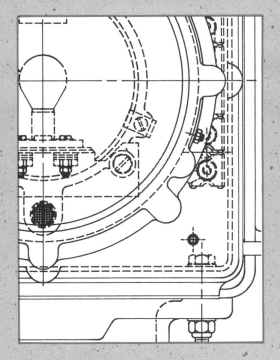

Although pictorial drawings are not used extensively in drafting and designing, the drafter should have the ability to create a pictorial drawing if necessary. A pictorial drawing is often needed in industry to convey an idea, present a new product, or to aid in the assembly of a completed object. This chapter describes and illustrates how to develop the major kinds of pictorial drawings, and discusses the techniques used to draw such features as threads, chamfers, and knurls. How to add dimensions is fully illustrated.

CHAPTER EIGHTEEN
Pictorial Drawings

Pictorial drawings are, as their name implies, pictorial views of an object. They are used in industry for sales presentations, to aid workers on complicated assemblies, to record new design ideas, for owner's manuals and parts catalogs, and for technical printed articles. In industry, a technical illustrator usually does most of this type of drawing, but all drafters should have a basic working knowledge in this area of drawing. This chapter touches on the various kinds of illustrations used, and basically how they are developed.

Types of Pictorial Drawings

Three types of pictorial drawings are used in industry today: axonometric, oblique, and perspective. Each type is illustrated using a simple cube in order that they may be compared and studied. Each is explained in full.

Axonometric Drawings

In *axonometric drawing*, an object is represented by its perpendicular projection on a surface so that it appears as inclined and shows three faces. Axonometric projection includes isometric, dimetric and trimetric projections, Figure 18-1. It is customary to consider the three edges of the basic shape that meet at the corner *nearest* the viewer as the *axonometric axes*, Figure 18-2.

Isometric Projection

Isometric means "equal measure"; all three principal edges or axes are projected with equal 120° angles, Figure 18-3. Any line on any surface that is either parallel to or perpendicular to one of the principal edges is called an *isometric line*. Any line on any surface that is *not* either parallel to or perpendicular to one of the principal edges is called a *nonisometric line*. In drawing an isometric projection, use the 30°-60° triangle to construct all isometric lines.

Technically, an isometric projection should be drawn approximately 80% of its true size, but, in actual practice, it is drawn full size. Isometric templates and grid paper are available to aid and speed up the drawing process.

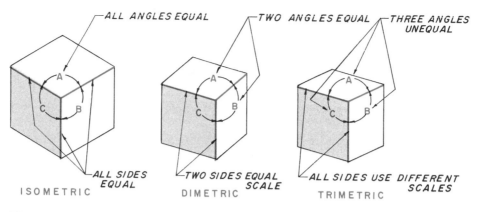

Figure 18-1 Axonometric projections

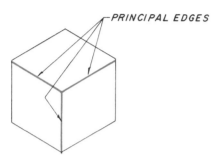

Figure 18-2 Axonometric axes

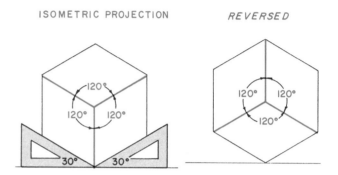

Figure 18-3 Isometric projection

Dimetric Projection

Dimetric projection differs from isometric projection in that only *two* angles are equal. Figure 18-4 illustrates some of the many combinations used to draw a dimetric projection.

In drawing a dimetric drawing, the object is turned so that two of the axes make the *same* angle with the plane of projection while the third is at a different angle. Edges that are parallel to the first two axes are drawn full size. The edge parallel to the third axis is drawn to a different scale. Because two different scales are used, less distortion is apparent and the object looks more natural.

A dimetric drawing is laid out in exactly the same way as an isometric drawing. All layout procedures to locate and draw nondimetric lines, circles, and arcs are exactly the same as those used in isometric projections.

Isometric and dimetric grid paper are available to aid and speed up the drawing process, Figure 18-5. Dimetric templates are also available.

Trimetric Projection

In *trimetric projection* the axes are rotated so that each of the three axes are drawn at different angles to the plane of projection. Each axis uses a different scale

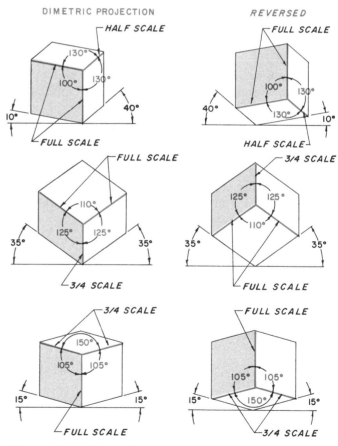

Figure 18-4 Dimetric projection

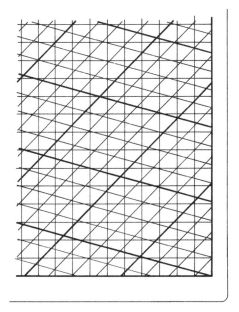

Figure 18-5 Isometric and dimetric grid

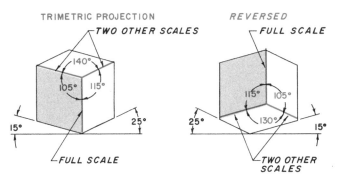

Figure 18-6 Trimetric projection

of reduction, Figure 18-6. Of the three kinds of axonometric projection, trimetric is the most complicated to draw, but it has by far the least amount of distortion, and truly appears as a picture of the object. Trimetric projections are rarely used in industry as they take too much drafting time.

All basic procedures used in laying out an isometric or dimetric projection are incorporated into drawing the trimetric projection.

As this text is primarily for drafters and not particularly for technical illustrators, concentration is given only to isometric projections. Once mastered, these same methods and basic procedures can be applied to dimetric and trimetric projections.

Oblique Drawings

The easiest type of pictorial drawing to develop is the oblique projection. In *oblique drawing*, one surface of the object, usually the most important view, is drawn exactly as it would be drawn in a multiview projection. It is drawn true size and shape.

Oblique drawings use three axes. Two at right angles to each other, as in multiview drawings; the other, the receding axis, is drawn at any convenient angle to the horizon.

There are three kinds of oblique drawings: cavalier, cabinet, and general oblique, Figure 18-7. In each of the three kinds of oblique drawings, its most important surface is drawn parallel to the plane of projection.

Cavalier Drawing

In the *cavalier drawing*, the receding axis is drawn from 30° to 60° to the horizon. As with the isometric projection, the receding distances are drawn full size,

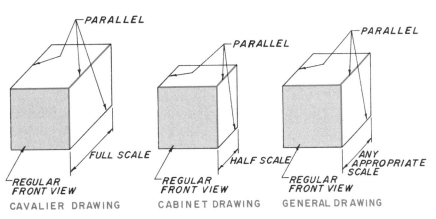

Figure 18-7 Oblique drawings

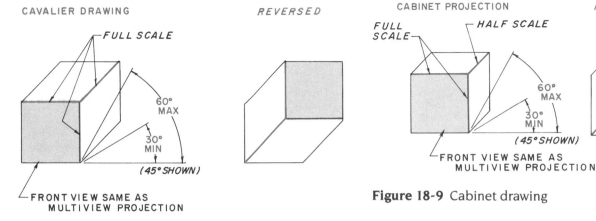

Figure 18-8 Cavalier drawing

Figure 18-9 Cabinet drawing

Figure 18-8. Notice how this creates much distortion; therefore, this type of drawing is seldom used.

The term *cavalier* originated from the drawing of medieval fortifications. The center area of these fortifications was much higher than the rest of the fortification, and was referred to as the *cavalier* because of its command position.

Cabinet Drawing

In the *cabinet drawing*, the receding axis is drawn from 30° to 60° to the horizon. The receding distances are drawn half size, Figure 18-9. This helps somewhat to eliminate the distortion associated with the oblique projection system. In past years, cabinet drawings were used to illustrate furniture and cabinets.

General Oblique

The *general oblique* drawing is very similar to the cabinet drawing, except that the receding distances are drawn to *any* scale that seems the most natural for that particular object. This could be from one-third full size to three-quarters full size, Figure 18-10.

The receding angle can vary from 30° to 60°, as with all oblique projections. In oblique drawings, it is best to choose the most complete shape as the front surface, and project all receding lines from that surface.

Perspective Drawing

Perspective drawings illustrate the object better than any other method. In a *perspective drawing*, there is little or no distortion, and it approximates the object as it would be seen by the human eye or as projected upon the film inside a camera, Figure 18-11.

A perspective drawing is used by architects, designers, and technical illustrators in order to convey their ideas. Architects often use perspectives to illustrate

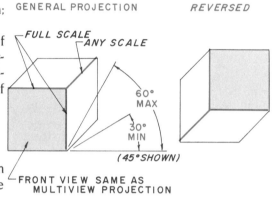

Figure 18-10 General oblique

how a proposed building will look when completed. As a general rule, however, perspective drawings are *not* used in the mechanical drafting field. Although they illustrate the object exactly as it will appear, they are too time-consuming and costly for drafters to construct. Therefore, drafters need only a working knowledge of perspective layout procedure.

Perspective drawings are discussed more fully later in this chapter.

Isometric Principles

An isometric drawing is drawn around three equally spaced principal edges or axes, Figure 18-12. The top illustration is the most commonly used. The object should be placed either in the position it is usually seen or in the position that best illustrates all the most important features.

How To Draw an Isometric Drawing

Given: A multiview drawing of an object, Figure 18-13A.

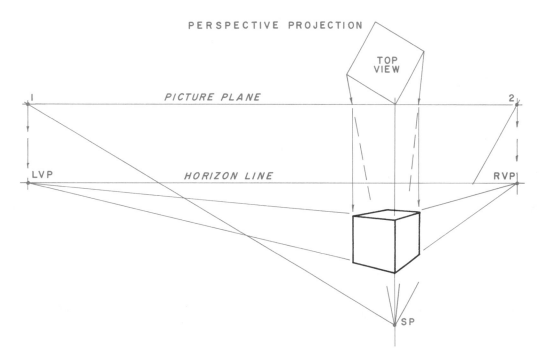

Figure 18-11 Perspective drawing

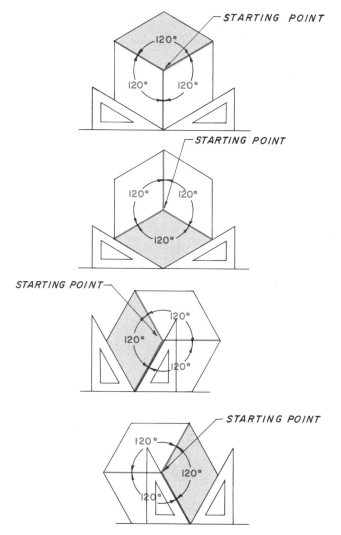

Figure 18-12 Isometric principles

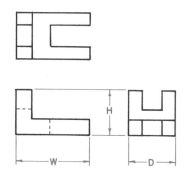

Figure 18-13A How to draw an isometric drawing

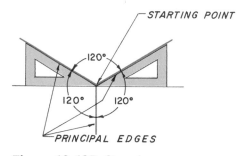

Figure 18-13B Step 1

Step 1. Locate the starting point and lightly draw the three principal edges 120°, as illustrated in Figure 18-13B. Use the 30°-60° triangle.

Step 2. Transfer the depth, height, and width directly from the multiview drawing, Figure 18-13A, to the three principal edges. Measure

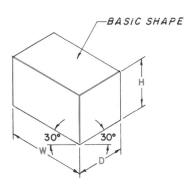

Figure 18-13C Step 2

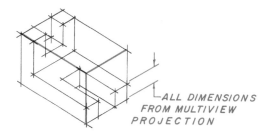

Figure 18-13D Step 3

Figure 18-13E
Completed isometric
view

full size directly along each edge, and lightly construct an isometric, basic shape of the object, Figure 18-13C.

Step 3. Transferring the full size lengths from the multiview drawing, fill in the various features of the object. Again, all full size measurements must be constructed along or parallel with one of the principal edges, Figure 18-13D.

Step 4. Check all work, and, if correct, darken in the object using correct line thickness, Figure 18-13E.

Nonisometric Lines

If a line is *not* parallel or perpendicular to any of the three principal edges, it is a *nonisometric* line. A nonisometric line must be reduced to two points: a point at each end of the line, as shown in the given multiview drawing, Figure 18-14A. Line a-b is *not* parallel or perpendicular to any of the three principal edges, thus it is a nonisometric line.

Referring to Figure 18-14B, the isometric drawing is developed exactly as described, except that points a and b must be located before the nonisometric line a-b can be drawn. Point a is located along one axis distance X from the back end; point b is located

Y distance up from the bottom surface. When points a and b are located, the nonisometric line a-b can be drawn. Notice, the 30° angular dimension given on the multiview drawing in Figure 18-14B has *no* bearing on the isometric layout and is *not* used. Also, line a-b is *not* the true length.

Hidden Lines

Hidden lines in pictorial drawings, regardless of which kind is used, are *not* drawn, unless needed to illustrate some important hidden feature that otherwise would not be seen. The object should be rotated or placed in such a position that no important feature is omitted.

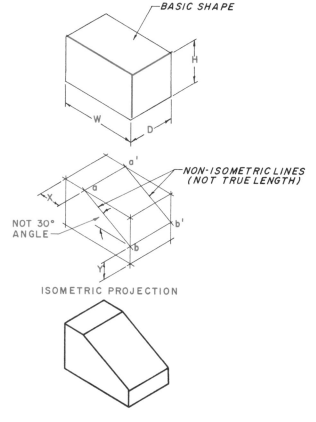

Figure 18-14B Locating points

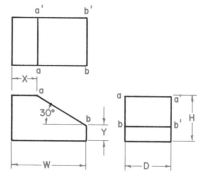

Figure 18-14A Nonisometric
lines

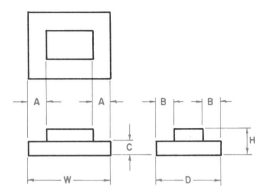

Figure 18-15A Multiview projection

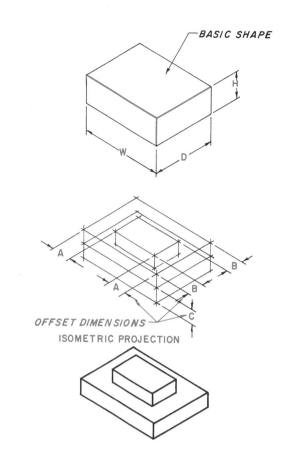

Figure 18-15B Offset measurements

Offset Measurements

Offset measurements are used to locate one feature in relationship to another. All offset measurements must be made either parallel to or perpendicular to any of the three principal surfaces.

A multiview drawing is given in Figure 18-15A. The isometric drawing is developed as outlined before, except that the offset measurements A, B, and C are measured parallel to and perpendicular to the three principal edges, Figure 18-15B. If a line or surface is *parallel* to another line or surface in the multiview drawing, the line or surface must be parallel, respectively, in the isometric drawing alşo.

Center Lines

Center lines are used to indicate symmetry and to aid in dimensioning. Center lines in an isometric drawing are drawn following the same drafting standards as in a multiview drawing. All holes must include the coordinates to indicate the exact center point, Figure 18-16. Center lines must extend outside the circle.

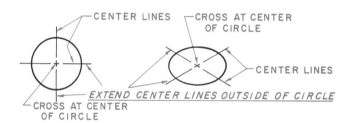

Figure 18-16 Center lines

Box Construction

Some objects do not conform to any of the three principal edges. If an object does not conform, a *box* or rectangle must be constructed around the object parallel to and perpendicular to the principal edges. This then becomes the basic shape of the object. Refer to the multiview in Figure 18-17A. A basic shape is constructed around the object so various points of the object touch, Figure 18-17B.

An isometric basic shape of the object is drawn as shown in Figure 18-17C, and the various points, A, B, C and 0, are located on or within the basic shape, Figure 18-17D. Points A, B, and C are located on the base of the basic shape, and point 0 is located, by offset measurements, on the top surface.

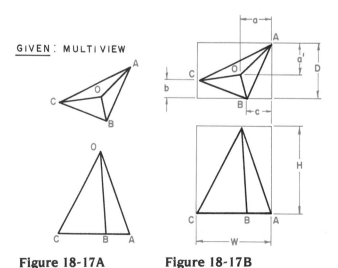

Figure 18-17A
Box construction

Figure 18-17B
Basic shape

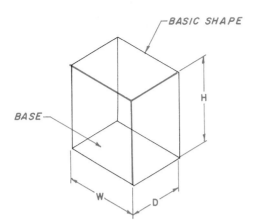

Figure 18-17C Isometric basic shape

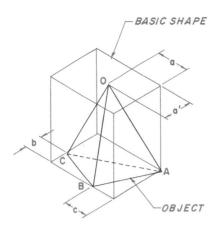

Figure 18-17D Locating points of object

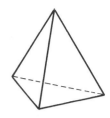

ISOMETRIC VIEW

Figure 18-17E isometric view Completed

After all points are found, lines are drawn from A to B, B to C, and C to A to form the triangular base. Lines from 0 to A, 0 to B and 0 to C are constructed and darkened in to complete the object, Figure 18-17E.

Irregularly Shaped Objects

An irregularly shaped object is drawn by breaking it into a series of sections. Carefully locate each section in relationship to the others, and draw each individual section as an isometric, Figure 18-18A. Illustrated is a multiview drawing of a simple boat. Divide the object into a series of sections, A through H. Starting from a baseline, in this example the center line, locate and draw each section A through H, Figure 18-18B. Connect all sections together and the object is completed, Figure 18-18C.

Isometric Curves

Curves are drawn either by a series of offset distances or by the use of a grid. In using the series of offset distances, use any desired number of evenly spaced lines parallel to one of the principal edges, Figure 18-19A. In this view, the curved surface is shown as a profile. In this example, the lines are drawn in the front view. These lines are all parallel to one another and parallel to the bottom surface. Transfer these lines to the isometric layout and transfer each distance, respectively, to find each offset distance, Figure 18-19B. To give depth, project a line 30° from

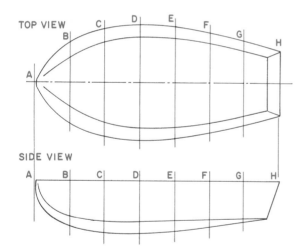

Figure 18-18A Irregularly shaped object

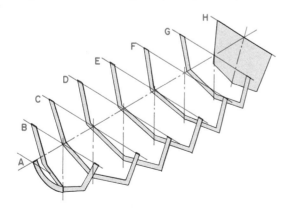

Figure 18-18B Drawing each section

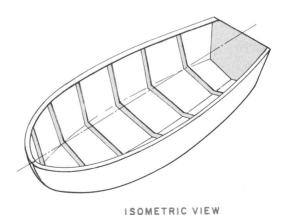

ISOMETRIC VIEW

Figure 18-18C Completed object

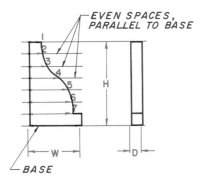

Figure 18-19A Drawing an isometric curve (offset method) distance

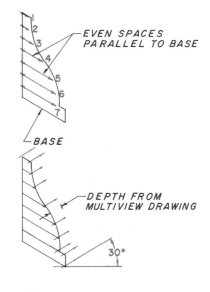

Figure 18-19B Drawing an isometric curve (finding the offset distance)

each point and transfer the distance of the depth of the object along each of the 30° projection lines. Check all work and, if correct, darken in the object using correct line thickness.

In using a grid, choose a grid of a size that will give the maximum number of points along the irregular curve in order to pick up as much detail as possible. See the given multiview in Figure 18-20A. A grid is drawn in the view where the irregular line appears; in this example the front view.

Draw the basic shape of the object as an isometric projection, and add the grid to the basic shape as constructed in the multiview. Transfer the irregular curve, square-by-square, from the multiview drawing to the isometric grid, Figure 18-20B. Check all work, and, if correct, darken in using correct line thickness.

Isometric Circles or Arcs

The drawing of circles or arcs in isometric drawing takes a lot of valuable drafting time. For small circles or arcs, use an isometric (circle) elliptical template, Figure 18-21. The actual angle of an isometric elliptical template is 35°16', which is slightly different from the standard 30° ellipse template. An isometric elliptical template has eight short hash marks

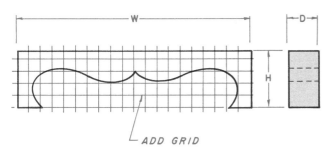

Figure 18-20A Drawing an isometric curve (grid method)

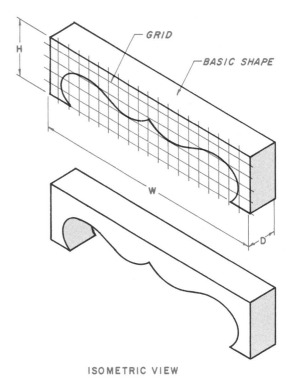

Figure 18-20B Isometric grid layout

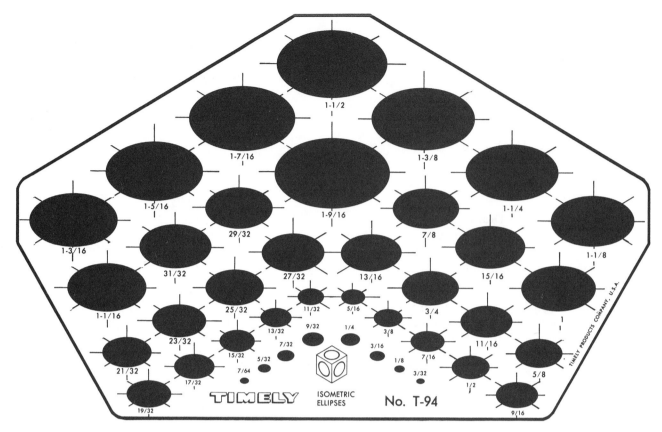

Figure 18-21 Isometric ellipse template (*Courtesy Timely Products Company*)

printed around the perimeter of each circle. The four longer hash marks indicate the *isometric* center lines of the circle or arc, and also indicate the four tangent points of the isometric circle with the isometric basic shape of the circle. The four short hash marks line up with the axis of the circle and are perpendicular to the axis of the circle, Figure 18-22A.

How To Use an Isometric Circle Arc Template

Step 1. Locate and draw the actual center lines of the circle or arc.

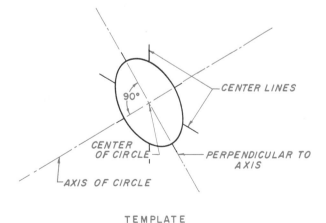

TEMPLATE

Figure 18-22A How to use an isometric ellipse template

Step 2. Draw the basic shape of the circle or arc, Figure 18-22B. The basic shape of the circle must be drawn parallel to and perpendicular to the principal edges where the circle or arc is located. *This is important.*

Step 3. Lightly extend the axis of the circle out as far as space will permit (refer again to Figure 18-22B).

Step 4. Choose the correct template size, and place it *within* the basic shape of the circle or arc. Align the longer center line hash marks on the template on the center lines of the circle or arc. These four hash marks should also be *tangent* to the basic shape of the circle, Figure

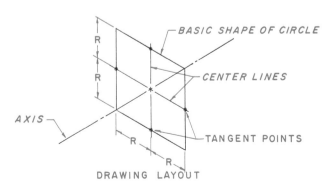

Figure 18-22B Draw the basic shape of the circle

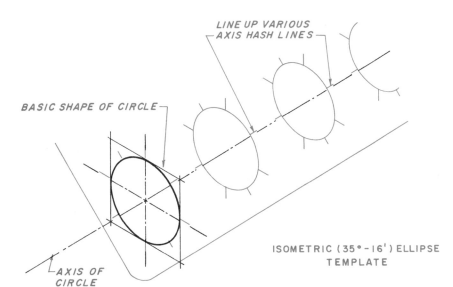

LINE UP VARIOUS AXIS HASH LINES

Figure 18-22C Drawing the ellipse from the basic shape

BASIC SHAPE OF CIRCLE

ISOMETRIC (35°-16') ELLIPSE TEMPLATE

AXIS OF CIRCLE

18-22C. To double check position for accuracy, line up *all* hash marks from other circles along the light axis line of the circle.

Step 5. If everything is in line and fits correctly, darken in the circle or arc using correct line thickness.

Regardless of the view in which the circle or arc is located, the foregoing steps apply. The basic shape of the circle or arc helps to eliminate any layout errors, Figure 18-23.

How To Draw a True Isometric Circle or Arc

Where accuracy is very important or where various points around a circle or arc must be established, the following method is used to draw an isometric circle or arc.

A true isometric circle or arc is drawn very much like any isometric circle using various offset measurements. Because the circle or arc is symmetrical, it is best to use evenly spaced offset measurements. In this example, 12 evenly spaced increments are used.

Before beginning, notice how a regular circle compares to an isometric circle, Figure 18-24A. Both have a basic shape with equal sides — each side being equal to the circle's diameter. Both are tangent to or touch the basic shape at four places, indicated by 3, 6, 9 and 12. Both are divided into four parts or quadrants.

Given: Regular circle in Figure 18-24A.

Step 1. In the view of the regular circle, divide it into 12 equal, even spaces, using the 30°-60° triangle, Figure 18-24B. Lightly label each point clockwise 1 through 12.

Step 2. Divide the circle into:
   1. A square, the basic shape, tangent to points 3, 6, 9 and 12.

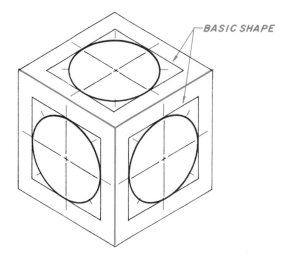

Figure 18-23 Isometric circles in various planes

GIVEN: REGULAR CIRCLE   ISOMETRIC CIRCLE

BASIC SHAPE

TANGENT POINTS 3,6,9,12

'X' RADIUS

TANGENT POINTS 3,6,9,12

'X' RADIUS

Figure 18-24A Regular circle and isometric circle

Figure 18-24B How to draw an isometric circle — Step 1

2. A rectangle tangent to points 1, 5, 7 and 11.
3. Another rectangle tangent to points 2, 4, 8 and 10. See Figure 18-24C.

Step 3. Locate the center of the isometric circle and draw the basic shape of the circle, Figure 18-24D. This automatically locates points 3, 6, 9 and 12. Lightly label these points.

Step 4. Using the offset measurement technique, lightly draw the rectangle 1, 5, 7 and 11. Transfer distances X and Y from the regular view of the circle, Figure 18-24E. Lightly label each point.

Step 5. Again, using the offset measurement technique, lightly draw the rectangle 2, 4, 8 and 10. Transfer distances X and Y from the regular view of the circle, Figure 18-24F.

Step 6. Check all work, and, if correct, darken in the isometric circle using an irregular or French curve, Figure 18-24G.

To draw an arc, use the foregoing steps, but use only that part of the circle as necessary, Figure 18-25.

How To Draw an Approximate Isometric Circle or Arc

For most drafting, an approximate isometric circle or arc is sufficient. This method takes much less time to draw, and is close enough for illustration where various points around the perimeter are not required. This method is sometimes referred to as the four-center isometric circle or arc.

Given: Regular circle with X radius (refer back to Figure 18-24A).

Step 1. Locate the center of the isometric circle and draw the basic shape of the circle, Figure 18-26A. Locate and label each of the four tangent points clockwise, 1, 2, 3 and 4.

Step 2. From each of the four tangent points, draw a line perpendicular (90°) as shown in Figure 18-26B. Where these lines *intersect* is the location of the four swing points, A, B, C, and D, Figure 18-26C.

Step 3. Set compass at swing point A and adjust lead to tangent point 1, Figure 18-26D. From swing point A, swing an arc to tangent point 2. From swing point B, swing this same arc from tangent point 3 to tangent point 4.

Step 4. Set compass at swing point C and adjust lead to tangent point 2, Figure 18-26E. From swing point C, swing an arc to tangent point 3. From swing point D, swing this same arc from tangent point 4 to tangent point 1.

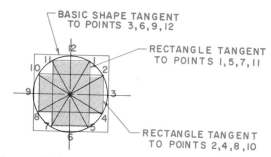

Figure 18-24C Step 2

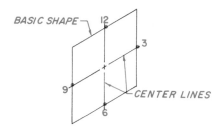

Figure 18-24D Step 3

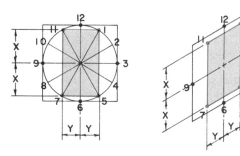

Figure 18-24E Step 4

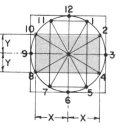

Figure 18-24F Step 5

Figure 18-24G Completed isometric circle

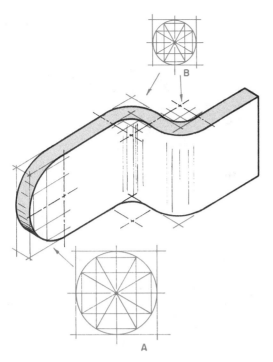

Figure 18-25 How to draw an isometric arc

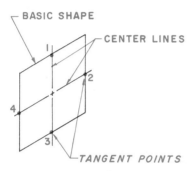

Figure 18-26A How to draw an approximate
isometric circle — Step 1

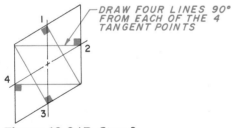

Figure 18-26B Step 2

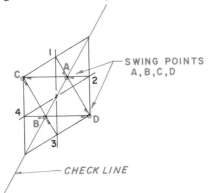

Figure 18-26C Step 2

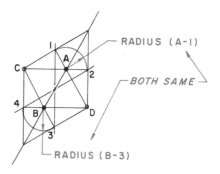

Figure 18-26D Step 3

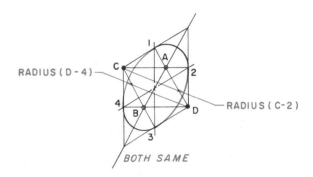

Figure 18-26E Step 4

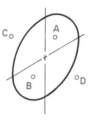

ISOMETRIC CIRCLE

Figure 18-26F Completed
isometric circle

Step 5. Check all work, and, if correct, darken in
the completed isometric circle using correct
line thickness; then, darken in the center lines,
Figure 18-26F.

Regardless of the position in which the isometric
circle is placed, the exact same steps are used (refer
back to Figure 18-23).

Isometric Arcs

Usually, the four-center isometric circle method is
used for drawing isometric arcs. In this example,
radius X is measured off from the projected corners
to locate the tangent points, Figure 18-27. From the
tangent points, construct perpendicular lines to locate
the swing points.

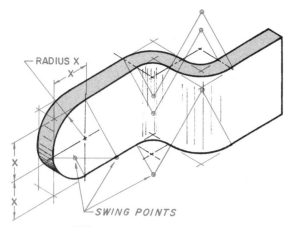

Figure 18-27 Drawing isometric arcs

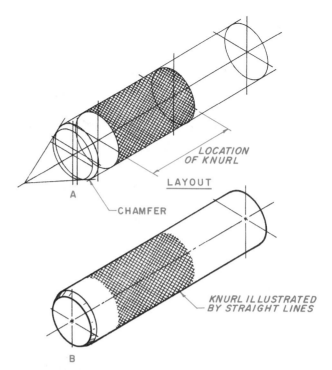

Figure 18-28 Isometric knurls

Isometric Knurls

Isometric knurls are usually drawn with straight lines, and are very seldom curved, Figure 18-28. Part A illustrates how to lay out the area for the knurl; Part B illustrates how it will look as a finished drawing. Notice the chamfer at the end of the object. This was drawn with an isometric template slightly smaller in size than the outside diameter of the object.

Isometric Screw Threads

In order to quickly draw thread representations, parallel partial ellipses, spaced approximately the thread pitch distance, are drawn, Figure 18-29. These ellipses represent the *crest* of the screw thread. These parallel partial ellipses can be drawn with either an ellipse template or by the four-center ellipse method as described.

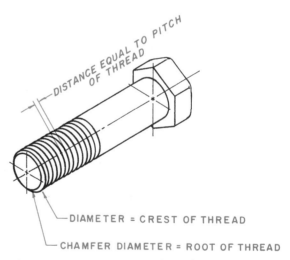

Figure 18-29 Isometric threads

Isometric Spheres

The layout and construction of an isometric sphere or various parts of a sphere are sometimes required. An *isometric sphere* is actually a true circle in three dimensions.

If a full sphere is needed, draw a circle equal in diameter to 1.22 times its actual diameter, Figure 18-30. If a half sphere is needed, draw an isometric circle of the required diameter and, from the center point, swing half a true circle, Figure 18-31. If a quarter sphere is needed, draw two half isometric circles at 90° to each other, and connect the two halves with an arc from the center point, Figure 18-32. If a three-quarter sphere is needed, draw two full isometric circles at 90° to each other, and connect the two arcs with an arc from the center point, Figure 18-33.

Figure 18-30
Isometric sphere

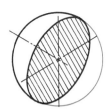

Figure 18-31
Isometric half
sphere

Figure 18-32
Isometric quarter
sphere

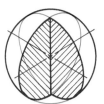

Figure 18-33
Isometric three-
quarter sphere

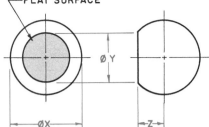

Figure 18-34 How to draw a flat
surface on a sphere

How To Draw a Flat Surface on a Sphere

Given: Multiview drawing in Figure 18-34 with X
  diameter/Y diameter, flat surface Z, and distance from center of sphere to flat surface.

Step 1. Draw two isometric circles with a diameter
  of X and at 90° to each other. Connect the two
  isometric circles with an arc from the center
  point, Figure 18-35.

Step 2. Measure distance Z to locate the center
  point of the flat surface. From this point, draw
  the required center lines of the flat surface,
  and draw an isometric circle of a diameter of Y
  (refer again to Figure 18-35).

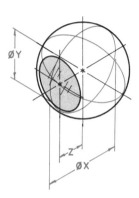

Figure 18-35
Steps 1 and 2

Isometric Intersections

To draw the intersection of a cylindrical hole in an
oblique plane, draw the isometric circle in a projected top surface, Figures 18-36A and 18-36B. Project the 12 points down to the oblique plane A, B, C,
and D, as shown. Each point is projected to the right-
side view and over to the corresponding point in the
front point. These points are then joined to form
the intersection.

To draw the generated curve between two cylinders, see Figures 18-37A and 18-37B. Divide one circle into 12 equally spaced points. Find the true length
of each of the 12 segments and transfer them to the
isometric view. Connect the points to draw the intersection line. This is exactly the process for developments, Chapter 8.

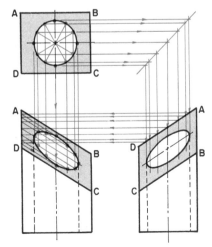

Figure 18-36A Isometric intersections – given:
three views

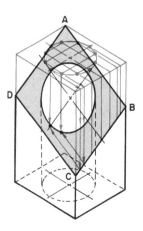

Figure 18-36B Isometric
intersections

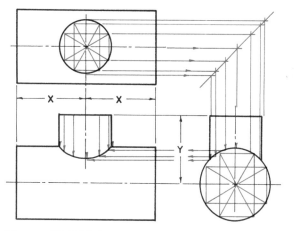

Figure 18-37A Isometric intersections – given: three views

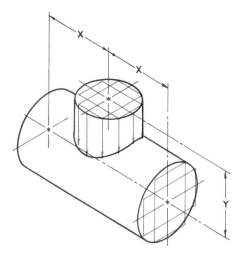

Figure 18-37B Isometric intersections

How To Draw an Ellipse on an Inclined Plane

Given: Multiview drawing in Figure 18-38A.

Step 1. Divide the top view into 12 equally spaced areas, 1 through 12. Project the 12 points into the front view to find the true length of each segment.

Step 2. Construct the isometric circle base using the offset measurement method in order to locate the 12 points around the base circle, Figure 18-38B.

Step 3. Project each of the 12 true length segments up from the base. Connect the ends of the segments to form the top surface – an ellipse on an inclined plane.

How To Find the True Length of Nonisometric Lines

The true length of a nonisometric line can be found by using an ellipse radius.

Given: The side view of a simple box with its lid at 60° and 135°, Figure 18-39, Part A.

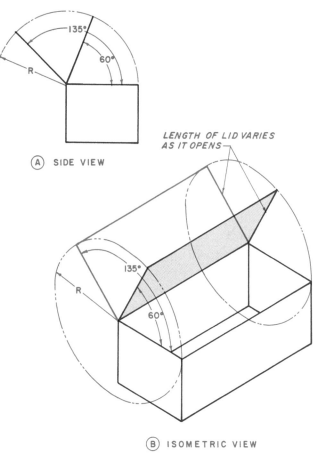

Figure 18-39 True length of nonisometric lines

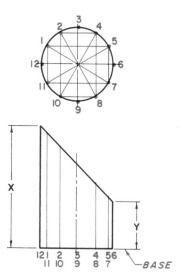

Figure 18-38A
How to draw an ellipse on an inclined plane – given: two views

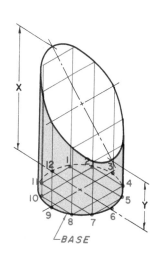

Figure 18-38B
An isometric ellipse on an inclined plane

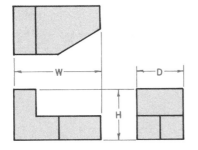

Figure 18-40A Centering an isometric view within the work area

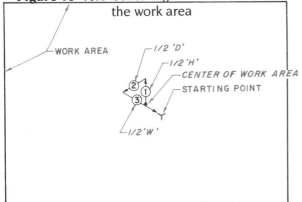

Figure 18-40B Step 1

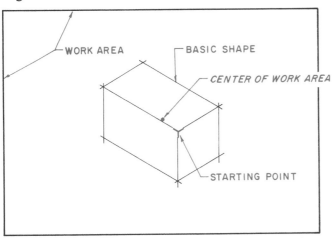

Figure 18-40C Step 2

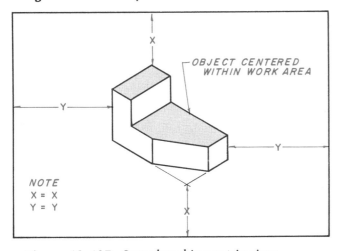

Figure 18-40D Completed isometric view

The true length of the lid can be found by drawing an ellipse with its center at the hinge point and its radius equal to the lid when *closed*, Figure 18-39, Part B. Notice how the length of the lid varies in relation to its position. At 60°, the lid is shorter than it is at 135°.

How To Center an Isometric Object

A quick method to center an isometric object within the work area is given in the following steps.

Given: Multiview drawing in Figure 18-40A.

Step 1. Locate the center point of the work area. From this center point, project straight upward *half the height* of the object. From this point, project to the left at 30° downward *half the depth* of the object. From this point, project to the right 30° downward *half the width* of the object. This is the *starting point* of the basic shape, Figure 18-40B.

Step 2. From the starting point as found in Step 1, lightly draw the basic shape of the object, Figure 18-40C.

Step 3. Construct the object within the basic shape, Figure 18-40D.

Notice that this method centers only the basic shape. If the object is irregularly shaped or has a part removed from the overall basic shape, it will not be 100% centered. Referring to Figure 18-40D, notice how a portion is cut off from the front surface; thus, this object is not 100% centered, but is considered to be close enough for most work.

Isometric Rounds and Fillets

Isometric rounds and fillets can be illustrated in various ways, Figure 18-41. Part A uses a solid line with solid arcs at the corners. Part B is similar, but uses a series of dashed lines. Part C uses a series of curved lines to indicate the rounds and fillets. The most important thing is to illustrate the object so that there is no question whatsoever as to where the rounded surfaces are.

Isometric Dimensioning

Generally, the exact same dimensioning standards, rules, and methods used to dimension a regular multiview drawing apply to an isometric drawing.

Today, the unidirectional system is used because it is much faster and easier to understand. Each dimension is placed at the center of the dimension line if possible.

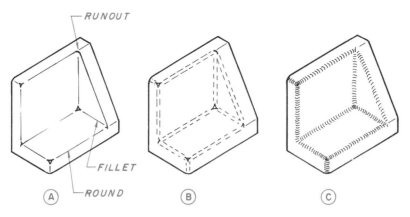

Figure 18-41 Isometric rounds and fillets

Extension lines are projected from the various surface features and at the same angle as the feature. All dimension lines are drawn parallel to the angle of the corresponding surface, Figure 18-42.

Isometric Templates

Various isometric templates are available to the drafter to speed up the drawing process. A unique template to help simplify isometric drawing is the Iso-Drafter, Figure 18-43. This template combines optical reference lines, 16 different isometric ellipses, and two scales. It also offers a right angle (90°) reference line, and isometric reference lines offset 30° on either side of it. These isometric reference lines are at precisely 120° angles to the main straight edges. Complete color instructions that clearly explain its many functions, operation, and uses are included with the template.

Perspective Drawing Procedures

A *perspective drawing* is a three-dimensional drawing of an object that shows it exactly as the eye views it from one particular viewing location. The perspective drawing is actually like a photograph of an object, see Figure 18-44. At the top of the figure is an outside view of a building; below it is an inside view of a room. Notice how all lines project to two points in space.

Perspective Terms

A full understanding of all terminology used in perspective drawing is very important in learning the layout procedures that follow. Refer to Figure 18-45.

Station point (SP) is the exact location from which the observer views the object.

Object to be viewed, as implied, is simply the object that is to be drawn.

Ground line (GL) is the edge view of the ground, or base upon which the object to be viewed rests.

Horizon (H) is a line at the level of the viewer; in this example, approximately 5'-0'' above the ground

Figure 18-42 Isometric dimensioning

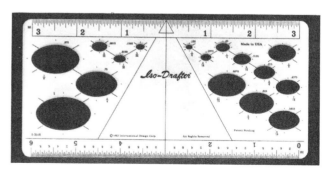

Figure 18-43 Isometric template (*Courtesy International Design Corporation*)

Outside view in two-point perspective

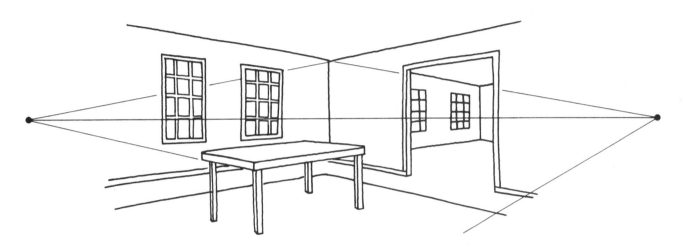

Inside view in two-point perspective

Figure 18-44 Outside and inside perspective view

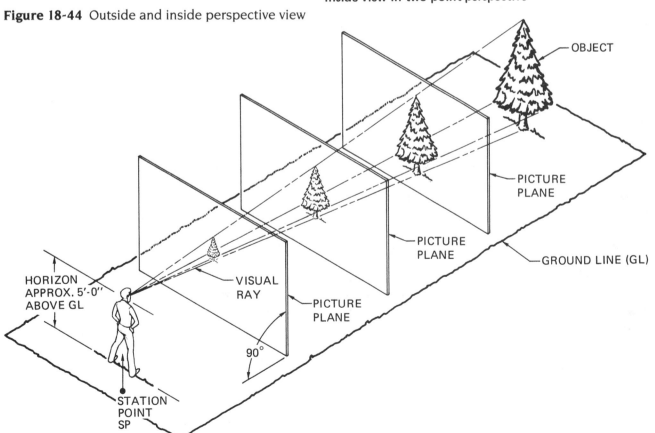

OBJECT

PICTURE
PLANE

PICTURE
PLANE

GROUND LINE (GL)

PICTURE
PLANE

HORIZON
APPROX. 5'-0''
ABOVE GL

VISUAL
RAY

PICTURE
PLANE

90°

STATION
POINT
SP

Perspective drawing: definition of terms

Figure 18-45 Perspective terms

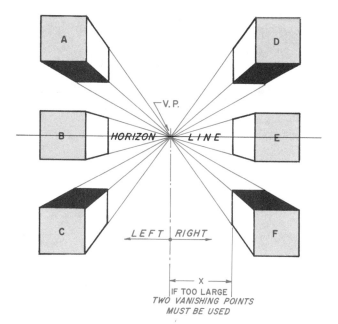

Figure 18-46 Vanishing point and horizon line

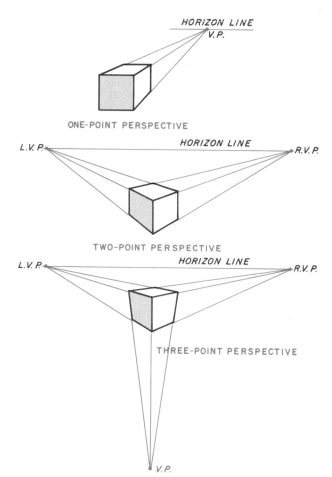

Figure 18-47 Kinds of perspective drawings

line. Note that the vanishing points are always located on the horizon line.

Picture plane line (PP) is an imaginary plane perpendicular to the ground line and located between the viewer and the object. Think of the picture plane line as the paper on which the object will be drawn.

Visual rays (VR) are lines projected from the eye of the viewer while standing at the station point to each and every point of the object. Notice how the visual rays project through the picture plane line. These rays, collectively, form the perspective view of the object upon the picture plane or paper.

Vanishing point (VP) the vanishing point(s) (not illustrated in the figure) is a point on the horizon line to which all other lines are projected, Figure 18-46. Cube A is *above* the horizon line and to the *left* of the vanishing point; thus, the bottom and right side of the cube are seen. Cube B is centered *on* the horizon line and to the *left* of the vanishing point; thus, only the right side of the cube is seen. Cube C is *below* the horizon line and to the *left* of the vanishing point; thus, the top and right side of the cube are seen. Cubes D, E, and F are opposite to A, B, and C. If dimension X is too great, a two-point perspective drawing must be used.

Measuring line (M) the measuring line is a vertical line 90° from the picture plane line, where all true length distances or heights are projected to and measured from. All true vertical heights must be projected to the measuring line and back into space to their respective vanishing points.

Types of Perspective Views

There are three types of perspective views: one-point, two-point, and three-point perspective, Fig-

ure 18-47. Only two-point perspective is discussed in this chapter, as it is the type most commonly used in industry.

Sketching

It is a good idea to make a simple, quick sketch of the object before actually starting the drawing in order to choose the position that best illustrates the object. This also gives an indication as to the approximate direction and location of the vanishing points.

How To Sketch a Two-Point Perspective

The multiview drawing in Figure 18-48A is given, and various sketches are made, Figure 18-48B.

In setting up a two-point perspective, two of the regular multiviews must be used; usually, the top view (plan view) and either a front or side view. The views are sometimes changed, if necessary, to draw the object in a particular position. Multiview drawing 18-48A, and sketch C of Figure 18-48B are used for describing the following steps:

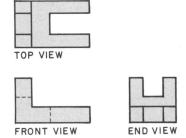

GIVEN : MULTIVIEW DRAWING

TOP VIEW

FRONT VIEW   END VIEW

Figure 18-48A How to draw a two-point perspective drawing

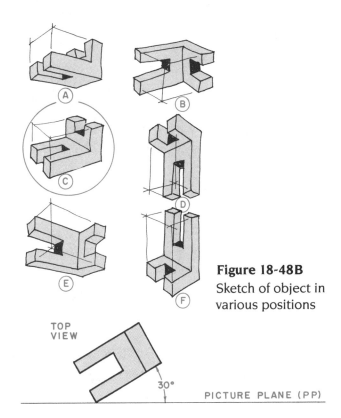

Figure 18-48B Sketch of object in various positions

TOP VIEW

30°

PICTURE PLANE (PP)

Figure 18-48C Step 1

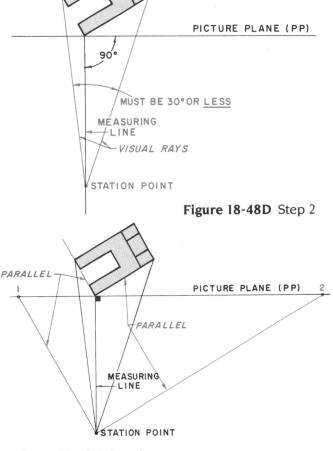

Step 1. Draw the top view at the top of a sheet of paper at any angle; in this example, 30° in order to come close to the sketch. Draw the picture plane line tangent to the front edge of the object, Figure 18-48C.

Step 2. From the front edge of the object, where it is tangent to the picture plane line, project the measuring line 90° to the picture plane line, Figure 18-48D. Locate the station point anywhere on the measuring line so that the maximum inclusive angle is 30° or less.

Step 3. From the station point, construct lines parallel to the two edges of the top view above, Figure 18-48E. Think of these three steps as 1) an overall top view of the object, 2) a picture plane line on edge, and 3) a station point. In a regular two-point perspective there are actually two views superimposed over each other. This is the first of the two views.

Step 4. Draw a line to represent the horizon line at any convenient location. It is best to draw this horizon line away from the other construction work, if possible. If space is limited, it can be drawn over the original work, Figure 18-48F. From points 1 and 2 on the picture plane line, project downward to the horizon line to locate the left and right vanishing points. Locate the ground line and draw the side or front view of the object on the ground line. In this example, the object is viewed from above. If the object is to be viewed from below, the ground line would have been located *above* the horizon line (refer back to Figure 18-46).

Step 5. From the right-side view, project the true height to the measuring line. From the height on the measuring line, project back into space

TOP VIEW

PICTURE PLANE (PP)

90°

MUST BE 30° OR <u>LESS</u>

MEASURING LINE

VISUAL RAYS

STATION POINT

Figure 18-48D Step 2

PARALLEL

1   PICTURE PLANE (PP)   2

PARALLEL

MEASURING LINE

STATION POINT

Figure 18-48E Step 3

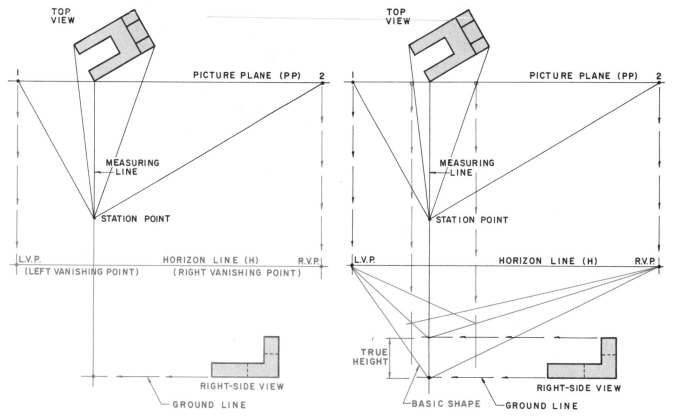

Figure 18-48F Step 4

Figure 18-48G Step 5

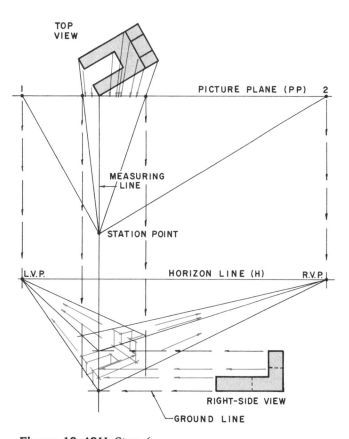

Figure 18-48H Step 6

PERSPECTIVE VIEW

Figure 18-48I
Completed two-point
perspective drawing

to the left and right vanishing points, Figure 18-48G. Construct the perspective basic shape of the object. The basic shape projects back into space until it intersects a line projected downward from the picture plane line.

Step 6. Project the true heights of each feature from the right-side view to the measuring line and back toward the appropriate vanishing point to a point where the same feature intersects the picture plane line, Figure 18-48H. Lightly develop the various features of the object as would be done in an isometric view.

Step 7. Check all work, and, if correct, darken in the object using correct line thickness, Figure 18-48I. As a general rule, hidden lines are *not* usually drawn except to illustrate an important feature that would otherwise not be seen.

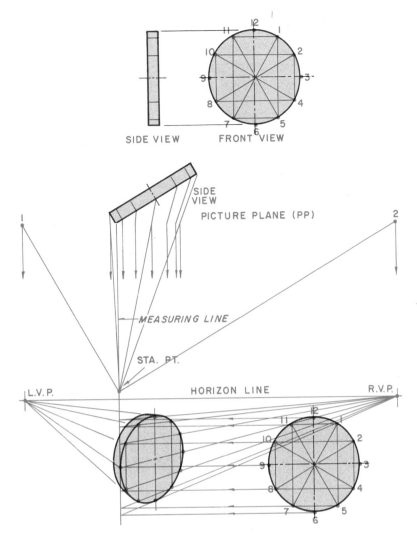

Figure 18-49 Perspective circles or arcs

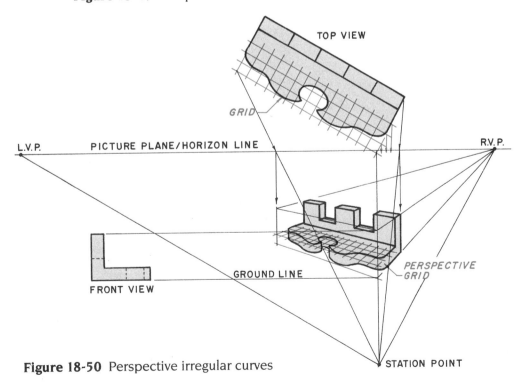

Figure 18-50 Perspective irregular curves

Perspective Circles or Arcs

The circle or arc is broken down into 12 equal spaces and, as with an isometric, reduced to a square and two rectangles. The square and two rectangles are now drawn as a simple perspective square and rectangle, Figure 18-49. The 12 points are connected, and the perspective circle or arc is drawn.

Perspective Irregular Curves

The perspective irregular curve is developed exactly as with an isometric irregular curve. A grid is drawn on the multiview drawing and transferred to the perspective layout. Each point is then transferred from the multiview grid to the perspective grid, Figure 18-50.

Review

1. Explain when the true isometric circle method must be used over the four-center isometric circle method even though it is much slower.

2. Explain how an isometric drawing is dimensioned.

3. Explain the use of box construction and why it is used.

4. Under what classification do isometric, dimetric, and trimetric projections fall?

5. List the three types of pictorial drawings used today in industry.

6. What are the principal edges of a pictorial drawing?

7. Which kind of pictorial drawing best illustrates the object as it would be viewed by the eye?

8. Describe two ways to draw an irregular curve in an isometric drawing.

9. What is meant by a nonisometric line?

10. How are pictorial drawings used today?

11. List three kinds of oblique drawings. Explain what a cabinet drawing is, and what it was originally used for.

12. Are hidden lines used in pictorial drawings? Explain.

Chapter Eighteen Problems

The following problems are intended to give the beginning drafter practice in laying out and developing isometric and perspective renderings of various objects.

The steps to follow in laying out the drawings in this chapter are:

Step 1. Study the problem carefully.

Step 2. Position the object so it *best* illustrates the most features.

Step 3. Make a pictorial sketch of the object.

Step 4. Calculate the basic shape size.

Step 5. Find the center of the work area.

Step 6. Lightly center and lay out the basic shape within the work area.

Step 7. Lightly develop the object *within* the basic shape.

Step 8. Check to see that the completed object is centered within the work area.

Step 9. Check all size dimensions, and check for accuracy of object.

Step 10. Darken in the object.

Step 11. Recheck all work, and, if correct, neatly fill out the title block using light guidelines and neat lettering.

Problems 18-1 through 18-27

Construct an isometric drawing of the object in Problems 18-1 through 18-27. (These problems have only *straight* lines.)

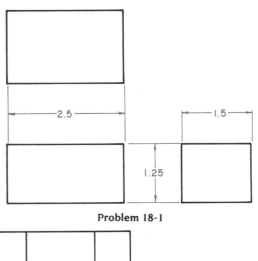

Problem 18-1

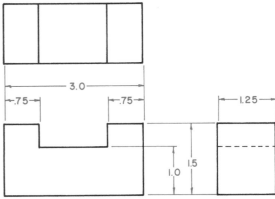

Problem 18-2

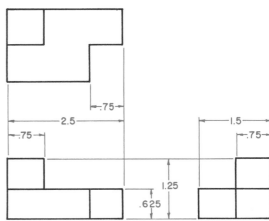

Problem 18-3

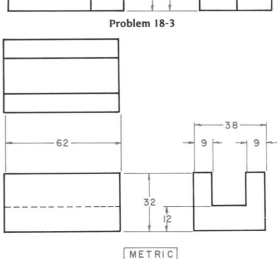

METRIC

Problem 18-4

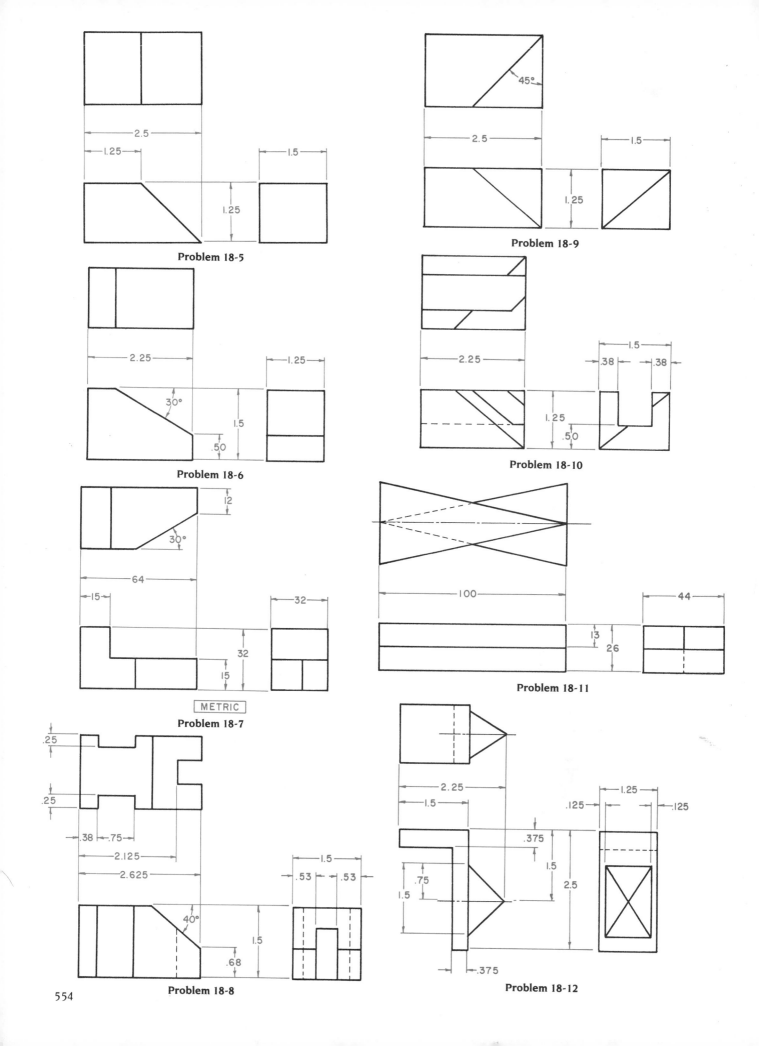

Problem 18-5

Problem 18-9

Problem 18-6

Problem 18-10

Problem 18-7

METRIC

Problem 18-11

Problem 18-8

Problem 18-12

554

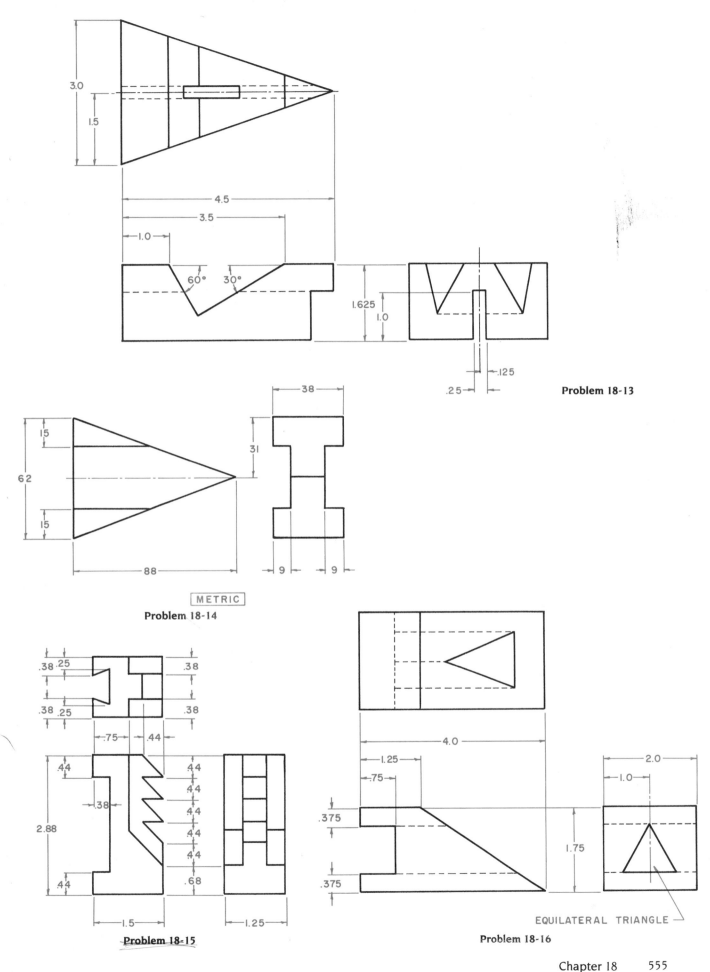

Problem 18-13

METRIC

Problem 18-14

Problem 18-15

Problem 18-16

EQUILATERAL TRIANGLE

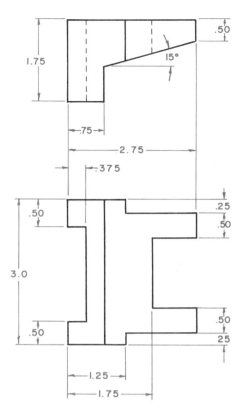

Problem 18-17

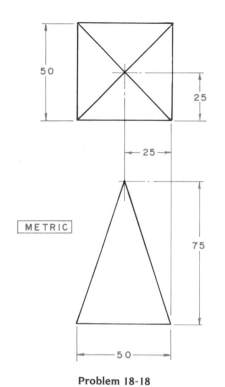

METRIC

Problem 18-18

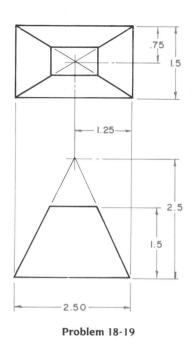

Problem 18-19

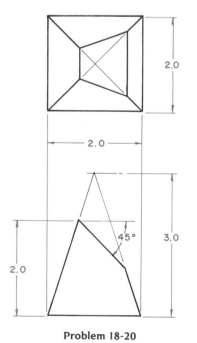

Problem 18-20

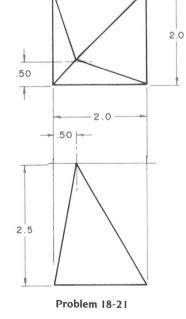

Problem 18-21

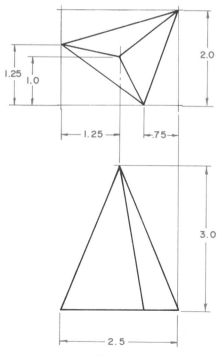

Problem 18-22

556   Section 4

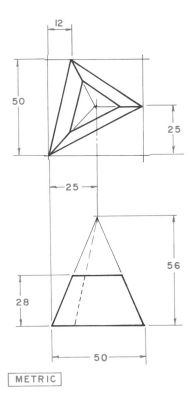

METRIC

Problem 18-23

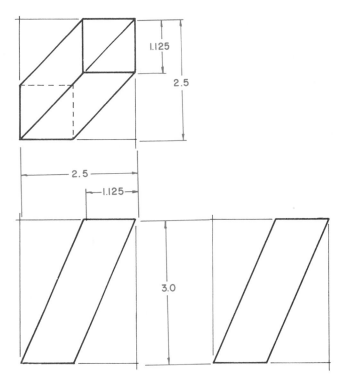

Problem 18-25

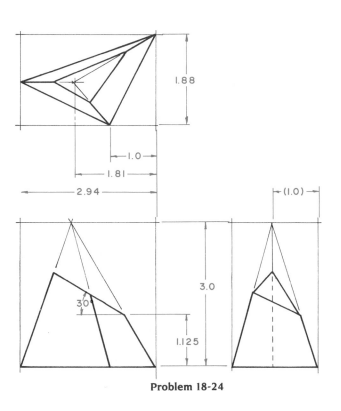

Problem 18-24

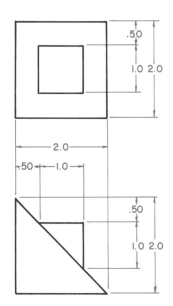

Problem 18-26

.25 SQUARES

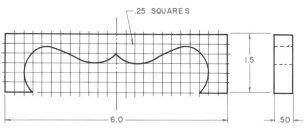

Problem 18-27

Problems 18-28 through 18-51

Construct an isometric drawing of the object in Problems 18-28 through 18-51. (These problems have *straight lines, arcs* and *circles*.)

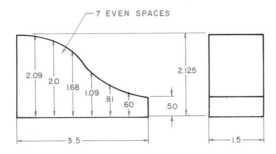

Problem 18-28

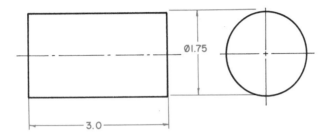

Problem 18-29

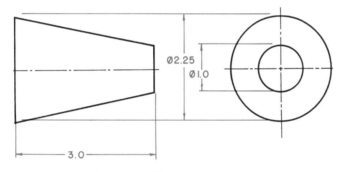

Problem 18-30

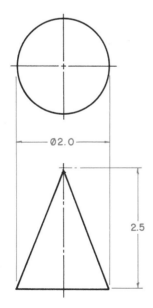

Problem 18-31

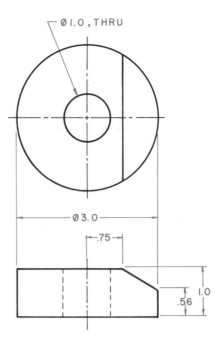

Problem 18-32

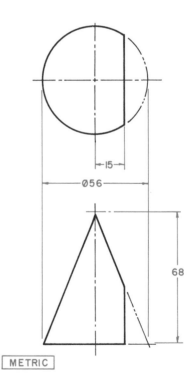

Problem 18-33

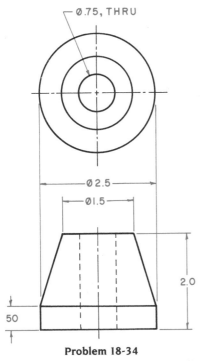

Ø.75, THRU
Ø2.5
Ø1.5
2.0
50

Problem 18-34

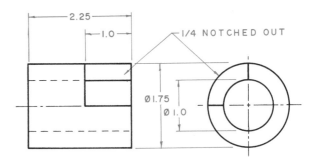

2.25
1.0
1/4 NOTCHED OUT
Ø1.75
Ø1.0

Problem 18-37

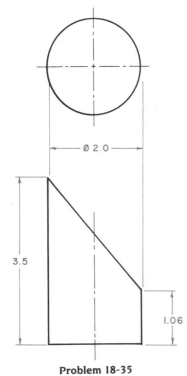

Ø2.0
3.5
1.06

Problem 18-35

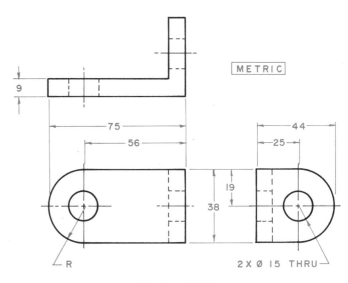

METRIC
9
75
56
44
25
19
38
R
2 X Ø 15 THRU

Problem 18-38

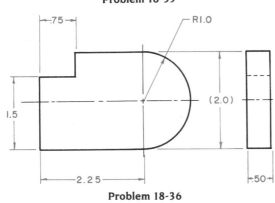

.75
R1.0
(2.0)
1.5
2.25
.50

Problem 18-36

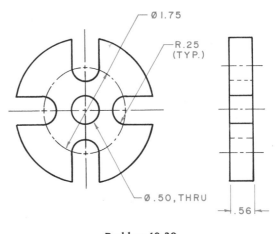

Ø1.75
R.25
(TYP.)
Ø.50, THRU
.56

Problem 18-39

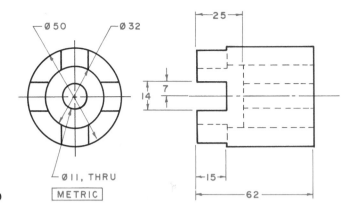

Problem 18-40

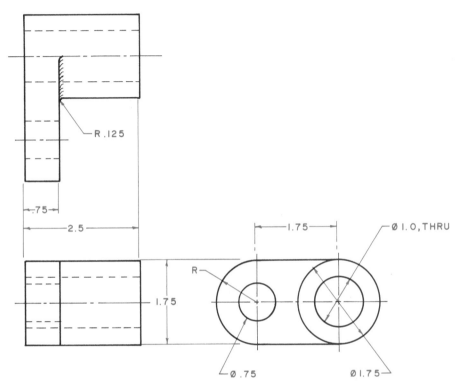

Problem 18-41

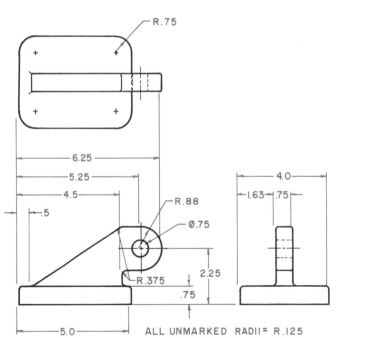

Problem 18-42

ALL UNMARKED RADII = R.125

Problem 18-43

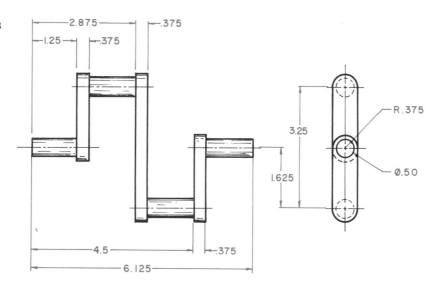

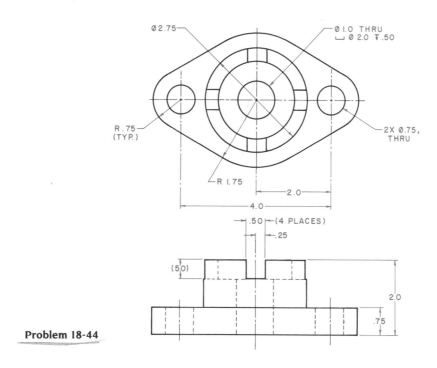

Problem 18-44

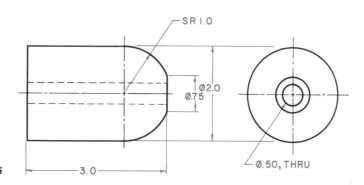

Problem 18-45

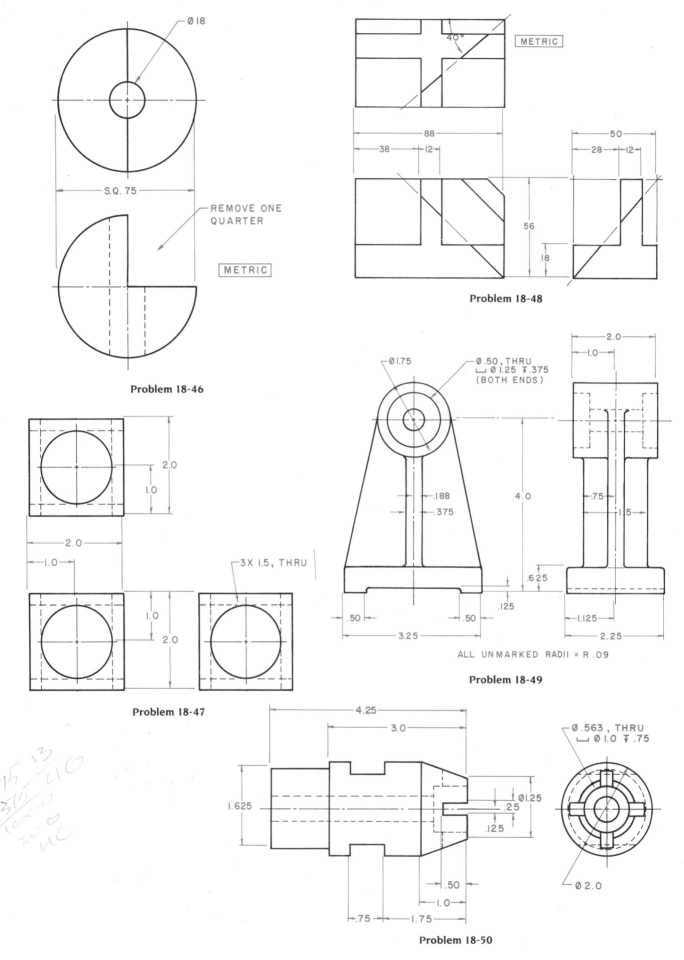

Ø 18

S.Q. 75

REMOVE ONE
QUARTER

METRIC

Problem 18-46

40°

METRIC

88

38

12

56

18

50

28

12

Problem 18-48

2.0

1.0

2.0

1.0

1.0

2.0

3X 1.5, THRU

Problem 18-47

Ø1.75

Ø.50, THRU
⌴ Ø 1.25 ⌵.375
(BOTH ENDS)

2.0

1.0

.188

.375

4.0

.75

1.5

.625

.125

.50

.50

.50

1.125

3.25

2.25

ALL UNMARKED RADII = R .09

Problem 18-49

4.25

3.0

1.625

Ø.563, THRU
⌴ Ø 1.0 ⌵.75

Ø1.25

.25

.125

.50

1.0

.75

1.75

Ø 2.0

Problem 18-50

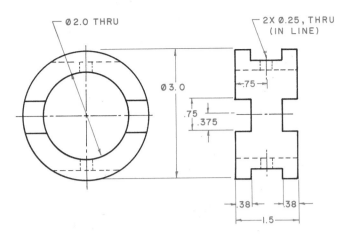

Problem 18-51

Problem 18-52

Construct an exploded isometric drawing of the assembly of a vise, Problem 18-52. The vise consists of 7 different parts and various fasteners, Problems 18-52-1 through 18-52-7. Illustrate the exploded view with the parts in relationship to its assembly. Space parts with an approximate 1" (25 mm) space between parts. Center the completed view within the work area.

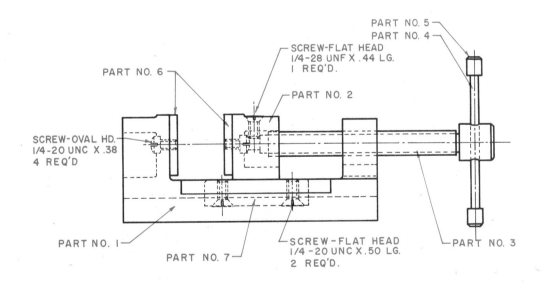

ASSEMBLY

Problem 18-52

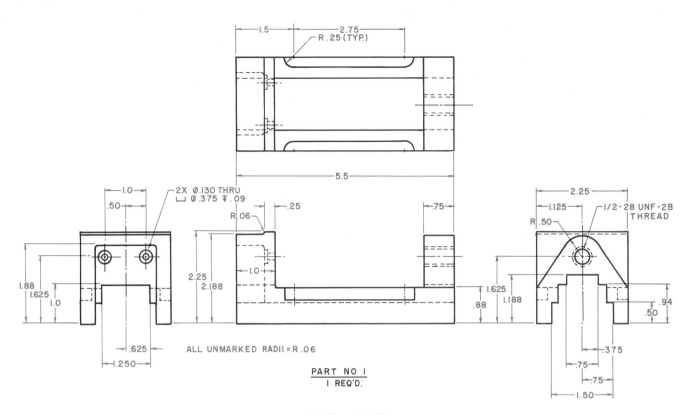

Problem 18-52-1

ALL UNMARKED RADII = R .06

PART NO I
1 REQ'D.

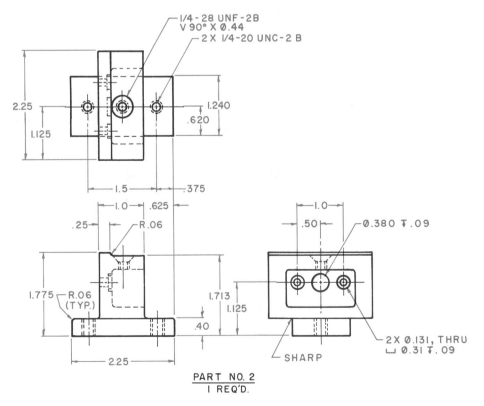

PART NO. 2
1 REQ'D.

Problem 18-52-2

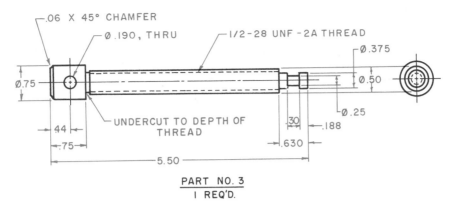

.06 X 45° CHAMFER

Ø .190, THRU

1/2-28 UNF -2A THREAD

Ø 0.375

Ø .50

Ø .25

UNDERCUT TO DEPTH OF THREAD

Ø.75

.44

.75

.30

.188

.630

5.50

PART NO. 3
1 REQ'D.

Problem 18-52-3

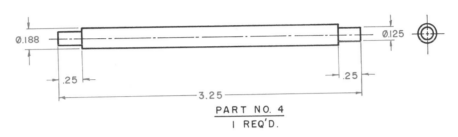

Ø .188

Ø .125

.25

3.25

.25

PART NO. 4
1 REQ'D.

Problem 18-52-4

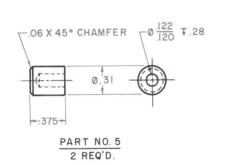

.06 X 45° CHAMFER

Ø .122 / .120 ▼.28

Ø .31

.375

PART NO. 5
2 REQ'D.

Problem 18-52-5

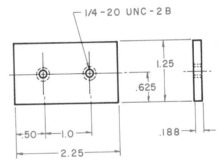

1/4 -20 UNC - 2 B

1.25

.625

.50

1.0

2.25

.188

Problem 18-52-6

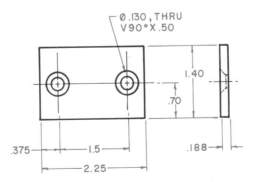

Ø .130 , THRU
⌵90° X .50

1.40

.70

.375

1.5

2.25

.188

Problem 18-52-7

Problem 18-53

Using the given multiview drawing in Problem 18-53, develop a perspective drawing in the positions as illustrated by sketches A, B, C, D, E and F.

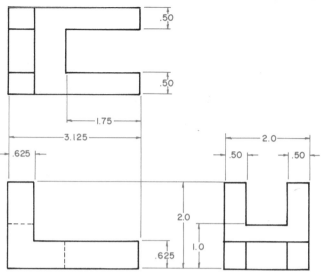

SKETCHES

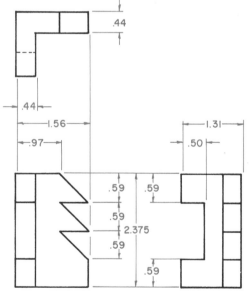

A   B   C

D   E

Problem 18-53

Problem 18-54

Using the given multiview drawing in Problem 18-54, develop a perspective drawing in the positions as illustrated by sketches A and B.

Problem 18-54

SKETCHES

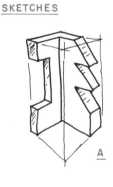

A

B

Problem 18-55

Choose various objects from Problems 18-1 through 18-51, and develop them into a perspective drawing. Compare the isometric view with the isometric drawing.

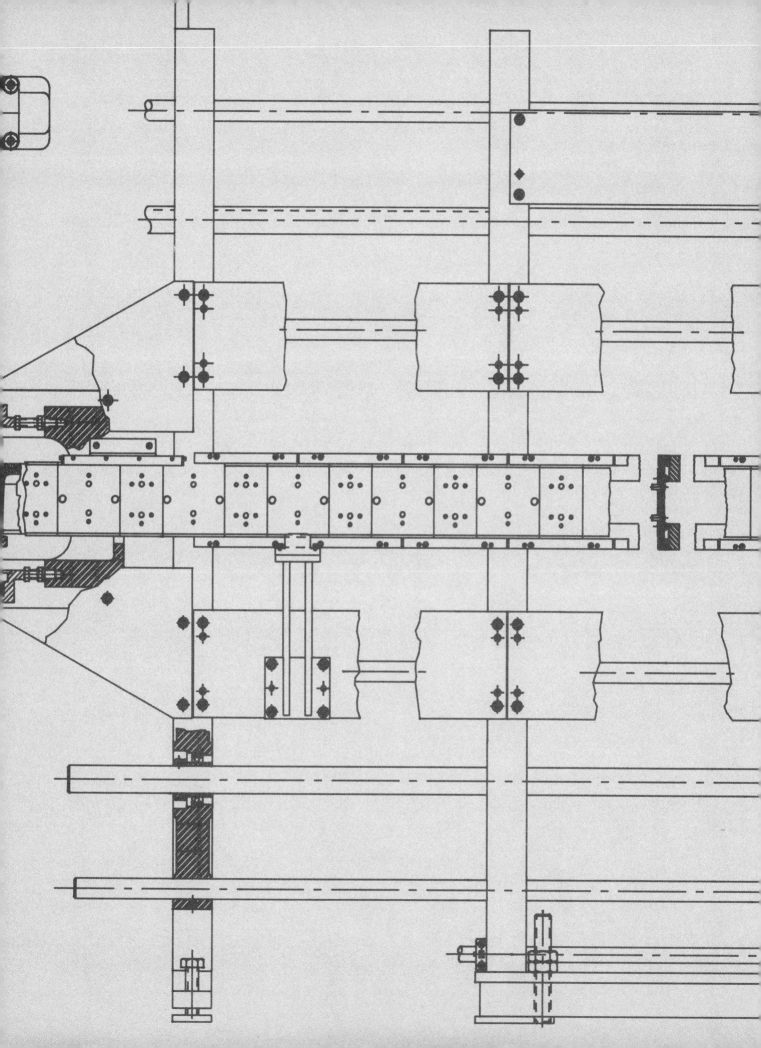

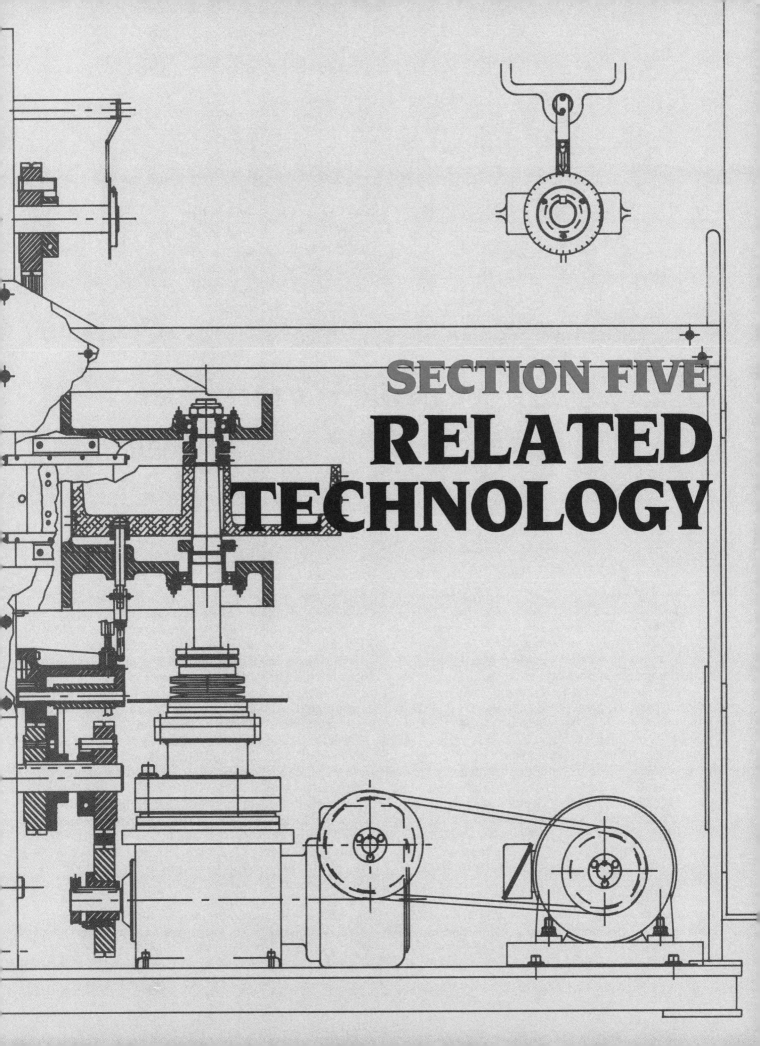

SECTION FIVE
RELATED TECHNOLOGY

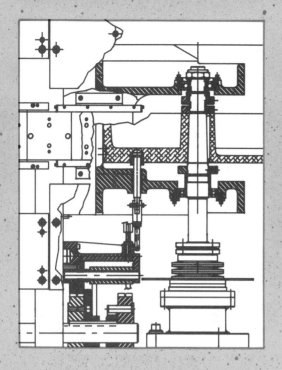

This *chapter discusses the various types of welds. All welding symbols and their meanings are studied and fully illustrated. Also discussed are the size, length, and placement of welds; welding processes; welding joints; depth and actual size of welds; multiple reference lines; dimensioning of welds; spot welds; projection welds; seam welds; and weld templates.*

CHAPTER NINETEEN

Welding

Welding Processes

Pieces of metal can be fastened together with mechanical fasteners as was discussed in Chapter 13, "Fasteners." They can also be held together by soldering or brazing, and some are fastened together by an adhesive. A permanent way to join pieces of metal together is by *welding*.

One method used to weld parts together is to heat the edges of the pieces to be joined until they melt and join or fuse together. When the pieces cool, they become one homogeneous mass, permanently joined together. A filler rod is sometimes used to mix with the molten metal to make the joint stronger. When a lot of heat is used to melt and fuse the joint, the weld is called a *fusion weld*. Heat for this process can come from a torch, burning gasses or with a high electric current. Because of the extremely high temperatures involved in welding, the part or parts may be distorted; thus, all machining is usually done after welding.

Another kind of weld used in years past was the pressure weld. In this process, the parts to be welded together are heated to a plastic state, and forced together by pressure or by hammering. This was done by the local blacksmith. This old method was called

forge welding. Today, faster and better methods are used.

Welding assemblies are usually built-up from stock forms, such as plate steel, square bars, tubing, angle iron and the like. These parts are cut to shape and welded together. The various welding processes are illustrated in Figure 19-1. Welded assemblies are much less expensive and more satisfactory than the casting process, especially in a case where only one or only a few identical parts are required.

Welding is used on large structures that would be difficult or impossible to build in a manufacturing plant. Large structures such as building frames, bridges, and ships are welded into one large assembly.

Major welding methods classified by the American Welding Society include: brazing, gas welding, arc welding, resistance welding, and fusion welding. *Brazing* is the process of joining metals together with a nonferrous filler rod. A temperature above 1000°F, but just below the melting point, is used to join the parts. The most commonly used *gas* welding is oxyacetylene welding. Resistance welding and fusion welding, the two major processes, are explained in this text. Detailed information regarding the other types of welding processes can be obtained by writ-

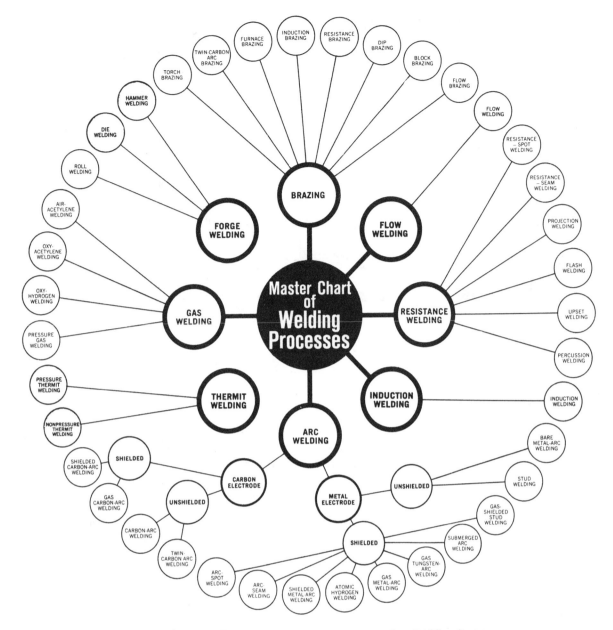

Figure 19-1 Master chart of welding processes *(Courtesy American Welding Society)*

ing to the American Welding Society, P.O. Box 351040, Miami, Florida 33125.

Resistance welding is the process of passing an electric current through the exact location where the parts are to be joined. This is usually done under pressure. The combination of the pressure and the heat generated by the electric current welds the parts together. This process is usually done on thin sheet metal parts.

Fusion welding is usually used for large parts. The fusion welding process uses many standard types of welds, Figure 19-2. Each is explained in full.

The next few figures use the fillet welding symbol as an example, but the same method applies, regardless of which welding symbol is used.

Basic Welding Symbol

The *basic welding symbol* consists of a reference line, a leader line and arrow, and, if needed, a tail, Figure 19-3. The tail is added for specific information or notes in regard to welding specifications, processes or reference information. The reference line of the basic welding symbol is usually drawn horizontally. Any welding symbol placed on the upper side of the reference line indicates WELD OPPOSITE SIDE, Figure 19-4. Any welding symbol placed on the lower side of the reference line indicates WELD ARROW SIDE, Figure 19-4. The direction of the leader line and arrow had no significance whatsoever to the reference line.

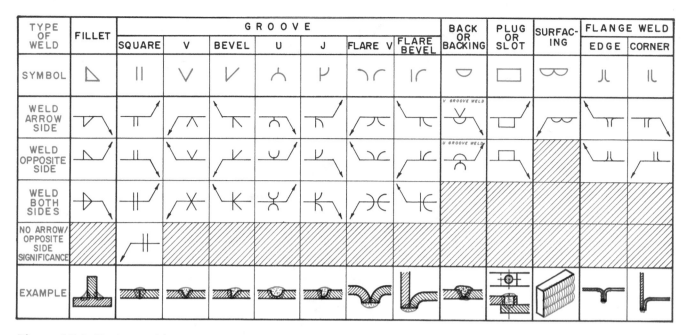

Figure 19-2 Fusion welding

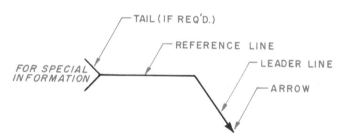

Figure 19-3 Basic welding symbol

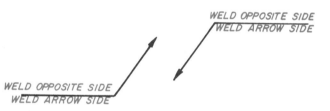

Figure 19-4 Any welding symbol placed on the upper side of the reference line indicates weld OPPOSITE side. Any welding symbol placed on the lower side of the reference line indicates weld ARROW side.

A fillet weld symbol (see Figure 19-2) is added above or below the reference line with the left leg always drawn vertical, Figure 19-5. A welding symbol placed *above* the reference line means to weld the OPPOSITE side, as illustrated in Figure 19-6. A weld symbol placed *below* the reference line means to weld ARROW side, as illustrated in Figure 19-6. The positions as drawn and as welded are shown in Figures 19-7 and 19-8. A weld symbol placed above and below the reference line means to weld both the OPPOSITE side and the ARROW side, as illustrated in Figure 19-9.

Figure 19-5 Left leg of welding symbol is always drawn vertical

Size of Weld

The weld must be fully dimensioned so that there is no question whatsoever as to its intended and designed size. The size of a weld refers to the length of the leg or side of the weld. The size is placed directly to the left of the welding symbol. The two

Figure 19-6 Welding symbol placed *above* the reference line means to weld OPPOSITE side. Welding symbol placed *below* the reference line means to weld ARROW side.

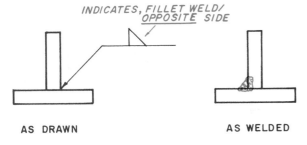

INDICATES, FILLET WELD/
<u>OPPOSITE</u> SIDE

AS DRAWN   AS WELDED

Figure 19-7 Position as drawn and as welded

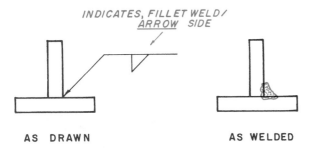

INDICATES, FILLET WELD/
<u>ARROW</u> SIDE

AS DRAWN   AS WELDED

Figure 19-8 Position as drawn and as welded

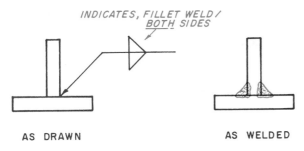

INDICATES, FILLET WELD/
<u>BOTH</u> SIDES

AS DRAWN   AS WELDED

Figure 19-9 Welding symbol placed above and below the reference line means to weld both sides

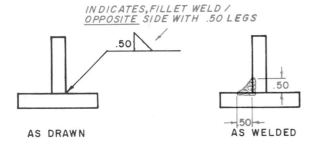

INDICATES, FILLET WELD /
<u>OPPOSITE</u> SIDE WITH .50 LEGS

.50

AS DRAWN   AS WELDED

.50

.50

Figure 19-10 Dimension to the left of welding symbol indicates length of leg or side of weld

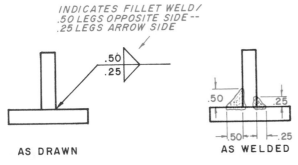

INDICATES FILLET WELD /
.50 LEGS OPPOSITE SIDE --
.25 LEGS ARROW SIDE

.50
.25

AS DRAWN   AS WELDED

.50   .25
.50   .25

Figure 19-11 Size of each weld is required

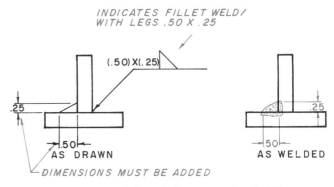

INDICATES FILLET WELD/
WITH LEGS .50 X .25

(.50) X (.25)

.25

.50

AS DRAWN

.25

.50

AS WELDED

DIMENSIONS MUST BE ADDED

Figure 19-12 Welds with legs or side of different sizes

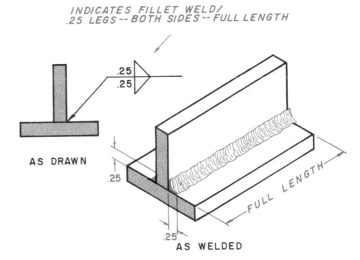

INDICATES FILLET WELD/
.25 LEGS -- BOTH SIDES -- FULL LENGTH

.25
.25

AS DRAWN

.25

.25

.25

FULL LENGTH

AS WELDED

Figure 19-13 If weld length is not specified, it is assumed to be continuous full length

legs are assumed to be equal in size unless otherwise dimensioned, Figure 19-10. If a double weld is needed, the size of each weld must be included as illustrated in Figure 19-11.

If the legs or sides of the weld are to be different, the dimensions of the sides must be indicated to the left of the welding symbol in parentheses, as reference only dimensions, Figure 19-12. Because the symbol does not indicate which is the .50 leg and which is the .25 leg, the detail drawing must include the dimensions.

Length of Weld

When a weld length is not specified it is assumed to be continuous the full length, Figure 19-13. If a weld must be made to a special length, it must be indicated. This is done by a dimension directly to the right of the weld symbol, Figure 19-14.

When a weld is to be made continuously all around an object, it is indicated on the welding symbol by adding a small circle between the reference line and the leader line, Figure 19-15.

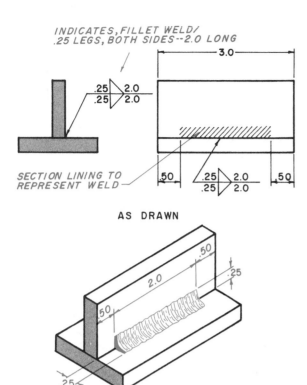

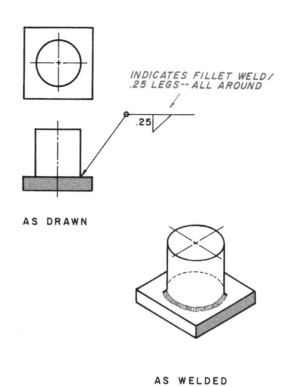

Figure 19-14 Length of weld must be noted to the right of the welding symbol

Figure 19-15 A small circle indicates weld continuously around an object

Placement of Weld

When a weld is not continuous and is needed in only a few areas, *section lining* is used to indicate where the weld is to be placed, Figure 19-16. If a weld is to be placed in a few areas, this is indicated by the use of multiple leader lines and arrows, Figure 19-17.

Intermittent Welds

Intermittent welds are a series of short welds. When this type of weld is needed, an extra dimension is added to the welding symbol, Figure 19-18. The usual size of leg is noted to the left of the symbol, and the length of the weld is indicated to the right of the welding symbol followed by a dash line and the pitch of the intermittent welds. See Figure 19-19. An *increment* is the length of the weld; the *pitch* is the center-to-center distance between increments.

Chain Intermittent Weld

A *chain intermittent weld* is a weld where each weld is applied directly opposite to each other on opposite sides of the joint, Figure 19-20. Note the welding symbol is applied above and below the reference line in line with each other.

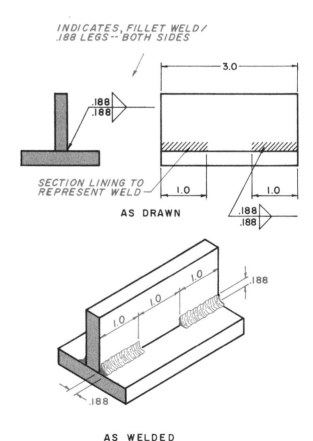

Figure 19-16 Section lining indicates location of weld

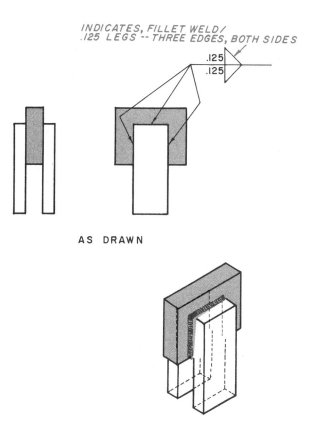

AS DRAWN

Figure 19-17 Leader line and arrows indicate location of weld

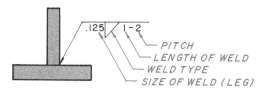

AS DRAWN

Figure 19-18 Intermittent welds are noted by dimensions to the right of the welding symbol

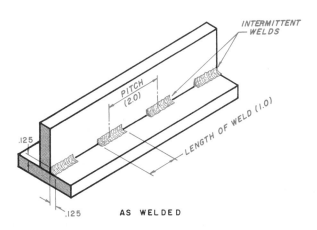

AS WELDED

Figure 19-19 Pitch is the center-to-center distance between welds

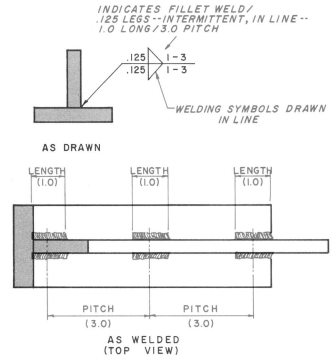

AS DRAWN

Figure 19-20 Chain intermittent welds are applied directly opposite to each other

Staggered Intermittent Weld

A *staggered intermittent weld* is similar to the chain intermittent weld except the welds are staggered directly opposite to each other on opposite sides, Figure 19-21. Note the welding symbol is applied above and below the reference line staggered to each other.

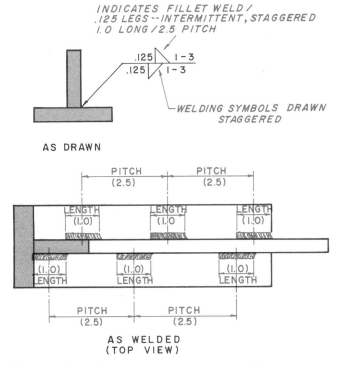

Figure 19-21 Staggered intermittent welds are placed staggered opposite each other

AAC	air carbon arc cutting
AAW	air acetylene welding
ABD	adhesive bonding
AB	arc brazing
AC	arc cutting
AHW	atomic hydrogen welding
AOC	oxygen arc cutting
AW	arc welding
B	brazing
BB	block brazing
BMAW	bare metal arc welding
CAC	carbon arc cutting
CAW	carbon arc welding
CAW-G	gas carbon arc welding
CAW-S	shielded carbon arc welding
CAW-T	twin carbon arc welding
CW	cold welding
DB	dip brazing
DFB	diffusion brazing
DFW	diffusion welding
DS	dip soldering
EASP	electric arc spraying
EBC	electron beam cutting
EBW	electron beam welding
ESW	electroslag welding
EXW	explosion welding
FB	furnace brazing
FCAW	flux cored arc welding
FCAW-EG	flux cored arc welding—electrogas
FLB	flow brazing
FLOW	flow welding
FLSP	flame spraying
FOC	chemical flux cutting
FOW	forge welding
FRW	friction welding
FS	furnace soldering
FW	flash welding
GMAC	gas metal arc cutting
GMAW	gas metal arc welding
GMAW-EG	gas metal arc welding—electrogas
GMAW-P	gas metal arc welding—pulsed arc
GMAW-S	gas metal arc welding—short circuiting arc
GTAC	gas tungsten arc cutting
GTAW	gas tungsten arc welding
GTAW-P	gas tungsten arc welding—pulsed arc
HFRW	high frequency resistance welding
HPW	hot pressure welding
IB	induction brazing
INS	iron soldering
IRB	infrared brazing
IRS	infrared soldering
IS	induction soldering
IW	induction welding
LBC	laser beam cutting
LBW	laser beam welding
LOC	oxygen lance cutting
MAC	metal arc cutting
OAW	oxyacetylene welding
OC	oxygen cutting
OFC	oxyfuel gas cutting
OFC-A	oxyacetylene cutting
OFC-H	oxyhydrogen cutting
OFC-N	oxynatural gas cutting
OFC-P	oxypropane cutting
OFW	oxyfuel gas welding
OHW	oxyhydrogen welding
PAC	plasma arc cutting
PAW	plasma arc welding
PEW	percussion welding
PGW	pressure gas welding
POC	metal powder cutting
PSP	plasma spraying
RB	resistance brazing

Figure 19-22 Designation of welding and allied processes by letters (abbreviations)

RPW	projection welding
RS	resistance soldering
RSEW	resistance seam welding
RSW	resistance spot welding
ROW	roll welding
RW	resistance welding
S	soldering
SAW	submerged arc welding
SAW-S	series submerged arc welding
SMAC	shielded metal arc cutting
SMAW	shielded metal arc welding
SSW	solid state welding
SW	stud arc welding
TB	torch brazing
TCAB	twin carbon arc brazing
TW	thermit welding
USW	ultrasonic welding
UW	upset welding

Figure 19-22 (continued)

LETTER	METHOD
C	CHIPPING
G	GRINDING
H	HAMMERING
R	ROLLING
M	MACHINING

Figure 19-23 Abbreviation of standard method used to obtain a particular contour

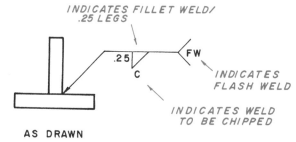

Figure 19-24 Process abbreviations are placed in the tail of the basic welding symbol

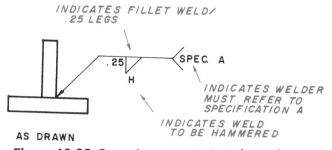

Figure 19-25 Sometimes, a note is indicated somewhere on the drawing.

Process Reference

There are many welding processes developed by the American Welding Society. Each process has a standard abbreviation designation, Figure 19-22. Letter designations are used to specify a method used to obtain a particular contour of a weld, Figure 19-23. These abbreviations are taken from the latest American Welding Society's standard #AWS-2.4-79,71.

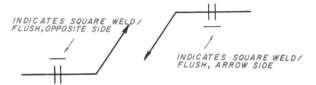

Figure 19-26 Contour symbol

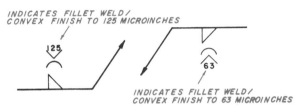

Figure 19-27 Contour symbol is added to the welding symbol

Figure 19-28 Finish number specifies finish required

The standard process abbreviations or letter designations are usually added to the tail of the basic welding symbol as required, Figure 19-24. Sometimes, they are called-off in a note somewhere on the drawing, Figure 19-25.

Contour Symbol

If a special finish must be made to a weld, a contour symbol must be added to the welding symbol. There are three kinds of contour symbols used: flush, convex, or concave, Figure 19-26. The contour symbol is added directly above or below the welding symbol. If the welding symbol is *above* the reference line, the contour symbol is placed *above* the welding symbol. If the welding symbol is placed *below* the reference line, the contour symbol is placed *below* the welding symbol, Figure 19-27. The degree of finish is not usually included, but, if a specific finish is desired, a finish number designation must be added, Figure 19-28.

Figure 19-29 represents that the finish weld must be concave and ground to a 125-microinch surface finish. Figure 19-30 represents that the finish weld must be machined flat on the top surface to a surface finish of 125 microinches. The weld must be convex on the bottom surface with a hammered finish.

Field Welds

Any weld not made in the factory, that is, it is to be made at a later date, perhaps at final assembly on site, is called a *field weld*. Its symbol is added to the

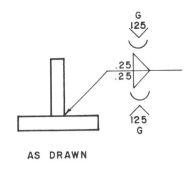

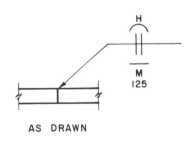

Figure 19-29 Finish weld is to be concave and ground to a 125 micro-inch surface finish

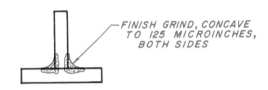

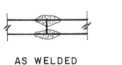

Figure 19-30 Finish weld is to be machined flat on top surface to 125-microinch surface and convex on the bottom surface with a hammered finish

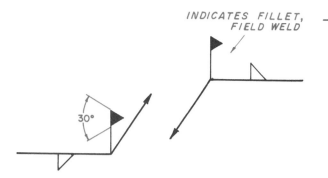

Figure 19-31 Field weld

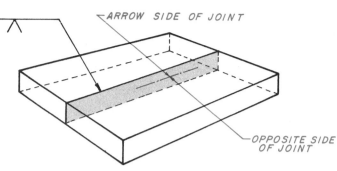

BUTT JOINT

Figure 19-32A Butt joint

basic weld symbol by a filled-in flag, located between the reference line and the leader line and drawn using a 30°-60° triangle, Figure 19-31.

Welding Joints

There are five basic kinds of welding joints and they are classified according to the position of the parts that are being joined, Figures 19-32A through 19-32E. There are many types of welds used to weld these joints together. Considerations as to which type of weld to use are based on the particular application, thickness of material, required strength of the joint, and available welding equipment, among others.

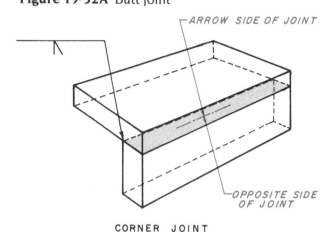

CORNER JOINT

Figure 19-32B Corner joint

Types of Welds

There are six major types of welds (refer back to Figure 19-2): fillet, groove, back or backing, plug or slot, surface, and flange welds.

More than one type of weld may be applied to a single joint — usually for strength or appearance. For example, a groove V weld may be used on one side, and a backing weld on the other side, Figure 19-33. This is known as a *multiweld*.

The same type of weld may be used on opposite sides of a single joint. Such a joint is called a *double weld*; for example, double V, double U or double J weld, Figure 19-34. These too, are used for strength and/or appearance.

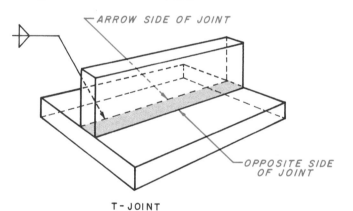

T-JOINT

Figure 19-32C T-joint

Fillet Welds

Up to this point, the fillet weld and its symbol have been used as an example. The other five major types of welding symbols used are very similar.

Groove Weld

Seven types of welds are considered to be *groove welds*: square, V, bevel groove, J, U, flare V, and flare

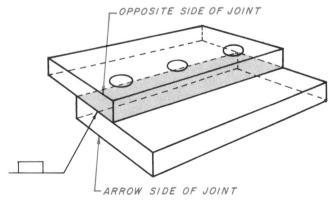

LAP JOINT

Figure 19-32D Lap joint

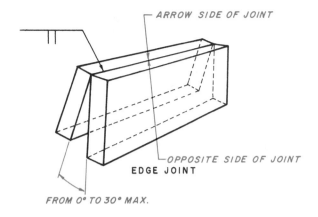

Figure 19-32E Edge joint

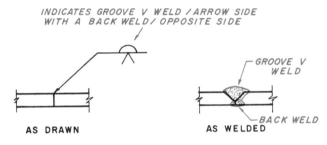

Figure 19-33 Multiwelds (different type of welds)

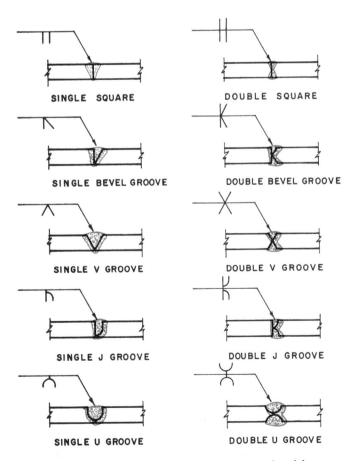

Figure 19-34 Double welds (same type of welds)

bevel. Each type has its own welding symbol, Figure 19-35. In the bevel groove and J welds, only one part is actually chamfered or grooved. To do this, the leader line and arrow point toward the part that has the bevel groove or J groove, Figure 19-36. Note that

GROOVE WELDS		
TYPE	SYMBOL	AS SEEN
SQUARE	||	
V	V	
BEVEL GROOVE	V	
U	⋂	
J	Ʋ	
FLARE V	)(	
FLARE BEVEL	I(	

Figure 19-35 Groove welds

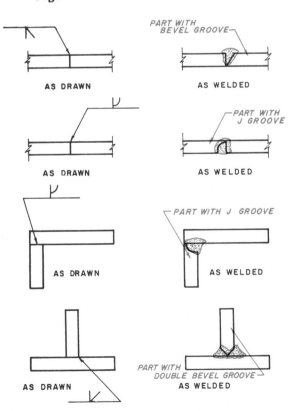

Figure 19-36 Groove weld illustrations

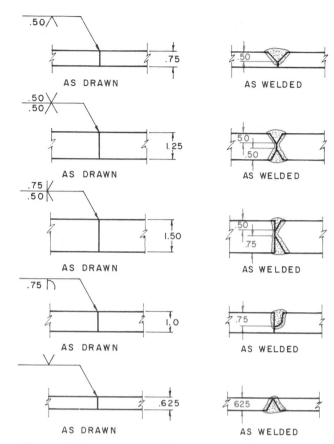

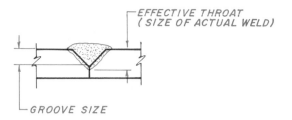

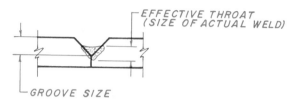

Figure 19-38 Depth or actual size of weld

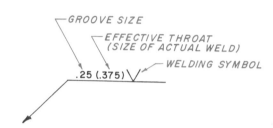

Figure 19-39 Size of weld only

Figure 19-37 Groove welds with dimensions

Figure 19-40 Size of weld is added to the left of the welding symbol in parentheses

the welding symbol is placed above or below the reference line, as usual, to indicate which side the chamfer or groove is located.

Size of Groove Weld. In the groove weld, the dimension refers to the actual depth of the chamfer or groove *not* the size of the weld. With the fillet weld, the size of the chamfer or groove is located directly to the left of the welding symbol and on the same side of the reference line, Figure 19-37. If a V groove weld symbol is shown without a size dimension, the size or depth is assumed to be equal to the thickness of the pieces.

Depth or Actual Size of Weld. The depth or actual size of the weld includes the penetration of the complete weld. This is illustrated in Figure 19-38. Note in this example that the depth or actual size of the weld is deeper than the depth of the chamfer. Occasionally, the depth or actual size of the weld is less than the size of the chamfer, Figure 19-39.

In order to call-off the depth or actual size of the weld *and* the size of the chamfer or groove, a dual dimension is used and both are added to the *left* of the welding symbol. In order to differentiate between the two, the size of the chamfer or groove is added to the extreme left of the welding symbol. The size of the depth or actual size of the weld is added directly

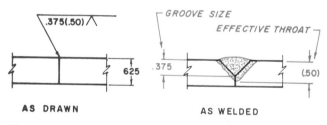

Figure 19-41 Example of effective throat-weld larger than groove

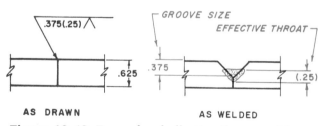

Figure 19-42 Example of effective throat-weld smaller than groove

to the left of the welding symbol, but in parentheses, Figure 19-40. Usually the weld is larger than the groove, Figure 19-41.

If the weld is less than the size of the chamfer, the dimensions are added to the left of the welding symbol in the exact same manner, Figure 19-42.

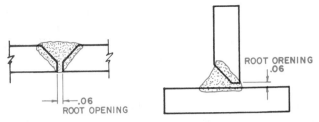

Figure 19-43 Root opening

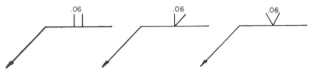

Figure 19-44 Size of root opening

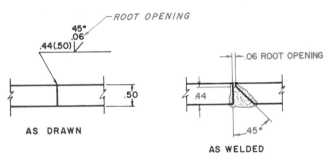

Figure 19-45 Chamfered angle

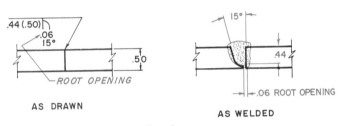

Figure 19-46 Required radius

Figure 19-47 Back weld symbol/melt-thru weld symbol

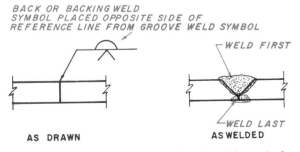

Figure 19-48 Illustration of back weld symbol

Root Opening. The allowed space between parts is called the *root opening*, Figure 19-43. The root opening is applied *inside* the welding symbol, Figure 19-44.

The root opening may be called-off as a general note. For example, "unless otherwise noted, root opening for all groove welds is .06." If there is *no* opening between parts, a zero is added inside the welding symbol.

Chamfer Angle. In a groove weld, a V or bevel groove must be held to tight tolerances. The angle must also be included with the size dimension in order to obtain the exact required geometric size and shape of all manufactured parts. The angle is also added inside the welding symbol, Figure 19-45. Note that the leader line and arrow point toward the part with the chamfer. Because the welding symbol is placed *above* the reference line, the chamfer and weld are on the side opposite from the arrow.

Groove Radius. If a J or U groove weld must be held to tight tolerances, the angle and radius must be included with the size dimensions in order to obtain the exact required geometric size and shape of all manufactured parts. The angle is also added inside the welding symbol and the required radius is called off as a note, Figure 19-46. Note that the leader line and arrow point toward the part with the groove. Because the welding symbol is placed *below* the reference line, the groove and weld is on the side of the arrow.

Back Weld — Backing Weld

A *back weld* is a weld applied to the opposite side of the joint *after* the major weld has been applied. A backing weld is a weld applied *first,* followed by the major weld. Both use the same welding symbol and both are used to strengthen a weld (refer back to Figure 19-2). The back weld is also used for appearance.

If a welding melt-through is required, the same symbol is used, except it is filled in solid, Figure 19-47. When the back weld or backing weld symbol is used, it is placed on the opposite side of the reference line from the groove weld symbol, Figures 19-48 and 19-49.

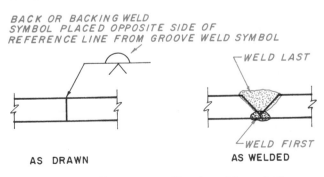

Figure 19-49 Illustration of back weld symbol

Plug and Slot Welds

The *plug* and *slot welds* both use the same welding symbol. The only difference between the two welds is the shape of the hole through which the weld is applied. A *plug weld* is made through a round hole. A *slot weld* is made through an elongated hole (refer back to Figure 19-2).

The plug or slot welding symbol is applied to the basic welding symbol and interpreted in exactly the same way. If the welding symbol is applied *above* the reference line, the hole or slot is on the *opposite* side of the arrow. If *below* the reference line, the hole or slot is on the *same side* as the arrow, Figure 19-50.

Size of Plug or Slot Weld. The given dimension for a plug or slot weld refers to the diameter of the plug or slot at the *base* of the weld. The required dimension is placed directly to the left of the plug or slot weld symbol, Figure 19-51. The size of the slot weld is drawn and dimensioned directly on the detail drawing.

Depth of Weld. Unless noted, the plug or slot weld hole is completely filled by the weld. If filling the hole is not necessary, the full depth of weld must be called off. The required depth of weld is added *inside* the welding symbol, Figure 19-52.

Tapered Plug or Slot. If the plug or slot hole is tapered, the taper inclusive angle is added directly above or below the welding symbol, Figure 19-53.

Surface Weld

If a surface must have material added to it or built-up, a *surface weld* is added (refer back to Figure 19-2). This welding symbol is *not* used to indicate the joining of parts together; therefore, the welding symbol is simply added below the reference line, Figure 19-54.

Size of Surface Weld. If the surface is to be built-up and a special height is *not* required, the welding sym-

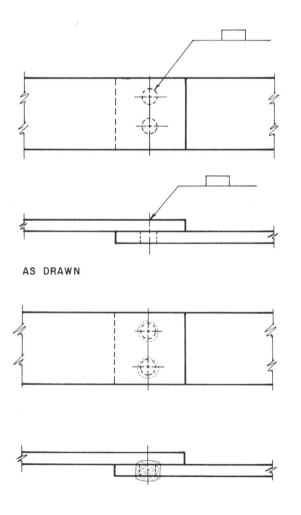

Figure 19-50 Plug and slot welds

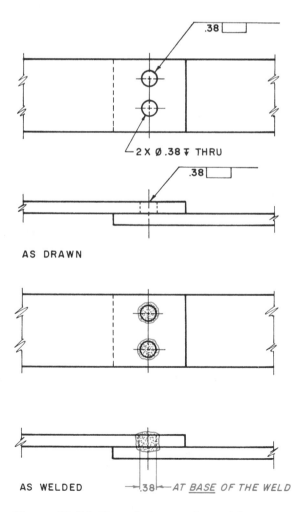

Figure 19-51 Size of plug or slot weld

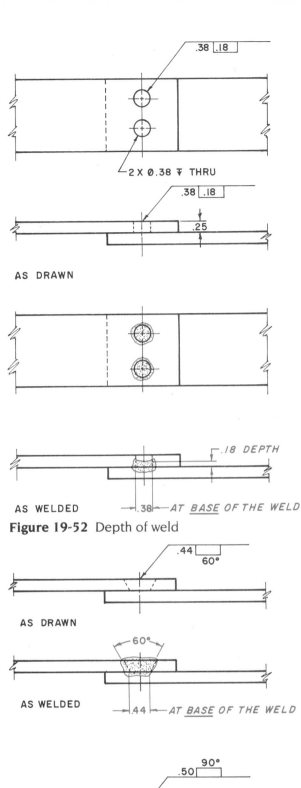

Figure 19-52 Depth of weld

Figure 19-53 Tapered plug or slot

Figure 19-54 Surface weld

INDICATES DEPTH OF WELD

Figure 19-55 Size of surface weld

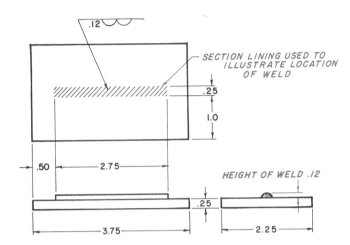

Figure 19-56 Illustration of a surface weld

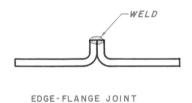

EDGE-FLANGE JOINT

Figure 19-57 Edge-flange weld

bol is drawn as illustrated in Figure 19-54. If the surface is to be built-up and a *specific* height is required, the required height is added directly to the left of the welding symbol, Figure 19-55. If only a *portion* of the surface is to be built-up, that portion must be illustrated and dimensioned on the detail drawing, Figure 19-56. The actual weld is represented by section lining.

Flange Weld

A *flange weld* is used to join thin metal parts together. Instead of welding to the surfaces of the parts to be joined, the weld is applied to the edges of thin material. A weld applied to thin metal could burn right through.

There are two distinct flange weld symbols (refer back to Figure 19-2): the first is called an edge-flange weld, Figure 19-57; the other is called a corner-flange

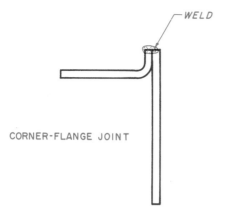

Figure 19-58 Corner-flange weld

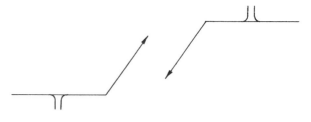

Figure 19-59 Edge-flange weld symbol

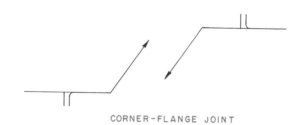

Figure 19-60 Corner-flange weld

weld, Figure 19-58. The straight line of this welding symbol is always drawn to the left of the partially curved line, regardless of the actual joint being illustrated. The edge-flange or corner-flange weld symbol is placed above or below the reference line to indicate OPPOSITE side or ARROW side, but never on both sides, Figure 19-59 and 19-60.

Dimensioning a Flange Weld. It requires three dimensions to fully dimension a flange weld: the radius of the flange, the height above the point of tangency between parts, and the size of the weld itself, Figures 19-61 and 19-62. The three dimensions are added to the welding symbol as indicated previously. The welding symbol should appear at the peak of the joint as illustrated. If there is to be a root opening, that is, a space between parts, the required space must be indicated on the drawing, *not* on the welding symbol, Figure 19-63.

Multiple Reference Line

If there is more than one operation on a particular joint, these operations are indicated on *multiple reference lines*. The first reference line closest to the arrow is for the first operation, the second reference line is for the second operation or supplemental data, and the third reference line is for the third operation or test information, Figure 19-64.

Spot Weld

Spot welding is a resistance welding process done by passing an electric current through the exact location where the parts are to be joined. Spot welding is usually done to hold thin sheet metal parts together. The resistance weld process uses three standard types of welds, Figure 19-65. Each is explained in full.

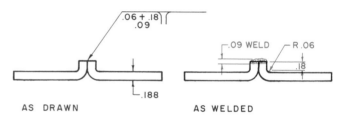

Figure 19-61 Dimensioning a flange weld

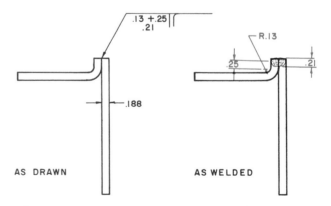

Figure 19-62 Dimensioning a corner-flange weld

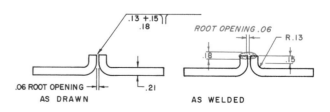

Figure 19-63 Root opening

The symbol for a spot weld, as illustrated in Figure 19-66, is located tangent to or touching the reference line, and uses the same added supplemental data symbols and dimensions as are illustrated with the fusion welding process, Figure 19-67. The tail is *always* included, in order to indicate the process required to make the weld (refer back to Figure 19-22).

Dimensioning the Spot Weld

The diameter size of the spot weld is added directly to the left of the weld symbol, Figure 19-68. Some-times the diameter size is omitted, and a required shear strength is inserted in its place. In this condition, an explanation of the value is added to the drawing by a note, Figure 19-69.

The spacing of the spot weld is called-off by a number directly to the right of the weld symbol. This number is the pitch, and refers to the center-to-center distance between spots, Figure 19-70. The number of welds for a particular joint is added above or below the welding symbol in parentheses.

SPOT WELD SYMBOL

Figure 19-66
Spot weld symbol

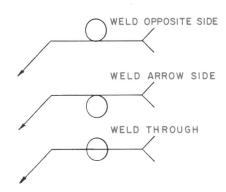

Figure 19-67 Spot weld symbol used with reference line

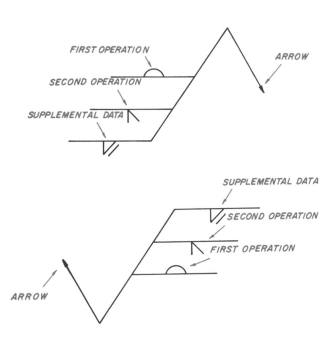

Figure 19-64 Multiple reference line

Figure 19-65 Spot welds

Figure 19-68 Diameter of spot weld

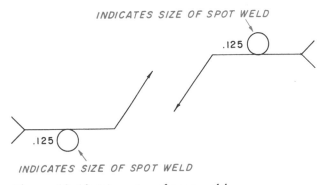

Figure 19-69 Shear strength of spot weld

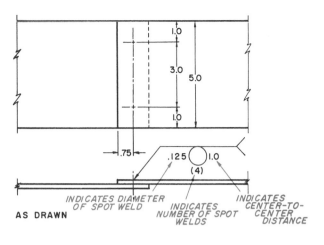

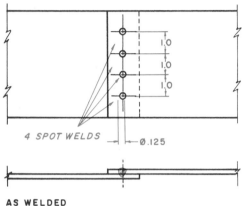

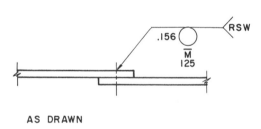

Figure 19-70 Pitch of spot welds

Contour and Finish Symbols

Contour and finish symbols are added to the welding symbol exactly as is done in the fusion welding symbol, Figure 19-71.

Projection Weld

A *projection weld* is similar to a spot weld and uses the exact same welding symbol. One of the two parts has a series of evenly spaced dimples stamped into it. Each stamped dimple is the location of an individual spot weld, Figure 19-72. A projection weld allows more penetration and, as a result, a much better weld. Note, the dimples are *not* illustrated on the detail drawing.

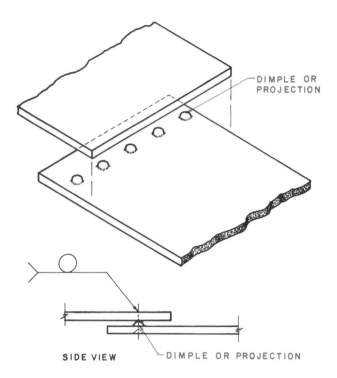

Figure 19-72 Projection weld

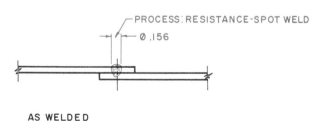

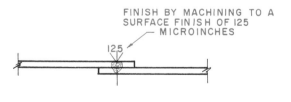

Figure 19-71 Contour and finish symbols

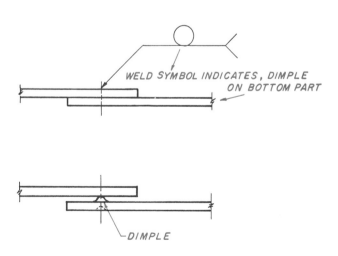

Figure 19-73 Projection weld symbol

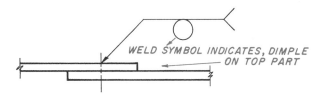

WELD SYMBOL INDICATES, DIMPLE
ON TOP PART

DIMPLE

Figure 19-74 Projection weld symbol

The reference line (OPPOSITE side-ARROW side significance) is changed slightly as used for a projection weld. The OPPOSITE side-ARROW side indicates which of the two parts to be joined has the dimples on it, Figures 19-73 and 19-74.

Dimensioning, Contour, and Finish Symbols

Dimensioning, contour, and finish symbols are applied to projection welds exactly as they are for spot welds.

Figure 19-75
Seam weld symbol

Seam Weld

A *seam weld* is like a spot weld, except that the weld is actually continuous from start to finish.

The welding symbol is modified slightly to indicate a seam weld, Figure 19-75. Figure 19-76 indicates how the seam weld symbol is applied to the reference line.

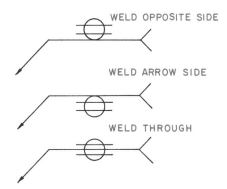

WELD OPPOSITE SIDE

WELD ARROW SIDE

WELD THROUGH

Figure 19-76 Seam weld symbol used with reference line

Welding Template

In order to simplify and speed up the drawing time for welding symbols, welding templates can be used. The welding template standardizes the size of each welding symbol and provides a quick, ready reference for welding symbols, Figure 19-77.

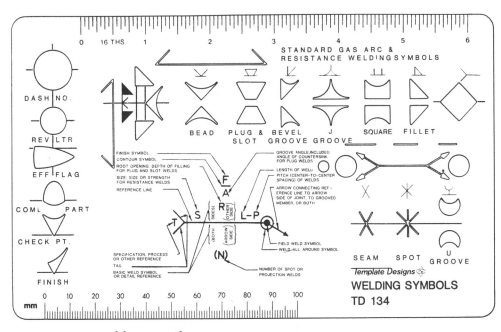

Figure 19-77 Welding template

Review

1. What is the difference between a plug weld and a slot weld?

2. Make a sketch of a welding symbol for a fillet weld with legs of .250 and a weld 4.5 long and weld on OPPOSITE side of arrowhead.

3. List six major types of welds.

4. Explain what the *root opening* is.

5. Describe an intermittent weld.

6. What are the two major welding processes and generally, where is each used?

7. List the five basic kinds of welding joints. How are they classified?

8. Describe the basic welding symbol and explain the significance of welding symbols being placed above and below the reference line.

9. Where and why is a flange weld used?

10. What is meant by a process reference and where is it located on the welding symbol?

11. What is the advantage of using a projection weld over a spot weld?

12. Explain in full the difference between a back weld and a backing weld. How is the melt-through designated to these symbols?

13. List the seven types of groove welds.

14. How is a field weld indicated on the welding symbol?

15. How is a weld that is to be continuously around an object indicated on the welding symbol?

Chapter Nineteen Problems

The following problems are intended to give the beginning drafter practice in studying the many factors involved in dimensioning and calling-off welding processes. The student will sharpen skills in centering the views within the work area, line weight, dimensioning, and speed.

The steps to follow in laying out the following problems are:

1. Study the problem carefully.

2. Choose the view with the most detail as the front view.

3. Make a sketch of the required views.

4. Add required dimensions to the above sketch per latest drafting standards.

5. Add welding symbols as required to the sketch per the latest welding standards.

6. Lightly lay out all required views and check that they are centered within the work area.

7. Check that all dimensions are added using the latest drafting standards.

8. Darken in all views using correct line thickness and add all dimensioning using light guidelines.

9. Recheck all work, and, if correct, neatly fill out the title block using light guidelines and neat lettering.

10. Recheck that all required welding symbols are added correctly.

Problems 19-1 through 19-11

Follow the listed steps, using the latest drafting standards.

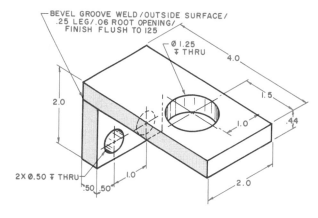

Problem 19-1

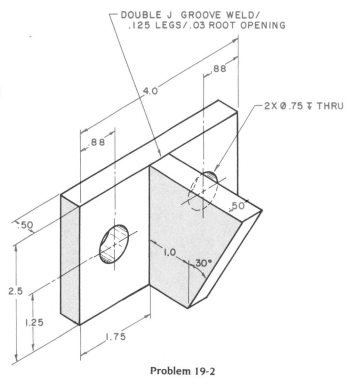

Problem 19-2

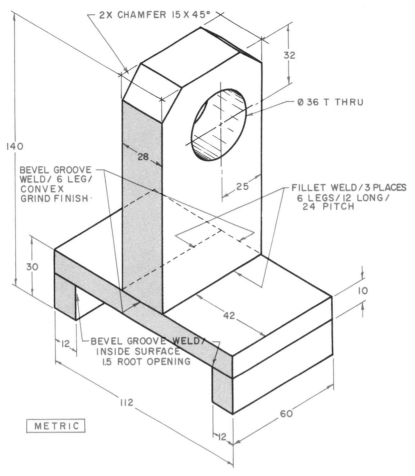

Problem 19-3

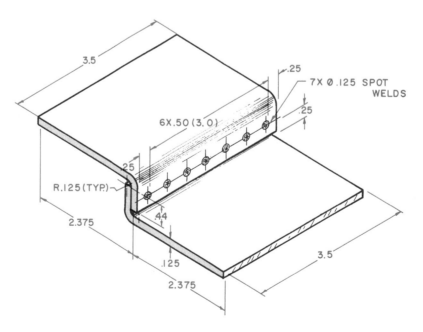

Problem 19-4

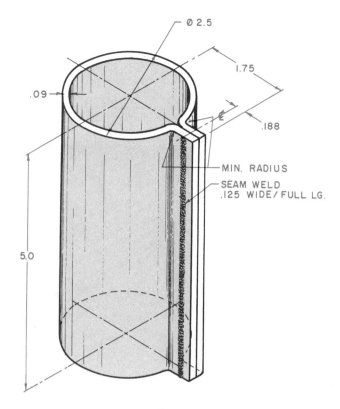

Ø 2.5

1.75

.09

¢

.188

MIN. RADIUS

SEAM WELD
.125 WIDE/ FULL LG.

5.0

Problem 19-5

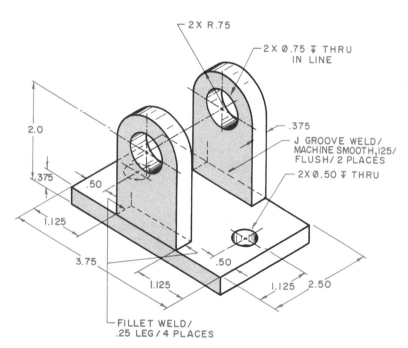

2X R.75

2X Ø.75 ⊽ THRU
IN LINE

.375

J GROOVE WELD/
MACHINE SMOOTH,125/
FLUSH/ 2 PLACES

2X Ø.50 ⊽ THRU

2.0

.375

.50

1.125

3.75

.50

1.125

1.125   2.50

FILLET WELD/
.25 LEG/4 PLACES

Problem 19-6

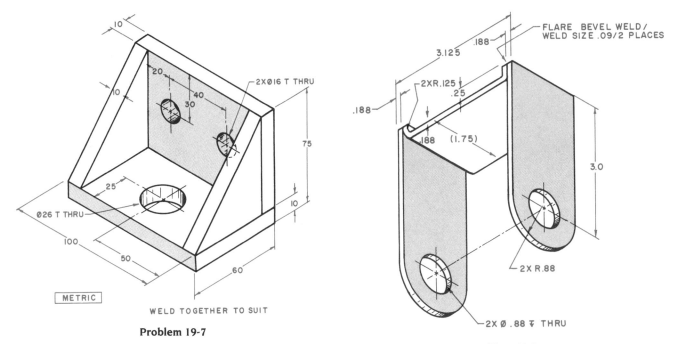

Problem 19-7

METRIC

WELD TOGETHER TO SUIT

FLARE BEVEL WELD/
WELD SIZE .09/2 PLACES

Problem 19-8

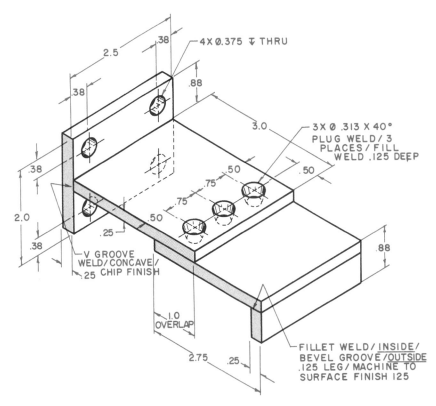

Problem 19-9

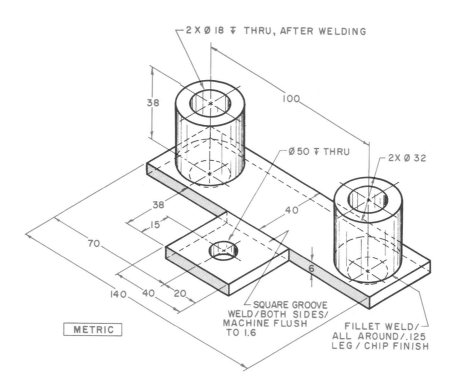

2 X Ø 18 ▼ THRU, AFTER WELDING

100

38

Ø 50 ▼ THRU

2X Ø 32

38

15

40

70

140   40   20

6

SQUARE GROOVE
WELD/BOTH SIDES/
MACHINE FLUSH
TO 1.6

FILLET WELD/
ALL AROUND/.125
LEG/CHIP FINISH

METRIC

Problem 19-10

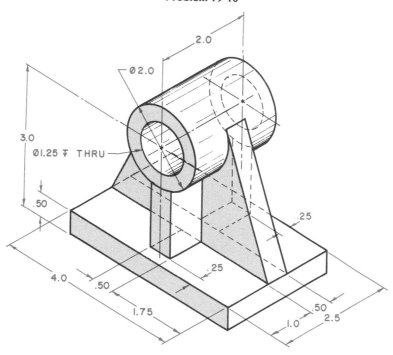

2.0

Ø2.0

3.0

Ø1.25 ▼ THRU

.50

.25

.25

4.0   .50

1.75

.50

1.0   2.5

ADD WELDS TO SUIT

Problem 19-11

Problems 19-12 through 19-15

Convert the following problems from Chapter 4 from casting drawings to weldment drawings:

Problem 19-12

Refer back to Problem 4-27.

Problem 19-13

Refer back to Problem 4-28.

Problem 19-14

Refer back to Problem 4-36.

Problem 19-15

Refer back to Problem 4-37.

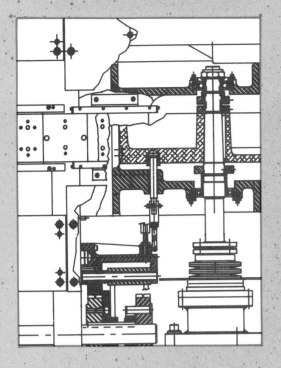

This chapter describes the methods used to make cast products. Covered are the basic parts of a casting mold, forging die, and stamping die; the processes used to forge and stamp metals; the process and tools used to extrude metals; machine tools and the processes they perform; workholding devices; and processes used to heat treat steels.

CHAPTER TWENTY

Shop Processes

The ultimate purpose of any engineering drawing is to provide the information necessary to make or fabricate an object. To achieve this purpose an engineering drawing must completely describe and detail the desired object. The drawing must show the specific size and geometric shape of the object as well as furnishing the related information concerning the material specifications, finish requirements, and any special treatments required.

To properly construct an engineering drawing, the designer must be throughly familiar with the methods used by the shop to transform the drawn image into the actual object specified in the drawing. By knowing the basic processes used to manufacture an object, the designer can design the object with the manufacturing processes in mind. This will not only make manufacture easier but will many times reduce the cost of a manufactured product. Therefore, the purpose of this chapter is to acquaint the new designer with the basic processes used to fabricate objects, as well as the capabilities and limitations of the machine tools used to machine these objects to their final form.

Shop Processes

The term *shop processes* refers to the basic methods used by the shop to make the object described in the engineering drawing. The specific processes used to make the object depend on the object itself. In some cases the part may be cast, while other parts may be forged, extruded, or stamped. In many instances the part, once fabricated into its basic form, must also be machined to maintain a specific degree of accuracy or to produce a feature not possible with other manufacturing processes.

When the shop receives the engineering drawing, in the form of a part print, it is first reviewed to ensure all pertinent data and information necessary to make the object are contained in the drawing. During this review, the shop personnel will consider several factors necessary to determine how the part must be made. These factors include: the type and condition of the material used to make the object, the overall size and shape of the part, the types of operations required, the required accuracy of the part, and number of parts to be made.

The type and condition of the material used to make the part are important considerations. Some parts may be made from solid bar stock while other parts must be extruded, cast, forged, or stamped. In most cases, parts made from bar stock require the least lead time. The *lead time* is the interval from the time the shop receives the drawing until production begins. Most shops maintain a sufficient supply of bar stock to begin production as soon as the drawing is received. However, when a part must be extruded, cast, forged, or stamped a longer lead time is required to make the necessary molds or dies to fabricate the object.

The size and shape of the object must be considered to determine if the object is within the capabilities of the shop to make. In addition, the size and shape may also determine the size of the machine tools required as well as the datum, or reference, surfaces used to locate the part during manufacture.

The types of operations required determine the types of equipment and machine tools needed to make the part. If, for example, the part requires holes, a drill press, vertical milling machine, or other machine tool may be used. The next consideration, the required accuracy of the part, also determines how the operations are performed. A hole with a required accuracy of .002″ (0.05 mm), for example, would require reaming, while a hole with a required accuracy of .020″ (0.5 mm) could be drilled.

The number of parts to be made will frequently determine if the part should be cast, forged, extruded, or machined from solid stock. Likewise, the number of parts will also determine if any special workholders are to be made. Larger production runs will normally justify more sophisticated tools and processes since the cost can be spread over a larger number of parts. Smaller production runs normally demand the parts be made at the lowest possible cost with little or no investment in special molds, dies, or workholders.

While each of these factors has been separated for the purpose of this discussion, in practice each is considered as part of the other. These factors are so closely related in production that they frequently overlap and one cannot be considered without the others.

The primary processes used to fabricate manufactured products are casting, forging, extruding, stamping, and machining. To use these processes to their best advantage, the designer must be familiar with the strengths and weaknesses of each process as well as the fundamental aspects of each process.

Casting

Casting is basically a process of pouring molten metal into a mold that contains the desired shape in the form of a cavity. The principal types of casting used in manufacturing today are sand casting, investment casting, centrifugal casting, and die casting.

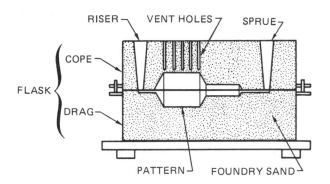

Figure 20-1 Components of a sand casting mold (From *Materials Processing*, B.R. Thode. Delmar Publishers Inc.)

Sand Casting

Sand casting is the most common type of casting method. The major components of the molds used to make sand castings are shown in Figure 20-1 and include the *flask, pattern,* and *green sand.* The flask is a two-part box, or frame, used to contain the sand. The top half of the flask is called the *cope* and the bottom half is called the *drag.* Occasionally, a third section may be installed between the cope and drag. This section is called a *cheek,* and is used where a deep or complex shape must be cast. In sand casting the mold is prepared by ramming the green sand around a model of a part, called the *pattern.* The model is then removed to form the cavity for the molten metal. The **plane** of division between the cope and drag is called the *parting line.* In many castings, the parting line occurs at the approximate middle of the part. The parting line can be seen on most casting by a ragged line which is usually ground off.

The molten metal enters the mold through the *sprue hole* and is directed to the cavity by one or more *gates.* The sprue hole is formed by installing a sprue peg in the cope. This peg is then removed after the final ramming of the cope. The *riser* is used to vent the mold and to allow gases to escape. The riser also acts as a small reservoir to keep the cavity full as the metal begins to shrink during the cooling process.

Once the cope and drag are rammed and the pattern is removed, a solid part could be poured in the mold. However, in some cases a hollow part, or one which has large holes, must be poured. To reduce the amount of material needed to fill the cavity and to reduce the time necessary to machine the part a *sand core* may be installed in the cavity. When sand cores are intended to be installed in a cavity, *core prints* must also be provided to locate and anchor the sand cores during the casting process. Core prints are normally a simple extension of the pattern.

Figure 20-2 Shrink rule (*Courtesy L.S. Starrett Co.*)

Making Patterns. When patterns are made for casting two important factors must always be considered. The first is the draft. *Draft* is the slope or taper of the sides of a pattern which permits it to be removed from the cope and drag without disturbing the cavity. The draft also permits the cast part to be removed easily from some molds. The amount of draft necessary will normally depend on the part being cast, but will normally be about one degree.

The second consideration is shrinkage. When metals are cast a certain amount of *shrinkage* occurs as the metal cools. The specific amount of shrinkage depends on the metal being cast. Steel, for example, shrinks at a rate of approximately three-sixteenths of an inch per foot, while cast iron shrinks about one-eighth of an inch per foot. To allow for this shrinkage and to make sure the final part is the correct size the pattern must be made slightly larger than the final size of the desired part.

When making a pattern, the pattern maker will normally use a shrink rule to compensate for the shrinkage. A *shrink rule*, Figure 20-2, is a standard steel rule which has the graduations marked for a specific amount of shrinkage. A shrink rule used for cast iron has an extra one-eighth inch added to each foot, while a shrink rule for steel has an additional three-sixteenths added.

In most instances, when a pattern must be made, the pattern maker will receive the engineering drawing which contains all the final sizes of the part. The pattern maker will then make all the necessary calculations needed to make the oversized pattern. But, occasionally, the pattern maker will be given a drawing with all the calculations already made to produce the pattern. In either case, the patterns must be made to suit the material being cast.

Investment Casting

Investment casting produces parts with great detail and accuracy, while at the same time allowing very thin cross sections to be cast. In this process, the pattern is made by casting wax in the desired form. The pattern is then placed in a sand mold and the mold is fired to melt out the wax pattern. This process is also referred to as *lost wax casting*. The molten metal is then fed into the mold to produce the cast part.

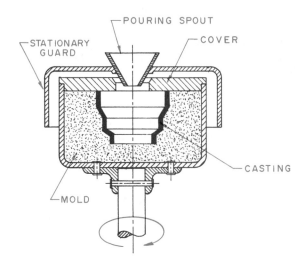

Figure 20-3 Centrifugal casting

Centrifugal Casting

Centrifugal casting, Figure 20-3, is a process of pouring a measured amount of molten metal into a rotating mold. This process can be used for a single mold or multiple molds. The centrifugal force created by the rotation forces the molten metal to fill the cavity. This same centrifugal force, along with the measured amount of material, also controls the wall thickness of the cast part and results in a less porous cast surface than is possible with sand casting. The molds used for this purpose are usually permanent molds made from metal rather than sand. These molds are used repeatedly and produce highly accurate parts requiring very little machining. However, due to their cost, permanent molds are normally used only for high-volume production.

Die Casting

Die casting, Figure 20-4, is a process where molten metal is forced into metal dies under pressure. This process is very well suited for such materials as zinc alloys, aluminum alloys, copper alloys, and magnesium alloys. Parts produced by die casting are superior in appearance and accuracy and require little or no machining to final size.

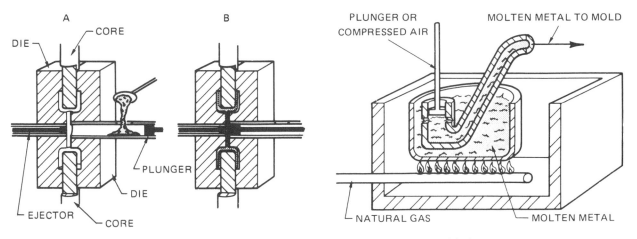

Figure 20-4 Die casting (From *Materials Processing*, B.R. Thode. Delmar Publishers Inc.)

Powder Metallurgy

Powder metallurgy, Figure 20-5, while not an actual casting process, does have some similarities to casting. In the *powder metallurgy* process, metal particles, or powder, are blended and mixed to achieve the desired composition. The powder is then forced into a die of the desired form under pressure from 15,000 to 100,000 pounds per square inch. The resulting heat fuses the powder into a solid piece which can be machined.

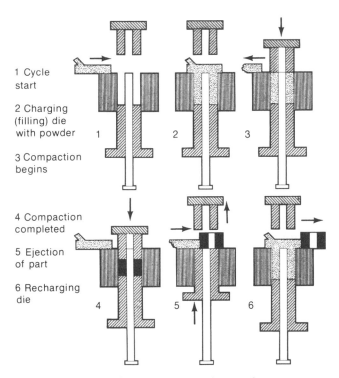

1 Cycle start

2 Charging (filling) die with powder

3 Compaction begins

4 Compaction completed

5 Ejection of part

6 Recharging die

Figure 20-5 Producing parts with powder metallurgy (*Courtesy Metal Powder Industry Foundation*)

Forging

Forging is the process of forming metals under pressure using a variety of different processes. The most common forms of forging are drop forging and press forging. Other variations of forging include rolling and upsetting.

Drop Forging

Drop forging is the process of forming a heated metal bar, or billet, in dies. In practice, the heated metal is placed on a lower portion of a forging die and struck repeatedly with the upper die portion. This forces the metal into the shape of the cavity of the dies. The pressure required to form the metal is provided by a drop hammer.

Most drop forge dies contain at least four different stations, or dies, to complete the part, Figure 20-6. The first station is called the *fuller*. This station is used to rough-form the part to fit the other cavities. The second station, when used, is called the *breakdown*, or *bender*. This station forms the part into any special contours required by the next station, the roughing impression. The roughing impression rough-forms the part to the desired form. The final station is called the *finishing impression* and is used to finish the part to the desired shape and size. Additional stations, such as a trimmer to remove the fins and a cutoff to sever the forged part from the bar, may also be included on the die if they are necessary.

Press Forging

Press forging is a single-step process in which the heated metal bar or billet is forced into a single die and is completed in a single press stroke. The press normally used for press forging is hydraulic.

DROP FORGING PROCESS

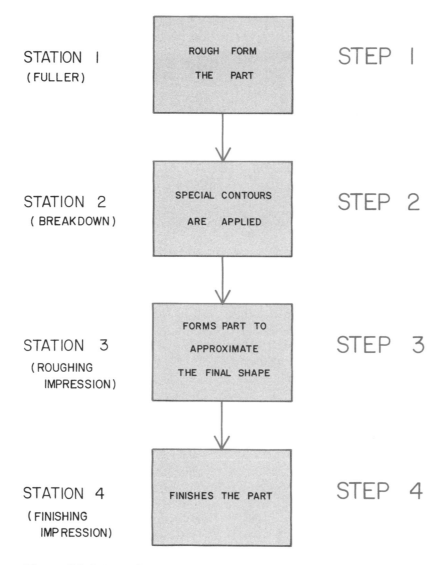

STATION 1
(FULLER)

ROUGH FORM
THE PART

STEP 1

STATION 2
(BREAKDOWN)

SPECIAL CONTOURS
ARE APPLIED

STEP 2

STATION 3
(ROUGHING
IMPRESSION)

FORMS PART TO
APPROXIMATE
THE FINAL SHAPE

STEP 3

STATION 4
(FINISHING
IMPRESSION)

FINISHES THE PART

STEP 4

Figure 20-6 Drop forging process

Rolling

Rolling is a process where the heated metal bar is formed by passing through rollers having the desired form impressed on their surfaces. Rolling is frequently used to flatten and thin out thick sections.

Upsetting

U*psetting*, Figure 20-7, is a process used to enlarge selected sections of a metal part. Bolt heads, for example, are normally produced by an upsetting process.

Extruding

Extruding is a process of forcing metal through a die of a desired form and cross section. The bars

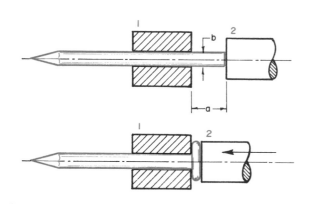

Figure 20-7 Upsetting

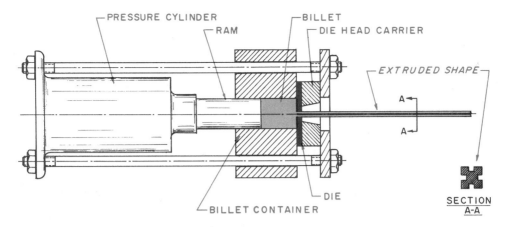

Figure 20-8 Extruding

produced are then cut to the required lengths. Typical examples of extruded parts include parts for aluminum windows and doors. Extrusions provide a near final shape, as shown in Figure 20-8 and, in many cases, only need to be cut off and machined slightly to complete a finished part.

Stamping

Stamping is a process of using dies to cut or form metal sheets or strips into a desired form. The main tool used for metal stamping is a die. The term *die* has a double meaning. It can be used to describe the entire assembled tool or the lower cutting part of the tool. The exact meaning of the term can usually be determined from the context of its use. The principal parts of a die are shown in Figure 20-9. The upper die shoe is used to mount the punch. The die is mounted on the lower die shoe. The guide pins and guide pin bushings are used to maintain the alignment between the punch and die. The stripper is designed to strip the stock material from the punch after the cutting stroke of the die.

When more than one operation takes place on a part during a single stroke, blanking and punching for example, the operation is said to be *compound*. However, when several operations take place sequentially in a die the die is called a *progressive die*.

The principal stamping operations normally performed are shearing, cutting off, parting, blanking, punching, piercing, perforating, trimming, slitting, shaving, bending, forming, drawing, coining, and embossing.

Operations that Produce Blanks

The operations that produce blanks are shearing, cutting off, parting, and blanking, Figure 20-10. *Shearing* is a cutting action performed along a straight line. *Cutting off* is a cutting operation which is performed on a part to produce an edge other than a straight edge. In many cases, cutting off is used to finish the

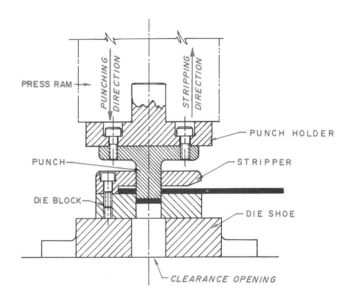

Figure 20-9 Metal stamping die

edges of a part. Like shearing, cutting off does not produce any scrap. *Parting*, on the other hand, is also an operation used to finish the edges of a part, but, unlike the other operations, parting does produce scrap.

Blanking is an operation that produces a part cut completely by the punch and die. In blanking operations, the piece that falls through the die is the desired part. The scrap is the skeleton left on the stock strip.

Operations that Produce Holes

The operations that produce holes are punching, piercing, and perforating. *Punching* is an operation that cuts a hole in a metal sheet or strip, Figure 20-11A. In punching, the piece that falls through the die is scrap, and the area on the strip is the desired part. Punching is normally performed on parts that are to be blanked to provide holes for bolts, screws, or similar parts.

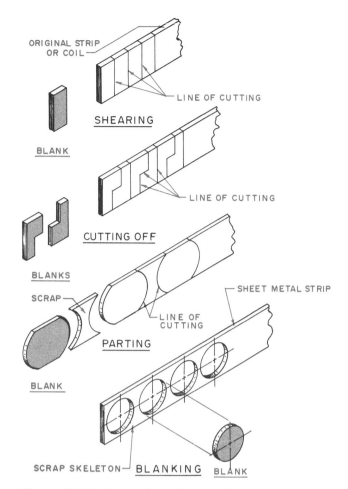

Figure 20-10 Operations that produce blanks

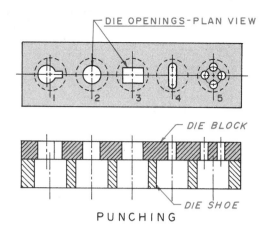

Figure 20-11A Punching operation to produce holes

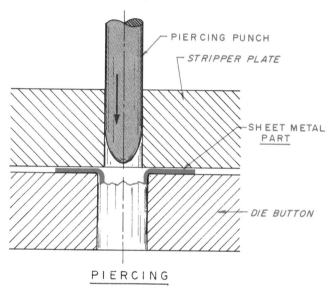

Figure 20-11B Piercing operation to produce holes

Piercing is an operation similar to punching, except that in piercing no scrap is produced, Figure 20-11B. A pierced hole is sometimes used to increase the thickness of metal around a hole for tapping threads in a stamped part. *Perforating* is simply a punching operation performed on sheets to produce either a uniform hole pattern or a decorative form in the sheet.

Operations that Control Size

The operations used to control size are trimming, slitting, and shaving, Figures 20-12A and 20-12B. *Trimming* is an operation performed on formed or drawn parts to remove the ragged edge of the blank. *Slitting* is performed on large sheets to produce thin stock strips. Slitting may be performed by rollers or by straight blades. *Shaving* is an operation performed on a blanked part to produce an exact dimensional size or to square a blanked edge. Shaving produces a part with a close dimensional size, and a cut edge which is almost perpendicular to the top and bottom surfaces of the blanked part.

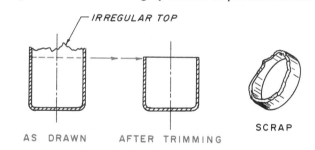

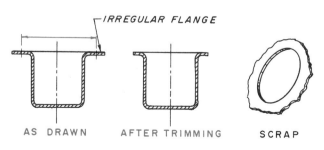

Figure 20-12A Operation that controls size – Trimming

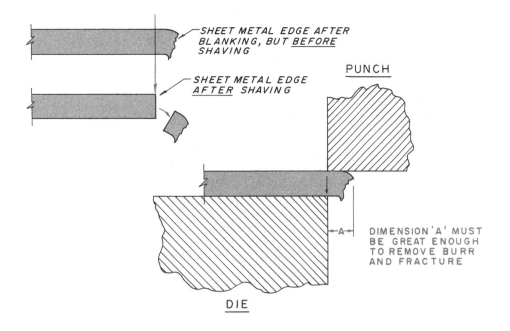

Figure 20-12B Operation that controls size – Shaving

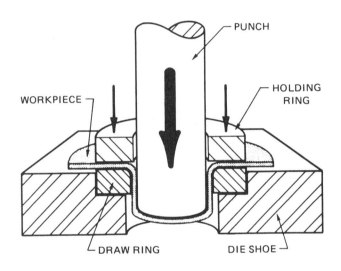

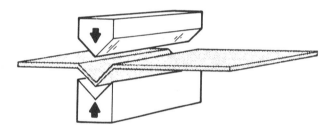

Figure 20-13 Operations that bend or form parts (From *Materials Processing*, B.R. Thode. Delmar Publishers Inc.)

Operations that Bend or Form Parts

The operations used to bend or form parts are bending, forming, and drawing, Figure 20-13. *Bending* is an operation in which a metal part is simply bent to a desired angle. *Forming*, on the other hand, is an operation in which a part is bent or formed into a complex shape. Forming is also a catchall word frequently used to describe any bending operation that does not fall into one of the other categories. *Drawing* is a process of stretching and forming a metal sheet into a shape similar to a cup or top hat.

Machining

Machining is the process of removing metal with machine tools and cutters to achieve a desired form or feature. The principal types of machine tools used to perform these operations are lathes, milling machines, drill presses, saws, and grinders. Other machines that are variations of the basic machines used in the machine shop include shapers and planers, boring mills, broaching machines, and numerically controlled machines. The specific type of machine used to perform a particular machining operation is determined by the type of operation and the degree of accuracy required.

Lathes

A lathe, Figure 20-14, is one of the most commonly used and versatile machine tools in the shop. Typically, a *lathe* performs such operations as straight and

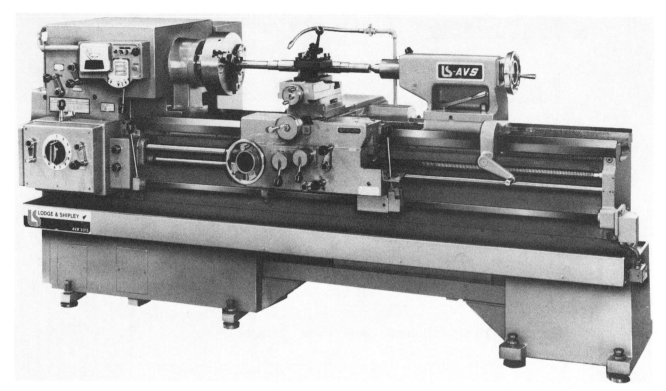

Figure 20-14 Lathe (*Courtesy Lodge & Shipley Co.*)

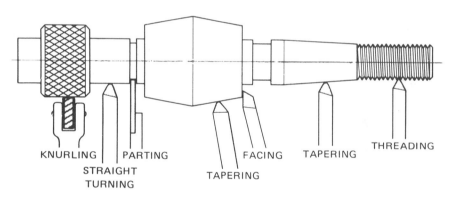

KNURLING   PARTING   FACING   TAPERING   THREADING

STRAIGHT   TAPERING
TURNING

Figure 20-15 Basic lathe operations (From *Materials Processing*, B.R. Thode. Delmar Publishers Inc.)

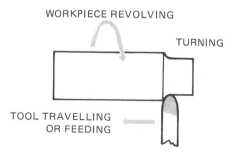

WORKPIECE REVOLVING

TURNING

TOOL TRAVELLING
OR FEEDING

Figure 20-16 Lathe setup for straight turning (From *Turning Technology*, S.F. Krar and J.W. Oswald. Delmar Publishers Inc.)

taper turning, facing, straight and taper boring, drilling, reaming, tapping, threading, and knurling, Figure 20-15.

In operation, the workpiece is held in the headstock and supported by the tailstock. The cutting tool is mounted in the tool post and is traversed past the rotating workpiece to machine the desired form, Figure 20-16. The most common method used to hold parts in a lathe is with a chuck. A *chuck* is a device much like a vise, having movable jaws that grip and hold the workpiece. The most popular type of chuck is the three-jaw chuck, but there are also two-jaw, four-jaw, and six-jaw chucks as well. Other types of devices used to hold and drive a workpiece in a lathe include collets, face plates, drive plates, and lathe dogs.

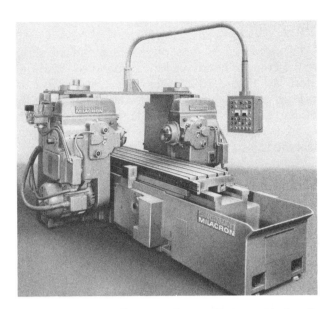

Figure 20-17 Milling machines (*Courtesy Cincinnati Milacron*)

CONVENTIONAL MILLING   CLIMB MILLING

Figure 20-18 Basic milling machine operations (From *Materials Processing*, B.R. Thode. Delmar Publishers Inc.)

Figure 20-19A Standard milling cutter (*Courtesy Cincinnati Milacron*)

Milling Machines

Milling machines are used to machine workpieces by feeding the part into a rotating cutter. The two basic variations of the milling machine are the vertical milling machine and the horizontal milling machine, Figure 20-17.

In operation, the milling cutter is mounted in either the spindle or arbor, and the workpiece is held on the machine table. The most commonly used device to mount the workpiece for milling is the milling machine vise, but the workpiece may also be held directly on the machine table or in other workholding devices. The operations most frequently performed on a milling machine include plain milling, face milling, end milling, straddle milling, gang milling, form milling, keyseat milling, and gear cutting, Figure 20-18. Figures 20-19A and 20-19B show two types of cutters commonly used for milling operations.

Drill Presses

A *drill press*, Figure 20-20, is a machine that is used mainly for producing holes. The principal types of drill presses used in the shop are the radial drill press and the sensitive drill press.

The operations normally performed on drill presses include drilling, reaming, tapping, chamfering, spotfacing, counterboring, countersinking, reverse countersinking, and reverse spotfacing, Figure 20-21.

When using a drill press, the workpiece is normally held in a vise or clamped directly to the machine table. The drill, or other cutting tool, is mounted in the machine spindle and is fed into the workpiece either by hand or with a mechanical feed unit.

Figure 20-19B Standard milling cutter (*Courtesy Cincinnati Milacron*)

Power Saws

The *power saws* used most often in the machine shop are the contour band machine or band saw and the cutoff saw, Figure 20-22. The *contour band machine* or *band saw* is used primarily for sawing intricate or detailed shapes; the *cutoff saw* is used to cut rough bar stock to lengths suitable for machining in other machine tools.

Precision Grinders

Precision grinders are available in several styles and types to suit their many and varied applications. The principal types of precision grinders used in the machine shop are the surface grinder and the cylindrical grinder, among others.

The *surface grinder*, Figure 20-23, is mainly used to produce flat, angular or special contours on flat workpieces. The most popular variation of this machine consists of a grinding wheel mounted on a horizontal spindle, and a reciprocating table that traverses back and forth under the grinding wheel. One other variation of the surface grinder frequently found in the machine shop uses a vertical spindle and a round, rotating table, Figure 20-24. The workpiece is generally mounted and held on a magnetic chuck during the grinding operation on both styles of surface grinders.

Cylindrical grinders are used to precisely grind cylindrical or conical workpieces. When grinding, the workpiece is mounted either between centers or in a precision chuck, much like a lathe. The grinding wheel is mounted behind the workpiece on a horizontal spindle. The table of the grinder traverses the workpiece past the grinding wheel in a reciprocating motion, while the workpiece rotates at a preset speed.

Figure 20-20 Drill press (*Courtesy Wilton Corporation*)

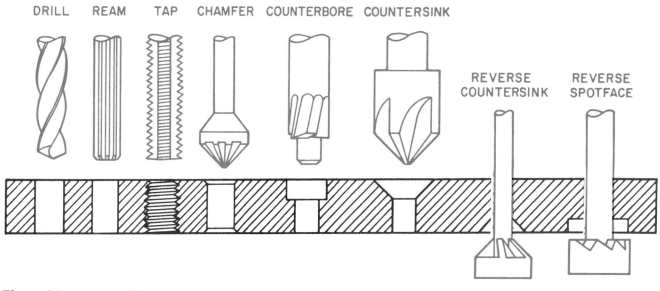

Figure 20-21 Basic drill press operations

Figure 20-22 Power saw (*Courtesy DoALL Co.*)

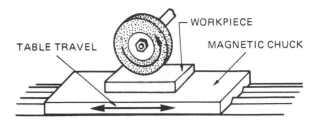

Figure 20-23 Surface grinder (From *Materials Processing*, B.R. Thode. Delmar Publishers Inc.)

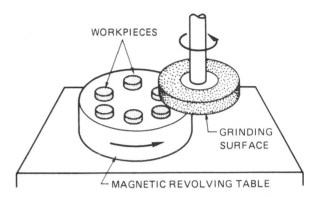

Figure 20-24 Vertical spindle — rotating table surface grinder (From *Materials Processing*, B.R. Thode. Delmar Publishers Inc.)

Shapers and Planers

Shapers and *planers* are machine tools that use single-point cutting tools to perform their cutting operations. The shaper uses a reciprocating ram to drive the cutter. The cutting tool on this machine tool is mounted on the end of the ram in a unit called the *clapper box*. The clapper box can be positioned vertically or at an angle, depending on the work to be performed. The depth of cut is adjusted by lowering the vertical slide of the clapper box or by raising the position of the table. The workpiece is mounted on the table or vise and is fed past the reciprocating cutter by the table feed. The length and position of the stroke are regulated by the position of the ram and are easily adjusted to suit the size of the workpiece.

Planers operate in a manner similar to the shaper. The major differences between these two machines are their size and method of cutting. The planer is normally much larger than the shaper and the work, rather than the cutter, moves on the planer. In operation, the workpiece is mounted on the table of the planer and the table reciprocates under the stationary tool. As the table reciprocates, the tool is moved across the workpiece by the feed unit. The depth of cut is determined by the height of the tool from the table and is adjusted by lowering or raising the cross beam. The length and position of the cut are determined by the position of the table.

These processes are seldom used today.

Boring Mills

A *boring mill*, Figure 20-25, is a machine tool normally used to machine large workpieces. The two common variations of these machines are horizontal and

Figure 20-25 Boring mill (*Courtesy Cincinnati Milacron*)

vertical. The distinction between the two is determined by the position of the spindle. *Horizontal boring mills* perform a wide variety of different machining tasks normally associated with a milling machine but on a much larger scale. *Vertical boring mills* are commonly used to turn, bore, and face large parts in much the same way as a lathe. Another variation of the horizontal boring mill sometimes found in the machine

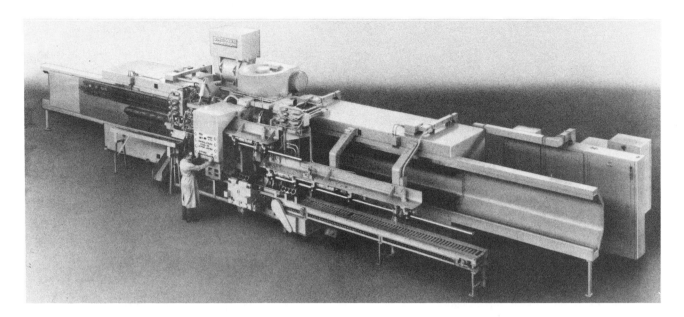

Figure 20-26 Broaching machine (*Courtesy Cincinnati Milacron*)

shop is the *vertical turret lathe*. This machine serves the same basic function as the vertical boring mill but, rather than using a single tool, a vertical turret lathe uses a turret arrangement to mount and position the cutting tools.

Broaching Machines

A *broaching machine*, Figure 20-26, is used to modify the shape of a workpiece by pulling tools called *broaches* across or through the part. Both internal and external forms can be broached. Internal broaching can produce holes with a wide variety of different forms,

Figure 20-27. *External broaching* is typically used to produce some gear teeth, plier jaws, and other similar details.

Another type of internal broaching operation is broached in an *arbor press* using a push-type, rather than a pull-type broach. This type of broaching is often used in the top to produce keyways or other simple shapes.

Numerically Controlled Machines

Numerically controlled machine tools represent one of the newer innovations in machine tool design in wide

Figure 20-27 Typical internal broached shapes (*Courtesy The DuMont Corporation*)

Figure 20-28 Machining center (*Courtesy Cincinnati Milacron*)

use today. These machine tools are operated by either punched tapes or by computers, and virtually eliminate the errors caused by human operators.

The two basic variations of these machines in use today are the point-to-point and the continuous path machines. The principal difference between the two is in the movements of the tool with reference to the workpiece. *Point-to-point* machines operate on a series of preprogrammed coordinates to locate the position of the tool. When the tool finishes at one point, it automatically goes to the next point by the shortest route. This type of control is very useful for drilling machines, but machines such as milling machines require greater control of the tool movement. So, *continuous path* controls are used to control the movement of the tool throughout the cutting cycle. For example, if a circle or radius were to be milled, the operator or programmer would need only to program a few points along the arc. The computer would compute the remaining points and guide the cutting tool throughout the complete cycle.

These controls are frequently used for a wide range of different machine tools, from drill presses and milling machines to lathes and grinders. Due to the advent

of these controls, a whole new form of machine tool has evolved: the machining center, Figure 20-28. These machines can take a rough, unmachined part and completely machine the whole part without removing it from the machine or changing a setup.

Special Workholding Devices

Special workholding devices are frequently used to produce parts in high-volume production runs. The major categories of special workholding devices commonly used in the shop are jigs and fixtures. The primary function of *jigs* and *fixtures* is to transfer the required accuracy and precision from the operator to the tool. This permits duplicate parts to be produced within the specified limits of size without error. Both jigs and fixtures hold, support and locate the workpiece; the principal difference between these tools is the method used to control the relationship of the tool to the workpiece. *Jigs* guide the cutting tool through a hardened drill bushing during the cutting cycle. *Fixtures*, on the other hand, reference the cutting tool by means of a set block, Figure 20-29.

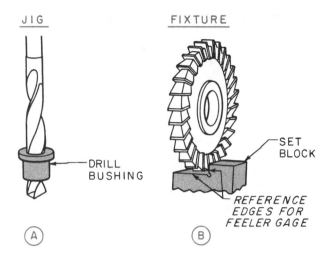

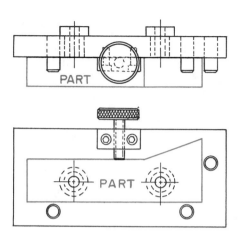

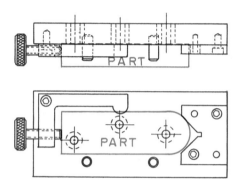

Figure 20-29 Referencing the tool to the workpiece

Figure 20-30 Plate jig

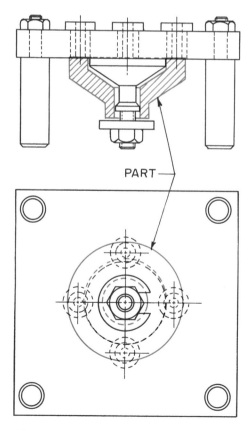

Classification of Jigs

Jigs are normally classified by the type of operation they perform and their basic construction. Typically, jigs are used to drill, ream, tap, countersink, chamfer, counterbore, and spotface. Jigs are also divided into two general construction categories: open and closed. *Open jigs* are jigs that cover only one side of the part, and are used for relatively simple operations. *Closed jigs* are jigs that enclose the part on more than one side, and are intended to machine the part on several sides without removing the part from the jig.

The most common types of jigs include plate jigs, angle plate jigs, box jigs, and indexing jigs. While there are several other distinct styles of jigs, these represent the most common forms.

Plate jigs, Figure 20-30, are the most common form of jig. These jigs consist of a simple plate which contains the required drill bushings, locators, and clamping elements. Typical variations of the basic plate jig include the *table jig*, Figure 20-31A, and the *sandwich jig*, Figure 20-31B. Another and even simpler version of the plate jig is the *template jig*, Figure 20-32. These jigs are used where accuracy and not speed is the prime consideration. Template jigs may or may not have drill bushings, and do not normally have a clamping device. This form of jig is frequently used for light machining or for layout work.

An *angle plate jig*, Figure 20-33, is a modified form of a plate jig in which the surface to be machined is perpendicular to the locating surface. This type of jig is often used to machine pulleys, gears, hubs or similar parts. Another variation of this type of jig is the *modified angle plate jig*. These jigs are used to machine parts at angles other than 90°.

Figure 20-31A Table jig

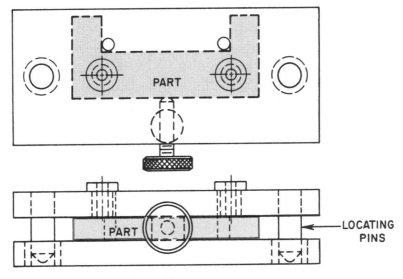

Figure 20-31B Sandwich jig

LOCATING PINS

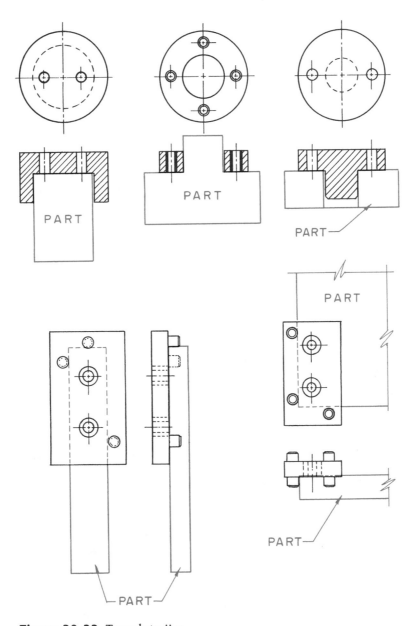

Figure 20-32 Template jigs

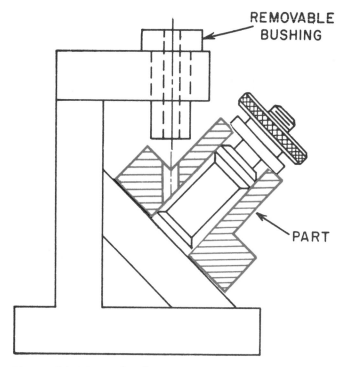

Figure 20-33 Angle plate jig

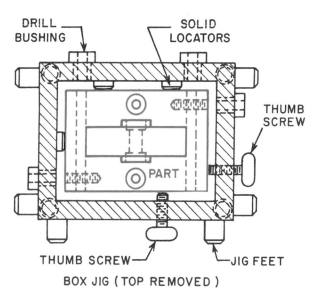

BOX JIG (TOP REMOVED)

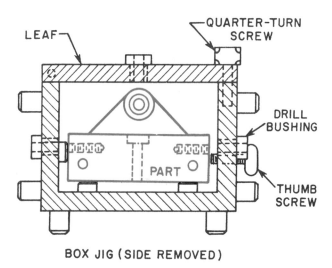

BOX JIG (SIDE REMOVED)

Figure 20-34 Box jig

A *box jig*, Figure 20-34, is designed to be used for parts that require machining on several sides. With these jigs, the part is mounted in the jig and clamped with a leaf or door. Other variations of the basic box jig include the *channel jig*, Figure 20-35A, and the *leaf jig*, Figure 20-35B. These jigs are similar in design to the box jig but they machine the part on only two or three sides, rather than all six sides.

An *indexing jig*, Figure 20-36, is used primarily to machine parts which have machined details at intervals around the part. Drilling four holes 90° apart is a typical example of the type of work this jig is best suited to perform. Another type of jig which uses an indexing arrangement to locate the jig, rather than the part, is the *multistation jig*, Figure 20-37. These jigs are used to machine several parts at one time.

Classification of Fixtures

Fixtures are classified by the type of machine they are used on, the type of operation performed, and by their basic construction features. The principal types of fixtures normally used in the shop include plate fixtures, angle plate fixtures, vise jaw fixtures, and indexing fixtures. Typically, fixtures are used for milling, turning, sawing, grinding, inspecting, and several other varied operations.

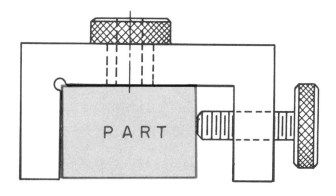

Figure 20-35A Channel jig

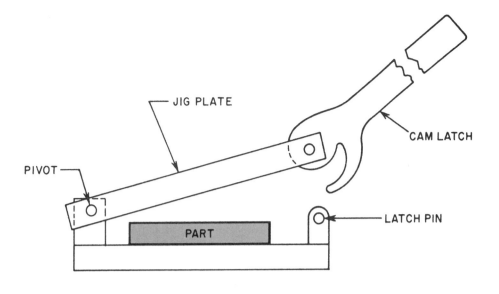

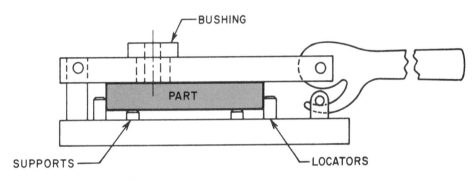

Figure 20-35B Leaf jig

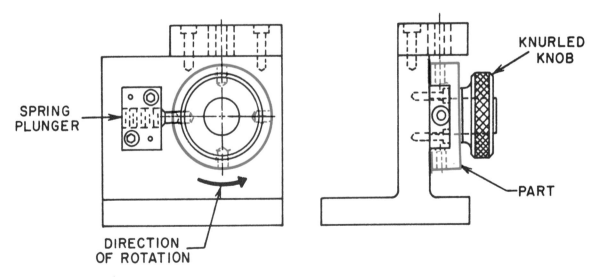

Figure 20-36 Indexing jig

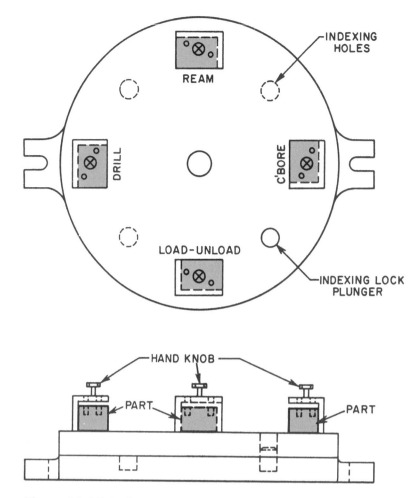

Figure 20-37 Multistation jig

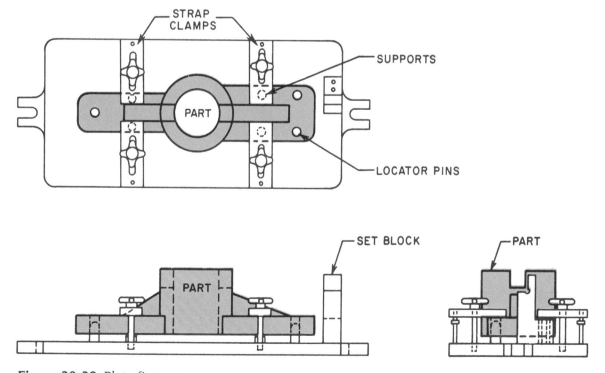

Figure 20-38 Plate fixture

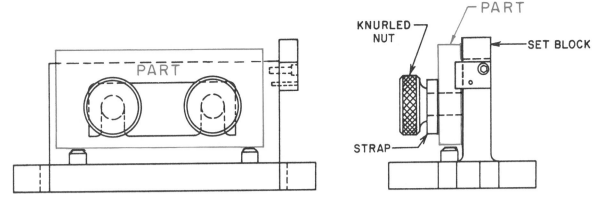

Figure 20-39A Angle plate fixture

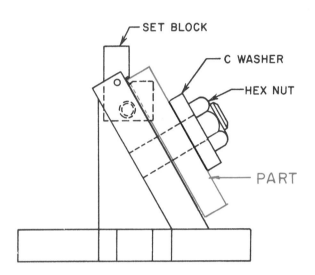

Figure 20-39B Modified angle plate fixture

A *plate fixture*, Figure 20-38, is the most common type of fixture. Like plate jigs, plate fixtures are simply a plate containing the locators, set blocks, and clamping devices necessary to locate and hold the workpiece and to reference the cutter. While similar in design to a plate jig, plate fixtures are normally made much heavier than plate jigs to resist the additional cutting forces.

An *angle plate fixture*, Figure 20-39A, and a *modified angle plate fixture*, Figure 20-39B, are simple modifications of the basic plate fixture design. These fixtures are used when the reference surface is at an angle to the surface to be machined.

A *vise jaw fixture*, Figure 20-40, is a useful modification to the standard milling machine vise. With this type of fixture, the standard jaws of a milling machine vise are replaced with specially shaped jaws to suit the part to be machined. The result is an accurate fixture which can be made at a minimal cost. Since the clamping devise is contained within the vise, this fixture is very cost effective. Likewise, one vise can

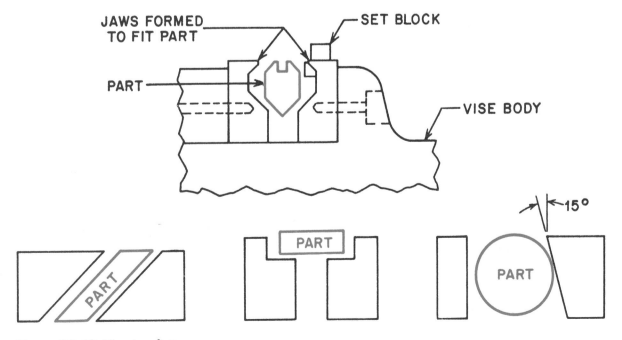

Figure 20-40 Vise jaw fixture

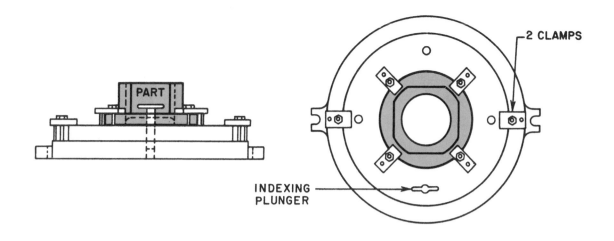

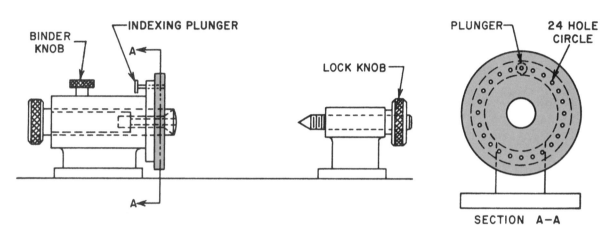

Figure 20-41 Indexing fixture

HEXAGON   SQUARE   GEAR   SPLINE   KEYWAYS

Figure 20-42 Typical parts machined in an indexing fixture

be used for a countless number of different fixtures by simply changing the jaws.

Indexing fixtures, Figure 20-41, are mainly used to machine parts which have a repeating part feature. Typical examples of the types of parts which are machined in an indexing fixture are shown in Figure 20-42. Here again, the indexing feature may also be used to position the tool as well as the part. The *duplex fixture* shown in Figure 20-43 shows a method of indexing the fixture to machine two parts. In use, the first part is machined while the second is loaded. The fixture is then rotated and the second is unloaded and a fresh part loaded. This process is continued throughout the production run.

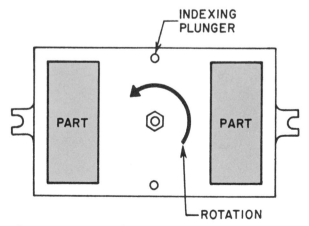

Figure 20-43 Duplex fixture

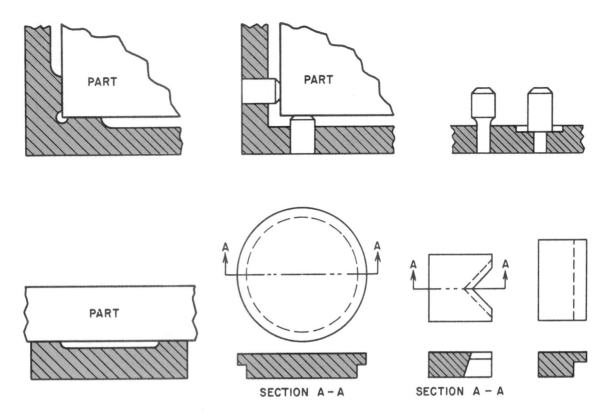

Figure 20-44 Relieving locators

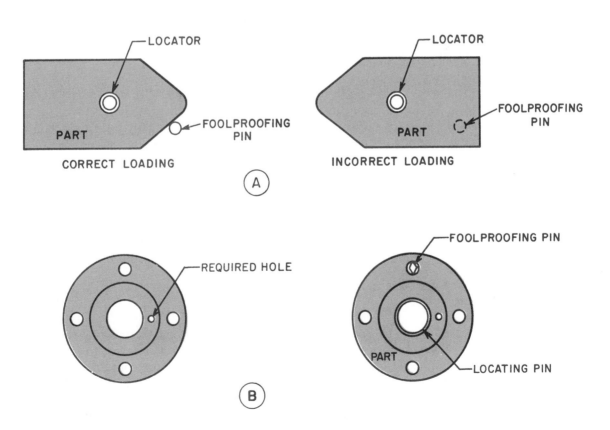

Figure 20-45 Foolproofing a workholder

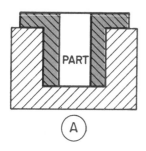
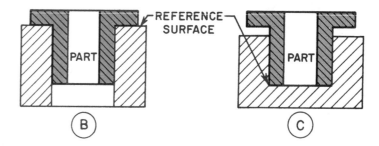

Figure 20-46 Duplicate locating

Locating Principles

To properly machine any part in a jig or fixture, the part must first be located correctly. The first rule of locating is repeatability. *Repeatability* is the feature of a jig or fixture which permits parts to be loaded in the tool in the same position part after part. When selecting locators for a part, the prime considerations must be position, foolproofing, tolerance, and elimination of duplicate or redundant locators.

Locators must be *positioned* as far apart as practical, and should contact the workpiece on a reliable surface to ensure repeatability. The locators must also be designed to minimize the effect of chips or dirt. Figure 20-44 shows a few methods generally used to relieve locators to prevent interference from chips.

Foolproofing is simply a method used to prevent the part from being loaded incorrectly. A simple pin, Figure 20-45, is normally enough to ensure proper loading of every part. The *tolerance* of a locator is determined by the specific size of the part to be machined. As a general rule, the tolerance of jigs and fixtures should be approximately 30% to 50% of the part tolerance. For example, if a hole were to be located within .010'' (0.25 mm) from an edge, the locators in the jig or fixture should be positioned within .003'' to .005'' (0.08 mm to 0.13 mm) to make sure the part is properly positioned. An overly tight tolerance only adds cost, not quality, to a tool.

Finally, locators should never duplicate any location. As shown in Figure 20-46, a part should be located on only one surface. Locating a part on two parallel surfaces only serves to reduce the effectiveness of the location, and could improperly locate the part. First determine the reference surface, and use only that surface for locating.

Restricting Movement. Every part is free to move in a limitless number of directions if left unrestricted. But, for the purpose of jig and fixture design, the number of directions in which a part can move has been limited to twelve: six axial and six radial, Figure 20-47. The methods used to restrict these twelve movements

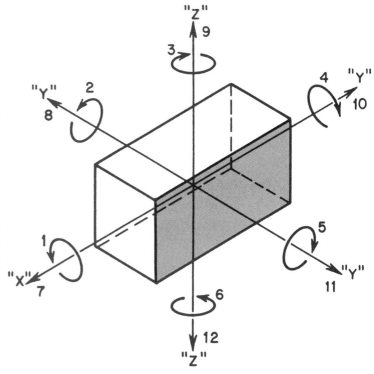

Figure 20-47 Planes of movement

normally depend on the part itself, but the examples in Figures 20-48A and 20-48B show two methods that are frequently used. In the first example, three-pin base restricts five directions of movement. In the second example, a five-pin base restricts eight directions of movement. In Figure 20-48C, a six-pin base restricts nine directions of movement.

Types of Locators. Locators are commercially available in many styles and types. Figures 20-49A through 20-49F show several of the most common types of locators.

Clamping Principles

In addition to locators, most jigs and fixtures use some type of clamping device to restrict the directions of movement not contained by the locators. When designing a clamping arrangement, several factors should be considered. These include position of the clamps, clamping forces, and tool forces.

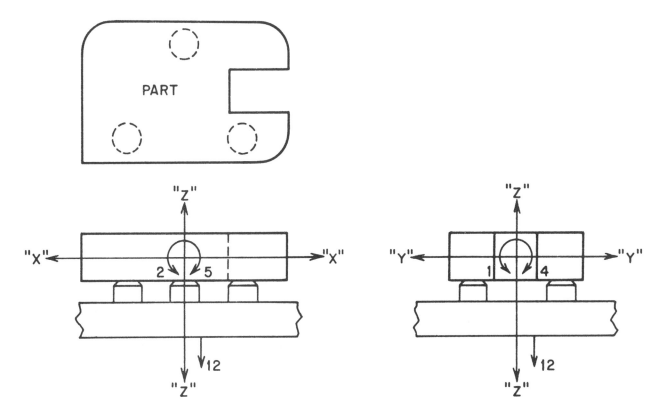

Figure 20-48A Restricting part movement

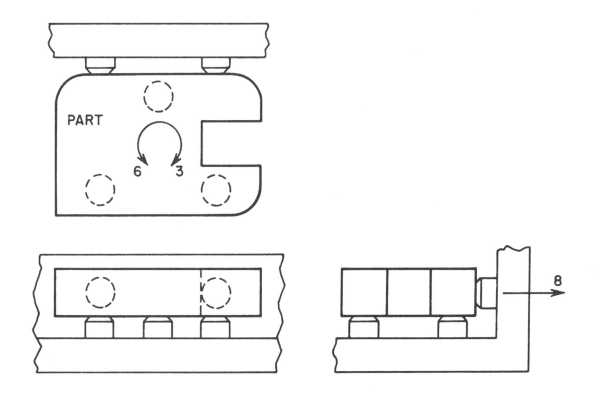

Figure 20-48B Restricting part movement

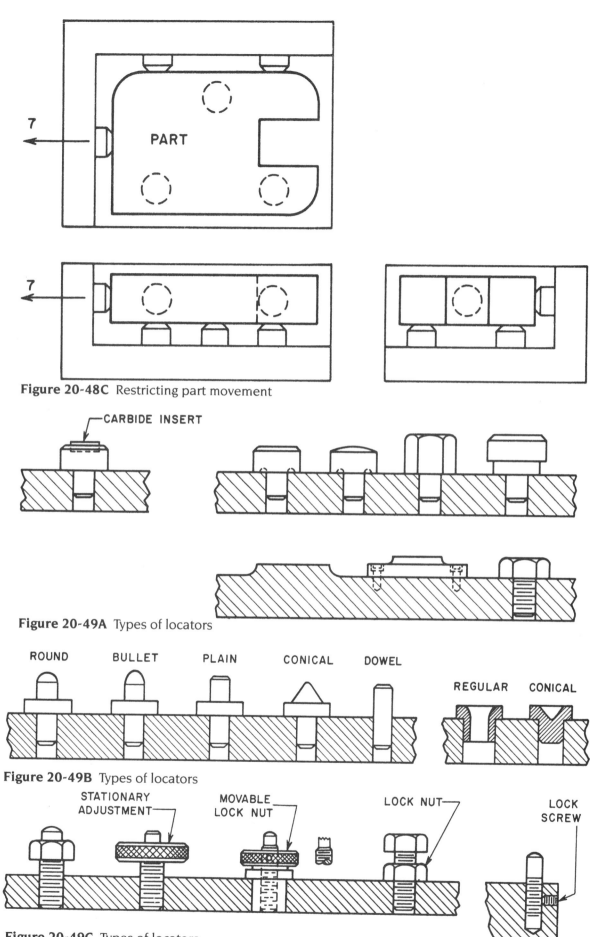

Figure 20-48C Restricting part movement

CARBIDE INSERT

Figure 20-49A Types of locators

ROUND   BULLET   PLAIN   CONICAL   DOWEL   REGULAR   CONICAL

Figure 20-49B Types of locators

STATIONARY ADJUSTMENT   MOVABLE LOCK NUT   LOCK NUT   LOCK SCREW

Figure 20-49C Types of locators

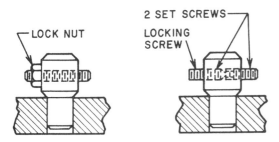

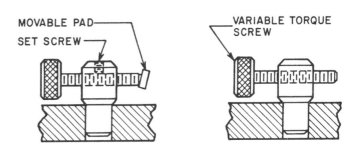

Figure 20-49D Types of locators

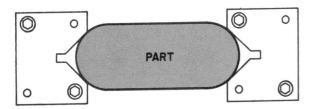

Figure 20-49E Types of locators

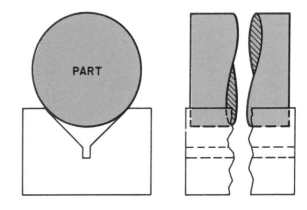

Figure 20-49F Types of locators

Clamps should always be positioned to contact the part at either its most rigid point or a supported point. Clamping a part at any other point could bend or distort the part, as shown in Figure 20-50. The clamping forces used to hold a part should always be directed toward the most solid part of the tool. Clamping a part as shown in Part A of Figure 20-51, will normally result in an egg-shaped part because the clamping forces are directed only toward each other. However, by clamping the part as shown in Part B of Figure 20-51, the part is not only held securely, but the chance of distortion is greatly reduced. As a rule, use only enough clamping force to hold the part against the locators. the locators should always resist the bulk of the tool thrust, not the clamps.

When designing a jig or fixture, remember to direct the tool forces toward the locators, not the clamps. *Tool forces* are those forces generated by the cutting tool during the machining cycle. In most cases, the tool forces can be used to advantage when holding any part. As shown in Figure 20-52, the downward tool forces are actually pushing the part into the tool. The rotational tool forces are contained by the locators. The only force the clamps need to hold are the forces generated by the drill as the point breaks out the opposite side of the part, and these forces are only a fraction of the cutting forces. So, when designing a jig or fixture, always direct the tool forces so that they act to hold the part in the tool. Use the clamps only to hold the part against the locators.

Types of Clamps. Clamps, like locators, are commercially available in many styles and types. Figures

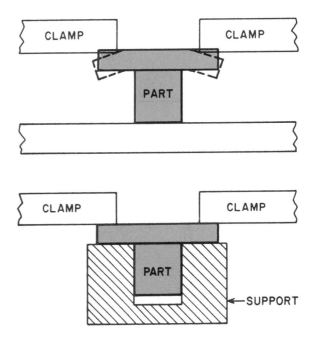

Figure 20-50 Always clamp a part at its most rigid point or a supported point

20-53A through 20-53D show some of the more common types of clamps used for jigs and fixtures. The specific type of clamp you should select for any workholding application will normally be determined by the part, and the type of holding force desired.

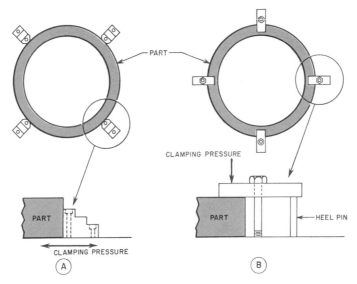

Figure 20-51 Clamping forces

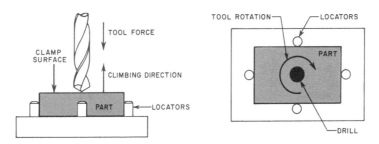

Figure 20-52 Tool forces

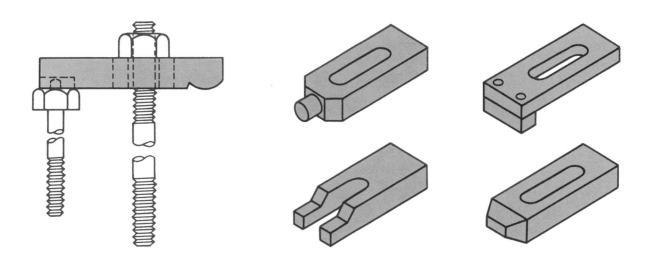

Figure 20-53A Types of clamps

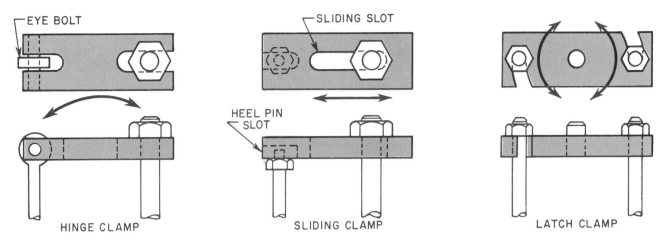

EYE BOLT

SLIDING SLOT

HEEL PIN SLOT

HINGE CLAMP   SLIDING CLAMP   LATCH CLAMP

Figure 20-53B Types of clamps

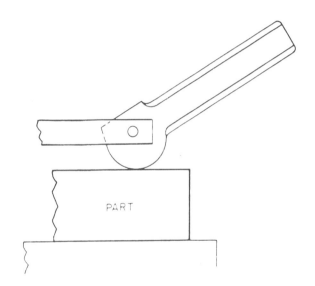

PART

Figure 20-53C Types of clamps

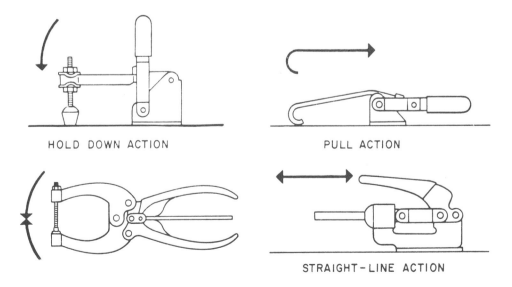

HOLD DOWN ACTION   PULL ACTION

STRAIGHT-LINE ACTION

Figure 20-53D Types of clamps

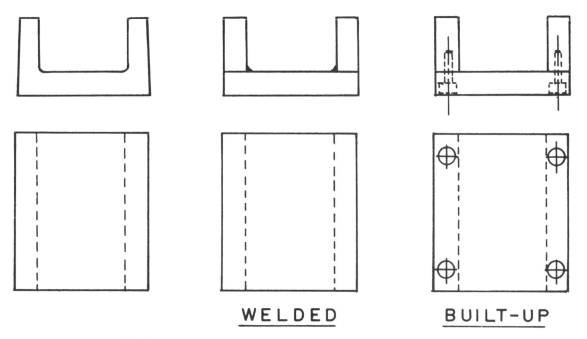

CAST

WELDED

BUILT-UP

Figure 20-54 Tool bodies

Basic Construction Principles

When designing any jig or fixture, the first consideration is normally the tool body. Tool bodies are generally made in any of three ways: cast, welded or built-up, Figure 20-54. *Cast tool* bodies are generally the most expensive, and require the longest lead time to make. They do, however, offer such advantages as good material distribution, good stability, and good vibration-damping qualities. *Welded tool* bodies offer a faster lead time, but must be machined frequently to remove any distortion caused by the heat of welding. The most popular and common type of tool body in use today is the built-up. *Built-up* tool bodies are made from pieces of preformed stock, such as precision-ground flat stock, ground rod or plain cold-rolled sections which are pinned and bolted together. The principle advantages of using this type of construction are fast lead time, minimal machining, and easy modification.

Drill Bushings. *Drill bushings* are used to position and guide the cutting tools used in jigs. The three basic types of drill bushing used for jigs are press fit bushings, renewable bushings, and liner bushings.

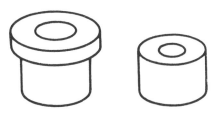

Figure 20-55 Press-fit drill bushings

Press-fit bushings, Figure 20-55, as their name implies, are pressed directly into the jig plate. These bushings are useful for short-run jigs or in applications where the bushings are not likely to be replaced frequently. The two principle types of press fit bushings are the head type and headless type.

Renewable bushings are used in high-volume applications or where the bushings must be changed to suit different cutting tools used in the same hole. The two types of renewable bushings that are commercially available are the slip-renewable type, Figure 20-56, Part A, and the fixed-renewable type, Figure 20-56, Part B.

Slip-renewable bushings, Figure 20-56, Part C, are used where the bushing must be changed quickly. Typical applications include holes which must be drilled, countersunk, and tapped. The bushings are held in place with a lock screw and need only to be turned counterclockwise and lifted out of the hole. The next bushing is inserted in the hole and turned clockwise to lock it in place.

Fixed-renewable bushings are used for applications where the bushings do not need to be changed often, but when they are changed more often than press fit bushings. High-volume production is a typical example where fixed renewable bushings are often used.

Liner bushings, Figure 20-56, Part D, are used in conjunction with renewable bushings and provide a hardened, wear-resistant mount for renewable bushings. Liner bushings are actually press-fit bushings with a large-diameter hole. Like press-fit bushings, liner bushings are available in both a head type and a headless style. The clamps normally used to hold the fixed-renewable bushing in the jig plate are shown in Figure 20-56, Part E. The screw is used to mount

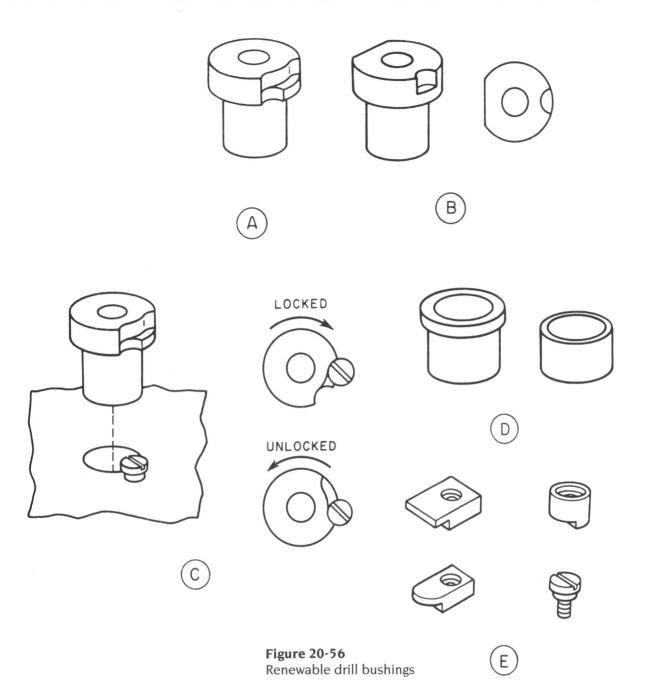

Figure 20-56
Renewable drill bushings

slip-renewable bushings, whereas the other three styles are used to secure the fixed-renewable bushings.

Several other variations of these basic bushing styles include special purpose bushings, serrated and knurled bushings, and oil-groove bushings. Oil-groove bushings are shown in Figure 20-57.

Mounting Drill Bushings. When installing a drill bushing in the jig plate, the specific size of the jig plate and spacing from the part are important factors the designer must consider. As shown in Figure 20-58A, the jig plate should be between one to two times the tool diameter. If a thinner jig plate must be used, a head-type bushing can be used to achieve the added thickness needed to support the cutting tool. The space between the jig plate and the workpiece should be one to one and one half the tool diameter for drilling, Figure 20-58B, and one quarter to one half the tool diameter for reaming, Figure 20-58C.

In those cases where both drilling and reaming operations are to be performed in the same hole, an arrangement similar to the one shown in Figure 20-58D may be used. Here the jig plate is properly positioned for drilling, and the bushing used for reaming is made longer to achieve the desired distance for reaming. When special shapes or contours must be drilled, the ends of the bushings can be modified to suit the contour and maintain the proper support for the cutting tool, Figures 20-58E and 20-58F.

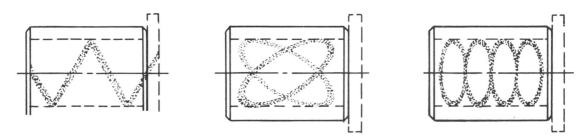

Figure 20-57 Oil-groove bushings

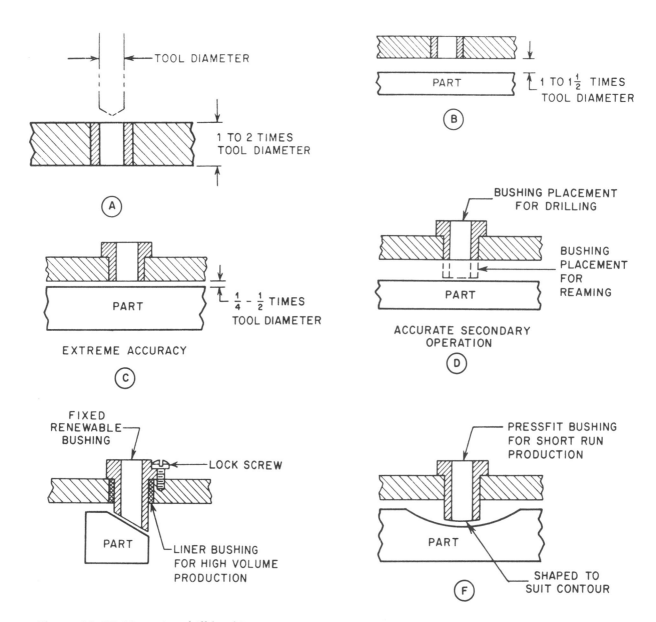

Figure 20-58 Mounting drill bushings

Set Blocks. Set blocks are used along with thickness gages to properly position the cutting tool with a fixture. The specific design of the set block is determined by the part and shape to be machined. The basic set block designs shown in Figure 20-59 are typical of the styles found on many fixtures.

Fastening Devices. The most common fastening devices used with jigs and fixtures are dowel pins and socket-head cap screws. Other fasteners sometimes used with these tools include the nuts and washers shown in Figure 20-60.

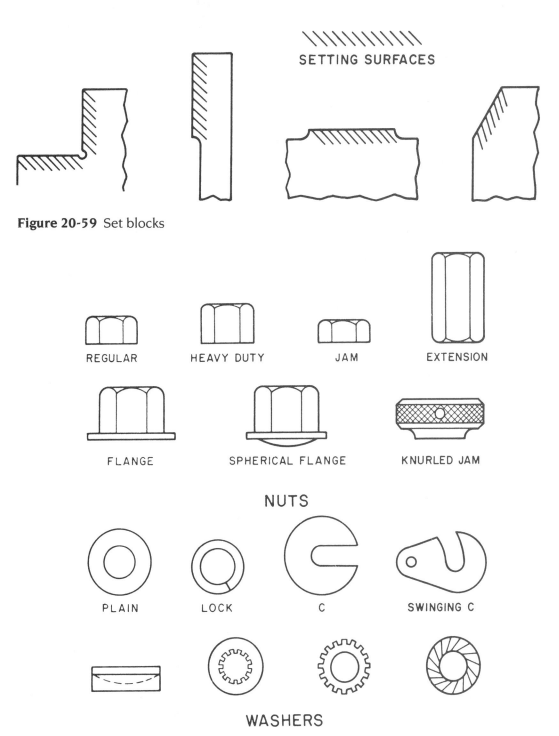

Figure 20-59 Set blocks

SETTING SURFACES

NUTS

REGULAR HEAVY DUTY JAM EXTENSION

FLANGE SPHERICAL FLANGE KNURLED JAM

WASHERS

PLAIN LOCK C SWINGING C

Figure 20-60 Nuts and washers frequently used with jigs and fixtures

Heat Treatment of Steels

Heat treatment is a series of processes used to alter or modify the existing properties in a metal to obtain a specific condition required for a workpiece. The specific properties normally changed by heat treatment include hardness, toughness, brittleness, malleability, ductility, wear resistance, tensile strength, and yield strength. The five standard heat treating operations normally performed on steel parts are hardening, tempering, annealing, normalizing, and case hardening.

Hardening is the process of heating a metal to a predetermined temperature, allowing the part to soak until throughly heated, and cooling rapidly in a cooling material called a *quench*. The most common quench media are air, oil, water, and a water and salt mixture called brine. Hardening is mainly used to increase the hardness, wear resistance, toughness, tensile strength, and yield strength of a workpiece.

Tempering is the process of heating a hardened workpiece to a temperature below the hardening temperature, allowing the part to soak for a specific time period, and cooling. Tempering is mainly used to reduce the hardness so the toughness is increased and brittleness is decreased. Almost every hardened part is tempered to control the amount of hardness in the finished part.

Annealing is the process of heating a part, soaking it until it is throughly heated, and slowly cooling it by turning off the furnace. Annealing is mainly used to completely remove all hardness in a metal. Frequently, hardened parts are annealed to be remachined or modified and then rehardened.

Normalizing is the process used to remove the effects of machining or cold-working metals. In this process, the metal is heated, allowed to soak, and cooled in still air. This process produces a uniform grain size, and eliminates almost all stresses in the metal.

Case hardening is the process of hardening the outside surface of a part to a reselected depth. In most case-hardening operations, carbon is added to the surface of the part by packing the part in a carbonous material, and then heating it so that the carbon is transferred into the surface of the part. Once the carbon is added to the surface of the part, the part is then hardened to produce a hardened shell around a normalized core.

Review

1. List four of the major factors to consider before any part is made in the shop.

2. What is meant by the term *lead time*?

3. List four primary methods used to fabricate manufactured products.

4. What terms are used to describe the top and bottom sections of the flask used for casting?

5. What two important factors must be considered when making a pattern for casting?

6. What is the approximate rate of shrinkage for steel? For cast iron?

7. Which casting process is also referred to as *lost wax casting*?

8. In which casting process is the molten metal forced into metal dies?

9. List three common forging processes.

10. How many stations are normally contained in a drop forge die?

11. Explain the extruding process.

12. What is the principal tool used for metal stamping?

13. List two stamping operations that produce blanks.

14. List two stamping operations that produce holes.

15. What do operations such as slitting, trimming, and shaving control?

16. What is the difference between notching and seminotching?

17. List four operations used to bend or form parts.

18. List four basic types of machine tools used for machining parts.

19. Explain the operation of a lathe.

20. What two variations of the milling machine are the most common?

21. List five operations normally performed on a drill press.

22. What are the two most common variations of saws used in the machine shop?

23. List three forms of precision grinders.

24. How is the cutter on a shaper driven?

25. On most planers, which element moves, the tool or the part?

26. What are the two variations of boring mills?

27. What type of operation is normally used to machine special shapes in holes?

28. What are the two variations of numerically controlled machines?

29. What are the two primary forms of workholding devices?

30. Which workholding device references the tool to the work? Which guides the tool?

31. Which form of workholder for drilling is the most common?

32. Which type of workholder is normally used to mill hexagonal or similar shapes?

33. What is the first rule of locating a part?

34. What is the general guide for the tolerance of workholders?

35. In how many directions is an unrestricted part free to move insofar as workholder design is concerned?

36. What devices are normally used to restrict the movement of a part?

37. List the three factors which must be considered when selecting a clamping device.

38. What is the principal purpose of clamps in a workholding device?

39. What are the three types of tool bodies normally used for workholding devices?

40. What are the three general variations of drill bushings?

41. How thick should the jig plate be?

42. How far should the bushing be from the work for drilling? For reaming?

43. What is the purpose of a set block?

44. What are the principal fasteners used for workholders?

45. List six of the properties of steels modified by heat treatment.

46. What are the five common heat treating operations?

47. Which heat treating operation is used to produce a hardened shell around a tough core?

48. Which heat treating operation is used to soften a hardened part?

49. Which heat treating operation should always follow the hardening process?

50. What effect does normalizing have on a part?

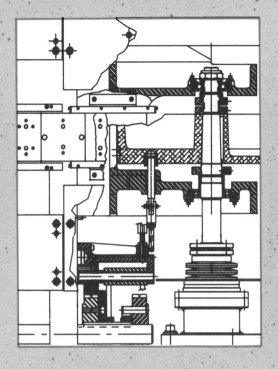

This chapter covers some of the more commonly used manual shortcuts. These shortcuts are time-saving tools and techniques that can be used to increase productivity in drafting settings where CAD systems are not present. Major topics covered include: inking techniques, use of appliques, use of burnishing plates, typewritten text, overlay drafting, and scissors drafting.

CHAPTER TWENTY-ONE
Shortcuts

Inking Techniques

Many old-timers in drafting still talk about the good old days when drawings were done in ink. Linen was the primary drafting medium at the time. However, in spite of the high quality of line work it produced, ink on linen was a slow, nonnproductive technique. Consequently, it was inevitable that the pressures to improve productivity would eventually push this technique aside; and, they did.

Pencil lines on vellum soon became the standard drafting technique. Line quality was, of course, diminished. However, improved reproduction techniques, coupled with increased productivity, served to compensate for this. That is, until drawing storage became a problem.

Pencil on vellum drawings were found to have several undesirable characteristics: 1) they become cracked and faded with age, 2) they are not dimensionally stable, and 3) over the years they collect and become a storage problem.

It was the storage problem that led to the introduction of microfilming in drafting. Through microfilming, drawings could be reduced so substantially that drawings that used to fill an entire set of filing

drawers could be stored in a container the size of a shoebox.

However, microfilming presented drafters with a new set of problems, the worst of which being that pencil lines on vellum tend to fade out and often will not reproduce during the microfilming process. This problem, coupled with the development of better inking tools, higher-quality vellum, and better polyester film led to a resurgence of inking in modern drafting rooms.

Inking Tools

Inking in drafting involves creating graphic and alphanumeric data. The primary tools used in inking are technical pens and mechanical inking sets, coupled with standard drafting tools.

A technical pen consists of a cap, point, needle retainer, pen body, spacer rink, ink container, and lock ring, Figure 21-1. Each pen point is manufactured to a specific width. Unlike ruling pens, technical pens cannot be adjusted to make lines of varying widths. Each pen point makes only one width of line. To compensate for this, technical pens come in sets,

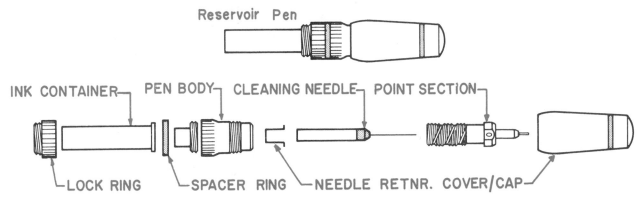

Figure 21-1 Technical pen parts

Figure 21-2. Each pen in the set has a different point width. Technical pens come in a variety of point sizes ranging from 0000 (the smallest) to 7 (the largest). Figure 21-3 is a line width chart with several commonly used technical pen point sizes.

Mechanical lettering in drafting has come to be known as "Leroy" lettering. However, this is not correct. Leroy is the brand name of one mechanical lettering set; other brands of mechanical lettering sets include Staedtler-Mars and Koh-I-Nor, for example. Therefore, the proper term for this inking technique is *mechanical lettering*.

Three basic components of a mechanical lettering set are: the scribe, the pen point, and the lettering guide. Scribes are hand-held devices especially designed to hold the pen point while the drafter follows the lettering guide, Figure 21-4. Scribes come in two basic configurations. One is designed to hold special pen point tips that are used only in mechanical lettering, Figure 21-5. The other holds the point mechanism for a regular technical pen with the holder removed, Figure 21-6.

The special pen point tips are used less and less frequently in modern drafting rooms because they hold less ink, are cumbersome to fill, and spill ink easily. However, they are sturdy and tend to last a long time. Consequently, they are still in evidence in drafting. These special tips come in sets with point sizes that match those of technical pens, Figure 21-7. The trend in most drafting rooms is toward using technical pens screwed into the scribe with the holders removed, resulting in lighter weight.

Filling a Technical Pen

The first step in using a technical pen is to fill it with ink. This is done by removing the holder and, if

Figure 21-2 Technical pen set (*Courtesy Keuffel & Esser*)

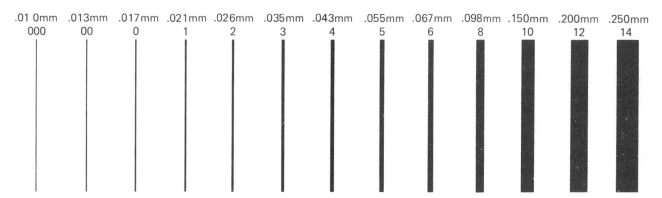

Figure 21-3 Technical pen point widths

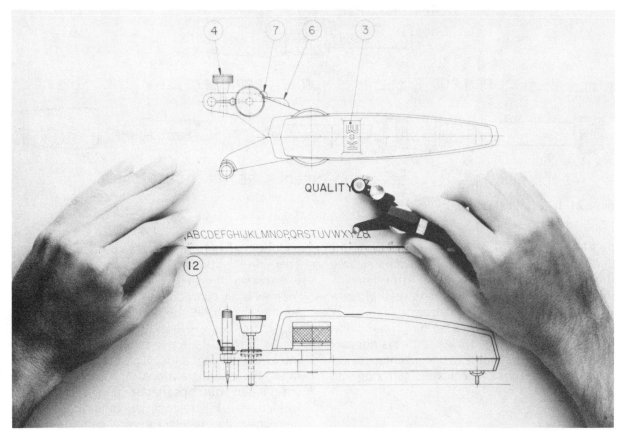

Figure 21-4 Lettering with a scribe (*Courtesy Keuffel & Esser*)

Figure 21-5 Hand-held scribe for mechanical lettering (*Courtesy Keuffel & Esser*)

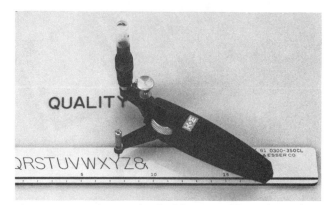

Figure 21-6 This scribe holds a regular technical penpoint (*Courtesy Keuffel & Esser*)

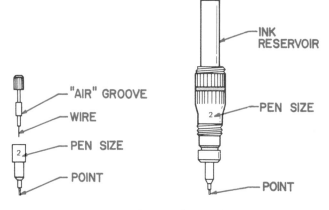

Figure 21-7 Pen tip (left) and matching technical pen point (right)

there is one, unscrewing the retainer ring holding the ink well on, and removing the ink well cartridge.

The ink well should be filled approximately 1/2 to 3/4 full. This will ensure enough pressure to keep the ink flowing evenly, and prevent an ink overflow. The filled ink well is reconnected to the point mechanism slowly and without applying any pressure. Pressure may cause ink to be forced through the point, creating an overflow and blotches. If this happens, the point should be wiped clean before using. The final step is reconnecting the holder by screwing it on.

Once the technical pen is filled, the ink flow is initiated by alternately shaking the pen in a sideways motion parallel to the floor, and trying out lines on a sheet of paper. Once the ink is flowing smoothly and evenly, any excess ink can be cleaned off of the pen using a soft cloth or paper towel. As a technical pen is used, its point should be constantly cleaned in this manner to eliminate accumulations of dust and dirt.

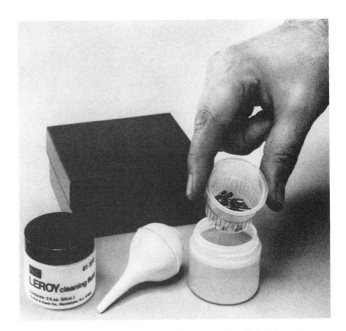

Figure 21-8 Cleaning kit (*Courtesy Keuffel & Esser*)

Cleaning a Technical Pen

Technical pens are designed so that they can be used continuously and with only occasional cleaning. A good rule of thumb is to clean technical pens each time they need to be refilled with ink. Another rule of thumb is to clean technical pens and leave them empty if they are to be stored without being used for more than three days. New humidifiers and rubber/neoprene caps are making this storage step unnecessary with some brands of pens.

A technical pen is cleaned by breaking it down and washing each part with clear, warm water. If ink has dried in the pen, you may need to soak the parts in warm water overnight. Cleaning kits can be purchased from manufacturers of technical pens. These kits are particularly helpful in cleaning pens in which the ink has dried. Figure 21-8 contains an example of a commercially produced cleaning kit for technical pens.

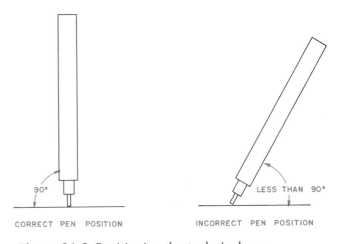

Figure 21-9 Positioning the technical pen

Using Technical Pens

Using a technical pen is much different from using a mechanical pencil. Several rules to remember when using technical pens are:

1. Grip the pen firmly, but gently. Apply a slight downward pressure and keep the pen perpendicular to the inking surface, Figure 21-9.

2. Use beveled or grooved straight edges to prevent ink from running and spreading, Figure 21-10.

3. Elevate templates, triangles, and other devices to prevent ink from running and spreading

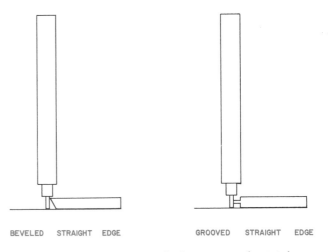

Figure 21-10 Use of beveled or grooved straight edges

underneath. Commercially produced riser pads are available for this.

4. Lay out inked lines in a sequence that reduces the amount of time spent waiting for ink to dry. For example, draw all horizontal lines beginning at the top of the drawing and working down. Then, draw all vertical lines beginning at the left side of the paper and moving to the right. Last, draw angular lines. If tangencies are called for, draw circular lines first and straight lines last.

Erasing Inked Lines

Several methods can be used for erasing inked lines. The best method to use depends on the drafting media to which the ink has been applied (i.e., vellum or polyester film).

Inked lines may be removed from polyester film very easily with a damp white vinyl eraser. Simply moisten the tip of the vinyl eraser lightly and erase the line using soft, short strokes. Remoisten the eraser as needed so as not to rub off the special coating on the polyester film.

Inked lines may be removed from vellum using a soft pink or green eraser in an electric erasing machine. Use light strokes being careful not to damage the surface of the vellum. Once the inked line has been removed, smooth down the grain of the vellum using a soft eraser before reinking the line. Other methods include special ink erasers that are impregnated with ink eradicaters and thinner solution on polyester film and acetate.

Use of Appliques

The word *applique* is a generic term used to describe a variety of shortcut products used in drafting. These products include such items as tapes, pads, and various other ready-made appliques for creating printed circuit board artwork, Figure 21-11. These same materials may also be used for a variety of tasks in other drafting fields, Figure 21-12. For example, architects use tapes for making lines and walls on floor plans.

Transfer cards are used primarily as substitutes for mechanical lettering, but any type of symbol or frequently used piece of graphic data can be placed on a transfer card. Transfer cards are especially designed to fit against a parallel bar, drafting rule or other straight edge for ease of alignment. Symbols are transferred from the card by rubbing them with a blunt point.

Dry transfer sheets are designed according to the same principles as transfer cards. The major differences are that transfer sheets are just that, sheets—not cards. Dry transfer sheets are used a great deal in architectural drafting and technical illustration. The transfer

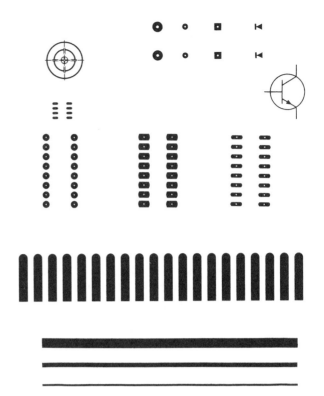

Figure 21-11 Tapes and pads for printed circuit board drafting

is made by rubbing the symbols on the sheet with a blunt, rounded point or a special burnisher.

Dry transfer materials do have some drawbacks. The heat of ammonia-developing print machines tends to lift dry transfer material from the sheet. In addition, the material may dry out and crack with age.

Use of Burnishing Plates

Burnishing is another shortcut for creating graphic symbology fast and easily. *Burnishing* involves placing an especially textured plate under the drafting medium (usually paper or vellum) and rubbing the drafting surface with a pencil. The pencil may be soft and dark or light and hard, depending on the amount of emphasis desired. Two of the most commonly used symbols on burnishing plates are bricks and stone, but any symbol could be made into a plate.

One weakness of burnishing plates is that the symbols they produce do not reproduce well.

Typewritten Text

Text on drawings and other types of documentation consists of dimensions, notes, and callouts. Creating text, or lettering, is one of the slowest, least productive manually performed tasks in drafting. Typewritten text is a shortcut for improving on lettering in manual drafting situations.

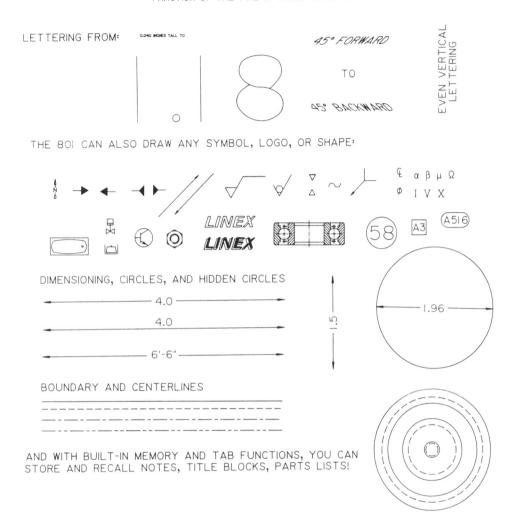

LINEX 801 SCRIBER

DIRECT INK LETTERING ON DRAFTING SURFACES

THE LINEX 801 SCRIBER DOES SCRIBER QUALITY LETTERING IN A
FRACTION OF THE TIME IT TAKES TO DO MANUALLY!

LETTERING FROM: 0.040 INCHES TALL TO

45° FORWARD

TO

45° BACKWARD

EVEN VERTICAL LETTERING

THE 801 CAN ALSO DRAW ANY SYMBOL, LOGO, OR SHAPE:

DIMENSIONING, CIRCLES, AND HIDDEN CIRCLES

BOUNDARY AND CENTERLINES

AND WITH BUILT-IN MEMORY AND TAB FUNCTIONS, YOU CAN
STORE AND RECALL NOTES, TITLE BLOCKS, PARTS LISTS!

Figure 21-12 Sample applique (*Courtesy A.D.S.Linex, Inc.*)

A number of different methods for typing text are used in drafting. The most frequently used are the open-carriage typewriter and the lettering machine. The *open-carriage typewriter* is any brand of typewriter that has been especially designed to hold larger-than-normal media, such as drawings, bills of material, and parts lists. Once a drawing is completed and ready for annotation, the drafter or engineering secretary rolls it into an open-carriage typewriter and types the required text. Typewritten text is fast, neat, and consistent, Figures 21-13 and 21-14.

Lettering machines provide another means for accomplishing typewritten text. Lettering machines are used primarily for titles on drawings, but may be used in any situation where Leroy lettering is used.

Lettering machines, such as the Kroy machine, output the letters on a clear tape that is pressed onto the drawing surface, Figure 21-15. Lettering machines are capable of producing text in a number of different sizes and styles. There are some drawbacks to this shortcut. The machine and its type fonts are expensive.

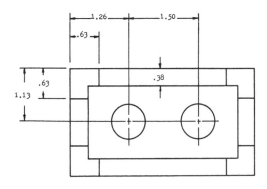

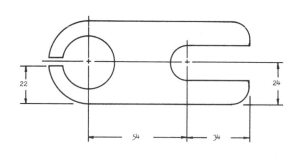

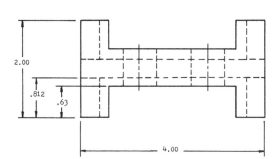

Figure 21-13 Typewritten text on drawings

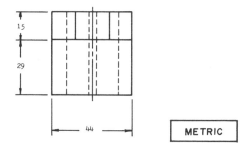

METRIC

Figure 21-14 Typewritten text on drawings

Another method used more and more frequently in modern drafting rooms for accomplishing typewritten text is the *computerized lettering machine*, Figures 21-16 and 21-17. This machine allows drafters to enter data through a keyboard and receive a simultaneous readout on a small display screen. If the text is correct as typed, an ENTER command will cause the lettering to be accomplished in ink automatically. If there are errors, they can be corrected before giving the ENTER command.

Overlay Drafting

Overlay drafting is a complete drafting process that uses advanced reproduction techniques and materials to reduce the amount of time spent in preparing drafting documentation. Actually more than a shortcut technique, the underlying principle of overlay drafting is nonrepetition. N*onrepetition* means that once any type of graphic data or symbology has been drawn the first time, it should never have to be drawn again.

The special tools and materials of overlay drafting include pin bars, registration tabs, prepunched polyester film or a punch to punch holes in standard film, a flatbed vacuum frame printer, and a diazo print machine (see Chapter 22).

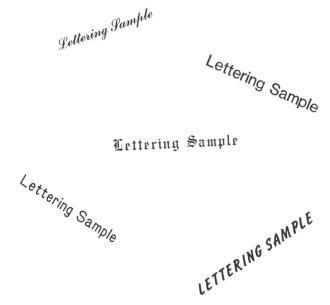

LETTERING SAMPLE

Figure 21-15 Kroy lettering samples

Overlay drafting involves placing a punched base sheet on a drafting table over a pin bar and placing various overlay sheets on top of it. Each successive overlay sheet is automatically lined up (registered) by the pin bar. The simplest example of how overlay

Figure 21-16
Computerized lettering
machine (*Courtesy Ozalid Corp.*)

Figure 21-17
Computerized lettering
machine (*Courtesy A.D.S.Linex, Inc.*)

drafting works is in preparing a set of commercial architectural plans.

In manual drafting, this preparation requires drawing the floor plan four separate times: once for the floor plan, once for the electrical plan, once for the plumbing plan, and once for the HVAC plan. Overlay drafting eliminates this time-consuming repetition.

In overlay drafting, the floor plan is drawn once. Then, before it is dimensioned, three additional originals are created from it, using a special sensitized polyester film, a flatbed vacuum frame printer, and a diazo print machine. (This process is treated in more depth in Chapter 22.) In general terms, this is how the process works:

Step 1. A sheet of punched polyester film is placed on the drafting board. The holes across the top of the film fit over the pin bar which is permanently attached to the top edge of the drafting board. The floor plan is drawn on this base sheet, but, for the moment, the dimensions and all other annotation are left off. These will be added later, after the base sheet is used to reproduce several other originals that do not require dimensions or annotation.

Step 2. The base sheet is taken from the drafting board. A sheet of punched sensitized polyester film is placed on top of it. The two sheets are fastened together with plastic registration pins. Together, they are placed in the flatbed vacuum frame exposure unit. The lights in the unit expose the sensitized film, "burning away" all of the special light-sensitive emulsion, except where the light is blocked out by lines from the base sheet. The exposed polyester film is then run through the ammonia developing section of a print machine producing what is called a slick. A *slick* is a polyester reproduction of the base sheet original. It is not drawn on; rather, it is used as a base sheet over which electrical, plumbing, and HVAC plans may be overlaid. The original base sheet of the floor plan may now be completed by adding dimensions and other annotation.

Step 3. The new slick may now be placed over the pin bar on a drafting board and a clear sheet of polyester film placed on top of it. The electrical symbols for the electrical plan are added on this new sheet. Only information for the electrical plan is entered on this overlay sheet. To get a print of the electrical plan superimposed on the floor plan, the slick base sheet containing the floor plan and the electrical plan overlay are placed in the flatbed vacuum frame printer, along with a sheet of ammonia developing print paper. All three sheets are secured together with plastic registration pins. When the exposure step is completed, the sheets are separated and the print paper is run through the ammonia developing section of a diazo print machine, thus producing a print of the electrical plan. The same process is repeated for the plumbing plan and the electrical plan. To save even more time, three slicks of the floor plan could have been made and given to three different drafters; one of whom would create the electrical plan overlay, one the plumbing plan overlay, and the other the HVAC plan overlay. This process is illustrated in Figure 21-18.

Scissors Drafting Techniques

Scissors drafting is an extension of overlay drafting. The two combined can bring substantial productivity benefits in manual drafting situations. In *scissors drafting*, if any part of a set of drawings has ever been drawn before, say a typical detail or sectional view, it

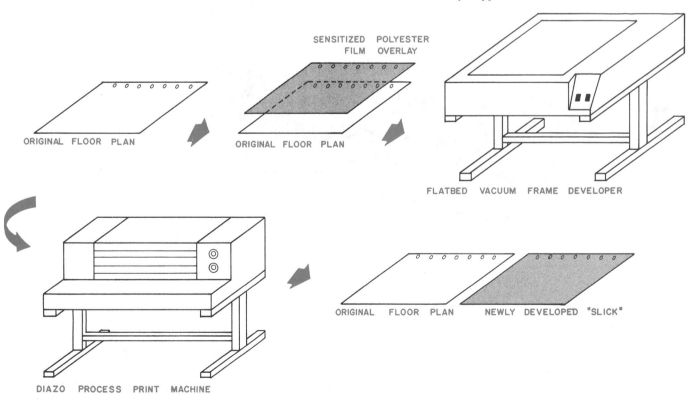

Figure 21-18 Overlay drafting process

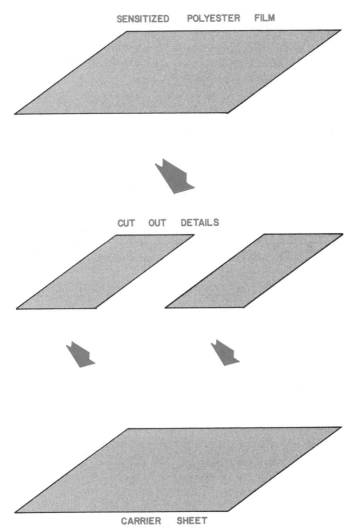

SENSITIZED POLYESTER FILM

CUT OUT DETAILS

CARRIER SHEET

Figure 21-19 Scissors drafting

need not be redrawn. Rather, the techniques outlined previously to create a slick are used, and the details and other data needed from the slick are cut out and taped to a carrier sheet, Figure 21-19. A new slick is created using the carrier sheet as the original.

The scissors drafting technique is best illustrated by example. A mechanical drafter must construct a sheet of typical details and sectional views for a documentation package. All of the needed details have been drawn before, but in different jobs. Some of the details needed are in one job, some in another, and so on. Rather than redrawing, the drafter decides to use scissors drafting techniques.

First, the drafter locates all of the details needed and pulls the required sheets from the filing drawers.

A slick of each sheet is then made. After cutting out the details from the slicks, the drafter assembles them on a carrier sheet. The first slicks allow the original drawings to be kept in case they are ever needed again.

Using the carrier sheet as an original, the drafter creates a slick containing all of the details. The slick may then be copied onto polyester film that accepts plastic lead or ink, and this new medium is used as any other original. The entire process takes about 20 minutes, whereas completely redrawing each detail would take hours.

Review

1. What was the most frequently used drafting medium prior to pencil and paper techniques?

2. Why did the switch from old methods to pencil lines on vellum have to eventually take place?

3. Pencil on vellum drawings have three undesirable characteristics. What are they?

4. What was found to be the biggest problem with the microfilming of drawings?

5. What is the range for technical pen point sizes?

6. What is Leroy lettering?

7. What are the three components of a mechanical lettering set?

8. Explain how to fill a technical pen with ink.

9. Explain how to clean a technical pen.

10. Why is the sequence of drawing lines so important in inking?

11. Explain how to erase an inked line from polyester film.

12. Define the term *applique*.

13. What is the difference between a transfer card and a dry transfer sheet?

14. What is a Kroy lettering machine?

15. What tools and materials are needed to do overlay drafting?

16. What is scissors drafting?

Chapter Twenty-One Problems

The following problems are intended to give beginning drafters practice in using drafting shortcuts that will make them more productive on the job.

The steps to follow in completing the problems are:

1. Study the problem carefully.

2. Make a checklist of tasks you will need to complete.

3. Center the required view or views in the work area.

4. Include all dimensions per the most recent drafting standard.

5. Recheck all work, and, if correct, neatly fill out the title block using light guidelines and neat lettering.

Problem 21-1

Using the techniques described in this chapter, make an inked copy of the drawing. Use freehand lettering.

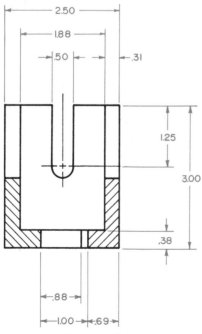

Problem 21-1

Problem 21-2

Using the techniques described in this chapter, make an inked copy of the drawing. Use a mechanical lettering technique, such as Leroy lettering.

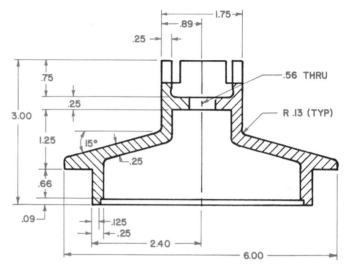

Problem 21-2

Problem 21-3

Using the techniques described in this chapter, make an inked copy of the drawing. Use Kroy lettering or an equivalent process.

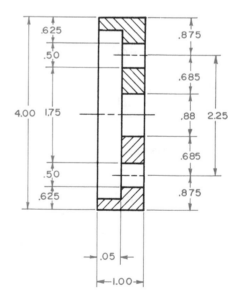

Problem 21-3

Problem 21-4

Using the techniques described in this chapter, make an inked copy of the drawing. Use mechanical lettering and omit the crosshatching. Cut the required crosshatching out of an applique sheet, and apply it to your original drawing.

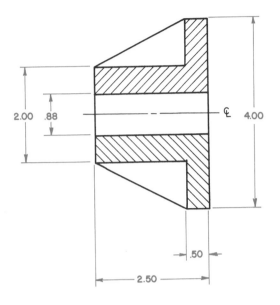

Problem 21-4

Problem 21-5

Using the techniques described in this chapter, make an inked copy of the drawing. Use Kroy lettering or an equivalent process and omit the crosshatching. Cut the required crosshatching out of an applique sheet, and apply it to your original drawing.

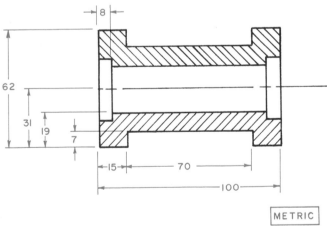

METRIC

Problem 21-5

Problem 21-6

Using the techniques described in this chapter, make an inked copy of the drawing. Use mechanical lettering and omit the crosshatching. Be especially careful to match all fillets, rounds, and radii when cutting the crosshatching out of an applique sheet.

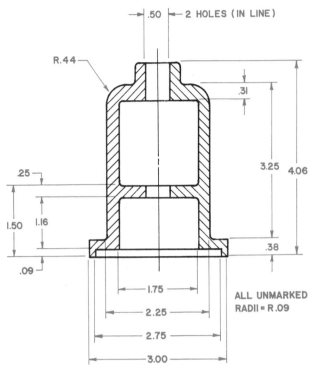

ALL UNMARKED
RADII = R.09

Problem 21-6

Problem 21-7

Using the techniques described in this chapter, make an inked copy of the drawing. Use typewritten notation for the text.

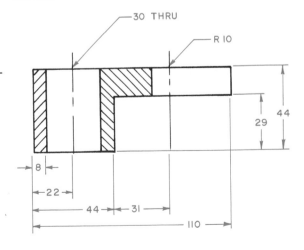

METRIC

Problem 21-7

Problem 21-8

Using the techniques described in this chapter, make an inked copy of the drawing. Use typewritten notation for the text.

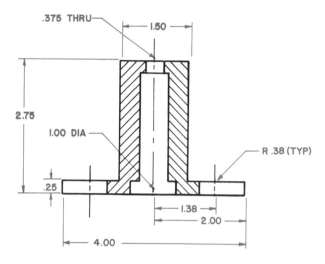

Problem 21-8

Problem 21-9

Using the techniques described in this chapter, make an inked copy of the drawing. Use typewritten notation for the text.

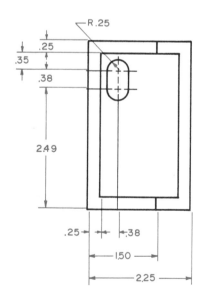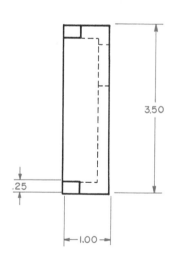

Problem 21-9

Problem 21-10

Using the techniques described in this chapter, make an inked copy of the drawing. Use typewritten notation.

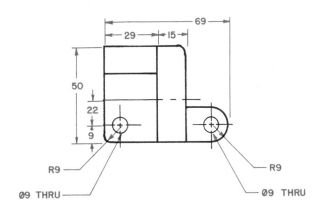

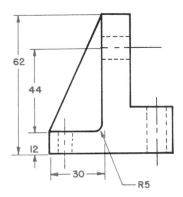

Problem 21-10

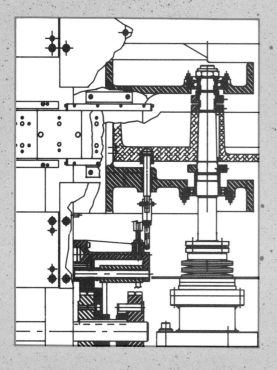

This chapter covers the various types of media used in producing technical drawings manually and on CAD systems, as well as the various processes used in reproducing them. Major topics covered include drafting vellum, polyester drafting films, sepia diazo intermediate paper, diazo print paper, sensitized polyester intermediate film, graphite pencil lead, plastic lead, erasers, diazo printing, and high-speed printing.

CHAPTER TWENTY-TWO
Media and Reproduction

The term "media" is being used more and more frequently in modern drafting rooms. It is the plural of the word medium. *Media* in the past referred to pencils and paper. Paper is a drafting medium, and so are pencils. However, the media used in modern drafting rooms include much more than just pencils and paper.

Modern drafting media include drafting vellum, polyester drafting film, sepia intermediate paper, diazo print paper, sensitized polyester intermediate film, graphite lead, plastic lead, and erasers of various types and textures. Drafters should be well informed as to the various types of media used in the modern drafting room.

Drafting Vellum

Often referred to as drafting paper, *drafting vellum* is a special, highly transparent medium that accepts graphite lead and ink. It is rated according to its *rag content*, the percentage of cotton cloth, as opposed to wood chips, used in making the vellum. The highest quality drafting vellum is 100% rag vellum.

Other factors that are important in terms of the quality of drafting vellum are erasability, translucency,

resistance to crease reprinting, resistance to ghosting after erasing, dimensional stability, and application flexibility.

Erasability is the characteristic which allows for frequent erasing without causing significant damage to the vellum. A vellum with high erasability can be erased on and drawn over several times with no negative effect.

Translucency is the quality of allowing light to pass. This is a critical characteristic when viewed from the perspective of drawing reproduction. The higher the translucency, the better the prints will be during the printing process.

Resistance to crease printing is also an important characteristic. It is a common occurrence in drafting to have a drawing become creased. This usually happens when trying to refile a drawing that is to be placed near the bottom of a full file drawer. It also happens at times during the reproduction process when a drawing becomes ensnarled in the roller mechanism of the print machine. Creased vellum has a tendency to print the crease as a ghost line — a fuzzy, indistinct line — during the diazo printing process. The more resistant a vellum is to showing creases on prints the better.

Resistance to ghosting after erasing is similar to crease reprinting resistance. Drafters must bear down on pencils in order to achieve lines that are dark enough to reprint clearly. Low-quality vellums leave a ghost line permanently indented even after erasing. This ghost line, if it is prominent enough, will reprint during the diazo process. The more resistant a vellum is to ghosting the better.

Dimensional stability is important in those situations where the drawings will be used as templates or masters for such manufacturing processes as those used in printed circuit board drafting. In printed circuit board drafting, the circuit is laid out in tape on the desired drafting medium and photographed. The negative or the positive of the photograph is used as a master in actually manufacturing the printed circuit board. Consequently, the medium used must be *dimensionally stable*, meaning it must not expand or contract; to do so would impair the dimensional accuracy of the master.

All vellum shrinks and expands to a certain extent. The less expansion and shrinkage the better. Because even high-quality vellums are relatively unstable, closely toleranced masters such as those done in printed circuit board drafting are usually done on polyester film.

Application flexibility is the final criterion for judging the quality of a given vellum. *Application flexibility,* in a vellum means that it can be used in a number of different drafting applications, such as manual drafting, computer-aided drafting, technical illustration, overlay drafting, and so forth. To be application flexible, a vellum must accept pencil, plastic lead, ink, and appliques. It must be translucent and strong enough to be placed on and taken off a pin bar repeatedly without fraying.

Drafting vellum and other types of drafting media come in several standard sizes. The American National Standards Institute (ANSI) sizes drafting media in inches: A size = 8½ x 11, B size = 11 x 17, C size = 17 x 22, D size = 22 x 34, E size = 34 x 44, and F size = 28 x 40. See Figure 22-1.

The International Standards Organization (ISO) sizes drafting media in millimetres: A 0 size = 841 X 1189, A 1 size = 594 x 841, A 2 size = 420 x 594, A 3 size = 297 x 420, and A 4 size = 210 x 297. See Figure 22-2.

Polyester Drafting Film

Polyester drafting film is frequently called Mylar, the trade name for the DuPont Company's brand of polyester film. DuPont is the leader in this field, but other brands are also available. Consequently, this medium should be referred to by its generic name — polyester film.

Polyester film is a base film coated with a special finish that allows it to accept graphite lead, ink, and plastic lead. It may have either a single- or a double-matte coating, and comes in various thicknesses which range from 1.5 mils to 7 mils (1.5 thousandths of an inch to 7 thousandths of an inch).

Polyester film, although more expensive than vellum, offers several advantages as a drafting medium: 1) it is more stable than vellum, 2) it is tougher than vellum, 3) it has high erasability characteristics, and 4) it has good reproducibility characteristics. High-quality polyester film does not expand and shrink like vellum. Consequently, it can be used in such situations as printed circuit board drafting where dimensional stability is important.

Polyester film is tough. It is not prone to ripping, tearing or fraying, and it maintains its durability even with age. Ink, graphite lead, and plastic lead are easily erased off polyester film. Finally, polyester film is highly reproducible. The trend in both manual drafting and computer-aided drafting is toward more and more use of ink on polyester film.

Sepia Diazo Intermediate Paper

It is very common in drafting to need a copy that can also serve as an original. That is, it can be drawn on, erased, and reproduced through the diazo process.

SIZE DESIGNATION	VERTICAL (WIDTH)	HORIZONTAL (LENGTH)
A	8½	11
B	11	17
C	17	22
D	22	34
E	34	44
F	28	40

Figure 22-1 ANSI sheet sizes

SIZE DESIGNATION	VERTICAL (WIDTH)	HORIZONTAL (LENGTH)
A 0	841	1189
A 1	594	841
A 2	421	594
A 3	297	421
A 4	210	297

Figure 22-2 ISO sheet sizes

One type of medium that allows for this is *sepia diazo intermediate paper.* Drawings reproduced using this medium are commonly referred to as "sepias." The word *sepia* means dark reddish brown. This is the color of lines on most sepias, although sepia paper is available that produces black lines.

A sepia is properly referred to as an *intermediate.* This is because, as a medium, a sepia falls between an original and a print. It has some of the characteristics of both, but is actually neither. It is produced as a diazo print, rather than by drawing. But, on the other hand, it can be drawn on, erased, and copied using the diazo process.

Early in the development of this medium, sepia paper was brittle, tear prone, and nonerasable. Removing lines required a special fluid that had a strong, offensive odor and left white splotches wherever a line was removed. Modern sepia intermediate paper is more durable and may be erased. However, none of its characteristics is equal to those of either polyester film or vellum.

Diazo Print Paper

Diazo print paper is a special paper coated with a light- and ammonia-sensitive emulsion. It is used for making reproductions of original drawings which have been produced on vellum or polyester film. In years past, diazo prints were called "blueprints." Actual blueprints, which have not been used in drafting for many years, were deep blue prints with white lines. Modern diazo prints are properly referred to as *blueline prints.* They have blue lines on a white (or light blue) background. Although still frequently referred to as blueprints, modern diazo prints should be called blueline prints.

Diazo print paper is a low-grade paper coated with a special yellow emulsion that is sensitive to both light and ammonia. The emulsion reacts to strong light by evaporating. It reacts to ammonia by turning blue. The longer it is exposed to ammonia, the deeper the blue color of the lines. These two characteristics of this emulsion are the basis of the diazo printing process.

In step one of the process, a translucent original drawn on vellum or polyester film is placed face up on top of a piece of diazo print paper with the yellow emulsion side up, Figure 22-3. The two sheets are then run through the exposure section of the diazo machine. In this section, pulleylike belts transport both the original and the print paper around a cylinder which contains several high-intensity light tubes. As the sheets travel around the cylinder, the light emitted "burns off" all of the yellow emulsion except that which is under pen or pencil lines on the original drawing.

In the next step, the exposed piece of print paper is run by itself through the developing section of the

diazo machine. Any yellow emulsion remaining on the print paper, as was protected by pen or pencil lines, reacts to the ammonia by turning blue.

Sensitized Polyester Intermediate Film

Sensitized polyester intermediate film is used in a way that is similar to sepia paper. It is referred to as intermediate film for two reasons: 1) it is a polyester film medium, and 2) like a sepia, it is between an original drawing and a print.

Intermediate film is used a great deal in overlay drafting. The product made by using sensitized intermediate film is called a "slick." A *slick* is a copy of an original from which other copies can be made. However, unlike a sepia, a slick is not drawn on, rather, it is used as one of the ingredients of a shortcut technique used in overlay drafting (see Chapter 21).

Graphite Pencil Lead

Graphite pencil leads are classified as hard, medium, and soft. A soft lead makes very dark lines, a medium lead medium weight lines, and a hard lead makes

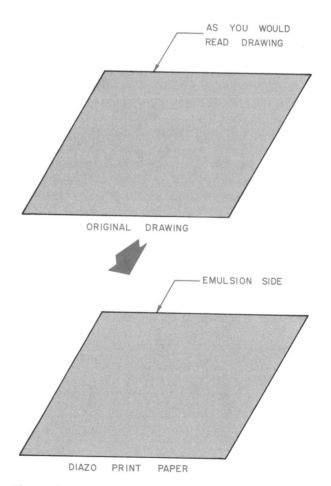

Figure 22-3 Diazo printing

light lines. All graphite leads have alphanumeric or letter designations. The hardest graphite pencil lead is the 9H. Hard leads range from 9H down to 4H (9H, 8H, 7H, 6H, 5H, and 4H). Medium leads range from 3H to B (3H, 2H, H, F, HB, and B). Soft leads range from 2B to 7B (2B, 3B, 4B, 5B, 6B, 7B).

Graphite leads used in drafting are either medium leads or hard leads. Soft leads are too smear prone for mechanical drawing. They are used mainly in such applications as freehand drawing and art.

Plastic Lead

Plastic lead was developed for use on polyester film. Drawing with plastic lead has been described as being similar to drawing with a crayon. And, in fact, it does require some practice and adjustment.

However, once a drafter masters the technique, plastic lead offers several advantages over graphite lead: 1) it produces dark, dense lines that reproduce well, 2) it can approximate the line quality of ink work and the speed of pencil work, 3) it does not smear or smudge as readily as graphite lead, and 4) it erases easily.

Plastic leads are coded with one of six alphanumeric designations: E0, E1, E2, E3, E4, and E5. E0 is the darkest and E5 is the lightest. An E0 plastic lead is the approximate equivalent of an F graphite lead. An E5 is equivalent, approximately, to a 5H graphite lead (refer to the chart in Figure 22-4).

Erasers

Erasers come in a number of different textures, shapes, and types, Figure 22-5. Textures range from rough, gritty erasers to soft, nonabrasive erasers. Shapes may be long and cylindrical, such as those ready-made for use in electric erasing machines. They may be rectangular and small enough for easy hand-held manipulation, or they may fit on the end of a pencil.

Erasers come in several types including soft rubber, vinyl, and plastic. The eraser used should be appropriate for the medium. Graphite lead on vellum is best erased with a soft rubber eraser. Ink or plastic lead on polyester film is easily erased with a white vinyl eraser. Seldom would a hard, gritty eraser be used in drafting because it would damage the medium — vellum or polyester film.

Reproduction of Drawings

A number of reproduction processes are used in drafting. All of them serve the same basic purpose: to produce copies from some type of original as rapidly and as economically as possible. The two most frequently used methods are diazo printing and high-speed printing.

Diazo Printing

In *diazo printing*, copies of original drawings are made by exposing special light-sensitive print paper to ammonia which brings on a chemical reaction that turns the yellow emulsion on the print paper blue. Three components are necessary to do diazo printing:

1. An original on some type of translucent medium.

2. Special light-sensitive print paper.

3. A diazo machine (contains a light source and an ammonia source).

The original can be graphite lead, plastic lead or ink on vellum, polyester film, or any other translucent medium. Line work, in order to reproduce well using the diazo process, must be dark and dense.

Print paper comes in several varieties including blue line, black line, and sepia or brown line. The paper can have a white, light blue, light green, light

EQUIVALENCY   CHART

PLASTIC LEAD	GRAPHITE LEAD
E0	F
E1	H
E2	2H
E3	3H
E4	4H
E5	5H

Figure 22-4 Equivalency chart for plastic and graphite leads

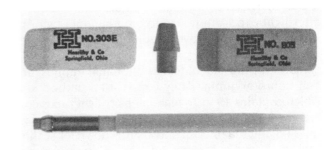

Figure 22-5 Erasers (*Courtesy Hearlihy & Co.*)

Figure 22-6 Rotary diazo print machine (*Courtesy Ozalid Corp.*)

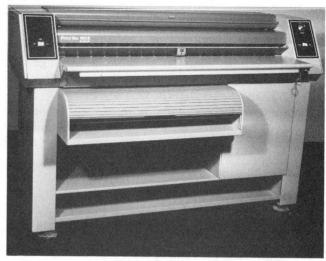

Figure 22-7 Rotary diazo print machine (*Courtesy Ozalid Corp.*)

yellow or light pink background. The various colors of backgrounds make print management and filing easier by color coding. For example, in some companies checkprint number one might always be yellow, checkprint number two blue, and checkprint number three green.

Print machines may be of the rotary or flatbed varieties. Rotary diazo machines have a bottom and a top component. The bottom component consists of a rotary system which pulls an original and a piece of print paper sandwiched together around a plexiglass cylinder which contains several high-intensity light tubes or a single lamp. The top component consists of a pulley system which pulls print paper which has already been exposed to light through a compartment filled with ammonia vapors. Figures 22-6, 22-7, and 22-8 are examples of diazo print machines of the rotary type.

In diazo printing, an original drawing is placed face up on top of a piece of print paper yellow emulsion coating side face up. This sandwiched combination is fed through the light exposure of the diazo machine. The light "burns off" all of the light-sensitive emulsion, except that which is covered by dark, dense lines from the original. Then, the exposed print paper is run through the ammonia developing section of the diazo machine by itself. The emulsion that remains after light exposure reacts to the ammonia by turning blue, black or brown, depending on the variety of paper used.

Flatbed print machines, as the name implies, are flat, nonrotary units. They consist of a flatbed unit which contains 10 to 16 high-intensity light tubes covered by a piece of supertranslucent plexiglass or plastic, a lid unit that lifts up and clamps down for sealing purposes, and a vacuum unit which creates a vacuum seal when activated, Figure 22-9.

Figure 22-8 Rotary diazo print machine (*Courtesy Ozalid Corp.*)

Figure 22-9 Flatbed print machine (*Courtesy Ozalid Corp.*)

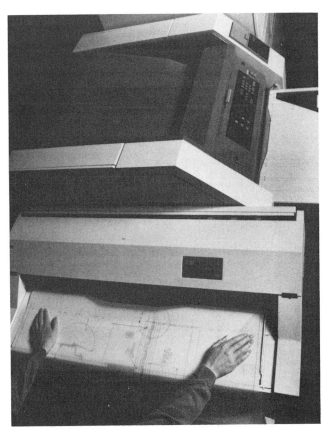

Figure 22-10 High-speed engineering printer (*Courtesy Xerox Corp.*)

Figure 22-11 Print expediter (*Courtesy Xerox Corp.*)

In flatbed printing, the original and print paper or sensitized polyester film are clamped together with plastic registration pins and placed on the flatbed unit. The lid is then closed and clamped shut. The timing device on the front panel of the machine is set for a specified exposure time and the vacuum unit is activated. The vacuum unit creates a vacuum inside the flatbed printer holding the original and printing medium tightly together so that perfect registration is achieved. As soon as a vacuum is achieved, the light tubes illuminate and "burn off" the light-sensitive coating on either the print paper or sensitized polyester film. When the lights go out, the original and exposed printing medium are removed from the flatbed machine. The exposed print paper or other medium is then run through the ammonia developing section of a diazo print machine to complete the process.

The advantage of the flatbed printer is that it gives a perfect one-to-one reproduction that because of the vacuum is dimensionally stable. Rotary print machines tend to stretch the media slightly as it is pulled around the light source.

High-Speed Printing

The always present pressures to improve productivity led to the development of high-speed printing. Diazo printing has improved continuously over the years, but it is still a relatively slow process. This causes "lag" time between the time a drawing is completed and the time it can be distributed. The *lag time* is the time that it takes to reproduce, sort, stamp, and fold prints of drawings.

Fast printing techniques can cut lag time by more than 70% as compared to conventional diazo printing. A typical fast printing machine can reproduce, sort, stamp, and fold as many as 41 prints per minute in sizes ranging from A to E. Figure 22-10 is an example of a high-speed engineering printer. This particular printer can produce prints at a rate of 58 per minute in sizes A, B or C. The originals from which the prints are made may be sized from A to E. Consequently, it follows that fast printers can make reductions of 50% to 65%. The fast printer in the figure automatically sorts, stamps, and folds copies so that they exit the machine ready for distribution.

Figure 22-11 shows how original drawings are fed into a high-speed printer. Notice that, unlike diazo printing, only the original is fed into a fast printer. The reproduction paper is contained in the fast print machine. Figure 22-12 illustrates the fast printer's capability of accepting different size originals intermixed. It automatically senses the sizes and adjusts accordingly.

Figure 22-12 This high-speed printer accepts different sized originals intermixed (*Courtesy Xerox Corp.*).

Review

1. What did the term *media* refer to in the past?

2. What is drafting vellum?

3. What does the term *rag content* mean in regard to vellum?

4. Define the following terms and phrases: *erasability, translucency, resistance to crease printing, resistance to ghosting, dimensional stability,* and *application flexibility.*

5. Give the ANSI sizes for the following sheets:
   A size =
   C size =
   E size =

6. Give the ISO sizes for the following sheets:
   A 0 size =
   A 2 size =
   A 4 size =

7. Define the term *polyester film.*

8. Give four advantages of polyester film as compared with vellum.

9. What is a *sepia?*

10. Outline step-by-step how a diazo reproduction of an original is made.

11. Define the term *slick.*

12. Give two reasons why sensitized polyester intermediate film is referred to as *intermediate film.*

13. Give the scope of the range for hard graphite leads.

14. Give the scope of the range for medium graphite leads.

15. Give the scope of the range for soft graphite leads.

16. List four advantages of plastic lead over graphite lead.

17. List the six code designations for plastic lead from darkest to lightest.

18. Three components are necessary in order to do diazo printing. Name them.

19. What is the primary advantage of a flatbed printer over a rotary diazo printer?

20. Explain the term *lag time* as it relates to printing. How does fast printing reduce lag time?

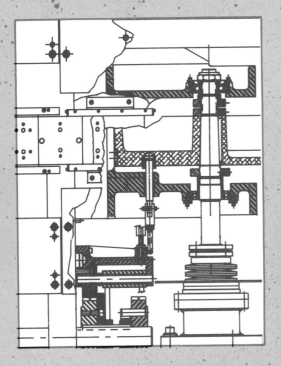

All technical occupations require the use of mathematics. The drafter is constantly involved with applied mathematics. Much of the mathematics is simply using the basic operations of addition, subtraction, multiplication, and division to compute drawing dimensions. However, a knowledge of the fundamentals of algebra, geometry, and trigonometry, as they relate to drafting applications, is essential. This chapter explains these essential functions in detail.

CHAPTER TWENTY-THREE
Mechanical Drafting Mathematics

Mathematics for Drafters

The drafter is often required to solve problems by using formulas obtained from technical handbooks. Geometry is fundamental to mechanical technology: An engineering drawing is an example of applied geometry. Geometric principles are the bases of many dimension calculations and problem solutions. Trigonometry is used to compute unknown angles and sides of triangles. The drafter and designer often work with triangular configurations. Working dimensions must be determined in which parts of complex shapes are broken down into a series of triangles.

Problems encountered in fully defining details of parts or assemblies to be manufactured are often solved by using a combination of algebra, geometry, and trigonometry.

It is essential that the drafter develop the ability to analyze a problem and relate the mathematical principles that are involved in its solution. Then the problem must be solved in clear, orderly steps based on mathematical fact.

Rounding Decimal Fractions

When working with decimals, the computations and answers may contain more decimal places than are required. The number of decimal places needed depends on the degree of precision required. The degree of precision depends on how the decimal value is going to be used. *Rounding a decimal* means expressing the decimal with fewer decimal places.

Procedure: To round a decimal fraction

- Determine the number of decimal places required in an answer.
- If the digit directly following the last decimal place required is less than 5, drop all digits that follow the required number of decimal places.
- If the digit directly following the last decimal place required is 5 or larger, add one to the last required digit and drop all digits that follow the required number of decimal places.

Example 1: Round 0.68247 in to three decimal places.
The digit following the third decimal place is 4.
0.684④7 in.
Because 4 is less than 5, drop all digits after the third decimal place.
0.682 in Ans

Example 2: Round 12.3876 mm to two decimal places.
The digit following the second decimal place is 7.
12.38⑦6 mm
Because 7 is greater than 5, add 1 to the 8.
12.39 mm Ans

Example 3: Round 3.918256 in to four decimal places.
The digit following the fourth decimal place is 5.
3.9182⑤6 in
Therefore, add 1 to the 2.
3.9183 in Ans

Expressing Common Fractions as Decimal Fractions

Given or computed common = fraction values are often converted to decimal = fraction dimensions.

A common fraction is an indicated division. For example, $7/16$ is the same as $7 \div 16$. Because the numerator and the denominator of a common fraction are both whole numbers, expressing a common fraction as a decimal fraction requires division with whole numbers.

Procedure: To express a common fraction as a decimal fraction divide the numerator by the denominator.

Example: Express $7/8$ in as a decimal.
Divide the numerator 7 by the denominator 8.
$$\begin{array}{r} 0.875 \\ 8\overline{)7.000} \end{array}$$

Decimal Equivalent Tables

Generally, fractional engineering drawing dimensions are given in multiples of 64ths of an inch. Decimal equivalent tables are widely used in the manufacturing industry. Many of the equivalents are memorized after using the tables for a period of time. Some of the decimals listed in the table are given to six places. A decimal is rounded to the degree of precision required for a particular application. The amount of computation and the chances of error can be reduced by using the decimal equivalent table shown in Figure 23-1.

The following examples illustrate the use of the decimal equivalent table.

Example 1: Find the decimal equivalent of $7/32$ in. The decimal equivalent is shown directly to the right of the common fraction.
$7/32$ in = 0.21875 in Ans

DECIMAL EQUIVALENT TABLE

Fraction	Decimal	Fraction	Decimal	Fraction	Decimal	Fraction	Decimal
1/64	.015625	17/64	.265625	33/64	.515625	49/64	.765625
1/32	.03125	9/32	.28125	17/32	.53125	25/32	.78125
3/64	.046875	19/64	.296875	35/64	.546875	51/64	.796875
1/16	.0625	5/16	.3125	9/16	.5625	13/16	.8125
5/64	.078125	21/64	.328125	37/64	.578125	53/64	.828125
3/32	.09375	11/32	.34375	19/32	.59375	27/32	.84375
7/64	.109375	23/64	.359375	39/64	.609375	55/64	.859375
1/8	.125	3/8	.375	5/8	.625	7/8	.875
9/64	.140625	25/64	.390625	41/64	.640625	57/64	.890625
5/32	.15625	13/32	.40625	21/32	.65625	29/32	.90625
11/64	.171875	27/64	.421875	43/64	.671875	59/64	.921875
3/16	.1875	7/16	.4375	11/16	.6875	15/16	.9375
13/64	.203125	29/64	.453125	45/64	.703125	61/64	.953125
7/32	.21875	15/32	.46875	23/32	.71875	31/32	.96875
15/64	.234375	31/64	.48438	47/64	.734375	63/64	.984375
1/4	.250	1/2	.500	3/4	.750	1	1.000

Figure 23-1 Decimal equivalent table

Example 2: Find the fractional equivalent of 0.8125 in. The fractional equivalent is shown directly to the left of the decimal fraction.
0.8125 in = $13/16$ in Ans

Example 3: Find the nearer fractional equivalent of 0.757 in. The decimal 0.757 lies between 0.750 and 0.765625. The difference between 0.757 and 0.750 is 0.007. The difference between 0.757 and 0.765625 is 0.008625. Since 0.007 is less than 0.008625, the 0.750 value is closer to 0.757. The nearer fractional equivalent of 0.757 in is therefore $3/4$ in.

Millimetre - Inch Equivalents (Conversion Factors)

Since the English and the metric systems are both used in this country, it is sometimes necessary to express equivalents between systems. Metric-English equivalents other than millimetre-inch are seldom used in mechanical technology.

The relationship between English decimal inch units and metric millimetre units is shown by comparing the scales in Figure 23-2.

1 inch (in) = 25.4 millimetres (mm)
1 millimetre (mm) = 0.03937 inch (in)

Procedure: To express a dimension given in inches as an equivalent dimension in millimetres, multiply the inch value by 25.4 mm.

Example: Express 1.278 in millimetres. Round the answer to two decimal places.
Multiply 1.278 by 25.4 mm.
1.278 × 25.4 mm = 32.46 mm Ans

Procedure: To express a dimension given in millimetres as an equivalent dimension in inches, multiply the millimetre value by 0.03937 inch.

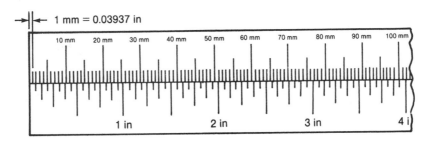

Figure 23-2 Decimal inch - millimetre scales

Example: Express 68.74 mm in inches. Round the answer to three decimal places.
Multiply 68.74 by 0.03937 in.
68.74 × 0.03937 in = 2.706 in Ans

Example: The template shown in Figure 23-3 is dimensioned in millimetres. Determine, in inches, the total length of the template. Round the answer to three decimal places. Do *not* express each of the dimensions as inches. Add the dimensions as they are given and express the sum in inches.

87.63 mm + 102.34 mm + 98.75 mm = 288.72 mm
Multiply 288.72 by 0.03937 in:
288.72 × 0.03937 in = 11.367 in Ans

Evaluating Formulas

By letters and other symbols, generalizations called *formulas* can be stated mathematically. A formula uses symbols to show the relationship between quantities. Problems are often solved by using formulas found in technical manuals and other reference materials.

Procedure: To solve (evaluate) a formula substitute the numerical values for the letter values and solve by following the order of operations of arithmetic. The order of operations follows:

• Do all operations within the grouping symbol first. Parentheses, the fraction bar, and the radical symbol are used to group numbers. If an expression contains parentheses within parentheses or brackets, do the work within the innermost parentheses first.
• Do powers and roots next. The operations are performed in the order in which they occur. If a root consists of two or more operations within the radical symbol, perform all the operations within the radical symbol, then extract the root.

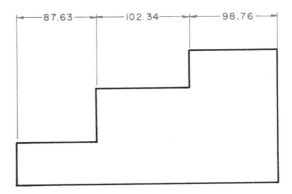

Figure 23-3 Template

• Do multiplication and division next in the order in which they occur.
• Do addition and subtraction last in the order in which they occur.

Example 1: Determine the outside diameter of a gear with 16 teeth and a circular pitch of 0.7854 in. Use the formula

$$D_O = \frac{P_c(N + 2)}{3.1416}$$

where D_O = outside diameter
P_c = circular pitch
N = number of teeth

Substitute the given numerical values for the letter values.

$$D_O = \frac{0.7854(16 + 2)}{3.1416}$$

Do the work in parentheses.

$$D_O = \frac{0.7854(18)}{3.1416}$$

Multiply.

$$D_O = \frac{14.1372}{3.1416}$$

Divide.

$$D_O = 4.5000 \text{ in Ans}$$

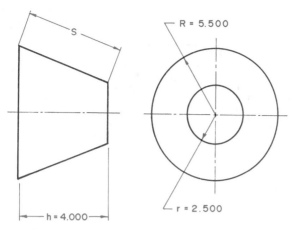

Figure 23-4

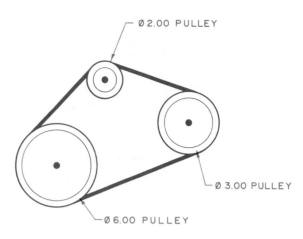

Figure 23-5 Pulley system

Example 2: The dimensions in Figure 23-4 are given in inches. Refer to the figure and determine slant height(s) by using the formula

$$S = \sqrt{(R - r)^2 + h^2}.$$

Substitute numerical values for letter values.
$S = \sqrt{(5.500 - 2.500)^2 + 4.000^2}$
Do the work within the grouping (radical) symbol.
Do the work in parentheses.
$S = \sqrt{3.000^2 + 4.000^2}$
Compute powers within the radical symbol.
$S = \sqrt{9.000 + 16.000}$
Add the values within the radical symbol.
$S = \sqrt{25.000}$
Extract the root.
$S = 5.000$ in Ans

Ratio and Proportion

Ratios

Ratio and proportion are widely used in manufacturing applications, such as computing gear sizes and speeds.

Ratio is the comparison of two *like* quantities. The terms of a ratio are the two numbers that are compared. Both terms of a ratio must be expressed in the same units of measure. Ratios are generally expressed with a colon between the two terms, such as 3:8 (which is read 3 to 8) or as a fraction: $\frac{3}{8}$.

The terms of a ratio must be compared in the order in which they are given. The first term is the numerator of a fraction, and the second is the denominator.

Examples:

1. The ratio 2 to $3 = 2 \div 3 = \frac{2}{3}$.
2. The ratio 3 to $2 = 3 \div 2 = \frac{3}{2}$.
   Generally, a ratio should be expressed in lowest fractional terms.

Examples: A pulley is shown in Figure 23-5.
  1. The ratio of the 2-in diameter pulley to the 6-in diameter pulley is 2 to $6 = \frac{2}{6} = \frac{1}{3}$.
  2. The ratio of the 6-in pulley to the 3-in pulley is 6 to $3 = \frac{6}{3} = \frac{2}{1}$.

Proportions

A *proportion* is an expression that states the equality of two ratios. Proportions are expressed in the following two ways:

  1. 3:4 :: 6:8, which is read: "3 is to 4 as 6 is to 8."
  2. $\frac{3}{4} = \frac{6}{8}$. This form is the usual way to write a proportion.

A proportion consists of four terms. The first and the fourth terms are called *extremes*, and the second and third terms are called the *means*.

Example: $\frac{7}{16} = \frac{14}{32}$. 7 and 32 are the extremes; 16 and 14 are the means.

In a proportion the product of the means equals the product of the extremes. If the terms are cross multiplied, their products are equal.

Example: $\frac{3}{4} = \frac{6}{8}$

Cross Multiply. $\frac{3}{4}$ ✕ $\frac{6}{8}$
$3 \times 8 = 4 \times 6$
$24 = 24$

The method of cross multiplying is used in solving proportions that have an unknown term. After the terms have been cross multiplied, the division principle of equality is applied.

Example 1: $\dfrac{X}{7.5} = \dfrac{23.4}{20}$. Solve for the value of X.
Cross multiply.
$20X = 7.5 \,(23.4)$
$20X = 175.5$

1 circle	= 360 degrees	1 degree	= $\frac{1}{360}$ circle
1 degree	= 60 minutes	1 minute	= $\frac{1}{60}$ degree
1 minute	= 60 seconds	1 second	= $\frac{1}{60}$ minute

Figure 23-6 Relationship of degrees, minutes, and seconds

Apply the division principle of equality. Divide both sides of the equation by 20.

$$\frac{20X}{20} = \frac{175.5}{20}$$

X = 8.775 Ans

Example 2: $\frac{80 \text{ mm}}{150 \text{ mm}} = \frac{70 \text{ mm}}{X}$. Solve for the value of X.

Cross multiply.

80X = 150(70 mm)

80X = 10,500 mm

Apply the division principle of equality. Divide both sides of the equation by 80.

$$\frac{80X}{80} = \frac{10,500 \text{ mm}}{80}$$

X = 131.25 mm Ans

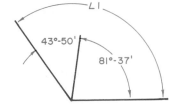

Figure 23-7

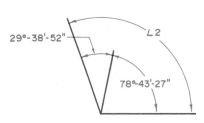

Figure 23-8

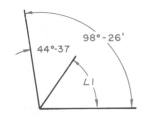

Figure 23-9

Arithmetic Operations on Angles Expressed in Degrees, Minutes, and Seconds

The *degree* is the basic unit of angular measure. The symbol ° means degree. A radius rotated one revolution makes a complete circle or 360°. In the English system, computations and measurements are in degrees and minutes. Degrees, minutes, and seconds are used for applications requiring precise angular measurement.

A degree is divided into 60 equal parts called *minutes*. The symbol for minute '. A minute is divided into 60 equal parts called *seconds*. The symbol for second is ''. The relationship of degrees, minutes, and seconds is shown in Figure 23-6.

Adding Angles Expressed in Degrees, Minutes, and Seconds

Example 1: Determine ∠1 in Figure 23-7.

∠1 = 43°50' + 81°37'

43°50'
+ 81°37'
124°87' = 125°27' Ans

Note: 87' = 60' + 27' = 1°27'; therefore, 124°87' = 125°27'.

Example 2: Determine ∠2 in Figure 23-8.

∠2 = 78°43'27'' + 29°38'52''

78°43'27''
+ 29°38'52''
107°81'79'' = 107°82'19'' = 108°22'19'' Ans

Note: 79'' = 60'' + 19'' = 1'19''; therefore, 107°81'79'' = 107°82'19''.

82' = 60' + 22' = 1°22'; therefore, 107°82'19'' = 108°22'19''.

Subtracting Angles Expressed in Degrees, Minutes, and Seconds

Example 1: Determine ∠1 in Figure 23-9.

∠1 = 98°26' − 44°37'

98°26' = 97°86'
− 44°37' = 44°37'
53°49' Ans

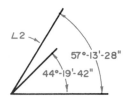

Figure 23-10

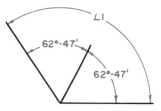

Figure 23-11

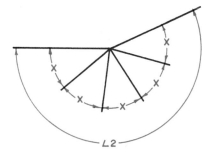

Figure 23-12

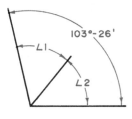

Figure 23-13

Note: Since 37' cannot be subtracted from 26', express
1° as 60'.
98°26' = 97° + 1° + 26' = 97° + 60' + 26' = 97°86'.

Example 2: Determine ∠2 in Figure 23-10.

$$\angle 2 = 57°13'28'' - 44°19'88''$$

$$
\begin{array}{rcccl}
57°13'28'' &=& 56°73'28'' &=& 56°72'88'' \\
-44°19'42'' &=& 44°19'42'' &=& 44°19'42'' \\
\hline
& & & & 12°53'46'' \text{ Ans}
\end{array}
$$

Note: Since 19' cannot be subtracted from 13', and
42'' cannot be subtracted from 28'', express 1° as 60'
and 1' as 60''.
57°13'28'' = 56° + 1° + 13' + 28'' = 56°73'28'' =
56°72' + 1' + 28'' = 56°72'88''.

Multiplying Angles Expressed in Degrees, Minutes, and Seconds

Example 1: Determine ∠1 in Figure 23-11.

$$\angle 1 = 2(62°47')$$

$$
\begin{array}{r}
62°47' \\
\times \ 2 \\
\hline
124°94' = 125°34' \text{ Ans}
\end{array}
$$

Note: 94' = 1°34'.

Example 2: Refer to Figure 23-12 . Determine ∠2
when $x = 41°27'42''$.

$$\angle 2 = 5x = 5(41°27'42'')$$

$$
\begin{array}{r}
41°27'42'' \\
\times \ 5 \\
\hline
205°135'2\,10'' = 205°138'30'' = 207°18'30'' \text{ Ans}
\end{array}
$$

Note: 210'' = 3'30'' and 138' = 2°18'.

Dividing Angles Expressed in Degrees, Minutes, and Seconds

Example 1: Determine ∠1 and ∠2 in Figure 23-13.

$$\angle 1 = \angle 2 = 103°26' \div 2$$

Divide 103° by 2.

$$
\begin{array}{r}
51° \\
2\overline{)103°} \\
102° \\
\hline
1°
\end{array}
$$

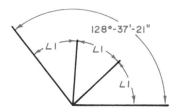

Figure 23-14

103 ÷ 2 = 51° plus a remainder of 1°

Add the 1° (60') to the 26'.
  60' + 26' = 86'.
Divide 86' by 2.

$$
\begin{array}{r}
43' \\
2\overline{)86'}
\end{array}
$$

  86' ÷ 2 = 43'
Combine degrees and minutes.
  51° + 43' = 51°43' Ans

Example 2: Determine ∠1, ∠2, and ∠3 in Figure
23-14.
  ∠1 = ∠2 − ∠3 = 128°37'21'' ÷ 3
Divide 128' by 3.

$$
\begin{array}{r}
42° \\
3\overline{)128°} \\
126° \\
\hline
2°
\end{array}
$$

  128 ÷ 3 = 42° plus a remainder of 2°.

Add the 2° (120') to the 37'.
  120' + 37' = 157'
Divide 157' by 3.

$$
\begin{array}{r}
52' \\
3\overline{)157'} \\
156' \\
\hline
1'
\end{array}
$$

157' ÷ 3 = 52' plus a remainder of 1'.

Add the 1' (60'') to the 21''.
  60'' + 21'' = 81''.
Divide 81'' by 3.

$$
\begin{array}{r}
27'' \\
3\overline{)81''}
\end{array}
$$

81'' ÷ 3 = 27''
Combine.
  42° + 52' + 27'' = 42°52'27'' Ans

Degrees, Minutes, Seconds — Decimal Degree Conversion

In the metric system the preferred method of angular measure is decimal degrees. However, degrees, minutes, and seconds are also used. It is important to be able to compute and express angular measurements in both the English and the metric systems and to express angular measure between systems.

Expressing Decimal Degrees as Degrees, Minutes, and Seconds

The measure of an angle given in the form of decimal degrees, such as 41.1938°, must often be expressed as degrees, minutes, and seconds.

Procedure: To express decimal degrees as degrees, minutes, and seconds.

- Multiply the decimal part of the degrees by 60' in order to obtain minutes.
- If the number of minutes obtained is not a whole number, multiply the decimal part of the minutes by 60'' in order to obtain seconds. Round to the nearer whole second if necessary.
- Combine degrees, minutes, and seconds.

Example: Express 47.1938° as degrees, minutes, and seconds.
Multiply 0.1938 by 60' to obtain minutes.
60'(0.1938) = 11.6280'
Multiply 0.6280 by 60'' to obtain seconds.
60''(0.6280) = 38''
Round to the nearer whole second.
Combine degrees, minutes, and seconds.
47° + 11' + 38'' = 47°11'38'' Ans

Expressing Degrees, Minutes, and Seconds as Decimal Degrees

Often an angle given in degrees and minutes is to be expressed as decimal degrees.

Procedure: To express degrees and minutes as decimal degrees

- Divide the minutes by 60.
- Combine whole degrees and decimal degrees. Round the answer to two places.

Example: Express 76°29' as decimal degrees.
Divide 29 by 60 to obtain decimal degrees.
29 ÷ 60 = 0.48°
Combine whole degrees with decimal degrees.
76° + 0.48 = 76.48° Ans

When working with English and metric units of measure, it may be necessary to express angles given in degrees, minutes, and seconds as angles in decimal degrees.

Procedure: To express degrees, minutes, and seconds as decimal degrees

- Divide the seconds by 60 in order to obtain decimal minutes.
- Combine whole minutes and decimal minutes.
- Divide the total minutes by 60 in order to obtain decimal degrees.
- Combine whole degrees and decimal degrees. Round the answer to four decimal places.

Example: Express 23°18'44'' as decimal degrees.
Divide 44 by 60 to obtain decimal minutes.
44 ÷ 60 = 0.7333'
Combine whole minutes and decimal minutes.
18' + 0.7333' = 18.7333'
Divide 18.7333 by 60 to obtain decimal degrees.
18.7333 ÷ 60 = 0.3122°
Combine whole degrees with decimal degrees.
23° + 0.3122° = 23.3122° Ans

Types of Angles

An *acute angle* is an angle that is less than 90°. Angle 1 in Figure 23-15 is acute.

A *right angle* is an angle of 90°. Angle A in Figure 23-16 is a right angle.

An *obtuse angle* is an angle greater than 90° but less than 180°. Angle ABC in Figure 23-17 is an obtuse angle.

A *straight angle* is an angle of 180°. A straight line is a straight angle. Line EFG in Figure 23-18 is a straight angle.

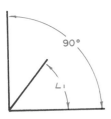

Figure 23-15
Acute angle

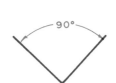

Figure 23-16
Right angle

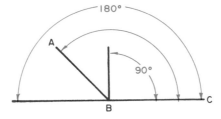

Figure 23-17
Obtuse angle

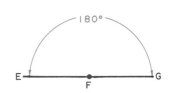

Figure 23-18
Straight angle

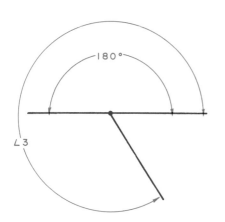

Figure 23-19 Reflex angle

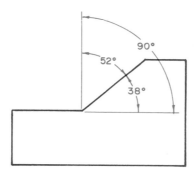

Figure 23-20
Complementary angles

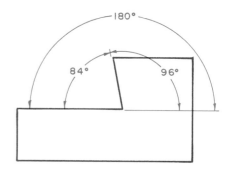

Figure 23-21
Supplementary angles

A *reflex angle* is an angle greater than 180° and less than 360°. Angle 3 in Figure 23-19 is a reflex angle.

Two angles are *complementary* when their sum is 90°. For example, 38° + 52° = 90°. Therefore, 38° is the complement of 52°, and 52° is the complement of 38°. Refer to Figure 23-20.

Two angles are *supplementary* when their sum is 180°. For example, 84° + 96° = 180°. Therefore, 84° is the supplement of 96°, and 96° is the supplement of 84°. Refer to Figure 23-21.

Two angles are *adjacent* if they have a common side and a common vertex. A *vertex* is the point where the two lines forming the angle meet. Angles 1 and 2 in Figure 23-22 are adjacent since they both contain the common side BC and the common vertex B. Angles 3 and 4 in Figure 23-23 are *not* adjacent because they do not have a common vertex.

Angles Formed by a Transversal

A *transversal* is a line that intersects (cuts) two or more lines. Refer to Figure 23-24. Line EF is a transversal since it cuts lines AB and CD.

Alternate interior angles are pairs of interior angles on opposite sides of the transversal. The angles have different vertices. For example, angles 3 and 5 and angles 4 and 6 are alternate interior angles.

Corresponding angles are pairs of angles, one interior and one exterior, on the same side of the transversal. The angles have different vertices. For example, angles 1 and 5, 2 and 6, 3 and 7, and 4 and 8 are corresponding angles.

If two parallel lines are intersected by a transversal, then the alternate interior angles are equal.

Refer to Figure 23-25. Given: AB ‖ CD. (The symbol // means parallel.) Conclusion: ∠3 = ∠5; therefore ∠5 = 62°. ∠4 = ∠6; therefore ∠6 = 118°.

If two lines are intersected by a transversal and a pair of alternate interior angles are equal, then the lines are parallel. Refer to Figure 23-26. Given: ∠1 = 70°, ∠2 = 70°; ∠1 = ∠2. Conclusion: AB//CD.

If two parallel lines are intersected by a transversal, then the corresponding angles are equal. Refer to Figure 23-27. Given: AB//CD. Conclusion: ∠1 = ∠5 = 57°, ∠2 = ∠6 = 123°, ∠3 = ∠7 = 57°, ∠4 = ∠8 = 123°.

If two lines are intersected by a transversal and a pair of corresponding angles are equal, then the lines are parallel. Refer to Figure 23-28. Given: ∠1 = 95°, ∠2 = 95°; ∠1 = ∠2. Conclusion: AB//CD.

Types of Triangles

A *polygon* is a closed plane figure formed by three or more straight-line segments.

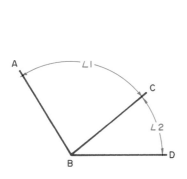

Figure 23-22

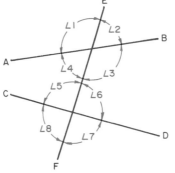

Figure 23-23

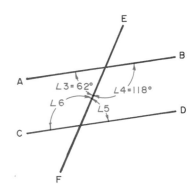

Figure 23-24

Figure 23-25

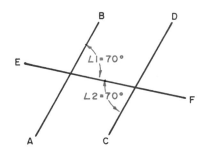

Figure 23-26

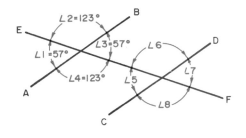

Figure 23-27

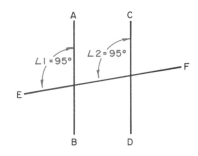

Figure 23-28

A *triangle* is a three-sided polygon; it is the simplest kind of polygon.

The sum of the three angles of any triangle equals 180°.

Scalene Triangle

A *scalene triangle* has three unequal sides. It also has three unequal angles. Triangle ABC shown in Figure 23-29 is scalene. Sides AB, AC, and BC are unequal, and angles A, B, and C are unequal.

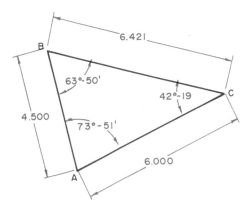

Figure 23-29 Scalene triangle

Isosceles Triangle

An *isosceles triangle* has two equal sides, called *legs*, and two equal *base angles* (the angles opposite the legs). In isosceles triangle EFG in Figure 23-30, leg EF = leg FG = 27.60 mm and base ∠E base ∠G = 40°.

In an isosceles triangle an altitude to the base bisects the base and the vertex angle.

An *altitude* is a line drawn from a vertex perpendicular to the opposite side.

To *bisect* means to divide into two equal parts.

Isosceles triangle ABC in Figure 23-31 is formed by the intersection of center lines between holes A, B, and C. Altitude CD bisects base AB; AD = DB = 6.840 ÷ 2 = 3.420 in. Altitude CD bisects vertex angle ACB; ∠1 = ∠2 = 71°38′ ÷ 2 = 35°49′.

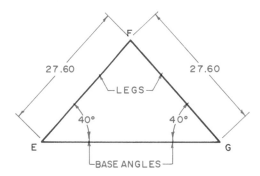

Figure 23-30 Isosceles triangle

Equilateral Triangle

An *equilateral triangle* has three equal sides and three equal angles, each equal to 60°. In equilateral triangle DEF in Figure 23-32, sides DE = DF = EF = 37.86 mm and ∠D = ∠E = ∠F = 60°.

In an equilateral triangle an altitude to any side bisects the side and the vertex angle.

Equilateral triangle HJK in Figure 23-33 is formed by the intersection of center lines between holes H, J, and K. Altitude MJ bisects side HK; HM = MK = 4.500 ÷ 2.250 in. Altitude MJ bisects vertex angle HJK; ∠1 = ∠2 = 60 ÷ 2 = 30°.

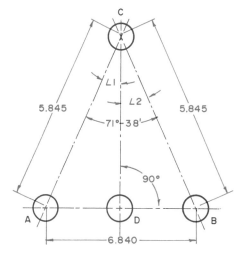

Figure 23-31

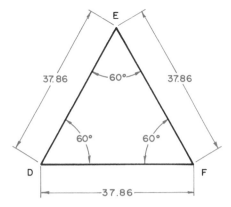

Figure 23-32 Equilateral triangle

Right Triangle

A *right triangle* has one right or 90° angle. The side opposite the right angle is called the *hypotenuse*. The other two sides are called *legs*. In right triangle EFG in Figure 23-34, $\angle E = 90°$, and FG is the hypotenuse.

In a right triangle the square of the hypotenuse is equal to the sum of the squares of the other two sides. This principle, called the *Pythagorean Theorem*, is often used for solving drafting problems. As applied to the right triangle with sides a, b, and c (hypotenuse) shown in Figure 23-35, the Pythagorean Theorem formula is expressed as a, b, and c (hypotenuse) shown in Figure 23-35, the Pythagorean Theorem formula is expressed as $c^2 = a^2 + b^2$.

Example 1: Solve for side c when side $a = 60.00$ mm and side $b = 80.00$ mm, as shown in Figure 23-35.

$c^2 = a^2 + b^2$

Substitute the given values and solve for c.

$c^2 = (60.00 \text{ mm})^2 + (80.00 \text{ mm})^2$
$c^2 = 3600 \text{ mm}^2 + 6400 \text{ mm}^2$
$c^2 = 10,000 \text{ mm}^2$

Add terms. Apply the root principle of equality. Extract the square root of both sides of the equation.

$\sqrt{c^2} = \sqrt{10,000} \text{ mm}^2$
$c^2 = 100.00 \text{ mm Ans}$

Example 2: In right triangle EFG in Figure 23-36, side $g = 5.800$ in and hypotenuse $e = 7.200$ in. Determine side f.

$e^2 = f^2 + g^2$

Substitute the given values and solve for f.

$(7.200 \text{ in})^2 = f^2 + (5.800 \text{ in})^2$
$51.840 \text{ sq in} = f^2 + 33.640 \text{ sq in}$

Apply the subtraction principle of equality. Subtract 33.640 sq in from both sides of the equation.

$51.840 \text{ sq in} = f^2 + 33.640 \text{ sq in}$
$\underline{- 33.640 \text{ sq in} \qquad - 33.640 \text{ sq in}}$
$18.200 \text{ sq in} = f^2$

Extract the square root.

$\sqrt{18.200} \text{ sq in} = \sqrt{f^2}$
$4.266 \text{ in} = f^2$
$f^2 = 4.266 \text{ in Ans}$

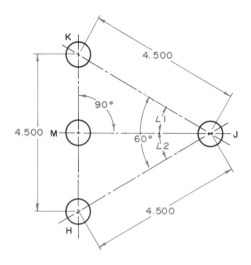

Figure 23-33

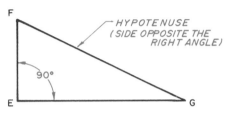

Figure 23-34 Right triangle

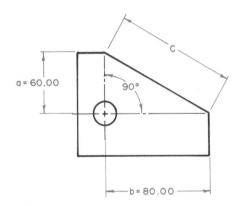

Figure 23-35

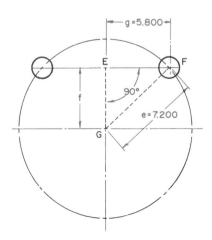

Figure 23-36

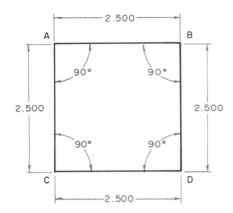

Figure 23-37 Square

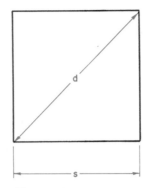

Figure 23-38

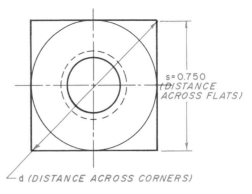

s = 0.750 (DISTANCE ACROSS FLATS)

d (DISTANCE ACROSS CORNERS)

Figure 23-39

Common Polygons

The types of polygons most widely used in drafting in addition to triangles are squares, rectangles, parallelograms, and regular hexagons. A *regular polygon* is one with equal sides and equal angles.

In addition to the description and properties of commonly used polygons, formulas are given that simplify computations when working with polygons.

Squares

A *square* is a regular four-sided polygon. Each angle equals 90°. In square ABCD in Figure 23-37, AB = BC = CD = AD = 2.500 in, and $\angle A = \angle B = \angle C = \angle D$ = 90°.

The following formulas are used with squares. Refer to Figure 23-38.

$a = 1.414s$ where d = diagonal
$s = 0.7071\,d$ s = side
$A = s^2$ or $0.5d^2$ A = area

Example: A square nut is shown in Figure 23-39. Given S (distance across flats) = 0.750 in, determine d (distance across corners). Substitute 0.750 for S in the formula $d = 1.414S$ and solve.

$d = 1.414S$
$d = 1.414(0.750 \text{ in})$
$d = 1.061$ in Ans

Rectangles

A rectangle is a four-sided polygon with opposite sides parallel and equal. Each angle is 90°. In rectangle EFGH in Figure 23-40, EF ∥ GH, EH ∥ FG; EF = GH = 40.85 mm, EH = FG = 75.20 mm; $\angle E = \angle F = \angle G = \angle H$ = 90°.

The following formulas are used with rectangles. Refer to Figure 22-41.

$d = \sqrt{l^2 + w^2}$ where d = diagonal
$l = \sqrt{d^2 - w^2}$ l = length
$w = \sqrt{d^2 - l^2}$ w = width
$A = lw$ A = area

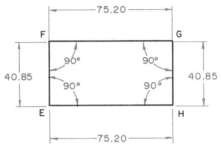

Figure 23-40 Rectangle

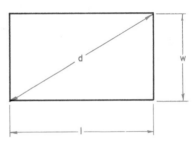

Figure 23-41

Example: A rectangular sheet of aluminum is 30 in wide and 42 in long.
Determine:
a. the distance across opposite corners, diagonal d.
b. the area or the sheet in square feet.
  a. Substitute the given values in the formula $d = \sqrt{l^2 + w^2}$ and solve.
    $d = \sqrt{l^2 + w^2}$
    $d = \sqrt{(42 \text{ in})^2 + (30 \text{ in})^2}$
    $d = \sqrt{1764 \text{ sq in} + 900 \text{ sq in}}$
    $d = \sqrt{2664 \text{ sq in}}$
    $d = 51.614$ in Ans
  b. Substitute the values in the formula $A = lw$ and solve. Since 1 sq ft = 144 sq in, 1260 sq in is divided by 144 to obtain square feet.
    $A = lw$
    $A = 42 \text{ in } (30 \text{ in})$
    $A = 1260$ sq in
    1260 sq in ÷ 144 = 8.75 sq ft Ans

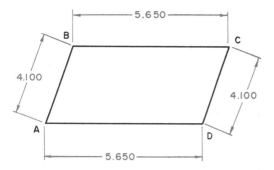

Figure 23-42 Parallelogram

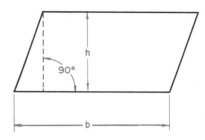

Figure 23-43

Parallelograms

A *parallelogram* is a four-sided polygon with opposite sides parallel and equal and opposite angles equal. In parallelogram ABCD in Figure 23-42, AB ∥ CD, AD ∥ BC; AB = CD = 4.100 in, AD = BC = 5.650 in; ∠A = ∠C = 70°, ∠B = ∠D = 110°.

The area of a parallelogram is equal to the product of the base and the height or altitude. Refer to Figure 23-43.

$$A = bh$$ where A = area
$$b = base$$
$$h = height \ or \ altitude.$$

Note that h is perpendicular to b.

Example: A sheet-metal detail in the shape of a parallelogram is shown in Figure 23-44. Determine the area in square feet. Substitute values in the formula A = bh and solve.

A = bh
A = 28 in (40 in)
A = 1120 sq in

Since 1 sq ft. = 144 sq in, 1120 sq in is divided by 144 to obtain square feet.

1120 sq in ÷ 144 = 7.78 sq ft Ans

Regular Hexagons

A *regular hexagon* is a six-sided figure with all sides equal and all angles equal. Each angle equals 120°.

In the regular hexagon ABCDEF shown in Figure 23-45, AB = BC = CD = DE = EF = AF = 50.25 mm, and ∠A = ∠B = ∠C = ∠D = ∠E = ∠F = 120°.

The following formulas are used with regular hexagons. Refer to Figure 23-46.

$c = 2s$ where s = side
$s = 0.5c$ c = distance across corners
$f = 1.732s$ f = distance across flats
$s = 0.577f$ A = area
$c = 1.155f$
$f = 0.866c$
$A = 2.598s^2 = 0.650c^2 = 0.866f^2$

Example: A hexagonal nut is shown in Figure 23-47. Given f (distance across flats) determine:

    a. s (length of side)
    b. c (distance across corners)
        a. Substitute 0.875 for f in the formula $s = 0.577f$ and solve.
        $s = 0.577f$
        $s = 0.577(0.875 \ in)$
        $s = 0.505 \ in$ Ans
        b. Substitute 0.875 for f in the formula $c = 1.555f$ and solve.
        $c = 1.155f$
        $c = 1.155(0.875 \ in)$
        $c = 1.011 \ in$ Ans
        $c = 1.155f$

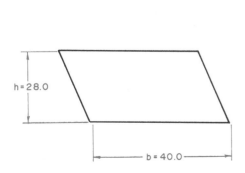

Figure 23-44

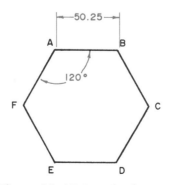

Figure 23-45 Regular hexagon

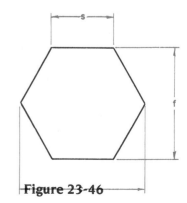

Figure 23-46

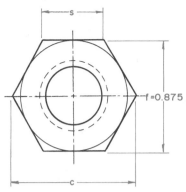

Figure 23-47

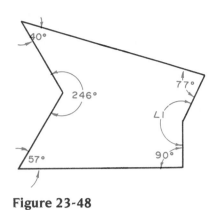

Figure 23-48

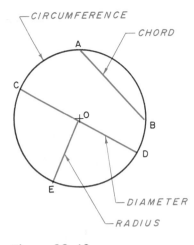

Figure 23-49

The following principle is widely applied when determining unknown angles in any type of polygon.

The sum of the interior angles of a polygon of N sides is equal to (N - 2) times 180°.

Example: Determine ∠1 in the polygon in Figure 23-48.
Count the number of sides. The polygon has six sides: N = 6. The sum of the six angles is (N-2) 180° = (6-2) 180° = 4 (180°) = 720°.
Add the five known interior angles and subtract from 720° to find ∠1.

∠1 = 720° − (57° + 246° + 40° + 77° + 90°)
∠1 = 720° − 510°
∠1 = 210° Ans

Definitions of Properties of Circles

The following terms are commonly used to describe the properties of circles. These properties are often applied in mechanical drafting and design situations.

A *circle* is a closed curve of which every point on the curve is equally distant from a fixed point called the center.

Refer to Figure 23-49 for the following definitions.

A *circumference* is the length of the curved line which forms the circle.

A *chord* is a straight-line segment that joins two points on the circle. AB is a chord.

A *diameter* is a chord that passes through the center of a circle. CD is a diameter.

A *radius* (plural radii) is a straight-line segment that connects the center of the circle with a point on the circle. The radius is one-half the diameter. OE is a radius.

Refer to Figure 23-50 for the following definitions.

An *arc* is that part of a circle between any two points on the circle. The symbol ⌒ written above the letters means arc. A͡B is an arc.

A *tangent* to a circle is a straight line that touches the circle at one point only. The point on the circle touched by the tangent is called the *point of tangency* or *tangent point*. CD is a tangent, P is a tangent point.

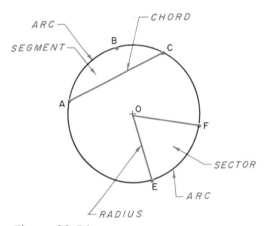

Figure 23-50

Figure 23-51

A *secant* is a straight line that passes through a circle and intersects the circle at two points. EF is a secant.

Refer to Figure 23-51 for the following definitions.

A *segment* is that part of a circle bounded by a chord and its arc. The shaded figure ABC is a segment.

A *sector* is that part of a circle bounded by two radii and the arc intercepted by the radii. The shaded figure EOF is a sector.

Refer to Figure 23-52 for the following definitions.

A *central angle* is an angle whose vertex is at the center of a circle and whose sides are radii. Angle MON is a central angle.

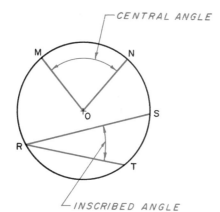

Figure 23-52

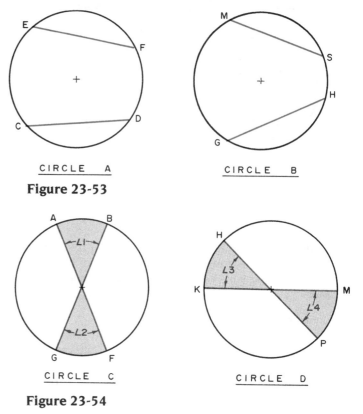

CIRCLE A   CIRCLE B

Figure 23-53

Figure 23-54

An *inscribed angle* is an angle whose vertex is on the circle and whose sides are chords. Angle SRT is an inscribed angle.

Geometric Principles of Circle Circumference, Central Angles, Arcs, and Tangents

The circumference of a circle is equal to pi (π) times the diameter. Generally, for the degree of precision required in machining applications, a value of 3.1416 is used for π.

$C = \pi d$ where C = circumference
or π = pi
$C = 2\pi r$ d = diameter
 r = radius

Example 1: Compute the circumference of a circle with a 50.70-mm diameter.
$C = \pi d = 3.1416(50.70 \text{ mm}) = 159.28 \text{ mm}$ Ans

Example 2: Determine the radius of a circle with a circumference of 14.860 in.

$$C = 2\pi r$$
$$14.860 \text{ in} = 2(3.1416)r$$
$$r = 2.365 \text{ in Ans}$$

In the same circle or in equal circles, equal chords cut off equal arcs.
Refer to Figure 22-53. Given: Circle A = Circle B and chords CD = EF = GH = MS. Conclusion: CD = EF = GH = MS.

In the same circle or in equal circles, equal central angles cut off equal arcs. Refer to Figure 22-54. Given: Circle D and ∠1 = ∠2 = ∠3 = ∠4. Conclusion: AB = FG = HK = MP.

In the same circle or in equal circles, two central angles have the same ratio as the arcs that are cut off by the angles.

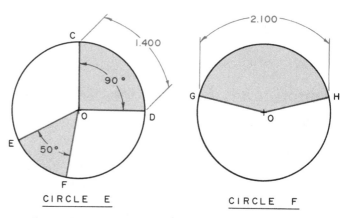

CIRCLE E   CIRCLE F

Figure 23-55

Example: Refer to Figure 23-55. Circle E = Circle F. If ∠COD = 90°, ∠EOF = 50°, CD = 1.400″, and GH = 2.100″, determine (a) the length of EF and (b) the size of ∠GOH. All dimensions are in inches.

a. Set up a proportion between CD and EF with their respective central angles. Solve for EF.

$$\frac{\angle COD}{\angle EOF} = \frac{CD}{EF}$$
$$\frac{90°}{50°} = \frac{1.400″}{EF}$$
$$9EF = 5(1.400″)$$
$$EF = \frac{5(1.400″)}{9}$$
$$EF = 0.778″ \text{ Ans}$$

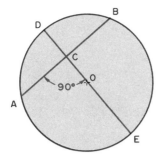

Figure 23-56

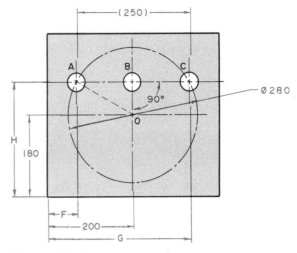

Figure 23-57

b. Set up a proportion between CD and GH with their central angles. Solve for ∠GOH.

$$\frac{\angle COD}{\angle GOH} = \frac{CD}{GH}$$
$$\frac{90°}{\angle GOH} = \frac{1.400''}{2.100''}$$
$$1.400\,(\angle GOH) = 90°(2.100)$$
$$\angle GOH = \frac{90°(2.100)}{1.400}$$
$$\angle GOH = 135° \text{ Ans}$$

A *line drawn from the center of a circle perpendicular to a chord bisects the chord and the arc cut off by the chord. The perpendicular bisector of a chord passes through the center of a circle.*

Refer to Figure 23-56. Given: Diameter DE is perpendicular to chord AB. Conclusion: AC = BC and AD = BD and AE = BE.

The use of this principle with the Pythagorean Theorem has wide practical application in mechanical technology.

Example: Holes A, B, C are to be drilled in the plate shown in Figure 23-57. The centers of holes A and C lie on a 280-mm diameter circle. The center of hole B lies on the intersection of chord AC and segment OB, which is perpendicular to AC. Compute working dimensions F, G, and H. All dimensions are in millimetres.

Compute dimension F: Applying the principle, AC is bisected by OB.

AB = BC = 250 mm ÷ 2 = 125 mm
F = 200 mm − 125 mm = 75 mm Ans

Compute dimension G.
G = 200 mm + 125 mm = 325 mm Ans

Compute dimension H. In right triangle ABO, AB = 125 mm, AO = 280 mm ÷ 2 = 140 mm. Compute OB by applying the Pythagorean Theorem.

$$AO^2 = OB^2 + AB^2$$
$$(140 \text{ mm})^2 = OB^2 + (125 \text{ mm})^2$$
$$OB = 63.05 \text{ mm}$$

Hence, H = 180 mm + 63.05 mm = 243.05 mm Ans

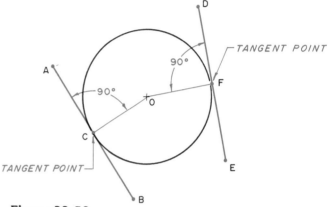

Figure 23-58

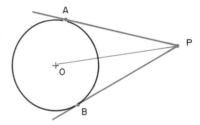

Figure 23-59

A *line perpendicular to a radius at its extremity is tangent to the circle. A tangent is perpendicular to a radius at its tangent point.*

Refer to Figure 23-58.

Example 1:

Given: Line AB is perpendicular to a radius CO at point C. Conclusion: Line AB is a tangent.

Example 2:

Given: Tangent DE passes through point F of radius FO. Conclusion: Tangent DE is perpendicular to radius FO.

Two tangents *drawn to a circle from a point outside the circle are equal. The angle at the outside point is bisected by a line drawn from the point to the center of the circle.*

Refer to Figure 23-59.

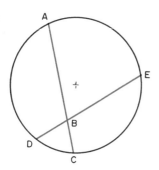

Figure 23-60

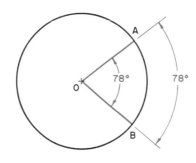

Figure 23-61

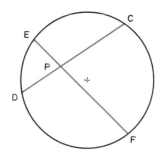

Figure 23-62

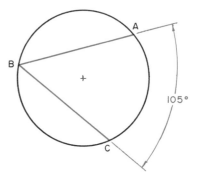

Figure 23-63

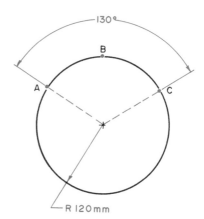

Figure 23-64

Example 1:

Given: Tangents AP and BP are drawn to the circle from point P. Conclusion: AP = BP

Example 2:

Given: Line OP extending from outside point P to center O. Conclusion: ∠APO = ∠BPO

If two chords intersect inside a circle, the product of the two segments of one chord is equal to the product of the two segments of the other chord.

Refer to Figure 23-60.

Example 1:

Given: Chords AC and DE intersect at point B. Conclusion: AB(BC) = BD(BE).

Example 2: If AB = 7.5 in, BC = 2.8 in, and BD = 2.1 in, determine the length of BE.

$$AB(BC) = BD(BE)$$
$$7.5(2.8) = 2.1(BE)$$
$$21.0 = 2.1BE$$
$$BE = 10.0 \text{ in Ans}$$

Geometric Principles of Angles Formed Inside a Circle

A central angle is equal to its intercepted arc. (An intercepted arc is an arc cut off by a central angle.) Refer to Figure 23-61. Given: AB = 78°. Conclusion: ∠AOB = 78°.

An angle formed by two chords intersecting inside a circle is equal to one-half the sum of its two intercepted arcs.

Refer to Figure 23-62.

Example 1: Given: Chords CD and EF intersect at point P.
Conclusion: ∠EPD = ½ (CF + DE).

Example 2: If CF = 106° and ED = 42°, determine ∠EPD. ∠EPD = ½ (106° + 42° = 74° Ans

An inscribed angle is equal to one-half of its intercepted arc. Refer to Figure 23-63. Given: AC = 105°. Conclusion: ∠ABC = ½AC = ½(105°) = 52°30'.

Arc Length Formula

Consider a complete circle as an arc of 360°. The ratio of the number of degrees of an arc to 360° is the fractional part of the circumference that is used to find the length of an arc. *The length of an arc equals the ratio of the number of degrees of the arc to 360° times the circumference.*

$$\text{Arc Length} = \frac{\text{Arc Degrees}}{360°}(2\pi r)$$
$$\text{or}$$
$$\text{Arc Length} = \frac{\text{Central Angle}}{360°}(2\pi r)$$

Example: Refer to Figure 23-64. ABC = 130° and the radius is 120 mm. Determine the arc length ABC.

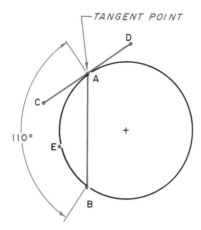

Figure 23-65

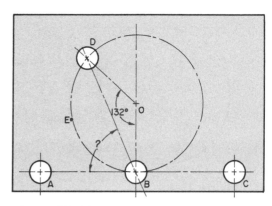

Figure 23-66

$$\text{Arc Length} = \frac{\text{Arc Degrees}}{360°}(2\pi r)$$
$$= \frac{130°}{360°}[2(3.1416)(120 \text{ mm})]$$
$$= 272.271 \text{ mm Ans}$$

Geometric Principles of Angles Formed on a Circle and Angles Formed Outside a Circle

An angle formed by a tangent and a chord at the tangent point is equal to one-half of its intercepted arc.

Example 1: Refer to Figure 23-65. Tangent CD meets chord AB at tangent point A and AEB = 110°. Determine $\angle$CAB.

$\angle$CAB = ½ AEB = ½ (110°) = 55° Ans

Example 2: Refer to Figure 23-66. The centers of three holes lie on line ABC. Line ABC is tangent to circle O at hole-center B. The hole-center D, of a fourth hole, lies on the circle. Determine $\angle$ABD. A central angle is equal to its intercepted arc.

DEB = $\angle$DOB = 132°

An angle formed by a tangent and a chord is equal to one-half of its intercepted arc.

$\angle$ABD = ½ DEB = ½ (132°) = 66° Ans

An angle formed at a point outside a circle by two secants, two tangents, or a secant and a tangent is equal to one-half the difference of the intercepted arcs.

Two Secants

Refer to Figure 23-67.

Example 1:

Given: Secants AP and DP meet at point P and intercept BC and AD. Conclusion: $\angle$P = ½ (AD − BC).

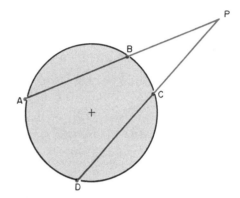

Figure 23-67

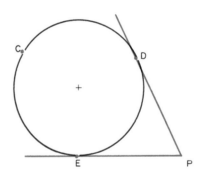

Figure 23-68

Example 2: If AD = 85°40′ and BC = 39°17′, find $\angle$P.

$\angle$P = ½ (AD − BC) = ½ (85°40′ − 39°17′) =
½ (46°23′) = 23°11′30′′ Ans

Two Tangents

Refer to Figure 23-68.

Example 1:
Given: Tangents DP and EP meet at point P and intercept DE and DCE. Conclusion: $\angle$P = ½ (DCE − DE).

Example 2: If DCE = 253°37′ and DE = 106°23′, determine $\angle$P.

$\angle$P = ½ (DCE − DE) = ½ (253°37′ − 106°23′) =
½ (147°14′) = 73°37′ Ans

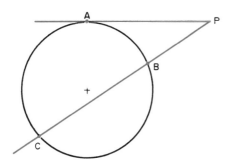

Figure 23-69

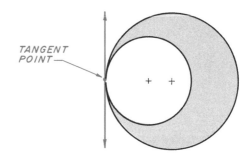

Figure 23-70 Internally tangent circles

A Tangent and a Secant

Example 1: Refer to Figure 23-69. Given: Tangent AP and secant CP meet at point P and intercept AC and AB. Conclusion: $\angle P = \frac{1}{2}(AC - AB)$.

Example 2: If AC − 126°38′ and AB = 68°58′, determine $\angle P$.

$$\angle P = \frac{1}{2}(AC - AB) = \frac{1}{2}(126°38′ - 68°58′)$$
$$= \frac{1}{2}(57°40′) = 28°50′ \text{ Ans}$$

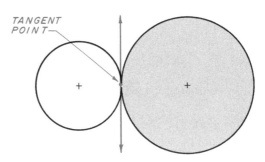

Figure 23-71 Externally tangent circles

Internally and Externally Tangent Circles

Two circles that are tangent to the same line at the same point are tangent to each other. Circles are either internally tangent or externally tangent.

Two circles are *internally tangent* if both circles are on the same side of the common tangent line. Refer to Figure 23-70.

Two circles are *externally tangent* if the circles are on opposite sides of the common tangent line. Refer to Figure 23-71.

If two circles are either internally tangent or externally tangent, a line connecting the centers of the circles passes through the point of tangency and is perpendicular to the tangent line.

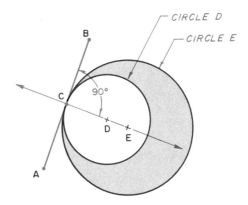

Figure 23-72

Internally Tangent Circles

Example 1: Refer to Figure 23-72. Given: Circle D and Circle E are internally tangent at point C. D is the center of Circle D, and E is the center of Circle E. Line AB is tangent to both circles at point C.
Conclusion: An extension of line DE passes through tangent point C and line CDE is perpendicular to tangent line AB.

This principle is often used as the basis for computing dimensions of parts on which two or more radii blend to give a smooth curved surface. This type of application is illustrated by the following example.

Example 2: A part is to be drawn as shown in Figure 23-73. The proper locations of the two radii will

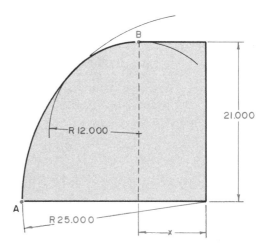

Figure 23-73

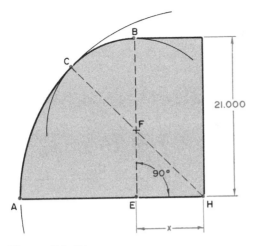

Figure 23-74

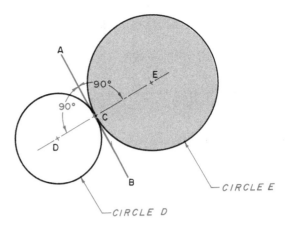

Figure 23-75

result in a smooth curve from point A to point B. Note that the curve from A to B is not an arc of one circle, but is made up of two circles of different sizes. In order to completely dimension the part the location to the center of the 12.000-in radius (dimension x) must be determined. Compute x. All dimensions are in inches. Refer to Figure 23-74. The 12.000-in radius arc and the 25.000-in radius arc are internally tangent. A line connecting arc centers F and H passes through tangent point C. Tangent point C is the endpoint of the 25.000-in radius: (FH = 25.000-in. Tangent point C is the endpoint of the 12.000-in radius: CF = 12.000-in.

FH = 25.000-in − 12.000-in = 13.000-in.

Since BFE is vertical and AEH is horizontal, $\angle$FEH = 90°. In right triangle FEH, FH = 13.000-in, FE = 21.000-in - BF = 9.000-in.
Apply the Pythagorean Theorem to compute EH.

$$FH^2 = EH^2 + FE^2$$
$$(13.000 \text{ in})^2 = EH^2 + (9.000 \text{ in})^2$$
$$169.000 \text{ sq in} = EH^2 + 81.000 \text{ sq in}$$
$$169.000 \text{ sq in} - 81.000 \text{ sq in} = EH^2$$
$$88.000 \text{ sq in} = EH^2$$
$$\sqrt{88.000 \text{ sq in}} = EH$$
$$9.381 \text{ in} = EH$$
$$x = EH,$$
$$x = 9.381 \text{ in Ans}$$

Externally Tangent Circles

Example 1: Refer to Figure 23-75. Given: Circle D and Circle E are externally tangent at point C. D is the center of Circle D, and E is the center of Circle E. Line AB is tangent to both circles at point C. Conclusion: Line DE passes through tangent point C, and line DE is perpendicular to tangent line AB at point C.

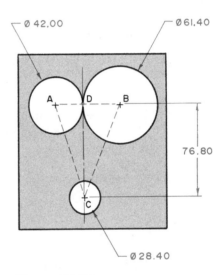

Figure 23-76

Example 2: Three holes are to be bored in the steel plate of Figure 23-76. The 42.00-mm and 61.40-mm diameter holes are tangent at point C. CD is the common tangent line. Determine the distances between hole centers (AB, AC, and BC). All dimensions are in millimetres. Compute AB. AB connects the centers of two tangent circles; AB passes through tangent point D.

AB = AD + DB = 21.00 mm + 30.70 mm = 51.70 mm Ans

Compute AC and BC. Since AB connects the centers of two tangent circles, AB is perpendicular to tangent line DC. Triangle ADC and triangle BDC are right triangles. Apply the Pythagorean Theorem.
In right triangle ADC, AD = 21.00 mm and DC = 76.80 mm.

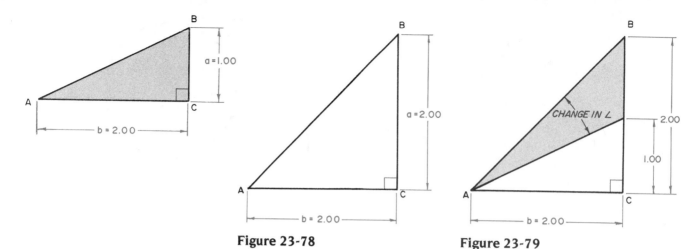

Figure 23-78 **Figure 23-79**

$AC^2 = AD^2 + DC^2$
$AC^2 = (21.00\ mm)^2 + (76.80\ mm)^2$
$AC^2 = 441.00\ mm^2 + 5898.24\ mm^2$
$AC^2 = 6339.24\ mm^2$
$AC\ = 79.62\ mm$ Ans

In right triangle BDC, DB = 30.70 mm and DC = 76.80 mm.

$BC^2 = DB^2 + DC^2$
$BC^2 = (30.70\ mm)^2 + (76.80\ mm)^2$
$BC^2 = 942.49\ mm^2 + 5898.24\ mm^2$
$BC^2 = 6840.73\ mm^2$
$BC\ = \sqrt{6840.73\ mm^2}$
$BC\ = 82.71\ mm$ Ans

Trigonometry: Trigonometric Functions

Trigonometry is the branch of mathematics used to compute unknown angles and sides of triangles. Many drafting problems that cannot be solved by the use of geometry alone are easily solved by trigonometry.

Ratio of Right Triangle Sides

In a right triangle the ratio of two sides of the triangle determines the sizes of the angles, and the angles determine the ratio of two sides. Refer to the triangles in Figures 23-77 - 23-79. The size of angle A is determined by the ratio of side a to side b. When side $a = 1$ in and side $b = 2$ in (Figure 23-77), the ratio of a to b is 1:2 or 1/2. If side a is increased to 2 in and side b remains 2 in (Figure 23-78), the ratio of a to b is 1:1 or 1/1. Observe the increase in angle A (Figure 23-79) as the ratio changed from 1/2 to 1/1.

Note: The symbol for a right angle is ∟, which is shown at the vertex of the angle.

Identifying Right Triangles by Name

The sides of a right triangle are named opposite side, adjacent side, and hypotenuse. The *hypotenuse* (hyp) is always the side opposite the right angle. It is always the longest side of a right triangle. The positions of the opposite and adjacent sides depend on the reference angle. The *opposite side* (opp side) is opposite the reference angle, and the *adjacent side* (adj side) is next to the reference angle.

In the triangle in Figure 23-80 showing ∠A as the reference angle, side b is the adjacent side, and side a is the opposite side. In the triangle in Figure 23-81

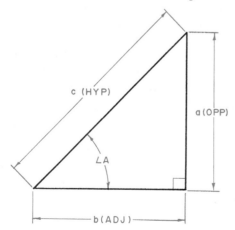

Figure 23-80

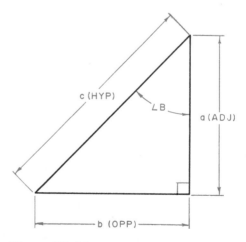

Figure 23-81

FUNCTION	SYMBOL	DEFINITION OF FUNCTION
sine of angle A	sin A	$\sin A = \dfrac{\text{opp side}}{\text{hyp}} = \dfrac{a}{c}$
cosine of angle A	cos A	$\cos A = \dfrac{\text{adj side}}{\text{hyp}} = \dfrac{b}{c}$
tangent of angle A	tan A	$\tan A = \dfrac{\text{opp side}}{\text{adj side}} = \dfrac{a}{b}$
cotangent of angle A	cot A	$\cot A = \dfrac{\text{adj side}}{\text{opp side}} = \dfrac{b}{a}$
secant of angle A	sec A	$\sec A = \dfrac{\text{hyp}}{\text{adj side}} = \dfrac{c}{b}$
cosecant of angle A	csc A	$\csc A = \dfrac{\text{hyp}}{\text{opp side}} = \dfrac{c}{a}$

Figure 23-82

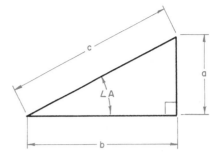

Figure 23-83

showing ∠B as the reference angle, side b is the opposite side, and side a is the adjacent side. It is important to be able to identify the opposite and adjacent sides of right triangles with reference to any angle regardless of the positions of the triangles.

Trigonometric Functions: Ratio Method

There are two methods of defining trigonometric functions: the unity or unit circle method, and the ratio method. Only the ratio method is presented here.

Since a triangle has three sides and a ratio is the comparison of any two sides, there are six different ratios. The names of the ratios are *sine, cosine, tangent, cotangent, secant,* and *cosecant.*

The six trigonometric functions are defined in the table in Figure 23-82 in relation to the triangle shown in Figure 23-83. The reference angle is A, and adjacent side is b, the opposite side is a, and the hypotenuse is c.

To properly use trigonometric functions it is essential to know that the function of an angle depends upon the ratio of the sides and not upon the size of the triangle. Similar triangles are alike in shape but different in size. The functions of similar triangles are the same regardless of the sizes of the triangles, since the sides of similar triangles are proportional. For example, in the similar triangles shown in Figure 23-84, the functions of angle A are the same for the three triangles. The equality of the tangent function is shown. Each of the other five functions has equal values for the three similar triangles.

Note: The symbol △ means triangle.

$$\text{In } \triangle ABC, \tan \angle A = \frac{0.500}{1.000} = 0.500$$

$$\text{In } \triangle ADE, \tan \angle A = \frac{0.800}{1.600} = 0.500$$

$$\text{In } \triangle AFG, \tan \angle A = \frac{1.200}{2.400} = 0.500$$

In a right triangle the reference angle is greater than 0° and less than 90°. Values for the six trigonometric functions from 0° to 90° are available in degree-minute and decimal-degree tables. Scientific elec-

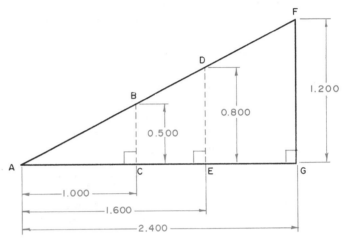

Figure 23-84

tronic calculators provide a quick and convenient determination of trigonometric functions. It is assumed that users of this text will use calculators; therefore, trigonometric function tables are not provided. However, if a calculator is not used, supplementary trigonometric function tables are generally readily available. Conversion between decimal-minutes and decimal-degree angular measure may be required. If necessary, refer to the section Degrees, Minutes, Seconds – Decimal Degree Conversion.

Trigonometry: Basic Calculations of Angles

In order to solve for an unknown angle of a right triangle when neither acute angle is shown, at least two sides must be known. An understanding of the procedures required for solving for unknown angles is essential to the drafter.

Procedure: *To determine an unknown angle when two sides of a right triangle are known*

- Identify two given sides as adjacent, opposite, or hypotenuse in relation to the desired angle.
- Determine the functions that are ratios of the sides identified in relation to the desired angle.

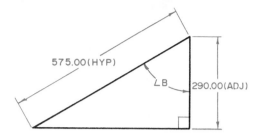

Figure 23-85

Note: Two of the six trigonometric functions are ratios of the two known sides. Either of the two functions can be used. Both produce the same value for the unknown, except for cotangents, secants, and cosecants of angles less than 15° and tangents, secants, and cosecants of angles greater than 75°.

- Choose one of the two functions, substitute the given sides in the ratio, and divide.
- Using either a calculator or a trigonometric function table, determine the angle that corresponds to the quotient obtained.

Example 1: Determine ∠B of the triangle in Figure 23-85. All dimensions are in millimetres.

In relation to ∠B, the 290.000-mm side is the adjacent side and the 575.00-mm side is the hypotenuse. Determine the two functions whose ratios consist of the adjacent side and the hypotenuse:

$$\text{secant } \angle B = \frac{\text{hypotenuse}}{\text{adjacent side}},$$

and

$$\text{cosine } \angle B = \frac{\text{adjacent side}}{\text{hypotenuse}}.$$

Either function can be used. Choosing the cosine function:

$$\cos \angle B = \frac{\text{adj side}}{\text{hyp}}$$
$$\cos \angle B = \frac{290.00}{575.00}$$
$$\cos \angle B = 0.50435$$
$$\angle B = 59.72° \text{ or } 59°43' \text{ Ans}$$

Example 2: Determine ∠1 and ∠2 of the triangle in Figure 23-86. All dimensions are in inches.

Calculate either ∠1 or ∠2. Choose any two of the three given sides. In relation to ∠1, the 3.420-in side is the opposite side. The 5.845-in side is the hypotenuse. Determine the two functions whose ratios consist of the opposite side and the hypotenuse:

$$\angle 1 = \frac{\text{opposite side}}{\text{hypotenuse}},$$

and

$$\text{cosecant } \angle 1 = \frac{\text{hypotenuse}}{\text{opposite side}}.$$

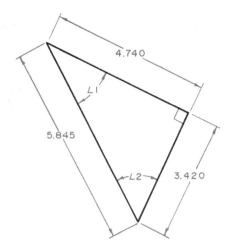

Figure 23-86

Either function can be used. Choosing the sine function:

$$\sin \angle 1 = \frac{\text{opp side}}{\text{hyp}}$$
$$\sin \angle 1 = \frac{3.420}{5.845}$$
$$\sin \angle 1 = 0.58512$$
$$\angle 1 = 35°49' \text{ Ans}$$
$$\angle 2 = 90° - 35°49'$$
$$\angle 2 = 54°11' \text{ Ans}$$

Trigonometry: Basic Calculations of Sides

Procedure: To determine an unknown side when an acute angle and one side of a right triangle are known

- Identify the given side and the unknown side as adjacent, opposite, or hypotenuse in relation to the given angle.
- Determine the trigonometric functions that are ratios of the sides identified in relation to the given angle.
  Note: Two of the six functions will be found as ratios of the two identified sides. Either of the two functions can be used. Both produce the same value for the unknown, except for cotangents, secants, and cosecants of angles less than 15° and tangents, secants, and cosecants of angles greater than 75°. If the unknown side is made the numerator of the ratio, the problem is solved by multiplication. If the unknown side is made the denominator of the ratio, the problem is solved by division.
- Using either a calculator or a trigonometric function table, find the function of the given angle and substitute this value.
- Solve as a proportion for the unknown side.

Example: Determine side of the triangle in Figure 23-87. All dimensions are in inches.

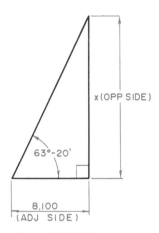

Figure 23-87

In relation to the 63°20' angle, the 8.100-in side is the adjacent side and side x is the opposite side. Determine the two functions whose ratios consist of the adjacent and opposite sides:

$$\text{tangent } 63°20' = \frac{\text{opposite side}}{\text{adjacent side}}$$

and

$$\text{cotangent } 63°20' = \frac{\text{adjacent side}}{\text{opposite side}}$$

Either function can be used. Choosing the tangent function:

$$\tan 63°20' = \frac{\text{opp side}}{\text{adj side}}$$

$$\tan 63°20' = \frac{x}{8.100}$$

$$1.9912 = \frac{x}{8.100}$$

Solve as a proportion.

$$\frac{1.9912}{1} = \frac{x}{8.100}$$
$$x = 1.9912 \,(8.100)$$
$$x = 16.129 \text{ in Ans}$$

Trigonometry: Common Drafting Applications

Method of Solution

The following examples are simple practical applications of right-angle trigonometry, although they may not be given directly in the form of right triangles. To solve the examples it is necessary to project auxiliary lines to produce a right triangle. The unknown, or a dimension required to compute the unknown, is part of the triangle. The auxiliary lines may be projected between given points or from given points. The lines may be projected parallel or perpendicular to center lines, tangents, or other reference lines.

It is important to study carefully the procedures and the use of auxiliary lines as they are applied to the following examples. The same basic method is

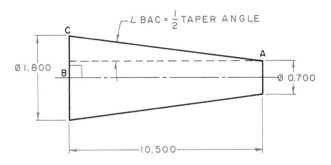

Figure 23-88

used in solving many similar drafting problems. A knowledge of both geometric principles and trigonometric functions and the ability to relate and apply them to specific situations are required in solving many applied problems.

Tapers and Bevels

Example 1: Determine the included taper angle of the shaft in Figure 23-88. All dimensions are in inches.

The problem must be solved by using a figure in the form of a right triangle. Therefore, project line AB from point A parallel to the center line. Right △ABC is formed in which ∠BAC is one-half the included taper angle.

Side AB = 10.500''

Side BC = $\frac{1.800'' - 0.700''}{2}$ = 0.550''

Using sides AB and BC, solve for ∠BAC.

$$\tan \angle BAC = \frac{BC}{AB} = \frac{0.500}{10.500} = 0.05238$$
$$\angle BAC = 3°0'$$

The included taper angle = 2(3°0') = 6°0' Ans

Example 2: Determine diameter x of the part shown in Figure 23-89. All dimensions are in millimetres. Project line DE from point D parallel to the centerline in order to form right △DEF.

Side DE = 21.80 mm − 7.50 mm = 14.30 mm.
∠EDF = 32.5°

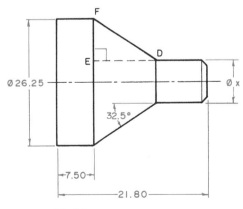

Figure 23-89

Using side DE and ∠EDF, solve for side EF.

$$\tan \angle EDF = \frac{EF}{DE}$$

$$\tan 32.5° = \frac{EF}{14.30}$$

$$EF = 0.63707(14.30)$$

$$EF = 9.11 \text{ mm}$$

Dia $x = 26.25$ mm $- 2(9.11$ mm$) = 8.03$ mm Ans

Isosceles Triangle Applications: Distance Between Holes and V-Slots

The solutions to many practical trigonometry problems are based on recognizing figures as isosceles triangles. In an isosceles triangle an altitude to the base bisects the base and the vertex angle.

Example 1: In Figure 23-90 five holes are equally spaced on a 5.200-in diameter circle. Determine the straight-line distance between centers of two consecutive holes.

Project radii from center O to hole centers A and B. Project a line from A to B. $\angle AOB = \frac{360°}{5} = 72°$.

Since OA = OB, △AOB is isosceles. Project line OC perpendicular to AB from center O. Line OC bisects ∠AOB and side AB. In right △AOC, ∠AOC $= \frac{72°}{2} = 36°$. AO $= \frac{5.200 \text{ in}}{2} = 2.600$ in.

Solve for side AC.

$$\sin \angle AOC = \frac{AC}{AO}$$

$$\sin 36° = \frac{AC}{2.600}$$

$$0.58779 = \frac{AC}{2.600}$$

$$AC = 0.58779(2.600)$$

$$AC = 1.528 \text{ in}$$

AB $= 2(1.528$ in$) = 3.056$ in Ans

Example 2: Determine dimension x of the V-slot in Figure 23-91. All dimensions are in inches. Connect a line between points R and T. Sides RS = TS; therefore, △RST is isosceles. Project line SM from point S perpendicular to RT. Side RT and ∠RST are bisected. In right △RSM,

$$\angle RSM = \frac{62°40'}{2} = 31°20'.$$

$$RM = \frac{3.800 \text{ in}}{2} = 1.900 \text{ in}.$$

Solve for distance MS.

$$\cot \angle RSM = \frac{MS}{RM}$$

$$\cot 31°20' = \frac{MS}{1.900}$$

$$1.6426 = \frac{MS}{1.900}$$

$$MS = 1.6426(1.900)$$

$$MS = 3.121 \text{ in}$$

$$x = MS = 3.121 \text{ in Ans}$$

Tangents to Circles Applications: Angle Cuts and Dovetails

Example: An internal dovetail is shown in Figure 23-92. Two pins or balls are used to check the dovetail for both location and angular accuracy. Calculate check dimension x. All dimensions are in inches.

Project line HO from point H to the pin center O; HO bisects the 72°20' angle. Project a radius from point O to the point of tangency K; ∠HKO is a right angle, since a radius is perpendicular to a tangent at the point of tangency. In right △HOK, ∠KHO $= \frac{72°20'}{2} = 36°10'$, KO $= \frac{1.000 \text{ in}}{2}$

$= 0.500$ in

Solve for side HK.

$$\cot \angle KHO = \frac{HK}{KO}$$

$$\cot 36°10' = \frac{HK}{0.500}$$

$$1.3680 = \frac{HK}{0.500}$$

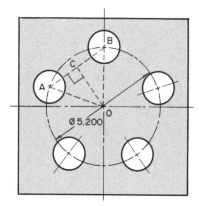

Figure 23-90

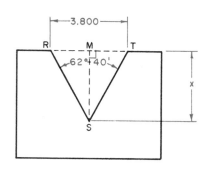

Figure 23-91

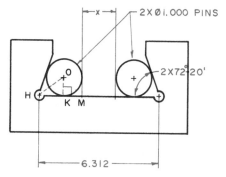

Figure 23-92

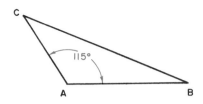

Figure 23-93

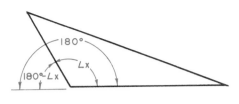

Figure 23-94

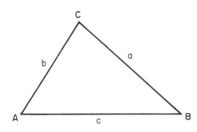

Figure 23-95

HK = 0.500(1.3680)
HK = 0.684 in
KM = pin radius = 0.500 in
HM = HK + KM = 0.684 in + 0.500 in = 1.184. in
x = 6.312 in − 2(HM) = 6.312 in − 2(1.184 in) =
3.944 in Ans

Figure 23-96

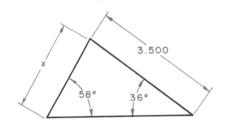

Figure 23-97

Trigonometry: Oblique Triangles — Law of Sines and Law of Cosines

An *oblique triangle* is one that does not contain a right angle. Drafting and design problems often involve oblique triangles. These problems can be reduced to a series of right triangles, but the process can be cumbersome and time consuming. Two formulas, the *law of sines* and the *law of cosines*, can be used to simplify such computations. In order to use either formula, three parts of an oblique triangle must be known; at least one part must be a side.

An oblique triangle can contain an angle greater than 90°, such as oblique triangle ABC in Figure 23-93. Therefore, sine and cosine functions of angles between 90° and 180° must be determined.

Sine Functions of Angles Between 90° and 180°

The sine of an angle greater than 90° and less than 180° equals the sine of the supplement of the angle. If the value of $\angle x$ is between 90° and 180°, as shown in Figure 23-94, $\sin \angle x = \sin(180° − \angle x)$.

Examples:
1. $\sin 115° = \sin(180° − 115°) = \sin 65° = 0.90631$ Ans
2. $\sin 90°42' = \sin(180° − 90°42') = \sin 89°18' = 0.99992$ Ans
3. $\sin 176.25° = \sin(180° − 176.25°) = \sin 3.75° = 0.06540$ Ans

Law of Sines

The law of sines states that, in any triangle, the sides are proportional to the sines of the opposite angles. That is (see Figure 23-95)

$$\frac{a}{\sin A} = \frac{b}{\sin B} = \frac{c}{\sin C}$$

The law of sines is used to solve the following kinds of problems:

- Problems where any two angles and any one side of an oblique triangle are known.
- Problems where any two sides and an angle opposite one of the given sides of an oblique triangle are known.

Solving Oblique Triangle Problems Given Two Angles and a Side

Example 1: Given two angles and a side, determine side x of the oblique triangle in Figure 23-96. (All dimensions are in inches.)

Set up a proportion and solve for x.

$$\frac{x}{\sin 36°} = \frac{3.500}{\sin 58°}$$

$$\frac{x}{0.58779} = \frac{3.500}{0.84805}$$

$$0.84805x = 3.500(0.58779)$$

$$x = \frac{3.500(0.58779)}{0.84805}$$

$$x = 2.426 \text{ in Ans}$$

Example 2: Given two angles and a side of the oblique triangle formed by the intersection of hole centerlines shown in Figure 23-97. All dimensions

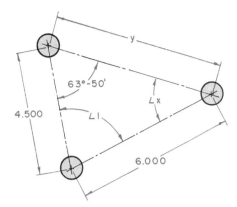

Figure 23-98

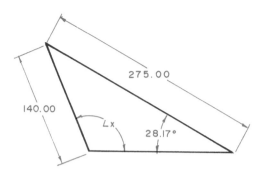

Figure 23-99

are in millimetres.
   a. Determine ∠A.
   b. Determine side *a*.
a. Determine ∠A.

   ∠A = 180° - (37.3° + 24.5°) = 118.2° Ans

   b. Determine side *a*. Set up a proportion and solve for side *a*.

$$\frac{a}{\sin 118.2°} = \frac{108.60}{\sin 37.3°}$$
$$\sin 118.2° = \sin(180° - 118.2°) = \sin 61.8° = 0.88130$$
$$\sin 37.3° = 0.60599$$
$$\frac{a}{0.88130} = \frac{108.60}{0.60599}$$
$$a = 157.94 \text{ mm Ans}$$

Solving Oblique Triangle Problems Given Two Sides and an Angle Opposite One of the Given Sides

Example 1: Given two sides and an opposite angle of the oblique triangle formed by the intersection of hole center lines shown in Figure 23-98. (All dimensions are in inches.)
   a. Determine ∠x.
   b. Determine side *y*.
a. Determine ∠x.

$$\frac{4.500}{\sin ∠x} = \frac{6.000}{\sin 63°50'}$$
$$\frac{4.500}{\sin ∠x} = \frac{6.000}{0.89752}$$
$$6.000(\sin ∠x) = 4.500(0.89752)$$
$$\sin ∠x = \frac{4.500(0.89752)}{6.000}$$
$$\sin ∠x = 0.67314$$
$$∠x = 42°18' \text{ Ans}$$

   b. Determine side *y*.

$$∠1 = 180° - (63°50' + ∠x) =$$
$$180° - (63°50' + 42°18') = 180° - 106°8' = 73°52'$$
$$\frac{6.000}{\sin 63°50'} = \frac{y}{\sin 73°52'}$$

$$\frac{6.000}{0.89752} = \frac{y}{0.96062}$$
$$0.89752y = 6.000(0.96054)$$
$$y = \frac{6.000(0.96054)}{0.89752}$$
$$y = 6.422 \text{ in Ans}$$

Example 2: Given two sides and an opposite angle, determine ∠x of the oblique triangle shown in Figure 23-99. (All dimensions are in millimetres.)

$$\frac{140.00}{\sin 28.17°} = \frac{275.00}{\sin ∠x}$$
$$\frac{140.00}{0.47209} = \frac{275.00}{\sin ∠x}$$
$$\sin ∠x = 0.92731$$

The angle that corresponds to the sine function 0.92731 is 68°. Because ∠x is greater than 90°, ∠x = the supplement of 68°. ∠x = 180° − 68° = 112°. Ans

Cosine Functions of Angles Between 90° and 180°

The cosine of an angle greater than 90° and less than 180° equals the negative cosine of the supplement of the angle. If the value of ∠x is between 90° and 180°, cos ∠x = −cos(180° − ∠x). A negative function of an angle does *not* mean that the angle is negative; it is a negative function of a positive angle. For example, − cos 65° does *not* mean cos (−65°).

Examples:

   1. cos 115° = −cos(180° − 115°) = −cos 65° = −0.42262.
   Note: Determine the cosine of 65° and prefix a negative sign:
      cos 65° = 0.42262; −cos 65° = 0.42262.
   2. cos 90°42' = −cos(180° − 90°42') = −cos 89°18' = -0.01222.
   3. cos 176.25° = −cos(180° - 176.25°) = −cos 3.75° = −0.99786.

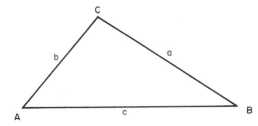

Figure 23-100

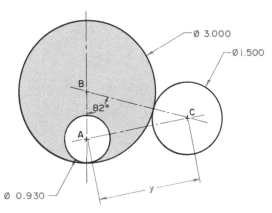

Figure 23-101

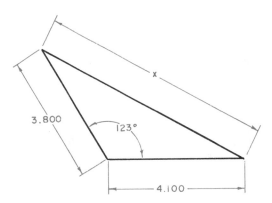

Figure 23-102

Law of Cosines

The law of cosines states that, in any triangle, the square of any side is equal to the sum of the squares of the other two sides minus twice the product of these two sides multiplied by the cosine of their included angle. That is (see Figure 23-100)

$$a^2 = b^2 + c^2 - 2bc(\cos A)$$
$$b^2 = a^2 + c^2 - 2ac(\cos B)$$
$$c^2 = a^2 + b^2 - 2ab(\cos C)$$

The equations can be rearranged in this form:

$$\cos A = \frac{b^2 + c^2 - a^2}{2bc}$$
$$\cos B = \frac{a^2 + c^2 - b^2}{2ac}$$
$$\cos C = \frac{a^2 + b^2 - c^2}{2ab}$$

The law of cosines is used to solve the following kinds of problems:

- Problems where two sides and the included angle of an oblique triangle are known.
- Problems where three sides of an oblique triangle are known.

Solving Oblique Triangle Problems Given Two Sides and the Included Angle

Example 1: Internally and externally tangent circles are shown in Figure 23-101. Determine dimension y, the distance between hole centers A and C.
All dimensions are in inches.
In oblique $\triangle ABC$:

$$AB = \frac{3.000}{2} - \frac{0.930}{2} = 1.035$$

$$BC = \frac{3.000}{2} + \frac{1.500}{2} = 2.250$$

The 82° angle is the included angle between AB and BC. Substitute values in their appropriate places in the formula and solve for dimension y.

$$y^2 = 1.035^2 + 2.250^2 - 2(1.035)(2.250)(\cos 82°)$$
$$y^2 = 1.0712 + 5.0625 - 2(1.035)(2.250)(0.13917)$$
$$y^2 = 6.1337 - 0.6482$$
$$y^2 = 5.4855$$
$$y = 2.342 \text{ in Ans}$$

Example 2: Given two sides and the included angle, determine side x of the oblique triangle in Figure 23-102. All dimensions are in inches. Substitute values to solve for x.

$$x^2 = 3.800^2 + 4.100^2 - 2(3.800)(4.100)(\cos 123°)$$

Since the given angle is greater than 90°, the cosine of the angle is equal to the negative cosine of its supplement. Therefore,

$$\cos 123° = -\cos(180° - 123°) = -\cos 57° = -0.54464$$

This negative value must be used in computing side x.

$$x^2 = 3.800^2 + 4.100^2 - 2(3.800)(4.100)(-0.54464)$$
$$x^2 = 31.250 - (-16.971)$$

Recall that subtracting a negative value is the same as adding a positive value.

$$x^2 = 31.250 + 16.971$$
$$x^2 = 48.221$$
$$x = 6.944 \text{ in Ans}$$

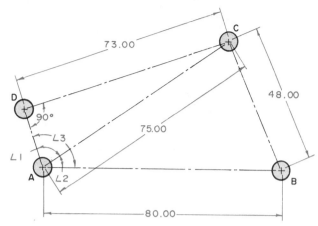

Figure 23-103

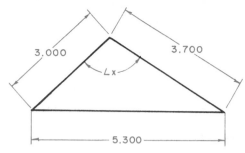

Figure 23-104

Solving Oblique Triangle Problems Given Three Sides

Example 1: Four pins are located on a fixture at A, B, C, and D as shown in Figure 23-103. Compute ∠3.

All dimensions are in millimetres.

∠3 = ∠1 + ∠2. Compute ∠1 in the right △ADC. Compute ∠2 in oblique △ABC. In right △ADC, AC = 75.00, DC = 73.00.

$$\sin \angle 1 = \frac{DC}{AC}$$

$$\sin \angle 1 = \frac{73.00}{75.00} = 0.97333$$

$$\angle 1 = 76.73°$$

In oblique △ABC, AC = 75.00, AB = 80.00, BC = 48.00.

Substitute values in their appropriate places in the formula.

$$\cos \angle 2 = \frac{75.00^2 + 80.00^2 - 48.00^2}{2(75.00)(80.00)}$$

$$\cos \angle 2 = \frac{5625 + 6400 - 2304}{12000} = 0.81008$$

$$\angle 2 = 35.90°$$
$$\angle 3 = \angle 1 + \angle 2 = 76.73° + 35.90° = 112.63° \text{ Ans}$$

Example 2: Given three sides, determine ∠x of the oblique triangle in Figure 23-104. All dimensions are in inches.

Substitute values in the formula.

$$\cos \angle x = \frac{3.000^2 + 3.700^2 - 5.300^2}{2(3.000)(3.700)}$$

$$\cos \angle x = \frac{9.000 + 13.690 - 28.090}{22.200}$$

$$\cos \angle x = \frac{22.690 - 28.090}{22.200}$$

$$\cos \angle x = \frac{-5.400}{22.200}$$

$$\cos \angle x = -0.24324$$

Since cos ∠x is a negative value, ∠x is equal to the supplement of the angle found. The angle corresponding to the cosine function 0.24324 is 75°55′. Therefore, the angle whose cosine function is −0.24324 is equal to 180° − 75°55′ = 104°05′. Therefore, ∠x = 104°05′ Ans

Review

Solutions to the following problems require the application of mathematical operations, procedures, and principles presented in Chapter 23.

1. Round 4.02463 inches to three decimal places.

2. Round 59.6841 mm to two decimal places.

3. Using the decimal equivalwent table on page 650,
   a. find the decimal equivalent of 57/64 in;
   b. find the fractional equivalent of 0.4375 in;
   c. find the nearer fractional equivalent of 0.706 in.

4. Express 0.983 in in millimetres. Round the answer to two decimal places.

5. Express 70.15 mm in inches. Round the answer to three decimal places.

Problems 6-11 involve working with formulas that have been obtained from technical handbooks. For each problem substitute the given numerical values for letters in the formula and solve for the unknown. Where necessary, round answers to two decimal places.

6. A = π (ab - cd). Solve for A when π = 3.1416, a = 5.000 in, b = 3.000 in, c = 4.000 in, and d = 2.000 in.

7. $A = \frac{\pi (R^2 - r^2)}{2}$. Solve for A when π = 3.1416, R = 7.000 in, and r = 3.000 in.

8. $V = \frac{(2a + c)bh}{6}$. Solve for V when a = 9.000 in, b = 7.000 in. c = 11.000 in, and h = 8.000 in.

9. All dimensions are in millimetres.
   a. Find the length of this arc (*l*).
   $$l = \frac{3.1416\, R\, \alpha}{180°}$$
   b. Find the area of this sector (A).
   $$A = \frac{1}{2}\, Rl$$

10. All dimensions are in millimetres.
    a. Find the radius of this circle.
    $$r = \frac{c^2 + 4h^2}{8h}$$
    b. Find the length of the arc (*l*).
    $$l = 0.0175\, ra$$

11. All dimensions are in inches. Find the belt length on the pulleys.
    $$\text{Length of belt} = 2C + \frac{11D + 11d}{7} + \frac{(D - d)^2}{4C}$$

12. Of two gears that mesh, the one which has the greater number of teeth is called the gear, and the one which has the fewer teeth is called the pinion. The proportion that expresses the relationship of gear teeth and speeds is

    $$\frac{\text{Number of teeth on gear}}{\text{Number of teeth on pinion}} = \frac{\text{Speed of pinion}}{\text{Speed of gear}}$$

    For each problem, a - e, determine the unknown value *x*. Where necessary, round answer to one decimal place.

	Number of Teeth on Gear	Number of Teeth on Pinion	Speed of Gear (rpm)	Speed of Pinion (rpm)
a	48	20	120	*x*
b	32	24	*x*	210
c	35	*x*	160	200
d	*x*	15	150	250
e	54	28	80	*x*

13. Add the angles in each of these problems.
    a. $23°43' + 71°12' + 19°27'$ b. $9°33'46'' + 37°12'19''$

14. Subtract the angles in each of these problems.
    a. $63°23' - 32°58'$ b. $54°32'13'' - 19°13'42''$

15. Multiply the angles in each of these problems.
    a. $2(43°43')$ b. $5(22°10'13'')$

16. Divide the angles in each of these problems.
    a. $105°20' \div 4$ b. $110°51'5'' \div 5$

17. In the figure,
    $\angle 1 = \angle 2 = \angle 3 = \angle 4 = \angle 5 = 53°41'$
    Determine $\angle 6$.

18. In the figure,
    $\angle 1 = \angle 2 = \angle 3 = \angle 4$
    Determine $\angle 1$.

19. Express each of the following decimal-degrees as degrees and minutes.
    a. $76.95°$ b. $117.70°$

20. Express the following decimal-degrees as degrees, minutes, and seconds.
    a $7.9250°$ b. $37.9365°$

21. Express each of the following degrees and minutes as decimal-degrees. Round the answer to two decimal places.
    a. $93°18'$ b. $79°59'$

22. Express each of the following degrees, minutes, and seconds as decimal-degrees. Round the answer to four decimal places.
    a. $53°10'45''$ b. $176°27'18''$

23. Centerlines AB ∥ CD and EF ∥ GH. Determine the values of $\angle 1 - \angle 2$.

24. Determine dimension *y*. All dimensions are in millimetres.

25. Determine dimension *x*. All dimensions are in inches.

26. Determine $\angle 1$.

27. Point A is a tangent point. All dimensions are in inches. Determine dimension *x*.

28. Determine dimension *y*. All dimensions are in millimetres.

29. Determine dimension *x*. All dimensions are in inches.

30. AC is a diameter. Determine $\angle 1$.

31. Determine the included taper angle *x*. All dimensions are in inches.

32. Determine diameter *x*. All dimensions are in inches.

33. Determine center distance *y*. All dimensions are in millimetres.

34. Determine depth of cut *x*. All dimension are in inches.

35. Determine $\angle x$. All dimension are in inches.

36. Determine gage dimension *x*. All dimensions are in inches.

37. Determine dimension *y*. All dimensions are in millimetres.

38. Determine $\angle x$. All dimensions are in millimetres.

39. Determine dimension *y*. All dimensions are in inches.

40. Determine dimension *x*. All dimensions are in inches.

41. Determine $\angle x$. All dimensions are in inches.

42. Determine dimension *y*. All dimensions are in millimetres.

Glossary

Acme   Screw thread form

Acute Angle   An angle less than 90°

Addendum   Radial distance from pitch circle to top of gear tooth

Allen Screw   Special set screw or cap screw with hexagon socket in head

Allowance   Minimum clearance between mating parts

Alloy   Two or more metals in combination, usually a fine metal with a baser metal

Alphanumeric   Refers to the totality of characters that are either alphabetic or numeric

Aluminum   A lightweight but relatively strong metal; often alloyed with copper to increase hardness and strength

Ammonia   A colorless gas used in the development process of diazo and sepia prints

Angle Iron   A structural shape whose section is a right angle

Anneal   To soften metals by heating to remove internal stresses caused by rolling and forging

Anodizing   The process of protecting aluminum by oxidizing in an acid bath using a direct current

Application   A definable set of drafting tasks to be accomplished in a given drafting area; may be accomplished partly through manual procedures and partly through computerized procedures

Arc   A part of a circle

Arc-weld   To weld by electric arc; the work is usually the positive terminal

Assembly Drawing   A drawing showing the working relationship of the various parts of a machine or structure as they fit together

Auxiliary View   An additional view of an object, usually of a surface inclined to the principal surfaces of the object to provide a true size and shape view

Axes   Plural of axis

Axis   An imaginary line around which parts rotate or are regularly arranged

Babbitt   A soft alloy for bearings, mostly of tin with small amounts of copper and antimony

Backlash   Lost motion between moving parts, such as threaded shaft and nut or the teeth of meshing gears

Baseline Dimensioning   A system of dimensioning where as many features of a part as are functionally practical are located from a common set of datums

Basic Dimension   A theoretically exact value used to describe the size, shape or location of a feature

Basic Size   The size from which the limits of size are derived by the application of allowances and tolerances

Bearing   A supporting member for a rotating shaft

Bend Allowance   The amount of sheet metal required to make a bend over a specific radius

Bevel   An inclined edge; not a right angle to joining surface

Bisect   To divide into two equal parts

Blanking   A stamping operation in which a press uses a die to cut blanks from flat sheets or strips of metal

Blueprint   A copy of a drawing

Bolt Circle   A circular center line on a drawing, containing the centers of holes about a common center

Bore   To enlarge a hole with a boring bar or tool in a lathe, drill press or boring mill

Boss   A cylindrical projection on a casting or a forging

Brass   An alloy of copper and zinc

Braze   To join with hard solder of brass or zinc

Broach   A tool for removing metal by pulling or pushing it across the work; the most common use is producing irregular hole shapes such as squares, hexagons, ovals or splines

Bronze   An alloy of eight or nine parts of copper and one part of tin

Buff   To finish or polish on a buffing wheel composed of fabric with abrasive powders

Burr   The ragged edge or ridge left on metal after a cutting operation

Burnish   To finish or polish by pressure upon a smooth rolling or sliding tool

Bushing   A metal lining which acts as a bearing between rotating parts such as a shaft and pulley; also used on jigs to guide cutting tool

CAD (Computer-aided drafting) The use of computers and peripheral devices to aid in the documentation for design projects

Callout   A note on the drawing giving a dimension, specification or machine process

Calipers   Instrument for measuring diameters

Cam A rotating shape for changing circular motion to reciprocating motion

Carburize To heat a low-carbon steel to approximately 2000°F in contact with material which adds carbon to the surface of the steel

Case harden To harden the outer surface of a carburized steel by heating and then quenching

Center Drill A special drill to produce bearing holes in the ends of a workpiece to be mounted between centers

Central Processing Unit (CPU) A unit of a computer that includes circuits controlling the interpretation and execution of instructions

Chain Dimensioning Successive dimensions that extend from one feature to another, rather than each originating at a datum

Chamfer A bevel on an external edge or corner, usually at 45°

Character A letter, digit or other symbol that is used as part of the organization, control, or representation of data

Chase To cut threads with an external cutting tool

Chill To harden the outer surface of cast iron by quick cooling, as in a metal mold

Chuck A mechanism for holding a rotating tool or workpiece

Circular Pitch The length of the arc along the pitch circle between the center of one gear tooth to the center of the next

Clockwise Rotation in the same direction as hands of a clock

Coin To form a part in one stamping operation

Cold Rolled Steel Bessemer steel containing .12% to .20% carbon that has been rolled while cold to produce a smooth, quite accurate stock

Collar A round flange or ring fitted on a shaft to prevent sliding

Concentric Having a common center as circles or diameters

Coordinate An ordered set of data values, either absolute or relative, that specifies a location

Coordinate Dimensioning A type of rectangular datum dimensioning in which all dimensions are measured from two or three mutually perpendicular datum planes; all dimensions originate at a datum and include regular extension and dimension lines and arrowheads

Cope The upper portion of a flask used in molding

Core To form a hollow portion in a casting by using a dry-sand core or a green-sand core in a mold

Counterbore The enlargement of the end of a hole to a specified diameter and depth

Countersink The chamfered end of a hole to receive a flat head screw

CPU Central processing unit

Cursor (1) On CRT, a movable marker that is visible on the viewing surface and is used to indicate a position at which an action is to take place or the position on which the next device operation would normally be directed. (2) On digitizers, a movable reference, usually optical cross hairs used by an operator to indicate manually the position of a reference point or line where an action is to take place.

Data A representation of facts, concepts or instructions in formalized manner suitable for communication, interpretation or processing by human or automatic means

Datum Points, lines, planes, cylinders and the like, assumed to be exact for purposes of computation from which the location or geometric relationship (form) of features of a part may be established

Dedendum The radial distance between the pitch circle and the bottom of the tooth

Delineation Pictorial representation: a chart, a diagram or a sketch

Design Size The size of a feature after an allowance for clearance has been applied and tolerances have been assigned

Detail Drawing A drawing of a single part that provides all the information necessary in the production of that part

Development Drawing of the surface of an object unfolded on a plane

Deviation The variance from a specified dimension or design requirement

Diagram A figure or drawing which is marked out by lines; a chart or outline

Diameter The length of a straight line passing through the center of a circle and terminating at the circumference on each end

Diazo Material that is either a film or paper sensitized by means of azo dyes used for photocopying

Die Hardened metal piece shaped to cut or form a required shape in a sheet or metal by pressing it against a mating die

Die Casting Process of forcing molten metal under pressure into metal dies or molds, producing a very accurate and smooth casting

Die Stamping Process of cutting or forming a piece of sheet metal with a die

Digitize To use numeric characters to express or represent data

Digitizer A device for converting positional information into digital signals; typically, a drawing or other graphic is placed on the measuring surface of the digitizer and traced by the operator using a cursor

Dimension Measurements given on a drawing such as size and location

Display A visual presentation of data on an output device

Dowel A cylindrical pin, commonly used to prevent sliding between two contacting flat surfaces

Draft The tapered shape of the parts of a pattern to permit it to be easily withdrawn from the sand or withdrawn from the dies

Drag   Lower portion of a flask used in molding

Draw   To temper steel

Drill   To cut a cylindrical hole with a drill; a blind hole does not go through the piece.

Drill Press   A machine for drilling and other hole-forming operations

Drop Forge   To form a piece while hot between dies in a drop hammer or with great pressure

Eccentric   Not having the same center; off center

Extrusion   Metal which has been shaped by forcing it in its hot or cold state through dies of the desired shape

Face   To finish a surface at right angles, or nearly so, to the center line of rotation on a lathe

FAO   Finish all over

Fastener   A mechanical device for holding two or more bodies in definite positions with respect to each other

Feature   A portion of a part, such as a diameter, hole, keyway or flat surface

Ferrous   Having iron as a metal's base material

File   To finish or smooth with a file

Fillet   An interior rounded intersection between two surfaces

Fin   A thin extrusion of metal at the intersection of dies or sand molds

Finish   General finish requirements such as paint, chemical or electroplating, rather than surface texture or roughness (see surface texture)

Fit   The clearance or interference between two mating parts

Fixture   A special device for holding the work in a machine tool

Flange   A relatively thin rim around a piece

Flask   A box made of two or more parts for holding the sand in sand molding

Flatbed Plotter   A plotter that draws an image on a data medium such as paper or film mounted on a table

Flat Pattern   A layout showing true dimensions of a part before bending; may be actual size pattern on polyester film for shop use

Flute   Groove, as on twist drills, reamers, and taps

Forge   To force metal while it is hot to take on a desired shape by hammering or pressing

Form Tolerancing   The permitted variation of a feature from the perfect form indicated on the drawing

Fusion Weld   The intimate mixing of molten metals

Gage   The thickness of sheet metal by number

Galvanize   To cover a surface with a thin layer of molten alloy, composed mainly of zinc

Gasket   A thin piece of rubber, metal or some other material, placed between surfaces to make a tight joint.

Gate   The opening in a sand mold at the bottom of the sprue through which the molten metal passes to enter the cavity or mold

Geometric Dimensioning and Tolerancing   A means of dimensioning and tolerancing a drawing with respect to the actual function or relationship or part features which can be most economically produced; it includes positional and form dimensioning and tolerancing

Grind   To remove metal by means of an abrasive wheel, often made of carborundum, used where accuracy is required

Gusset   A small plate used in reinforcing assemblies

Hard Copy   A copy of output in a visually readable form (e.g., printed reports, listings, documents, summaries, and drawings)

Harden   To heat steel above a critical temperature and then quench in water, oil or air

Hardness Test   Techniques used to measure the degree of hardness of heat-treated materials

Heat Treat   To change the properties of metals by heating and then cooling

Hexagon   A polygon having six angles and six sides

Hone   A method of finishing a hole or other surface to a precise tolerance by using a spring-loaded abrasive block and rotary motion

Horizontal   Parallel to the horizon

Inclined   A line or plane at an angle to a horizontal line or plane

Incremental System   A system of numerically controlled machining that always refers to the preceding point when making the next movement; also known as continuous path or contouring method of NC machining

Input   The transfer of information into a computer or machine control unit (see "output")

Interchangeable   Refers to a part made to limit dimensions so that it will fit any mating part similarly manufactured

Intermediate   A translucent reproduction made on vellum, cloth or film from an original drawing to serve in place of the original for making other prints

Involute   A spiral curve generated by a point on a chord as it unwinds from a circle or a polygon

Isometric Drawing   A pictorial drawing of an object so positioned that all three axes make equal angles with the picture plane, and measurements on all three axes are made to the same scale

Jig   A device for guiding a tool in cutting a piece

Journal   Portion of a rotating shaft supported by a bearing

Kerf   Groove or cut made by a saw

Key   A small piece of metal sunk partly into both shaft and hub to prevent rotation of mating parts

Keyseat A slot in a shaft to hold a key

Keyslot The slot machined in a shaft for Woodruff-type keys

Keyway A slot in a hub or portion surrounding a shaft to receive a key

Lathe A machine used to shape materials by rotating against a tool

Light Pen A hand-held data-entry device used only with refresh displays. It consists of an optical lens and photocell, with associative circuitry mounted in a wand. Most light pens have a switch allowing the pen to be sensitive to light from the screen.

Limits The extreme permissible dimensions of a part resulting from the application of a tolerance

Lug An irregular projection of metal, but not round as in the case of a boss; usually with a hole in it for a bolt or screw

Magnaflux A nondestructive inspection technique that makes use of a magnetic field and magnetic particles to locate internal flaws in ferrous metal parts

Main Storage The general-purpose storage of a computer; usually, main storage can be accessed directly by the operating registers

Malleable Casting A casting that has been made less brittle and tougher by annealing

Maximum Material Condition (MMC) When a feature contains the maximum amount of material; that is, minimum hole diameter and maximum shaft diameter

Menu A list of options which are displayed on the CRT or a plastic or paper overlay

Microcomputer A computer that is constructed using a microprocessor as a basic element of the CPU; all electronic components are arranged on one printed circuit board

Mill To remove material by means of a rotating cutter on a milling machine

Mold The mass of sand or other material that forms the cavity into which molten metal is poured

Neck To cut a groove called a neck around a cylindrical piece

Nonferrous A description of metals not derived from an iron base or an iron alloy base; nonferrous metals include aluminum, magnesium, and copper, among others

Normalize To heat steel above its critical temperature, and then cooling it in air

Normalizing A process in which ferrous alloys are heated and then cooled in still air to room temperatures to restore the uniform grain structure free of strains caused by cold working or welding

Numerical Control A system of controlling a machine or tool by means of numeric codes that direct commands to control devices attached or built into the machine or tool

Oblique Drawing A pictorial drawing of an object so drawn that one of its principal faces is parallel to the plane of projection, and is projected in its true size and shape; the third set of edges is oblique to the plane of projection at some convenient angle

Obtuse Angle An angle larger than 90°

Octagon A polygon having eight angles and eight sides

Ordinate The Y coordinate of a point; i.e., its vertical distance from the X axis measured parallel to the Y axis; the vertical axis of a graph or chart

Orthographic Projection A projection on a picture plane formed by perpendicular projectors from the object to the picture plane; third-angle projection is used in the United States and first-angle projection is used in most countries outside the United States

Output Data that have been processed from an internal storage to an external storage; opposite of "input"

Parallel Having the same direction, such as two lines which, if extended, would never meet

Pattern A model, usually of wood, used in forming a mold for a casting

Peen To hammer into shape with a ballpeen hammer

Pentagon A polygon having five angles and five sides

Perpendicular Lines or planes at a right angle to a given line or plane

Perspective Drawing A pictorial drawing in which receding lines converge at vanishing points on the horizon; the most natural of all pictorial drawings.

Pickle To clean forgings or castings in dilute sulphuric acid

Pilot A piece that guides a tool or machine part

Pilot Hole A small hole used to guide a cutting tool for making a larger hole

Pinion The smaller of two mating gear

Pitch The distance from a point on one thread to a corresponding point on the next thread; the slope of a surface

Pitch Circle An imaginary circle corresponding to the circumference of the friction gear from which the spur gear is derived

Plane To remove material by means of a planer

Plate To coat a metal piece with another metal, such as chrome or nickel, by electrochemical methods

Polish To produce a highly finished or polished surface by friction, using a very fine abrasive

Polygon A plane geometric figure with three or more sides

Positional Tolerancing The permitted variation of a feature from the exact or true position indicated on the drawing

Precision The quality or state of being precise or accurate; mechanical exactness

Printer A device that prints the output of a computer

Prism A solid whose bases or ends are any congruent and parallel polygons, and whose sides are parallelograms

Prismatic Pertaining to or like a prism

Processor The computer in a CAD system

Program A set of step-by-step instructions telling the computer to solve a problem with the information input to it or contained in memory

Punch To cut an opening of a desired shape with a rigid tool having the same shape

Quench To immerse a heated piece of metal in water or oil in order to harden it

Rack A flat bar with gear teeth in a straight line to engage with teeth in a gear

Raster The coordinate grid dividing the display area of a graphics display

Reference Dimension Used only for information purposes; does not govern production or inspection operations

Refresh Display A CRT device that requires the refreshing of its screen presentation at a high rate so that the image will not fade or flicker

Regardless of Feature Size (RFS) The condition where tolerance of position or form must be met irrespective of where the feature lies within its size tolerance

Relief An offset of surfaces to provide clearance

Rendering Finishing a drawing to give it a realistic appearance; a representation

Resistance Welding The process of welding metals by using the resistance of the metals to the flow of electricity to produce the heat for fusion of the metals

Rib A relatively thin flat member acting as a brace or support

Rivet To connect with rivets or to clench over the end of a pin by hammering

Rotate To revolve through an angle relative to an origin

Round An exterior rounded intersection of two surfaces

Sandblast To blow sand at high velocity with compressed air against castings or forgings to clean them

Scrape To remove metal by scraping with a hand scraper, usually to fit a bearing

Sectional View A view of an object obtained by the imaginary cutting away of the front portion of the object to show the interior detail

Shape To remove metal from a piece with a shaper

Shear To cut metal by means of shearing with two blades in sliding contact

Shim A thin piece of metal or other material used as a spacer in adjusting two parts

Software A set of programs, procedures, rules, and possibly associated documentation concerned with the operation of a CAD system

Solder To join with solder, usually composed of lead and tin

Spin To form a rotating piece of sheet metal into a desired shape by pressing it with a smooth tool against a rotating form

Spline A keyway, usually one of a series cut around a shaft or hole

Spotface To produce a round spot or bearing surface around a hole, usually with a spotfacer; similar to a counterbore

Spot Weld A resistance-type weld that joins pieces of metal by welding separate spots rather than a continuous weld

Sprue A hole in the sand leading to the gate which leads to the mold, through which the metal enters

Stress Relieving To heat a metal part to a suitable temperature and holding that temperature for a determined time, then gradually cooling it in air; this treatment reduces the internal stresses induced by casting, quenching, machining, cold working or welding

Stretchout A flat pattern development for use in laying out, cutting and folding lines on flat stock, such as paper or sheet metal, to be formed into an object

Stylus A penlike device that provides input or output of coordinate data usually for the purpose of indicating where the next entered character will be displayed.

Surface Texture The roughness, waviness, lay, and flaws of a surface

Swage To hammer metal into shape while it is held over a swage or die that fits in a hole in the swage block or anvil

Sweat To fasten metal together by the use of solder between the pieces and by the application of heat and pressure

Symbol A letter, character or schematic design representing a unit or component

Tabular Dimension A type of rectangular datum dimensioning in which dimensions from mutually perpendicular datum planes are listed in a table on the drawing instead of on the pictorial portion

Tap A tool used to produce internal threads by hand or machine

Taper Pin A small tapered pin for fastening, usually to prevent a collar or hub from rotating on a shaft

Taper Reamer Produces accurately tapered holes, as for a taper pin

Temper To reheat hardened steel to bring it to a desired degree of hardness

Template A guide or pattern used to mark out the work

Tensile Strength The maximum load (pull) a piece supports without breaking or failing

682

Tin A silvery metal used in alloys and for coating other metals with tin plate

Tolerance Total amount of variation permitted in limit dimension of a part

Torque The rotational or twisting force in a turning shaft

Trammel An instrument consisting of a straightedge with two adjustable fixed points for drawing curves and ellipses

Translucent A quality of material that passes light but diffuses it so that objects are not identifiable

True Position The basic or theoretically exact position of a feature

Truncated Having the apex, vertex or end cut off by a plane

Turn To produce on a lathe a cylindrical surface parallel to the center line

Typical This term, when associated with any dimension or feature, means the dimension or feature applies to the locations that appear to be identical in size and shape

Uniform Having the same form or character; unvarying

Upset To form a head or enlarged end on a bar by pressure or by hammering between dies

User A person who operates a CAD system

Vector A directed line segment which, in computer graphics, is always defined by its two end points

Vernier Scale A small movable scale, attached to a larger fixed scale, for obtaining fractional subdivisions of the fixed scale

Vertex The highest point of something; the top; the summit; plural: vertices

Vertical Perpendicular to the horizon

Web A thin flat part joining larger parts; also known as a rib

Welding Uniting metal pieces by pressure or fusion-welding processes

Woodruff Key A semicircular flat key

Working Drawings A set of drawings which provides details for the production of each part, and information for the correct assembly of the finished product

Workstation The assigned location where a worker performs his or her job (i.e., the keyboard and the system display)

Wrought Iron Iron of low carbon content; useful because of its toughness, ductility, and malleability

Appendix
Contents

Table		Page
1	Inches to Millimetres/Millimetres to Inches	685
2	Inch/Metric - Equivalents	686
3	Metric Equivalents	687
4	Multipliers for Drafters	688
5	Circumferences and Areas of Circles	690
6	Trigonometric Formulas	691
7	Right Triangle Formulas	692
8	Oblique-Angled Triangle Formulas	693
9	Running and Sliding Fits	694
10	Locational Clearance Fits	695
11	Locational Transition Fits	696
12	Locational Interference Fits	697
13	Force and Shrink Fits	698
14	Unified Standard Screw Thread Series	699
15	Drill and Tap Sizes	700
16	Drilled Hole Tolerance	701
17	Inch-Metric Thread Comparison	704
18	I.S.O. Basic Metric Thread Information	705
19	Hex Bolts	706
20	Finished Hex Bolts	707
21	Square Bolts	708
22	Slotted Flat Head Cap Screws	709
23	Slotted Round Head Cap Screws	710
24	Fillister Head Cap Screws	711
25	Cross Recessed Flat Head Machine Screws	712

Table		Page
26	Slotted Oval Head Machine Screws	713
27	Slotted Pan Head Machine Screws	714
28	Slotted Headless Set Screws	715
29	Square Head Set Screws	716
30	Plain Washers	718
31	Helical Spring Lock Washers	719
32	External Tooth Lock Washers	720
33	Internal Tooth Lock Washers	721
34	Hex Nuts and Hex Jam Nuts	722
35	Square Nuts	723
36	Keyway Size/Key Size Charts	724
37	Woodruff Keys	725
38	Woodruff Keyseats	727
39	Woodruff Key Sizes for Shaft Diameters	728
40	Parallel and Taper Keys	729
41	Gib Head Keys	730
42	Sheet Metal and Wire Gage Designation	731
43	Bend Allowance for 90° Bends	732
44	Bend Allowance for 1° Bends	733
45	Spur/Pinion Gear Tooth Parts	734
46	Surface Texture Roughness	735
47	Properties, Grade Numbers, and Usage of Steel Alloys	736
48	American Standards	737

INCHES TO MILLIMETRES

in.	mm	in.	mm	in.	mm	in.	mm
1	25.4	26	660.4	51	1295.4	76	1930.4
2	50.8	27	685.8	52	1320.8	77	1955.8
3	76.2	28	711.2	53	1346.2	78	1981.2
4	101.6	29	736.6	54	1371.6	79	2006.6
5	127.0	30	762.0	55	1397.0	80	2032.0
6	152.4	31	787.4	56	1422.4	81	2057.4
7	177.8	32	812.8	57	1447.8	82	2082.8
8	203.2	33	838.2	58	1473.2	83	2108.2
9	228.6	34	863.6	59	1498.6	84	2133.6
10	254.0	35	889.0	60	1524.0	85	2159.0
11	279.4	36	914.4	61	1549.4	86	2184.4
12	304.8	37	939.8	62	1574.8	87	2209.8
13	330.2	38	965.2	63	1600.2	88	2235.2
14	355.6	39	990.6	64	1625.6	89	2260.6
15	381.0	40	1016.0	65	1651.0	90	2286.0
16	406.4	41	1041.4	66	1676.4	91	2311.4
17	431.8	42	1066.8	67	1701.8	92	2336.8
18	457.2	43	1092.2	68	1727.2	93	2362.2
19	482.6	44	1117.6	69	1752.6	94	2387.6
20	508.0	45	1143.0	70	1778.0	95	2413.0
21	533.4	46	1168.4	71	1803.4	96	2438.4
22	558.8	47	1193.8	72	1828.8	97	2463.8
23	584.2	48	1219.2	73	1854.2	98	2489.2
24	609.6	49	1244.6	74	1879.6	99	2514.6
25	635.0	50	1270.0	75	1905.0	100	2540.0

The above table is exact on the basis: 1 in. = 25.4 mm

MILLIMETRES TO INCHES

mm	in.	mm	in.	mm	in.	mm	in.
1	0.039370	26	1.023622	51	2.007874	76	2.992126
2	0.078740	27	1.062992	52	2.047244	77	3.031496
3	0.118110	28	1.102362	53	2.086614	78	3.070866
4	0.157480	29	1.141732	54	2.125984	79	3.110236
5	0.196850	30	1.181102	55	2.165354	80	3.149606
6	0.236220	31	1.220472	56	2.204724	81	3.188976
7	0.275591	32	1.259843	57	2.244094	82	3.228346
8	0.314961	33	1.299213	58	2.283465	83	3.267717
9	0.354331	34	1.338583	59	2.322835	84	3.307087
10	0.393701	35	1.377953	60	2.362205	85	3.346457
11	0.433071	36	1.417323	61	2.401575	86	3.385827
12	0.472441	37	1.456693	62	2.440945	87	3.425197
13	0.511811	38	1.496063	63	2.480315	88	3.464567
14	0.551181	39	1.535433	64	2.519685	89	3.503937
15	0.590551	40	1.574803	65	2.559055	90	3.543307
16	0.629921	41	1.614173	66	2.598425	91	3.582677
17	0.669291	42	1.653543	67	2.637795	92	3.622047
18	0.708661	43	1.692913	68	2.677165	93	3.661417
19	0.748031	44	1.732283	69	2.716535	94	3.700787
20	0.787402	45	1.771654	70	2.755906	95	3.740157
21	0.826772	46	1.811024	71	2.795276	96	3.779528
22	0.866142	47	1.850394	72	2.834646	97	3.818898
23	0.905512	48	1.889764	73	2.874016	98	3.858268
24	0.944882	49	1.929134	74	2.913386	99	3.897638
25	0.984252	50	1.968504	75	2.952756	100	3.937008

The above table is approximate on the basis: 1 in. = 25.4 mm, 1/25.4 = 0.039370078740+

Table 1

INCH/METRIC – EQUIVALENTS

Fraction	Decimal Equivalent Customary (in.)	Metric (mm)	Fraction	Decimal Equivalent Customary (in.)	Metric (mm)
1/64	.015625	0.3969	33/64	.515625	13.0969
1/32	.03125	0.7938	17/32	.53125	13.4938
3/64	.046875	1.1906	35/64	.546875	13.8906
1/16	.0625	1.5875	9/16	.5625	14.2875
5/64	.078125	1.9844	37/64	.578125	14.6844
3/32	.09375	2.3813	19/32	.59375	15.0813
7/64	.109375	2.7781	39/64	.609375	15.4781
1/8	.1250	3.1750	5/8	.6250	15.8750
9/64	.140625	3.5719	41/64	.640625	16.2719
5/32	.15625	3.9688	21/32	.65625	16.6688
11/64	.171875	4.3656	43/64	.671875	17.0656
3/16	.1875	4.7625	11/16	.6875	17.4625
13/64	.203125	5.1594	45/64	.703125	17.8594
7/32	.21875	5.5563	23/32	.71875	18.2563
15/64	.234375	5.9531	47/64	.734375	18.6531
1/4	.250	6.3500	3/4	.750	19.0500
17/64	.265625	6.7469	49/64	.765625	19.4469
9/32	.28125	7.1438	25/32	.78125	19.8438
19/64	.296875	7.5406	51/64	.796875	20.2406
5/16	.3125	7.9375	13/16	.8125	20.6375
21/64	.328125	8.3384	53/64	.828125	21.0344
11/32	.34375	8.7313	27/32	.84375	21.4313
23/64	.359375	9.1281	55/64	.859375	21.8281
3/8	.3750	9.5250	7/8	.8750	22.2250
25/64	.390625	9.9219	57/64	.890625	22.6219
13/32	.40625	10.3188	29/32	.90625	23.0188
27/64	.421875	10.7156	59/64	.921875	23.4156
7/16	.4375	11.1125	15/16	.9375	23.8125
29/64	.453125	11.5094	61/64	.953125	24.2094
15/32	.46875	11.9063	31/32	.96875	24.6063
31/64	.484375	12.3031	63/64	.984375	25.0031
1/2	.500	12.7000	1	1.000	25.4000

Table 2

From Drafting for Trades and Industry—Basic Skills. Nelson. Delmar Publishers Inc.

METRIC EQUIVALENTS

LENGTH

U.S. to Metric

1 inch = 2.540 centimetres
1 foot = .305 metre
1 yard = .914 metre
1 mile = 1.609 kilometres

Metric to U.S.

1 millimetre = .039 inch
1 centimetre = .394 inch
1 metre = 3.281 feet or 1.094 yards
1 kilometre = .621 mile

AREA

1 inch^2 = 6.451 centimetre^2
1 foot^2 = .093 metre^2
1 yard^2 = .836 metre^2
1 acre^2 = 4,046.873 metre^2

1 millimetre^2 = .00155 inch^2
1 centimetre^2 = .155 inch^2
1 metre^2 = 10.764 foot^2 or 1.196 yard^2
1 kilometre^2 = .386 mile^2 or 247.04 acre^2

VOLUME

1 inch^3 = 16.387 centimetre^3
1 foot^3 = .028 metre^3
1 yard^3 = .764 metre^3
1 quart = .946 litre
1 gallon = .003785 metre^3

1 centimetre^3 = 0.61 inch^3
1 metre^3 = 35.314 foot^3 or 1.308 yard^3
1 litre = .2642 gallons
1 litre = 1.057 quarts
1 metre^3 = 264.02 gallons

WEIGHT

1 ounce = 28.349 grams
1 pound = .454 kilogram
1 ton = .907 metric ton

1 gram = .035 ounce
1 kilogram = 2.205 pounds
1 metric ton = 1.102 tons

VELOCITY

1 foot/second = .305 metre/second
1 mile/hour = .447 metre/second

1 metre/second = 3.281 feet/second
1 kilometre/hour = .621 mile/second

ACCELERATION

1 inch/second^2 = .0254 metre/second^2
1 foot/second^2 = .305 metre/second^2

1 metre/second^2 = 3.278 feet/second^2

FORCE

N (newton) = basic unit of force, kg-m/s^2. A mass of one kilogram (1 kg) exerts a gravitational force of 9.8 N (theoretically 9.80665 N) at mean sea level.

Table 3

MULTIPLIERS FOR DRAFTERS

Multiply	By	To Obtain	Multiply	By	To Obtain
Acres	43,560	Square feet	Degrees/sec.	0.002778	Revolutions/sec.
Acres	4047	Square metres	Fathoms	6	Feet
Acres	1.562×10^{-3}	Square miles	Feet	30.48	Centimetres
Acres	4840	Square yards	Feet	12	Inches
Acre–feet	43,560	Cubic feet	Feet	0.3048	Metres
Atmospheres	76.0	Cms. of mercury	Foot–pounds	1.286×10^{-3}	British Thermal Units
Atmospheres	29.92	Inches of mercury	Foot–pounds	5.050×10^{-7}	Horsepower–hrs.
Atmospheres	33.90	Feet of water	Foot–pounds	3.241×10^{-4}	Kilogram–calories
Atmospheres	10,333	Kgs./sq. metre	Foot–pounds	0.1383	Kilogram–metres
Atmospheres	14.70	Lbs./sq. inch	Foot–pounds	3.766×10^{-7}	Kilowatt–hrs.
Atmospheres	1.058	Tons/sq. ft.	Foot–pounds/min.	1.286×10^{-3}	B.T.U./min.
Board feet	144 sq. in. $\times$ 1 in.	Cubic inches	Foot–pounds/min.	0.01667	Foot–pounds/sec.
British Thermal Units	0.2520	Kilogram–calories	Foot–pounds/min.	3.030×10^{-5}	Horsepower
British Thermal Units	777.5	Foot–lbs.	Foot–pounds/min.	3.241×10^{-4}	Kg.–calories/min.
British Thermal Units	3.927×10^{-4}	Horsepower–hrs.	Foot–pounds/min.	2.260×10^{-5}	Kilowatts
British Thermal Units	107.5	Kilogram–metres	Foot–pounds/sec.	7.717×10^{-2}	B.T.U./min.
British Thermal Units	2.928×10^{-4}	Kilowatt–hrs.	Foot–pounds/sec.	1.818×10^{-3}	Horsepower
B.T.U./min.	12.96	Foot–lbs./sec.	Foot–pounds/sec.	1.945×10^{-2}	Kg.–calories/min.
B.T.U./min.	0.02356	Horsepower	Foot–pounds/sec.	1.356×10^{-3}	Kilowatts
B.T.U./min.	0.01757	Kilowatts	Gallons	3785	Cubic centimetres
B.T.U./min.	17.57	Watts	Gallons	0.1337	Cubic feet
Cubic centimetres	3.531×10^{-5}	Cubic feet	Gallons	231	Cubic inches
Cubic centimetres	6.102×10^{-2}	Cubic inches	Gallons	3.785×10^{-3}	Cubic metres
Cubic centimetres	10^{-6}	Cubic metres	Gallons	4.951×10^{-3}	Cubic yards
Cubic centimetres	1.308×10^{-6}	Cubic yards	Gallons	3.785	Litres
Cubic centimetres	2.642×10^{-4}	Gallons	Gallons	8	Pints (liq.)
Cubic centimetres	10^{-3}	Litres	Gallons	4	Quarts (liq.)
Cubic centimetres	2.113×10^{-3}	Pints (liq.)	Gallons–Imperial	1.20095	U.S. gallons
Cubic centimetres	1.057×10^{-3}	Quarts (liq.)	Gallons–U.S.	0.83267	Imperial gallons
Cubic feet	2.832×10^{4}	Cubic cms.	Gallons water	8.3453	Pounds of water
Cubic feet	1728	Cubic inches	Horsepower	42.44	B.T.U./min.
Cubic feet	0.02832	Cubic metres	Horsepower	33,000	Foot–lbs./min.
Cubic feet	0.03704	Cubic yards	Horsepower	550	Foot–lbs./sec.
Cubic feet	7.48052	Gallons	Horsepower	1.014	Horsepower (metric)
Cubic feet	28.32	Litres	Horsepower	10.70	Kg.–calories/min.
Cubic feet	59.84	Pints (liq.)	Horsepower	0.7457	Kilowatts
Cubic feet	29.92	Quarts (liq.)	Horsepower	745.7	Watts
Cubic feet/min.	472.0	Cubic cms./sec.	Horsepower–hours	2547	B.T.U.
Cubic feet/min.	0.1247	Gallons/sec.	Horsepower–hours	1.98×10^{6}	Foot–lbs.
Cubic feet/min.	0.4720	Litres/sec.	Horsepower–hours	641.7	Kilogram–calories
Cubic feet/min.	62.43	Pounds of water/min.	Horsepower–hours	2.737×10^{5}	Kilogram–metres
Cubic feet/sec.	0.646317	Millions gals./day	Horsepower–hours	0.7457	Kilowatt–hours
Cubic feet/sec.	448.831	Gallons/min.	Kilometres	10^{5}	Centimetres
Cubic inches	16.39	Cubic centimetres	Kilometres	3281	Feet
Cubic inches	5.787×10^{-4}	Cubic feet	Kilometres	10^{3}	Metres
Cubic inches	1.639×10^{-5}	Cubic metres	Kilometres	0.6214	Miles
Cubic inches	2.143×10^{-5}	Cubic yards	Kilometres	1094	Yards
Cubic inches	4.329×10^{-3}	Gallons	Kilowatts	56.92	B.T.U./min.
Cubic inches	1.639×10^{-2}	Litres	Kilowatts	4.425×10^{4}	Foot–lbs./min.
Cubic inches	0.03463	Pints (liq.)	Kilowatts	737.6	Foot–lbs./sec.
Cubic inches	0.01732	Quarts (liq.)	Kilowatts	1.341	Horsepower
Cubic metres	10^{6}	Cubic centimetres	Kilowatts	14.34	Kg.–calories/min.
Cubic metres	35.31	Cubic feet	Kilowatts	10^{3}	Watts
Cubic metres	61.023	Cubic inches	Kilowatt–hours	3415	B.T.U.
Cubic metres	1.308	Cubic yards	Kilowatt–hours	2.655×10^{6}	Foot–lbs.
Cubic metres	264.2	Gallons	Kilowatt–hours	1.341	Horsepower–hrs.
Cubic metres	10^{3}	Litres	Kilowatt–hours	860.5	Kilogram–calories
Cubic metres	2113	Pints (liq.)	Kilowatt–hours	3.671×10^{5}	Kilogram–metres
Cubic metres	1057	Quarts (liq.)	Lumber Width (in.) $\times$		
Degrees (angle)	60	Minutes	$\dfrac{\text{Thickness (in.)}}{12}$	Length (ft.)	Board feet
Degrees (angle)	0.01745	Radians			
Degrees (angle)	3600	Seconds	Metres	100	Centimetres
Degrees/sec.	0.01745	Radians/sec.	Metres	3.281	Feet
Degrees/sec.	0.1667	Revolutions/min.	Metres	39.37	Inches

Table 4

MULTIPLIERS FOR DRAFTERS (cont'd)

Multiply	By	To Obtain	Multiply	By	To Obtain
Metres	10^{-3}	Kilometres	Pounds (troy)	373.24177	Grams
Metres	10^3	Millimetres	Pounds (troy)	0.822857	Pounds (avoir.)
Metres	1.094	Yards	Pounds (troy)	13.1657	Ounces (avoir.)
Metres/min.	1.667	Centimetres/sec.	Pounds (troy)	3.6735×10^{-4}	Tons (long)
Metres/min.	3.281	Feet/min.	Pounds (troy)	4.1143×10^{-4}	Tons (short)
Metres/min.	0.05468	Feet/sec.	Pounds (troy)	3.7324×10^{-4}	Tons (metric)
Metres/min.	0.06	Kilometres/hr.	Quadrants (angle)	90	Degrees
Metres/min.	0.03728	Miles/hr.	Quadrants (angle)	5400	Minutes
Metres/sec.	196.8	Feet/min.	Quadrants (angle)	1.571	Radians
Metres/sec.	3.281	Feet/sec.	Radians	57.30	Degrees
Metres/sec.	3.6	Kilometres/hr.	Radians	3438	Minutes
Metres/sec.	0.06	Kilometres/min.	Radians	0.637	Quadrants
Metres/sec.	2.237	Miles/hr.	Radians/sec.	57.30	Degrees/sec.
Metres/sec.	0.03728	Miles/min.	Radians/sec.	0.1592	Revolutions/sec.
Microns	10^{-6}	Metres	Radians/sec.	9.549	Revolutions/min.
Miles	5280	Feet	Radians/sec./sec.	573.0	Revs./min./min.
Miles	1.609	Kilometres	Radians/sec./sec.	0.1592	Revs./sec./sec.
Miles	1760	Yards	Reams	500	Sheets
Miles/hr.	1.609	Kilometres/hr.	Revolutions	360	Degrees
Miles/hr.	0.8684	Knots	Revolutions	4	Quadrants
Minutes (angle)	2.909×10^{-4}	Radians	Revolutions	6.283	Radians
Ounces	16	Drams	Revolutions/min.	6	Degrees/sec.
Ounces	437.5	Grains	Square yards	2.066×10^{-4}	Acres
Ounces	0.0625	Pounds	Square yards	9	Square feet
Ounces	28.349527	Grams	Square yards	0.8361	Square metres
Ounces	0.9115	Ounces (troy)	Square yards	3.228×10^{-7}	Square miles
Ounces	2.790×10^{-5}	Tons (long)	Temp. (°C.) + 273	1	Abs. temp. (°C.)
Ounces	2.835×10^{-5}	Tons (metric)	Temp. (°C.) + 17.78	1.8	Temp. (°F.)
Ounces (troy)	480	Grains	Temp. (°F.) + 460	1	Abs. temp. (°F.)
Ounces (troy)	20	Pennyweights (troy)	Temp. (°F.) − 32	5/9	Temp. (°C.)
Ounces (troy)	0.08333	Pounds (troy)	Watts	0.05692	B.T.U./min.
Ounces (troy)	31.103481	Grams	Watts	44.26	Foot−pounds/min.
Ounces (troy)	1.09714	Ounces (avoir.)	Watts	0.7376	Foot−pounds/sec.
Ounces (fluid)	1.805	Cubic inches	Watts	1.341×10^{-3}	Horsepower
Ounces (fluid)	0.02957	Litres	Watts	0.01434	Kg.−calories/min.
Ounces/sq. inch	0.0625	Lbs./sq. inch	Watts	10^{-3}	Kilowatts
Pounds	16	Ounces	Watt−hours	3.415	B.T.U.
Pounds	256	Drams	Watt−hours	2655	Foot−pounds
Pounds	7000	Grains	Watt−hours	1.341×10^{-3}	Horsepower−hrs.
Pounds	0.0005	Tons (short)	Watt−hours	0.8605	Kilogram−calories
Pounds	453.5924	Grams	Watt−hours	367.1	Kilogram−metres
Pounds	1.21528	Pounds (troy)	Watt−hours	10^{-3}	Kilowatt−hours
Pounds	14.5833	Ounces (troy)	Yards	91.44	Centimetres
Pounds (troy)	5760	Grains	Yards	3	Feet
Pounds (troy)	240	Pennyweights (troy)	Yards	36	Inches
Pounds (troy)	12	Ounces (troy)	Yards	0.9144	Metres

Table 4 (Cont'd)

Table 5

CIRCUMFERENCES AND AREAS OF CIRCLES
From 1/64 to 50, Diameter

Dia.	Circum.	Area	Dia.	Circum.	Area	Dia.	Circum.	Area	Dia.	Circum.	Area
1/64	.04909	.00019	8	25.1327	50.2655	17	53.4071	226.980	26	81.6814	530.929
1/32	.09818	.00077	8⅛	25.5254	51.8485	17⅛	53.7998	230.330	26⅛	82.0741	536.047
1/16	.19635	.00307	8¼	25.9181	53.4562	17¼	54.1925	233.705	26¼	82.4668	541.188
⅛	.39270	.01227	8⅜	26.3108	55.0883	17⅜	54.5852	237.104	26⅜	82.8595	546.355
3/16	.58905	.02761	8½	26.7035	56.7450	17½	54.9779	240.528	26½	83.2522	551.546
¼	.78540	.04909	8⅝	27.0962	58.4262	17⅝	55.3706	243.977	26⅝	83.6449	556.761
5/16	.98175	.07670	8¾	27.4889	60.1321	17¾	55.7633	247.450	26¾	84.0376	562.002
⅜	1.1781	.11045	8⅞	27.8816	61.8624	17⅞	56.1560	250.947	26⅞	84.4303	567.266
7/16	1.3744	.15033	9	28.2743	63.6173	18	56.5487	254.469	27	84.8230	572.555
½	1.5708	.19635	9⅛	28.6670	65.3967	18⅛	56.9414	258.016	27⅛	85.2157	577.869
9/16	1.7671	.24850	9¼	29.0597	67.2007	18¼	57.3341	261.587	27¼	85.6084	583.207
⅝	1.9635	.30680	9⅜	29.4524	69.0292	18⅜	57.7268	265.182	27⅜	86.0011	588.570
11/16	2.1598	.37122	9½	29.8451	70.8822	18½	58.1195	268.803	27½	86.3938	593.957
¾	2.3562	.44179	9⅝	30.2378	72.7597	18⅝	58.5122	272.447	27⅝	86.7865	599.369
13/16	2.5525	.51849	9¾	30.6305	74.6619	18¾	58.9049	276.117	27¾	87.1792	604.806
⅞	2.7489	.60132	9⅞	31.0232	76.5886	18⅞	59.2976	279.810	27⅞	87.5719	610.267
15/16	2.9452	.69029				19	59.6903	283.529	28	87.9646	615.752
1	3.1416	.78540	10	31.4159	78.5398	19⅛	60.0830	287.272	28⅛	88.3573	621.262
1⅛	3.5343	.99402	10⅛	31.8086	80.5156	19¼	60.4757	291.039	28¼	88.7500	626.797
1¼	3.9270	1.2272	10¼	32.2013	82.5159	19⅜	60.8684	294.831	28⅜	89.1427	632.356
1⅜	4.3197	1.4849	10⅜	32.5940	84.5408	19½	61.2611	298.648	28½	89.5354	637.940
1½	4.7124	1.7671	10½	32.9867	86.5902	19⅝	61.6538	302.489	28⅝	89.9281	643.548
1⅝	5.1051	2.0739	10⅝	33.3794	88.6641	19¾	62.0465	306.354	28¾	90.3208	649.181
1¾	5.4978	2.4053	10¾	33.7721	90.7626	19⅞	62.4392	310.245	28⅞	90.7135	654.838
1⅞	5.8905	2.7612	10⅞	34.1648	92.8856	20	62.8319	314.159	29	91.1062	660.520
2	6.2832	3.1416	11	34.5575	95.0332	20⅛	63.2246	318.099	29⅛	91.4989	666.226
2⅛	6.6759	3.5466	11⅛	34.9502	97.2053	20¼	63.6173	322.062	29¼	91.8916	671.957
2¼	7.0686	3.9761	11¼	35.3429	99.4020	20⅜	64.0100	326.051	29⅜	92.2843	677.713
2⅜	7.4613	4.4301	11⅜	35.7356	101.623	20½	64.4027	330.064	29½	92.6770	683.493
2½	7.8540	4.9087	11½	36.1283	103.869	20⅝	64.7954	334.101	29⅝	93.0697	689.297
2⅝	8.2467	5.4119	11⅝	36.5210	106.139	20¾	65.1881	338.163	29¾	93.4624	695.127
2¾	8.6394	5.9396	11¾	36.9137	108.434	20⅞	65.5808	342.250	29⅞	93.8551	700.980
2⅞	9.0321	6.4918	11⅞	37.3064	110.753				30	94.2478	706.858
3	9.4248	7.0686	12	37.6991	113.097	21	65.9735	346.361	30⅛	94.6405	712.761
3⅛	9.8175	7.6699	12⅛	38.0918	115.466	21⅛	66.3662	350.496	30¼	95.0332	718.689
3¼	10.2102	8.2958	12¼	38.4845	117.859	21¼	66.7589	354.656	30⅜	95.4259	724.640
3⅜	10.6029	8.9462	12⅜	38.8772	120.276	21⅜	67.1516	358.841	30½	95.8186	730.617
3½	10.9956	9.6211	12½	39.2699	122.718	21½	67.5442	363.050	30⅝	96.2113	736.618
3⅝	11.3883	10.3206	12⅝	39.6626	125.185	21⅝	67.9369	367.284	30¾	96.6040	742.643
3¾	11.7810	11.0447	12¾	40.0553	127.676	21¾	68.3296	371.542	30⅞	96.9967	748.693
3⅞	12.1737	11.7932	12⅞	40.4480	130.191	21⅞	68.7223	375.825	31	97.3894	754.768
4	12.5664	12.5664	13	40.8407	132.732	22	69.1150	380.133	31⅛	97.7821	760.867
4⅛	12.9591	13.3640	13⅛	41.2334	135.297	22⅛	69.5077	384.465	31¼	98.1748	766.990
4¼	13.3518	14.1863	13¼	41.6261	137.886	22¼	69.9004	388.821	31⅜	98.5675	773.139
4⅜	13.7445	15.0330	13⅜	42.0188	140.500	22⅜	70.2931	393.203	31½	98.9602	779.311
4½	14.1372	15.9043	13½	42.4115	143.139	22½	70.6858	397.608	31⅝	99.3529	785.509
4⅝	14.5299	16.8002	13⅝	42.8042	145.802	22⅝	71.0785	402.038	31¾	99.7456	791.731
4¾	14.9226	17.7206	13¾	43.1969	148.489	22¾	71.4712	406.493	31⅞	100.1383	797.977
4⅞	15.3153	18.6655	13⅞	43.5896	151.201	22⅞	71.8639	410.972			
5	15.7080	19.6350	14	43.9823	153.938	23	72.2566	415.476	32	100.5310	804.248
5⅛	16.1007	20.6290	14⅛	44.3750	156.699	23⅛	72.6493	420.004	32⅛	100.9237	810.543
5¼	16.4934	21.6476	14¼	44.7677	159.485	23¼	73.0420	424.557	32¼	101.3164	816.863
5⅜	16.8861	22.6906	14⅜	45.1604	162.295	23⅜	73.4347	429.134	32⅜	101.7091	823.208
5½	17.2788	23.7583	14½	45.5531	165.130	23½	73.8274	433.736	32½	102.1018	829.577
5⅝	17.6715	24.8505	14⅝	45.9458	167.989	23⅝	74.2201	438.363	32⅝	102.4945	835.971
5¾	18.0642	25.9672	14¾	46.3385	170.873	23¾	74.6128	443.014	32¾	102.8872	842.389
5⅞	18.4569	27.1085	14⅞	46.7312	173.782	23⅞	75.0055	447.689	32⅞	103.2799	848.831
6	18.8496	28.2743	15	47.1239	176.715	24	75.3982	452.389	33	103.6726	855.299
6⅛	19.2423	29.4647	15⅛	47.5166	179.672	24⅛	75.7909	457.114	33⅛	104.0653	861.791
6¼	19.6350	30.6796	15¼	47.9094	182.654	24¼	76.1836	461.863	33¼	104.4580	868.307
6⅜	20.0277	31.9191	15⅜	48.3020	185.661	24⅜	76.5763	466.637	33⅜	104.8507	874.848
6½	20.4204	33.1831	15½	48.6947	188.692	24½	76.9690	471.435	33½	105.2434	881.413
6⅝	20.8131	34.4716	15⅝	49.0874	191.748	24⅝	77.3617	476.258	33⅝	105.6361	888.003
6¾	21.2058	35.7847	15¾	49.4801	194.828	24¾	77.7544	481.106	33¾	106.0288	894.618
6⅞	21.5985	37.1223	15⅞	49.8728	197.933	24⅞	78.1471	485.977	33⅞	106.4215	901.257
7	21.9912	38.4845	16	50.2655	201.062	25	78.5398	490.874	34	106.8142	907.920
7⅛	22.3839	39.8712	16⅛	50.6582	204.216	25⅛	78.9325	495.795	34⅛	107.2069	914.609
7¼	22.7765	41.2825	16¼	51.0509	207.394	25¼	79.3252	500.740	34¼	107.5996	921.321
7⅜	23.1692	42.7183	16⅜	51.4436	210.597	25⅜	79.7179	505.711	34⅜	107.9923	928.058
7½	23.5619	44.1787	16½	51.8363	213.825	25½	80.1106	510.705	34½	108.3850	934.820
7⅝	23.9546	45.6636	16⅝	52.2290	217.077	25⅝	80.5033	515.724	34⅝	108.7777	941.607
7¾	24.3473	47.1730	16¾	52.6217	220.353	25¾	80.8960	520.768	34¾	109.1704	948.417
7⅞	24.7400	48.7069	16⅞	53.0144	223.654	25⅞	81.2887	525.836	34⅞	109.5631	955.253

TRIGONOMETRIC FORMULAS

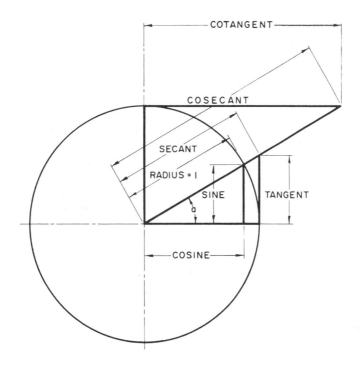

FORMULAS FOR FINDING FUNCTIONS OF ANGLES	
$\dfrac{\text{Side opposite}}{\text{Hypotenuse}}$ = SINE	
$\dfrac{\text{Side adjacent}}{\text{Hypotenuse}}$ = COSINE	
$\dfrac{\text{Side opposite}}{\text{Side adjacent}}$ = TANGENT	
$\dfrac{\text{Side adjacent}}{\text{Side opposite}}$ = COTANGENT	
$\dfrac{\text{Hypotenuse}}{\text{Side adjacent}}$ = SECANT	
$\dfrac{\text{Hypotenuse}}{\text{Side opposite}}$ = COSECANT	
FORMULAS FOR FINDING THE LENGTH OF SIDES FOR RIGHT-ANGLE TRIANGLES WHEN AN ANGLE AND SIDE ARE KNOWN	
Length of side opposite	Hypotenuse × Sine Hypotenuse ÷ Cosecant Side adjacent × Tangent Side adjacent ÷ Cotangent
Length of side adjacent	Hypotenuse × Cosine Hypotenuse ÷ Secant Side opposite × Cotangent Side opposite ÷ Tangent
Length of Hypotenuse	Side opposite × Cosecant Side opposite ÷ Sine Side adjacent × Secant Side adjacent ÷ Cosine

Table 6

RIGHT-TRIANGLE FORMULAS

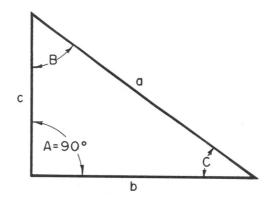

TO FIND ANGLES	FORMULAS	
C	$\frac{c}{a}$ = Sine C	90°−B
C	$\frac{b}{a}$ = Cosine C	90°−B
C	$\frac{c}{b}$ = Tan. C	90°−B
C	$\frac{b}{c}$ = Cotan. C	90°−B
C	$\frac{a}{b}$ = Secant C	90°−B
C	$\frac{a}{c}$ = Cosec. C	90°−B
B	$\frac{b}{a}$ = Sine B	90°−C
B	$\frac{c}{a}$ = Cosine B	90°−C
B	$\frac{b}{c}$ = Tan. B	90°−C
B	$\frac{c}{b}$ = Cotan. B	90°−C
B	$\frac{a}{c}$ = Secant B	90°−C
B	$\frac{a}{b}$ = Cosec. B	90°−C

TO FIND SIDES	FORMULAS	
a	$\sqrt{b^2 + c^2}$	
a	c × Cosec. C	$\frac{c}{\text{sine C}}$
a	c × Secant B	$\frac{c}{\text{Cosine B}}$
a	b × Cosec. B	$\frac{b}{\text{Sine B}}$
a	b × Secant C	$\frac{b}{\text{Cosine C}}$
b	$\sqrt{a^2 - c^2}$	
b	a × Sine B	$\frac{a}{\text{Cosecant B}}$
b	a × Cos. C	$\frac{a}{\text{Secant C}}$
b	c × Tan. B	$\frac{c}{\text{Cotangent B}}$
b	c × Cot. C	$\frac{c}{\text{Tangent C}}$
c	$\sqrt{a^2 - b^2}$	
c	a × Cos. B	$\frac{a}{\text{Secant B}}$
c	a × Sine C	$\frac{a}{\text{Cosecant C}}$
c	b × Cot. B	$\frac{b}{\text{Tangent B}}$
c	b × Tan C	$\frac{b}{\text{Cotangent C}}$

Table 7

OBLIQUE-ANGLED TRIANGLE FORMULAS

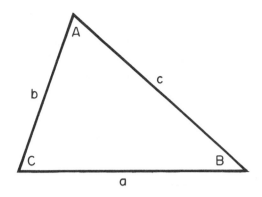

TO FIND	KNOWN	SOLUTION
C	A-B	$180° - (A + B)$
b	a-B-A	$\dfrac{a \times \text{Sin. B}}{\text{Sin. A}}$
c	a-A-C	$\dfrac{a \times \text{Sin. C}}{\text{Sin. A}}$
Tan. A	a-C-b	$\dfrac{a \times \text{Sin. C}}{b - (a \times \text{Cos. C})}$
B	A-C	$180° - (A + C)$
Sin. B	b-A-a	$\dfrac{b \times \text{Sin. A}}{a}$
A	B-C	$180° - (B + C)$
Cos. A	a-b-c	$\dfrac{b^2 + C^2 - a^2}{2bc}$
Sin. C	c-A-a	$\dfrac{c \times \text{Sin. A}}{a}$
Cot. B	a-C-b	$\dfrac{a \times \text{cscC}}{b} - \text{Cot. C}$
c	b-C-B	$b \times \text{Sin. C} \times \text{cscB}$

Table 8

RUNNING AND SLIDING FITS

VALUES IN THOUSANDTHS OF AN INCH

Nominal Size Range Inches		Class RC1 Precision Sliding			Class RC2 Sliding Fit			Class RC3 Precision Running			Class RC4 Close Running			Class RC5 Medium Running		
		Hole Tol. GR5	Minimum Clearance	Shaft Tol. GR4	Hole Tol. GR6	Minimum Clearance	Shaft Tol. GR5	Hole Tol. GR7	Minimum Clearance	Shaft Tol. GR6	Hole Tol. GR8	Minimum Clearance	Shaft Tol. GR7	Hole Tol. GR8	Minimum Clearance	Shaft Tol. GR7
Over	To	-0		+0	-0		+0	-0		+0	-0		+0	-0		+0
0	.12	+0.15	0.10	-0.12	+0.25	0.10	-0.15	+0.40	0.30	-0.25	+0.60	0.30	-0.40	+0.60	0.60	-0.40
.12	.24	+0.20	0.15	-0.15	+0.30	0.15	-0.20	+0.50	0.40	-0.30	+0.70	0.40	-0.50	+0.70	0.80	-0.50
.24	.40	+0.25	0.20	-0.15	+0.40	0.20	-0.25	+0.60	0.50	-0.40	+0.90	0.50	-0.60	+0.90	1.00	-0.60
.40	.71	+0.30	0.25	-0.20	+0.40	0.25	-0.30	+0.70	0.60	-0.40	+1.00	0.60	-0.70	+1.00	1.20	-0.70
.71	1.19	+0.40	0.30	-0.25	+0.50	0.30	-0.40	+0.80	0.80	-0.50	+1.20	0.80	-0.80	+1.20	1.60	-0.80
1.19	1.97	+0.40	0.40	-0.30	+0.60	0.40	-0.40	+1.00	1.00	-0.60	+1.60	1.00	-1.00	+1.60	2.00	-1.00
1.97	3.15	+0.50	0.40	-0.30	+0.70	0.40	-0.50	+1.20	1.20	-0.70	+1.80	1.20	-1.20	+1.80	2.50	-1.20
3.15	4.73	+0.60	0.50	-0.40	+0.90	0.50	-0.60	+1.40	1.40	-0.90	+2.20	1.40	-1.40	+2.20	3.00	-1.40
4.73	7.09	+0.70	0.60	-0.50	+1.00	0.60	-0.70	+1.60	1.60	-1.00	+2.50	1.60	-1.60	+2.50	3.50	-1.60
7.09	9.85	+0.80	0.60	-0.60	+1.20	0.60	-0.80	+1.80	2.00	-1.20	+2.80	2.00	-1.80	+2.80	4.50	-1.80
9.85	12.41	+0.90	0.80	-0.60	+1.20	0.80	-0.90	+2.20	2.50	-1.20	+3.00	2.50	-2.00	+3.00	5.00	-2.00
12.41	15.75	+1.00	1.00	-0.70	+1.40	1.00	-1.00	+2.20	3.00	-1.40	+3.50	3.00	-2.20	+3.50	6.00	-2.20

Nominal Size Range Inches		Class RC6 Medium Running			Class RC7 Free Running			Class RC8 Loose Running			Class RC9 Loose Running		
		Hole Tol. GR9	Minimum Clearance	Shaft Tol. GR8	Hole Tol. GR9	Minimum Clearance	Shaft Tol. GR8	Hole Tol. GR10	Minimum Clearance	Shaft Tol. GR9	Hole Tol. GR11	Minimum Clearance	Shaft Tol. GR10
Over	To	-0		+0	-0		+0	-0		+0	-0		+0
0	.12	+1.00	0.60	-0.60	+1.00	1.00	-0.60	+1.60	2.50	-1.00	+2.50	4.00	-1.60
.12	.24	+1.20	0.80	-0.70	+1.20	1.20	-0.70	+1.80	2.80	-1.20	+3.00	4.50	-1.80
.24	.40	+1.40	1.00	-0.90	+1.40	1.60	-0.90	+2.20	3.00	-1.40	+3.50	6.00	-2.20
.40	.71	+1.60	1.20	-1.00	+1.60	2.00	-1.00	+2.80	3.50	-1.60	+4.00	6.00	-2.80
.71	1.19	+2.00	1.60	-1.20	+2.00	2.50	-1.20	+3.50	4.50	-2.00	+5.00	7.00	-3.50
1.19	1.97	+2.50	2.00	-1.60	+2.50	3.00	-1.60	+4.00	5.00	-2.50	+6.00	8.00	-4.00
1.97	3.15	+3.00	2.50	-1.80	+3.00	4.00	-1.80	+4.50	6.00	-3.00	+7.00	9.00	-4.50
3.15	4.73	+3.50	3.00	-2.20	+3.50	5.00	-2.20	+5.00	7.00	-3.50	+9.00	10.00	-5.00
4.73	7.09	+4.00	3.50	-2.50	+4.00	6.00	-2.50	+6.00	8.00	-4.00	+10.00	12.00	-6.00
7.09	9.85	+4.50	4.00	-2.80	+4.50	7.00	-2.80	+7.00	10.00	-4.50	+12.00	15.00	-7.00
9.85	12.41	+5.00	5.00	-3.00	+5.00	8.00	-3.00	+8.00	12.00	-5.00	+12.00	18.00	-8.00
12.41	15.75	+6.00	6.00	-3.50	+6.00	10.00	-3.50	+9.00	14.00	-6.00	+14.00	22.00	-9.00

VALUES IN MILLIMETRES

Nominal Size Range Millimetres		Class RC1 Precision Sliding			Class RC2 Sliding Fit			Class RC3 Precision Running			Class RC4 Close Running			Class RC5 Medium Running		
		Hole Tol. H5	Minimum Clearance	Shaft Tol. g4	Hole Tol. H6	Minimum Clearance	Shaft Tol. g5	Hole Tol. H7	Minimum Clearance	Shaft Tol. f6	Hole Tol. H8	Minimum Clearance	Shaft Tol. f7	Hole Tol. H8	Minimum Clearance	Shaft Tol. e7
Over	To	-0		+0	-0		+0	-0		+0	-0		+0	-0		+0
0	3	+0.004	0.003	-0.003	+0.006	0.003	-0.004	+0.010	0.008	-0.006	+0.015	0.008	-0.010	+0.015	0.015	-0.010
3	6	+0.005	0.004	-0.004	+0.008	0.004	-0.005	+0.013	0.010	-0.008	+0.018	0.010	-0.013	+0.018	0.020	-0.013
6	10	+0.006	0.005	-0.004	+0.010	0.005	-0.006	+0.015	0.013	-0.010	+0.023	0.013	-0.015	+0.023	0.025	-0.015
10	18	+0.008	0.006	-0.005	+0.010	0.006	-0.008	+0.018	0.015	-0.010	+0.025	0.015	-0.018	+0.025	0.030	-0.018
18	30	+0.010	0.008	-0.006	+0.013	0.008	-0.010	+0.020	0.020	-0.013	+0.030	0.020	-0.020	+0.030	0.040	-0.020
30	50	+0.010	0.010	-0.008	+0.015	0.010	-0.010	+0.030	0.030	-0.015	+0.040	0.030	-0.030	+0.040	0.050	-0.030
50	80	+0.013	0.010	-0.008	+0.018	0.010	-0.013	+0.030	0.030	-0.020	+0.050	0.030	-0.030	+0.050	0.060	-0.030
80	120	+0.015	0.013	-0.010	+0.023	0.013	-0.015	+0.040	0.040	-0.020	+0.060	0.040	-0.040	+0.060	0.080	-0.040
120	180	+0.018	0.015	-0.013	+0.025	0.015	-0.018	+0.040	0.040	-0.030	+0.060	0.040	-0.040	+0.060	0.090	-0.040
180	250	+0.020	0.015	-0.015	+0.030	0.015	-0.020	+0.050	0.050	-0.030	+0.070	0.050	-0.050	+0.070	0.110	-0.050
250	315	+0.023	0.020	-0.015	+0.030	0.020	-0.023	+0.050	0.060	-0.030	+0.080	0.060	-0.050	+0.080	0.130	-0.050
315	400	+0.025	0.025	-0.018	+0.036	0.025	-0.025	+0.060	0.080	-0.040	+0.090	0.080	-0.060	+0.090	0.150	-0.060

Nominal Size Range Millimetres		Class RC6 Medium Running			Class RC7 Free Running			Class RC8 Loose Running			Class RC9 Loose Running		
		Hole Tol. H9	Minimum Clearance	Shaft Tol. e8	Hole Tol. H9	Minimum Clearance	Shaft Tol. d8	Hole Tol. H10	Minimum Clearance	Shaft Tol. e9	Hole Tol. GR11	Minimum Clearance	Shaft Tol. gr10
Over	To	-0		+0	-0		+0	-0		+0	-0		+0
0	3	+0.025	0.015	-0.015	+0.025	0.025	-0.015	+0.041	0.064	-0.025	+0.060	0.100	-0.040
3	6	+0.030	0.015	-0.018	+0.030	0.030	-0.018	+0.046	0.071	-0.030	+0.080	0.110	-0.050
6	10	+0.036	0.025	-0.023	+0.030	0.040	-0.023	+0.056	0.076	-0.036	+0.070	0.130	-0.060
10	18	+0.040	0.030	-0.025	+0.040	0.050	-0.025	+0.070	0.090	-0.040	+0.100	0.150	-0.070
18	30	+0.050	0.040	-0.030	+0.050	0.060	-0.030	+0.090	0.110	-0.050	+0.130	0.180	-0.090
30	50	+0.060	0.050	-0.040	+0.060	0.080	-0.040	+0.100	0.130	-0.060	+0.150	0.200	-0.100
50	80	+0.080	0.060	-0.050	+0.080	0.100	-0.050	+0.110	0.150	-0.080	+0.180	0.230	-0.120
80	120	+0.090	0.080	-0.060	+0.090	0.130	-0.060	+0.130	0.180	-0.090	+0.230	0.250	-0.130
120	180	+0.100	0.090	-0.060	+0.100	0.150	-0.060	+0.150	0.200	-0.100	+0.250	0.300	-0.150
180	250	+0.110	0.100	-0.070	+0.110	0.180	-0.070	+0.180	0.250	-0.110	+0.300	0.380	-0.180
250	315	+0.130	0.130	-0.080	+0.130	0.200	-0.080	+0.200	0.300	-0.130	+0.300	0.460	-0.200
315	400	+0.150	0.150	-0.090	+0.150	0.250	-0.090	+0.230	0.360	-0.150	+0.360	0.560	-0.230

Table 9

From Drafting for Trades and Industry—Mechanical and Electronic. Nelson. Delmar Publishers Inc.

694

LOCATIONAL CLEARANCE FITS

VALUES IN THOUSANDTHS OF AN INCH

Nominal Size Range Inches		Class LC1			Class LC2			Class LC3			Class LC4			Class LC5			Class LC6		
		Hole Tol. GR6	Minimum Clearance	Shaft Tol. GR5	Hole Tol. GR8	Minimum Clearance	Shaft Tol. GR7	Hole Tol. GR10	Minimum Clearance	Shaft Tol. GR9	Hole Tol. GR7	Minimum Clearance	Shaft Tol. GR6	Hole Tol. GR9	Minimum Clearance	Shaft Tol. GR8	Hole Tol. GR9	Minimum Clearance	Shaft Tol. GR8
Over	To	-0		+0	-0		+0	-0		+0	-0		+0	-0		+0	-0		+0
0	.12	+0.25	0	-0.15	+0.4	0	-0.25	+0.6	0	-0.4	+1.6	0	-1.0	+0.4	0.10	-0.25	+1.0	0.3	-0.6
.12	.24	+0.30	0	-0.20	+0.5	0	-0.30	+0.7	0	-0.5	+1.8	0	-1.2	+0.5	0.15	-0.30	+1.2	0.4	-0.7
.24	.40	+0.40	0	-0.25	+0.6	0	-0.40	+0.9	0	-0.6	+2.2	0	-1.4	+0.6	0.20	-0.40	+1.4	0.5	-0.9
.40	.71	+0.40	0	-0.30	+0.7	0	-0.40	+1.0	0	-0.7	+2.8	0	-1.6	+0.7	0.25	-0.40	+1.6	0.6	-1.0
.71	1.19	+0.50	0	-0.40	+0.8	0	-0.50	+1.2	0	-0.8	+3.5	0	-2.0	+0.8	0.30	-0.50	+2.0	0.8	-1.2
1.19	1.97	+0.60	0	-0.40	+1.0	0	-0.60	+1.6	0	-1.0	+4.0	0	-2.5	+1.0	0.40	-0.60	+2.5	1.0	-1.6
1.97	3.15	+0.70	0	-0.50	+1.2	0	-0.70	+1.8	0	-1.2	+4.5	0	-3.0	+1.2	0.40	-0.70	+3.0	1.2	-1.8
3.15	4.73	+0.90	0	-0.60	+1.4	0	-0.90	+2.7	0	-1.4	+5.0	0	-3.5	+1.4	0.50	-0.90	+3.5	1.4	-2.2
4.73	7.09	+1.00	0	-0.70	+1.6	0	-1.00	+2.5	0	-1.6	+6.0	0	-4.0	+1.6	0.60	-1.00	+4.0	1.6	-2.5
7.09	9.85	+1.20	0	-0.80	+1.8	0	-1.20	+2.8	0	-1.8	+7.0	0	-4.5	+1.8	0.60	-1.20	+4.5	2.0	-2.8
9.85	12.41	+1.20	0	-0.90	+2.0	0	-1.20	+3.0	0	-2.0	+8.0	0	-5.0	+2.0	0.70	-1.20	+5.0	2.2	-3.0
12.41	15.75	+1.40	0	-1.00	+2.2	0	-1.40	+3.5	0	-2.2	+9.0	0	-6.0	+2.2	0.70	-1.40	+6.0	2.5	-3.5

Nominal Size Range Inches		Class LC7			Class LC8			Class LC9			Class LC10			Class LC11		
		Hole Tol. GR10	Minimum Clearance	Shaft Tol. GR9	Hole Tol. GR10	Minimum Clearance	Shaft Tol. GR9	Hole Tol. GR11	Minimum Clearance	Shaft Tol. GR10	Hole Tol. GR12	Minimum Clearance	Shaft Tol. GR11	Hole Tol. GR13	Minimum Clearance	Shaft Tol. GR12
Over	To	-0		+0	-0		+0	-0		+0	-0		+0	-0		+0
0	.12	+1.6	0.6	-1.0	+1.6	1.0	-1.0	+2.5	2.5	-1.6	+1.0	4.0	-2.5	+6.0	5.0	-4.0
.12	.24	+1.8	0.8	-1.2	+1.8	1.2	-1.2	+3.0	2.8	-1.8	+5.0	4.5	-3.0	+7.0	6.0	-5.0
.24	.40	+2.2	1.0	-1.4	+2.2	1.6	-1.4	+3.5	3.0	-2.2	+6.0	5.0	-3.5	+9.0	7.0	-6.0
.40	.71	+2.8	1.2	-1.6	+2.8	2.0	-1.6	+4.0	3.5	-2.8	+7.0	6.0	-4.0	+10.0	8.0	-7.0
.71	1.19	+3.5	1.6	-2.0	+3.5	2.5	-2.0	+5.0	4.5	-3.5	+8.0	7.0	-5.0	+12.0	10.0	-8.0
1.19	1.97	+4.0	2.0	-2.5	+4.0	3.6	-2.5	+6.0	5.0	-4.0	+10.0	8.0	-6.0	+16.0	12.0	-10.0
1.97	3.15	+4.5	2.5	-3.0	+4.5	4.0	-3.0	+7.0	6.0	-4.5	+12.0	10.0	-7.0	+18.0	14.0	-12.0
3.15	4.73	+5.0	3.0	-3.5	+5.0	5.0	-3.5	+9.0	7.0	-5.0	+14.0	11.0	-9.0	+22.0	16.0	-14.0
4.73	7.09	+6.0	3.5	-4.0	+6.0	6.0	-4.0	+10.0	8.0	-6.0	+16.0	12.0	-10.0	+25.0	18.0	-16.0
7.09	9.85	+7.0	4.0	-4.5	+7.0	7.0	-4.5	+12.0	10.0	-7.0	+18.0	16.0	-12.0	+28.0	22.0	-18.0
9.85	12.41	+8.0	4.5	-5.0	+8.0	7.0	-5.0	+12.0	12.0	-8.0	+20.0	20.0	-12.0	+30.0	28.0	-20.0
12.41	15.75	+9.0	5.0	-6.0	+9.0	8.0	-6.0	+14.0	14.0	-9.0	+22.0	22.0	-14.0	+35.0	30.0	-22.0

VALUES IN MILLIMETRES

Nominal Size Range Millimetres		Class LC1			Class LC2			Class LC3			Class LC4			Class LC5			Class LC6		
		Hole Tol. H6	Minimum Clearance	Shaft Tol. h5	Hole Tol. H7	Minimum Clearance	Shaft Tol. h6	Hole Tol. H8	Minimum Clearance	Shaft Tol. h7	Hole Tol. H10	Minimum Clearance	Shaft Tol. h9	Hole Tol. H7	Minimum Clearance	Shaft Tol. g6	Hole Tol. H9	Minimum Clearance	Shaft Tol. f8
Over	To	-0		+0	-0		+0	-0		+0	-0		+0	-0		+0	-0		+0
0	3	+0.006	0	-0.004	+0.010	0	-0.006	+0.015	0	-0.010	+0.041	0	-0.025	+0.010	0.002	-0.006	+0.025	0.008	-0.015
3	6	+0.008	0	-0.005	+0.013	0	-0.008	+0.018	0	-0.013	+0.046	0	-0.030	+0.013	0.004	-0.008	+0.030	0.010	-0.018
6	10	+0.010	0	-0.006	+0.015	0	-0.010	+0.023	0	-0.015	+0.056	0	-0.036	+0.015	0.005	-0.010	+0.036	0.013	-0.023
10	18	+0.010	0	-0.008	+0.018	0	-0.010	+0.025	0	-0.018	+0.070	0	-0.040	+0.018	0.006	-0.010	+0.041	0.015	-0.025
18	30	+0.013	0	-0.010	+0.020	0	-0.013	+0.030	0	-0.020	+0.090	0	-0.050	+0.020	0.008	-0.013	+0.050	0.020	-0.030
30	50	+0.015	0	-0.010	+0.025	0	-0.015	+0.041	0	-0.025	+0.100	0	-0.060	+0.025	0.010	-0.015	+0.060	0.030	-0.040
50	80	+0.018	0	-0.013	+0.030	0	-0.018	+0.046	0	-0.030	+0.110	0	-0.080	+0.030	0.010	-0.018	+0.080	0.030	-0.050
80	120	+0.023	0	-0.015	+0.036	0	-0.023	+0.056	0	-0.036	+0.130	0	-0.080	+0.036	0.013	-0.023	+0.090	0.040	-0.060
120	180	+0.025	0	-0.018	+0.041	0	-0.025	+0.064	0	-0.041	+0.150	0	-0.100	+0.041	0.015	-0.025	+0.100	0.040	-0.060
180	250	+0.030	0	-0.020	+0.046	0	-0.030	+0.071	0	-0.046	+0.180	0	-0.110	+0.046	0.015	-0.030	+0.110	0.050	-0.070
250	315	+0.020	0	-0.023	+0.051	0	-0.030	+0.076	0	-0.051	+0.200	0	-0.130	+0.051	0.018	-0.030	+0.130	0.060	-0.080
315	400	+0.036	0	-0.025	+0.056	0	-0.036	+0.089	0	-0.056	+0.230	0	-0.150	+0.056	0.018	-0.036	+0.150	0.060	-0.090

Nominal Size Range Millimetres		Class LC7			Class LC8			Class LC9			Class LC10			Class LC11		
		Hole Tol. H10	Minimum Clearance	Shaft Tol. e9	Hole Tol. H10	Minimum Clearance	Shaft Tol. d9	Hole Tol. H11	Minimum Clearance	Shaft Tol. c10	Hole Tol. GR12	Minimum Clearance	Shaft Tol. gr11	Hole Tol. GR13	Minimum Clearance	Shaft Tol. gr12
Over	To	-0		+0	-0		+0	-0		+0	-0		+0	-0		+0
0	3	+0.041	0.015	-0.025	+0.041	0.025	-0.025	+0.064	0.06	-0.041	+0.10	0.10	-0.06	+0.15	0.13	-0.10
3	6	+0.046	0.020	-0.030	+0.046	0.030	-0.030	+0.076	0.07	-0.046	+0.13	0.11	-0.08	+0.18	0.15	-0.13
6	10	+0.056	0.025	-0.036	+0.056	0.041	-0.036	+0.089	0.08	-0.056	+0.15	0.13	-0.09	+0.23	0.18	-0.15
10	18	+0.070	0.030	-0.040	+0.070	0.050	-0.040	+0.100	0.09	-0.070	+0.18	0.15	-0.10	+0.25	0.20	-0.18
18	30	+0.090	0.040	-0.050	+0.090	0.060	-0.050	+0.130	0.11	-0.090	+0.20	0.18	-0.13	+0.31	0.25	-0.20
30	50	+0.100	0.050	-0.060	+0.100	0.090	-0.060	+0.150	0.13	-0.100	+0.25	0.20	-0.15	+0.41	0.31	-0.25
50	80	+0.110	0.060	-0.080	+0.110	0.100	-0.080	+0.180	0.15	-0.110	+0.31	0.25	-0.18	+0.46	0.36	-0.31
80	120	+0.130	0.080	-0.090	+0.130	0.130	-0.090	+0.230	0.18	-0.130	+0.36	0.28	-0.23	+0.56	0.41	-0.36
120	180	+0.150	0.090	-0.100	+0.150	0.150	-0.100	+0.250	0.20	-0.150	+0.41	0.31	-0.25	+0.64	0.46	-0.41
180	250	+0.180	0.100	-0.110	+0.180	0.180	-0.110	+0.310	0.25	-0.180	+0.46	0.41	-0.31	+0.71	0.56	-0.46
250	315	+0.200	0.110	-0.130	+0.200	0.180	-0.130	+0.310	0.31	-0.200	+0.51	0.51	-0.31	+0.76	0.71	-0.51
315	400	+0.230	0.130	-0.150	+0.230	0.200	-0.150	+0.360	0.36	-0.230	+0.56	0.56	-0.36	+0.89	0.76	-0.56

Table 10

From Drafting for Trades and Industry—Mechanical and Electronic. Nelson. Delmar Publishers Inc.

LOCATIONAL TRANSITION FITS

VALUES IN THOUSANDTHS OF AN INCH

Nominal Size Range Inches		Class LT1			Class LT2			Class LT3			Class LT4			Class LT5			Class LT6		
		Hole Tol. GR7	Maximum Interference	Shaft Tol. GR6	Hole Tol. GR8	Maximum Interference	Shaft Tol. GR7	Hole Tol. GR7	Maximum Interference	Shaft Tol. GR6	Hole Tol. GR8	Maximum Interference	Shaft Tol. GR7	Hole Tol. GR7	Maximum Interference	Shaft Tol. GR6	Hole Tol. GR8	Maximum Interference	Shaft Tol. GR7
Over	To	-0		+0	-0		+0	-0		+0	-0		+0	-0		+0	-0		+0
0	.12	+0.4	0.10	-0.25	+0.6	0.20	-0.4	+0.4	0.25	-0.25	+0.6	0.4	-0.4	+0.4	0.5	-0.25	+0.6	0.65	-0.4
.12	.24	+0.5	0.15	-0.30	+0.7	0.25	-0.5	+0.5	0.40	-0.30	+0.7	0.6	-0.5	+0.5	0.6	-0.30	+0.7	0.80	-0.5
.24	.40	+0.6	0.20	-0.40	+0.9	0.30	-0.6	+0.6	0.50	-0.40	+0.9	0.7	-0.6	+0.6	0.8	-0.40	+0.9	1.00	-0.6
.40	.71	+0.7	0.20	-0.40	+1.0	0.30	-0.7	+0.7	0.50	-0.40	+1.0	0.8	-0.7	+0.7	0.9	-0.40	+1.0	1.20	-0.7
.71	1.19	+0.8	0.25	-0.50	+1.2	0.40	-0.8	+0.8	0.60	-0.50	+1.2	0.9	-0.8	+0.8	1.1	-0.50	+1.2	1.40	-0.8
1.19	1.97	+1.0	0.30	-0.60	+1.6	0.50	-1.0	+1.0	0.70	-0.60	+1.6	1.1	-1.0	+1.0	1.3	-0.60	+1.6	1.70	-1.0
1.97	3.15	+1.2	0.30	-0.70	+1.8	0.60	-1.2	+1.2	0.80	-0.70	+1.8	1.3	-1.2	+1.2	1.5	-0.70	+1.8	2.00	-1.2
3.15	4.73	+1.4	0.40	-0.90	+2.2	0.70	-1.4	+1.4	1.00	-0.90	+2.2	1.5	-1.4	+1.4	1.9	-0.90	+2.2	2.40	-1.4
4.73	7.09	+1.6	0.50	-1.00	+2.5	0.80	-1.6	+1.6	1.10	-1.00	+2.5	1.7	-1.6	+1.6	2.2	-1.00	+2.5	2.80	-1.6
7.09	9.85	+1.8	0.60	-1.20	+2.8	0.90	-1.8	+1.8	1.40	-1.20	+2.8	2.0	-1.8	+1.8	2.6	-1.20	+2.8	3.20	-1.8
9.85	12.41	+2.0	0.60	-1.20	+3.0	1.00	-2.0	+2.0	1.40	-1.20	+3.0	2.2	-2.0	+2.0	2.6	-1.20	+3.0	3.40	-2.0
12.41	15.75	+2.2	0.70	-1.40	+3.5	1.00	-2.2	+2.2	1.60	-1.40	+3.5	2.4	-2.2	+2.2	3.0	-1.40	+3.5	3.80	-2.2

VALUES IN MILLIMETRES

Nominal Size Range Millimetres		Class LT1			Class LT2			Class LT3			Class LT4			Class LT5			Class LT6		
		Hole Tol. H7	Maximum Interference	Shaft Tol. js6	Hole Tol. H8	Maximum Interference	Shaft Tol. js7	Hole Tol. H7	Maximum Interference	Shaft Tol. k6	Hole Tol. H8	Maximum Interference	Shaft Tol. k7	Hole Tol. H7	Maximum Interference	Shaft Tol. n6	Hole Tol. H8	Maximum Interference	Shaft Tol. n7
Over	To	-0		+0	-0		+0	-0		+0	-0		+0	-0		+0	-0		+0
0	3	+0.010	0.002	-0.006	+0.015	0.005	-0.010	+0.010	0.006	-0.006	+0.015	0.010	-0.010	+0.010	0.013	-0.006	+0.015	0.016	-0.010
3	6	+0.013	0.004	-0.008	+0.018	0.006	-0.013	+0.013	0.010	-0.008	+0.018	0.015	-0.013	+0.013	0.015	-0.008	+0.018	0.020	-0.013
6	10	+0.015	0.005	-0.010	+0.023	0.008	-0.015	+0.015	0.013	-0.010	+0.023	0.018	-0.015	+0.015	0.020	-0.010	+0.023	0.025	-0.015
10	18	+0.018	0.005	-0.010	+0.025	0.008	-0.018	+0.018	0.013	-0.010	+0.025	0.020	-0.018	+0.018	0.023	-0.010	+0.025	0.030	-0.018
18	30	+0.020	0.006	-0.013	+0.030	0.010	-0.020	+0.020	0.015	-0.013	+0.030	0.023	-0.020	+0.020	0.028	-0.013	+0.030	0.036	-0.020
30	50	+0.025	0.008	-0.015	+0.041	0.013	-0.025	+0.025	0.018	-0.015	+0.041	0.028	-0.025	+0.025	0.033	-0.015	+0.041	0.044	-0.025
50	80	+0.030	0.008	-0.018	+0.046	0.015	-0.030	+0.030	0.020	-0.018	+0.046	0.033	-0.030	+0.030	0.038	-0.018	+0.046	0.051	-0.030
80	120	+0.036	0.010	-0.023	+0.056	0.018	-0.036	+0.036	0.025	-0.023	+0.056	0.038	-0.036	+0.036	0.048	-0.023	+0.056	0.062	-0.036
120	180	+0.041	0.013	-0.025	+0.064	0.020	-0.041	+0.041	0.028	-0.025	+0.064	0.044	-0.041	+0.041	0.056	-0.025	+0.064	0.071	-0.041
180	250	+0.046	0.015	-0.030	+0.071	0.023	-0.046	+0.046	0.036	-0.030	+0.071	0.051	-0.046	+0.046	0.066	-0.030	+0.071	0.081	-0.046
250	315	+0.051	0.015	-0.030	+0.076	0.025	-0.051	+0.051	0.036	-0.030	+0.076	0.056	-0.051	+0.051	0.066	-0.030	+0.076	0.086	-0.051
315	400	+0.056	0.018	-0.036	+0.089	0.025	-0.056	+0.056	0.041	-0.036	+0.089	0.062	-0.056	+0.056	0.076	-0.036	+0.089	0.096	-0.056

From Drafting for Trades and Industry—Mechanical and Electronic. Nelson. Delmar Publishers Inc.

Table 11

LOCATIONAL INTERFERENCE FITS

VALUES IN THOUSANDTHS OF AN INCH

Nominal Size Range Inches		Class LN1 Light Press Fit			Class LN2 Medium Press Fit			Class LN3 Heavy Press Fit			Class LN4			Class LN5			Class LN6		
		Hole Tol. GR6	Maximum Interference	Shaft Tol. GR5	Hole Tol. GR7	Maximum Interference	Shaft Tol. GR6	Hole Tol. GR7	Maximum Interference	Shaft Tol. GR6	Hole Tol. GR8	Maximum Interference	Shaft Tol. GR7	Hole Tol. GR9	Maximum Interference	Shaft Tol. GR8	Hole Tol. GR10	Maximum Interference	Shaft Tol. GR9
Over	To	-0		+0	-0		+0	-0		+0	-0		+0	-0		+0	-0		+0
0	.12	+0.25	0.40	-0.15	+0.4	0.65	-0.25	+0.4	0.75	-0.25	+0.6	1.2	-0.4	+1.0	1.8	-0.6	+1.6	3.0	-1.0
.12	.24	+0.30	0.50	-0.20	+0.5	0.80	-0.30	+0.5	0.90	-0.30	+0.7	1.5	-0.5	+1.2	2.3	-0.7	+1.8	3.6	-1.2
.24	.40	+0.40	0.65	-0.25	+0.6	1.00	-0.40	+0.6	1.20	-0.40	+0.9	1.8	-0.6	+1.4	2.8	-0.9	+2.2	4.4	-1.4
.40	.71	+0.40	0.70	-0.30	+0.7	1.10	-0.40	+0.7	1.40	-0.40	+1.0	2.2	-0.7	+1.6	3.4	-1.0	+2.8	5.6	-1.6
.71	1.19	+0.50	0.90	-0.40	+0.8	1.30	-0.50	+0.8	1.70	-0.50	+1.2	2.6	-0.8	+2.0	4.2	-1.2	+3.5	7.0	-2.0
1.19	1.97	+0.60	1.00	-0.40	+1.0	1.60	-0.60	+1.0	2.00	-0.60	+1.6	3.4	-1.0	+2.5	5.3	-1.6	+4.0	8.5	-2.5
1.97	3.15	+0.70	1.30	-0.50	+1.2	2.10	-0.70	+1.2	2.30	-0.70	+1.8	4.0	-1.2	+3.0	6.3	-1.8	+4.5	10.0	-3.0
3.15	4.73	+0.90	1.60	-0.60	+1.4	2.50	-0.90	+1.4	2.90	-0.90	+2.2	4.8	-1.4	+4.0	7.7	-2.2	+5.0	11.5	-3.5
4.73	7.09	+1.00	1.90	-0.70	+1.6	2.80	-1.00	+1.6	3.50	-1.00	+2.5	5.6	-1.6	+4.5	8.7	-2.5	+6.0	13.5	-4.0
7.09	9.85	+1.20	2.20	-0.80	+1.8	3.20	-1.20	+1.8	4.20	-1.20	+2.8	6.6	-1.8	+5.0	10.3	-2.8	+7.0	16.5	-4.5
9.85	12.41	+1.20	2.30	-0.90	+2.0	3.40	-1.20	+2.0	4.70	-1.20	+3.0	7.5	-2.0	+6.0	12.0	-3.0	+8.0	19.0	-5.0
12.41	15.75	+1.40	2.60	-1.00	+2.2	3.90	-1.40	+2.2	5.90	-1.40	+3.5	8.7	-2.2	+6.0	14.5	-3.5	+9.0	23.0	-6.0

VALUES IN MILLIMETRES

Nominal Size Range Millimetres		Class LN1 Light Press Fit			Class LN2 Medium Press Fit			Class LN3 Heavy Press Fit			Class LN4			Class LN5			Class LN6		
		Hole Tol. GR6	Maximum Interference	Shaft Tol. gr5	Hole Tol. H7	Maximum Interference	Shaft Tol. p6	Hole Tol. H7	Maximum Interference	Shaft Tol. t6	Hole Tol. GR8	Maximum Interference	Shaft Tol. gr7	Hole Tol. GR9	Maximum Interference	Shaft Tol. gr8	Hole Tol. GR10	Maximum Interference	Shaft Tol. gr9
Over	To	-0		+0	-0		+0	-0		+0	-0		+0	-0		+0	-0		+0
0	3	+0.006		-0.004	+0.010	0.016	-0.006	+0.010	0.019	-0.006	+0.015	0.030	-0.010	+0.025	0.046	-0.015	+0.041	0.076	-0.025
3	6	+0.008		-0.005	+0.013	0.020	-0.008	+0.013	0.023	-0.008	+0.018	0.038	-0.013	+0.030	0.059	-0.018	+0.046	0.091	-0.030
6	10	+0.010		-0.006	+0.015	0.025	-0.010	+0.015	0.030	-0.010	+0.023	0.046	-0.015	+0.036	0.071	-0.023	+0.056	0.112	-0.036
10	18	+0.010		-0.008	+0.018	0.028	-0.010	+0.018	0.036	-0.010	+0.025	0.056	-0.018	+0.041	0.086	-0.025	+0.071	0.142	-0.041
18	30	+0.013		-0.010	+0.020	0.033	-0.013	+0.020	0.044	-0.013	+0.030	0.066	-0.020	+0.051	0.107	-0.030	+0.089	0.178	-0.051
30	50	+0.015		-0.010	+0.025	0.041	-0.015	+0.025	0.051	-0.015	+0.041	0.086	-0.025	+0.064	0.135	-0.041	+0.102	0.216	-0.064
50	80	+0.018		-0.013	+0.030	0.054	-0.018	+0.030	0.059	-0.018	+0.046	0.102	-0.030	+0.076	0.160	-0.046	+0.114	0.254	-0.076
80	120	+0.023		-0.015	+0.036	0.064	-0.023	+0.036	0.074	-0.023	+0.056	0.122	-0.036	+0.102	0.196	-0.056	+0.127	0.292	-0.102
120	180	+0.025		-0.018	+0.041	0.071	-0.025	+0.041	0.089	-0.025	+0.064	0.142	-0.041	+0.114	0.221	-0.064	+0.152	0.343	-0.114
180	250	+0.030		-0.020	+0.046	0.081	-0.030	+0.046	0.107	-0.030	+0.071	0.168	-0.046	+0.127	0.262	-0.071	+0.178	0.419	-0.127
250	315	+0.030		-0.023	+0.051	0.086	-0.030	+0.051	0.119	-0.030	+0.076	0.191	-0.051	+0.152	0.305	-0.076	+0.203	0.483	-0.152
315	400	+0.036		-0.025	+0.056	0.099	-0.036	+0.056	0.150	-0.036	+0.089	0.221	-0.056	+0.152	0.368	-0.089	+0.229	0.584	-0.152

From Drafting for Trades and Industry—Mechanical and Electronic. Nelson. Delmar Publishers Inc.

Table 12

FORCE AND SHRINK FITS

VALUES IN THOUSANDTHS OF AN INCH

Nominal Size Range Inches		Class FN1 Light Drive Fit			Class FN2 Medium Drive Fit			Class FN3 Heavy Drive Fit			Class FN4 Shrink Fit			Class FN5 Heavy Shrink Fit		
		Hole Tol. GR6	Maximum Interference	Shaft Tol. GR5	Hole Tol. GR7	Maximum Interference	Shaft Tol. GR6	Hole Tol. GR7	Maximum Interference	Shaft Tol. GR6	Hole Tol. GR7	Maximum Interference	Shaft Tol. GR6	Hole Tol. GR8	Maximum Interference	Shaft Tol. GR7
Over	To	-0		+0	-0		+0	-0		+0	-0		+0	-0		+0
0	.12	+0.25	0.50	-0.15	+0.40	0.85	-0.25				+0.40	0.95	-0.25	+0.60	1.30	-0.40
.12	.24	+0.30	0.60	-0.20	+0.50	1.00	-0.30				+0.50	1.20	-0.30	+0.70	1.70	-0.50
.24	.40	+0.40	0.75	-0.25	+0.60	1.40	-0.40				+0.60	1.60	-0.40	+0.90	2.00	-0.60
.40	.56	+0.40	0.80	-0.30	+0.70	1.60	-0.40				+0.70	1.80	-0.40	+1.00	2.30	-0.70
.56	.71	+0.40	0.90	-0.30	+0.70	1.60	-0.40				+0.70	1.80	-0.40	+1.00	2.50	-0.70
.71	.95	+0.50	1.10	-0.40	+0.80	1.90	-0.50				+0.80	2.10	-0.50	+1.20	3.00	-0.80
.95	1.19	+0.50	1.20	-0.40	+0.80	1.90	-0.50	+0.80	2.10	-0.50	+0.80	2.30	-0.50	+1.20	3.30	-0.80
1.19	1.58	+0.60	1.30	-0.40	+1.00	2.40	-0.60	+1.00	2.60	-0.60	+1.00	3.10	-0.60	+1.60	4.00	-1.00
1.58	1.97	+0.60	1.40	-0.40	+1.00	2.40	-0.60	+1.00	2.80	-0.60	+1.00	3.40	-0.60	+1.60	5.00	-1.00
1.97	2.56	+0.70	1.80	-0.50	+1.20	2.70	-0.70	+1.20	3.20	-0.70	+1.20	4.20	-0.70	+1.80	6.20	-1.20
2.56	3.15	+0.70	1.90	-0.50	+1.20	2.90	-0.70	+1.20	3.70	-0.70	+1.20	4.70	-0.70	+1.80	7.20	-1.20
3.15	3.94	+0.90	2.40	-0.60	+1.40	3.70	-0.90	+1.40	4.40	-0.70	+1.40	5.90	-0.90	+2.20	8.40	-1.40

VALUES IN MILLIMETRES

Nominal Size Range Millimetres		Class FN1 Light Drive Fit			Class FN2 Medium Drive Fit			Class FN3 Heavy Drive Fit			Class FN4 Shrink Fit			Class FN5 Heavy Shrink Fit		
		Hole Tol. GR6	Maximum Interference	Shaft Tol. gr5	Hole Tol. H7	Maximum Interference	Shaft Tol. s6	Hole Tol. H7	Maximum Interference	Shaft Tol. t6	Hole Tol. GR8	Maximum Interference	Shaft Tol. gr7	Hole Tol. H8	Maximum Interference	Shaft Tol. t7
Over	To	-0		+0	-0		+0	-0		+0	-0		+0	-0		+0
0	3	+0.006	0.013	-0.004	+0.010	0.216	-0.006				+0.010	0.024	-0.006	+0.015	0.033	-0.010
3	6	+0.007	0.015	-0.005	+0.013	0.025	-0.007				+0.013	0.030	-0.007	+0.018	0.043	-0.013
6	10	+0.010	0.019	-0.006	+0.015	0.036	-0.010				+0.015	0.041	-0.010	+0.023	0.051	-0.015
10	14	+0.010	0.020	-0.008	+0.018	0.041	-0.010				+0.018	0.046	-0.010	+0.025	0.058	-0.018
14	18	+0.010	0.023	-0.008	+0.018	0.041	-0.010				+0.018	0.046	-0.010	+0.025	0.064	-0.018
18	24	+0.013	0.028	-0.010	+0.020	0.048	-0.013				+0.020	0.053	-0.013	+0.030	0.076	-0.020
24	30	+0.013	0.030	-0.010	+0.020	0.048	-0.013	+0.020	0.053	-0.013	+0.020	0.058	-0.013	+0.030	0.084	-0.020
30	40	+0.015	0.033	-0.010	+0.025	0.061	-0.015	+0.025	0.066	-0.015	+0.025	0.079	-0.015	+0.041	0.102	-0.025
40	50	+0.015	0.036	-0.010	+0.025	0.061	-0.015	+0.025	0.071	-0.015	+0.025	0.086	-0.015	+0.041	0.127	-0.025
50	65	+0.018	0.046	-0.013	+0.030	0.069	-0.018	+0.030	0.082	-0.018	+0.030	0.107	-0.018	+0.046	0.157	-0.030
65	80	+0.018	0.048	-0.013	+0.030	0.074	-0.018	+0.030	0.094	-0.018	+0.030	0.119	-0.018	+0.046	0.183	-0.030
80	100	+0.023	0.061	-0.015	+0.035	0.094	-0.023	+0.035	0.112	-0.023	+0.036	0.150	-0.023	+0.056	0.213	-0.036

From Drafting for Trades and Industry—Mechanical and Electronic. Nelson. Delmar Publishers Inc.

Table 13

UNIFIED STANDARD SCREW THREAD SERIES

Sizes Primary	Sizes Secondary	Basic Major Diameter	Coarse UNC	Fine UNF	Extra fine UNEF	4UN	6UN	8UN	12UN	16UN	20UN	28UN	32UN	Sizes
0		0.0600	—	80	—	—	—	—	—	—	—	—	—	0
	1	0.0730	64	72	—	—	—	—	—	—	—	—	—	1
2		0.0860	56	64	—	—	—	—	—	—	—	—	—	2
	3	0.0990	48	56	—	—	—	—	—	—	—	—	—	3
4		0.1120	40	48	—	—	—	—	—	—	—	—	—	4
5		0.1250	40	44	—	—	—	—	—	—	—	—	—	5
6		0.1380	32	40	—	—	—	—	—	—	—	—	UNC	6
8		0.1640	32	36	—	—	—	—	—	—	—	—	UNC	8
10		0.1900	24	32	—	—	—	—	—	—	—	—	UNF	10
	12	0.2160	24	28	32	—	—	—	—	—	—	UNF	UNEF	12
¼		0.2500	20	28	32	—	—	—	—	—	UNC	UNF	UNEF	¼
5⁄16		0.3125	18	24	32	—	—	—	—	—	20	28	UNEF	5⁄16
3⁄8		0.3750	16	24	32	—	—	—	—	UNC	20	28	UNEF	3⁄8
7⁄16		0.4375	14	20	28	—	—	—	—	16	UNF	UNEF	32	7⁄16
½		0.5000	13	20	28	—	—	—	—	16	UNF	UNEF	32	½
9⁄16		0.5625	12	18	24	—	—	—	UNC	16	20	28	32	9⁄16
5⁄8		0.6250	11	18	24	—	—	—	12	16	20	28	32	5⁄8
	11⁄16	0.6875	—	—	24	—	—	—	12	16	20	28	32	11⁄16
¾		0.7500	10	16	20	—	—	—	12	UNF	UNEF	28	32	¾
	13⁄16	0.8125	—	—	20	—	—	—	12	16	UNEF	28	32	13⁄16
7⁄8		0.8750	9	14	20	—	—	—	12	16	UNEF	28	32	7⁄8
	15⁄16	0.9375	—	—	20	—	—	—	12	16	UNEF	28	32	15⁄16
1		1.0000	8	12	20	—	—	UNC	UNF	16	UNEF	28	32	1
	1 1⁄16	1.0625	—	—	18	—	—	8	12	16	20	28	—	1 1⁄16
1 1⁄8		1.1250	7	12	18	—	—	8	UNF	16	20	28	—	1 1⁄8
	1 3⁄16	1.1875	—	—	18	—	—	8	12	16	20	28	—	1 3⁄16
1 ¼		1.2500	7	12	18	—	—	8	UNF	16	20	28	—	1 ¼
	1 5⁄16	1.3125	—	—	18	—	—	8	12	16	20	28	—	1 5⁄16
1 3⁄8		1.3750	6	12	18	—	UNC	8	UNF	16	20	28	—	1 3⁄8
	1 7⁄16	1.4375	—	—	18	—	6	8	12	16	20	28	—	1 7⁄16
1 ½		1.5000	6	12	18	—	UNC	8	UNF	16	20	28	—	1 ½
	1 9⁄16	1.5625	—	—	18	—	6	8	12	16	20	—	—	1 9⁄16
1 5⁄8		1.6250	—	—	18	—	6	8	12	16	20	—	—	1 5⁄8
	1 11⁄16	1.6875	—	—	18	—	6	8	12	16	20	—	—	1 11⁄16
1 ¾		1.7500	5	—	—	—	6	8	12	16	20	—	—	1 ¾
	1 13⁄16	1.8125	—	—	—	—	6	8	12	16	20	—	—	1 13⁄16
1 7⁄8		1.8750	—	—	—	—	6	8	12	16	20	—	—	1 7⁄8
	1 15⁄16	1.9375	—	—	—	—	6	8	12	16	20	—	—	1 15⁄16
2		2.0000	4½	—	—	—	6	8	12	16	20	—	—	2
	2 1⁄8	2.1250	—	—	—	—	6	8	12	16	20	—	—	2 1⁄8
2 ¼		2.2500	4½	—	—	—	6	8	12	16	20	—	—	2 ¼
	2 3⁄8	2.3750	—	—	—	—	6	8	12	16	20	—	—	2 3⁄8
2 ½		2.5000	4	—	—	UNC	6	8	12	16	20	—	—	2 ½
	2 5⁄8	2.6250	—	—	—	4	6	8	12	16	20	—	—	2 5⁄8
2 ¾		2.7500	4	—	—	UNC	6	8	12	16	20	—	—	2 ¾
	2 7⁄8	2.8750	—	—	—	4	6	8	12	16	20	—	—	2 7⁄8
3		3.0000	4	—	—	UNC	6	8	12	16	20	—	—	3
	3 1⁄8	3.1250	—	—	—	4	6	8	12	16	—	—	—	3 1⁄8
3 ¼		3.2500	4	—	—	UNC	6	8	12	16	—	—	—	3 ¼
	3 3⁄8	3.3750	—	—	—	4	6	8	12	16	—	—	—	3 3⁄8
3 ½		3.5000	4	—	—	UNC	6	8	12	16	—	—	—	3 ½
	3 5⁄8	3.6250	—	—	—	4	6	8	12	16	—	—	—	3 5⁄8
3 ¾		3.7500	4	—	—	UNC	6	8	12	16	—	—	—	3 ¾
	3 7⁄8	3.8750	—	—	—	4	6	8	12	16	—	—	—	3 7⁄8
4		4.0000	4	—	—	UNC	6	8	12	16	—	—	—	4
	4 1⁄8	4.1250	—	—	—	4	6	8	12	16	—	—	—	4 1⁄8
4 ¼		4.2500	—	—	—	4	6	8	12	16	—	—	—	4 ¼
	4 3⁄8	4.3750	—	—	—	4	6	8	12	16	—	—	—	4 3⁄8
4 ½		4.5000	—	—	—	4	6	8	12	16	—	—	—	4 ½
	4 5⁄8	4.6250	—	—	—	4	6	8	12	16	—	—	—	4 5⁄8
4 ¾		4.7500	—	—	—	4	6	8	12	16	—	—	—	4 ¾
	4 7⁄8	4.8750	—	—	—	4	6	8	12	16	—	—	—	4 7⁄8
5		5.0000	—	—	—	4	6	8	12	16	—	—	—	5
	5 1⁄8	5.1250	—	—	—	4	6	8	12	16	—	—	—	5 1⁄8
5 ¼		5.2500	—	—	—	4	6	8	12	16	—	—	—	5 ¼
	5 3⁄8	5.3750	—	—	—	4	6	8	12	16	—	—	—	5 3⁄8
5 ½		5.5000	—	—	—	4	6	8	12	16	—	—	—	5 ½
	5 5⁄8	5.6250	—	—	—	4	6	8	12	16	—	—	—	5 5⁄8
5 ¾		5.7500	—	—	—	4	6	8	12	16	—	—	—	5 ¾
	5 7⁄8	5.8750	—	—	—	4	6	8	12	16	—	—	—	5 7⁄8
6		6.0000	—	—	—	4	6	8	12	16	—	—	—	6

Table 14

DECIMAL EQUIVALENTS AND TAP DRILL SIZES
OF WIRE GAGE LETTER AND FRACTIONAL SIZE DRILLS
(TAP DRILL SIZES BASED ON 75% MAXIMUM THREAD)

FRACTIONAL SIZE DRILLS INCHES	WIRE GAGE DRILLS	DECIMAL EQUIVALENT INCHES	TAP SIZE OF THREAD	TAP THREADS PER INCH
	80	.0135		
	79	.0145		
1/64	.....	.0156		
	78	.0160		
	77	.0180		
	76	.0200		
	75	.0210		
	74	.0225		
	73	.0240		
	72	.0250		
	71	.0260		
	70	.0280		
	69	.0292		
	68	.0310		
1/32	.....	.0312		
	67	.0320		
	66	.0330		
	65	.0350		
	64	.0360		
	63	.0370		
	62	.0380		
	61	.0390		
	60	.0400		
	59	.0410		
	58	.0420		
	57	.0430		
	56	.0465		
3/64	.....	.0469	0	80
	55	.0520		
	54	.0550		
	53	.0595	1	64
1/16	.....	.0625		72
	52	.0635		
	51	.0670		
	50	.0700	2	56
	49	.0730		64
	48	.0760		
5/64	.....	.0781		
	47	.0785	3	48
	46	.0810		
	45	.0820	3	56
	44	.0860		
	43	.0890	4	40
	42	.0935	4	48
3/32	.....	.0937		
	41	.0960		
	40	.0980		
	39	.0995		
	38	.1015	5	40
	37	.1040	5	44
	36	.1065	6	32
7/64	.....	.1094		
	35	.1100		
	34	.1110		
	33	.1130	6	40
	32	.1160		
	31	.1200		
1/8	.....	.1250		
	30	.1285		
	29	.1360	8	32
	28	.1405		36

FRACTIONAL SIZE DRILLS INCHES	WIRE GAGE DRILLS	DECIMAL EQUIVALENT INCHES	TAP SIZE OF THREAD	TAP THREADS PER INCH
9/64	.....	.1406		
	27	.1440		
	26	.1470		
	25	.1495	10	24
	24	.1520		
	23	.1540		
5/32	.....	.1562		
	22	.1570		
	21	.1590	10	32
	20	.1610		
	19	.1660		
	18	.1695		
11/64	.....	.1719		
	17	.1730		
	16	.1770	12	24
	15	.1800		
	14	.1820	12	28
	13	.1850		
3/16	.....	.1875		
	12	.1890		
	11	.1910		
	10	.1935		
	9	.1960		
	8	.1990		
	7	.2010	1/4	20
13/64	.....	.2031		
	6	.2040		
	5	.2055		
	4	.2090		
	3	.2130	1/4	28
7/32	.....	.2187		
	2	.2210		
	1	.2280		
	A	.2340		
15/64	.....	.2344		
	B	.2380		
	C	.2420		
	D	.2460		
1/4	E	.2500		
	F	.2570	5/16	18
	G	.2610		
17/64	.....	.2656		
	H	.2660		
	I	.2720	5/16	24
	J	.2770		
	K	.2810		
9/32	.....	.2812		
	L	.2900		
	M	.2950		
19/64	.....	.2969		
	N	.3020		
5/16	.....	.3125	3/8	16
	O	.3160		
	P	.3230		
21/64	.....	.3281		
	Q	.3320	3/8	24
	R	.3390		
11/32	.....	.3437		
	S	.3480		
	T	.3580		

FRACTIONAL SIZE DRILLS INCHES	WIRE GAGE DRILLS	DECIMAL EQUIVALENT INCHES	TAP SIZE OF THREAD	TAP THREADS PER INCH
23/64	.....	.3594		
	U	.3680	7/16	14
3/8		.3750		
	V	.3770		
	W	.3860		
25/64	.....	.3906	7/16	20
	X	.3970		
	Y	.4040		
13/32	.....	.4062		
	Z	.4130		
27/64	.....	.4219	1/2	13
7/16	.....	.4375		
29/64	.....	.4531	1/2	20
15/32	.....	.4687		
31/64	.....	.4844	9/16	12
1/2	.....	.5000		
33/64	.....	.5156	9/16	18
17/32	.....	.5312	5/8	11
35/64	.....	.5469		
9/16	.....	.5625		
37/64	.....	.5781	5/8	18
19/32	.....	.5937		
39/64	.....	.6094		
5/8	.....	.6250		
41/64	.....	.6406		
21/32	.....	.6562	3/4	10
43/64	.....	.6719		
11/16	.....	.6875	3/4	16
45/64	.....	.7031		
23/32	.....	.7187		
47/64	.....	.7344		
3/4	.....	.7500		
49/64	.....	.7656	7/8	9
25/32	.....	.7812		
51/64	.....	.7969		
13/16	.....	.8125	7/8	14
53/64	.....	.8281		
27/32	.....	.8437		
55/64	.....	.8594		
7/8	.....	.8750	1	8
57/64	.....	.8906		
29/32	.....	.9062		
59/64	.....	.9219		
15/16	.....	.9375	1	14
61/64	.....	.9531		
31/32	.....	.9687		
63/64	.....	.9844	1 1/8	7
1	.....	1.0000		

Table 15

DRILLED HOLE TOLERANCE (UNDER NORMAL SHOP CONDITIONS)

STANDARD DRILL SIZE				TOLERANCE IN DECIMALS	
DRILL SIZE				PLUS	MINUS
Number	Fraction	Decimal	Metric (MM)		
80		0.0135	0.3412	0.0023	
79		0.0145	0.3788	0.0024	
—	1/64	0.0156	0.3969	0.0025	
78		0.0160	0.4064	0.0025	
77		0.0180	0.4572	0.0026	
76		0.0200	0.5080	0.0027	
75		0.0210	0.5334	0.0027	
74		0.0225	0.5631	0.0028	
73		0.0240	0.6096	0.0028	
72		0.0250	0.6350	0.0029	
71		0.0260	0.6604	0.0029	
70		0.0280	0.7112	0.0030	.0005
69		0.0292	0.7483	0.0030	
68		0.0310	0.7874	0.0031	
—	1/32	0.0312	0.7937	0.0031	
67		0.0320	0.8128	0.0031	
66		0.0330	0.8382	0.0032	
65		0.0350	0.8890	0.0032	
64		0.0360	0.9144	0.0033	
63		0.0370	0.9398	0.0033	
62		0.0380	0.9652	0.0033	
61		0.0390	0.9906	0.0033	
60		0.0400	1.0160	0.0034	
59		0.0410	1.0414	0.0034	
58		0.0420	1.0668	0.0034	
57		0.0430	1.0922	0.0035	
56		0.0465	1.1684	0.0035	
—	3/64	0.0469	1.1906	0.0036	
55		0.0520	1.3208	0.0037	
54		0.0550	1.3970	0.0038	
53		0.0595	1.5122	0.0039	
—	1/16	0.0625	1.5875	0.0039	
52		0.0635	1.6002	0.0039	
51		0.0670	1.7018	0.0040	
50		0.0700	1.7780	0.0041	
49		0.0730	1.8542	0.0041	.001
48		0.0760	1.9304	0.0042	
—	5/64	0.0781	1.9844	0.0042	
47		0.0785	2.0001	0.0042	
46		0.0810	2.0574	0.0043	
45		0.0820	2.0828	0.0043	
44		0.0860	2.1844	0.0044	
43		0.0890	2.2606	0.0044	
42		0.0935	2.3622	0.0045	
—	3/32	0.0937	2.3812	0.0045	
41		0.0960	2.4384	0.0045	
40		0.0980	2.4892	0.0046	
39		0.0995	2.5377	0.0046	
38		0.1015	2.5908	0.0046	
37		0.1040	2.6416	0.0047	
36		0.1065	2.6924	0.0047	
—	7/64	0.1094	2.7781	0.0047	

Table 16

DRILLED HOLE TOLERANCE (UNDER NORMAL SHOP CONDITIONS)

STANDARD DRILL SIZE				TOLERANCE IN DECIMALS	
DRILL SIZE				PLUS	MINUS
No./Letter	Fraction	Decimal	Metric (MM)		
35		0.1100	2.7490	0.0047	
34		0.1110	2.8194	0.0048	
33		0.1130	2.8702	0.0048	
32		0.1160	2.9464	0.0048	
31		0.1200	3.0480	0.0049	
—	1/8	0.1250	3.1750	0.0050	
30		0.1285	3.2766	0.0050	
29		0.1360	3.4544	0.0051	
28		0.1405	3.5560	0.0052	
—	9/64	0.1406	3.5719	0.0052	
27		0.1440	3.6576	0.0052	
26		0.1470	3.7338	0.0052	
25		0.1495	3.7886	0.0053	
24		0.1520	3.8608	0.0053	
23		0.1540	3.9116	0.0053	
—	5/32	0.1562	3.9687	0.0053	
22		0.1570	3.9878	0.0053	
21		0.1590	4.0386	0.0054	
20		0.1610	4.0894	0.0054	
19		0.1660	4.2164	0.0055	
18		0.1695	4.3180	0.0055	
—	11/64	0.1719	4.3656	0.0055	
17		0.1730	4.3942	0.0055	
16		0.1770	4.4958	0.0056	.001
15		0.1800	4.5720	0.0056	
14		0.1820	4.6228	0.0057	
13		0.1850	4.6990	0.0057	
—	3/16	0.1875	4.7625	0.0057	
12		0.1890	4.8006	0.0057	
11		0.1910	4.8514	0.0057	
10		0.1935	4.9276	0.0058	
9		0.1960	4.9784	0.0058	
8		0.1990	5.0800	0.0058	
7		0.2010	5.1054	0.0058	
—	13/64	0.2031	5.1594	0.0058	
6		0.2040	5.1816	0.0058	
5		0.2055	5.2070	0.0059	
4		0.2090	5.3086	0.0059	
3		0.2130	5.4102	0.0059	
—	7/32	0.2187	5.5562	0.0060	
2		0.2210	5.6134	0.0060	
1		0.2280	5.7912	0.0061	
A		0.2340	5.9436	0.0061	
—	15/64	0.2344	5.9531	0.0061	
B		0.2380	6.0452	0.0061	
C		0.2420	6.1468	0.0062	
D		0.2460	6.2484	0.0062	
E	1/4	0.2500	6.3500	0.0063	
F		0.2570	6.5278	0.0063	
G		0.2610	6.6294	0.0063	
—	17/64	0.2656	6.7469	0.0064	
H		0.2660	6.7564	0.0064	
I		0.2720	6.9088	0.0064	.002
J		0.2770	7.0358	0.0065	
K		0.2810	7.1374	0.0065	
—	9/32	0.2812	7.1437	0.0065	
L		0.2900	7.3660	0.0066	
M		0.2950	7.4930	0.0066	
—	19/64	0.2969	7.5406	0.0066	

Table 16 (Cont'd)

DRILLED HOLE TOLERANCE (UNDER NORMAL SHOP CONDITIONS)

STANDARD DRILL SIZE				TOLERANCE IN DECIMALS	
DRILL SIZE				PLUS	MINUS
Letter	Fraction	Decimal	Metric (MM)		
N		0.3020	7.6708	0.0067	
—	5/16	0.3125	7.9375	0.0067	
O		0.3160	8.0264	0.0068	
P		0.3230	8.2042	0.0068	
—	21/64	0.3281	8.3344	0.0068	
Q		0.3320	8.4328	0.0069	
R		0.3390	8.6106	0.0069	
—	11/32	0.3437	8.7312	0.0070	
S		0.3480	8.8392	0.0070	
T		0.3580	9.0932	0.0071	
—	23/64	0.3594	9.1281	0.0071	
U		0.3680	9.3472	0.0071	
—	3/8	0.3750	9.5250	0.0072	
V		0.3770	9.5758	0.0072	
W		0.3860	9.8044	0.0072	
—	25/64	0.3906	9.9219	0.0073	
X		0.3970	10.0838	0.0073	
Y		0.4040	10.2616	0.0073	
—	13/32	0.4062	10.3187	0.0074	
Z		0.4130	10.4902	0.0074	.002
	27/64	0.4219	10.7156	0.0075	
	7/16	0.4375	10.1125	0.0075	
	29/64	0.4531	11.5094	0.0076	
	15/32	0.4687	11.9062	0.0077	
	31/64	0.4844	12.3031	0.0078	
	1/2	0.5000	12.7000	0.0079	
	33/64	0.5156	13.0968	0.0080	
	17/32	0.5312	13.4937	0.0081	
	35/64	0.5469	13.8906	0.0081	
	9/16	0.5625	14.2875	0.0082	
	37/64	0.5781	14.6844	0.0083	
	19/32	0.5927	15.0812	0.0084	
	39/64	0.6094	15.4781	0.0084	
	5/8	0.6250	15.8750	0.0085	
	41/64	0.6406	16.2719	0.0086	
	21/32	0.6562	16.6687	0.0086	
	43/64	0.6719	17.0656	0.0087	
	11/16	0.6875	17.4625	0.0088	
	45/64	0.7031	17.8594	0.0088	
	23/32	0.7187	18.2562	0.0089	
	47/64	0.7344	18.6532	0.0090	
	3/4	0.7500	19.0500	0.0090	
	49/64	0.7656	19.4469	0.0091	
	25/32	0.7812	19.8433	0.0092	
	51/64	0.7969	20.2402	0.0092	
	13/16	0.8125	20.6375	0.0093	
	53/64	0.8281	21.0344	0.0093	
	27/32	0.8437	21.4312	0.0094	
	55/64	0.8594	21.8281	0.0095	
	7/8	0.8750	22.2250	0.0095	.003
	57/64	0.8906	22.6219	0.0096	
	29/32	0.9062	23.0187	0.0096	
	59/64	0.9219	23.4156	0.0097	
	15/16	0.9375	23.8125	0.0097	
	61/64	0.9531	24.2094	0.0098	
	31/32	0.9687	24.6062	0.0098	
	63/64	0.9844	25.0031	0.0099	
	1	1.0000	25.4000	0.0100	

From Drafting for Trades and Industry—Mechanical and Electronic. Nelson. Delmar Publishers Inc.

Table 16
(Cont'd)

INCH — METRIC THREAD COMPARISON

INCH SERIES			METRIC			
Size	Dia.(In.)	TPI	Size	Dia. (In.)	Pitch (MM)	TPI (Approx)
			M1.4	.055	.3 .2	85 127
#0	.060	80				
			M1.6	.063	.35 .2	74 127
#1	.073	64 72				
			M2	.079	.4 .25	64 101
#2	.086	56 64				
			M2.5	.098	.45 .35	56 74
#3	.099	48 56				
#4	.112	40 48				
			M3	.118	.5 .35	51 74
#5	.125	40 44				
#6	.138	32 40				
			M4	.157	.7 .5	36 51
#8	.164	32 36				
#10	.190	24 32				
			M5	.196	.8 .5	32 51
			M6	.236	1.0 .75	25 34
¼	.250	20 28				
⁵⁄₁₆	.312	18 24				
			M8	.315	1.25 1.0	20 25
³⁄₈	.375	16 24				
			M10	.393	1.5 1.25	17 20
⁷⁄₁₆	.437	14 20				
			M12	.472	1.75 1.25	14.5 20
½	.500	13 20				
			M14	.551	2 1.5	12.5 17
⁵⁄₈	.625	11 18				
			M16	.630	2 1.5	12.5 17
			M18	.709	2.5 1.5	10 17
¾	.750	10 16				
			M20	.787	2.5 1.5	10 17
			M22	.866	2.5 1.5	10 17
⅞	.875	9 14				
			M24	.945	3 2	8.5 12.5
1"	1.000	8 12				
			M27	1.063	3 2	8.5 12.5

Table 17

I.S.O. BASIC METRIC THREAD INFORMATION

Basic Major DIA & Pitch	Tap Drill DIA	INTERNAL THREADS		EXTERNAL THREADS		Clearance Hole
		Minor DIA MAX	Minor DIA MIN	Major DIA MAX	Major DIA MIN	
M1.6 × 0.35	1.25	1.321	1.221	1.576	1.491	1.9
M2 × 0.4	1.60	1.679	1.567	1.976	1.881	2.4
M2.5 × 0.45	2.05	2.138	2.013	2.476	2.013	2.9
M3 × 0.5	2.50	2.599	2.459	2.976	2.870	3.4
M3.5 × 0.6	2.90	3.010	2.850	3.476	3.351	4.0
M4 × 0.7	3.30	3.422	3.242	3.976	3.836	4.5
M5 × 0.8	4.20	4.334	4.134	4.976	4.826	5.5
M6 × 1	5.00	5.153	4.917	5.974	5.794	6.6
M8 × 1.25	6.80	6.912	6.647	7.972	7.760	9.0
M10 × 1.5	8.50	8.676	8.376	9.968	9.732	11.0
M12 × 1.75	10.20	10.441	10.106	11.966	11.701	13.5
M14 × 2	12.00	12.210	11.835	13.962	13.682	15.5
M16 × 2	14.00	14.210	13.835	15.962	15.682	17.5
M20 × 2.5	17.50	17.744	17.294	19.958	19.623	22.0
M24 × 3	21.00	21.252	20.752	23.952	23.577	26.0
M30 × 3.5	26.50	26.771	26.211	29.947	29.522	33.0
M36 × 4	32.00	32.270	31.670	35.940	35.465	39.0
M42 × 4.5	37.50	37.799	37.129	41.937	41.437	45.0
M48 × 5	43.00	43.297	42.587	47.929	47.399	52.0
M56 × 5.5	50.50	50.796	50.046	55.925	55.365	62.0
M64 × 6	58.00	58.305	57.505	63.920	63.320	70.0
M72 × 6	66.00	66.305	65.505	71.920	71.320	78.0
M80 × 6	74.00	74.305	73.505	79.920	79.320	86.0
M90 × 6	84.00	84.305	83.505	89.920	89.320	96.0
M100 × 6	94.00	94.305	93.505	99.920	99.320	107.0

Table 18

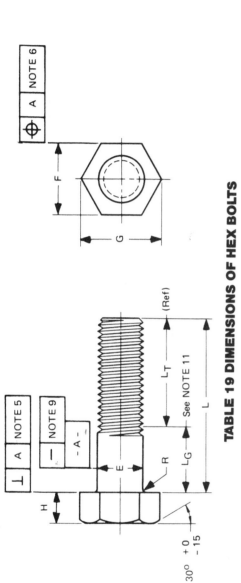

TABLE 19 DIMENSIONS OF HEX BOLTS

Nominal Size or Basic Product Dia (17)		E Body Dia (7) Max	F Width Across Flats (4)			G Width Across Corners		H Height			R Radius of Fillet		L_T Thread Length For Bolt Lengths (11) 6 in. and shorter Basic	L_T over 6 in. Basic
			Basic	Max	Min	Max	Min	Basic	Max	Min	Max	Min		
1/4	0.2500	0.260	7/16	0.438	0.425	0.505	0.484	11/64	0.188	0.150	0.03	0.01	0.750	1.000
5/16	0.3125	0.324	1/2	0.500	0.484	0.577	0.552	7/32	0.235	0.195	0.03	0.01	0.875	1.125
3/8	0.3750	0.388	9/16	0.562	0.544	0.650	0.620	1/4	0.268	0.226	0.03	0.01	1.000	1.250
7/16	0.4375	0.452	5/8	0.625	0.603	0.722	0.687	19/64	0.316	0.272	0.03	0.01	1.125	1.375
1/2	0.5000	0.515	3/4	0.750	0.725	0.866	0.826	11/32	0.364	0.302	0.03	0.01	1.250	1.500
5/8	0.6250	0.642	15/16	0.928	0.906	1.083	1.033	27/64	0.444	0.378	0.06	0.02	1.500	1.750
3/4	0.7500	0.768	1 1/8	1.125	1.088	1.299	1.240	1/2	0.524	0.455	0.06	0.02	1.750	2.000
7/8	0.8750	0.895	1 5/16	1.312	1.269	1.516	1.447	37/64	0.604	0.531	0.06	0.02	2.000	2.250
1	1.0000	1.022	1 1/2	1.500	1.450	1.732	1.653	43/64	0.700	0.591	0.09	0.03	2.250	2.500
1 1/8	1.1250	1.149	1 11/16	1.688	1.631	1.949	1.859	3/4	0.780	0.658	0.09	0.03	2.500	2.750
1 1/4	1.2500	1.277	1 7/8	1.875	1.812	2.165	2.066	27/32	0.876	0.749	0.09	0.03	2.750	3.000
1 3/8	1.3750	1.404	2 1/16	2.062	1.994	2.382	2.273	29/32	0.940	0.810	0.09	0.03	3.000	3.250
1 1/2	1.5000	1.531	2 1/4	2.250	2.175	2.598	2.480	1	1.036	0.902	0.09	0.03	3.250	3.500
1 3/4	1.7500	1.785	2 5/8	2.625	2.538	3.031	2.893	1 5/32	1.196	1.054	0.12	0.04	3.750	4.000
2	2.0000	2.039	3	3.000	2.900	3.464	3.306	1 11/32	1.388	1.175	0.12	0.04	4.250	4.500
2 1/4	2.2500	2.305	3 3/8	3.375	3.262	3.897	3.719	1 1/2	1.548	1.327	0.19	0.06	4.750	5.000
2 1/2	2.5000	2.559	3 3/4	3.750	3.625	4.330	4.133	1 21/32	1.708	1.479	0.19	0.06	5.250	5.500
2 3/4	2.7500	2.827	4 1/8	4.125	3.988	4.763	4.546	1 13/16	1.869	1.632	0.19	0.06	5.750	6.000
3	3.0000	3.081	4 1/2	4.500	4.350	5.196	4.959	2	2.060	1.815	0.19	0.06	6.250	6.500
3 1/4	3.2500	3.335	4 7/8	4.875	4.712	5.629	5.372	2 3/16	2.251	1.936	0.19	0.06	6.750	7.000
3 1/2	3.5000	3.589	5 1/4	5.250	5.075	6.062	5.786	2 5/16	2.380	2.057	0.19	0.06	7.250	7.500
3 3/4	3.7500	3.858	5 5/8	5.625	5.437	6.495	6.198	2 1/2	2.572	2.241	0.19	0.06	7.750	8.000
4	4.0000	4.111	6	6.000	5.800	6.928	6.612	2 11/16	2.764	2.424	0.19	0.06	8.250	8.500

30° + 0 − 15

From The American Society of Mechanical Engineers - ANSI B18.2.1 - 1981

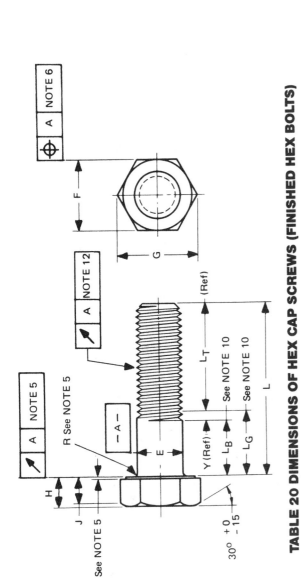

TABLE 20 DIMENSIONS OF HEX CAP SCREWS (FINISHED HEX BOLTS)

Nominal Size or Basic Product Dia (18)	E Body Dia (8) Max	E Body Dia (8) Min	F Width Across Flats Basic	F Width Across Flats Max	F Width Across Flats Min	G Width Across Corners (4) Max	G Width Across Corners (4) Min	H Height Basic	H Height Max	H Height Min	J Wrenching Height (4) Min	L_T Thread Length For Screw Lengths (10) 6 in. and Shorter Basic	L_T Thread Length For Screw Lengths (10) Over 6 in. Basic	Y Transition Thread Length (10) Max	Runout of Bearing Surface FIM (5) Max
1/4 0.2500	0.2500	0.2450	7/16	0.438	0.428	0.505	0.488	5/32	0.163	0.150	0.106	0.750	1.000	0.250	0.010
5/16 0.3125	0.3125	0.3065	1/2	0.500	0.489	0.577	0.557	13/64	0.211	0.195	0.140	0.875	1.125	0.278	0.011
3/8 0.3750	0.3750	0.3690	9/16	0.562	0.551	0.650	0.628	15/64	0.243	0.226	0.160	1.000	1.250	0.312	0.012
7/16 0.4375	0.4375	0.4305	5/8	0.625	0.612	0.722	0.698	9/32	0.291	0.272	0.195	1.125	1.375	0.357	0.013
1/2 0.5000	0.5000	0.4930	3/4	0.750	0.736	0.866	0.840	5/16	0.323	0.302	0.215	1.250	1.500	0.385	0.014
9/16 0.5625	0.5625	0.5545	13/16	0.812	0.798	0.938	0.910	23/64	0.371	0.348	0.250	1.375	1.625	0.417	0.015
5/8 0.6250	0.6250	0.6170	15/16	0.938	0.922	1.083	1.051	25/64	0.403	0.378	0.269	1.500	1.750	0.455	0.017
3/4 0.7500	0.7500	0.7410	1 1/8	1.125	1.100	1.299	1.254	15/32	0.483	0.455	0.324	1.750	2.000	0.500	0.020
7/8 0.8750	0.8750	0.8660	1 5/16	1.312	1.285	1.516	1.465	35/64	0.563	0.531	0.378	2.000	2.250	0.556	0.023
1 1.0000	1.0000	0.9900	1 1/2	1.500	1.469	1.732	1.675	39/64	0.627	0.591	0.416	2.250	2.500	0.625	0.026
1 1/8 1.1250	1.1250	1.1140	1 11/16	1.688	1.631	1.949	1.859	11/16	0.718	0.658	0.461	2.500	2.750	0.714	0.029
1 1/4 1.2500	1.2500	1.2390	1 7/8	1.875	1.812	2.165	2.066	25/32	0.813	0.749	0.530	2.750	3.000	0.714	0.033
1 3/8 1.3750	1.3750	1.3630	2 1/16	2.062	1.994	2.382	2.273	27/32	0.878	0.810	0.569	3.000	3.250	0.833	0.036
1 1/2 1.5000	1.5000	1.4880	2 1/4	2.230	2.175	2.598	2.480	15/16	0.974	0.902	0.640	3.250	3.500	0.833	0.039
1 3/4 1.7500	1.7500	1.7380	2 5/8	2.625	2.538	3.031	2.893	1 3/32	1.134	1.054	0.748	3.750	4.000	1.000	0.046
2 2.0000	2.0000	1.9880	3	3.000	2.900	3.464	3.306	1 7/32	1.263	1.175	0.825	4.250	4.500	1.111	0.052
2 1/4 2.2500	2.2500	2.2380	3 3/8	3.375	3.262	3.897	3.719	1 3/8	1.423	1.327	0.933	4.750	5.000	1.111	0.059
2 1/2 2.5000	2.5000	2.4880	3 3/4	3.750	3.625	4.330	4.133	1 17/32	1.583	1.479	1.042	5.250	5.500	1.250	0.065
2 3/4 2.7500	2.7500	2.7380	4 1/8	4.125	3.988	4.763	4.546	1 11/16	1.744	1.632	1.151	5.750	6.000	1.250	0.072
3 3.0000	3.0000	2.9880	4 1/2	4.500	4.350	5.196	4.959	1 7/8	1.935	1.815	1.290	6.250	6.500	1.250	0.079

From The American Society of Mechanical Engineers - ANSI B18.2.1 - 1981

30° +0 −15

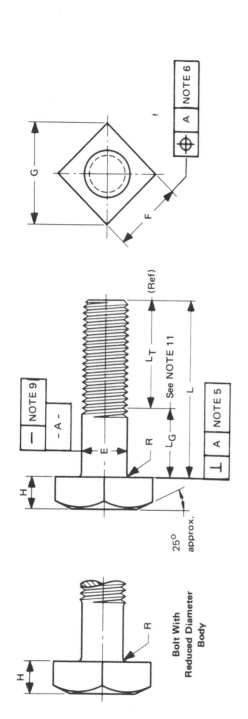

Bolt With Reduced Diameter Body

TABLE 21 DIMENSIONS OF SQUARE BOLTS

Nominal Size or Basic Product Dia (17)		E Body Dia (7), (14) Max	F Width Across Flats (4) Basic	F Max	F Min	G Width Across Corners Max	G Min	H Height Basic	H Max	H Min	R Radius of Fillet Max	R Min	L_T Thread Length For Bolt Lengths (11) 6 in. and shorter Basic	L_T over 6 in. Basic
1/4	0.2500	0.260	3/8	0.375	0.362	0.530	0.498	11/64	0.188	0.156	0.03	0.01	0.750	1.000
5/16	0.3125	0.324	1/2	0.500	0.484	0.707	0.665	13/64	0.220	0.186	0.03	0.01	0.875	1.125
3/8	0.3750	0.388	9/16	0.562	0.544	0.795	0.747	1/4	0.268	0.232	0.03	0.01	1.000	1.250
7/16	0.4375	0.452	5/8	0.625	0.603	0.884	0.828	19/64	0.316	0.278	0.03	0.01	1.125	1.375
1/2	0.5000	0.515	3/4	0.750	0.725	1.061	0.995	21/64	0.348	0.308	0.03	0.01	1.250	1.500
5/8	0.6250	0.642	15/16	0.938	0.906	1.326	1.244	27/64	0.444	0.400	0.06	0.02	1.500	1.750
3/4	0.7500	0.768	1 1/8	1.125	1.088	1.591	1.494	1/2	0.524	0.476	0.06	0.02	1.750	2.000
7/8	0.8750	0.895	1 5/16	1.312	1.269	1.856	1.742	19/32	0.620	0.568	0.06	0.02	2.000	2.250
1	1.0000	1.022	1 1/2	1.500	1.450	2.121	1.991	21/32	0.684	0.628	0.09	0.03	2.250	2.500
1 1/8	1.1250	1.149	1 11/16	1.688	1.631	2.386	2.239	3/4	0.780	0.720	0.09	0.03	2.500	2.750
1 1/4	1.2500	1.277	1 7/8	1.875	1.812	2.652	2.489	27/32	0.876	0.812	0.09	0.03	2.750	3.000
1 3/8	1.3750	1.404	2 1/16	2.062	1.994	2.917	2.738	29/32	0.940	0.872	0.09	0.03	3.000	3.250
1 1/2	1.5000	1.531	2 1/4	2.250	2.175	3.182	2.986	1	1.036	0.964	0.09	0.03	3.250	3.500

From The American Society of Mechanical Engineers - ANSI B18.2.1 - 1981

SLOTTED

FLAT

Type of Head

AMERICAN NATIONAL STANDARD
MACHINE SCREWS AND MACHINE SCREW NUTS

ANSI B18.6.3—1972

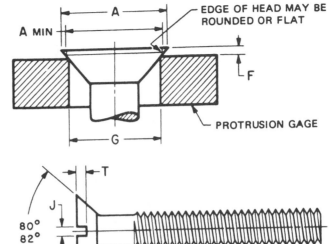

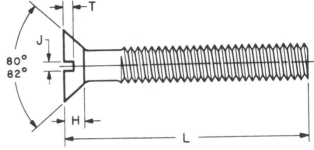

TABLE 22 DIMENSIONS OF SLOTTED FLAT COUNTERSUNK HEAD MACHINE SCREWS

Nominal Size[1] or Basic Screw Diameter		L[2] These Lengths or Shorter are Undercut.	A Head Diameter		H[3] Head Height	J Slot Width		T Slot Depth		F[4] Protrusion Above Gaging Diameter		G[4] Gaging Diameter
			Max, Edge Sharp	Min, Edge Rounded or Flat	Ref	Max	Min	Max	Min	Max	Min	
0000	0.0210	—	0.043	0.037	0.011	0.008	0.004	0.007	0.003	*	*	*
000	0.0340	—	0.064	0.058	0.016	0.011	0.007	0.009	0.005	*	*	*
00	0.0470	—	0.093	0.085	0.028	0.017	0.010	0.014	0.009	*	*	*
0	0.0600	1/8	0.119	0.099	0.035	0.023	0.016	0.015	0.010	0.026	0.016	0.078
1	0.0730	1/8	0.146	0.123	0.043	0.026	0.019	0.019	0.012	0.028	0.016	0.101
2	0.0860	1/8	0.172	0.147	0.051	0.031	0.023	0.023	0.015	0.029	0.017	0.124
3	0.0990	1/8	0.199	0.171	0.059	0.035	0.027	0.027	0.017	0.031	0.018	0.148
4	0.1120	3/16	0.225	0.195	0.067	0.039	0.031	0.030	0.020	0.032	0.019	0.172
5	0.1250	3/16	0.252	0.220	0.075	0.043	0.035	0.034	0.022	0.034	0.020	0.196
6	0.1380	3/16	0.279	0.244	0.083	0.048	0.039	0.038	0.024	0.036	0.021	0.220
8	0.1640	1/4	0.332	0.292	0.100	0.054	0.045	0.045	0.029	0.039	0.023	0.267
10	0.1900	5/16	0.385	0.340	0.116	0.060	0.050	0.053	0.034	0.042	0.025	0.313
12	0.2160	3/8	0.438	0.389	0.132	0.067	0.056	0.060	0.039	0.045	0.027	0.362
1/4	0.2500	7/16	0.507	0.452	0.153	0.075	0.064	0.070	0.046	0.050	0.029	0.424
5/16	0.3125	1/2	0.635	0.568	0.191	0.084	0.072	0.088	0.058	0.057	0.034	0.539
3/8	0.3750	9/16	0.762	0.685	0.230	0.094	0.081	0.106	0.070	0.065	0.039	0.653
7/16	0.4375	5/8	0.812	0.723	0.223	0.094	0.081	0.103	0.066	0.073	0.044	0.690
1/2	0.5000	3/4	0.875	0.775	0.223	0.106	0.091	0.103	0.065	0.081	0.049	0.739
9/16	0.5625	—	1.000	0.889	0.260	0.118	0.102	0.120	0.077	0.089	0.053	0.851
5/8	0.6250	—	1.125	1.002	0.298	0.133	0.116	0.137	0.088	0.097	0.058	0.962
3/4	0.7500	—	1.375	1.230	0.372	0.149	0.131	0.171	0.111	0.112	0.067	1.186

[1] Where specifying nominal size in decimals, zeros preceding decimal and in the fourth decimal place shall be omitted.
[2] Screws of these lengths and shorter shall have undercut heads as shown in Table 5.
[3] Tabulated values determined from formula for maximum H, Appendix V.
[4] No tolerance for gaging diameter is given. If the gaging diameter of the gage used differs from tabulated value, the protrusion will be affected accordingly and the proper protrusion values must be recalculated using the formulas shown in Appendix I.

*Not practical to gage.

From The American Society of Mechanical Engineers - ANSI B18.6.3 - 1972

AMERICAN NATIONAL STANDARD — SLOTTED HEAD CAP SCREWS,
SQUARE HEAD SET SCREWS, AND SLOTTED HEADLESS SET SCREWS

ANSI B18.6.2—1972

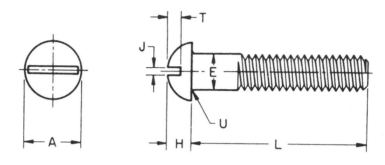

TABLE 23 DIMENSIONS OF SLOTTED ROUND HEAD CAP SCREWS

Nominal Size[1] or Basic Screw Diameter		E Body Diameter		A Head Diameter		H Head Height		J Slot Width		T Slot Depth		U Fillet Radius	
		Max	Min	Max	Min	Max	Min	Max	Min	Max	Min	Max	Min
1/4	0.2500	0.2500	0.2450	0.437	0.418	0.191	0.175	0.075	0.064	0.117	0.097	0.031	0.016
5/16	0.3125	0.3125	0.3070	0.562	0.540	0.245	0.226	0.084	0.072	0.151	0.126	0.031	0.016
3/8	0.3750	0.3750	0.3690	0.625	0.603	0.273	0.252	0.094	0.081	0.168	0.138	0.031	0.016
7/16	0.4375	0.4375	0.4310	0.750	0.725	0.328	0.302	0.094	0.081	0.202	0.167	0.047	0.016
1/2	0.5000	0.5000	0.4930	0.812	0.786	0.354	0.327	0.106	0.091	0.218	0.178	0.047	0.016
9/16	0.5625	0.5625	0.5550	0.937	0.909	0.409	0.378	0.118	0.102	0.252	0.207	0.047	0.016
5/8	0.6250	0.6250	0.6170	1.000	0.970	0.437	0.405	0.133	0.116	0.270	0.220	0.062	0.031
3/4	0.7500	0.7500	0.7420	1.250	1.215	0.546	0.507	0.149	0.131	0.338	0.278	0.062	0.031

[1] Where specifying nominal size in decimals, zeros preceding decimal and in the fourth decimal place shall be omitted.

From The American Society of Mechanical Engineers - ANSI B18.6.2 - 1972

AMERICAN NATIONAL STANDARD – SLOTTED HEAD CAP SCREWS,
SQUARE HEAD SET SCREWS, AND SLOTTED HEADLESS SET SCREWS

ANSI B18.6.2–1972

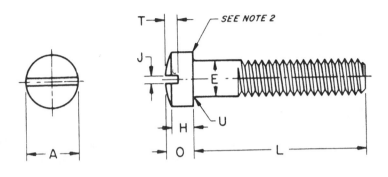

TABLE 24 DIMENSIONS OF SLOTTED FILLISTER HEAD CAP SCREWS

Nominal Size[1] or Basic Screw Diameter		E Body Diameter		A Head Diameter		H Head Side Height		O Total Head Height		J Slot Width		T Slot Depth		U Fillet Radius	
		Max	Min	Max	Min	Max	Min	Max	Min	Max	Min	Max	Min	Max	Min
1/4	0.2500	0.2500	0.2450	0.375	0.363	0.172	0.157	0.216	0.194	0.075	0.064	0.097	0.077	0.031	0.016
5/16	0.3125	0.3125	0.3070	0.437	0.424	0.203	0.186	0.253	0.230	0.084	0.072	0.115	0.090	0.031	0.016
3/8	0.3750	0.3750	0.3690	0.562	0.547	0.250	0.229	0.314	0.284	0.094	0.081	0.142	0.112	0.031	0.016
7/16	0.4375	0.4375	0.4310	0.625	0.608	0.297	0.274	0.368	0.336	0.094	0.081	0.168	0.133	0.047	0.016
1/2	0.5000	0.5000	0.4930	0.750	0.731	0.328	0.301	0.413	0.376	0.106	0.091	0.193	0.153	0.047	0.016
9/16	0.5625	0.5625	0.5550	0.812	0.792	0.375	0.346	0.467	0.427	0.118	0.102	0.213	0.168	0.047	0.016
5/8	0.6250	0.6250	0.6170	0.875	0.853	0.422	0.391	0.521	0.478	0.133	0.116	0.239	0.189	0.062	0.031
3/4	0.7500	0.7500	0.7420	1.000	0.976	0.500	0.466	0.612	0.566	0.149	0.131	0.283	0.223	0.062	0.031
7/8	0.8750	0.8750	0.8660	1.125	1.098	0.594	0.556	0.720	0.668	0.167	0.147	0.334	0.264	0.062	0.031
1	1.0000	1.0000	0.9900	1.312	1.282	0.656	0.612	0.803	0.743	0.188	0.166	0.371	0.291	0.062	0.031

[1] Where specifying nominal size in decimals, zeros preceding decimal and in the fourth decimal place shall be omitted.

[2] A slight rounding of the edges at periphery of head shall be permissible provided the diameter of the bearing circle is equal to no less than 90 per cent of the specified minimum head diameter.

From The American Society of Mechanical Engineers - ANSI B18.6.2 - 1972

AMERICAN NATIONAL STANDARD
MACHINE SCREWS AND MACHINE SCREW NUTS

ANSI B18.6.3–1972

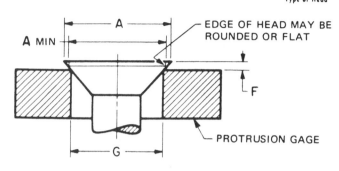

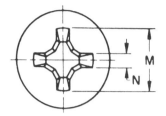

This type of recess has a large center opening, tapered wings, and blunt bottom, with all edges relieved or rounded.

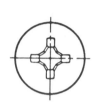

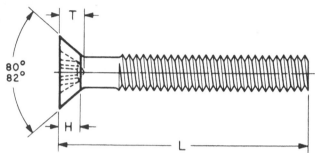

TABLE 25 DIMENSIONS OF TYPE I CROSS RECESSED FLAT COUNTERSUNK HEAD MACHINE SCREWS

Nominal Size[1] or Basic Screw Diameter		L[2] These Lengths or Shorter are Undercut	A Head Diameter Max, Edge Sharp	A Head Diameter Min, Edge Rounded or Flat	H[3] Head Height Ref	M Recess Diameter Max	M Recess Diameter Min	T Recess Depth Max	T Recess Depth Min	N Recess Width Min	Driver Size	Recess Penetration Gaging Depth Max	Recess Penetration Gaging Depth Min	F[4] Protrusion Above Gaging Diameter Max	F[4] Protrusion Above Gaging Diameter Min	G[4] Gaging Diameter
0	0.0600	1/8	0.119	0.099	0.035	0.069	0.056	0.043	0.027	0.014	0	0.036	0.020	0.026	0.016	0.078
1	0.0730	1/8	0.146	0.123	0.043	0.077	0.064	0.051	0.035	0.015	0	0.044	0.028	0.028	0.016	0.101
2	0.0860	1/8	0.172	0.147	0.051	0.102	0.089	0.063	0.047	0.017	1	0.056	0.040	0.029	0.017	0.124
3	0.0990	1/8	0.199	0.171	0.059	0.107	0.094	0.068	0.052	0.018	1	0.061	0.045	0.031	0.018	0.148
4	0.1120	3/16	0.225	0.195	0.067	0.128	0.115	0.089	0.073	0.018	1	0.082	0.066	0.032	0.019	0.172
5	0.1250	3/16	0.252	0.220	0.075	0.154	0.141	0.086	0.063	0.027	2	0.075	0.052	0.034	0.020	0.196
6	0.1380	3/16	0.279	0.244	0.083	0.174	0.161	0.106	0.083	0.029	2	0.095	0.072	0.036	0.021	0.220
8	0.1640	1/4	0.332	0.292	0.100	0.189	0.176	0.121	0.098	0.030	2	0.110	0.087	0.039	0.023	0.267
10	0.1900	5/16	0.385	0.340	0.116	0.204	0.191	0.136	0.113	0.032	2	0.125	0.102	0.042	0.025	0.313
12	0.2160	3/8	0.438	0.389	0.132	0.268	0.255	0.156	0.133	0.035	3	0.139	0.116	0.045	0.027	0.362
1/4	0.2500	7/16	0.507	0.452	0.153	0.283	0.270	0.171	0.148	0.036	3	0.154	0.131	0.050	0.029	0.424
5/16	0.3125	1/2	0.635	0.568	0.191	0.365	0.352	0.216	0.194	0.061	4	0.196	0.174	0.057	0.034	0.539
3/8	0.3750	9/16	0.762	0.685	0.230	0.393	0.380	0.245	0.223	0.065	4	0.225	0.203	0.065	0.039	0.653
7/16	0.4375	5/8	0.812	0.723	0.223	0.409	0.396	0.261	0.239	0.068	4	0.241	0.219	0.073	0.044	0.690
1/2	0.5000	3/4	0.875	0.775	0.223	0.424	0.411	0.276	0.254	0.069	4	0.256	0.234	0.081	0.049	0.739
9/16	0.5625	—	1.000	0.889	0.260	0.454	0.431	0.300	0.278	0.073	4	0.280	0.258	0.089	0.053	0.851
5/8	0.6250	—	1.125	1.002	0.298	0.576	0.553	0.342	0.316	0.079	5	0.309	0.283	0.097	0.058	0.962
3/4	0.7500	—	1.375	1.230	0.372	0.640	0.617	0.406	0.380	0.087	5	0.373	0.347	0.112	0.067	1.186

[1]Where specifying nominal size in decimals, zeros preceding decimal and in the fourth decimal place shall be omitted.
[2]Screws of these lengths and shorter shall have undercut heads.
[3]Tabulated values determined from formula for maximum H, Appendix V, ANSI B18.6.3-1972.
[4]No tolerance for gaging diameter is given. If the gaging diameter of the gage used differs from tabulated value, the protrusion will be affected accordingly and the proper protrusion values must be recalculated using the formulas shown in Appendix I, ANSI B18.6.3-1972.

From The American Society of Mechanical Engineers - ANSI B18.6.3 - 1972

SLOTTED

OVAL

Type of Head

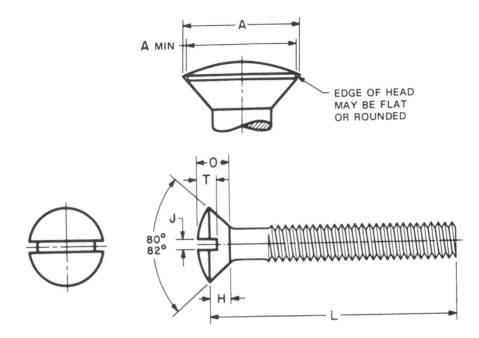

TABLE 26 DIMENSIONS OF SLOTTED OVAL COUNTERSUNK HEAD MACHINE SCREWS

Nominal Size[1] or Basic Screw Diameter		L[2] These Lengths or Shorter are Undercut	A Head Diameter Max, Edge Sharp	A Head Diameter Min, Edge Rounded or Flat	H[3] Head Side Height Ref	O Total Head Height Max	O Total Head Height Min	J Slot Width Max	J Slot Width Min	T Slot Depth Max	T Slot Depth Min
00	0.0470	—	0.093	0.085	0.028	0.042	0.034	0.017	0.010	0.023	0.016
0	0.0600	1/8	0.119	0.099	0.035	0.056	0.041	0.023	0.016	0.038	0.031
1	0.0730	1/8	0.146	0.123	0.043	0.068	0.052	0.026	0.019	0.045	0.037
2	0.0860	1/8	0.172	0.147	0.051	0.080	0.063	0.031	0.023	0.052	0.043
3	0.0990	1/8	0.199	0.171	0.059	0.092	0.073	0.035	0.027	0.059	0.049
4	0.1120	3/16	0.225	0.195	0.067	0.104	0.084	0.039	0.031	0.067	0.055
5	0.1250	3/16	0.252	0.220	0.075	0.116	0.095	0.043	0.035	0.074	0.060
6	0.1380	3/16	0.279	0.244	0.083	0.128	0.105	0.048	0.039	0.088	0.072
8	0.1640	1/4	0.332	0.292	0.100	0.152	0.126	0.054	0.045	0.103	0.084
10	0.1900	5/16	0.385	0.340	0.116	0.176	0.148	0.060	0.050	0.117	0.096
12	0.2160	3/8	0.438	0.389	0.132	0.200	0.169	0.067	0.056	0.136	0.112
1/4	0.2500	7/16	0.507	0.452	0.153	0.232	0.197	0.075	0.064	0.171	0.141
5/16	0.3125	1/2	0.635	0.568	0.191	0.290	0.249	0.084	0.072	0.206	0.170
3/8	0.3750	9/16	0.762	0.685	0.230	0.347	0.300	0.094	0.081	0.210	0.174
7/16	0.4375	5/8	0.812	0.723	0.223	0.345	0.295	0.094	0.081	0.216	0.176
1/2	0.5000	3/4	0.875	0.775	0.223	0.354	0.299	0.106	0.091	0.250	0.207
9/16	0.5625	—	1.000	0.889	0.260	0.410	0.350	0.118	0.102	0.285	0.235
5/8	0.6250	—	1.125	1.002	0.298	0.467	0.399	0.133	0.116	0.353	0.293
3/4	0.7500	—	1.375	1.230	0.372	0.578	0.497	0.149	0.131		

[1]Where specifying nominal size in decimals, zeros preceding decimal and in the fourth decimal place shall be omitted.
[2]Screws of these lengths and shorter shall have undercut heads.
[3]Tabulated values determined from formula for maximum H, Appendix V, ANSI B18.6.3-1972.

From The American Society of Mechanical Engineers - ANSI B18.6.3 - 1972

AMERICAN NATIONAL STANDARD
MACHINE SCREWS AND MACHINE SCREW NUTS

ANSI B18.6.3—1972

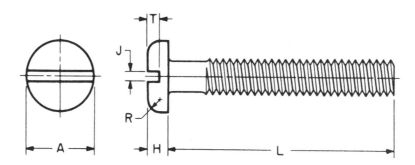

TABLE 27 DIMENSIONS OF SLOTTED PAN HEAD MACHINE SCREWS

Nominal Size[1] or Basic Screw Diameter		A Head Diameter		H Head Height		R Head Radius	J Slot Width		T Slot Depth	
		Max	Min	Max	Min	Max	Max	Min	Max	Min
0000	0.0210	0.042	0.036	0.016	0.010	0.007	0.008	0.004	0.008	0.004
000	0.0340	0.066	0.060	0.023	0.017	0.010	0.012	0.008	0.012	0.008
00	0.0470	0.090	0.082	0.032	0.025	0.015	0.017	0.010	0.016	0.010
0	0.0600	0.116	0.104	0.039	0.031	0.020	0.023	0.016	0.022	0.014
1	0.0730	0.142	0.130	0.046	0.038	0.025	0.026	0.019	0.027	0.018
2	0.0860	0.167	0.155	0.053	0.045	0.035	0.031	0.023	0.031	0.022
3	0.0990	0.193	0.180	0.060	0.051	0.037	0.035	0.027	0.036	0.026
4	0.1120	0.219	0.205	0.068	0.058	0.042	0.039	0.031	0.040	0.030
5	0.1250	0.245	0.231	0.075	0.065	0.044	0.043	0.035	0.045	0.034
6	0.1380	0.270	0.256	0.082	0.072	0.046	0.048	0.039	0.050	0.037
8	0.1640	0.322	0.306	0.096	0.085	0.052	0.054	0.045	0.058	0.045
10	0.1900	0.373	0.357	0.110	0.099	0.061	0.060	0.050	0.068	0.053
12	0.2160	0.425	0.407	0.125	0.112	0.078	0.067	0.056	0.077	0.061
1/4	0.2500	0.492	0.473	0.144	0.130	0.087	0.075	0.064	0.087	0.070
5/16	0.3125	0.615	0.594	0.178	0.162	0.099	0.084	0.072	0.106	0.085
3/8	0.3750	0.740	0.716	0.212	0.195	0.143	0.094	0.081	0.124	0.100
7/16	0.4375	0.863	0.837	0.247	0.228	0.153	0.094	0.081	0.142	0.116
1/2	0.5000	0.987	0.958	0.281	0.260	0.175	0.106	0.091	0.161	0.131
9/16	0.5625	1.041	1.000	0.315	0.293	0.197	0.118	0.102	0.179	0.146
5/8	0.6250	1.172	1.125	0.350	0.325	0.219	0.133	0.116	0.197	0.162
3/4	0.7500	1.435	1.375	0.419	0.390	0.263	0.149	0.131	0.234	0.192

[1] Where specifying nominal size in decimals, zeros preceding decimal and in the fourth decimal place shall be omitted.

From The American Society of Mechanical Engineers - ANSI B18.6.3 - 1972

AMERICAN NATIONAL STANDARD — SLOTTED HEAD CAP SCREWS,
SQUARE HEAD SET SCREWS, AND SLOTTED HEADLESS SET SCREWS

ANSI B18.6.2–1972

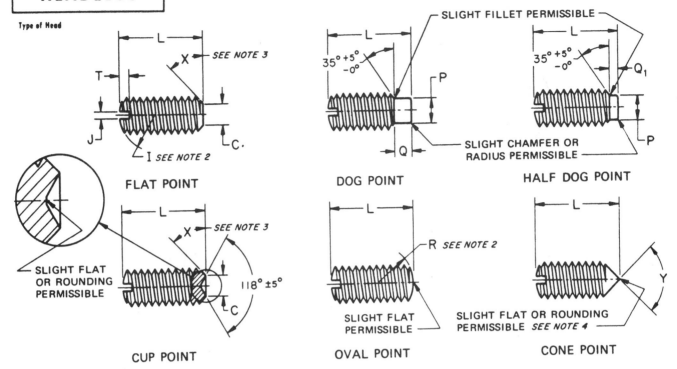

FLAT POINT DOG POINT HALF DOG POINT

CUP POINT OVAL POINT CONE POINT

TABLE 28 DIMENSIONS OF SLOTTED HEADLESS SET SCREWS

Nominal Size[1] or Basic Screw Diameter		I^2 Crown Radius Basic	J Slot Width Max	Min	T Slot Depth Max	Min	C Cup and Flat Point Diameters Max	Min	P Dog Point Diameters Max	Min	Q Point Length Dog Max	Min	Q_1 Half Dog Max	Min	R^2 Oval Point Radius Basic	Y Cone Point Angle 90° ±2° For These Nominal Lengths or Longer; 118° ±2° For Shorter Screws
0	0.0600	0.060	0.014	0.010	0.020	0.016	0.033	0.027	0.040	0.037	0.032	0.028	0.017	0.013	0.045	5/64
1	0.0730	0.073	0.016	0.012	0.020	0.016	0.040	0.033	0.049	0.045	0.040	0.036	0.021	0.017	0.055	3/32
2	0.0860	0.086	0.018	0.014	0.025	0.019	0.047	0.039	0.057	0.053	0.046	0.042	0.024	0.020	0.064	7/64
3	0.0990	0.099	0.020	0.016	0.028	0.022	0.054	0.045	0.066	0.062	0.052	0.048	0.027	0.023	0.074	1/8
4	0.1120	0.112	0.024	0.018	0.031	0.025	0.061	0.051	0.075	0.070	0.058	0.054	0.030	0.026	0.084	5/32
5	0.1250	0.125	0.026	0.020	0.036	0.026	0.067	0.057	0.083	0.078	0.063	0.057	0.033	0.027	0.094	3/16
6	0.1380	0.138	0.028	0.022	0.040	0.030	0.074	0.064	0.092	0.087	0.073	0.067	0.038	0.032	0.104	3/16
8	0.1640	0.164	0.032	0.026	0.046	0.036	0.087	0.076	0.109	0.103	0.083	0.077	0.043	0.037	0.123	1/4
10	0.1900	0.190	0.035	0.029	0.053	0.043	0.102	0.088	0.127	0.120	0.095	0.085	0.050	0.040	0.142	1/4
12	0.2160	0.216	0.042	0.035	0.061	0.051	0.115	0.101	0.144	0.137	0.115	0.105	0.060	0.050	0.162	5/16
1/4	0.2500	0.250	0.049	0.041	0.068	0.058	0.132	0.118	0.156	0.149	0.130	0.120	0.068	0.058	0.188	5/16
5/16	0.3125	0.312	0.055	0.047	0.083	0.073	0.172	0.156	0.203	0.195	0.161	0.151	0.083	0.073	0.234	3/8
3/8	0.3750	0.375	0.068	0.060	0.099	0.089	0.212	0.194	0.250	0.241	0.193	0.183	0.099	0.089	0.281	7/16
7/16	0.4375	0.438	0.076	0.068	0.114	0.104	0.252	0.232	0.297	0.287	0.224	0.214	0.114	0.104	0.328	1/2
1/2	0.5000	0.500	0.086	0.076	0.130	0.120	0.291	0.270	0.344	0.334	0.255	0.245	0.130	0.120	0.375	9/16
9/16	0.5625	0.562	0.096	0.086	0.146	0.136	0.332	0.309	0.391	0.379	0.287	0.275	0.146	0.134	0.422	5/8
5/8	0.6250	0.625	0.107	0.097	0.161	0.151	0.371	0.347	0.469	0.456	0.321	0.305	0.164	0.148	0.469	3/4
3/4	0.7500	0.750	0.134	0.124	0.193	0.183	0.450	0.425	0.562	0.549	0.383	0.367	0.196	0.180	0.562	7/8

[1] Where specifying nominal size in decimals, zeros preceding decimal and in the fourth decimal place shall be omitted.

[2] Tolerance on radius for nominal sizes up to and including 5 (0.125 in.) shall be plus 0.015 in. and minus 0.000 in. and for larger sizes, plus 0.031 in. and minus 0.000. Slotted ends on screws may be flat at option of manufacturer.

[3] Point angle X shall be 45° plus 5°, minus 0°, for screws of nominal lengths equal to or longer than those listed in Column Y, and 30° minimum for screws of shorter nominal lengths.

[4] The extent of rounding or flat at apex of cone point shall not exceed an amount equivalent to 10 per cent of the basic screw diameter.

From The American Society of Mechanical Engineers - ANSI B18.6.2 - 1972

AMERICAN NATIONAL STANDARD – SLOTTED HEAD CAP SCREWS,
SQUARE HEAD SET SCREWS, AND SLOTTED HEADLESS SET SCREWS   ANSI B 18.6.2–1972

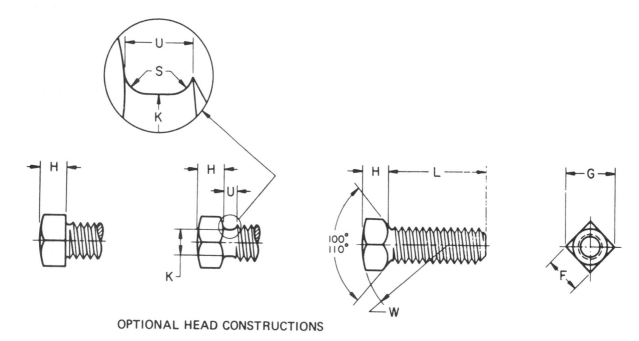

OPTIONAL HEAD CONSTRUCTIONS

TABLE 29 DIMENSIONS OF SQUARE HEAD SET SCREWS

Nominal Size[1] or Basic Screw Diameter		F Width Across Flats		G Width Across Corners		H Head Height		K Neck Relief Diameter		S Neck Relief Fillet Radius	U Neck Relief Width	W Head Radius
		Max	Min	Max	Min	Max	Min	Max	Min	Max	Min	Min
10	0.1900	0.188	0.180	0.265	0.247	0.148	0.134	0.145	0.140	0.027	0.083	0.48
1/4	0.2500	0.250	0.241	0.354	0.331	0.196	0.178	0.185	0.170	0.032	0.100	0.62
5/16	0.3125	0.312	0.302	0.442	0.415	0.245	0.224	0.240	0.225	0.036	0.111	0.78
3/8	0.3750	0.375	0.362	0.530	0.497	0.293	0.270	0.294	0.279	0.041	0.125	0.94
7/16	0.4375	0.438	0.423	0.619	0.581	0.341	0.315	0.345	0.330	0.046	0.143	1.09
1/2	0.5000	0.500	0.484	0.707	0.665	0.389	0.361	0.400	0.385	0.050	0.154	1.25
9/16	0.5625	0.562	0.545	0.795	0.748	0.437	0.407	0.454	0.439	0.054	0.167	1.41
5/8	0.6250	0.625	0.606	0.884	0.833	0.485	0.452	0.507	0.492	0.059	0.182	1.56
3/4	0.7500	0.750	0.729	1.060	1.001	0.582	0.544	0.620	0.605	0.065	0.200	1.88
7/8	0.8750	0.875	0.852	1.237	1.170	0.678	0.635	0.731	0.716	0.072	0.222	2.19
1	1.0000	1.000	0.974	1.414	1.337	0.774	0.726	0.838	0.823	0.081	0.250	2.50
1 1/8	1.1250	1.125	1.096	1.591	1.505	0.870	0.817	0.939	0.914	0.092	0.283	2.81
1 1/4	1.2500	1.250	1.219	1.768	1.674	0.966	0.908	1.064	1.039	0.092	0.283	3.12
1 3/8	1.3750	1.375	1.342	1.945	1.843	1.063	1.000	1.159	1.134	0.109	0.333	3.44
1 1/2	1.5000	1.500	1.464	2.121	2.010	1.159	1.091	1.284	1.259	0.109	0.333	3.75

[1] Where specifying nominal size in decimals, zeros preceding decimal and in the fourth decimal place shall be omitted.

AMERICAN NATIONAL STANDARD – SLOTTED HEAD CAP SCREWS, SQUARE HEAD SET SCREWS, AND SLOTTED HEADLESS SET SCREWS   ANSI B18.6.2–1972

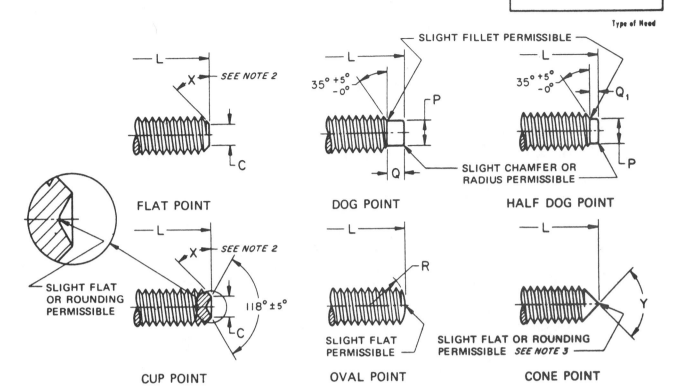

FLAT POINT   DOG POINT   HALF DOG POINT

CUP POINT   OVAL POINT   CONE POINT

TABLE 29 DIMENSIONS OF SQUARE HEAD SET SCREWS (CONTINUED)

Nominal Size[1] or Basic Screw Diameter		C		P		Q		Q₁		R	Y
		Cup and Flat Point Diameters		Dog and Half Dog Point Diameters		Point Length				Oval Point Radius +0.031 -0.000	Cone Point Angle 90° ±2° For These Nominal Lengths or Longer; 118° ±2° For Shorter Screws
						Dog		Half Dog			
		Max	Min	Max	Min	Max	Min	Max	Min		
10	0.1900	0.102	0.088	0.127	0.120	0.095	0.085	0.050	0.040	0.142	1/4
1/4	0.2500	0.132	0.118	0.156	0.149	0.130	0.120	0.068	0.058	0.188	5/15
5/16	0.3125	0.172	0.156	0.203	0.195	0.161	0.151	0.083	0.073	0.234	3/8
3/8	0.3750	0.212	0.194	0.250	0.241	0.193	0.183	0.099	0.089	0.281	7/16
7/16	0.4375	0.252	0.232	0.297	0.287	0.224	0.214	0.114	0.104	0.328	1/2
1/2	0.5000	0.291	0.270	0.344	0.334	0.255	0.245	0.130	0.120	0.375	9/16
9/16	0.5625	0.332	0.309	0.391	0.379	0.287	0.275	0.146	0.134	0.422	5/8
5/8	0.6250	0.371	0.347	0.469	0.456	0.321	0.305	0.164	0.148	0.469	3/4
3/4	0.7500	0.450	0.425	0.562	0.549	0.383	0.367	0.196	0.180	0.562	7/8
7/8	0.8750	0.530	0.502	0.656	0.642	0.446	0.430	0.227	0.211	0.656	1
1	1.0000	0.609	0.579	0.750	0.734	0.510	0.490	0.260	0.240	0.750	1 1/8
1 1/8	1.1250	0.689	0.655	0.844	0.826	0.572	0.552	0.291	0.271	0.844	1 1/4
1 1/4	1.2500	0.767	0.733	0.938	0.920	0.635	0.615	0.323	0.303	0.938	1 1/2
1 3/8	1.3750	0.848	0.808	1.031	1.011	0.698	0.678	0.354	0.334	1.031	1 5/8
1 1/2	1.5000	0.926	0.886	1.125	1.105	0.760	0.740	0.385	0.365	1.125	1 3/4

[1] Where specifying nominal size in decimals, zeros preceding decimal and in the fourth decimal place shall be omitted.
[2] Point angle X shall be 45° plus 5°, minus 0°, for screws of nominal lengths equal to or longer than those listed in Column Y, and 30° minimum for screws of shorter nominal lengths.
[3] The extent of rounding or flat at apex of cone point shall not exceed an amount equivalent to 10 per cent of the basic screw diameter.

From The American Society of Mechanical Engineers - ANSI B18.6.2 - 1972

AMERICAN STANDARD

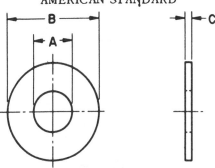

TABLE 30 DIMENSIONS OF PREFERRED SIZES OF TYPE A PLAIN WASHERS **

Nominal Washer Size***			Inside Diameter A			Outside Diameter B			Thickness C		
			Basic	Plus	Minus	Basic	Plus	Minus	Basic	Max	Min
—	—		0.078	0.000	0.005	0.188	0.000	0.005	0.020	0.025	0.016
—	—		0.094	0.000	0.005	0.250	0.000	0.005	0.020	0.025	0.016
—	—		0.125	0.008	0.005	0.312	0.008	0.005	0.032	0.040	0.025
No. 6	0.138		0.156	0.008	0.005	0.375	0.015	0.005	0.049	0.065	0.036
No. 8	0.164		0.188	0.008	0.005	0.438	0.015	0.005	0.049	0.065	0.036
No. 10	0.190		0.219	0.008	0.005	0.500	0.015	0.005	0.049	0.065	0.036
3/16	0.188		0.250	0.015	0.005	0.562	0.015	0.005	0.049	0.065	0.036
No. 12	0.216		0.250	0.015	0.005	0.562	0.015	0.005	0.065	0.080	0.051
1/4	0.250	N	0.281	0.015	0.005	0.625	0.015	0.005	0.065	0.080	0.051
1/4	0.250	W	0.312	0.015	0.005	0.734*	0.015	0.007	0.065	0.080	0.051
5/16	0.312	N	0.344	0.015	0.005	0.688	0.015	0.007	0.065	0.080	0.051
5/16	0.312	W	0.375	0.015	0.005	0.875	0.030	0.007	0.083	0.104	0.064
3/8	0.375	N	0.406	0.015	0.005	0.812	0.015	0.007	0.065	0.080	0.051
3/8	0.375	W	0.438	0.015	0.005	1.000	0.030	0.007	0.083	0.104	0.064
7/16	0.438	N	0.469	0.015	0.005	0.922	0.015	0.007	0.065	0.080	0.051
7/16	0.438	W	0.500	0.015	0.005	1.250	0.030	0.007	0.083	0.104	0.064
1/2	0.500	N	0.531	0.015	0.005	1.062	0.030	0.007	0.095	0.121	0.074
1/2	0.500	W	0.562	0.015	0.005	1.375	0.030	0.007	0.109	0.132	0.086
9/16	0.562	N	0.594	0.015	0.005	1.156*	0.030	0.007	0.095	0.121	0.074
9/16	0.562	W	0.625	0.015	0.005	1.469*	0.030	0.007	0.109	0.132	0.086
5/8	0.625	N	0.656	0.030	0.007	1.312	0.030	0.007	0.095	0.121	0.074
5/8	0.625	W	0.688	0.030	0.007	1.750	0.030	0.007	0.134	0.160	0.108
3/4	0.750	N	0.812	0.030	0.007	1.469	0.030	0.007	0.134	0.160	0.108
3/4	0.750	W	0.812	0.030	0.007	2.000	0.030	0.007	0.148	0.177	0.122
7/8	0.875	N	0.938	0.030	0.007	1.750	0.030	0.007	0.134	0.160	0.108
7/8	0.875	W	0.938	0.030	0.007	2.250	0.030	0.007	0.165	0.192	0.136
1	1.000	N	1.062	0.030	0.007	2.000	0.030	0.007	0.134	0.160	0.108
1	1.000	W	1.062	0.030	0.007	2.500	0.030	0.007	0.165	0.192	0.136
1 1/8	1.125	N	1.250	0.030	0.007	2.250	0.030	0.007	0.134	0.160	0.108
1 1/8	1.125	W	1.250	0.030	0.007	2.750	0.030	0.007	0.165	0.192	0.136
1 1/4	1.250	N	1.375	0.030	0.007	2.500	0.030	0.007	0.165	0.192	0.136
1 1/4	1.250	W	1.375	0.030	0.007	3.000	0.030	0.007	0.165	0.192	0.136
1 3/8	1.375	N	1.500	0.030	0.007	2.750	0.030	0.007	0.165	0.192	0.136
1 3/8	1.375	W	1.500	0.045	0.010	3.250	0.045	0.010	0.180	0.213	0.153
1 1/2	1.500	N	1.625	0.030	0.007	3.000	0.030	0.007	0.165	0.192	0.136
1 1/2	1.500	W	1.625	0.045	0.010	3.500	0.045	0.010	0.180	0.213	0.153
1 5/8	1.625		1.750	0.045	0.010	3.750	0.045	0.010	0.180	0.213	0.153
1 3/4	1.750		1.875	0.045	0.010	4.000	0.045	0.010	0.180	0.213	0.153
1 7/8	1.875		2.000	0.045	0.010	4.250	0.045	0.010	0.180	0.213	0.153
2	2.000		2.125	0.045	0.010	4.500	0.045	0.010	0.180	0.213	0.153
2 1/4	2.250		2.375	0.045	0.010	4.750	0.045	0.010	0.220	0.248	0.193
2 1/2	2.500		2.625	0.045	0.010	5.000	0.045	0.010	0.238	0.280	0.210
2 3/4	2.750		2.875	0.065	0.010	5.250	0.065	0.010	0.259	0.310	0.228
3	3.000		3.125	0.065	0.010	5.500	0.065	0.010	0.284	0.327	0.249

*The 0.734 in., 1.156 in., and 1.469 in. outside diameters avoid washers which could be used in coin operated devices.
**Preferred sizes are for the most part from series previously designated "Standard Plate" and "SAE." Where common sizes existed in the two series, the SAE size is designated "N" (narrow) and the Standard Plate "W" (wide). These sizes as well as all other sizes of Type A Plain Washers are to be ordered by ID, OD, and thickness dimensions.
***Nominal washer sizes are intended for use with comparable nominal screw or bolt sizes.

From The American Society of Mechanical Engineers - ANSI B18.22.1 - 1965 (R1975)

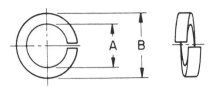 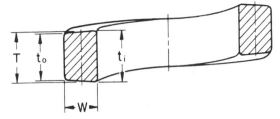

ENLARGED SECTION

TABLE 31 DIMENSIONS OF REGULAR HELICAL SPRING LOCK WASHERS[1]

Nominal Washer Size		A Inside Diameter		B Outside Diameter	T Mean Section Thickness $\left(\frac{t_i + t_o}{2} \right)$	W Section Width
		Max	Min	Max[2]	Min	Min
No. 4	0.112	0.120	0.114	0.173	0.022	0.022
No. 5	0.125	0.133	0.127	0.202	0.030	0.030
No. 6	0.138	0.148	0.141	0.216	0.030	0.030
No. 8	0.164	0.174	0.167	0.267	0.047	0.042
No. 10	0.190	0.200	0.193	0.294	0.047	0.042
¼	0.250	0.262	0.254	0.365	0.078	0.047
5/16	0.312	0.326	0.317	0.460	0.093	0.062
3/8	0.375	0.390	0.380	0.553	0.125	0.076
7/16	0.438	0.455	0.443	0.647	0.140	0.090
½	0.500	0.518	0.506	0.737	0.172	0.103
5/8	0.625	0.650	0.635	0.923	0.203	0.125
¾	0.750	0.775	0.760	1.111	0.218	0.154
7/8	0.875	0.905	0.887	1.296	0.234	0.182
1	1.000	1.042	1.017	1.483	0.250	0.208
1⅛	1.125	1.172	1.144	1.669	0.313	0.236
1¼	1.250	1.302	1.271	1.799	0.313	0.236
1⅜	1.375	1.432	1.398	2.041	0.375	0.292
1½	1.500	1.561	1.525	2.170	0.375	0.292
1¾	1.750	1.811	1.775	2.602	0.469	0.383
2	2.000	2.061	2.025	2.852	0.469	0.383
2¼	2.250	2.311	2.275	3.352	0.508	0.508
2½	2.500	2.561	2.525	3.602	0.508	0.508
2¾	2.750	2.811	2.775	4.102	0.633	0.633
3	3.000	3.061	3.025	4.352	0.633	0.633

[1] For use with 1960 Series Socket Head Cap Screws specified in American National Standard, ANSI B18.3.
[2] The maximum outside diameters specified allow for the commercial tolerances on cold drawn wire.

From The American Society of Mechanical Engineers—ANSI B18.21.1—1972

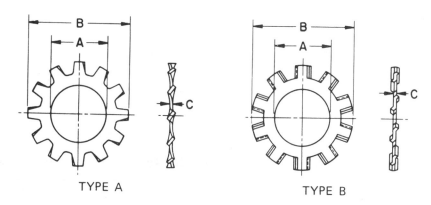

TYPE A   TYPE B

TABLE 32 DIMENSIONS OF EXTERNAL TOOTH LOCK WASHERS

Nominal Washer Size		A Inside Diameter		B Outside Diameter		C Thickness	
		Max	Min	Max	Min	Max	Min
No. 3	0.099	0.109	0.102	0.235	0.220	0.015	0.012
No. 4	0.112	0.123	0.115	0.260	0.245	0.019	0.015
No. 5	0.125	0.136	0.129	0.285	0.270	0.019	0.014
No. 6	0.138	0.150	0.141	0.320	0.305	0.022	0.016
No. 8	0.164	0.176	0.168	0.381	0.365	0.023	0.018
No. 10	0.190	0.204	0.195	0.410	0.395	0.025	0.020
No. 12	0.216	0.231	0.221	0.475	0.460	0.028	0.023
$\frac{1}{4}$	0.250	0.267	0.256	0.510	0.494	0.028	0.023
$\frac{5}{16}$	0.312	0.332	0.320	0.610	0.588	0.034	0.028
$\frac{3}{8}$	0.375	0.398	0.384	0.694	0.670	0.040	0.032
$\frac{7}{16}$	0.438	0.464	0.448	0.760	0.740	0.040	0.032
$\frac{1}{2}$	0.500	0.530	0.513	0.900	0.880	0.045	0.037
$\frac{9}{16}$	0.562	0.596	0.576	0.985	0.960	0.045	0.037
$\frac{5}{8}$	0.625	0.663	0.641	1.070	1.045	0.050	0.042
$\frac{11}{16}$	0.688	0.728	0.704	1.155	1.130	0.050	0.042
$\frac{3}{4}$	0.750	0.795	0.768	1.260	1.220	0.055	0.047
$\frac{13}{16}$	0.812	0.861	0.833	1.315	1.290	0.055	0.047
$\frac{7}{8}$	0.875	0.927	0.897	1.410	1.380	0.060	0.052
1	1.000	1.060	1.025	1.620	1.590	0.067	0.059

From The American Society of Mechanical Engineers—ANSI B18.21.1—1972

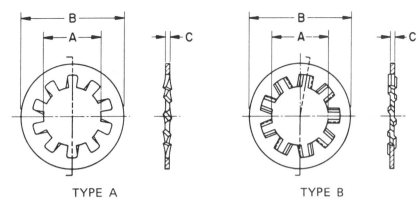

TYPE A   TYPE B

TABLE 33 DIMENSIONS OF INTERNAL TOOTH LOCK WASHERS

Nominal Washer Size		A Inside Diameter		B Outside Diameter		C Thickness	
		Max	Min	Max	Min	Max	Min
No. 2	0.086	0.095	0.089	0.200	0.175	0.015	0.010
No. 3	0.099	0.109	0.102	0.232	0.215	0.019	0.012
No. 4	0.112	0.123	0.115	0.270	0.255	0.019	0.015
No. 5	0.125	0.136	0.129	0.280	0.245	0.021	0.017
No. 6	0.138	0.150	0.141	0.295	0.275	0.021	0.017
No. 8	0.164	0.176	0.168	0.340	0.325	0.023	0.018
No. 10	0.190	0.204	0.195	0.381	0.365	0.025	0.020
No. 12	0.216	0.231	0.221	0.410	0.394	0.025	0.020
$\frac{1}{4}$	0.250	0.267	0.256	0.478	0.460	0.028	0.023
$\frac{5}{16}$	0.312	0.332	0.320	0.610	0.594	0.034	0.028
$\frac{3}{8}$	0.375	0.398	0.384	0.692	0.670	0.040	0.032
$\frac{7}{16}$	0.438	0.464	0.448	0.789	0.740	0.040	0.032
$\frac{1}{2}$	0.500	0.530	0.512	0.900	0.867	0.045	0.037
$\frac{9}{16}$	0.562	0.596	0.576	0.985	0.957	0.045	0.037
$\frac{5}{8}$	0.625	0.663	0.640	1.071	1.045	0.050	0.042
$\frac{11}{16}$	0.688	0.728	0.704	1.166	1.130	0.050	0.042
$\frac{3}{4}$	0.750	0.795	0.769	1.245	1.220	0.055	0.047
$\frac{13}{16}$	0.812	0.861	0.832	1.315	1.290	0.055	0.047
$\frac{7}{8}$	0.875	0.927	0.894	1.410	1.364	0.060	0.052
1	1.000	1.060	1.019	1.637	1.590	0.067	0.059
$1\frac{1}{8}$	1.125	1.192	1.144	1.830	1.799	0.067	0.059
$1\frac{1}{4}$	1.250	1.325	1.275	1.975	1.921	0.067	0.059

From The American Society of Mechanical Engineers—ANSI B18.21.1—1972

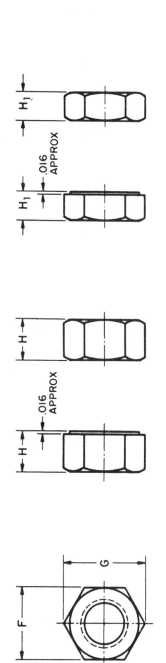

TABLE 34 DIMENSIONS OF HEX NUTS AND HEX JAM NUTS

Nominal Size or Basic Major Dia of Thread		F Width Across Flats			G Width Across Corners		H Thickness Hex Nuts			H_1 Thickness Hex Jam Nuts			Hex Nuts Specified Proof Load Up to 150,000 psi	150,000 psi and Greater	Jam Nuts All Strength Levels
		Basic	Max	Min	Max	Min	Basic	Max	Min	Basic	Max	Min	Runout of Bearing Face, FIR Max		
1/4	0.2500	7/16	0.438	0.428	0.505	0.488	7/32	0.226	0.212	5/32	0.163	0.150	0.015	0.010	0.015
5/16	0.3125	1/2	0.500	0.489	0.577	0.557	17/64	0.273	0.258	3/16	0.195	0.180	0.016	0.011	0.016
3/8	0.3750	9/16	0.562	0.551	0.650	0.628	21/64	0.337	0.320	7/32	0.227	0.210	0.017	0.012	0.017
7/16	0.4375	11/16	0.688	0.675	0.794	0.768	3/8	0.385	0.365	1/4	0.260	0.240	0.018	0.013	0.018
1/2	0.5000	3/4	0.750	0.736	0.866	0.840	7/16	0.448	0.427	5/16	0.323	0.302	0.019	0.014	0.019
9/16	0.5625	7/8	0.875	0.861	1.010	0.982	31/64	0.496	0.473	5/16	0.324	0.301	0.020	0.015	0.020
5/8	0.6250	15/16	0.938	0.922	1.083	1.051	35/64	0.559	0.535	3/8	0.387	0.363	0.021	0.016	0.021
3/4	0.7500	1 1/8	1.125	1.088	1.299	1.240	41/64	0.665	0.617	27/64	0.446	0.398	0.023	0.018	0.023
7/8	0.8750	1 5/16	1.312	1.269	1.516	1.447	3/4	0.776	0.724	31/64	0.510	0.458	0.025	0.020	0.025
1	1.0000	1 1/2	1.500	1.450	1.732	1.653	55/64	0.887	0.831	35/64	0.575	0.519	0.027	0.022	0.027
1 1/8	1.1250	1 11/16	1.688	1.631	1.949	1.859	31/32	0.999	0.939	39/64	0.639	0.579	0.030	0.025	0.030
1 1/4	1.2500	1 7/8	1.875	1.812	2.165	2.066	1 1/16	1.094	1.030	23/32	0.751	0.687	0.033	0.028	0.033
1 3/8	1.3750	2 1/16	2.062	1.994	2.382	2.273	1 11/64	1.206	1.138	25/32	0.815	0.747	0.036	0.031	0.036
1 1/2	1.5000	2 1/4	2.250	2.175	2.598	2.480	1 9/32	1.317	1.245	27/32	0.880	0.808	0.039	0.034	0.039
See Notes	9	3			4								2		

From The American Society of Mechanical Engineers—ANSI B18.2.2–1972

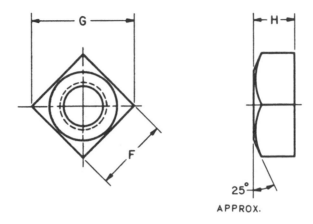

TABLE 35 DIMENSIONS OF SQUARE NUTS

Nominal Size or Basic Major Dia of Thread		F Width Across Flats			G Width Across Corners		H Thickness		
		Basic	Max	Min	Max	Min	Basic	Max	Min
1/4	0.2500	7/16	0.438	0.425	0.619	0.584	7/32	0.235	0.203
5/16	0.3125	9/16	0.562	0.547	0.795	0.751	17/64	0.283	0.249
3/8	0.3750	5/8	0.625	0.606	0.884	0.832	21/64	0.346	0.310
7/16	0.4375	3/4	0.750	0.728	1.061	1.000	3/8	0.394	0.356
1/2	0.5000	13/16	0.812	0.788	1.149	1.082	7/16	0.458	0.418
5/8	0.6250	1	1.000	0.969	1.414	1.330	35/64	0.569	0.525
3/4	0.7500	1 1/8	1.125	1.088	1.591	1.494	21/32	0.680	0.632
7/8	0.8750	1 5/16	1.312	1.269	1.856	1.742	49/64	0.792	0.740
1	1.0000	1 1/2	1.500	1.450	2.121	1.991	7/8	0.903	0.847
1 1/8	1.1250	1 11/16	1.688	1.631	2.386	2.239	1	1.030	0.970
1 1/4	1.2500	1 7/8	1.875	1.812	2.652	2.489	1 3/32	1.126	1.062
1 3/8	1.3750	2 1/16	2.062	1.994	2.917	2.738	1 13/64	1.237	1.169
1 1/2	1.5000	2 1/4	2.250	2.175	3.182	2.986	1 5/16	1.348	1.276
See Notes	8	3							

From The American Society of Mechanical Engineers—ANSI B18.2.2—1972

TABLE 36 KEY & KEYWAY SIZES

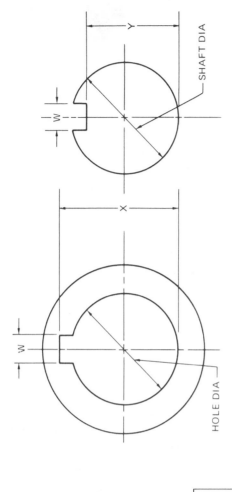

KEY SIZE = W × H
LENGTH (L) TO SUIT

Nom. Size (Inch)	DIA. (Shaft) Inch	mm	'X' (Collar) Inch	mm	'Y' (Shaft) Inch	mm
1/2	.500	12.700	.560	14.224	.430	10.922
9/16	.562	14.290	.623	15.824	.493	12.522
5/8	.625	15.875	.709	18.008	.517	13.132
11/16	.688	17.470	.773	18.618	.581	14.757
3/4	.750	19.050	.837	21.259	.644	16.357
13/16	.812	20.640	.900	22.860	.708	17.983
7/8	.875	22.225	.964	24.485	.771	19.583
15/16	.938	23.820	1.051	26.695	.791	20.091
1	1.000	25.400	1.114	28.295	.859	21.818
1 1/16	1.062	26.985	1.178	29.921	.923	23.444
1 1/8	1.125	28.575	1.241	31.521	.986	25.044
1 3/16	1.188	30.165	1.304	33.121	1.049	26.644
1 1/4	1.250	31.750	1.367	34.722	1.112	28.244
1 5/16	1.312	33.340	1.455	36.957	1.137	28.879
1 3/8	1.375	34.923	1.518	38.557	1.201	30.505

Shaft Nom. Size – DIA. – From	To & Incl.	Square (W = H)
5/16 (8)	7/16 (11)	3/32 (2.38)
7/16 (11)	9/16 (14)	1/8 (3.175)
9/16 (14)	7/8 (22)	3/16 (4.76)
7/8 (22)	1 1/4 (32)	1/4 (6.35)
1 1/4 (32)	1 3/8 (35)	5/16 (7.94)
1 3/8 (35)	1 3/4 (44)	3/8 (9.53)
1 3/4 (44)	2 1/4 (57)	1/2 (12.7)
2 1/4 (57)	2 3/4 (70)	5/8 (15.88)
2 3/4 (70)	3 1/4 (82)	3/4 (19.05)
3 1/4 (82)	3 3/4 (95)	7/8 (22.23)

Type	Square Key From	To & Incl.	Tolerance
Bar Stock	—	3/4 (19.05)	+.000 – .002 (+.0000 – .0254)
Bar Stock	—	1 1/2 (38.1)	+.000 – .003 (+.0000 – .0762)
Bar Stock	3/4 (19.05)	2 1/2 (63.5)	+.000 – .004 (+.0000 – .1016)
Bar Stock	1 1/2 (38.1)	3 1/2 (88.9)	+.000 – .006 (+.0000 – .1524)
Keystock	—	1 1/4 (31.75)	+.001 – .000 (+.0254 – .0000)
Keystock	1 1/4 (31.75)	3 (76.2)	+.002 – .000 (+.0508 – .0000)
Keystock	3 (76.2)	3 1/2 (88.9)	+.003 – .000 (+.0762 – .0000)

(Figures in parenthesis = mm)

From Drafting for Trades and Industry—Mechanical and Electronic. Nelson. Delmar Publishers Inc.

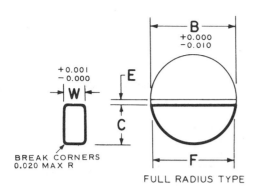

FULL RADIUS TYPE

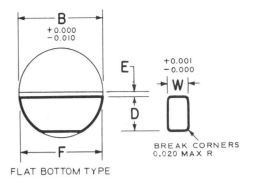

FLAT BOTTOM TYPE

TABLE 37 WOODRUFF KEYS

Key No.	Nominal Key Size W × B	Actual Length F +0.000-0.010	Height of Key				Distance Below Center E
			C		D		
			Max	Min	Max	Min	
202	1/16 × 1/4	0.248	0.109	0.104	0.109	0.104	1/64
202.5	1/16 × 5/16	0.311	0.140	0.135	0.140	0.135	1/64
302.5	3/32 × 5/16	0.311	0.140	0.135	0.140	0.135	1/64
203	1/16 × 3/8	0.374	0.172	0.167	0.172	0.167	1/64
303	3/32 × 3/8	0.374	0.172	0.167	0.172	0.167	1/64
403	1/8 × 3/8	0.374	0.172	0.167	0.172	0.167	1/64
204	1/16 × 1/2	0.491	0.203	0.198	0.194	0.188	3/64
304	3/32 × 1/2	0.491	0.203	0.198	0.194	0.188	3/64
404	1/8 × 1/2	0.491	0.203	0.198	0.194	0.188	3/64
305	3/32 × 5/8	0.612	0.250	0.245	0.240	0.234	1/16
405	1/8 × 5/8	0.612	0.250	0.245	0.240	0.234	1/16
505	5/32 × 5/8	0.612	0.250	0.245	0.240	0.234	1/16
605	3/16 × 5/8	0.612	0.250	0.245	0.240	0.234	1/16
406	1/8 × 3/4	0.740	0.313	0.308	0.303	0.297	1/16
506	5/32 × 3/4	0.740	0.313	0.308	0.303	0.297	1/16
606	3/16 × 3/4	0.740	0.313	0.308	0.303	0.297	1/16
806	1/4 × 3/4	0.740	0.313	0.308	0.303	0.297	1/16
507	5/32 × 7/8	0.866	0.375	0.370	0.365	0.359	1/16
607	3/16 × 7/8	0.866	0.375	0.370	0.365	0.359	1/16
707	7/32 × 7/8	0.866	0.375	0.370	0.365	0.359	1/16
807	1/4 × 7/8	0.866	0.375	0.370	0.365	0.359	1/16
608	3/16 × 1	0.992	0.438	0.433	0.428	0.422	1/16
708	7/32 × 1	0.992	0.438	0.433	0.428	0.422	1/16
808	1/4 × 1	0.992	0.438	0.433	0.428	0.422	1/16
1008	5/16 × 1	0.992	0.438	0.433	0.428	0.422	1/16
1208	3/8 × 1	0.992	0.438	0.433	0.428	0.422	1/16
609	3/16 × 1 1/8	1.114	0.484	0.479	0.475	0.469	5/64
709	7/32 × 1 1/8	1.114	0.484	0.479	0.475	0.469	5/64
809	1/4 × 1 1/8	1.114	0.484	0.479	0.475	0.469	5/64
1009	5/16 × 1 1/8	1.114	0.484	0.479	0.475	0.469	5/64

TABLE 37 WOODRUFF KEYS **(Concluded)**

Key No.	Nominal Key Size W × B	Actual Length F +0.000-0.010	Height of Key				Distance Below Center E
			C		D		
			Max	Min	Max	Min	
610	3/16 × 1¼	1.240	0.547	0.542	0.537	0.531	5/64
710	7/32 × 1¼	1.240	0.547	0.542	0.537	0.531	5/64
810	¼ × 1¼	1.240	0.547	0.542	0.537	0.531	5/64
1010	5/16 × 1¼	1.240	0.547	0.542	0.537	0.531	5/64
1210	3/8 × 1¼	1.240	0.547	0.542	0.537	0.531	5/64
811	¼ × 1⅜	1.362	0.594	0.589	0.584	0.578	3/32
1011	5/16 × 1⅜	1.362	0.594	0.589	0.584	0.578	3/32
1211	3/8 × 1⅜	1.362	0.594	0.589	0.584	0.578	3/32
812	¼ × 1½	1.484	0.641	0.636	0.631	0.625	7/64
1012	5/16 × 1½	1.484	0.641	0.636	0.631	0.625	7/64
1212	3/8 × 1½	1.484	0.641	0.636	0.631	0.625	7/64

All dimensions given are in inches.

The key numbers indicate nominal key dimensions. The last two digits give the nominal diameter B in eighths of an inch and the digits preceding the last two give the nominal width W in thirty-seconds of an inch.

Example:
No. 204 indicates a key 2/32 × 4/8 or 1/16 × ½.
No. 808 indicates a key 8/32 × 8/8 or ¼ × 1.
No. 1212 indicates a key 12/32 × 12/8 or 3/8 × 1½.

From The American Society of Mechanical Engineers—ANSI B17.2—1967

WOODRUFF KEYS AND KEYSEATS

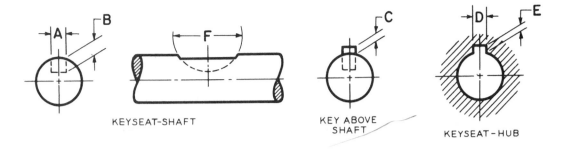

KEYSEAT-SHAFT KEY ABOVE SHAFT KEYSEAT-HUB

| Key Number | Nominal Size Key | Keyseat — Shaft | | | | | Key Above Shaft | Keyseat — Hub | |
| | | Width A* | | Depth B | Diameter F | | Height C | Width D | Depth E |
		Min	Max	+0.005 −0.000	Min	Max	+0.005 −0.005	+0.002 −0.000	+0.005 −0.000
202	1/16 × 1/4	0.0615	0.0630	0.0728	0.250	0.268	0.0312	0.0635	0.0372
202.5	1/16 × 5/16	0.0615	0.0630	0.1038	0.312	0.330	0.0312	0.0635	0.0372
302.5	3/32 × 5/16	0.0928	0.0943	0.0882	0.312	0.330	0.0469	0.0948	0.0529
203	1/16 × 3/8	0.0615	0.0630	0.1358	0.375	0.393	0.0312	0.0635	0.0372
303	3/32 × 3/8	0.0928	0.0943	0.1202	0.375	0.393	0.0469	0.0948	0.0529
403	1/8 × 3/8	0.1240	0.1255	0.1045	0.375	0.393	0.0625	0.1260	0.0685
204	1/16 × 1/2	0.0615	0.0630	0.1668	0.500	0.518	0.0312	0.0635	0.0372
304	3/32 × 1/2	0.0928	0.0943	0.1511	0.500	0.518	0.0469	0.0948	0.0529
404	1/8 × 1/2	0.1240	0.1255	0.1355	0.500	0.518	0.0625	0.1260	0.0685
305	3/32 × 5/8	0.0928	0.0943	0.1981	0.625	0.643	0.0469	0.0948	0.0529
405	1/8 × 5/8	0.1240	0.1255	0.1825	0.625	0.643	0.0625	0.1260	0.0685
505	5/32 × 5/8	0.1553	0.1568	0.1669	0.625	0.643	0.0781	0.1573	0.0841
605	3/16 × 5/8	0.1863	0.1880	0.1513	0.625	0.643	0.0937	0.1885	0.0997
406	1/8 × 3/4	0.1240	0.1255	0.2455	0.750	0.768	0.0625	0.1260	0.0685
506	5/32 × 3/4	0.1553	0.1568	0.2299	0.750	0.768	0.0781	0.1573	0.0841
606	3/16 × 3/4	0.1863	0.1880	0.2143	0.750	0.768	0.0937	0.1885	0.0997
806	1/4 × 3/4	0.2487	0.2505	0.1830	0.750	0.768	0.1250	0.2510	0.1310
507	5/32 × 7/8	0.1553	0.1568	0.2919	0.875	0.895	0.0781	0.1573	0.0841
607	3/16 × 7/8	0.1863	0.1880	0.2763	0.875	0.895	0.0937	0.1885	0.0997
707	7/32 × 7/8	0.2175	0.2193	0.2607	0.875	0.895	0.1093	0.2198	0.1153
807	1/4 × 7/8	0.2487	0.2505	0.2450	0.875	0.895	0.1250	0.2510	0.1310
608	3/16 × 1	0.1863	0.1880	0.3393	1.000	1.020	0.0937	0.1885	0.0997
708	7/32 × 1	0.2175	0.2193	0.3237	1.000	1.020	0.1093	0.2198	0.1153
808	1/4 × 1	0.2487	0.2505	0.3080	1.000	1.020	0.1250	0.2510	0.1310
1008	5/16 × 1	0.3111	0.3130	0.2768	1.000	1.020	0.1562	0.3135	0.1622
1208	3/8 × 1	0.3735	0.3755	0.2455	1.000	1.020	0.1875	0.3760	0.1935
609	3/16 × 1 1/8	0.1863	0.1880	0.3853	1.125	1.145	0.0937	0.1885	0.0997
709	7/32 × 1 1/8	0.2175	0.2193	0.3697	1.125	1.145	0.1093	0.2198	0.1153
809	1/4 × 1 1/8	0.2487	0.2505	0.3540	1.125	1.145	0.1250	0.2510	0.1310
1009	5/16 × 1 1/8	0.3111	0.3130	0.3228	1.125	1.145	0.1562	0.3135	0.1622

From The American Society of Mechanical Engineers—ANSI B17.2—1967

Table 38

727

WOODRUFF KEY SIZES FOR DIFFERENT SHAFT DIAMETERS

Shaft Diameter	5/16 to 3/8	7/16 to 1/2	9/16 to 3/4	13/16 to 15/16	1 to 1 3/16	1 1/4 to 1 7/16	1 1/2 to 1 3/4	1 13/16 to 2 1/8	2 3/16 to 2 1/2
Key Numbers	204	304 305	404 405 406	505 506 507	606 607 608 609	807 808 809	810 811 812	1011 1012	1211 1212

Table 39

4. KEY DIMENSIONS AND TOLERANCES

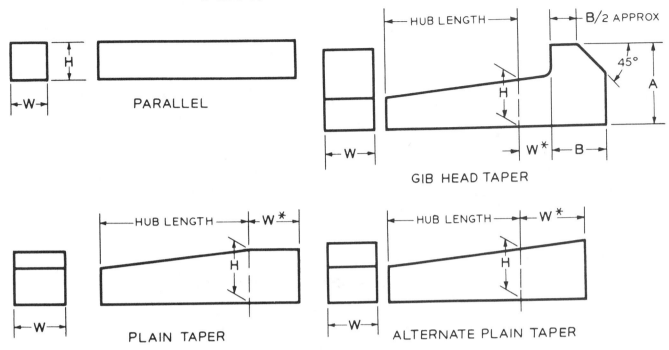

PARALLEL

GIB HEAD TAPER

PLAIN TAPER

ALTERNATE PLAIN TAPER

Plain and Gib Head Taper Keys Have a 1/8″ Taper in 12″

KEY			NOMINAL KEY SIZE		TOLERANCE	
			Width, W		Width, W	Height, H
			Over	To (Incl)		
Parallel	Square	Bar Stock	— 3/4 1-1/2 2-1/2	3/4 1-1/2 2-1/2 3-1/2	+0.000 −0.002 +0.000 −0.003 +0.000 −0.004 +0.000 −0.006	+0.000 −0.002 +0.000 −0.003 +0.000 −0.004 +0.000 −0.006
		Keystock	— 1-1/4 3	1-1/4 3 3-1/2	+0.001 −0.000 +0.002 −0.000 +0.003 −0.000	+0.001 −0.000 +0.002 −0.000 +0.003 −0.000
	Rectangular	Bar Stock	— 3/4 1-1/2 3 4 6	3/4 1-1/2 3 4 6 7	+0.000 −0.003 +0.000 −0.004 +0.000 −0.005 +0.000 −0.006 +0.000 −0.008 +0.000 −0.013	+0.000 −0.003 +0.000 −0.004 +0.000 −0.005 +0.000 −0.006 +0.000 −0.008 +0.000 −0.013
		Keystock	— 1-1/4 3	1-1/4 3 7	+0.001 −0.000 +0.002 −0.000 +0.003 −0.000	+0.005 −0.005 +0.005 −0.005 +0.005 −0.005
Taper	Plain or Gib Head Square or Rectangular		— 1-1/4 3	1-1/4 3 7	+0.001 −0.000 +0.002 −0.000 +0.003 −0.000	+0.005 −0.000 +0.005 −0.000 +0.005 −0.000

From The American Society of Mechanical Engineers—ANSI B17.1—1967

*For locating position of dimension H. Tolerance does not apply.
See Table 41 for dimensions on gib heads.
All dimensions given in inches.

TABLE 40 KEY DIMENSIONS AND TOLERANCES

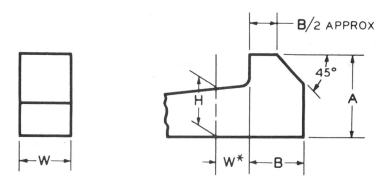

TABLE 41 GIB HEAD NOMINAL DIMENSIONS

Nominal Key Size Width, W	SQUARE			RECTANGULAR		
	H	A	B	H	A	B
1/8	1/8	1/4	1/4	3/32	3/16	1/8
3/16	3/16	5/16	5/16	1/8	1/4	1/4
1/4	1/4	7/16	3/8	3/16	5/16	5/16
5/16	5/16	1/2	7/16	1/4	7/16	3/8
3/8	3/8	5/8	1/2	1/4	7/16	3/8
1/2	1/2	7/8	5/8	3/8	5/8	1/2
5/8	5/8	1	3/4	7/16	3/4	9/16
3/4	3/4	1-1/4	7/8	1/2	7/8	5/8
7/8	7/8	1-3/8	1	5/8	1	3/4
1	1	1-5/8	1-1/8	3/4	1-1/4	7/8
1-1/4	1-1/4	2	1-7/16	7/8	1-3/8	1
1-1/2	1-1/2	2-3/8	1-3/4	1	1-5/8	1-1/8
1-3/4	1-3/4	2-3/4	2	1-1/2	2-3/8	1-3/4
2	2	3-1/2	2-1/4	1-1/2	2-3/8	1-3/4
2-1/2	2-1/2	4	3	1-3/4	2-3/4	2
3	3	5	3-1/2	2	3-1/2	2-1/4
3-1/2	3-1/2	6	4	2-1/2	4	3

*For locating position of dimension H.

For larger sizes the following relationships are suggested as guides for establishing A and B.

$$A = 1.8 H \qquad B = 1.2 H$$

All dimensions given in inches.

From The American Society of Mechanical Engineers—ANSI B17.1—1967

SHEET METAL AND WIRE GAGE DESIGNATION

GAGE NO.	AMERICAN OR BROWN & SHARPE'S A.W.G. OR B. & S.	UNITED STATES STANDARD	MANU- FACTURERS' STANDARD FOR SHEET STEEL	GAGE NO.
0000000	. . . .	.500	. . . .	0000000
000000	.5800	.469	. . . .	000000
00000	.5165	.438	. . . .	00000
0000	.4600	.406	. . . .	0000
000	.4096	.375	. . . .	000
00	.3648	.344	. . . .	00
0	.3249	.312	. . . .	0
1	.2893	.281	. . . .	1
2	.2576	.266	. . . .	2
3	.2294	.250	.2391	3
4	.2043	.234	.2242	4
5	.1819	.219	.2092	5
6	.1620	.203	.1943	6
7	.1443	.188	.1793	7
8	.1285	.172	.1644	8
9	.1144	.156	.1495	9
10	.1019	.141	.1345	10
11	.0907	.125	.1196	11
12	.0808	.109	.1046	12
13	.0720	.0938	.0897	13
14	.0642	.0781	.0747	14
15	.0571	.0703	.0673	15
16	.0508	.0625	.0598	16
17	.0453	.0562	.0538	17
18	.0403	.0500	.0478	18
19	.0359	.0438	.0418	19
20	.0320	.0375	.0359	20
21	.0285	.0344	.0329	21
22	.0253	.0312	.0299	22
23	.0226	.0281	.0269	23
24	.0201	.0250	.0239	24
25	.0179	.0219	.0209	25
26	.0159	.0188	.0179	26
27	.0142	.0172	.0164	27
28	.0126	.0156	.0149	28
29	.0113	.0141	.0135	29
30	.0100	.0125	.0120	30
31	.0089	.0109	.0105	31
32	.0080	.0102	.0097	32
33	.0071	.00938	.0090	33
34	.0063	.00859	.0082	34
35	.0056	.00781	.0075	35
36	.0050	.00703	.0067	36

Table 42

BEND ALLOWANCE FOR 90° BENDS (INCH)

Radii / Thickness	.031	.063	.094	.125	.156	.188	.219	.250	.281	.313	.344	.375	.438	.500	.531	.625
.013	.058	.108	.157	.205	.254	.304	.353	.402	.450	.501	.549	.598	.697	.794	.843	.991
.016	.060	.110	.159	.208	.256	.307	.355	.404	.453	.503	.552	.600	.699	.796	.845	.993
.020	.062	.113	.161	.210	.259	.309	.358	.406	.455	.505	.554	.603	.702	.799	.848	.995
.022	.064	.114	.163	.212	.260	.311	.359	.408	.457	.507	.556	.604	.703	.801	.849	.997
.025	.066	.116	.165	.214	.263	.313	.362	.410	.459	.509	.558	.607	.705	.803	.851	.999
.028	.068	.119	.167	.216	.265	.315	.364	.412	.461	.511	.560	.609	.708	.805	.854	1.001
.032	.071	.121	.170	.218	.267	.317	.366	.415	.463	.514	.562	.611	.710	.807	.856	1.004
.038	.075	.126	.174	.223	.272	.322	.371	.419	.468	.518	.567	.616	.715	.812	.861	1.008
.040	.077	.127	.176	.224	.273	.323	.372	.421	.469	.520	.568	.617	.716	.813	.862	1.010
.050		.134	.183	.232	.280	.331	.379	.428	.477	.527	.576	.624	.723	.821	.869	1.017
.064		.144	.192	.241	.290	.340	.389	.437	.486	.536	.585	.634	.732	.830	.878	1.026
.072			.198	.247	.296	.346	.394	.443	.492	.542	.591	.639	.738	.836	.885	1.032
.078			.202	.251	.300	.350	.399	.447	.496	.546	.595	.644	.743	.840	.889	1.036
.081			.204	.253	.302	.252	.401	.449	.498	.548	.598	.646	.745	.842	.891	1.038
.091			.212	.260	.309	.359	.408	.456	.505	.555	.604	.653	.752	.849	.898	1.045
.094			.214	.262	.311	.361	.410	.459	.507	.558	.606	.655	.754	.851	.900	1.048
.102				.268	.317	.367	.416	.464	.513	.563	.612	.661	.760	857	.906	1.053
.109				.273	.321	.372	.420	.469	.518	.568	.617	.665	.764	.862	.910	1.058
.125				.284	.333	.383	.432	.480	.529	.579	.628	.677	.776	.873	.922	1.069
.156					.355	.405	.453	.502	.551	.601	.650	.698	.797	.895	.943	1.091
.188						.427	.476	.525	.573	.624	.672	.721	.820	.917	.966	1.114
.203								.535	.584	.634	.683	.731	.830	.928	.976	1.124
.218								.546	.594	.645	.693	.742	.841	.938	.987	1.135
.234								.557	.606	.656	.705	.753	.852	.950	.998	1.146
.250								.568	.617	.667	.716	.764	.863	.961	1.009	1.157

EXAMPLE: MATERIAL THICKNESS = 1/8″, INSIDE RADII = 1/4″ R. WHERE THESE CROSS = .284″ BEND ALLOWANCE (B/A) PLUS TOTAL OF STRAIGHT LENGTHS = DEVELOPED LENGTH.

BEND ALLOWANCE FOR 90° BENDS (MILLIMETRE)

Radii / Thickness	0.80	1.58	2.38	3.18	3.96	4.76	5.56	6.35	7.15	7.94	8.74	9.52	11.12	12.70	13.50	15.85
.330	1.46	2.74	3.98	5.22	6.45	7.73	8.96	10.20	11.44	12.71	13.95	15.19	17.70	20.18	21.42	25.16
.406	1.52	2.80	4.04	5.27	6.51	7.78	9.02	10.26	11.50	12.77	14.01	15.24	17.76	20.23	21.47	25.22
.508	1.59	2.87	4.11	5.34	6.58	7.86	9.09	10.33	11.57	12.84	14.08	15.32	17.83	20.30	21.54	25.29
.559	1.63	2.91	4.14	5.38	6.62	7.89	9.13	10.37	11.60	12.88	14.12	15.35	17.86	20.34	21.57	25.32
.635	1.68	2.96	4.20	5.43	6.67	7.95	9.18	10.42	11.66	12.93	14.17	15.40	17.92	20.39	21.63	25.38
.711	1.74	3.02	4.25	5.49	6.72	8.00	9.24	10.47	11.71	12.99	14.22	15.46	17.98	20.44	21.68	25.43
.813	1.81	3.08	4.32	5.56	6.79	8.07	9.30	10.54	11.78	13.05	14.29	15.53	18.04	20.52	21.75	25.50
.965	1.91	3.19	4.42	5.66	6.90	8.18	9.41	10.65	11.89	13.16	14.40	15.64	18.15	20.62	21.86	25.61
1.016	1.95	2.23	4.46	5.70	6.94	8.21	9.45	10.69	11.92	13.20	14.44	15.67	18.19	20.66	21.90	25.64
1.270		3.40	4.64	5.88	7.11	8.39	9.63	10.86	12.10	13.38	14.61	15.85	18.36	20.84	22.07	25.82
1.625		3.65	4.89	6.13	7.36	8.64	9.88	11.11	12.35	13.63	14.86	16.10	18.61	21.08	22.32	26.07
1.829			5.03	6.27	7.51	8.78	10.02	11.26	12.49	13.77	15.01	16.24	18.76	21.23	22.47	26.22
1.981			5.14	6.38	7.61	8.89	10.13	11.36	12.60	13.88	15.11	16.35	18.86	21.34	22.57	26.32
2.058			5.19	6.43	7.67	8.94	10.18	11.42	12.65	13.93	15.17	16.40	18.92	21.39	22.63	26.38
2.311			5.37	6.60	7.85	9.12	10.36	11.60	12.83	14.11	15.35	16.58	19.09	21.57	22.80	26.55
2.388			5.43	6.66	7.90	9.17	10.41	11.65	12.89	14.16	15.40	16.64	19.15	21.62	22.86	26.61
2.591				6.81	8.04	9.31	10.55	11.79	13.03	14.30	15.54	16.78	19.29	21.76	23.00	26.75
2.769				6.93	8.17	9.44	10.68	11.92	13.15	14.43	15.67	16.90	19.42	21.89	23.13	26.88
3.175				7.22	8.45	9.73	10.96	12.20	13.44	14.71	15.95	17.19	19.70	22.17	23.41	27.16
3.962					9.00	10.28	11.52	12.75	13.99	15.27	16.74	17.74	20.25	22.72	23.96	27.71
4.775					9.58	10.85	12.09	13.32	14.56	15.84	17.07	18.31	20.82	23.30	24.53	28.28
5.156						11.12	12.36	13.59	14.83	16.11	17.34	18.58	21.09	23.57	24.80	28.55
5.537								13.85	15.10	16.37	17.61	18.85	21.36	23.83	25.07	28.82
5.944								14.15	15.38	16.66	17.89	19.13	21.64	24.12	25.35	29.10
6.350								14.43	15.67	16.94	18.18	19.42	21.93	24.40	25.64	29.39

EXAMPLE: MATERIAL THICKNESS = 3.18 MM, INSIDE RADII = 6.35 MM R. WHERE THESE CROSS = 12.20 MM BEND ALLOWANCE (B/A) PLUS TOTAL OF STRAIGHT LENGTHS = DEVELOPED LENGTH.

Table 43

From Drafting for Trades and Industry—Basic Skills, Nelson. Delmar Publishers Inc.

BEND ALLOWANCE FOR EACH 1° OF BEND (INCH)

Radii / Thickness	.031	.063	.094	.125	.156	.188	.219	.250	.281	.313	.344	.375	.438	.500	.531	.625
.013	.00064	.00120	.00174	.00228	.00282	.00338	.00392	.00446	.00500	.00556	.00610	.00664	.00774	.00883	.00937	.01101
.016	.00067	.00122	.00176	.00231	.00285	.00342	.00395	.00449	.00503	.00559	.00613	.00667	.00777	.00885	.00939	.01103
.020	.00069	.00125	.00179	.00233	.00287	.00343	.00397	.00452	.00506	.00561	.00616	.00670	.00780	.00888	.00942	.01106
.022	.00071	.00127	.00181	.00235	.00289	.00345	.00399	.00453	.00508	.00563	.00617	.00672	.00782	.00890	.00944	.01108
.025	.00074	.00129	.00184	.00238	.00292	.00348	.00402	.00456	.00510	.00566	.00610	.00674	.00784	.00892	.00946	.01110
.028	.00076	.00132	.00186	.00240	.00294	.00350	.00404	.00458	.00512	.00568	.00622	.00676	.00786	.00894	.00948	.01112
.032	.00079	.00134	.00189	.00243	.00297	.00353	.00407	.00461	.00515	.00571	.00625	.00679	.00789	.00897	.00951	.01115
.038	.00084	.00140	.00194	.00248	.00302	.00358	.00412	.00466	.00520	.00576	.00630	.00684	.00794	.00902	.00946	.01120
.040	.00085	.00141	.00195	.00249	.00303	.00359	.00413	.00468	.00522	.00577	.00632	.00686	.00796	.00904	.00958	.01122
.050		.00149	.00203	.00258	.00312	.00368	.00422	.00476	.00530	.00586	.00640	.00694	.00804	.00912	.00966	.01130
.064		.00160	.00214	.00268	.00322	.00378	.00432	.00486	.00540	.00596	.00650	.00704	.00814	.00922	.00976	.01140
.072			.00220	.00274	.00328	.00384	.00438	.00492	.00546	.00602	.00656	.00710	.00820	.00929	.00983	.01147
.078			.00225	.00279	.00333	.00389	.00443	.00497	.00551	.00607	.00661	.00715	.00825	.00933	.00987	.01152
.081			.00227	.00281	.00335	.00391	.00445	.00499	.00554	.00609	.00664	.00718	.00828	.00936	.00990	.01154
.091			.00235	.00289	.00343	.00399	.00453	.00507	.00561	.00617	.00671	.00725	.00835	.00944	.00998	.01162
.094			.00237	.00291	.00346	.00401	.00456	.00510	.00564	.00620	.00674	.00728	.00838	.00946	.00999	.01164
.102				.00298	.00352	.00408	.00462	.00516	.00570	.00626	.00680	.00734	.00844	.00952	.01006	.01170
.109				.00303	.00357	.00413	.00467	.00521	.00575	.00631	.00685	.00739	.00849	.00958	.01012	.01176
.125				.00316	.00370	.00426	.00480	.00534	.00588	.00644	.00698	.00752	.00862	.00970	.01024	.01188
.156					.00394	.00450	.00504	.00558	.00612	.00668	.00722	.00776	.00886	.00994	.01048	.01212
.188						.00475	.00529	.00583	.00637	.00693	.00747	.00802	.00911	.01019	.01073	.01237
.203								.00595	.00649	.00704	.00759	.00813	.00923	.01031	.01085	.01249
.218								.00606	.00660	.00716	.00770	.00824	.00934	.01042	.01097	.01261
.234								.00619	.00673	.00729	.00783	.00837	.00947	.01055	.01109	.01273
.250								.00631	.00685	.00741	.00795	.00849	.00959	.01068	.01122	.01286

EXAMPLE: MATERIAL THICKNESS = 1/8", INSIDE RADII = 1/4" R. WHERE THESE CROSS = .00534" IF THE INSIDE OF YOUR BEND IS 20° THE BEND ALLOWANCE (B/A) = .1068 (.00534 × 20°) BEND ALLOWANCE (B/A) PLUS TOTAL OF STRAIGHT LENGTHS = DEVELOPED LENGTH.

BEND ALLOWANCE FOR EACH 1° OF BEND (MILLIMETRE)

Radii / Thickness	0.80	1.58	2.38	3.18	3.96	4.76	5.56	6.35	7.15	7.94	8.74	9.52	11.12	12.70	13.50	15.85
.330	.0161	.0305	.0442	.0580	.0717	.0859	.0996	.1134	.1271	.1413	.1550	.1668	.1967	.2242	.2379	.2796
.406	.0169	.0311	.0448	.0586	.0723	.0865	.1002	.1140	.1277	.1419	.1556	.1694	.1973	.2248	.2385	.2802
.508	.0177	.0319	.0456	.0594	.0731	.0873	.1010	.1148	.1285	.1427	.1564	.1702	.1981	.2256	.2393	.2810
.533	.0181	.0323	.0460	.0598	.0735	.0877	.1014	.1152	.1289	.1431	.1568	.1706	.1985	.2260	.2397	.2814
.635	.0187	.0329	.0466	.0604	.0741	.0883	.1020	.1158	.1295	.1437	.1574	.1712	.1991	.2266	.2403	.2820
.711	.0193	.0335	.0472	.0610	.0747	.0889	.1026	.1164	.1301	.1443	.1580	.1718	.1997	.2272	.2409	.2826
.813	.0201	.0343	.0480	.0617	.0755	.0897	.1034	.1174	.1309	.1451	.1588	.1726	.2005	.2280	.2417	.2834
.965	.0213	.0355	.0492	.0629	.0767	.0909	.1046	.1183	.1321	.1463	.1600	.1737	.2017	.2291	.2429	.2845
1.016	.0217	.0358	.0496	.0633	.0771	.0913	.1050	.1187	.1325	.1467	.1604	.1741	.2021	.2295	.2433	.2849
1.270		.0378	.0516	.0653	.0790	.0932	.1070	.1207	.1345	.1486	.1624	.1761	.2040	.2315	.2453	.2869
1.625		.0406	.0543	.0681	.0818	.0960	.1097	.1235	.1372	.1514	.1652	.1789	.2066	.2343	.2480	.2897
1.829			.0559	.0697	.0834	.0976	.1113	.1251	.1388	.1530	.1667	.1805	.2084	.2359	.2496	.2913
1.981			.0571	.0709	.0846	.0988	.1125	.1263	.1400	.1542	.1679	.1817	.2096	.2371	.2508	.2925
2.058			.0577	.0715	.0852	.0994	.1131	.1269	.1406	.1548	.1685	.1823	.2102	.2377	.2514	.2931
2.311			.0597	.0734	.0872	.1014	.1151	.1288	.1426	.1568	.1705	.1842	.2122	.2396	.2534	.2950
2.388			.0603	.0740	.0878	.1020	.1157	.1294	.1432	.1574	.1711	.1848	.2128	.2402	.2540	.2956
2.591				.0756	.0894	.1035	.1173	.1310	.1448	.1589	.1727	.1864	.2143	.2418	.2556	.2972
2.769				.0770	.0907	.1049	.1187	.1324	.1461	.1603	.1741	.1878	.2157	.2432	.2570	.2986
3.175				.0802	.0939	.1081	.1218	.1356	.1493	.1635	.1772	.1910	.2189	.2464	.2601	.3018
3.962					.1001	.1142	.1280	.1417	.1555	.1696	.1834	.1971	.2250	.2525	.2663	.3079
4.775					.1064	.1206	.1343	.1481	.1618	.1760	.1897	.2035	.2314	.2589	.2726	.3143
5.156						.1235	.1373	.1510	.1648	.1789	.1927	.2064	.2344	.2618	.2756	.3172
5.537								.1540	.1677	.1819	.1957	.2094	.2373	.2648	.2785	.3202
5.944								.1572	.1709	.1851	.1988	.2126	.2405	.2680	.2817	.3234
6.350								.1603	.1741	.1883	.2020	.2157	.2437	.2711	.2849	.3265

EXAMPLE: MATERIAL THICKNESS = 3.175 MM, INSIDE RADII = 6.35 MM R. WHERE THESE CROSS = .1356 MM IF THE INSIDE OF YOUR BEND IS 20° THE BEND ALLOWANCE (B/A) = 2.712 MM (.1356 × 20°) BEND ALLOWANCE (B/A) PLUS TOTAL OF STRAIGHT LENGTHS = DEVELOPED LENGTH

From Drafting for Trades and Industry—Basic Skills, Nelson. Delmar Publishers Inc.

Table 44

SPUR/PINION GEAR TOOTH PARTS

20 Degree Pressure Angles

Diametral Pitch	Circular Pitch	Circular Thickness	Addend.	Dedend.	Standard Fillet Radius
D.P.	C.P.	C.T.	A	D	
1	3.1416	1.5708	1.0000	1.2500	0.3000
2	1.5708	0.7854	0.5000	0.6250	0.1500
2.5	1.2566	0.6283	0.4000	0.5000	0.1200
3	1.0472	0.5236	0.3333	0.4167	0.1000
3.5	0.8976	0.4488	0.2857	0.3571	0.0857
4	0.7854	0.3927	0.2500	0.3125	0.0750
4.5	0.6981	0.3491	0.2222	0.2778	0.0667
5	0.6283	0.3142	0.2000	0.2500	0.0600
5.5	0.5712	0.2856	0.1818	0.2273	0.0545
6	0.5236	0.2618	0.1667	0.2083	0.0500
6.5	0.4833	0.2417	0.1538	0.1923	0.0462
7	0.4488	0.2244	0.1429	0.1786	0.0429
7.5	0.4189	0.2094	0.1333	0.1667	0.0400
8	0.3927	0.1963	0.1250	0.1563	0.0375
8.5	0.3696	0.1848	0.1176	0.1471	0.0353
9	0.3491	0.1745	0.1111	0.1389	0.0333
9.5	0.3307	0.1653	0.1053	0.1316	0.0316
10	0.3142	0.1571	0.1000	0.1250	0.0300
11	0.2856	0.1428	0.0909	0.1136	0.0273
12	0.2618	0.1309	0.0833	0.1042	0.0250
13	0.2417	0.1208	0.0769	0.0962	0.0231
14	0.2244	0.1122	0.0714	0.0893	0.0214
15	0.2094	0.1047	0.0667	0.0833	0.0200

Table 45

SURFACE TEXTURE ROUGHNESS

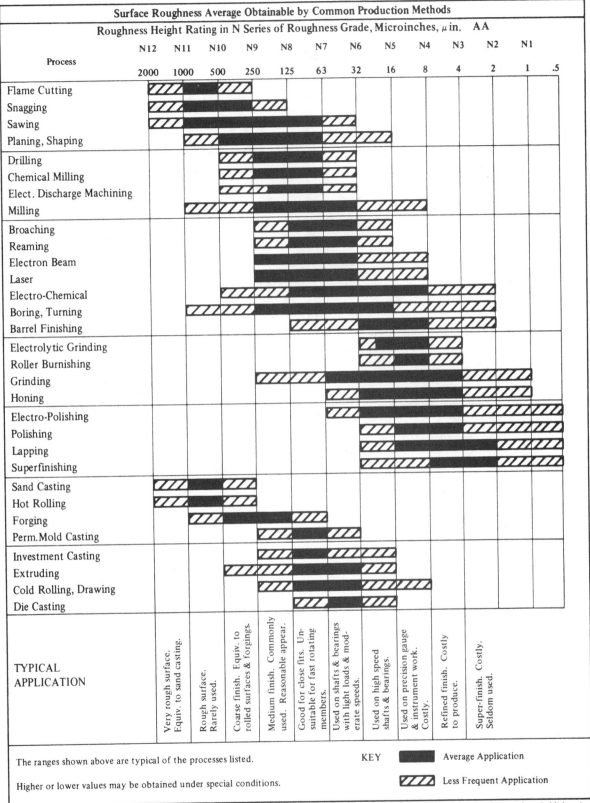

Surface Roughness Average Obtainable by Common Production Methods												
Roughness Height Rating in N Series of Roughness Grade, Microinches, μ in. AA												

Table 46

From Interpreting Engineering Drawings. Jensen. Delmar Publishers Inc.

PROPERTIES, GRADE NUMBERS, AND USAGE OF STEEL ALLOYS

Class of Steel	*Grade Number	Properties	Uses
Carbon - Mild 0.3% carbon	10xx	Tough - Less Strength	Rivets - Hooks - Chains - Shafts - Pressed Steel Products
Carbon - Medium 0.3% to 0.6% carbon	10xx	Tough & Strong	Gears - Shafts - Studs - Various Machine Parts
Carbon - Hard 1.6% to 1.7%	10xx	Less Tough - Much Harder	Drills - Knives - Saws
Nickel	20xx	Tough & Strong	Axles - Connecting Rods - Crank Shafts
Nickel Chromium	30xx	Tough & Strong	Rings Gears - Shafts - Piston Pins - Bolts - Studs - Screws
Molybdenum	40xx	Very Strong	Forgings - Shafts - Gears - Cams
Chromium	50xx	Hard W/Strength & Toughness	Ball Bearings - Roller Bearing - Springs - Gears - Shafts
Chromium Vanadium	60xx	Hard & Strong	Shafts - Axles -Gears - Dies - Punches - Drills
Chromium Nickel Stainless	60xx	Rust Resistance	Food Containers - Medical/Dental Surgical Instruments
Silicon - Manganese	90xx	Springiness	Large Springs

*The first two numbers indicate type of steel, the last two numbers indicate the approx. average carbon content — 1010 steel indicates, carbon steel w/approx. 0.10% carbon.

From Drafting for Trades and Industry—Mechanical and Electronic. Nelson. Delmar Publishers Inc.

Table 47

AMERICAN NATIONAL STANDARDS OF INTEREST TO DESIGNERS, ARCHITECTS, AND DRAFTERS

TITLE OF STANDARD

Abbreviations Y1.1-1972
American National Standard Drafting Practices
   Size and Format Y14.1-1980
   Line Conventions and Lettering Y14.2M-1979
   Multi and Sectional View Drawings . . . . Y14.3-1975(R1980)
   Pictorial Drawing. Y14.4-1957
   Dimensioning and Tolerancing Y14.5M-1982
   Screw Threads Y14.6-1978
   Screw Threads (Metric Supplement). Y14.6aM-1981
   Gears and Splines
      Spur, Helical, and Racks Y14.7.1-1971
      Bevel and Hyphoid. Y14.7.2-1978
   Forgings Y14.9-1958
   Springs Y14.13M-1981
   Electrical and Electronic Diagram . . . . Y14.15-1966(R1973)
   Interconnection Diagrams Y14.15a-1971
   Information Sheet Y14.15b-1973
   Fluid Power Diagrams . . . . Y14.17-1966(R1980)
   Digital Representation for Communication of Product Definition Data Y14.26M-1981
   Computer-Aided Preparation of Product Definition Data Dictionary of Terms. Y14.26.3-1975
   Digital Representation of Physical Object Shapes . . . . Y14 Report
   Guideline — User Instructions . . . . Y14 Report No. 2
   Guideline — Design Requirements. . . . . Y14 Report No. 3
   Ground Vehicle Drawing Practices . . . . In Preparation
   Chasis Frames Y14.32.1-1974
   Parts Lists, Data Lists, and Index Lists. Y14.34M-1982
   Surface Texture Symbols Y14.36-1978
Illustrations for Publication and Projection Y15.1M-1979
Time Series Charts Y15.2M-1979
Process Charts Y15.3M-1979
Graphic Symbols for:
   Electrical and Electronics Diagrams. Y32.2-1975
   Plumbing Y32.4-1977
   Use on Railroad Maps and Profiles . . . . Y32.7-1972(R1979)
   Fluid Power Diagrams . . . . Y32.10-1967(R1974)
   Process Flow Diagrams in Petroleum and Chemical Industries. Y32.11-1961
   Mechanical and Acoustical Element as Used in Schematic Diagrams . . . . Y32.18-1972(R1978)
   Pipe Fittings, Valves, and Piping . . . . Z32.2.3-1949(R1953)
   Heating, Ventilating, and Air Conditioning. . . . . Z32.2.4-1949(R1953)
   Heat Power Apparatus . . . . Z32.2.6-1950(R1956)
Letter Symbols for:
   Glossary of Terms Concerning Letter Symbols Y10.1-1972
   Hydraulics Y10.2-1958
   Quantities Used in Mechanics for Solid Bodies Y10.3-1968
   Heat and Thermodynamics. Y10.4-1982
   Quantities Used in Electrical Science and Electrical Engineering Y10.5-1968
   Aeronautical Sciences Y10.7-1954
   Structural Analysis Y10.8-1962
   Meteorology . . . . Y10.10-1953(R1973)
   Acoustics . . . . Y10.11-1953(R1959)
   Chemical Engineering . . . . Y10.12-1955(R1973)
   Rocket Propulsion Y10.14-1959
   Petroleum Reservoir Engineering and Electric Logging . . . . Y10.15-1958(R1973)
   Shell Theory. . . . . Y10.16-1964(R1973)
   Guide for Selecting Greek Letters Used as Symbols for Engineering Mathematics. . . . . Y10.17-1961(R1973)
   Illuminating Engineering . . . . Y10.18-1967(R1977)
   Mathematical Signs and Symbols for Use in Physical Sciences and Technology. Y10.20-1975

Table 48

Index

Acme thread form, 405
Acute angle, 97, 655
  construct an arc tangent to, 116
Adjustable curve, 31, 32
Adjustable triangle, 33
Airbrush, 54
American National Standards, list of, 737
Angle
  bisecting, 102
  defined, 97
  gears, 484
  transfer an, 107
Angles
  convert decimal degrees to degrees, minutes, and seconds, 97
  convert minutes and seconds to decimal degrees, 97
  dimensioning, 307-8
  transversal formed, 656
  types of, 655-56
Arc
  drawing through three given points, 103
  outlines, dimension of, 309
Arc tangent
  construct to a straight line and a curve, 118
  construct to an acute angle, 116
  construct to an obtuse angle, 117
  construct to two radii or diameters, 119
  constructing to a right angle, 115
Architect's scale, 41
Arcs, dimensioning, 307-8
Arrowheads, 304
Artistic drawings, 2
Assembly drawings, 507-8
Assembly section view, 198
Auxiliary section, 225
  view, 198
Auxiliary views, 219-34
  defined, 219-20
  drawing a round surface, 223
  drawing of, 220-21
  half, 226
  partial, 225

  plotting an irregular curved surface, 223-25
  projecting a round surface from an inclined surface, 221-22
  secondary, 225
Axonometric drawings, 529-31
Axonometric projections, 7, 8
Axonometric sketching, 78, 79, 83

Back welding, 581
Beam compass, 31
Bends in sheet metal, 282, 284-85
Bevel gears, 484, 485, 493
Bolt, defined, 412
Bolts, 415
  hex, dimensions of, 706, 707
  square, dimensions of, 708
Boring mills, 605-6
Bow compass with lead clutch, 31
Broaching machine, 606
Broken-out section view, 195-96
Butterfly-type scriber, 52
Buttress thread form, 406

Cabinet projections, 7
Camfers, dimensioning, 312, 313
Cams, 468-81
  dimensioning, 475-76
  drawing, 472-74
  motion of, 470-71
  terminology of, 469-70
  timing diagram of, 475
Cap screws, 412, 413
  dimensions for, 710-11
Cavalier drawings, 531-32
Cavalier projections, 7
Center angle, 99
Center lines, 159
Centrifugal casting, 596
Chords
  defined, 99
  dimensioning, 307-8
Circle
  defined, 99

drawing an involute of a, 138
drawing through three given points, 102
locating the center, 113
Circles
angles formed inside of, 664-65
angles formed on or outside of, 665-66
circumferences and areas of, 690
definitions of properties of, 661-62
geometric principles of, 662-64
tangents of, 666-67
Circular construction, 113-34
Circumference, defined, 99
Civil engineer's scale, 41
Compasses, 28-31
Compression springs, 452, 453, 454
drawing, 456-59
Computer-aided drafting (CAD)
advantages of, 352-53
difference between manual and CAD,
342, 344
input commands, 359-61, 363, 365
manipulation commands, 365, 367-68
operations, 357-75
output commands, 368, 373
software, 351-52
system configurations, 352
systems, 344-51
digitizer, 347-48
functions menu, 348-49
graphics display, 345-47
keyboard, 347
plotter, 349-50
processor, 350-51
text display, 347
technology of, 340-56
Concentric circles, 99
Cone
defined, 100
develop a truncated, using radial line
development, 277-79
Conic sections, 121-22
Conversion
degrees, minutes, seconds and decimal
degrees, 655
millimetre to inch, 650-51
Corners, rounded, dimensioning, 308
Curve
adjustable, 31, 32
cycloidal, drawing a, 138-39
ogee, drawing an, 120
plotting, 165-66
Cutting-plane line, 189
Cycloidal curve, drawing a, 138-39
Cylinder

develop a truncated, using parallel line
development, 273-75
finding intersection with a plane surface
by line projection, 252
Cylindrical intersections, 166-68

Decimal-inch dimensioning, 300
Descriptive geometry, 235-65
construct an edge view of a plane surface, 243-44
constructing a point view of a line, 239-40
determining piercing point
by construction, 247-49
by inspection, 247
by line projection, 249-51
determining visibility of lines, 246-47
finding the intersection
of a cylinder and plane surface by line
projection, 252
of a sphere and a plane surface, 252-53
of two planes by line projection, 251
of two prisms, 253
finding the true
angle between two planes, 245-46
distance between a line and a point
in space, 240-41
distance between a plane surface and
a point in space, 244-45
distance between two parallel lines,
241-42
length of a line, 238-39
fold lines and, 236-37
locating a point in space (right view), 238
notations and, 236
projecting a line into other views, 237-38
projecting a plane into another view, 242
projection, 235
Design layout, 507
Design procedure, 506-7
Design process, 8-9
Detail drawings, 509
Developments, 266-82
defined, 266
laps and seams, 267-68
parallel lines, 266, 269-75
radial line, 266, 275-79
surface, 266
triangulation, 266, 279-82
Diameter, defined, 99
Diameters, dimensioning, 306
Diazo printing, 645-47
Die casting, 596
Dihedral angle, finding the true angle
between two, 245-46

Dimension lines, 302-3
Dimensioning, 299-316
  arc outlines, 309
  camfers, 312-13
  chords, arcs, and angles, 307-8
  components, 301-5
    arrowheads, 304
    dimension lines, 302-3
    extension lines, 301-2
    leader lines, 303-4
  counterbored holes, 310-11
  countersunk holes, 311
  diameters, 306
  general rules for, 305-6
  geometric shapes, 313-14
  isometric, 545-46
  keyseats, 313
  locational systems, 314-16
  polar coordinate, 316
  radii, 307
  rectangular coordinate, 315
  round holes, 309-310
  rounded corners, 308
  rounded ends, 308
  slotted holes, 310, 311
  spotfaces, 312
  systems, 300-301
    decimal-inch, 300
    fractional, 301
    metric, 300, 301
  techniques, 306-14
Dimensioning and tolerancing. *See*
    Geometric dimensioning and
    tolerancing
Dimetric projection, 530
Direction of sight, 189
Dividers, 31
Drafting
  brush, 38-39
  defined, 9
  instruments, 23-65
    airbrush, 54
    butterfly-type scriber, 52
    care of, 60-61
    drawing sets, 27-39
    ink tools, 44-49
    list of, 23-24
    mechanical lettering sets, 49-52
    multipliers for, 688-89
    polyester film, 643
    scales used, 39-42
  machine, 26-27
  mathematics. *See* Mathematics of
    drafting

media, 642-45
sets, 27-39
vellum, 642-43
Drawing surface, 25
Drawing table, 24
Drawings
  artistic, 2
  assembly, 507-8
  auxiliary section, 225
  auxiliary views, 219-34
    defined, 219-20
    drawing of, 220-21
    half, 226
    partial, 225
    plotting an irregular curved surface,
      223-25
    projecting a round surface from an
      inclined surface, 221-22
    round surface, 223
    secondary, 225
  axonometric, 529-31
  cavalier, 531-32
  detail, 509
  drafter's checklist, 503, 505
  multiview, 148-87
    centering of, 162-63
    conventional breaks, 172-73
    curve plotting, 165-66
    cylindrical intersections, 166-68
    first-angle projection, 175
    incomplete views, 168-69
    intersecting surfaces, 165
    numbering of, 163-64
    orthographic projections, 149-60
    planning the, 160-61
    runouts, 164
    rounds and fillets, 164
    sketching procedure, 161-62
    visualization, 173-74
  notations on, 317-27
    rules for applying, 318, 320, 325-27
  numbering system, 505
  oblique, 531-32
  parts list, 505
  pattern, 510
  perspective, 532, 546-52
  pictorial, 529-66
  reproduction of, 645-48
  revisions of, 501-3
  sectional views, 189-218
    aligned, 201
    fasteners and shafts in, 201-2
    holes and, 199-200
    keyways in, 201

kinds of, 189-98
    ribs or webs, 199, 200-201
specifying the scale, 299-300
spokes in, 201
subassembly, 508-9
technical, 2-3, 5
    application of, 10-12
    purpose of, 7-9
    regulation of, 12, 16
    types of, 5-7
title block, 503, 504
types of, 2-7
working, 507-9
Drill and tap sizes, 700
Drill presses, 603
Drilled hole tolerance, 701-3
Drop bow compass, 30
Drop forging, 597
Dry cleaning powder, 39

Eccentric circles, 99
Ellipse
    drawing
        using concentric circle method, 122-24
        a foci, 125-26
        on an inclined plane, 544
        a tangent to an, 127-29
        a trammel, 124-25
    locating the major or minor axes of an, 127
Ends, rounded, dimensioning, 308
Engineering change order (ECO), 501-3
Engineering department organization, 501, 502
Equilateral triangle, 98, 657
    drawing, 110
Erasers, 38, 645
Extension line, 301-2
Extension springs, 454
    drawing, 459-60

Fasteners, 403-51
    classifications of, 403
    grooved, 420-33
        pins, types of, 421-30
        spring pins, 432-33
        studs, 431-32
    keys and keyseats, 418-19
    pitch, 406-7
    rivets, 416-18
    screws, bolts, and studs, 411-16
    in sectional views, 201-2
    single and multiple threads, 407
    tap and die, 406

threads, 404
    per inch (TPI), 406
    screws, 404-6
    single and multiple, 407
Fastening systems, 434
    retaining rings, 434-44
Fillet welding, 578
Fillets, 164
First-angle projection, 175
Fixtures, 610-24
    clamping principles, 616, 618-21
    locating principles, 616, 617-18
Flange welding, 583-84
Flat springs, 454-55
Flat-plane surface, 267
Foci ellipse, drawing a, 125-26
Force and shrink fits, 698
Foreshortening, 149
Fractional dimensioning, 301
French curves, 33, 36
    using, 36
Full section view, 191, 193

Gears, 483-500
    backlash, 488
    blank, 487
    center-to-center distance, 490
    design and layout of, 496
    diameter, 487
    diametral pitch, 489
    gear tooth caliper, use of, 490
    kinds of, 483-86
    pressure angle, 490
    rack, 492
    ratio, 486-87
    tooth-cutting data, 490-92
        worm gear, 495
    train, 494, 496
Geometric construction, 96-143
    bisect a line, 101
    bisect an angle, 102
    circular construction, 113-34
    dividing a line
        into equal parts, 105-6
        into proportional parts, 106
    drawing
        circle through three given points, 102
        line parallel to a curved line, 103
        line parallel to a straight line, 103
        perpendicular to a line at a point, 104-5
        perpendicular to a line from a point not on the line, 105
    enlarge or reduce a shape, 109

polygon, 109-113
transfer complex shapes, 108
transfer of, 107
odd shapes, 107-8
Geometric dimensioning and tolerancing,
378-402
angularity, 387
circularity (roundness), 386-87
cylindricity, 387
feature control symbol, 383-84
flatness, 385-86
modifiers, 380-83
parallelism, 387
perpendicularity, 388
profile, 388-89
runout, 389-90
straightness, 386
table of key, 729
true position, 384-85
Geometric dimensioning, defined, 379
Geometric nomenclature, 96-100
Geometry, descriptive. See Descriptive
geometry
Gib head nominal dimensions, 730
Glossary, 678-83
Groove welding, 578-80
Grooved pins
spring, 432-33
types of, 421-30

Half-section view, 194-95
Helical springs, 452-54
Helix, drawing a, 135-37
Hemispheres, defined, 100
Hex bolts, 415
dimensions of, 706-7
Hex jam nuts, dimensions of, 722
Hexagon, drawing a, 112-13
Hidden lines, 158-59, 534
Holes
counterbored, dimensioning of, 310-11
countersunk, dimensioning of, 311
round, dimensioning of, 309-10
slotted, dimensioning of, 310, 311
Hyperbola, drawing a, 132-34

Inch/metric equivalents, 686
Inches to millimetres table, 685
Incomplete view, 168-69
Intersecting surfaces treatment of, 165
Intersectings, cylindrical, 166-68
Invention agreements, 503
Investment casting, 596
ISO metric thread form, 404, 405

Isometric
dimensioning, 545-46
drawing principles, 532-46
box construction, 535
circles and arches, 537-42
curves, 536-37
irregularly shaped objects, 536
knurls, 542
screw threads, 542
spheres, 542-43
intersections, 543-45
projection, 529
drawing, 532-34
rounds and fillets, 545
sketching, 78, 79
steps in, 84-85
templates, 546
triangle, 98, 657

Jigs, 608-10

Key and keyways, sizes of, 724-27
Keys, 418-19
Keyseats, dimensioning, 313
Keysets, 418-19
Keyways in sectional views, 201
Knuckle thread form, 405-6

Laps and seams, 267-68
Lathes, 601-3
Lead
graphite pencils, 644-45
holders, 37-38
Leader lines, 303-4
Leads, 37
Lettering
Freehand, 66-71
characteristics of good, 69, 71
six basic strokes used for, 72
techniques, 71-73
Lettering sets, mechanical, 49-52
Line
bisecting, 101
constructing a point view of, 239-40
defined, 97
dividing into equal parts, 105-6
dividing into proportional parts, 106
drawing a perpendicular to or from a
point not on the line, 105
drawing an involute of a, 137
drawing parallel,
to a curved line, 103
to a straight line, 103
drawing perpendicular to a point, 104-5

finding true length of, 238-39
projecting onto other views, 237-38
work, 74-77
Lines
  angular, 74
  characteristics of, 74
  determining visibility of, 246-47
  dimensioning, 302-3
  extension, 301-2
  hidden, 158-59
  horizontal and vertical, 74
  leader, 303-4
  parallel, 75-76
    finding true distance between two, 241-42
  perpendicular, 76-77
Locational
  clearance fits, 695
  interference fits, 697
  transition fits, 696
Lock washers, dimensions for, 719-21

Machine screws, 412, 413
  dimensions of, 709, 712-14
Mathematics of drafting, 649-77
  angles
    formed by a transversal, 656
    formed inside a circle, 664-65
    formed outside a circle, 665-66
    types of, 655-56
  circles
    angles formed inside of, 664-65
    angles formed outside of, 665-66
    definitions of properties of, 661-62
    geometric principles of, 662-64
    tangents of, 666-67
  degrees, minutes, and seconds
    angles expressed in, 653-54
    decimal degree conversion, 655
  evaluating formulas, 651-52
  fractions
    expressing common as decimal, 650
    rounding decimal, 649-50
  millimetre-inch equivalents, 650-51
  polygons, 659-61
  ratio and proportion, 652-53
  triangles, 656-58
  trigonometric functions, 668-77
    drafting applications, 671-73
    oblique triangles, 673-76
    sides, calculations of, 670-71
Mechanical engineer's scale, 39-41
Mechanical lettering set, 49-52
Media, 642-45

Metallurgy, powder, 597
Metric
  dimensioning, 300, 301
  equivalents, 687
  system, 42
Micrometer, 43-44
Millimetres to inch table, 685
Milling machines, 603
Miter gears, 484, 485
Multipliers for drafters, 688-89
Multiview drawings, 148-87
  centering of, 162-63
  conventional breaks, 172-73
  curve plotting, 165-66
  cylindrical intersections, 166-68
  first-angle projection, 175
  incomplete views, 168-69
  intersecting surfaces, 165
  numbering of, 163-64
  orthographic projections, 149-60
  planning of, 160-61
  rounds and fillets, 164
  runouts, 164
  sketching procedure, 161-62
  visualization, 173-74

Notation on drawings, 317-27
  rules for applying, 318, 320, 325-27
Notches in sheet metal, 282, 283
Numerically controlled machines, 606-7
Nuts, dimensions for, 722-23

Oblique
  drawings, 531-32
  projections, 7
  sketching, 78, 79, 84
  triangles, 693
    formulas for, 693
    trigonometric functions and, 673-76
Obtuse angle, 97, 655
  construct an arc tangent to, 117
Obtuse angle triangle, 98
Offset section view, 193-94
Ogee curve, drawing an, 120
Orthographic projections, 5-7, 149-60
  center lines, 159
  dimension transfer methods, 157-58
  hidden lines, 158-59
  multiple views, 153-56
  normal surfaces, 149-50
  object description requirements, 156
  surface categories, 159-60
  two views, 150-51
    labeling of, 151-53

Orthographic sketching, 78, 79, 83

Paper sizes, 55-57
Parabola
    curve, joining two parts of, 134
    drawing a, 129-30
    finding the focus point of a, 130-31
Parallel line development, 266, 269-75
Parallel projections, 5
Parallel straightedge, 26
Parallelograms, 660-61
Pattern drawings, 510
Pencils, 37
Pens, technical, 44-47
    cleaning, 48-49
    sizes, 47
Pentagon, drawing a, 111
Perspective drawing, 532
    procedures, 546-52
    terminology of, 546, 548
Perspective projections, 7, 8
Perspective sketching, 78, 80
Pictorial drawings, 529-66
    axonometric, 529-31
    oblique, 531-32
    perspective, 546-52
Piercing point
    determining by construction, 247-49
    determining by inspection, 247
    determining by line projection, 249-51
Pinion gears, 483, 484
Plain washers, dimensions for, 718
Plane
    construct an edge view of, 243-44
    find true distance between a point in
        space and, 244-45
    finding the intersection of two by line
        projection, 251
    projecting into another view, 242
    surface, 267
Plug and slot welding, 582
Points
    defined, 96
    location of in space (right view), 238
    in space, 96
    tangent
        defined, 114
        locating, 114
Polyester drafting film, 643
Polygons, 659-61
    construction of, 109-113
    defined, 98, 656
Polyhedron, defined, 100
Power metallurgy, 597

Power saws, 604
Precision grinders, 604
Press forging, 597-98
Printing
    diazo, 645-47
    high-speed, 647-48
Prisms
    defined, 100
    developing a truncated, using parallel
        line development, 271-73
    finding the intersection of two, 253
Projection, 149
    descriptive geometry and, 235
    plane, 149
Projections
    axonometric, 7, 8
    cabinet, 7
    cavalier, 7
    dimetric, 530
    isometric, 529-30
    oblique, 7
    orthographic, 5-7, 149-60
        center lines, 159
        dimension transfer methods, 157-58
        hidden lines, 158-59
        multiple views, 153-56
        normal surfaces, 149-50
        object description requirements, 156
        surface categories, 159-60
        two views, 150-51
    parallel, 5
    perspective, 7, 8
    trimetric, 530-31
Proportional dividers, 31
Protractor, 36
Pyramid
    defined, 100
    develop a truncated, using radical line
        development, 276-77

Quadrant, defined, 99
Quadrilateral, defined, 98

Rack gears, 483, 484
Radical line developments, 266, 275-79
Radii, dimensioning, 307
Radius, defined, 99
Rectangles, defined, 659
Reflex angle, 656
Removed section view, 197
Retaining rings, 434-44
Revolved section view, 196-97
Ribs in sectional views, 199, 200-201
Right angle, 97, 655

constructing an arc tangent to a, 115
Right triangles, 98, 658
    formulas, 692
Ring gears, 483, 484
Rivets, 416-18
Rounds, 164
Rounds and fillets, isometric, 545
Running and sliding fits, 694
Runouts, 164

Sand casting, 595-96
Scalene triangle, 98, 657
Scales
    architect's, 41
    civil engineer's, 41
    mechanical engineer's, 39-41
    specifying, 299-300
Screw heads, types of, 416
Screw thread forms
    pitch, 406-7
    right-hand and left-hand, 308
    single and multiple threads, 407
    thread relief (undercut), 409-11
    thread representation, 408-9
    threads per inch (TPI), 406
Screw threads, 404-6
    inch-metric comparison, 704
    ISO basic metric thread information, 705
    unified standard series, 699
Screws, 412-14
    dimensions of, 709-17
Scriber
    butterfly-type, 52
    templates, 49-51
Seams and laps, 267-68
Section lining, 189-91
Sectional views, 189-218
    aligned, 201
    assembly section, 198
    auxiliary section, 198
    broken-out section, 195-96
    fasteners in, 201-2
    full, 191, 193
    half section, 194-95
    holes, 199-200
    keyways in, 201
    kinds of, 191-98
    offset, 193-94
    removal section, 197
    revolved section, 196-97
    ribs and webs, 199, 200-201
    shafts in, 201-2
    spokes in, 201
    thinwall section, 198

Sections
    defined, 99
    kinds of, 191-98
Segment, defined, 99
Semicircle, defined, 99
Sepia diazo paper, 643-44
Set screws, 413-14
    dimensions for, 715-17
Shafts in sectional views, 201-2
Shapers and planers, 605
Sheet metal
    laps and seams, 267-68
    thickness of, 268-69
    wire gage designation and, 731-33
Shop processes, 594-627
    castings, 595-97
    extruding, 598-99
    forging, 597-98
    heat treatment of steels, 625-26
    machining, 601-7
    stamping, 599-601
    workholding devices, 607-24
        fixtures, 610-24
        jigs, 608-10
Sketching, 77-88
    axonometric, 83-84
    circles, 81
    ellipses, 81
    lines and curves, 80-81
    materials, 80
    oblique, 78, 79, 84
    orthographic, 78, 79, 83
    perspectives, 84, 86-88
        steps in, 86-88
    proportion in, 81-82
    techniques, 80-88
    types of, 77-80.
    See also Technical drawing
Sphere, defined, 100
    finding intersection of a plane surface
        and, 252-53
Spiral, drawing a, 135
Spokes in sectional views, 201
Spotfaces, dimensioning, 312
Springs, 452-68
    drawing, 456-60
    flat, 454-55
    helical, 452-54
    terminology of, 455-56
    views of, 460-61
Spur gears, 483, 484
Spur pinion gear tooth parts, 734
Square bolt, 415
    dimensions, 708

Square nuts, dimensions of, 723
Square thread form, 405
Squares
    defined, 659
    drawing a, 110
    drawing an involute of a, 137-38
Steel alloys, table of, 736
Straight angle, 655
Straightedge, parallel, 26
Studs
    defined, 412
    sizes and specifications, 431-32
Subassembly drawings, 508-9
Surface quality, control of, 316
Surface texture roughness, 735

T-square, 25-26
Tangent, arc
    construct to
        an acute angle, 116
        a line and a curve, 118
        an obtuse angle, 117
        a right angle, 115
        two radii or diameters, 119
Tangent points
    defined, 114
    locating, 114
Tap and die, 406
Technical drawings, 2-3, 5
    applications of, 10-12
    purpose of, 7-9
    regulation of, 12, 16
    types of, 5-7.
    See also Sketching
Technical pens, 45-47
    cleaning, 48-49
    sizes, 47
Templates, 33, 34-35
    isometric, 546
    scriber, 49-51
    standard, 51-52
    welding, 587
Thinwall section view, 198
Threads per inch (TPI), 406
Threads, screw. See Screw thread forms
Tolerancing, 378-79
    positional, defined, 379
Torsion springs, 454
Trammel ellipse, drawing a, 124-25
Triangles, 31, 33
    defined, 97-98, 657
    drawing
        an equilateral, 110
        an involute of a, 137

with known length of sides, 109-110
    types of, 656-58
Triangular developments, 266, 267, 279-82
    develop, a transitional piece
        with no folded lines, 280, 282
        with a round end, 280, 281
        using, 279-80
Trigonometric formulas, 691
Trigonometric functions, 668-77
    drafting applications of, 671-73
    oblique triangles, 673-76
    sides, calculations of, 670-71
Trimetric projection, 530-31
Truncated cone, developed by using a
        radical line development, 277-79
Truncated cylinder, developed by using
        parallel line development, 273-75

Uniform national thread form, 404, 405

Vellum, drafting, 642-43
Views, incomplete, 168-69
Visualization, 173-74

Washers, dimensions for, 718-21
Webs in sectional views, 199, 200-201
Welding, 570-93
    double, 579
    field, 577-78
    intermittent, 574
    joints, 578
    length of, 573-74
    placement of, 574
    projection, 586-87
    process, 570-71
    seam, 587
    size of, 572-73
    spot, 584-86
    symbols, 571-72
        contour, 577
    template, 587
    types of, 578-84
        back, 581
        fillet, 578
        flange, 583-84
        groove, 578-80
        plug and slot, 582
        surface, 582-83
Whiteprinter, 57-59
Woodruff keys, table of, 725-27
Working drawings, 507-9
Worm gears, 484, 485, 493-94
    required tooth-cutting data, 495
Worm thread form, 405